The Earth Inside and Out: Some Major Contributions to Geology in the Twentieth Century

Geological Society Special Publications
Society Book Editors
A. J. Fleet (Chief Editor)
P. Doyle
F. J. Gregory
J. S. Griffiths
A. J. Hartley
R. E. Holdsworth
A. C. Morton
N. S. Robins
M. S. Stoker
J. P. Turner

Special Publication reviewing procedures

The Society makes every effort to ensure that the scientific and production quality of its books matches that of its journals. Since 1997, all book proposals have been refereed by specialist reviewers as well as by the Society's Books Editorial Committee. If the referees identify weaknesses in the proposal, these must be addressed before the proposal is accepted.

Once the book is accepted, the Society has a team of Book Editors (listed above) who ensure that the volume editors follow strict guidelines on refereeing and quality control. We insist that individual papers can only be accepted after satisfactory review by two independent referees. The questions on the review forms are similar to those for *Journal of the Geological Society*. The referees' forms and comments must be available to the Society's Book Editors on request.

Although many of the books result from meetings, the editors are expected to commission papers that were not presented at the meeting to ensure that the book provides a balanced coverage of the subject. Being accepted for presentation at the meeting does not guarantee inclusion in the book.

Geological Society Special Publications are included in the ISI Index of Scientific Book Contents, but they do not have an impact factor, the latter being applicable only to journals.

More information about submitting a proposal and producing a Special Publication can be found on the Society's web site: www.geolsoc.org.uk.

It is recommended that reference to all or part of this book should be made in one of the following ways:

Oldroyd, D. R. (ed.) 2002. *The Earth Inside and Out: Some Major Contributions to Geology in the Twentieth Century.* Geological Society, London, Special Publications, **192.**

Young, D. A. 2002. Norman Levi Bowen (1887–1956) and igneous rock diversity *In*: Oldroyd, D. R. (ed.) 2002. *The Earth Inside and Out: Some Major Contributions to Geology in the Twentieth Century.* Geological Society, London, Special Publications, **192,** 99–111.

GEOLOGICAL SOCIETY SPECIAL PUBLICATION NO. 192

The Earth Inside and Out: Some Major Contributions to Geology in the Twentieth Century

EDITED BY

DAVID R. OLDROYD
The University of New South Wales, Sydney, Australia

2002
Published by
The Geological Society
London

THE GEOLOGICAL SOCIETY

The Geological Society of London (GSL) was founded in 1807. It is the oldest national geological society in the world and the largest in Europe. It was incorporated under Royal Charter in 1825 and is Registered Charity 210161.

The Society is the UK national learned and professional society for geology with a worldwide Fellowship (FGS) of 9000. The Society has the power to confer Chartered status on suitably qualified Fellows, and about 2000 of the Fellowship carry the title (CGeol). Chartered Geologists may also obtain the equivalent European title, European Geologist (EurGeol). One fifth of the Society's fellowship resides outside the UK. To find out more about the Society, log on to www.geolsoc.org.uk.

The Geological Society Publishing House (Bath, UK) produces the Society's international journals and books, and acts as European distributor for selected publications of the American Association of Petroleum Geologists (AAPG), the American Geological Institute (AGI), the Indonesian Petroleum Association (IPA), the Geological Society of America (GSA), the Society for Sedimentary Geology (SEPM) and the Geologists' Association (GA). Joint marketing agreements ensure that GSL Fellows may purchase these societies' publications at a discount. The Society's online bookshop (accessible from www.geolsoc.org.uk) offers secure book purchasing with your credit or debit card.

To find out about joining the Society and benefiting from substantial discounts on publications of GSL and other societies worldwide, consult www.geolsoc.org.uk, or contact the Fellowship Department at: The Geological Society, Burlington House, Piccadilly, London W1J 0BG: Tel. +44 (0)20 7434 9944; Fax +44 (0)20 7439 8975; Email: enquiries@geolsoc.org.uk.

For information about the Society's meetings, consult *Events* on www.geolsoc.org.uk. To find out more about the Society's Corporate Affiliates Scheme, write to enquiries@geolsoc.org.uk.

Published by The Geological Society from:
The Geological Society Publishing House
Unit 7, Brassmill Enterprise Centre
Brassmill Lane
Bath BA1 3JN
UK

(*Orders*: Tel. +44 (0)1225 445046
Fax +44 (0)1225 442836)
Online bookshop: http://bookshop.geolsoc.org.uk

British Library Cataloguing in Publication Data
A catalogue record for this book is available from the British Library.

ISBN 1–86239–096–7

Typeset by Type Study, Scarborough, UK
Printed by The Alden Press, Oxford, UK

Distributors

USA
AAPG Bookstore
PO Box 979
Tulsa
OK 74101–0979
USA
Orders: Tel. +1 918 584-2555
Fax +1 918 560-2652
E-mail *bookstore@aapg.org*

Australia
Australian Mineral Foundation Bookshop
63 Conyngham Street
Glenside
South Australia 5065
Australia
Orders: Tel. +61 88 379-0444
Fax +61 88 379-4634
E-mail *bookshop@amf.com.au*

India
Affiliated East–West Press PVT Ltd
G-1/16 Ansari Road, Daryaganj,
New Delhi 110 002
India
Orders: Tel. +91 11 327-9113
Fax +91 11 326-0538
E-mail *affiliat@nda.vsnl.net.in*

Japan
Kanda Book Trading Co.
Cityhouse Tama 204
Tsurumaki 1-3-10
Tama-shi
Tokyo 206–0034
Japan
Orders: Tel. +81 (0)423 57-7650
Fax +81 (0)423 57-7651

Contents

Preface

The essays in this volume have developed from the proceedings of Section 27 of the International Geological Congress, held at Rio de Janeiro in August 2000. At that meeting – with a view to the arrival of the end of the second millennium – a symposium was held on 'Major Contributions to Geology in the Twentieth Century', organized by Dr Silvia Figueirôa, Professor Hugh Torrens, and myself, in our capacity as Members of the IUGS's International Commission on the History of Geological Sciences (INHIGEO), which was responsible for organizing the symposium.

Established in 1967, INHIGEO has about 170 Members representing 37 countries. Its role is to promote studies on the history of geological sciences and stimulate and coordinate the activities of national and regional organizations having the same purpose. It seeks to bring together, or facilitate communication between, persons working on the history of the geosciences worldwide. To this end, it holds annual conferences in different countries, and its *Proceedings* appear in various forms, according to the publication opportunities that may be available.

It was, then, with pleasure that INHIGEO received an invitation from The Geological Society to offer its papers from the Rio meeting as one of the Society's Special Publications. Evidently, the time was ripe for a retrospective look at some of the major 20th-century contributions to geology. The present volume follows three other recent Special Publications dealing with historical matters: Blundell & Scott (1998), Craig & Hull (1999), and Lewis & Knell (2001).

The Rio symposium had eight invited papers, and, by invitation, the number has been increased to fourteen, thereby adding to the international character of the present publication as well as the number of papers.

I am most grateful to all those who have contributed to the present collection, to the referees, and to Martyn Stoker for overseeing the volume.

David Oldroyd

References

Blundell, D. J. & Scott, A. C. (eds). 1998. *Lyell: The Past is the Key to the Present.* Geological Society, London, Special Publications, **143**.

Craig, G. Y. & Hull, J. H. (eds). 1999. *James Hutton – Present and Future.* Geological Society, London, Special Publications, **150**.

Lewis, C. L. E. & Knell, S. (eds). 2001. *The Age of the Earth: 4004 BC–AD 2002*. Geological Society, London, Special Publications, **190**.

Introduction: writing about twentieth century geology

DAVID OLDROYD

School of Science and Technology Studies, The University of New South Wales, Sydney, New South Wales 2052, Australia

(e-mail: D.Oldroyd@unsw.edu.au)

In a classic paper by the late Yale historian of science, Derek De Solla Price (1965), based mainly on the study of citations in a single scientific research field, it was shown how citations in a developing research area have a strong 'immediacy effect'.[1] Citation was found to be at a maximum for papers about two-and-a-half years old, and the 'major work of a paper . . . [is] finished after 10 years', as judged by citations. There were, however, some 'classic' papers that continue to be cited over long periods of time, and review papers specifically discussing the earlier literature. There appears to be a need for such review papers after the publication of about thirty to forty research papers in a field. And the knowledge is synthesized in book form from time to time.

De Solla Price saw citations as the means whereby activities at the research front are linked to what has gone before. He wrote:

> [E]ach group of new papers is 'knitted' to a small select part of the existing scientific literature but connected rather weakly and randomly to a much greater part. Since only a small part of the earlier literature is knitted together by the new year's crop of papers, we may look upon this small part as a sort of growing tip or epidermal layer, an active research front.

He continued:

> The total research front has never . . . been a single row of knitting. It is, instead, divided by dropped stitches into quite small segments and strips . . . most of these strips correspond[ing] to the work of, at most, a few hundred men [*sic*] at any one time.

So we may imagine the research front of science being a multitude of partly interconnected fields, each growing like the shoot or branch of a plant. The research progress occurs at the 'tip' of each 'shoot', and its lower part consists largely of 'dead wood' – though not wholly dead as occasional reference back to classical papers continues. Obviously, the 'shoots' are loosely interconnected, as references may sometimes be from one research field to another.

I represent some of De Solla Price's findings diagrammatically in Fig. 1; and in this diagram I have also indicated what may be the range of interest of historians of science. It will be seen that while the scientists' interest in the earlier literature declines quite rapidly with time the historians' interest is focused on the earlier work and falls off towards the present.

It is an interesting question whether the study of the history of science generally, or geology in particular, is part of science. Some think it is, and in some cases they are obviously right. For example, old data are of importance in earthquake prediction or studies of geomagnetism. Field mappers may use old field-slips to help locate outcrops. Mining records are important to economic geologists. Palaeontologists need to know the early literature to avoid problems of synonymy. And so on.

On the other hand, one could hardly claim that study of, say, the work of Arthur Holmes is advancing any modern scientific research front. Historians of science usually have other motivations than the direct advancement of science. They are interested in the past 'for its own sake', the history of ideas, correct attributions of credit, understanding the philosophy and sociology of science, 'ancestor worship', and so on and so forth. Such historical work can be called

[1] In fact, the field selected by De Solla Price turned out to be an illusory one – the study of 'N-rays'. But the practitioners of the field were not aware at the time that they were investigating a spurious phenomenon. The field selected by Price for his analysis was well suited to his purpose as it had a clearly defined beginning; and its literature 'behaved' like that of other research programmes. That it had an ignominious end was not relevant to Price's findings. It is true, however, that some fields such as palaeontology make much greater use of early literature than do others such as geochemistry. Palaeontologists and stratigraphers have to observe the principle of priority of nomenclature and so are always involved with the early literature of their fields.

From: OLDROYD, D. R. (ed.) 2002. *The Earth Inside and Out: Some Major Contributions to Geology in the Twentieth Century*. Geological Society, London, Special Publications, **192,** 1–16. 0305-8719/02/$15.00

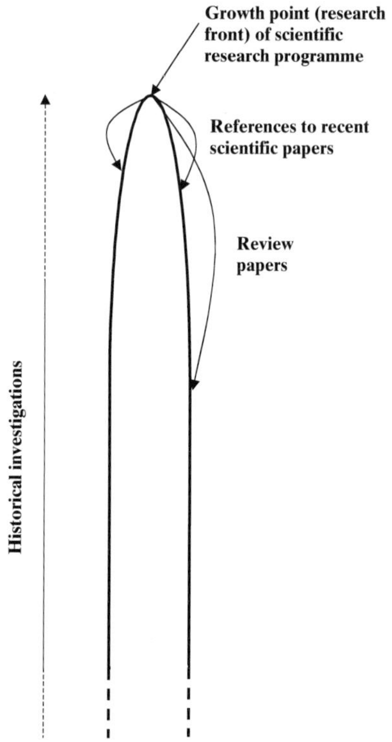

Fig. 1. Representation of the growth of a scientific sub-field, specialty, or research programme, based on the scientometric study of D. J. De Solla Price (1965), representing also the respective temporal interests of scientists and historians of science.

'metascientific'. It is different from what motivates scientists, as working scientists, to study the earlier stages of their fields of inquiry – to further the technical progress of science.

If we regard the study of the history of science as a 'metascientific' activity, then it too has some of the characteristics of a scientific research programme, as described by De Solla Price. But there are differences. The 'knitting' of, say, the history of geology literature into past work, via citations, tends to be more diffuse than is the case for scientific research programmes – though in some areas of the history of science (e.g. the study of Darwin or Lyell) there is a discernible 'research programme' with a developing research front not unlike that of a programme in science. In addition, if they are interested in recent science, historians of science have to scrutinize a target that does not remain fixed, as do the laws of the physical world, but expands indefinitely through time. However, most historians of science do not attend much to the very recent past. Such metascientific attention is the domain of the reviewer or the science journalist.

Studies of the history of geology were almost non-existent before the nineteenth century. Early contributions were 'part of' science (e.g. d'Archiac 1847–1860). Even Lyell's history (Lyell 1830–1833, **1**, pp. 5–74) served, for him, the polemical purpose of garnering support for his geo-philosophy. When studies of history of geology got going in a serious and professional way after the Second World War, most attention was given to the geoscience of the seventeenth, eighteenth, and nineteenth centuries (e.g. Gillispie 1956; Davies 1969; Ospovat 1971; Rudwick 1972; Porter 1977; Greene 1982). Such writings were different in character from the earlier efforts of scientist-historians (e.g. Geikie 1897; Zittel 1901; Woodward 1908). They were not necessarily concerned chiefly with the 'internal' history of science, and offered 'critical' historiography, attending in some cases to the social context of geology.

It was, of course, natural that historians should attend to earlier matters first. Remote events could be viewed with 'perspective' and without treading on the toes of people still alive. The foundations had to be established first, rather than the recent superstructure. Moreover, so far as the twentieth century is concerned, it is only just completed, so we can hardly expect to see much in the way of general synthetic overviews of twentieth century geology at the present juncture. Nevertheless, much more geology has been done in the twentieth century than in the whole of previous human history, and the task of trying to form an overview of it cannot be delayed long. So while the task of studying twentieth-century geology cannot be completed here and now, it can at least be started – or contributions made towards future syntheses.

If we look for generalizations, we immediately remark the development of specialization, with the division of science into research programmes, such as those perceived by De Solla Price. Such specialization, accompanied by a growing divide between the humanities and the sciences, has long been deplored, at least from the 1950s, when C. P. Snow's essay on the 'two cultures' (Snow 1964) caused heads to shake in disapproval, and remedies for the supposed problem were sought – including the study of the history of science by students of the humanities. The philosopher Nicholas Maxwell (1980) deplored the supposed departure from enlightenment arising from specialization.

However, in one of the best books that I know on the *sociology* of science, the geologist and oceanographer Henry William Menard (1971) argued that the pressure towards specialization is irresistible. Influenced by De Solla Price (1961, 1963), he likened the development of science to that of a bean sprout, which eventually, however, inevitably loses growth and withers. The growth of science is like that of water lilies on a pool of finite size, following the pattern of the S-shaped 'logistic curve'. But this applies to specialisms or research programmes rather than science as a whole, which keeps 'alive' by constant divisions into new specialisms. Why does this specialization occur?

The 'explosive' nature of the growth of scientific literature is well known, and science itself has ways to try to cope with the problem, through the production of review papers, bibliographies, and text-books (and perhaps ultimately retrospective histories), and the storage of data in computers as well as libraries. How do people keep on top of it all? The answer, for most, is through specialization. There are new 'hot' fields, and old ones with slowing growth that are becoming ossified almost by virtue of their age and size. Menard considers the case of a new field. There may only be a handful of people in it, and a young person can get a handle on its literature relatively easily and advance to a position of influence when young. By contrast, for a person joining an old field it may take years to gain a commanding position, and all the 'positions and perquisites of academic, professional, and economic power are out of his [*sic*] reach for 20 to 40 years' (Menard 1971, p. 18).

Menard estimates that a person entering a really new field might become '*au courant*' with its literature in perhaps two months. For an 'average' field it might take three years. But someone entering a mature field might be faced with a literature of nearly 30,000 items! The newcomer may be near retirement before he or she has a grip on the literature. In any case, positions in an old field are very likely filled, keeping out new aspirants. Or, if the field is declining, vacancies that may occur are not filled by people in that field but by neighbouring predators. The trick, then, is to get into a new field, but not one that is a bad risk because of shaky foundations. Menard recommends that the optimum time to enter a field is at about its third period of doubling. Then the risks are at a minimum and opportunities at their maximum. However, if one has invested a lifetime's work in a research programme or in working according to some paradigm, and if one has, despite the problems of old research fields, made a successful career therein, then one may be exceedingly disinclined to abandon it and try something new.

Leaving aside such questions of career tactics, it can be seen that pressure towards specialization is intense, the concerns of the likes of Maxwell or Snow notwithstanding. By way of illustration, we see the field of ammonite studies in decline in the latter part of the twentieth century; and one of the authors of the papers in the present volume decided to leave it to all intents and purposes, to become an authority on the history of geology, particularly in the early nineteenth century. Such a career response is one way for a person to respond to changing circumstances. The commoner response is to seek to become an administrator, university teacher (as opposed to researcher), or go in for university politics. Becoming an historian seems to me a more attractive proposition – though one may be hard pressed to find the necessary funding!

Be that as it may, we should note that Menard regarded geology as somewhat moribund in the first half of the twentieth century. It had, so to speak, run out of puff: it was, as a whole, becoming a 'mature' or even 'elderly' science. During the nineteenth century (as, for example, was the case in the State Surveys in the US), it had been a rapidly expanding enterprize, with rather few bureaucratic accessories. There was a large and successful research programme, based on primary or reconnaissance surveys. But such work was limited to the Earth's surface rocks. There was little technology to explore within the Earth by geophysical methods, or (obviously) from without by aerial survey or space travel.

Further, much of the Earth was covered by oceans and inaccessible. Conditions within the Earth could not be simulated in the laboratory. In addition, the overarching framework of geological theory was (as it now appears) unsatisfactory in important respects. It embraced vertical movements as the prime type (though Charles Lapworth had demonstrated the importance of lateral movements in the NW Highlands of Scotland; earlier, geologists in Switzerland such as Albert Heim had done likewise with the idea of nappes; and in America James Hall and the brothers Henry and William Rogers had envisaged significant lateral movements). Besides, geological research was seriously impeded by the two world wars, though geologists contributed their services to both (Underwood & Guth 1998; Rose & Nathanail 2000). In Britain, an ill-advised reorganization of science education before the First World War tended to separate geology from biology, physics, and chemistry at the secondary level. The subject was not taught at elementary schools, and at university it was not seen as a relevant study for engineering students. According to Percy Boswell, in a Presidential Address to the Geological Society, 'while our science was suffering these reverses, the Geological Society stood magnificently and gerontically aloof' (Boswell 1941, p. xli)!

Menard distinguished fields of science that are in a steady state or decline, in transition, or in a state of real (perhaps super-exponential) growth. In the last case, the literature may double in as little as five years. Under such circumstances, papers are brief and published rapidly. Often communication by word of mouth or by pre-prints (or now by e-mail) is more important than by journal communication. The literature of 'hot' fields is not burdened with reviews, and citations are rather few in number. The field's practioners do not concern themselves unduly with bureaucratic or stylistic niceties. Bibliographic work is put aside. By contrast, in old fields many practitioners may have been diverted into administrative functions. Publication delays are considerable. The literature has copious bibliographies, and arcane terminological distinctions are devised, as, for example, in Marshall Kay's baroque taxonomy for different kinds of geosynclines (Kay 1963). In severe cases, papers spend more time discussing other papers than the subject matter of the fields. (Such a state of affairs is found hyperdeveloped in Classics, which has rather little new empirical nutriment.)

As is well known, geological sciences as a whole became re-invigorated in the 1960s and '70s through the plate tectonics revolution. This came about through the application of new technical methods (such as the use of computers in geology) and the partial fusion of two previously distinct fields: geology and oceanography. Submersibles and aeroplanes became useful tools in the progress of geology, complementing the hammer, microscope, field survey instruments, etc. One might say, with Darwin: '[h]ere then I [or, in the case now under discussion, geologists as a whole] had at last got a theory by which to work' (Darwin F. 1887, **1**, p. 83). Several authors (e.g. Hallam 1973) have, appropriately I think, seen the revolution as 'Kuhnian' in character (cf. Kuhn 1962), which implies in a way – at least according to the earlier exposition of Kuhn's views – a revolution in 'world-view'. In this case, it entailed a shift from seeing tectonic movements of the Earth's crust as primarily vertical to lateral also. (Of course, the movement of plumes – part of modern tectonic theory – is essentially vertical.)

The transformation in theory associated with the plate tectonics revolution also led to significant changes in geology as a discipline. In many universities, departments were re-organized, involving fusion with, or incorporation of, studies in geophysics, and they were re-named as schools of 'Earth Science', or similar. In Australia, the changes occurred at about the same time as a notable expansion of prospecting and mining, and there was a 'boom' in geology as well as in mining shares. I am not sure whether that boom was linked to the plate tectonics revolution, but certainly geology began to be seen as an intellectually exciting, and (perhaps better) a lucrative field. There was a rush of students into the earth sciences, in parallel with the famous Poseidon Company (nickel) stock-market bubble. This story had an unhappy ending. The nickel market crashed and many geologists fell out of work or graduates failed to find jobs in the field in which they had trained. Thus the linkage of geology with the capitalist system may be remarked, though such links were nothing new in applied geology.

While important parts of geology became inextricably linked with physics, partly as a result of the plate tectonics revolution, it also became entwined in the latter part of the twentieth century with space science and aeronomy, so that we now find congresses in which the participants are partly earth scientists (seismologists, geomagneticians, tectonics specialists, etc.) and partly space scientists and space engineers (IAGA–IASPEI 2001), or even astronomers. The study of the Earth is now enriched by investigations of the Moon and planets. Geomagnetic studies (so important in the plate tectonics revolution) are linked to investigations of the Sun, the ionosphere, etc. Studies of movements of faults and plates are facilitated by the use of new techniques such as GPS, themselves made possible only by the work of artificial satellite engineers. Well before the end of the twentieth century, one of the leading journals for geologists was *Earth and Planetary Science Letters*. On the other hand, it should be emphasized that the effect of plate tectonic theory on the day-to-day activities of many geologists, particularly applied geologists, was often quite small.

In any case, much had gone on before the plate tectonics revolution actually occurred, both in theory and in technological development. Alfred Wegener (1915) and Alexander Du Toit (1937) had long before found much geological evidence for 'drift'. Arthur Holmes (1929) had upheld the idea of convection in the Earth's interior to account for 'drift'. Felix Vening Meinesz (1929 and other publications) had undertaken a series of underwater gravimetric investigations aboard a US submarine.

But mobilist theory was not generally accepted, meeting opposition in both dominant post-war powers: the US and the USSR. The reasons for the tardy acceptance of mobilist doctrine have been analyzed by Robert Muir Wood (1985) and Naomi Oreskes (1999).

Muir Wood suggests that Soviet scientists' opposition to new ideas was due to the conservative nature of society and the scientific community in the USSR, and the fact that Soviet scientists worked on a huge continental mass, had limited contacts with Western scientists, and lacked the oceanographic data available to the Americans. Oreskes argues that American opposition arose from several factors. First, American geology in the first half of the twentieth century had a certain style, exemplified by the grand collaborative effort of the US Coast and Geodetic Survey, begun in the nineteenth century, to determine the form of the geoid. For simpler calculation, this work assumed the Pratt (as opposed to the Airy) model for isostasy. A uniform global depth of isostatic compensation was assumed, and it appeared that the crust and mantle were generally in a state of isostatic equilibrium. Lateral movements, insofar as they occurred, were thought to be relatively small-scale, occurring in response to erosion of mountains and deposition of sediments in the oceans. The thinking was in accord with long-standing American ideas about the permanence of oceans and continental cratons, derived particularly from the work of James Dwight Dana. Americans such as Charles Schuchert and Bailey Willis attempted to account for faunal similarities across oceans by postulating various 'isthmian links'.

Second, there was the American delight in T. C. Chamberlin's (1897) 'method of multiple working hypotheses'. This was supposed to guard geologists against the uncritical adherence to grand theoretical systems, but in practice, according to Oreskes, it led to the overzealous collection of 'facts'. For William Bowie, the chief spokesperson on matters to do with isostasy, isostatic adjustment and balance was a 'fact', whereas continental drift was an 'interesting hypothesis'. Also, according to Oreskes, Lyellian uniformitarianism impeded acceptance of 'drift' theory. Schuchert believed that knowledge of present faunal distributions could not be applied to the past if there had been latitudinal changes in the positions of continents. It seemed to him that were this so, the present would no longer be the key to the past.

Such geological arguments may seem implausible, but the fact that they attracted favour can perhaps be explained by the hypothesis that geology was indeed in the doldrums before the plate tectonics revolution. Senior geologists were overly committed to an old paradigm and found it difficult to change their opinions. In the context of the 1960s, with the US as the dominant power in the West, it was unlikely that there could be a scientific revolution in geology unless the North Americans joined the revolutionaries. This they eventually did, with the work of J. Tuzo Wilson and the classic paper of Isacks, Oliver & Sykes (1968), in which it was shown, by seismological evidence, that there was movement along the fault planes postulated by theorists such as Wilson (1965*a*, *b*). But the transition was not easy.

The literature on the history of plate tectonics revolution is substantial, even if that on twentieth century geology as a whole is sparse. Besides the volumes by Hallam, Muir Wood, and Oreskes, one should mention particularly the earlier 'straight' account by Marvin (1973) and the later one by Le Grand (1988), which interprets the revolution in terms of the ideas of philosopher of science Larry Laudan rather than those of Kuhn. Henry Frankel (1978, 1979), by contrast, has seen the revolution through the eyes of the philosopher of science Imre Lakatos (which addresses the idea of competing research programmes, either 'progressive' or 'degenerating') than through those of Kuhn. For the oceanographical aspects, see Menard (1986) and Hsü (1992); and for the seismological aspects, see Oliver (1996). Geomagnetic issues are admirably treated by Glen (1982).

Away from the plate tectonics revolution, there are biographies of a few notable individuals, such as Alfred Wegener (Schwarzbach 1986; Milanovsky 2000), Johannes Walther (Seibold 1992), and Arthur Holmes (Lewis 2000); and in connection with work on the study of the age of the Earth, and radiometric dating more generally, the volume of Dalrymple (1991) holds the field. There are useful collections of classic papers from the first half of the century edited by Mather (1967) and Cloud (1970). A set of essays on the history of sedimentology (Ginsburg 1973) is interesting for an essay by Roger Walker (1973), which proposes that the coming of the idea of turbidity currents (Kuenen & Migliorini 1950) constituted a scientific revolution of Kuhnian dimensions in sedimentology. A volume by Peter Westbroek (1991) takes one in the direction of the 'Gaia hypothesis', discussing, as the title *Life as a Geological Force* suggests, ways in which living organisms are involved in geological processes. It also contains material of an historical nature, such as discussion of Robert Garrels' ideas on the cycling of elements through the oceans, atmosphere, and lithosphere. A related topic – controversial over many years – has been that of eustasy, which takes one into the domain of sequence stratigraphy. A collection of papers edited by Robert Dott (1992) gives much useful detail, and includes an essay by one of the main protagonists in the eustasy debate, Peter Vail. There are various institutional histories (e.g., Eckel 1982; Bachl-Hofmann *et al.* 1999), but not much 'critical history' in this area. A two-volume encyclopedia edited by Gregory Good (1998) contains interesting essays on twentieth century geology.

One of the oldest geological topics has been the problem of the causes of the formation of mountains and ocean basins, and interest in the issue has been sustained through the twentieth century. Few have made a concerted effort to view the wood, as distinct from all the trees in the literature. However, in a collection of papers on geological controversies, mostly on sedimentological topics (Müller *et al.* 1991), the Turkish geologist and historian of geology Celâl Şengör (1991) gives one of his several accounts of his interpretation of the 'taxonomy' of the history of tectonic theories. He proposes a general model for the history of tectonics, there being, he suggests, two different tectonic *Leitbilder* (e.g., Şengör 1982, 1999). He drew the notion of *Leitbilder* from Wegmann (1958).

Şengör's 'Manichean' dichotomy of tectonic theorists proposes that two broad ways of thinking were established as far back as the eighteenth century (in the ideas of Hutton and Werner) and, in a sense, have been ongoing ever since. He further traces the philosophical (but obviously not the geological or tectonic) roots of the eighteenth century thinking back to the atomists and Aristotelians in Antiquity. In the nineteenth century, the two modes of interpretation were, he suggests, manifest in uniformitarian and catastrophist geologies respectively. Şengör (1991, p. 417) lays out his dichotomy as summarized in Table 1.

Table 1. *Classification of tectonic theorists, according to A. M. C. Şengör*

Atomists (e.g. Democritus)	Aristotle
Hutton	Werner
Lyell	Cuvier Élie de Beaumont
Suess	Dana Chamberlin
Wegener Argand	Kober Stille

Followers in the two traditions were, suggests Şengör (1982, 1991):

Wegener–Argand ('mobilism')	**Kober–Stille (episodic, world-wide orogenies)**
du Toit	Haug
Daly	Willis
Holmes	Schuchert
Salomon-Calvi	Bucher
Staub	Haarmann
Griggs	van Bemmelen
Ketin	Hans Cloos
[Wilson]	Kay
	Tatyayev
	Beloussov

Şengör sees the members of the Wegener–Argand school as tending to recognize irregularities in Nature and as being in accord with the falsificationist philosophy of science of Karl Popper – of which he strongly approves. By contrast, he regards the members of the Kober–Stille school as tending to look for and see regularities, both geometrical and temporal, in Nature. These two ways of looking at, or thinking about, the world can be seen in the ancient atomists and in the Artistotelians.

I am not aware that many have adopted Şengör's schema, one obvious reason being that today hardly anyone (or no anglophone) has the necessary knowledge of the early Continental and Russian tectonic literature to be able to evaluate his dichotomy satisfactorily. (Of course, even if one accepts Şengör's dichotomy of tectonic theorists one need not agree with his parallel division along methodological and metaphysical approaches or attitudes; and some may doubt that Lyell and Wegener should be situated in the same geological tradition.) Be this as it may, it is evident that Şengör offers a view of the history of twentieth century tectonics quite different from the 'before and after the plate tectonics revolution' account of most English language texts. It proposes a fresh pattern, to make sense of the 'bloomin-buzzin-confusion' of the tectonics literature. It is probably not a pattern that professional historians of ideas would find attractive, but it is undoubtedly an interesting schema; and to my knowledge no other author has tried to identify the common factors in the tectonic theories that have been proposed over the years. Şengör sees conceptual continuity, and Popperian piecemeal change, in the history of tectonics. By contrast, the Kuhnian 'anglophone' theorists such as Hallam have seen conceptual discontinuities.

It should be noted that Şengör's modern theoretical work is typically grounded in all the early literature relevant to his given theme. The same was true of the French geologist and historian of geology, François Ellenberger (1915–2000), but such levels of scholarship are becoming rarer. A recent study by Şengör (1998 for 1996) traces the lengthy history of the concept of the Tethys Ocean (a topic he was worrying about in the middle of the night when he was about ten!).

If tectonics is a major theme in, or branch of, geology, so too is petrology, but to date little has been written on the history of twentieth century petrology, experimental or otherwise. Davis Young (1998) has written a biography of the petrologist Norman Bowen, and Young's paper in the present collection is in a sense a digest of that book. Sergei Tomkeieff's (1983) posthumous *Dictionary of Petrology* contains valuable terminological information, with copious references to the early literature, and an older volume by Loewinson-Lessing (1954) is still useful. Yoder (1993) has published a set of 'annals' of petrology, which provides a chronological framework for a synthetic study of twentieth century igneous and metamorphic petrology. Such a volume will probably first appear from Davis Young's hand.

While the plate tectonics revolution stands out in most people's minds when thinking about the history of twentieth century geology, the re-emergence of 'catastrophism' has also been a noteworth phenomenon. It has chiefly taken the form of the theory – put forward with increasing confidence in the last quarter of the twentieth century – that impacts from extra-terrestrial bodies (bolides) have had a substantial influence on the Earth's geological history, especially in the realms of stratigraphy, palaeoclimatology, and evolutionary palaeontology (see e.g., Albritton 1989; Huggett 1989; Clube & Napier 1990). It has been an uphill task for 'bolide theorists' in that the very notion of extra-terrestrial contacts and attendant catastrophes smacks of nineteenth century 'catastrophism', or even earlier 'theories of the Earth' such as those of Buffon or Whiston. It runs counter to what geologists have long been taught: uniformitarianism and the virtue of the methodological principle that 'the present is the key to the past'. So 'neo-catastrophism' has perhaps had an even more complex history than that to do with the plate tectonics revolution in that there has been no swift and successful 'coup' or scientific revolution, but a long-drawn-out series of battles. Its proponents have had to produce and justify the empirical evidence, and also show that their theory is metaphysically or methodologically sound.

The history of the shift of opinion on the question of neo-catastrophism has been complex in that it has involved different fields in geology (stratigraphy, palaeontology, geochemistry, planetary geology, mineralogy, etc.) with, broadly speaking, a debate between geologists chiefly involved with the life sciences and those associated more with the physical sciences. William Glen (1994) has edited an interesting collection, the papers of which examined the dynamics of the debate while still in progress – before the battle was over and one could see who had 'won'. Since the publication of that book, the conflict seems to have shifted in favour of the 'catastrophists', and recently, a neo-catastrophist, Charles Frankel (1999), has argued that the major subdivisions of the Cenozoic can all be matched with impacts, the 'smoking gun' for the K–T boundary being found at the Chicxulub Crater, by the edge of the Yucatán Peninsula, Mexico (as others had earlier suggested). The arguments of some stratigraphers and palaeontologists that the great change of flora and fauna at the end of the Cretaceous, including the demise of ammonites and dinosaurs, does not coincide in time with the layer of iridium-enriched sediment, thought by the bolide theorists to have been caused by some catastrophic impact, seems to have less appeal – at least to the public imagination – than the notion of an apocalyptic termination of the Cretaceous.

It is interesting that the nineteenth century (Cuvierian) catastrophists were looking to some such event to *explain* the discontinuities in the stratigraphic record; and it was discontinuities in the fossil record that made the establishment of stratigraphy by William Smith, Alcide d'Orbigny, Albert Oppel, and the like, possible. It is, therefore, a little ironic that, in the twentieth century, it has been chiefly biostratigraphers who have opposed the idea of extra-terrestrial impacts being responsible for fundamental features of the stratigraphic column. Be this as it may, the controversy is by no means over at the beginning of the twenty-first century. For example, one of the contributors to the present collection has recently co-authored a paper that argues with considerable cogency that the case for the Chicxulub event being responsible for the demise of the dinosaurs and other extinction events at about the end of the Cretaceous is anything but conclusive (Sarjeant & Currie 2001). It is, for example, not a little startling to read of the discovery of seemingly unreworked dinosaur egg remains (ornithoid theropod types) above the famous iridium horizon (Bajpai & Prasad 2000). It is not claimed that these fossils are Palaeocene, but it is suggested that the iridium layer does not mark the top of the Cretaceous (at least in India). It may well be some time, therefore, before Glen will be able to write a book recounting the closure of this controversy. The controversy may, in fact, eventually be resolved by some sort of compromise. Sarjeant and Currie certainly do not contest the occurrence of the Chicxulub impact event.

From what has been said above, it will be evident that any attempt to provide a synthetic overview of the history of twentieth century geology, as Zittel provided a summation of the geological endeavours of the nineteenth century, is at present premature. The story is infinitely more complex than that for the nineteenth century. The chapter of the twentieth century is only recently closed. Historians have not yet done the necessary analysis, which should precede the synthesis. A recent publication by Edward Young & Margaret Carruthers (2001) is interesting, however, in that it provides a kind of 'annals' or preliminary chronology of twentieth century geology – a 'year-by-year account of important advances since 1900'. The authors mention a deep 'crisis of identity' among those who study the Earth and the rocky bodies of the solar system. Even departmental names are 'doubtful'. The authors suggest that: '[i]n some quarters . . . the activities of scientists studying the Earth can no longer be described as belonging to a single discipline, and . . . just as it is rare to find the life sciences under a single roof in most universities today, so too will go the earth sciences'.

It is too soon to say whether the field of geology or earth sciences will eventually disappear as such, but it is true that it has been troubled, after the rush of adrenalin in the 1970s, by declining student interest, in some parts of the world at least. For example, in New South Wales, the decline in secondary-student enrolments in geology was so great that it appeared at one stage that the subject would vanish from the Higher School Certificate curriculum. The response was, in a sense, to 'disguise' geology in the clothing of 'environmental science'. This change was implemented in the late 1990s, and it is too soon at present to know whether it will prove effective in the long term from the point of view of those interested in the well-being of geology or the earth sciences, but I understand that enrolments have picked up. Clearly, students have been looking for a more 'holistic' approach to geoscience, and it is interesting therefore that in their 'annals' Young & Carruthers (2001) include a good deal of material on environmental issues and space science. For example, the publication of Rachel Carson's *Silent Spring* (1962) is seen as a milestone – along with Harry Hess's 'History of ocean basins' (Hess 1962). The authors' 'annals' of twentieth century earth science thus refer to issues traditionally categorized under the heads of geographic exploration (including satellite imaging), meteorology, environmental science, 'conservation' (such as the Rio summit of 1992), aeronomy, space science, etc.

Young & Carruthers' (2001) headings for the major branches of modern earth science are therefore interesting. They offer:

Understanding Earth's materials
Earth's deep interior
Geological time
Chemistry of Earth's near surface
Climate and global warming
Life on Earth
Plate tectonics
Beyond plate tectonics
Hazard assessment
Remote sensing
Planetary geology

These heads may strike the reader as somewhat whimsical, failing to cover the field adequately, or cutting the cake of geoscience inappropriately. They are, nonetheless, suggestive, and show the way the wind has begun to blow at the beginning of the twenty-first century. A register at the beginning of the twentieth century would surely have included stratigraphy or palaeontology as separate items. In the middle of the century, we would expected to see petrology, structural geology, and sedimentology in such a list. Now at the turn of the new century we remark the interest in the Earth, both 'inside and out'. To that extent, at least, the present collection of essays has common cause with the overview of twentieth century earth science sketched by Young & Curruthers. So far as I am concerned, it's not clear how geology could or would be geology if it were bereft of biostratigraphy. But perhaps that is to be the 'shape of things to come'.

When planning the Rio symposium we decided not to devote excessive attention to the history of plate tectonics. Despite the fact that the emergence of that theory has been the most important theoretical development in twentieth century geoscience (or at least it caused the greatest excitement in the earth science community), it has already been the object of substantial historical investigations, some of which are mentioned above. Nevertheless, the topic was unavoidable. So in the present collection we find that the papers by Lewis, Le Grand, Khain & Ryabukhin, and Barton deal with the question to a greater or lesser extent; and it appears in some of the other papers too.

Khain & Ryabukhin's paper should be of special interest. There has, I suggest, been a perception in 'the West', reinforced by Muir Wood's (1985) stimulating book, that Russian geology was reluctant to embrace plate tectonics. It is true that Russian geology adopted plate tectonics somewhat later than in the West, and one of her most influential geologists, Vladimir Beloussov, was antagonistic towards the theory, at least initially. However, Beloussov's opposition was not just 'perverse' or 'political'. His views were based on ideas developed by Nikolay Shatsky, based on seismic evidence for deep faults, apparently crossing the crust and upper mantle. In fact, as Khain and Ryabukhin reveal, there was intense discussion at Moscow State University, with, in effect, two contradictory theories being taught in the same institution. Khain was one of the main protagonists and actively promoted plate tectonic theory.

The tectonics theorist Khain is, of course, writing about the events of the 1970s from the perspective of the 'winning' side; and it might be said that, having lived longest, he now has the opportunity to write the history the way it appeared to him. Be this as may, there was evidently no monolithic anti-mobilist theory in Russia in the 1970s, and by the end of that decade immense efforts were being made to apply plate tectonics within the Russian domains, as is evident, for example, from the arduous work undertaken in the Urals (Zonenshain *et al.* 1984). Incidentally, it may be mentioned that geological theory at Moscow State University remains 'un-monolithic' to this day, as I understand, with some classes teaching expanding (or pulsating)-Earth theory, while the majority offer standard plate tectonic doctrine. The Russian paper also refers to some theoretical notions not well known in the West.

Some of the contributors to the present volume are scientists interested in the history of geology; some are historians of geology. Homer **Le Grand** is one of the latter. His paper utilizes oral history, providing some reminiscences about the extension of plate tectonic theory to 'terrane theory'. It is well that such reminiscences be captured for posterity. Le Grand, of course, has been an observer of events, rather than a participant.

The same may be said of the historian Cathy **Barton**. Her paper is partly based on interviews with Marie Tharp, well known for her contributions to the mapping of the ocean floors – a necessary empirical first step towards the plate tectonics revolution. There is currently considerable interest in the part played by women in science, and it is sometimes said that women have had a hard time in 'getting on' in geology. Barton's paper shows that Tharp was not much hindered because of her gender; but she had the advantage of working at a time when there were vacancies in civilian science due to the Second World War; and she also had Bruce Heezen's patronage. Interestingly, though Heezen and Tharp's work (or that like it) was, I think, necessary for the emergence of plate tectonics, it was not sufficient, for they adopted the now-rejected expanding-Earth theory.[2] Barton suggests that they were the geological equivalents of Tycho Brahe in the Copernican Revolution. They provided essential empirical information, but for them it led to what is (according to the present consensus) an erroneous theory.

Cherry **Lewis**, known among geologists for her work on fission-track estimates of the 'lost overburden' of some of the older rocks in Britain, has for some time been studying the work of Arthur Holmes, on whom she has published a biography (Lewis 2000). Lewis's paper raises the problem of the age of the Earth, which was for many years a major issue in geology and beyond, but was eventually solved in principle by Holmes, regarded by some as the outstanding geologist of the twentieth century. He was also one of those who accepted mobilist doctrines well before the plate tectonics revolution proper, and he advocated (but did not originate) the idea of a convectional mechanism for continental movement that still stands in essence.

Readers picking up this book will immediately notice its famous cover illustration, and the title. Two of the papers (those of Good and Marvin) deal respectively with the Earth's interior and with entities external to the Earth. Thus we are taken into the realms of geophysics and astronomy – where geology overlaps with physics and with planetary science (or even cosmology).

Ursula **Marvin**, geologist, meteoritics expert,

---

[2] But Ursula Marvin (pers. comm., 25 Sept. 2001) informs me that she heard Heezen say at a meeting in 1966 that some calculations he had made suggested that the Earth expansion required just to account for the opening of the Atlantic was unreasonably large. Heezen is generally regarded as an 'expansionist' but the matter perhaps deserves closer historical scrutiny.

and authority on the history of meteoritics, takes the reader into the world of outer space and what it can tell us about the geology of our Earth. As discussed above, one of the main trends in twentieth century earth science has been the extent to which it has been integrated with planetary science (and aeronomy). Marvin's paper is a perhaps unlikely, but also a good, place to start this collection of essays. Meteorites provide some of the most useful empirical evidence we have about ways in which the Earth may have formed. Also, the study of craters on the Moon and elsewhere has thrown light on terrestrial impacts, and their possible role in the history of life on Earth, which, as mentioned above, has been hotly debated over the last twenty years or so by 'astrogeologists' and traditional palaeontologists and stratigraphers (see e.g. the paper by Torrens in this volume).

Marvin takes us through the story of the efforts to find meteorites and discover whence they came, particularly those that seem to have come from the Moon and from Mars. I was particularly struck by two points she made in her Rio paper. She remarked that the 'vision' of our Earth, seen from space and depicted on the cover of this book, had a substantial impact on the way we now think about the Earth; and the 'vision' did wonders for the 'holistic' environmentalist movement. This is the planet where we live, which we can now 'see' as a whole from the outside; and this is where we shall likely perish as a species if we do not act sensibly as its stewards. Marvin also observed that the summary geological time-chart, which delegates received in their conference-bags at Rio (REPSOL: YPF 2000), listed the lunar names for the epochs of the Hadean Period (Cryptic, Basin Groups 1–9, Nectarian, and Early Imbrian) obtained by mapping of the Moon, which preserves a stratigraphic record that is keyed to dated samples reaching back to that time. Direct stratigraphic evidence on Earth for those remote times has long since been lost, so insofar as we have a 'stratigraphy' for the very early Earth it is inferred from entities *outside* our planet.

Incidentally, though the present collection does not have a paper specifically devoted to the question of bolide impacts and their implications for Earth history, Marvin addresses some aspects of the question, even though she does not discuss it in detail. (It was treated by her in a previous Special Publication: Marvin 1999).

The historian of geology, Gregory **Good**, takes us *inside* the Earth. He is interested in the changes that have taken place through the twentieth century in studies of the Earth's magnetic properties. The early work developed from the many observations of its magnetic field that go back to the beginnings of geomagnetic investigation. By the first half of the twentieth century, the subject had progressed well beyond Baconian (or Humboldtian) data-collecting, and attempts were made to develop theories about the causes of the existence of, and changes in, the Earth's magnetic field. This work, Good argues, lay within the domain of 'terrestrial magnetism'. It was related to problems in navigation, for example, rather than geological theories *per se*. But as time passed, more information became available about the Earth's interior and it became possible to produce theories about the origin of the Earth's field and its changes. After palaeomagnetic studies' substantial contributions to the plate tectonics revolution, much attention is now bestowed on palaeomagnetics, as geologists seek evidence about former positions of the poles in reconstructing the geological histories of different parts of the Earth (a matter also intimately related to terrane theory). Good argues that the very nature of geomagnetics has changed; and he holds that the view of earlier work has become distorted because it is seen through the lens of the later.

The paper by Richard **Howarth**, is authored by someone who assisted in the development of the use of the computer in geological studies. He has also made much use of statistical analyses for the purpose of geological research. It might not be obvious that there is a coherent field of 'mathematical geology', but in this paper and in his other historical publications Howarth has demonstrated the coherence of the field as a branch of geology appropriate to historical investigation (e.g. Howarth 1999). He has also been much interested in the use of figures such as 'rose diagrams' or stereograms in geological analysis, and for understanding geological ideas and making them comprehensible to others (cf. Rudwick 1976). Such representations did not begin *ex nihilo* in the twentieth century, though they are characteristic of the work of that period.

As mentioned, there has long been a dearth of studies in the history of petrology, perhaps the most basic of the geosciences, yet neglected by historians of science, especially for the twentieth century. For this reason I am gratified that the present collection contains four petrological papers. The field is, of course, enormous, and we cannot expect an author to cover the whole in a paper such as might fit into the present collection. In the contribution of Eugen and (his wife) Ilse **Seibold** we are provided with a straightforward survey of twentieth century sedimentological writings, extending into sedimentary petrology. It identifies the main themes in the field, and provides an entree to its vast literature. It will be particularly useful in that, written by German authors, it is not focused on English-language writings (this may well become appropriate for the twenty-first century, but it is not so for the twentieth century), but discusses English, French, German, and Russian publications. I am particularly grateful to Professor Eugen Seibold for completing this work in a year when he had to undergo an eye operation. He has been President of the International Union of Geological Sciences and participated in voyages undertaken for the purpose of ocean-floor surveys. Ilse Seibold is a foraminifera specialist and author of a book on Johannes Walther (1992).

As to igneous petrology, one of the most important topics for the twentieth century has been the problem of understanding the changes that occur during magma crystallization. Amongst those who worked on this topic, one of the most important figures was Norman Bowen. He came from the research institution where arguably the most important work in experimental petrology was done, at least in the first half of the twentieth century: the Geophysical Laboratory of the Carnegie Institution, Washington. The petrologist and historian of petrology Davis **Young** argues that this particular institution provided the ideal framework for Bowen's work in igneous petrology, most of it experimentally based, utilizing the apparatus for the study of rocks and rock melts at high temperatures and pressures available at the geophysical laboratory. The issue of what happens when melts cool and differentiate is fundamental to igneous petrology. For Bowen, it was essentially a laboratory problem, but his work led to fundamental progress in the understanding of rocks as they are present in the field, as discussed, for example, in the classic work of Wager & Brown (1968).

Eventually Bowen's work (in conjunction with Orville Frank Tuttle) led to a resolution of one of the great debates of twentieth century geology: the battle between the 'migmatists' and the 'magmatists' regarding the origin of granite, Tuttle & Bowen (1958) declaring in favour of the latter (see Read 1957). Consideration of this topic leads us into the intricacies of metamorphic petrology, discussed by Jacques **Touret** and Timo **Nijland**. The authors have undertaken the massive task of 'picking the eyes' out of twentieth century metamorphic petrology, to which field they have themselves contributed, having worked together in Scandinavia. The history of metamorphic geology still requires detailed analysis, but the Touret and Nijland paper should serve as a starting-point for all future studies. Like several other essays in the present collection, the authors have found it necessary to trace the roots of twentieth century debates in earlier ways of thinking – in this case even back to the eighteenth century. They also travel as far afield as the work of Miyashiro in Japan. Regretfully, this is the only paper in the collection that attends to ideas developed in the Far East.

Studies of metamorphic petrology are naturally associated with Scandinavian geology, for metamorphic rocks are particularly well exposed in the Baltic Shield, where they have led to new ideas about their production. In the essay by the historian of geosciences, Bernhard **Fritscher**, we look more closely at one of the Scandinavians mentioned in the Touret and Nijland paper: Victor Goldschmidt. He was a petrologist but is chiefly associated with geochemistry, being one of that discipline's founders, especially through his *Geochemistry* (Goldschmidt 1958). He also listed the abundances of elements in the solar system, on the basis of analyses of meteorites. So he too was interested in the Earth 'inside and out'. Here, however, Fritscher focuses on the application of the phase-rule to petrology, and debates about the development of petrology based on fundamental chemical principles – as opposed to the approach via fieldwork and the study of thin-sections favoured by British petrologists like Alfred Harker. Fritscher sees important differences between British and Continental workers and offers some socio-political explanation for the differences.

One of the points made *en passant* by Touret and Nijland is that they find metamorphic petrology in decline (at least in The Netherlands, admittedly a country lacking metamorphic rocks), with posts in the field disappearing, whereas it was formerly a leading area of research. This decline – matched in their country in some other fields such as mineralogy – may reflect changes in public concerns, such as a heightened awareness of environmental problems or dislike of fields regarded as being associated with mining and mineral exploration. It meshes with the broad shifts in emphasis in the second half of the twentieth century that were discussed above, but, I suggest, the current contraction of the field in some parts of the world should not be taken to imply that metamorphic petrology is shrinking for want of interesting and important problems. Indeed, new instruments used in well-funded institutions such as Edinburgh University are being used for exciting work on space material, oil-field metamorphism studies, and so on.

Be that as may, metamorphic petrology is not the only branch of geology whose fortunes have changed in the twentieth century. Hugh **Torrens** is (or formerly was) an ammonite specialist and stratigrapher, but now chiefly studies the history of geology in relation to technology. He too has seen his field contract during the span of his career, so that whereas biostratigraphy was once king it is now being 'squeezed' by specialties such as magnetostratigraphy or sequence stratigraphy. When presenting his Rio paper, Torrens sought (at my request!) to do the impossible, namely discuss stratigraphy as a whole during the twentieth century. In his revised version, he has focused on the question of precision and the extent to which measurements of time by various stratigraphic criteria are more or less precise, and well founded. He takes his starting-point in the nineteenth century, considering the work of the American Henry Shaler Williams and the English stratigrapher, Sydney Savory Buckman. They showed how fossils allowed the correlation of different rock units in different localities and how different thicknesses and types of rock can represent equal amounts of time. Particular lithologies may cross time-lines. For Torrens, the notion of correlation implies determination, or knowledge, of time. And the question he addresses in his paper is what measures are available for the determination of time, so that stratigraphy can make increasingly precise determination of time-intervals.

Torrens argues that biostratigraphy, where changes of fossil types are able to be calibrated by radiometric determinations (cf. Holmes's work), still provides the best way for stratigraphers to proceed, and in consequence he deplores the loss of 'ammonite lore', for example, that has begun to afflict stratigraphy. Torrens also has, with others, doubts about the efficacy of sequence stratigraphy, fearing that it may be prone to arguing in circles; for the 'packets' of sediments identified by seismic investigation are not always dated (calibrated) by palaeontological methods. However, he does not actually deal with sequence stratigraphy, its extensive successful use in (say) the oil industry notwithstanding. Rather, he discusses the question of the chronological precision of the events claimed to be associated with the impacts of meteorites.

In considering potential papers for the present collection, it was evidently impossible to have one that covered the whole of palaeontology, which would have been as unrealistic as a paper that might cover stratigraphy as a whole. So for palaeontology I invited William **Sarjeant** to write a paper on the history of one of his numerous fields of interest (e.g., palynology, ichnology, bibliography, writing novels, folk singing, . . .): namely palynology. He responded with enthusiasm but in so doing he found it necessary and appropriate to trace the historical roots of the field, so that, with its worldwide coverage, and considering the several branches of palynology, his paper starts before and does not reach the end of the twentieth century. Yet, as Sarjeant remarks, palynology has grown from 'a scientific backwater into a mainstream of research'. For example, in my own recent investigations of the history of geology in the English Lake District, I have been forcibly struck by the significance of acritarchs for making progress in the understanding of the stratigraphy of rocks such as the Skiddaw Slates, which have few macrofossils. To a significant extent, it has been acritarchs that have promoted major revisions in structural understanding, helping, for example, to reveal the presence of olistostrome structures in the Lakes. Palynology is also making major contributions to palaeoclimatology and Quaternary geology, not to mention the oil industry.

Palynologists (and palaeontologists more generally) are much concerned to inter-relate their knowledge of fossils by having knowledge of the literature – which may sometimes be published in disconcertingly obscure places. Sarjeant's paper does not pretend to offer a guide to the literature of palynology as a whole, even to his approximate closing date of the 1970s. He says he is writing a 'short history'. Nevertheless, his bibliography is massive, and should be of considerable value to palynologists, or to 'outsiders' who may become involved in the field from time to time. Sarjeant's paper is partly autobiographical, for he has himself played his part in twentieth century palynology. It is pleasing to have his own account of some of his contributions, and his recollections of encounters with colleagues. Whether the interest in matters bibliographical is a sign of the 'old age' of a discipline, as Menard's arguments might lead one to imagine, I leave others to figure out. Naturally, palynology has extended its influences considerably, subsequent to Professor Sarjeant's self-imposed cut-off date of 1970.

Microfossils are, of course, never likely to 'run out', but it is not obvious that the same holds true for macrofossils in a small country like Britain, where collectors from schoolchildren to professors have long been active. To what extent should collecting be open to all, and what regulations (if any) should apply to collecting and conservation? This became an acute problem in Britain and some other countries in the late twentieth century. Ideas on the matter – and the appropriate regulations – have varied considerably. The problem is treated historically, largely for Britain but also with reference to America, by the museologist Simon **Knell**. In his paper, we encounter the cultural, social, and political framework within which geology operates. Through his study of the recent history of collecting, Knell examines the issue of geology's changing social context, thereby showing the way that science operates in practice. He is interested in the public perception of geology and the way geology presents itself to the world, as well as its 'internal' workings.

There is no simple answer to the question: to have or not to have unregulated collection? But questions that have no simple answers are always worth asking. Knell concerns himself with fossils, but what he says applies equally to mineral collection and conservation, or even rocks. I think his paper sufficiently reveals the nature of the question, which is part of the much broader problem of the conservation of objects, whether they be buildings, archives, . . . or the environment as a whole. Knell focuses on geological collecting in one country in the late twentieth century. But his paper raises larger issues; and so far as geology is concerned it may prompt questions about policies in countries where problems of collection and conservation are not yet as acute as in Britain. It may be, as Touret and Nijland suggest, that metamorphic petrology is now in 'retreat'. But Knell's kinds of questions will necessarily become more acute in the twenty-first century and beyond. They link the present selection of papers with the trends towards the increasing interest on the part of earth scientists in environmental issues and conservation issues, previously noted.

It may also be mentioned that Knell's paper signals important changes that have occurred in the very nature of science, as a whole, towards the end of the twentieth century. When De Solla Price wrote, his 'growing-shoot' analogy was perhaps more apt than it is today. In the 1960s, the advancing fronts for different geological research programmes could be approximated by the metaphor of more or less discrete growing shoots – extending towards the light chiefly in the favourable environments of university departments, research institutes, or national government-funded geological surveys. But things became substantially different in the second half of the century. Tax-sourced funding declined. Problems came to be addressed, not just in the context of research programmes but in the context of particular technical applications or goals, which are diffused through society. Problems like the extraction of oil from beneath the North Sea could not be solved by expertise within a single discipline. We have, then, what has been called 'transdisciplinarity' (Gibbons *et al.* 1997). For science in this 'mode' (so-called 'Mode 2'), results are communicated, not primarily by publicly accessible journals, but by 'internal' reports and personal contacts. Knowledge may move with the practitioners as they transfer to new problems when old ones are solved. New kinds of sites for the production of knowledge emerge – in consultancies, think-tanks, industrial laboratories, etc. – side by side the traditional ones to be found in universities and research institutes. Funding is garnered from numerous different sources, according to what may be available and the nature of the problems in hand.

Concomitantly, the network of interested parties increases: we may find natural scientists, social scientists, lawyers, business people, engineers – a heterogeneous mix – all involved in developing solutions to problems. Those who are involved may find themselves embroiled in politics and have to be increasingly aware of the social, political, and economic implications of what they are doing. They must take account of the values and interests of groups normally regarded as outside the system of science and technology: solutions to problems have to be socially, politically, and economically acceptable. The fact that this came to be so increasingly in the late twentieth century is illustrated by Knell's paper. The science he discusses does not grow like a free shoot in a hot-house (or ivory tower). It has to interact with all the forces of the society in which it finds itself and negotiate its activities accordingly. It is, in consequence, a rather different *kind* of science from that which De Solla Price analyzed three decades earlier ('Mode 1') – which was based on the study of a scientific field from the first half of the century.

Regretfully, the present collection can only scratch the surface of the history of twentieth century geology. *How* and *why* the changes to science referred to in the preceding paragraph came about are problems too large to be entered into here. But, as said, analysis must precede synthesis. So without claiming to have achieved a synthesis, it is hoped nevertheless that the present collection will prove useful to those who may subsequently tackle the heroic task of furnishing an historical synthesis of twentieth century geology, earth science, planetary science, environmental science, conservation, . . .

I am most grateful to Gordon Craig, Gregory Good, Richard Howarth, Simon Knell, Cherry Lewis, Ursula Marvin, David Miller, Timo Nijland, Martyn Stoker, William Sarjeant, Hugh Torrens, Jacques Touret, and Davis Young for their helpful comments on drafts of this Introduction.

References

ALBRITTON, C. C. 1989. *Catastrophic Episodes in Earth History*, Chapman & Hall, London & New York.

BACHL-HOFMANN, CERNAJSEK, T., HOFMANN, T. & SCHEDL, A. 1999. *Die Geologie Bundesanstalt in Wien: 150 Jahre Geologie im Dienste Österreichs (1849–1999)*, In Kommission bei Böhlau Verlag Ges. m. b. H. & KG, Vienna.

BAJPAI, S. & PRASAD, G. V. R. 2000. Cretaceous age for Ir-rich Deccan intertrappean deposits: palaeontological evidence from Anjar, western India. *Journal of the Geological Society*, **157**, 257–260.

BOSWELL, P. G. H. 1941. Presidential Address. *The Quarterly Journal of the Geological Society*, **97**, xxxvi–lv.

CARSON, R. 1962. *Silent Spring*, Hamish Hamilton, London.

CHAMBERLIN, T. C. 1897. The method of multiple working hypotheses. *Journal of Geology*, **5**, 837–848.

CLOUD, C. (ed.) 1970. *Adventures in Earth History*, W. H. Freeman & Co., San Francisco & Reading.

CLUBE, V. & NAPIER, B. 1990. *The Cosmic Winter*, Blackwell, Oxford & Cambridge (Mass).

DALRYMPLE, G. B., 1991. *The Age of the Earth*, Stanford University Press, Stanford.

D'ARCHIAC, A. 1847–1860. *Histoire des progrés de la géologie*, Société Géologique de France, Paris.

DARWIN, F. (ed.). 1887. *The Life and Letters of Charles Darwin: Including an Autobiographical Sketch*, John Murray, London.

DAVIES, G. H[ERRIES]. 1969. *The Earth in Decay: A History of British Geomorphology 1578–1878*, Macdonald Technical and Scientific, London.

DE SOLLA PRICE, D. J. 1961. *Science Since Babylon*, Yale University Press.

DE SOLLA PRICE, D. J. 1963. *Little Science, Big Science*, Columbia University Press, New York.

DE SOLLA PRICE, D. J. 1965. Networks of scientific papers. *Science*, **149**, 510–515.

DOTT, R. H. (ed.). 1992. *Eustasy: The Historical Ups and Downs of a Major Geological Concept*, The Geological Society of America, Memoir No. **180**, Boulder.

DU TOIT, A. 1937. *Our Wandering Continents: An Hypothesis of Continental Wandering*, Oliver and Boyd, Edinburgh & London.

ECKEL, E. B. 1982. *The Geological Society of America: Life History of a Learned Society*, The Geological Society of America, Memoir No. **155**, Boulder.

FRANKEL, C. 1999. *The End of the Dinosaurs: Chicxulub Crater and Mass Extinctions*, Cambridge University Press, Cambridge, New York, Melbourne & Madrid (1st French edn 1996).

FRANKEL, H. 1978. The non-Kuhnian nature of the recent revolution in the earth-sciences. *In*: ASQUITH, P. L. & HACKING, I. (eds), *Proceedings of the 1978 Meeting of the Philosophy of Science Association*, The Philosophy Association, East Lancing, 197–214.

FRANKEL, H. 1979. The career of continental drift theory: an application of Imre Lakatos' analysis of scientific growth to the rise of drift theory. *Studies in the History and Philosophy of Science*, **10**, 10–66.

GEIKIE, A. 1897. *The Founders of Geology*, Macmillan & Co., New York (2nd edn 1905, reprinted 1962 by Dover Publications, Inc., New York).

GIBBONS, M., LIMOGES, C., NOWOTNY, H., SCHWARTZMAN, S., SCOTT, P. & TROW, M. 1997. *The New Production of Knowledge: The Dynamics of Science and Research in Contemporary Societies*, Sage Publications, London, Thousand Oaks & New Delhi (1st edn, Forskningsråfdämnden, Stockholm, 1994).

GILLISPIE, C. C. 1956. *Genesis and Geology, A Study in the Relations of Scientific Thought, Natural Theology, and Social Opinion in Great Britain, 1790–1850*, Harper Torchbooks, New York (reprint of Harvard Historical Studies, **58**, 1951).

GINSBURG, R. N. (ed.). 1973. *Evolving Concepts in Sedimentology*, The Johns Hopkins University Press, Baltimore & London.

GLEN, W. 1982. *The Road to Jaramillo: Critical Years of the Revolution in Earth Science*, Stanford University Press, Stanford.

GLEN, W. (ed.). 1994. *The Mass-Extinction Debates: How Science Works in a Crisis*, Stanford University Press, Stanford.

GOLDSCHMIDT, V. M. 1958. *Geochemistry*. Clarendon Press, Oxford.

GOOD, G. (ed.). 1998. *Sciences of the Earth: An Encyclopedia of Events, People, and Phenomena*, Garland Publishing, Inc., New York & London, 2 vols.

GREENE, M. T. 1982. *Geology in the Nineteenth Century: Changing Views of a Changing World*, Cornell University Press, Ithaca & London.

HALLAM, A. 1973. *A Revolution in the Earth Sciences: From Continental Drift to Plate Tectonics*, Clarendon Press, Oxford & London.

HESS, H. 1962. History of ocean basins. *In*: ENGLE, E. A. J. (ed.), *Petrologic Studies: A Volume to Honor A. F. Buddington*, The Geological Society of America, Denver, 599–602.

HOLMES, A. 1929. Radioactivity and earth movements. *Transactions of the Geological Society of Glasgow*, **18**, 559–606.

HOWARTH, R. J. 1999. Measurement, portrayal and analysis of orientation data and the origins of early modern structural geology (1670–1967), *Proceedings of the Geologists' Association*, **110**, 273–309.

HSÜ, K. J. 1992. *Challenger at Sea: A Ship that Revolutionized Earth Science*, Princeton University Press, Princeton.

HUGGETT, R. 1989. *Cataclysms and Earth History: The Development of Diluvialism*, Clarendon Press, Oxford.

INTERNATIONAL ASSOCIATION OF GEOMAGNETISM AND AERONOMY (IAGA) AND INTERNATIONAL ASSOCIATION OF SEISMOLOGY AND PHYSICS OF THE EARTH'S INTERIOR (IASPEI) OF THE INTERNATIONAL UNION OF GEODESY AND GEOPHYSICS (IUGG). 2001. *IAGA–IASPEI Joint Scientific Assembly 19–31 August 2001 Hanoi, Vietnam: Abstracts*, Institute of Geophysics, Hanoi.

ISACKS, B., OLIVER, J. & SYKES, L. R. 1968. Seismology and the new global tectonics. *Journal of Geophysical Research*, **73**, 5,855–5,899.

KAY, M. 1963. *North American Geosynclines*, Geological Society of America Memoir No. **48**, New York (1st edn 1951).

KUENEN, P. H. & MIGLIORINI, C. I. 1950. Turbidity currents as a cause of graded bedding. *Journal of Geology*, **58**, 91–127.

KUHN, T. S. 1962. *The Structure of Scientific Revolutions*. The University of Chicago Press, Chicago & London.

LE GRAND, H. E. 1988. *Drifting Continents and Shifting Theories*, Cambridge University Press, Cambridge.

LEWIS, C. 2000. *The Dating Game: One Man's Search for the Age of the Earth*, Cambridge University Press, Cambridge, New York, Oakleigh (Victoria), Madrid & Cape Town.

LOEWINSON-LESSING, F. Y. 1954. *A Historical Survey of Petrology*, translated from the Russian edition of 1936 by S. I. TOMKEIEFF, Oliver & Boyd, Edinburgh & London.

LYELL, C. 1830–1833. *Principles of Geology: An Attempt to Explain the Former Changes of the Earth's Surface, by Reference to Causes now in Operation*, John Murray, London (reprinted in facsimile with an introduction by M. J. S. Rudwick), The University of Chicago Press, Chicago & London, 1990).

MARVIN, U. B. 1973. *Continental Drift: The Evolution of a Concept*, Smithsonian Institution Press, Washington, D. C.

MARVIN, U. B. 1999. Impacts from space: the implications for uniformitarian geology. *In*: CRAIG, G. Y. & HULL, J. H. (eds), *James Hutton – Present and Future*. The Geological Society, London, Special Publications **150**, 89–117.

MATHER, K. F. 1967. *Source Book in Geology 1900–1950*, Harvard University Press, Cambridge (Mass).

MAXWELL, N. 1980. Science, reason, knowledge and wisdom: a critique of specialism. *Inquiry*, **23**, 19–81.

MENARD, H. W. 1971. *Science: Growth and Change*, Harvard University Press, Cambridge (Mass).

MENARD, H. W. 1986. *The Ocean of Truth: A Personal History of Global Tectonics*, Princeton University Press, Princeton.

MILANOVSKY, E. E. 2000. *Alfred Wegener: 1880–1930*, Nauka, Moscow [in Russian].

MÜLLER, D. W., J. A. MCKENZIE & WEISSERT, H. (eds). 1991. *Controversies in Modern Geology: Evolution of Geological Theories in Sedimentology, Earth History and Tectonics*, Academic Press, London, San Diego, New York, Boston, Sydney, Tokyo & Toronto.

MUIR WOOD, R. 1985. *The Dark Side of the Earth*, Allen & Unwin Ltd, London, Winchester (Mass) & Sydney.

OLIVER, J. E. 1996. *Shocks and Rocks: Seismology in the Plate Tectonics Revolution. The Story of Earthquakes and the Great Earth Science Revolution in the 1960's*, American Geophysical Union, Washington, D. C.

ORESKES, N. 1999. *The Rejection of Continental Drift: Theory and Method in American Earth Science*, Oxford University Press, New York & London.

OSPOVAT, A. 1971, *Abraham Gottlob Werner: Short Classification and Description of the Various Rocks Translated with an Introduction and Notes by Alexander M. Ospovat*, Hafner Publishing Company, New York.

PORTER, R. 1977. *The Making of Geology: Earth Science in Britain 1660–1815*, Cambridge University Press, Cambridge, London, New York & Melbourne.

READ, H. H. 1957. *The Granite Controversy: Geological Addresses Illustrating the Evolution of a Disputant*, Interscience Publishers, Inc., New York.

REPSOL: YPF 2000. *Geological Time-scale*. REPSOL, Rio de Janeiro.

ROSE, E. P. F. & NATHANAIL, C. P. (eds). 2000. *Geology and Warfare: Examples of the Influence of Terrain and Geologists on Military Operations*, The Geological Society, London.

RUDWICK, M. J. S. 1972. *The Meaning of Fossils: Episodes in the History of Palaeontology*, Macdonald, London & American Elsevier, New York.

RUDWICK, M. J. S. 1976. The emergence of a visual language for geological science, 1760–1840. *History of Science*, **14**, 149–195.

SARJEANT, W. A. S. & CURRIE, P. J. 2001. The "Great Extinction" that never happened: the demise of the dinosaurs considered. *Canadian Journal of Earth Sciences*, **38**, 239–247.

SCHWARZBACH, M. 1986. *Alfred Wegener The Father of Continental Drift*, Science Tech Publishers, Madison & Springer-Verlag, Berlin, Heidelberg, New York, London, Paris & Tokyo (1st German edn 1980).

SEIBOLD, I. 1992. *Der Weg zur Biogeologie: Johannes Walther (1860–1937)*, Springer-Verlag, Berlin, Heidelberg, London, Paris, Tokyo, Hong Kong, Barcelona & Budapest.

ŞENGÖR, A. M. C. 1982. Classical theories of orogenesis. *In*: MIYASHIRO, A., AKI, K. & SENGÖR, A. M. C. (eds) *Orogeny*, John Wiley & Sons, Chichester, New York, Brisbane, Toronto & Singapore, 1–48.

ŞENGÖR, A. M. C. 1991. Timing of orogenic events: a persistent geological controversy. *In*: MÜLLER, D. W., MCKENZIE, J. A. & WEISSERT, H. (eds) *Controversies in Modern Geology: Evolution of Geological Theories in Sedimentology, Earth History and Tectonics*, Academic Press, London, San Diego, New York, Boston, Sydney, Tokyo & Toronto, 405–473.

ŞENGÖR, A. M. C. 1998 for 1996. Tethys: vor hundert Jahren und heute. *Mitteilungen der Österreichischen Geologischen Gesellschaft*, 1–177.

ŞENGÖR, A. M. C. 1999. Continental interiors and cratons: any relation? *Tectonophysics*, **305**, 1–42.

SNOW, C. P. 1964. *The Two Cultures and A Second Look*, Cambridge University Press, Cambridge.

TOMKEIEFF, S. I. 1983. *Dictionary of Petrology*, WALTON, E. K., RANDALL, A. O., BATTEY, M. H. & TOMKEIEF, O. (eds) John Wiley & Sons, Chichester, New York, Brisbane, Toronto & Singapore.

TUTTLE, O. F. & BOWEN, N. J. 1958. *Origin of Granite in the Light of the Experimental System* $NaAlSi_3O_8$–$KAlSi_3O_8$–SiO_2–H_2O. The Geological Society of America Memoir No. **74**. The Geological Society of America, New York.

UNDERWOOD, J. R. & GUTH, P. L. (eds). 1998. *Military Geology in War and Peace. Reviews in Engineering Geology*, **13**. The Geological Society of America, Boulder, Colorado.

VENING MEINESZ, F. 1929. Gravity expedition of the US Navy. *Nature*, **123**, 473–475.

WAGER, L. R. & BROWN, G. M. 1968. *Layered Igneous Rocks*. Oliver & Boyd, Edinburgh & London.

WALKER, R. 1973. Mopping up the turbidite mess. *In*: GINSBURG, R. N. (ed.) *Evolving Concepts in Sedimentology*, The Johns Hopkins University Press, Baltimore & London, 1–37.

WEGENER, A. 1915. *Die Entstehung der Kontinente und Ozeane*, Friedrich Viewig, Braunschweig.

WEGMANN, E. 1958. Das Erbe Werners und Huttons, *Geologie*, **7**, 531–559.

WESTBROEK, P. 1991. *Life as a Geological Force: Dynamics of the Earth*, W. W. Norton & Co., New York & London.

WILSON, J. T. 1965*a*. A new class of faults and their bearing on continental drift. *Nature*, **207**, 343–347.

WILSON, J. T. 1965*b*. Transform faults, oceanic ridges, and magnetic anomalies southwest of Vancouver Island. *Science*, **150**, 482–485.

WOOD, R. M. *See* MUIR WOOD, R.

WOODWARD, H. B. 1908. *The History of the Geological Society of London*, Longmans, Green, & Co., London, New York, Bombay & Calcutta.

YODER, H. S. 1993. Timetable of petrology. *Journal of Geological Education*, **41**, 447–489.

YOUNG, D. A. 1998. *N. L. Bowen and Crystallization-Differentiation: The Evolution of A Theory*. Mineralogical Society of America's Monograph Series, Publication No. **4**. Mineralogical Society of America, Washington, D. C.

YOUNG, E. D. & CARRUTHERS, M. 2001. *Trends in Earth Sciences*. Helicon Publishing, Oxford.

ZITTEL, K. VON. 1901. *History of Geology and Palaeontology to the End of the Nineteenth Century, Translated by Maria M. Ogilvie-Gordon*, Walter Scott, London.

ZONENSHAIN, L. P., KORINEVSKY, V. G., KAZMIN, V. G., PECHERSKY, D. M., KHAIN, V. V. & MATVEENKOV, V. V. 1984. Plate tectonics model of the South Urals development. *Tectonophysics*, **109**, 95–135.

Geology: from an Earth to a planetary science in the twentieth century

URSULA B. MARVIN

Harvard–Smithsonian Center for Astrophysics, 60 Garden Street, Cambridge, MA 02138, USA

Abstract: Since the opening of the Space Age, images from spacecraft have enabled us to map the surfaces of all the rocky planets and satellites in the Solar System, thus transforming them from astronomical to geological objects. This progression of geology from being a strictly Earth-centred science to one that is planetary-wide has provided us with a wealth of information on the evolutionary histories of other bodies and has supplied valuable new insights on the Earth itself. We have learned, for example, that the Earth–Moon system most likely formed as a result of a collision in space between the protoearth and a large impactor, and that the Moon subsequently accreted largely from debris of Earth's mantle. The airless, waterless Moon still preserves a record of the impact events that have scarred its surface from the time its crust first formed. The much larger, volcanic Earth underwent a similar bombardment but most of the evidence was lost during the earliest 550 million years or so that elapsed before its first surviving systems of crustal rocks formed. Therefore, we decipher Earth's earliest history by investigating the record on the Moon. Lunar samples collected by the *Apollo* astronauts of the USA and the robotic *Luna* missions of the former USSR linked the Earth and Moon by their oxygen isotopic compositions and enabled us to construct a timescale of lunar events keyed to dated samples. They also permitted us to identify certain meteorites as fragments of the lunar crust that were projected to the Earth by impacts on the Moon. Similarly, analyses of the Martian surface soils and atmosphere by the *Viking* and *Pathfinder* missions led to the identification of meteorite fragments ejected by hypervelocity impacts on Mars. Images of Mars displayed landforms wrought in the past by voluminous floodwaters, similar to those of the long-controversial Channeled Scablands of Washington State, USA. The record on Mars confirmed catastrophic flooding as a significant geomorphic process on at least one other planet. The first views of the Earth photographed by the crew of *Apollo 8* gave us the concept of 'Spaceship Earth' and heightened international concern for protection of the global environment.

Until the latter years of the twentieth century, planets were the night-time 'stars that moved', seen as points of light or viewed through telescopes. Since then, images from spacecraft have transformed all those planetary bodies with solid surfaces from astronomical to geological objects, each one with its own unique evolutionary history. Manned and instrumented missions to the Moon have sampled its surface and probed its interior. And meteorites, sometimes called 'poor man's space probes', have furnished us with samples of a wide variety of asteroids, and of our Moon and Mars.

We also have learned about dangers to the Earth posed by bodies in space. Far from existing in isolation and subject only to processes of change that are intrinsic to it, the Earth hurtles around the Sun along a path that is gritty with interplanetary dust and rubble and bathed in solar and galactic radiation. Without its shielding atmosphere, the Earth would be as barren and lifeless as the Moon.

Every day, approximately 40 tons of interplanetary dust and debris, including about 50 meteorites weighing at least 100 grams, fall to the Earth. In historic times, all freshly fallen meteorites have been small objects of no urgent concern to us. But larger bodies have pockmarked Earth's surface with more than 160 impact craters, and every hundred million years or so a comet or asteroid, at least ten kilometres in diameter, has struck with great violence, sometimes triggering mass extinctions and terminating geological periods (Melosh 1997). Thus, geoscientists have learned to view the Earth as a very different place from the uniformitarian realm we inherited from the nineteenth century. However, the topic of impacts from space and the implications for uniformitarian geology has been reviewed elsewhere (Marvin 1999) and will not be pursued here.

This paper will review some of the insights we have gained since the opening of the Space Age from studies of meteorites, asteroids, the Moon and Mars and how we have applied this knowledge to gain a better understanding of the

From: OLDROYD, D. R. (ed.) 2002. *The Earth Inside and Out: Some Major Contributions to Geology in the Twentieth Century*. Geological Society, London, Special Publications, **192,** 17–57. 0305-8719/02/$15.00

Earth. It also will recount the efforts of a few individuals who helped to persuade the USA space agency to add planetary geology to its agenda. We will argue that the change in geology from being entirely Earth-centred to planetary-wide has been one of the truly outstanding advances of the twentieth century. Indeed, one scientist has compared its importance to the change in astronomy in the sixteenth century from the Ptolemaic to the Copernican system (Head 1999, p. 158).

The Earth in space

Meteorites: natural space probes

Meteorites have provided us with invaluable data on the geochemistry and petrology of planetary bodies, chiefly asteroids. Meteorites also carry a record of their bombardment by cosmic rays that produce short-lived cosmogenic isotopes and minute tracks of radiation damage in them as they orbit through space. When they plunge to the Earth, the atmosphere shields the meteorites from further bombardment and the isotopes begin to decay. They do so at a known rate and this makes it possible for us to calculate how much time has passed since they fell. Isotopically, each meteorite serves as a timekeeper of at least three important dates in its history: the date when its parent body originally formed (its formation age), the length of time it has orbited through space (its cosmic-ray exposure age), and the time since it fell to Earth (its terrestrial age). In some instances it also indicates the time that has elapsed since one or more shock events have reset certain atomic clocks in the meteorite. All of these isotopic techniques for measuring ages have been developed since the 1950s, as ever more sensitive analytical instruments have come into use.

Most meteorites are fragments of asteroids (also called minor planets), thousands of which populate a wide belt between Mars and Jupiter. Asteroids are small bodies mostly less than 200 kilometres in diameter although the four largest ones range from 400 to 935 kilometres in diameter. Collisions between asteroids send debris around the Sun in elliptical orbits, some of which cross that of the Earth. If the Earth happens to be at the intersection at just the right moment, a 'meteoroid' will enter the atmosphere. During its very brief passage through the atmosphere the falling body may burst into pieces and fall as a shower. All shower fragments are counted as a single meteorite and named for the nearest post office or for a prominent local landmark.

The very idea of solid rocks literally falling out of the sky is so counterintuitive that it was rejected utterly by savants of the Age of Enlightenment until a succession of four witnessed and widely publicized falls occurred between 1794 and 1798. Chemical analyses of these and other allegedly fallen stones and irons, published by E. C. Howard in 1802, demonstrated their differences with Earth's crustal rocks and finally convinced the most hardened skeptics of the authenticity of meteorites (e.g. Marvin 1996). As of December 1999, a total of 1005 meteorites had been catalogued from witnessed falls, and an additional 21 500 meteorite fragments had been found in all parts of the world (Grady 2000, p. 8).

Meteorites come in three main varieties with the descriptive names stony, iron, and stony-iron. Stony meteorites make up 93%, irons make up about 6%, and the rare stony-irons, make up less than 1% of all meteorites that have been collected after being seen to fall. We calculate percentages only on those from witnessed falls because these provide the best available evidence of their relative abundance in their parent bodies.

Ordinary chondrites

The overwhelming majority (87%) of stony meteorites seen to fall are chondrites. These meteorites are, in effect, cosmic sediments, widely viewed as aggregates of particles that existed in the primeval solar nebula. They consist of minute mineral grains and chondrules, which are rounded, millimetre-sized silicate bodies containing crystallites of one or more minerals (Fig. 1a). Chondrules were first seen in thin-sections by Henry Clifton Sorby (1826–1918), who had invented the technique of slicing rocks into transparent wafers and looking at them through a microscope. Sorby (1864) wrote that chondrules looked like droplets of a fiery rain. Indeed they do; many of them are partially glassy and clearly have been molten. However, the chondrites in which they occur never have been heated to melting temperatures, although most of them have been recrystallized by thermal metamorphism and so have lost their primitive textures. Spirited debates continue to this day on whether chondrules are quenched droplets from short-lived heating events within the primeval solar nebula or were formed by processes that occurred on or within the earliest planetary bodies.

Carbonaceous chondrites

Although ordinary chondrites are anhydrous, a few rare meteorites called carbonaceous

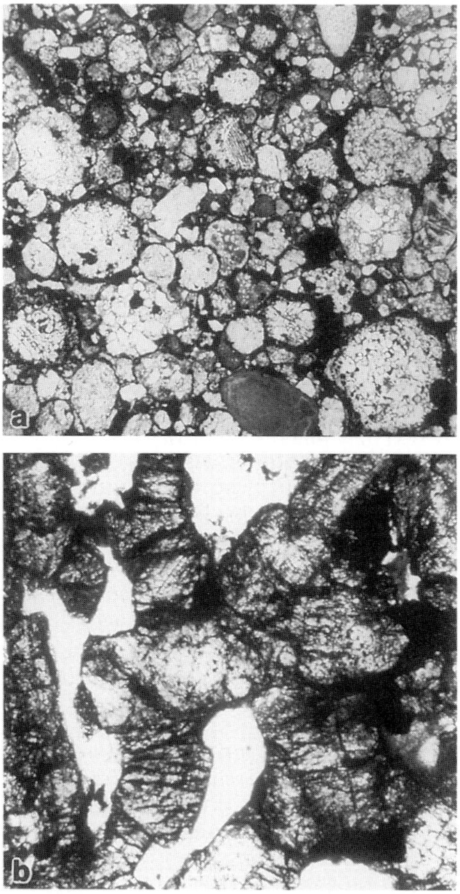

Fig. 1. Two classes of stony meteorites. (**a**) The Tieschitz chondrite is a cosmic sediment consisting of chondrules in a dark matrix of minute mineral grains. It fell on 15 July 1878, at Tieschitz, now in the Czech Republic (thin-section photograph by the author, in transmitted light. Width of field, 6 mm). (**b**) The Shergotty basaltic achondrite that fell on 26 August 1865 at Shergotty, India. It consists mainly of two pyroxenes, augite and pigeonite (grey), and maskelynite (white). The maskelynite was plagioclase feldspar that has been transformed *in situ* to glass by impact shockwaves (thin-section photograph by the author, in transmitted light; width of field 1.6 cm).

chondrites (even though they contain no chondrules) contain up to 20 wt% of water and 3.5 wt% of carbon. Only 36 such meteorites have been seen to fall. Their carbonaceous compounds include amino acids and all of the chemical building blocks of life. None of these compounds, however, are linked into proteins, the essential basis of living things. Some of them contain molecular hydrocarbons that are unknown on the Earth and so provide us with knowledge of organic (but non-biologic) compounds that occur in space. In bulk composition, carbonaceous chondrites closely match the content of non-gaseous elements in the Sun. Thus, they are tangible samples of the primitive matter from which all the bodies of the Solar System originated. Some carbonaceous chondrites contain so-called CAIs (calcium aluminium-rich inclusions) consisting of unusual assemblages of silicate and oxide minerals. Isotopic age determinations show that CAIs are *c.* 4566 million years old, making them the oldest dated materials in the Solar System. This supports the hypothesis that the CAIs formed in the solar nebula previous to their accretion into their carbonaceous parent bodies, which occurred *c.* 4560 million years ago (Clayton *et al.* 1973) .

The Earth accreted only a few million years later than the meteorite parent bodies. After more than a century of attempts, the age of the Earth was finally established in 1956 when Claire Patterson (1922–1995), at the California Institute of Technology, compared the ratios of primeval to radiogenic lead in five meteorites with those in deep sea sediments and calculated that the Earth formed $4.55 \pm 0.07 \times 10^9$ years ago. (Herein, such ages are expressed in aeons, or Ae, with 1 aeon = 10^9 years). However, the Earth is so large and warm and chemically active that none of its earliest crustal materials survived the early bombardment by impacting bodies and subsequent chemical recycling. The oldest dated terrestrial minerals are detrital zircons, *c.* 4.4 Ae old, found in Australia. They have been eroded out of their parent rocks and incorporated into younger sandstones. At present, the world's oldest dated rock outcrops are the Acasta gneisses in the northwestern corner of the Canadian shield. They formed 3.9 to 4.0 Ae ago (e.g. Bowring & Housh 1995). Rocks nearly as old occur in Wyoming, Greenland, western Australia and northeastern China. In the absence of older crustal rocks, petrologists now look to the Moon for evidence of what occurred during the first 550 million years or so of the Earth's Precambrian Era.

Presolar grains

In the early 1980s, primitive chondritic meteorites were found to contain minute traces of matter that formed in the atmospheres of distant stars long before the formation of our Solar System. These presolar grains occur as crystals, a few nanometres across, of highly refractory minerals, including diamond (C), lonsdaleite (C), graphite (C), corundum (Al_20_3), spinel ($MgAl_20_4$), carborundum (SiC), silicon nitride (Si_3N_4) and carbides of zirconium, titanium and molybdenum. Their isotopic compositions indicate that some of these grains were shot into space by the collapse of red giant stars and others are vapour condensates from supernovae (e.g. Anders & Zinner 1993; Sanford 1996). Eventually they mixed into the solar nebula and accreted into meteorite parent bodies, fragments of which have brought 'stardust' into our laboratories.

More recently, equally minute diamonds and intergrowths of diamonds with other minerals have been found on the Earth at impact sites. The melt rocks at the Ries Kessel (23 km across), near Nördlingen in Germany, contain minute intergrowths of cubic with hexagonal diamonds, and also of diamond with carborundum, a composite unknown elsewhere on the Earth. Both of these diamond-bearing intergrowths are ascribed to condensation not in stellar atmospheres but in impact-generated fireballs (Hough *et al.* 1995).

Traces of microdiamonds also have been found along with grains of shocked quartz and other exotic minerals in clays at the Cretaceous–Tertiary boundary, the horizon that marks a massive extinction coincident with the excavation of the Chicxulub Crater in Yucatán. These diamonds carry carbon and nitrogen isotopes indicative of an origin either from the shockwaves of the impact or, more likely, in the immense fireball it generated. These occurrences of microdiamonds that were not formed under high pressures in the Earth's mantle have revived interest in the possible origin of the well-known carbonados, or black diamonds, found in placer deposits in Brazil and the Central African Republic. Carbonados are extremely hard polycrystalline aggregates with a porous matrix of minute (0.5–20 μm) crystallites often enclosing octahedral and cubic diamonds and sometimes other minerals. Since carbonados never are found *in situ* in bedrock, an impact-related origin by vapour condensation is under investigation (Koeberl 1995).

Achondrites

A small proportion of meteorites – the achondrites, irons and stony-irons – come from parent bodies that melted and differentiated. About 8% of stony meteorites are igneous rocks called 'achondrites' because they lack chondrules. Many achondrites look so much like terrestrial basalts that they rarely have been collected unless they were seen to fall (Fig. 1b). Two associations of achondrites of special interest to us are the howardites, eucrites and diogenites (HEDs), and the shergottites, nakhlites and chassignites (SNCs).

Eucrites are basaltic lavas consisting of Ca-rich plagioclase and Ca-poor pyroxene; diogenites are more deep-seated, plutonic rocks consisting almost entirely of Ca-poor pyroxene. Howardites are breccias consisting of eucrite and diogenite fragments that have been broken up by impacts, mixed, and cemented together. Both eucrites and diogenites appear to have crystallized from the same magma as early as 4557 million years ago, showing that igneous processes occurred in some asteroids well within the first ten million years of their existence. Their rapid cooling rates suggest that their parent body(or bodies) was only partially melted. Today, about 200 eucrites, 95 diogenites and 93 howardites are catalogued in the world's collections.

Since the 1970s, comparisons have been made between the reflectance spectra, in visible to infrared wave lengths, of crushed meteorite samples and the surface compositions of asteroids, as determined by remote sensing. This technique has distinguished at least 14 compositional classes of asteroids, nine of which present fair matches with meteorites, although sometimes with more than one type of meteorites.

Remote sensing has led to a definitive match between the HED achondrites and the unique spectra of the large (*c.* 510 km diameter) asteroid, Vesta. This was the first instance in which meteorites were identified with a specific parent body. Most of Vesta reflects the spectra of basaltic eucrites, but the floors of at least two broad, deep craters reflect that of the deeper-seated diogenites, and one even deeper basin, exposes the underlying peridotite of Vesta's mantle. Like most asteroids, Vesta is irregular in shape due to high-energy collisions in orbit that have cratered its surface and split off large pieces of it. Its south polar region is occupied by a huge basin 460 km wide and 12 km deep. Some collisional fragments undoubtedly were lost into space; others were perturbed into Earth-crossing orbits and fell as the HED basaltic achondrites; still others, nicknamed 'Vestoids', remain in the asteroid belt. Several of these have been observed bridging the orbital space between Vesta and one of the gaps in the belt which serves as an escape hatch into interplanetary space (Binzell & Xu Shui 1993).

The first of the SNC meteorites fell in 1815 at Chassigny, France; the second one fell in 1869 at Shergotty in India; and the third fell in 1911 at

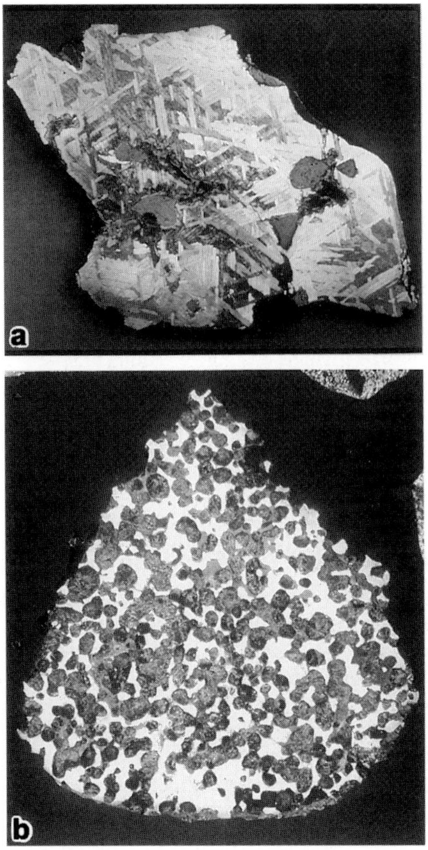

Fig. 2. An iron meteorite and a pallasite. (**a**) A fragment of the Gibeon iron meteorite, a Class IVA fine octahedrite, from the world's largest strewn field in central Namibia. The sawn surface has been polished and etched with acid to show the Widmanstätten pattern (the slice is 16 cm across. Smithsonian Institution photograph). (**b**) A slice of the Springwater pallasite found in 1931 in Saskatchewan, Canada. Transluscent olivine crystals are scattered through Ni–Fe metal with Widmanstätten patterns too fine to be visible in this view (the slab is 14 × 12 cm; courtesy of P. Lafaite, Muséum National d'Histoire Naturelle, Paris).

Nakhla in Egypt. Shergotty (see Fig. 1b) is a medium-textured pyroxene–plagioclase basalt in which the feldspar has been shocked *in situ* to glass (maskelynite). Nakhla is a clinopyroxenite, and Chassigny is a dunite, consisting almost entirely of olivine. These three meteorites contain small quantities of hydrous silicates and are more oxidized than other achondrites. Despite their individual differences they share enough similarities to be classed together as a group. Since 1911, 14 more SNCs have fallen or been found. We will discuss them below in more detail.

There are seven additional classes of achondrites, one of which, the ureilites, are of special interest to planetary geologists because they consist chiefly of olivine indicative of an origin in the mantle of an asteroid. Ureilites also contain graphite and clumps of microdiamonds which have been ascribed to shock transformation of graphite during collisions in space. The first ureilite fell in three pieces near Novo Urei in Russia in 1886. Two of the pieces are listed as lost, although a story persists that they were broken up and eaten by the local people, presumably for their medicinal value or magical powers. We may hope the people did not chew on them because the tiny diamonds could have done immense damage to their teeth. Ninety-two ureilites are known today.

Iron meteorites

Needless to say, the cores of completely melted asteroids are represented by iron meteorites, although some cored bodies may have coalesced with others thus producing larger bodies with irons distributed in a 'raisin-bread' texture. A few irons may even have accreted directly from grains of metal in the solar nebula. Iron meteorites resist weathering better than stones do and are more easily recognized in the field, hence they make up a large proportion of museum collections. However, irons make up only *c.* 7% of meteorites seen to fall, a figure that yields a truer picture of their abundance in their parent bodies.

All iron meteorites contain nickel, their chief defining characteristic which was discovered by E. C. Howard (1802). They range from 5 to 35 wt% Ni, but most of them carry no more than 20 wt% Ni. The most abundant irons, the so-called octahedrites, acquire a unique metallurgical texture while the metal cools in the solid state between *c.* 900 and 450°C. Nickel, diffusing through the hot metal, separates into two phases, Ni-rich taenite and Ni-poor kamacite, which form alternating lamellae oriented parallel to the eight faces of an octahedron. Any slice through an octahedrite, once the surfaces are polished and etched with acid, will show the lamellae in a handsome 'Widmanstätten' pattern (Fig. 2a).

This name is a historical accident. In 1804, William (Guiglielmo) Thompson (1761–1806) in Naples described this metallurgical pattern and published a drawing he made (severely straining his eyesight) in the journal *Bibliothèque Britannique*. However, the patterns were independently discovered in 1808 by Count Alois von Widmanstätten (1753–1849), Director of the Imperial Industrial Products Cabinet in Vienna, who inked the surfaces and printed the patterns directly on paper. He never published his 'nature prints,' but his friends, who evidently knew nothing of Thompson's article, called them 'Widmanstätten patterns' and the name has survived.

These patterns have been used as indicators of the cooling rates of the irons in their parent bodies. Computer calculations based on the diffusion rate of nickel and the widths of lamellae in different irons have yielded rates ranging from less than one degree to several thousand degrees centigrade per million years. Such cooling rates are in the range to be expected in the cores of asteroidal sized bodies (in contrast with the core of the Earth which still is partly molten). Nearly 870 iron meteorites are catalogued today, of which only 48 were seen to fall. The irons have been classified into 13 main chemical groups plus a number of lesser ones chiefly on the basis of their contents of nickel and the minor elements gallium and germanium. Additional irons are chemically anomalous or remain unclassified. At most, the iron meteorites are thought to have originated in about 60 different parent asteroids (Wasson 1985).

In 1891, diamonds were discovered in the Canyon Diablo iron meteorite from the vicinity of Coon Butte (Meteor Crater), Arizona. These hard masses stopped the saw blades as the metal was being sliced open. At that time, diamonds were believed to have formed at high pressures deep within the Earth. Quite naturally, therefore, it was assumed that the Canyon Diablo diamonds had formed at high confining pressures within a large parent body. This lent credibility to an old idea that a single planet once occupied the region between Mars and Jupiter but was shattered into asteroids by an explosion from within or a collision from without.

By 1963, however, artificial diamonds in clumps of nanometre-sized crystals had been produced by shockwave experiments on graphite, and X-ray diffraction films showed that the diamonds in the Canyon Diablo iron occurred in similar clumps of minute grains (Lipschutz & Anders 1961). This strongly suggested that the Canyon Diablo diamonds were shock-produced from inclusions of graphite during the hypervelocity impact that excavated the crater. Evidence suggestive of this was reported by H. H. Nininger (1965) who found diamonds only in shocked, shrapnel-like fragments of the iron that were concentrated on the crater's northeastern rim.

In 1982, however, typical shockwave diamonds were discovered in a small iron meteorite lying on the ice in Antarctica, where it clearly had not collided violently with the Earth. The diamonds in this small iron formed during a collision that occurred in space – the same explanation that had been applied to the microdiamonds in ureilites. Assuming (as scientists commonly do) that what is true for one must be true for all, many meteoriticists immediately concluded that the Canyon Diablo diamonds also were formed during collisions in space. Not all of us agree, however: surely the shockwaves that blasted open the 1.2 km wide Meteor Crater were capable of transforming some of the graphite to diamond in the most severely shocked fragments.

Stony-iron meteorites

These meteorites come in two varieties: pallasites and mesosiderites. The pallasites are extraordinarily beautiful meteorites consisting of nickel–iron metal, displaying delicate Widmanstätten patterns studded with large, translucent crystals of yellow-green olivine (Fig. 2b). Presumably, rocks of such a composition could form only at the core–mantle boundaries within their parent bodies. However, olivine, being much lighter in density, would float up and out of the molten metal unless it were somehow held in place. Most likely the odd mix occurred in each case when molten metal rose from below and invaded a mush of olivine crystals that had collected at the base of the mantle and were trapped there.

Mesosiderites are among the most puzzling of meteorites. They are coarse breccias of Ni Fe metal mixed with fragments of silicate rocks having the compositions of eucrites and diogenites. Olivine is present only in small traces. Thus, mesosiderites seem to record the violent breakup and selective reassembly of pieces of the core and the crust of a differentiated parent body, but with no samples from the mantle! As if this were not strange enough, the metallic iron in mesosiderites tends to be of an unusually uniform composition unlike that of any class of iron meteorites. Possibly this iron is a product of impact melting. A series of six epochs has been worked out to produce a plausible scenario for the formation of mesosiderites, every stage of which requires further investigation (Rubin 1997).

Once diamonds were dismissed as indicators of high pressures the case collapsed for large meteorite parent bodies and a consensus formed, for a variety of reasons, that the asteroids originated as small bodies. The nearby presence of the giant planet, Jupiter, with its powerful gravitational field, evidently prevented the formation of a single planet in that region, and probably contributed to maintaining Mars as a relatively small planet. In 1847 A. A. Boisse (1810–1896), in Paris, drew a diagram of a meteorite parent body with a nickel–iron core overlain concentrically by pallasitic stony-irons and stony meteorites of increasingly silica-rich compositions. Boisse's 'onion shell' meteoritic planet was widely taken as a model for the interior of the Earth.

Research on meteorites began to advance during the 1950s with the introduction of mass spectrometers and other analytical techniques. Meteorite studies did not assume importance as part of a national effort, however, until after the opening of the Space Age on 4 October 1957, with the orbiting of *Sputnik 1* by the Soviet Union.

The Space Age: the US programme

Sputnik I sent shockwaves through America. By that time, preparations for space missions were well along but with no sense of urgency about them. No one in the USA had any idea of how advanced the programme was in the Soviet Union. This is well illustrated by the following notice in the 'Science and the Citizen' section of *The Scientific American* for October 1957:

> Early Returns from the International Geophysical Year.
>
> But the satellite projects were not going too well. Scientists of the U. S. S. R. had not yet made laboratory models of their satellites, or even decided on their size or weight. In the United States, workers on Project Vanguard have built 20-pound models, but the propulsion problem is still so formidable that they think they may have to begin with projectiles no bigger than a softball, carrying no instruments except possibly a radio transmitter for tracking purposes.

This message reached many readers while *Sputnik 1* was beep-beeping overhead.

The following summer, on 29 July 1958, congress voted to establish the National Aeronautics and Space Administration (NASA). Until then almost everyone had assumed that space exploration would be a military project, but President Dwight D. Eisenhower, despite his rank as a five-star general of the United States Army (or possibly because of his rank), insisted that the space effort was to be strictly civilian. On 3 August, a Space Sciences Board was formed to recommend what types of science projects NASA should conduct. That very question aroused immense controversy. Neither NASA administrators nor NASA engineers, responsible for the flawless performance of rockets and capsules, were interested in diverting their time and attention and cluttering up their elegant machines to accommodate science projects (Wilhelms 1993).

A few physicists already had arranged to fly experiments, however, so on 31 January 1958, when *Explorer I*, the first US spacecraft, lifted off, it carried a Geiger counter designed by the physicist James A. Van Allen of the University of Iowa. It was designed to measure cosmic ray intensities above the atmosphere, but, as the capsule rose to an altitude of 2500 km, the Geiger counter roared into action as it passed through two belts of highly charged particles surrounding the Earth, trapped there by the magnetic field. These 'Van Allen belts' were totally unexpected, and their discovery was such a triumph for American science that it might well have led to a space programme focused entirely on measurements of interplanetary particles and fields. That would have produced major advances in space physics, but none at all in planetary geology. At that time, nobody in charge had the slightest interest in the Moon. Indeed, most scientists viewed the Moon as an inert body with a history of long-dead volcanism of no conceivable scientific value.

This view was described as early as 1935 by Frank E. Wright (1877–1953), director of the Carnegie Institution's Geophysical Laboratory in Washington. Wright himself took a strong interest in the Moon and lamented the prevailing attitudes (Wright 1935, p. 101):

> [the Moon's] presence in the night sky is resented by the modern astronomer, especially the astrophysicist. Its light interferes with the photography and analysis of far distant, faint celestial objects, such as stars, clusters and nebulae. . . . To him [the Moon] is a lifeless, inert mass, shining only by reflected sunlight and held by gravitation in its orbit about the Earth.

Wright's assessment still applied in the 1950s. After *Sputnik I*, however, geologists, geochemists and geophysicists, who were interested in the Moon and planets, worked hard to make themselves heard in conferences and on planning committees. Perhaps, at this juncture, we should ask why these scientists were interested in the Moon.

Earth's unaccountable Moon

Why does the Earth have its moon? How and when did the Moon begin to orbit the Earth? These are not trivial questions. Of the four rocky planets of the inner Solar System, Mercury has no moon and Venus has none. Mars has two moons but they are miniscule, misshapen bodies, only 12 and 22 km in their longest dimensions, that clearly are captured asteroids.

The Earth, however, has a gigantic moon that is nearly one-quarter the size of the Earth itself. No other planet has a moon, or even a family of moons, that adds up to such a proportion of the primary planet. (This excludes Pluto, a small icy body which no longer enjoys an uncontested status as a planet. Pluto follows a steeply tilted orbit among a crowd of smaller bodies in similar orbits that make up the Edgeworth–Kuiper comet belt just beyond Neptune.) So, the Earth–Moon pair is unique. Dynamically, it behaves like a double planet, lurching around the Sun like mismatched knobs on a dumb-bell. Yet our knowledge of our partner has been so fragmentary and in some ways so contradictory that the Moon always has been more mysterious, by far, to scientists than it ever was to poets.

The Moon before Apollo

What did we know about the Moon in the 1950s? We knew that the Moon turns only one face toward us as it circles the Earth once each month. That face shows us bright highlands surrounding dark plains that spread over about one-third of the visible surface. In 1610 Galileo called these features terrae and maria (lands and seas), and the Latin names are still in use. Galileo observed that the terrae stand higher than the maria, and he also described and sketched what he called circular 'spots' or 'cavities' in the surface.

From its size and mass, we had learned that the Moon's density (3.3 g cm^{-3}) is markedly lower than that of the whole Earth (5.4 g cm^{-3} but a close match to that of Earth's mantle. This told us that the Moon must be poor in iron and have only a small metal core or none at all. We also knew that the Moon has no atmosphere and no surface water, although there were speculations that internal waters might have deposited mineral veins in lunar bedrock and congealed its soils with permafrost. In any case, it was clear that the Moon is far from being a mini-Earth.

The Earth–Moon system has an extraordinarily high degree of angular momentum. Furthermore, the Earth's spin axis (and hence its equator) is tilted 23.45° from the plane of the ecliptic, in which the Earth and all the other planets revolve around the Sun. The Moon's spin axis is almost vertical to the ecliptic, but its orbital path around the Earth is tilted *c.* 7° to the ecliptic and *c.* 28° to Earth's equatorial plane. How could two such closely linked bodies become so out of kilter?

From its orbital characteristics we had deduced that the Moon is not quite a sphere. It appeared to have a gravitational bulge toward the Earth that is out of equilibrium with its present orbital distance. Thus the bulge was thought to have been 'frozen-in' at a much earlier date when the Moon was more plastic and closer to the Earth. As the Moon raises the tides, the drag of the waters on shelving ocean floors slows the Earth's rate of rotation, causing the Moon to retreat from the Earth at the rate of *c.* 3 cm each year.

However, if we were to spin its orbit backward, the Moon would come closer and closer to Earth until, only about 1500 million years ago, it would arrive at the Roche limit, just 2.89 earth radii away. What a dramatic sight a late Precambrian moonrise would have been as that huge body appeared over the horizon! The Moon never could have come closer because any object passing inside the Roche limit is doomed to break up and shed its debris on the Earth. Did Earth have a Moon before 1500 million years ago? Were there no ocean tides before then? We had no ready answers for these questions. In fact, we knew enough in the 1950s about the Moon's composition, shape, density and orbit to be thoroughly mystified by it. We did not know when or where the Moon formed, or when and how the Earth and Moon became linked together.

Hypotheses of the Moon's origin

Until 1984, the four most widely discussed modes of lunar origin were fission of the Earth, capture of the Moon from the asteroid belt or from a location near the Earth, simultaneous accretion in Earth orbit, or the accretion of a ring of Earth-orbiting moonlets.

Earth fission. The fission theory, first proposed in 1879 by George H. Darwin (1845–1912), son of Charles Darwin, pictured the Moon spinning out of the mantle of the rapidly rotating Earth due to resonance between its free oscillations and the solar tides. Diagrams of this process show a large bulge at Earth's equator stretching into a long, narrow neck until the tip finally breaks off to form the Moon. Darwin calculated that this should have taken place 3560 million years ago, but on remembering Lord Kelvin's low age for the Earth, he recalculated it to 57 million years ago. In either case the Moon and Earth would be roughly the same age. But the dynamic problems were extreme: why does the fissioned-off Moon not revolve around the Earth's equator? What tilted the Moon's orbit and the Earth's axis by strikingly different amounts? Above all, what force kept the Moon moving outward in full retreat from the Earth?

An interesting variation was proposed in 1881 by the Reverend Osmond Fisher (1817–1914) in England in his book, *Physics of the Earth's Crust*, that earned him the title 'The Father of Geophysics'. Fisher argued that the Moon was ripped out of the essentially solid Earth, leaving behind the vast basin of the Pacific Ocean as the unhealed scar. Subsequently, the Earth's remaining granitic crust split into fragments that started drifting toward the magma-lined hollow until they were grounded in their present positions. Fisher's theory, an early version of continental drift which accounted for the Moon, the Pacific basin and the fit of the continental shorelines across the Atlantic, had immense popular appeal that persisted into the middle of the twentieth century. However, geophysicists could imagine no force capable of tearing the Moon bodily out of the solid Earth, and keeping this huge mass moving outward. And, in hindsight, we know that the timing of events was wrong. The current episode of continental break-up and separation did not begin until the Jurassic Period *c.* 180 million years ago, when the Moon already was very old.

A captured moon. The Moon's low density closely matches that of chondritic meteorites, and this, together with its tilted orbit, led to hypotheses that it formed in the asteroid belt from which it was perturbed into an Earth-crossing orbit and captured. The asteroid belt does seem oddly deficient in mass: all the present asteroids put together are equivalent to only *c.* 3% the mass of the Moon. The escape of a Moon-sized asteroid and its capture by the Earth seemed statistically unlikely, but this mode of lunar origin attracted a large following. Perhaps the most extreme version of the capture hypothesis, posited in 1963 by Hannes Alfvén, held that the Moon approached Earth in a retrograde orbit and the force of the encounter ripped off the outer portion of the Moon and showered debris over the Earth, making the entire crust above the Mohorovičic discontinuity of 'Moon-stuff' – a dramatic reversal of the fission hypothesis that made the Moon of 'Earth-stuff'.

A more plausible hypothesis, strongly supported in the 1950s and 1960s by Harold C. Urey and others, was that the Moon was captured not from the asteroid belt but from the vicinity of the Earth, where many sizeable primitive objects must have accreted. After the other bodies coalesced to form the Earth, this remaining one was captured into Earth orbit. Urey, as will be discussed below, was influential in persuading NASA to conduct a scientific study of the Moon.

Simultaneous accretion of the Earth and Moon. The hypothesis of simultaneous accretion proposed that the Moon accreted in Earth orbit from an atmosphere of lighter elements left aloft when most of the heavier elements accreted into the Earth. If so, the growing Moon had (somehow) to maintain a steady retreat to avoid falling into the Earth. Once again, this fails to account for the Earth's tilted axis, the Moon's tilted orbital plane, and for the large amount of angular momentum possessed by the Earth–Moon system.

A ring of moonlets. A fourth hypothesis, proposed in the 1960s by Gordon J. F. MacDonald (b. 1929) envisioned the Moon accreting just beyond the Roche limit from a ring of Earth-orbiting moonlets, analogous to Saturn's ring. He calculated that the accretion would have been essentially complete *c.* 1500 million years ago, and that the final falling bodies marked the lunar surface with its multitudes of craters. Thereupon the Moon began its slow retreat (MacDonald 1965). When MacDonald's Moon was born close to the Earth, it suddenly would have initiated enormous ocean tides and tectonic disruptions that should be visible in the geological record. But no conclusive evidence of them has been found.

Each hypothesis of lunar origin presented such serious difficulties that in 1968 Harold Urey remarked: 'we might say that no method for the origin of the Moon is possible and the Moon simply cannot exist – but there it is, just the same'.

Three advocates of a lunar geology programme

A close look at the history of any enterprise generally reveals that one person, or a few individuals, played decisive roles in influencing the directions things took. Many people contributed significantly to persuading NASA to undertake planetary missions rather than to confine their scientific activities to space physics. But three men, each working separately, played roles of such crucial importance that it is doubtful if the decision to go to the Moon and to conduct science there would have been made without them. They were Ralph B. Baldwin (b. 1912), an astronomer who made his career in private industry, Harold C. Urey (1893–1981), a chemist, and Eugene M. Shoemaker (1928–1997), a geologist.

Ralph B. Baldwin

Baldwin's interest in the Moon was sparked in 1941 by a series of telescopic lunar photographs he examined while waiting to give evening lectures at the Adler Planetarium in Chicago. No one had ever mentioned the Moon as an object of serious interest during his student years at the University of Michigan, where he earned his PhD in astronomy in 1937. Now, as he puzzled over certain grooves and ridges, he tested the idea of projecting them along great circles and found that they crossed one another in the central region of Mare Imbrium. To him, this suggested that Imbrium was the site of an enormous explosion, and that the deep grooves were excavated by the forceful ejection of impact debris along radial lines. Baldwin continued his analyses of lunar surface forms and concluded that impacts from space have excavated most, if not all, of the lunar craters, large and small. He also detected stratigraphic evidence that the large craters such as Imbrium had stood open for considerable periods of time before the mare lavas entered them.

Baldwin soon learned that his views were too radical for publication and that his fellow astronomers were not interested in hearing him talk about them. They still saw the Moon as a relict, of extinct volcanism, unworthy of scientific inquiry. Thus, Baldwin's first paper, written in 1942, was rejected by three leading astronomical and astrophysical journals in succession before it was accepted by *Popular Astronomy*. Meanwhile, he had begun work on classified military research at the Applied Physics Laboratory in Washington, where he had access to facts and figures on the magnitudes of craters excavated by bombs, shells and high explosives. From this trove of data he confirmed earlier reports that explosive impacts create circular craters almost regardless of their angle of entry, a fact that had received scant attention.

Three years after the end of World War II, Baldwin decided to leave the laboratory and move to Grand Rapids, Michigan, to help in the running of his family's business, the Oliver Machinery Company. Thus, instead of leading research efforts in a federal laboratory or entering academia and mentoring a succession of students, Ralph Baldwin opted for a career in private industry. Nevertheless, recognizing that he had a fascinating research project all to himself, he continued his investigations of the Moon as time permitted, and ultimately assembled his evidence and insights into a 239-page book, *The Face of the Moon,* published in 1949. It has been called 'probably the most influential book ever written in lunar science (Wilhelms 1993, p. 15).

To support his argument for the impact origin of lunar craters Baldwin plotted the depth-to-diameter ratios of the freshest and least altered (Class 1) lunar craters on a log–log diagram which showed that these craters cluster along the upper end of a smooth curve, and that bomb and shell craters cluster along the lower end of the same curve, with a gap between them due to the lack of known craters in intermediate sizes. In addition he plotted the d/di (depth/diameter) ratios of the four freshest of the terrestrial meteorite craters known at that time, and they, too, lay on the curve. Two of the craters, Henbury-13 in Australia and Odessa-2 in Texas, plotted amid the bomb and shell craters; Odessa-1 was the largest of the explosion pits; Meteor Crater, in Arizona, plotted at the lower end of the lunar craters. No such regularity applies to volcanic craters or calderas. In later years, Baldwin's diagram would be a key factor in persuading many a geologist of the importance of impact cratering.

Baldwin's book sold poorly, but it exerted enormous influence by catching the attention of a few prominent scientists. Perhaps the first was Peter M. Millman (1909–1999), then of the Stellar Physics Division of the Dominion Observatory at Ottawa. Millman read the book and showed it to the Dominion Astronomer, Carlyle S. Beals (1899–1979) with whom he discussed the possibilities of finding impact craters on the Precambrian Canadian shield. Beals instituted a systematic search by air and on the ground. Within two decades, this effort had yielded 12 proven impact craters plus several probable ones. Today the number for all of Canada stands at 26, and counting. In 1981, the Royal Astronomical Society of Canada, presented Baldwin with an honorary membership. In his citation, Ian Halliday (1981) stated that it was in recognition of:

> *The Face of the Moon* (1949), which may properly be considered the generating force behind modern research on both terrestrial impact craters and lunar surface features.

Fig. 3. Ralph B. Baldwin (left) and Donald E. Gault in 1986 at an entrance to the American Museum of Natural History in New York. That year, The Meteoritical Society presented Baldwin with its Leonard Medal and Gault with its Barringer Medal (photograph courtesy of John A. Wood).

> Seldom has a single book had such far-reaching consequences in the progress of science as those which followed in the [next] three decades.

Another copy of Baldwin's book came into the hands of Harold Urey, then a distinguished service professor at the University of Chicago. There are at least four different stories of how this came about but a favoured one is that Urey was shown the newly published book by his host at a party one evening and he promptly dropped out of the festivities. When time came to go home Urey was found sitting alone totally absorbed in Baldwin's book. '[Urey's] reading of *The Face of the Moon* started a chain of events that eventually led to the choice of the Moon as America's main goal in space', wrote Don Wilhelms (1993, p. 19). In 1963, Baldwin published a second book, *The Measure of the Moon,* a 488-page work that provided a wealth of new data on terrestrial as well as lunar craters.

We are not accustomed in this day and age to the idea of one person working alone outside the academic and scientific communities and still wielding enormous influence. Baldwin continued to attend meetings, give talks and publish papers, but by the 1980s few of the young scientists conducting research in meteoritics and planetary science had any idea of his seminal contributions to the field. Their elders did, however, and in time the community honoured Baldwin with its three most prestigious awards. Figure 3 shows Baldwin with Donald E. Gault (1923–1999) at the 1986 meeting of The Meteoritical Society in New York where both men were presented with medals. Baldwin received the Society's Leonard Medal for his outstanding achievements in original research in meteoritics and closely allied fields; Gault received the Barringer Medal for the path-breaking experimental studies of cratering that he carried out at the NASA Ames Research Center in Sunnyvale, California, using a gun that fired gas-propelled pellets into various substances at a large range of velocities and angles. Later that same year, 1986, Baldwin received the G. K. Gilbert Award of the Planetary Sciences Division of the Geological Society of America for his outstanding contributions to the planetary sciences. In 2000 Baldwin received The Meteoritical Society's Barringer Medal for 'outstanding work in the field of terrestrial impact cratering or work that has led to a better understanding of impact phenomena'. Baldwin is only the second scientist to receive both the Leonard and Barringer Medals from The Meteoritical Society. The other one was Eugene Shoemaker.

Harold C. Urey

In 1934, Urey was awarded the Nobel Prize in Chemistry for his discovery of deuterium. During World War II he conducted research on the separation of isotopes of uranium, and afterward, in an effort to explore new fields, he took an interest in chemical abundances in meteorites and the origin and evolution of the Solar System. Urey worked out a hypothesis that all planetary bodies formed by the accretion of cold particles, and that some of them never heated to melting temperatures. Impressed by geophysical evidence that the Moon possesses sufficient internal strength to maintain its figure out of hydrostatic equilibrium and to support high mountain ranges, Urey concluded that the Moon is a prime example of a such a cold, primeval object that has survived from the earliest epoch of the Solar System. Baldwin's evidence of the impact origin of the Moon's craters fitted perfectly with this idea. In 1952, after Urey finished writing his own book, *The Planets*, he visited Baldwin in Michigan to discuss the Moon and planets. The two men agreed on some fundamental issues and strongly disagreed on others, but they always remained friends. Baldwin (pers. comm. 2000) wrote: 'I had a great liking and respect for [Urey]. He was often wrong in matters concerning the Moon, but he aided greatly in making it a proper body to study'.

Throughout his book, Baldwin referred to the mare fillings as lava flows, but he was not thinking of them as volcanic in origin. At that time,

Fig. 4. Harold C. Urey at the 1968 meeting of The Meteoritical Society in Cambridge, Massachusetts (photograph by the author).

both Baldwin and Urey believed that the 'lavas' consisted of impact-generated melt rock. Baldwin had a problem however: he had found clear stratigraphic evidence that a substantial delay had taken place between the excavation of large craters and their partial filling with mare flows. Urey saw no delay at all. He believed that impacts excavated craters and generated the flows of molten rock simultaneously. Indeed, for years Urey scorned geologists for not grasping what he saw as the essential fact that the impacts caused the mare flows. To account for the delay, Baldwin (1949, pp. 210–214) constructed a scenario, based on his knowledge of the behaviour of materials, in which a rebound from the Imbrium impact formed a massive structural dome that stayed in place for an appreciable time and then collapsed. In so doing, it displaced an immense volume of superheated melt rock that welled up the ring faults, filled Imbrium, and then flowed out across the surface into the other large open craters. Later on, Baldwin concluded that the mare basalts were, indeed, volcanic flows erupted from depth into craters that had stood open for a long time.

With his towering stature among scientists, high-level politicians and administrators, Urey (Fig. 4), was immensely influential in persuading NASA of the importance of going to the Moon and conducting scientific experiments there. No doubt, another factor helped to tip the scales: on 12 September 1959, the Soviet Union crash-landed a vehicle on the Moon showing that they had the guidance system to do it. Then, on 4–7 October 1959, the USSR's *Luna 3* mission orbited the Moon and sent back the first images of the Moon's far side. Although the pictures themselves were rather fuzzy, what they revealed was breathtaking: the Moon's faces, like those of the Earth, are asymmetrical. Bright highlands, interrupted by only a few small maria, occupy all of the lunar far side.

Eugene M. Shoemaker

In 1948, Eugene Shoemaker, a 20-year-old geologist working for the US Geological Survey on the uranium–vanadium deposits of the Colorado Plateau, read in a California Institute of Technology newspaper about certain rocket experiments at White Sands, New Mexico. In his own words Shoemaker described his thoughts (Wilhelms 1993, p. 20): 'Why, we're going to explore space, and I want to be part of it! The Moon is made of rock, so geologists are the logical ones to go there – me, for example!'

From then on, Shoemaker pursued every avenue that possibly could lead to the US exploration of the Moon with himself as one of the explorers. He examined volcanic maars, nuclear bomb test craters, and Meteor Crater in northern Arizona. Meanwhile, in 1949, he combed the literature and found, 'nothing but nonsense' about the Moon with the conspicuous exceptions of a paper published in 1893 by Grove Karl Gilbert and the newly published book by Ralph Baldwin. Both Baldwin and Shoemaker were so impressed with Gilbert's analysis of lunar features that it is worthwhile for us to examine his paper.

Grove Karl Gilbert (1843–1918), chief geologist of the US Geological Survey, spent 18 nights in September 1892, making telescopic observations of the Moon at the US Naval Observatory in Washington. 'The face the moon turns toward us is a territory as large as North America, and, on the whole, it is probably better mapped', he wrote in 1893. Clearly, he valued the remarkably detailed lunar maps and charts that had been published in Europe during the eighteenth and nineteenth centuries, but he felt there was much more to be done.

Gilbert studied the system of grooves radiating outward from Mare Imbrium and (as Baldwin would do again nearly 50 years later) he saw that they converged at the site of an explosion, immense beyond comprehension. He concluded that the grooves were cut by blocks of ejecta that were sent scouring through the surrounding highlands. He also noted the huge numbers of lunar craters, with their marked circularity, enormous size range and random distribution, and concluded that they could not be volcanic – they were too different in many respects from volcanic features on the Earth. Neither, he thought, could they have been excavated by meteorites. He needed a larger, more focused supply of impactors than meteorites which fall sporadically.

To account for the craters' predominantly circular form, which he believed could result only from direct hits, Gilbert hypothesized (50 years ahead of Gordon MacDonald) that the Moon had coalesced from a ring of moonlets in orbit around the Earth. Their bombardment caused the growing body to tilt this way and that until the entire surface was pockmarked with circular craters. He explained the dark maria (much as Baldwin would do at first) as resulting from a great flood of liquified, solid and plastic debris, generated by the heat of impact, that poured out of the Imbrium crater and flooded one-third of the near side of the Moon.

Observing that these dark plains are sparsely cratered and hence younger than the densely cratered uplands, he took them as a time marker separating 'antediluvial' from 'postdiluvial' lunar topography. Gilbert then constructed a stratigraphic chronology of the Moon based on crater counts and the geological principles of overlapping formations, cross-cutting relationships and degrees of preservation. For this path-breaking work Gilbert (belatedly) has been called the 'father of lunar stratigraphy'.

In December 1892, Gilbert presented his findings in an address he gave as the retiring president of the Philosophical Society of Washington. His 50-page paper, 'The Moon's face: a study of the origin of its features', appeared in that Society's *Bulletin* in January 1893. Then his ideas effectively passed into oblivion. Evidently, Gilbert was asking too much of his fellow geologists by trying to interest them in the Moon, and by introducing an entirely new cratering process to account for its features which were universally assumed to be volcanic. Nor is Gilbert's study highly valued by everyone today. Gilbert's biographer, Stephen J. Pyne, referred to 'The Moon's face' as 'a masterly investigation but also a magnificent distraction' (Pyne 1980, p. 160). A magnificent distraction from what, he did not say. Presumably it was from his field work, for which Gilbert was justly famous.

Ralph Baldwin sees the situation differently. Baldwin was unaware of Gilbert's paper until his own book was nearly completed. Then Reginald A. Daly (1871–1957), Professor Emeritus of Geology at Harvard University, recommended it to him. Baldwin has remarked that he feels most fortunate that Gilbert published in such an obscure journal, else there would have been nothing left for Baldwin himself to do 50 years later.

Perhaps the journal was comparatively obscure, but condensations of the text were presented to the National Academy of Science and the New York Academy of Science, and abstracts were published in their journals and several others. Furthermore, we may safely assume that the hall in Washington in which Gilbert delivered his presidential address was packed with a fair sampling of the nation's leading geologists from the nearby headquarters of the US Geological Survey. If he failed to strike a spark of interest it may have been because he was flouting two of the basic tenets of uniformitarianism, which dictated that all processes of change must originate within the Earth, and must be observed in operation. Crater-forming impacts would originate outside the Earth, and such things never had been observed in operation. Volcanism was the standard crater-forming process. Indeed, in 1791, when J. H. Schröter (1745–1816) first called the lunar features 'craters', the term was understood as volcanic, just as 'lava' was, and is today. For this reason, Daly (1946) enclosed 'craters' in quotation marks throughout a paper in which he raised his lonely voice favouring an impact origin of the lunar features. That paper is much admired today for Daly's remarkably prescient insights, particularly his argument that the Moon resulted from a glancing collision of the Earth with another planet-sized body (Baldwin & Wilhelms 1992).

Gilbert himself was not in the least hesitant to introduce a new process into geology. In 1891, when he heard of a rimmed bowl, nearly three-quarters of a mile wide excavated in the sandstone and limestone of the northern Arizona plateau with iron meteorites strewn on the surrounding plains, he immediately envisioned a possible impact origin. Once again, Gilbert was thinking of an impact not of a meteorite but of a late-falling planetesimal that struck the Earth after accretion of the planet was substantially complete. With this in mind he set out 'to hunt a star' at Coon Butte and conducted the first investigation in history to determine whether a crater could be of impact origin.

Gilbert assumed that the main mass of the iron would have buried itself beneath the crater floor where it would cause a magnetic anomaly. It also would occupy so much extra space that if the rim were packed back into the bowl the feature would form a mound. And he assumed that an impact crater should be elliptical, since most bodies must strike the Earth at an oblique angle. The Arizona crater failed all of Gilbert's tests: he detected no magnetic anomaly, he found the crater to be essentially circular, and he measured identical volumes of the rim and bowl. He found no trace of volcanic rock in or near the crater, but from the rim he could see the cones of the youthful San Francisco volcanic field on the northwestern horizon. So Gilbert yielded his 'fallen star' hypothesis with good grace and described the crater as a volcanic maar, caused by a deep-seated steam explosion when magma, migrating at depth, encountered ground water at this site. Gilbert concluded that the iron meteorites were coincidental.

Although Gilbert examined the crater in 1891 he did not publish his report, 'The origin of hypotheses, illustrated by a discussion of a topographic problem', until 1896. This paper cast a pall on the subject of impact craters – terrestrial and lunar – for the next 50 years, during which a large majority of scientists accepted as decisive Gilbert's volcanic explanation for Meteor Crater. It is ironic that Gilbert's paper on the Moon, which is much admired today, was ignored during his lifetime, and the one on Meteor Crater, which is rather an embarrassment today, wielded a strong, albeit decidedly negative, influence on cratering studies.

This was true even though Daniel Moreau Barringer (1860–1929), a mining entrepreneur who believed, as Gilbert had, that a large iron lay buried beneath the floor, staked a mining claim on the crater in 1903. In 1905, Barringer and his partner, B. C. Tilghman, published separate reports describing an impressive array of evidence (fully accepted today) for an impact origin, including tilted and overturned sedimentary strata on the rim, tonnages of pulverized quartz grains suggestive of shock, and nuggets of Ni–Fe oxide shale mixed into the rim, clearly relating the impact of the irons to the crater. Barringer went on to sink shafts and drill holes that revealed 21 m of Pleistocene lake sediments in the crater and Ni–Fe-rich sludge lying at a depth of 396 m, but no sign of volcanic rocks or of a large iron meteorite. As early as 1908, George P. Merrill (1854–1929) of the Smithsonian Institution, argued that the incoming iron largely vaporized itself due to the heat of impact – an unwelcome thought to Barringer but equally unwelcome to those favouring volcanism. No matter: the US Geological Survey staunchly refused to concede that Gilbert could have been wrong, and all but a few American geologists were closely enough linked to the Survey to shy away from the subject of meteorite craters (e.g. Marvin 1986). Consequently, members of the US Geological Survey were *persona non grata* at the crater.

Eugene Shoemaker changed all that. In 1957 he sought permission to study the crater through the good offices of one 'Major' Brady, who had run a school for boys in Mesa, Arizona, attended by two of Daniel M. Barringer's sons. Given such an introduction, D. Moreau Barringer, Jr, the director of the crater company, welcomed Shoemaker to the site. Shoemaker studied the crater in exquisite detail and, taking advantage of his federal security clearance, he compared it with the Teapot Ess Crater of Yucca Flat, Nevada, that had been formed by the explosion of an underground nuclear device. By then, Shoemaker had two hypotheses of origin to refute: the standard volcanic one and a second one, published in 1953 by Dorsey Hager, a petroleum geologist, who argued that the crater was, in fact, a sinkhole, resulting from the dissolution of 200 000-year-old beds of salt, gypsum and limestone at depth. Hager stated that this was the opinion of numerous geologists interested in an objective explanation of the crater.

By the time Shoemaker had finished his fieldwork in 1959, he had identified several lines of evidence for impact and calculated that Meteor Crater could have been formed by the impact of a 63 000-ton iron meteorite, 25 m in diameter, which struck Earth at a velocity of 15 km s^{-1} and triggered a 1.7 megaton explosion. Shoemaker detailed his observations and calculations in a dissertation that earned him his PhD from Princeton University in 1960. He also submitted a short version to the International Geological Congress (IGC) for presentation at its meeting in Copenhagen in August 1960. Today, the long version of his report is ranked as a landmark paper in cratering studies (Shoemaker 1963).

His work on Meteor Crater redoubled Shoemaker's interest in the Moon. In 1956 he had urged the director of the US Geological Survey to set up a studies group on lunar geology. But the idea was tabled for the time being. In 1958, he drew up a plan for proposed lunar research but that, too, was tabled. By then, however, interest in the Moon was developing. In December, 1959, NASA, in co-operation with the Jet Propulsion Laboratory (JPL) in Pasadena, California, launched the Lunar Ranger project that would crash-land a series of vehicles on the Moon to send back pictures and test the nature of the lunar surface. In addition, the US Air Force Chart and Information Center began a lunar mapping programme and in February 1960 published an airbrushed chart of the Copernicus region of the Moon, the first of a series at a scale of 1: 1 000 000. The following month Shoemaker visited JPL and happened to see the chart. He obtained a copy and, within a week, he had plotted five geological units on it and coloured them by hand. His colleague, R. J. Hackman (1923–1980) added lineaments and they had their geological map printed in time for Shoemaker to show it in August at the IGC in Copenhagen (Shoemaker & Hackman 1962).

Meanwhile, Edward T. C. Chao (b. 1919), at the US Geological Survey in Washington, discovered coesite by its X-ray diffraction pattern in a sample of the quartz sandstone from Meteor Crater. Coesite is a high-pressure polymorph of silica that was known at that time only as a product of shockwave experiments in laboratories. This first natural occurrence of coesite was reported by Chao, Shoemaker and Madsen in 1960.

En route to Copenhagen, Shoemaker stopped in Germany to examine the Ries Kessel, which was still considered to be a volcanic caldera although both Ralph Baldwin (1949) and Robert S. Dietz (1959), who found shatter cones there, had identified it as an impact crater. Robert Dietz (1914–1995) was a world leader in establishing studies of impact craters and structures as a new branch of geoscience. He authored an early paper (Dietz 1946) on the impact origin of lunar craters (of which Baldwin was unaware while writing *The Face of the Moon*), but Dietz devoted most of his energies to studies of terrestrial craters and global tectonics, for which he coined the term 'sea-floor spreading' in 1961. He did not play a role in persuading NASA to include geology in the *Apollo* programme.

At the Ries Kessel, Shoemaker collected samples of suevite (partially glassy impactite) and sent them to Chao in Washington. Chao found coesite in them and telephoned the news to Shoemaker in time for him to insert this new evidence into his talk in Copenhagen. Shoemaker and Chao published their results in 1961. These discoveries of coesite prompted searches for impact effects in other minerals and soon gave rise to shock metamorphism, a new branch of petrology (French & Short 1968).

Shoemaker had kept on urging the Survey to set up an Astrogeology Studies Group and, on 25 August 1960, the same day that he presented his talk and showed his Copernicus map at the IGC, the US Geological Survey officially established the studies group, with Shoemaker as the director. In 1961 it would become a branch of the Geological Survey with some members serving in Washington and others in Menlo Park, California. Four years later many members moved to Flagstaff, Arizona, where a new astrogeology branch headquarters was dedicated in October 1965.

Fig. 5. Eugene Shoemaker in his office at the California Institute of Technology, *c.* 1980 (courtesy of Carolyn S. Shoemaker).

The name 'astrogeology' was Shoemaker's choice. He wanted something connoting space science and the proposed name, 'photogeology', simply would not do. Purists objected that 'astro' means 'star' and stars have no geology, but Shoemaker prevailed. We already had so-called 'astronauts', although they do not fly to stars. Terminology can be light-hearted, but it also can have fateful consequences, as we have seen in the example of 'craters' with its strong volcanic connotations.

On 25 May 1961, President John F. Kennedy announced the national purpose to send a man to the Moon and return him safely to Earth within that decade. In 1962 and 1963, Shoemaker served at NASA headquarters in Washington where he lobbied hard for the addition of scientific investigations, particularly geological ones, to the *Apollo* missions. It was a difficult, uphill battle as none of the top administrators had any interest in science. President Kennedy had instructed NASA to send a man to the Moon and bring him safely back – he said nothing about taking pictures or, worse yet, collecting rocks. Observers are convinced that if Shoemaker had not been at headquarters at that time, it is more than likely that no geology would have been included in the *Apollo* programme.

Shoemaker (Fig. 5) played a major role in planning and analysing the results of the unmanned vehicles – the Lunar Rangers and Surveyors that imaged and tested the lunar surface before the *Apollo* landings. And he initiated a strong programme of geological training for the astronauts, including trips to volcanic fields and impact craters to learn about their distinguishing characteristics. He even created an artificial crater field on his property near Flagstaff by detonating charges of various magnitudes. One of the greatest disappointments in Shoemaker's life came with the realization that he would be unable to pass the rigorous physical tests required for astronauts because he had contracted Addison's disease, a life-threatening condition which he, fortunately, was able to keep under control by using cortisone. Nevertheless, he continued to devote his energies to extending geology from a strictly terrestrial enterprise to one that encompassed the geological mapping of the Moon and, subsequently, of all the rocky and icy planets and satellites of the Solar System.

In 1969, when the *Apollo 11* samples arrived at the Lunar Receiving Laboratory in Houston, Texas, Shoemaker cleared the way for the simplified use of their terminology. The committee responsible for the preliminary examination of the samples had agreed that, to avoid false connotations, they would avoid using terrestrial names for the lunar rocks and minerals. Some already had replaced 'geology' with 'selenology', along with 'selenodesy' 'selenochemistry', 'selenophysics' and so on. Thus, as the rock boxes were unsealed, the committee members dutifully intoned: if it were on Earth we would call it such and such. Finally, when he heard about a yellow-green mineral which 'If it were on Earth we would call it olivine', Shoemaker had had enough: 'Aw, come on then,' he said, 'let's call it olivine'. From that moment, discussions of the lunar samples and lunar geology were briefer and more informative with no perceived damage to the quality of lunar science (Brett 1999).

Perhaps the emphasis we have placed on the influence of Baldwin, Urey and Shoemaker seems to imply that most scientists favoured impact over volcanism at the time of the *Apollo* missions. Nothing could be farther from the truth. Many of the astrogeologists at Flagstaff and Menlo Park believed not only in mare but also in highland volcanism. Indeed, the *Apollo 16* landing site in the Descartes region of the highlands was chosen because the mountains there are so precipitous and the intermontane plains are so smooth that astrogeologists concluded the peaks must consist of youthful volcanic rhyolites or andesites, and the plains of fresh pyroclastic flows.

The Apollo missions

We could think of the *Apollo* missions as the greatest geological field excursion in history. On 20 July 1969, two astronauts climbed out of the *Apollo 11* module and stepped onto the Moon. The *Apollo 11* mission fulfilled President Kennedy's stated purposes to the letter: it was on time, it was on target, it returned three astronauts safely to Earth, and *mirabile dictu* it kept within its original budget!

Between then and December 1972, 12 astronauts landed on the Moon. One of them was a geologist, Harrison (Jack) Schmitt of New Mexico. The 11 others, all fighter pilots, had received the geological training initiated by Eugene Shoemaker. The astronauts photographed and described the moonscape, and set out instruments to measure details of the Moon's interior and of radiation from space. Seismometers revealed that the lunar crust ranges in thickness from *c.* 20 km on the near side to more than 100 km on the far side; the mantle is 1100–1300 km thick, and there is a small core 300–400 km in radius. Seismometers also showed that Moon's gravitational bulge is not literally a bulge but a reflection of the fact that, due to the greater abundance of denser basaltic rocks on the near side and the greater thickness of the feldspathic crust on the far side, the Moon's centre of mass is offset toward the Earth by 1.8 km from its centre of figure. Passive seismometers left on the Moon recorded about 1700 very weak moonquakes each year, most of which originated in the lower mantle due to stresses and strains from the monthly lunar body tides. Their total energy release would scarcely be noticed on the Earth even if they all occurred at once. Meteorite impacts also were recorded, including a very large one that struck the back of the Moon on 7 July 1972 (Nakamura *et al.* 1973). We shall have no news of another one: in 1977, to save on expenses, NASA switched off all the instruments still operating on the Moon.

Five passive laser ranging reflectors are still in use, however. The reflectors were emplaced on the Moon by three *Apollo* missions and two of the robotic Soviet *Lunakhod* Rovers. They reflect laser pulses from telescopes on Earth directly back to the same telescope, thus allowing accurate measurements of the Earth–Moon distance. Over time, the measurements have improved our knowledge of the Moon's orbit, its rotation, and its physical properties by more than two orders of magnitude. They also have shown evidence of a small, dense lunar core, detected free librations indicative of a recent large impact, and confirmed the 'equivalence principle' of Einstein's theory of general relativity as applied to a celestial body (Mulholland 1980).

The astronauts explored six landing sites (see Fig. 6) and brought back 841 kg of lunar rocks

Fig. 6. The sites on the Moon sampled by the USA *Apollo* and the USSR *Luna* missions (NASA photograph labelled by John A. Wood).

and soils. In addition, the Soviet Union sent up three unmanned sample-return missions that collected 321 g of soil samples. The USA shared *Apollo* samples with Russian scientists and they shared their *Luna* samples with the USA, to the great advantage of all. Every sampling site yielded new and interesting rock types to the general inventory.

In 1970, the mineralogists, Brian Mason and William Melson wrote:

> the lunar rocks represent a unique scientific adventure and an intellectual challenge of the first magnitude . . . they are certainly the most intensively and extensively studied materials in the history of science.

This is true beyond a doubt: every year since 1969, as increasingly sensitive techniques of microanalysis have been developed, samples have been allocated in repsonse to new requests from research laboratories around the world,

Fig. 7. Lunar rock samples. (**a**) A lunar anorthosite metamorphosed to a granulitic texture (thin-section photograph in cross-polarized light by John A. Wood; width of field, 6 mm). (**b**) Seven grains handpicked from a 1 to 4 mm fraction of *Apollo 12* soil sample 12033. The three lower grains are anorthositic gabbros. The other four grains are coarse-grained norites (photograph by the author).

including 355 allocations made between March 2000 and March 2001.

The *Apollo* missions also sent us our first images of the whole Earth taken by men in space. These views of our fragile-looking home planet in the blackness of space have been credited with unleashing the world-wide torrent of concern we are now experiencing for preserving our environment, an issue we will discuss presently.

The lunar highland samples

The *Apollo* samples provided us with several profound surprises. We learned, for example, that the bright highlands of the Moon consist in large part of fine-grained igneous rocks rich in plagioclase feldspar, specifically anorthite ($CaAl_2Si_2O_8$) (see Fig. 7a). Geologists had expected the highlands to consist of granites or rhyolites, the stuff of chondritic meteorites, or of basaltic achondrites. No one imagined that the lunar crust would be made in large part of feldspathic rocks of a type with no direct counterpart on the Earth.

The most feldspar-rich terrestrial rocks, called anorthosites, are very different from those of the lunar crust. They are much coarser grained, they occur not in igneous but in ancient metamorphic terranes, and they consist mainly of labradorite, a variety of plagioclase significantly poorer in calcium than anorthite. Despite the mismatches in texture and composition, the first small particles of white, feldspathic rocks found in samples of the dark soils of the *Apollo 11* landing site on Mare Tranquilitatis were called anorthosites, to denote that they consisted predominantly of plagioclase, by Wood *et al.* (1970) and Smith *et al.* (1970).

In addition to anorthosites and closely related anorthositic gabbros, the highlands yielded a suite of Mg-rich rocks, mainly norites and troctolites, that were derived from a separate but almost equally ancient parent magma that intruded the anorthositic crust. The *Apollo 16* mission to the Descartes highlands encountered none of the youthful volcanics anticipated by some astrogrologists. On the contrary, these mountains proved to be heaps of impact breccias and melt rocks ejected from the ancient basins nearby. In composition, they constituted an average sample of the highland crust. A highland component of special interest is KREEP, so named because it is rich in potassium (K), rare earth elements and phosphorus (P). It also contains traces of uranium and thorium, which render it weakly radioactive. A KREEP component was first identified by trace element analyses in glasses and impact breccias in the *Apollo 12* and *14* samples. Crystalline KREEP-rich rocks with basaltic textures occur in the *Apollo 15* and *17* samples. They consist of plagioclase (more sodic than anorthite), Fe-rich pyroxenes, and accessory minerals such as ilmenite, cristobalite, whitlockite, apatite, zircon and baddeleyite. Some petrologists view them as authentically igneous rocks while others argue that they are crystallized impact melts.

The distribution of KREEP, determined first by orbital gamma-ray measurements (Metzger *et al.* 1974), and most recently by neutron spectrometry (Elphic *et al.* 1998), show it to be concentrated in patches forming a large ring around the Imbrium basin. This distribution suggests that a KREEP-rich residual magma, which formed as the final product of the differentiation of the crust and mantle, rose close to the surface of the lunar near side where it was excavated, pulverized and distributed radially by the Imbrium impact.

Isotopic dating of the lunar anorthosites tells us that the Moon is very old. The average age of dated anorthosite samples is 4.4 ± 0.02 Ae. The highland norites and troctolites range from 4.3 to 4.4 Ae. KREEP samples range from a relatively youthful age of 3.8 to 4.3 Ae. Their ages overlap with those of the earliest mare basalts.

The maria

Moments after the *Apollo 11* astronauts stepped onto Mare Tranquilitatis, they declared that the main rock in the regolith was basalt, a volcanic lava. That put an end to fringe speculations on basins filled with electrostatically pooled dust or with black carbonaceous sediments. The mare basalts display a wide range in titanium content, colour and age. Typically they were of such low viscosity that some of them flowed across the surface for hundreds of kilometres. The visible flows of mare basalt range in age between 3.85 and 3.10 Ae, but older basalts existed as evidenced by clasts of them, *c.* 4.3 Ae old, found in highland breccias. By 3.0 Ae, mare volcanism had dwindled to a trickle, although minor eruptions continued until as recently as *c.* 0.8 Ae ago. The mare basalts were derived by partial melting of the mantle at depths of 200–400 km. They reached the surface of the near side, where they cover about one-third of the surface, much more readily than on the far side where the anorthositic crust is so much thicker. The mare flows are rather thin, however, and so mare basalts make up a trivial proportion of the lunar crust.

A few new minerals were identified in the lunar rocks. The first example, found in the *Apollo 11* basalt samples, is a titanium oxide [$(Fe,Mg)Ti_2O_5$] that was named armalcolite in honour of the three *Apollo 11* astronauts: *Arm*strong, *Al*drin and *Col*lins. The mineral remained unknown on the Earth until recently when it was found in the impact rocks of the Ries Kessel, and subsequently at other impact sites.

A most puzzling problem arose with the discovery that rock samples from both the highlands and maria display remanent magnetism acquired some 3.6 to 3.8 Ae ago. The Moon does not, however, possess an internally generated dipolar magnetic field. The *Lunar Prospector* mission of 1998 and 1999, which repeatedly circled the Moon in a polar orbit, detected regions of relatively strong magnetization lying on the lunar far side at the antipodal points directly opposite the huge Imbrium and Serenetatis basins. Lin *et al.* (1998) suggested that the force of basin-forming impacts sent expanding fireballs of ionized plasma racing around the Moon, pushing magnetic field lines ahead of them until they reached the antipodal points where the surface rocks were heated by seismic shock waves and magnetized as they cooled.

The lunar regolith

Previous to the *Apollo* missions, the airless lunar surface was known to be blanketed by a layer of impact debris that Eugene Shoemaker named the 'lunar regolith', a term he borrowed from terrestrial geology. It also is commonly called the 'lunar soil', although it is totally lacking in organic components. The regolith is 20–30 m thick in the ancient highlands, 2–8 m on the younger maria, and perhaps only a few centimetres thick on impact melt-sheets that floor youthful rayed craters such as Tycho (e.g. Hörz *et al.* 1991). All of the lunar samples were taken from the regolith, or from boulders lying in it; none were taken from bedrock, which was inaccessible to the astronauts. Regolith samples are treasure troves of rock types projected to the sampling site by impacts from many sources on the Moon.

The most common materials in the regolith are 'soil breccias', angular agglutinates of minute rock fragments welded together by impact glasses. Grains of individual rock types also occur in the regolith. Fortunately, most of the lunar rocks are so fine-grained that particles only 1 mm across often consist of two or more minerals, and particles 2–4 mm across are veritable boulders (Fig. 7b). All the lunar rocks originally were igneous and their most common mineral constituents are plagioclase feldspar, pyroxene, olivine and ilmenite. Rare components include silica minerals, zircon, phosphates and other accessory minerals. The most common lunar rocks are varieties of anorthosites, basalts, gabbros, norites and troctolites with minor dunites, quartz monzodiorites and 'granites'. A few unique lithologies include a Mg-rich cordierite–spinel troctolite (see Fig. 8), derived from deep within the lunar crust (Marvin *et al.* 1989).

The absence of water and volatiles

Petrologists received one more profound surprise when analyses showed all of the lunar rocks and minerals to be utterly dry. No lunar

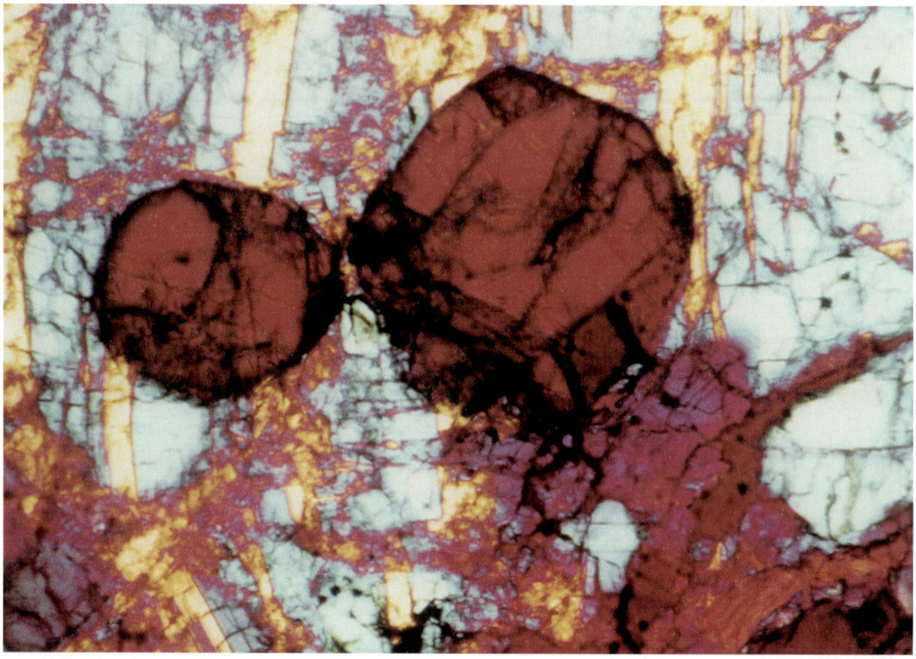

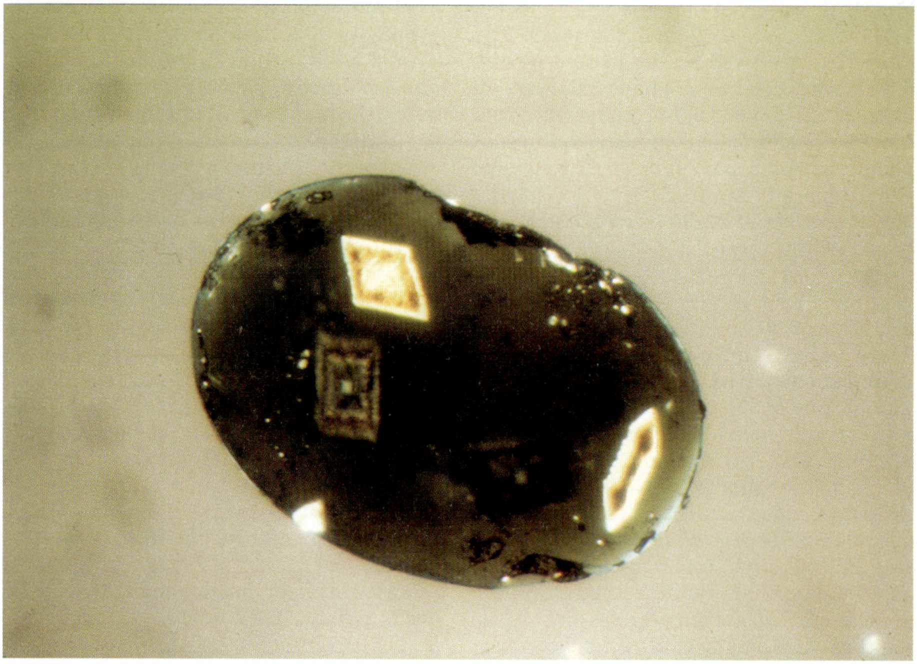

Fig. 8. A fragment of cordierite–spinel troctolite from *Apollo 15* regolith breccia 15295. Two spinel crystals (red-brown), and an adjacent grain of cordierite (pinkish-purple, lower right) are included in a large grain of twinned feldspar (blue and yellow). The crackled textures, with offset twin lamellae, and web-like patterns of finely crushed feldspar (pink and yellow), are shock features (false colour photomicrograph of thin section taken by the author, in partially cross-polarized light with gypsum accessory plate; width of field, 0.53 mm).

Fig. 9. An oval bead of green glass from the regolith at the *Apollo 15* landing site. The glass was formed by fire-fountaining of mare basalt 3400 million years ago. It quenched just after crystallites of olivine began to form (the bead is 8 mm from end to end; photograph by the author, in cross-polarized light).

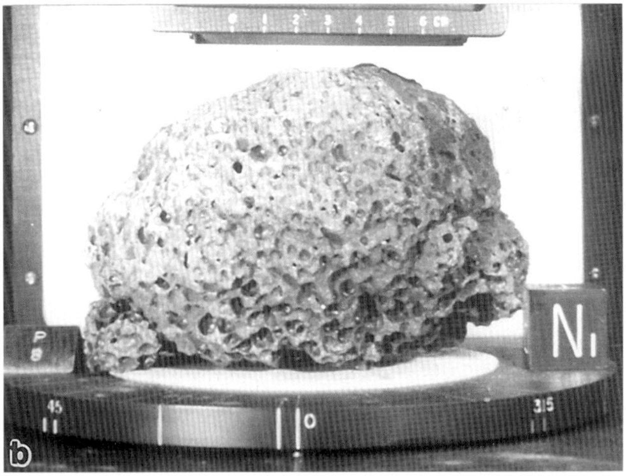

Fig. 10. A hand specimen of highly vesicular olivine basalt, No. 15556, from the *Apollo 15* mission shows that gas (most likely CO) escaped from the molten lava despite the absence of water (NASA photograph at the Lunar Receiving Laboratory in Houston; cube is 1 cm on an edge).

minerals contain water (H_2O), hydroxyl radicals (OH) or hydrogen bonds (H+). Therefore, there are no lunar micas, amphiboles, clay minerals or oxidation products. Besides the absence of water, the rocks are severely depleted in oxygen and other volatile elements. As a result, the lunar minerals are as fresh and gleaming as the day they crystallized. There are even lunar glasses, most of which were formed by impacts but some by mare-related fire-fountaining, that are perfectly transparent and undevitrified after thousands of millions of years (Fig. 9). On the warm, moist Earth glasses rarely survive for as long as 100 million years.

Despite the lack of water, many of the basaltic lavas of the maria are highly vesicular, so some gas or other escaped during crystallization (see Fig. 10). Carbon monoxide (CO) is the one favoured by experimental petrologists (Sato 1978; Head & Wilson 1979). Is the Moon, then, a completely waterless realm? This question was raised in 1994 when the US *Clementine* orbital mission detected abnormal amounts of hydrogen suggestive of ice in the immense, permanently shaded, South Polar–Aitken basin that is *c.* 2600 km across and more than 12 km deep. Early in 2000, however, a sensitive probe was crash-landed into the basin to test whether the impact would release H_2O. Not a trace was detected.

Oxygen isotopes

The ratios of the three isotopes of oxygen in the lunar rocks proved to be identical to those of the Earth (Clayton & Mayeda 1975). Petrologically, therefore, the Earth and Moon *do* belong together. This does not imply that the Moon fissioned off from the Earth; it simply means that the Earth and Moon formed in the same neighbourhood, defined as one astronomical unit (AU) from the Sun (1 AU = 150×10^6 km). Clayton and his colleagues had already shown that the ratios of oxygen isotopes in meteorites differ markedly depending on how far from Sun their parent bodies originated. This made it possible to group meteorites genetically, and disposed of all the 'onion skin' models, designed to derive all meteorites from a single body, that were still being designed in the 1960s, 120 years after Adolphe Boisse constructed the first one in 1847. It also disposed of the old theory that the Moon is a captured asteroid, but did not tell us whether the Moon accreted in orbit around the Sun or around the Earth.

The Russian Luna samples

The three automated sample-return missions sent to the Moon by the Soviet Union – *Luna 16* in September 1970, *Luna 20* in February 1972, and *Luna 24* in August 1976 – obtained highland and mare samples from the vicinity of Mare Crisium. They took drill cores, about 1 cm across, in the regolith and stowed them in the return capsules. These samples added significantly to the range of known lunar rock types. One notable example was a basalt from Mare Crisium with a very low content of titanium. Unquestionably, there are many more rock

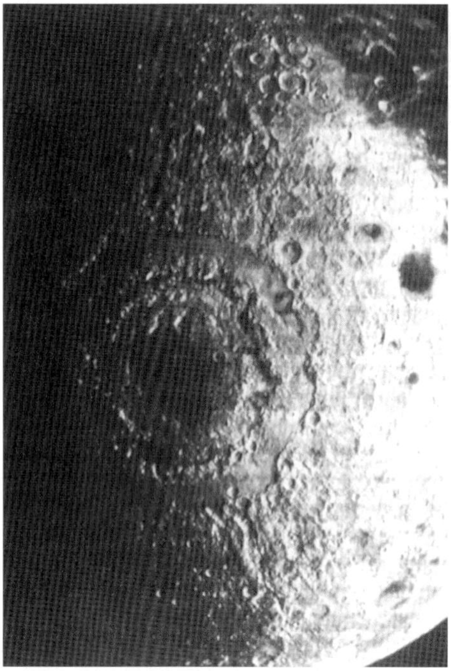

Fig. 11. Two lunar impact craters of contrasting form and magnitude. (**a**) The multiringed Oriental Basin on the west limb of the Moon. In this view, four rings are visible: the innermost ring is 320 km, and the next two are 480 and 620 km in diameter. The outermost, Cordillera Ring, which is taken as the basin rim, is 930 km across. Two outer rings, 1300 and 1900 km across, are not visible (NASA photo, *Lunar Orbiter* IV–187M).

types to be found on the Moon. Samples have been collected from within an area equaling only about 5% of the lunar near side. Intriguing (possibly volcanic) sites on the near side, as well as the entire expanse of the lunar far side, remain to be explored.

The lunar magma ocean

No one has proposed any method of constructing a planetary crust chiefly of anorthite except by crystal flotation. This calls for a molten (but water-free) lunar magma of just the right composition and density to allow plagioclase feldspar to crystallize early and float to the surface while most of the denser ferromagnesian silicate minerals sink to the lower crust or mantle. This melt had to be deep enough to yield massive volumes of feldspar, but shallow enough to cool and form a solid crust within only 150–200 million years. Calculations show that an ocean of magma 200 to 500 km deep could supply the feldspar (e.g. Wood 1971), but only the lower estimate, or an even shallower depth, would accommodate the rapid cooling rate. Details of the magma ocean are in dispute–how deep it was and whether it encompassed the entire Moon all at once or portions of it at different times – but the broad concept is widely accepted.

Heavy bombardment of the Moon and Earth: multiring basins

The lunar highlands show evidence of intense bombardment from the time the crust solidified *c.* 4.45 Ae ago until *c.* 3.8 Ae ago. The large impacting bodies that fell during that early period may simply have been late-falling successors of those that had coalesced to form the Moon. However, an abundance of highland samples with ages clustering between *c.* 3.95 and 3.85 Ae, and a dearth of impact glasses older than that, led to a hypothesis that a cataclysmic terminal bombardment occurred during that interval after a lull in the rate of impacts (Tera *et al.* 1974). Eminent scientists still argue each side of this question. Those who favour a continuous bombardment at a declining rate note a lack of hard evidence for the delay and ask where the large impactors were stored during the interlude. They argue that the abundance of relatively youthful samples reflects the fact that most of the *Apollo* samples are rich in ejecta from two large basins, Serenitatis and Imbrium, that formed at *c.* 3.9 Ae and 3.85 Ae, respectively. To test this explanation, Cohen *et al.* (2000) obtained dates on fragments of impact glass they picked out of four lunar meteorites and obtained ages ranging from 2.76 to a maximum of 3.92 Ae. Inasmuch as lunar meteorites (a topic discussed below) represent a broader sampling of the lunar crust than the returned samples do, this maximum age of impact glasses supports (but does not positively confirm) the hypothesis of a delayed terminal bombardment of the Moon.

Basins were not recognized as a special class of lunar features until 1961, when rectified lunar photographs were projected onto a large lunar globe at the University of Arizona's Lunar and Planetary Laboratory. This laboratory had been founded by Gerard P. Kuiper (1905–1973), another scientist with a long-term dedication to lunar studies. Suddenly, direct views of several huge craters with concentric rings and radial grooves or lineaments were visible to astonished viewers. The most dramatic example was the giant Oriental structure (Fig. 11a), which is 930

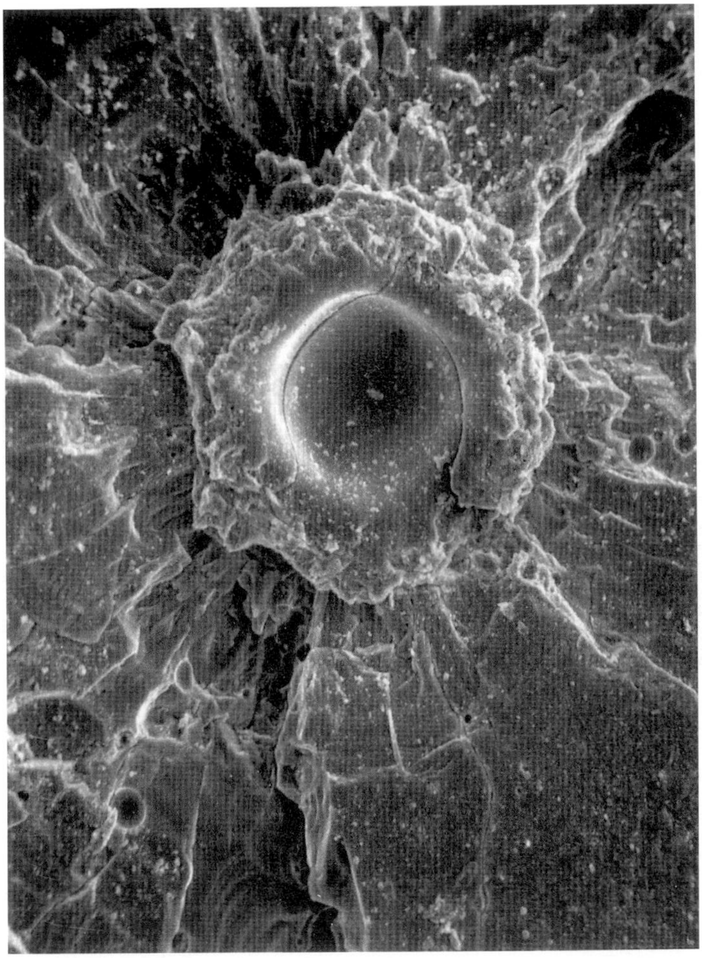

Fig. 11. (**b**) A zap-pit, caused by the impact of a micrometeorite into the impact-glass coating of Rock No. 15286 from the *Apollo 15* mission. The central pit is 0.07 mm across (courtesy of Donald E. Brownlee).

km across and has three inner rings that are 320, 480 and 620 km across and two outer rings that are 1300 and 1900 km across. Thin mare flows occur at its centre and in patches between the inner rings. In their initial description of these enormous ringed features, Hartmann & Kuiper (1962) introduced the term 'basins' to distinguish them from large, complex craters.

For comparison, extremely small craters also occur all over the Moon. Micrometre-sized 'zap-pits' (Fig. 11b) mark every rocky surface that is exposed to the lunar sky. Most zap-pits are lined with glass melted by the heat of impact. Zap-pits document the Moon's continual bombardment by tiny particles from space – a type of erosion from which the Earth is fully protected by its atmosphere.

Many lunar basins have at least one ring but to qualify as multiringed a basin must have at least two, and some investigators demand at least three rings (e.g. Spudis 1993 and references therein). Lunar multiring basins range in diameter from *c.* 300 to well over 1000 km and have up to six concentric rings. Six such basins are visible from the Earth and have been since the 1600s to anyone using a small telescope, not to mention those using the high magnification telescopes in the world's observatories (Hartmann 1981). Why did multiring basins go unrecognized until 1961?

No doubt the Oriental basin remained unnoticed for so long because it lies on the Moon's western limb where all but the edge of a prominent ring is out of sight from the Earth. (It was named 'Oriental' because early maps and pictures showed a glimpse of its rings on the eastern limb when lunar images were printed upside down – the way they look in telescopes.) Ralph Baldwin promptly pointed out that he had described rings in Imbrium and other large craters in *The Face of the Moon*; and, indeed, he had (Baldwin 1949, pp. 40–44). But somehow their special significance had been lost amid the plethora of new data and ideas in his text. Hartmann (1981) suggested that the shifting positions of the terminator on such large features tend to reveal arcs and to obscure rings. He also noted that since the eighteenth century a strong emphasis had been placed on mapping finer and finer details at the expense of broad views of the Moon; thus the gestalt was lacking for the recognition of this whole system of major features. We might also recall that very few astronomers using telescopes had spent any time looking at the Moon.

In 1965, the USSR's *Zond* mission provided detailed images of the lunar far side that revealed numerous ringed basins, most of them with no mare filling. By 1971 Hartmann and Wood had counted 27 multiring basins on the Moon; today the count is closer to 50. In 1971, *Mariner 9* imaged multiring basins on Mars, and a year later *Mariner 10* did so on Mercury. By 1980, *Voyagers 1* and *2*, on their grand tour of the Solar System, had imaged multiring basins on the rocky and icy satellites of Jupiter and Saturn. Clearly, they were features of planetary-wide importance. But, were there any multiring basins on the Earth?

While the Moon was being heavily bombarded, so, too, was the Earth; indeed the bombardment of the Earth may have been even more intense due to our planet's more powerful gravitational field. Yet, in 1961 when they were discovered on the Moon, multiring basins were unknown on the Earth and the prospects of finding them seemed bleak. The earliest and presumably the largest of Earth's impactors fell during the first 550 million years before the crust solidified, and were lost in the hot, volcanic mantle. Possibly, they did not vanish without leaving a trace, however; the plunging of large impactors into the deep mantle may have set up the physical and chemical reactions that determined the location of the earliest ocean basins, or perhaps more likely initiated the formation of continents (e.g. Spudis 1993, p. 229).

Originally it was assumed that multiring basins were to be expected only in the Earth's most ancient Precambrian terranes. Unfortunately, these terranes are deeply eroded and often severely deformed. Nevertheless, two of Earth's more than 160 known impact structures are Precambrian and show evidence of having formed as multiring basins: the Vredefort Dome of South Africa (250–300 km in diameter; *c.* 2.02 Ae old); and the Sudbury basin in Ontario, Canada (250 km in diameter; 1.85 Ae old). Each of these features lies at a site where erosion has lowered the land surface by 5 to 10 km and removed all topographic evidence of any rings they may have had. However, a good case for initial rings can be made at Sudbury. Although it has been deformed from a circular to a roughly elliptical structure, a radial succession of rock types strongly suggests that the Sudbury Basin originally had five concentric rings (Deutsch *et al.* 1995; Ivanov & Deutsch 1999). Isotopic investigations have revealed that the 'noritic' Sudbury Igneous Complex is, in fact, an impact melt of crustal rocks – granites, greenstones, and sediments – of such a huge volume that it differentiated in situ after being covered by a blanket of ejecta (Grieve *et al.* 1991). The Sudbury complex is the first known example of a petrologically differentiated impact melt, and it casts some doubt on the distinctions that have been made between crystallized impact melts and endogenous igneous lithologies found in the lunar highlands.

The immense Vredefort Dome surely must have originated as a multiringed impact basin. Although erosion has removed an 8 km thickness of material, including the crater itself and the impact melt, remnants of shocked and brecciated target rocks that lay at or beneath the basin floor display radial faults and a subtle concentric pattern of anticlines and synclines (Theriault *et al.* 1993). We may get a clearer idea of what the Vredefort Dome originally looked like by comparing it with the much younger multiring basin, Klenova on Venus, which is 150 km in diameter (Spudis 1993, p. 220). Venus is nearly as large as the Earth but it is so very dry that its evolution has followed a completely different course. The *Magellan* mission, which mapped Venus between September 1992 and October 1994, showed that basaltic volcanism has resurfaced the entire planet within the past 250 to 450 million years – since mid-Ordovician time on the Earth. Nevertheless, Venus has around 1000 randomly distributed impact craters, ranging in diameter from 3 to 150 km. The lower limit of 3 km indicates that Venus' thick atmosphere prevents crater-forming impacts by bodies less than 30 m across.

By 1985 four basins with at least three rings (Manicouagan, Quebec, 100 km diameter, 214 Ma; Wanapitei, Ontario, 7.5 km diameter, 37 Ma; the Ries Kessel, Germany, 24 km diameter, 15 Ma; and Popigai, Russia, 100 km diameter, 35 Ma) had been identified on the Earth (Pike 1985), in addition to the suspect ones at Vredefort and Sudbury. These sizes and ages show that a multiring structure need not be either huge or Precambrian. Then, in 1991, came the discovery of the deeply buried Chicxulub Crater in Yucatán.

Finding a multiring basin in Yucatán: twice

The 65 million year old Chicxulub structure of Yucatán, Mexico, which is covered by a 1 km thickness of Tertiary sediments, displays gravitational and magnetic evidence of rings. This crater, which is so famous today as the impact structure at the Cretaceous–Tertiary boundary suspected of having triggered the extinction of the dinosaurs, had to be 'discovered' twice before it gained the attention of the cratering community.

In 1950, a team of geophysicists employed by Petroleos Mexicanos (PEMEX), who were using gravity and magnetic surveys to explore for oil, located an enormous circular structure beneath the tip of the Yucatán peninsula, partly under land and partly under the waters of the Gulf of Mexico. In the 1960s and 1970s, PEMEX took drill cores that encountered crystalline basement rock beneath the Tertiary sediments. After that, no oil was expected, but Glen Penfield, an American consultant, and Antonio Camargo, a PEMEX geologist, made a detailed study of the maps and cores and began to think of the structure as a possible impact crater. Then, in 1980, Penfield read the landmark paper in *Science* in which Luis Alvarez (1911–1988) and his colleagues published findings of anomalously high iridium values in the Cretaceous–-Tertiary (K/T) boundary clay in Italy, Denmark and New Zealand, and speculated that the iridium was fallout from the hypervelocity impact of a meteorite that triggered global climate changes resulting in the extinctions.

In 1981, Penfield and Camargo described the crater at a meeting of the Society of Exploration Geophysicists in Houston. They suggested an impact origin, pointed to the crater's location at the K/T boundary, and invited investigations in the light of the hypothesis that the extinctions had resulted from a climate change consequent on a major impact. No members of the cratering community heard their talk – they all were at Snowbird, Utah, at the first international conference called to discuss the Alvarez hypothesis. A science reporter, Carlos Byars, heard it, however, and on 13 December 1981 he published an article in the *Houston Chronicle* titled 'Mexican site may be link to dinosaur's disappearance'. No one in Houston paid the slightest attention, although Houston is the site of both the NASA Johnson Space Center, where solar system samples are curated and studied, and the Lunar and Planetary Institute in which several young scientists were working on the 'cutting edge' of cratering studies.

Nor did any cratering specialist read and remember an account in the March 1982 issue of *Sky & Telescope* saying: 'Penfield . . . believes the feature, which lies within rocks dating to late Cretaceous times, may be the scar from a collision with an asteroid roughly 10 km across'. *Sky & Telescope* aims to interest amateurs but it diligently checks its facts to make its articles acceptable to professionals. Nevertheless, the Yucatán crater lapsed into oblivion for the next ten years while a worldwide search continued for the K/T impact crater (e.g. Powell 1998).

In 1989, Alan Hildebrand, a doctoral student at the University of Arizona, conducted a literature search that yielded a report of a bed, 50 cm thick, of volcanic glasses at the K/T boundary in Haiti. Hildebrand and a colleague, David Kring, visited the site in Haiti and found a thick deposit of large glass spherules and fragments of shocked quartz, both of which had been found, although in much smaller abundances, in samples from widespread locations at the boundary. Hildebrand's literature search also turned up a report of a circular feature under 2–3 km of sediments in the Gulf of Mexico north of Colombia, and the Penfield–Camargo abstract of 1981. By 1990, tsunami deposits, which proimpactors predicted from impacts in water, had been identified on top of Cretaceous beds at various sites around the Gulf of Mexico. These led Hildebrand to favour the crater near Colombia until it proved to be the wrong age. Then, he turned to the one in Yucatán. In 1991 Hildebrand and his thesis advisor, William Boynton, published an article in *Natural History* saying that they, together with Penfield and others, had identified 'Cretaceous Ground Zero', a deeply buried impact crater in Yucatán. They named it Chicxulub, for the village of Puerto Chicxulub near its centre.

Penfield (1991) immediately responded with a letter to *Natural History* stating that *he* had identified the structure in 1978, and he quoted the passage from the paper he had written with Camargo saying that the crater might be responsible for the worldwide distribution of iridium and the extinctions at the K/T boundary. Subsequently, Penfield and Camargo received their well-deserved recognition for their discovery of the crater, ten years after they first published a description of it.

When G. K. Gilbert spoke and wrote on lunar impact craters in 1891, his ideas were ignored because no one was interested in the Moon and impact was not accepted as a geological process. Nearly a century later, when Penfield and Camargo spoke and wrote on the Yucatán crater and presented evidence based on geophysical measurements and core samples, they, too, were ignored even though there was an eager international community of scientists searching the world for the Cretaceous–Tertiary impact crater. Does one, these days, have to belong to an 'in' group to gain an audience? Perhaps. But many scientists, meteoriticists in particular, receive so many erroneous reports of meteorite falls and finds that they hesitate to credit information from persons or institutions unknown to them. Happily, in this case, once the scientists recognized their oversight they listened with rapt attention to Camargo when he described the Penfield–Camargo data to the third Snowbird Conference in 1990, and they included Penfield and Camargo as co-authors of the first major report on the Chicxulub crater (Hildebrand *et al.* 1991). Currently, Camargo and other Mexican geologists are collaborating on the continuing research on the crater. Walter Alvarez (1997, p. 114) wrote that the ten-year waiting period was a blessing because it forced pro-impactors like himself to confront the repeated challenges of the opposition and to learn much more about the K–T boundary. However, it seems hard to concede that an extra ten years of research on the Chicxulub crater could have been anything but beneficial in the effort to understand the effects of large cratering events.

In 1997, the use of deep seismic reflection sounding at Chicxulub produced the first three-dimensional picture of the structure revealing that it is indeed a multiring basin albeit a somewhat asymmetrical one. The rim is *c.* 145 km across; inside the rim there is a rounded 'peak ring' of coarse breccia averaging 120 km across, and outside the rim are two exterior rings *c.* 169 and 250 km across (Morgan & Warner 1999). The outermost ring consists of a fault scarp that plunges beneath the crater at an angle of 30–40° and cuts all the way through the crust into Earth's mantle at a depth of 35 km. This is the first known example of such a deep fault, and it raises many questions about our beliefs on the nature of the crust–mantle boundary (Melosh 1997). Although Chicxulub was the first terrestrial impact structure to be suspected of triggering a mass extinction, evidence is accumulating that other impact events may have caused extinctions, including the most lethal extinction of all that wiped out about 96% of the world's species 250 million years ago at the end of the Permian.

One more multiring basin lies at Morokweng, South Africa. It appears to be at least 200 km across and 145 million years old (Reimold *et al.* 1999). More recently, gravity anomalies and core drilling have revealed that the deeply buried Woodleigh structure in Western Australia is a multiring basin. It is *c.* 120 km across and appears to be late Triassic in age, i.e. 206 to 210 million years old (Mory *et al.* 2000). Studies are under way of several other possible multiring basins.

Lunar stratigraphy and timescale: the Hadean period on Earth

Geological history began when Earth's first patches of surviving crust formed, which occurred approximately 4.0 Ae ago. Even then it was not until *c.* 3.80 Ae ago that the first permanent systems of rocks formed, such as those at Isua in Greenland, that preserve a decipherable record of geological events. The 750-million-year interval previous to that is called the Hadean Period (e.g. Harland *et al.* 1989, p. 22). We picture the Hadean as a time when impacting bodies broke up the earliest slabs of crust and mixed them back into the molten interior time and time again. Some geologists refer to this as 'pregeologic time' and it is true that only by examining the face of the Moon can we decipher the history of impacts and volcanic eruptions that took place on the Earth during that earliest interval.

In the early 1960s, astrogeologists worked out the first system of lunar geologic periods since that of G. K. Gilbert in 1893. They established a relative timescale of five major stratigraphic periods based on detailed mapping of formations linked to dated samples (Wilhelms 1993). The earliest period, designated as pre-Nectarian, was one of heavy bombardment and basin formation. The excavation of the Nectaris basin 3.92 Ae ago opened the Nectarian Period during which ten more large basins were formed. That period ended when a giant impact excavated the Imbrium Basin and opened the Imbrian Period 3.85 Ae ago. The oldest mare basalt flows erupted early in this period, which is divided into Early and Late Sub-Periods. Great volumes of basalt continued to flow into the

Table 1. *Timescale of lunar history*

Time (Ae (10^9 years) BP)	Periods and events
1.10 to Present	*Copernican Period* Copernicus Crater formed *c.* 0.81 Ae BP, minor basalt flows intercepted its rays; Tycho Crater formed *c.* 0.11 Ae BP; sporadic impacts and rains of small particles continue
3.20 to 1.10	*Eratosthenian Period* Eratosthenes Crater formed *c.* 3.20 Ae BP. New impact craters formed and old ones degraded; major mare volcanism ceased *c.* 3.10 Ae BP.
3.80 to 3.20	*Late Imbrian Sub-Period* Large craters formed while rate of impacts diminished; voluminous eruptions of mare flows, including Mare Tranquillitatis 3.84–3.57, Mare Serenetatis *c.* 3.72, Mare Fecunditatis *c.* 3.40, and Mare Crisium *c.* 3.3 Ae BP.
3.85 to 3.80	*Early Imbrian Sub-Period* Oriental Basin formed *c.* 3.80 (end of Hadean Era on Earth). KREEP-rich volcanic basalt erupted *c.* 3.85 Ae BP.
3.85 to 3.20	*Imbrian Period* Imbrium Basin formed *c.* 3.85 Ae BP.
3.92 to 3.85	*Nectarian Period* Nectaris basin formed *c.* 3.92 Ae BP.; heavy bombardment continued; ten more basins formed including Serenitatis *c.* 3.87, and Crisium *c.* 3.86 Ae BP.
4.15 to 3.92	*Pre-Nectarian Period* Heavy bombardment excavated *c.* 30 large basins, including Procellarum *c.* 4.15 and South Pole–Aitken *c.* 4.10 Ae BP, and 3400 craters >30 km diameter.
4.55 to 4.15	*Cryptic Division* Moon achieved present mass *c.* 4.55 Ae BP; lunar magma ocean formed and differentiated; lunar crust solidified *c.* 4.47 to 4.2 Ae BP; extensive volcanism, plutonism, impact melting, mixing and cratering.

Lunar dates are from Wilhelms (1993), Harland *et al.* (1989) and S. R. Taylor (pers. comm. 2001).

basins of the lunar near side until *c.* 3.2 Ae ago when the crater Eratosthenes was excavated. Eratosthenes is not a large crater; nevertheless, it serves well as a post-Imbrium stratigraphic time-marker. During the Eratosthenian Period, which lasted until *c.* 1.1 Ae ago, the rate of impacts declined dramatically and basaltic eruptions ceased, except for small events detectable only by mappers. The final, Copernican Period, which began *c.* 1.1 Ae ago, has been a time of sporadic impacts and a continual rain of small particles forming zap-pits. Its two most dramatic events were the impacts that formed the bright-rayed crater, Copernicus, *c.* 0.81 Ae ago, on the southern margin of the Imbrium Basin, and the even brighter-rayed crater, Tycho, *c.* 0.11 Ae ago. Tycho is the spectacular crater that dominates the Moon's southern highlands and has long rays that fan out in every direction except toward the SW. All lunar geology is Precambrian, as indicated in the comparative lunar and terrestrial timescales in Table 1. Astrogeologists have applied the lunar timescale to the other terrestrial planets and satellites, all of which show evidence of having been subjected to the initial heavy bombardment simultaneously.

Origin of the Moon: the giant impact hypothesis

Did the *Apollo* missions tell us the true story of how the Moon originated? No. But they furnished us with an abundance of new information that prompted a major rethinking of the problem. In the mid-1970s two research teams independently proposed a new hypothesis that the Moon resulted from a giant impact on the protoearth by another large body (Hartmann & Davis 1975; Cameron & Ward 1976). This idea was the principal topic discussed throughout a meeting in Kona, Hawaii, in 1984, from which it emerged as the most favoured modern hypothesis.

In its original form, the hypothesis assumed that the protoearth and a body at least the size of Mars had accreted in heliocentric orbits in our neighborhood of the Solar System. After both bodies had formed their cores, the smaller one struck the Earth a glancing blow in which the impactor itself and a sizeable portion of the Earth's mantle were ejected into space, largely as a long plume of white-hot vapour that went into orbit around the Earth. Most of the heavy metals of the impactor's core fell to the Earth, sank through the mantle, and coalesced with Earth's core. The rest of the ejected matter that was not lost into space collapsed into an orbiting disc that aggregated into the Moon.

The problems lay in the details. Did this process create the Moon in a few hours or a few million years? How much of the Moon consists of Earth's mantle and how much came from the impactor? Did all the debris vaporize or did some of it remain in sizeable chunks? Did accretion of solar material continue adding mass to either or both of the two bodies after the event? A. G. W. Cameron (b. 1925) has tested this hypothesis in long-term computer runs that have established important new constraints. In his current version, Cameron (2001) requires the impactor to have been at least three times the size of Mars, and the collision to have taken place when accretion of the Earth was only about half complete. A solid remnant of the impactor then looped back and struck Earth a second glancing blow. Most of the matter that went into orbit and accumulated into the Moon came from Earth's mantle. The whole process was rapid: only about 50 million years passed between the formation of the primitive solar nebula and formation of the Moon.

Cameron emphasizes that the Giant Impact still remains a hypothesis – not a theory – because it lacks a secure mechanism. However, he is convinced that this is the right approach because it explains better than any other model most of the puzzling aspects of the Earth–Moon pair. It accounts for the Earth's tilted axis and the Moon's tilted orbit; the Moon's lack of water and volatiles, and of any sizeable metal core; the matching of the Moon's oxygen isotopes with those of the Earth; the energy source for formation of the magma ocean that enabled the flotation of anorthite to form the lunar crust and the derivation, an aeon later, of basaltic lava flows by partial melting of the lunar mantle. Most convincing of all, it accounts for the anomalously high angular momentum of the Earth–Moon system. This work-in-progress provides us with the most promising key to understanding why the Earth has its Moon.

It does not, however, necessarily tell us about the moons of other planets. The tiny moons of Mars are generally believed to be captured asteroids; those of the giant planets may well include captured moons, moons accreted from planetary rings, and moons formed by simultaneous accretion. Images from space have shown us that every planet and satellite in the Solar System has had a different evolutionary history from every other.

Rocks from the Moon and Mars: meteorite recovery expeditions on the Earth

In December 1969, the same year the *Apollo 11* mission landed on the Moon, Japanese geologists in Antarctica discovered nine stony meteorite fragments on a small patch of bare ice. They returned the specimens to Tokyo assuming they were all pieces of the same meteorite. However, in 1973 two Japanese cosmochemists published petrographic and isotopic analyses of four of the specimens showing that, far from being shower fragments, they were samples of four different classes of meteorites: a carbonaceous chondrite, an enstatite chondrite, an olivine–bronzite chondrite, and a Ca-poor achondrite (Shima & Shima 1973).

As news of this announcement spread abroad, it created a sensation. Clearly, stones from different falls had, in some way, been concentrated into a 'placer deposit', presumably by ice motion. Up until then extraordinary good luck had been required either to see a meteorite fall or to recognize one in the field, but the Japanese discovery opened up the possibility of conducting systematic searches for them. In 1973, a team from Tokyo went to Antarctica to look for meteorites, and in 1976 the first team went there from the USA. Japanese and American members searched together for three years. Since then USA-led expeditions, most often including members from other countries, have gone to Antarctica annually, and Japanese teams and European consortiums have gone frequently.

The Antarctic situation is unique: meteorites fall onto the great dome-shaped ice sheet and are frozen-in and carried slowly downslope, roughly northward in all directions. Wherever the ice reaches the sea, its cargo of meteorites will be lost into the water. Occasionally, large masses of ice temporarily stagnate behind mountain ranges where they are worn down by wind ablation. As the winds cut deeper and deeper, meteorites of all types and ages are exposed at the surface. In three or four instances, meteorites with only their upper tips visible have been collected in blocks of ice and kept frozen until studies can be made of the crystalline structure of the ice.

During the past three decades, nearly 24 000

meteorite fragments have been collected by all parties (Grady 2000). No one can be certain how many individual falls this number represents, but a rough estimate of four fragments per fall would suggest that Antarctica has yielded samples of approximately 6000 individual meteorites – almost three times the number that were catalogued in the world collections in 1975. However, the large numbers of fragments are not nearly so important as the fact that they include new types of meteorites from asteroids and meteorites from the Moon and Mars. They also include a few stony meteorites that fell to the ice sheet two million years ago – long before most of those found on other continents.

Inasmuch as Antarctica is subject to an international treaty that allows the taking of geological or biological specimens solely for purposes of scientific research, all of the meteorites are collected untouched (see Fig. 12), by techniques designed to prevent contamination, and are shipped, still frozen, to receiving laboratories in the countries of the expedition leaders. There, they are processed under standard clean-room conditions and, following certain protocols, documented samples are sent to research laboratories around the world. Expeditions to Antarctica are a most elegant means of adding to our store of samples while we await future collecting missions to planetary bodies.

Fig. 12. On the USA meteorite collecting expedition of 1981–1982 to Antarctica, the author inspects a small meteoritic stone while John Schutt, the team's alpinist, takes out a sterile plastic bag in which to collect it (photograph courtesy of Ghislaine Crozaz).

Arid regions

Searches have also been conducted in arid regions where the dry climates and sparsity of population make the survival rate of meteorites relatively high. The most successful ones have been in Roosevelt County, New Mexico, USA, the Nullarbor plain of southwestern Australia, the Sahara Desert of North Africa from Morocco to Libya, and in Oman in the south-eastern portion of the Arabian peninsula.

In the high plains of New Mexico, gentle depressions in the surface collect temporary pools of rain water and are worn deeper and deeper as winds blow away the dry soils after the water evaporates. This process is hastened wherever buffalo or range cattle wallow in the waters. Eventually, the hollows expose the regional layer of 'caliche', a hard subsoil cemented by calcium carbonate. Meteorites from within the topsoil are left on the hard-pan floors of the hollows. From 1968 to 1979, 101 meteorites were found in Roosevelt County.

In most deserts, the winds continually blow away the fine-grained materials deflating the surfaces and leaving meteorites among the rocks in so-called desert pavements. Since the early 1960s, systematic searches of Australia's Nullarbor plain, conducted by the Western Australia Museum at Perth, have recovered more than 280 meteorites. Beginning in 1990, private individuals have collected 300 meteorites in Oman and more than 1500 fragments from reaches of the Sahara in northern Africa (Schultz *et al.* 2000). Portions of these meteorites are sent to universities and to private collections, and some of them reach the open market. Meteorites from hot deserts also include specimens from the Moon and Mars. How do these meteorites get to Earth, and how do we identify them when we find them?

Mars: the planet of phantasy

For centuries, Mars with its red colour and variable dark markings, has sparked the imaginations of storytellers and science fiction writers. One of the unintentional masters of this genre was Percival Lowell (1855–1916) who built an observatory at Flagstaff, Arizona, specifically to observe Mars. Lowell assumed that an 1877 report of 'canali' on Mars by the astronomer Giovanni Schiaparelli in Sicily, referred to canals, although the word properly translates to channels or lines. From 1895 onward Lowell published a succession of well-received popular books and articles on the technically advanced inhabitants of Mars with their great cities and monumental systems of canals. The facts are very different.

Mars is slightly over half the size of the Earth, lower in density (3.94 compared to 5.52 g cm^{-3}) and is half as far again from the Sun, making it a small, cold place with an average temperature of

–50°C. Our information on Mars comes from fly-by and orbital missions of the USA and USSR that began in the 1960s, particularly those from the USA *Mariner 9* orbiter of 1971, the *Viking Landers* of 1976, *Mars Pathfinder* of 1997 with its small rover, *Sojourner*, and the *Mars Global Surveyor* which, in the summer of 2001, was approaching the end of a long programme of orbiting and imaging the planet.

Mars has a very thin atmosphere, mainly of carbon dioxide, with high winds and fierce annual dust storms that sometimes last for months. Its two polar caps, which expand and contract seasonally, consist mainly of dry ice (CO_2), although the northern one (and perhaps also the southern one) has an underlying shield of water ice. Mars has the highest volcanoes, the longest system of rift valleys, and one of the largest multiringed impact basins (Hellas, more than 2000 km across) in the Solar System.

In 1965, *Mariner 4* provided our first view of Martian impact craters, and in 1972 the *Mariner 9* orbiter sent us clear images of two entirely different hemispheres. The southern two-thirds of Mars consists of heavily cratered, and hence older, highlands standing 1 to 4 km above Martian base level (the average radius). More than 20 multiring basins and a large number of craters, 20 to 200 km across, have been mapped there. All of them are degraded; none have fresh profiles or rays. Unlike those on the waterless Moon, many Martian craters are surrounded by lobes of smooth mud-like ejecta; they very likely are muddy, due to ice or permafrost in the soils.

The northern third of Mars consists mainly of sparsely cratered plains, lying below base level. Along the equator is a huge upland, the Tharsis bulge, 10 km high and 4000 km across, with three immense shield volcanoes along its crest. Nearby is Olympus Mons, 26 km high, the largest Martian volcano. The great system of subparallel rifts, called Valles Marineris, occurs along the bulge. It is 4000 km long, and would stretch across the USA from Boston to San Francisco. Mars has dramatic scenery for a small, cold planet.

Our first chemical data on Martian soils were obtained in 1976 by the *Viking* missions. That year, two *Landers*, loaded with instruments, put down about 6500 km apart and analysed the Martian soils and atmosphere. The soils proved to be basaltic in composition with a strong indication of permafrost at a shallow depth. The atmosphere is strikingly different from that of the Earth. The Martian atmosphere consists of 95% CO_2, 2.6% N_2, 1.4% Ar and 1% other gases; the Earth's, of 78% N_2, 21% O_2 and 1% other gases. Each of the *Viking Landers* carried out three separate experiments in search of living organisms and found no positive evidence of them.

In 1979, the suggestion was made for the first time in print that the Shergotty meteorite and at least some of its siblings in the SNC suite might have come from Mars (Wasson & Wetherill 1979, p. 164). These meteorites share several characteristics suggestive of an origin in a body larger than the Moon and much larger than an asteroid. They are igneous lavas and cumulate rocks that formed under oxidizing conditions and are richer in volatiles, including water, than any samples of the Moon or of other achondrites. And they are astonishingly youthful: they all crystallized less than 1.3 Ae ago and some of them only 100 to 300 million years ago. Where, besides Mars, can we find a parent body large and well insulated enough to have sustained magmatic activity throughout most of the history of the Solar System?

One problem, pointed out by Wasson and Wetherill, is that the SNC meteorites have very short cosmic ray exposure ages. Shergotty, for example, had its radiometric (Rb–Sr) clock reset by an impact event that transformed its feldspar to glass *c.* 165 million years ago. Yet it shows a record of being bombarded by cosmic rays for only two million years. Possibly the meteorite began as a well-shielded piece from the interior of a much larger mass that was ejected into space 165 million years ago and then broke into pieces only two million years ago. But from the standpoint of celestial dynamics, the ejection of so large a mass from Mars seemed almost out of the question. Nevertheless, David Walker and his colleagues at Harvard University (1979) argued that only well-insulated bodies at least as large as Mars could sustain volcanism throughout most of the age of the solar system. They referred favourably to the Wasson–Wetherill suggestion of Martian origin and supported the idea with a table showing a very close match in chemical compositions of the Martian soil and the Shergotty meteorite.

That same year a US team of Antarctic meteorite searchers at a site called Elephant Moraine found a large (7.5 kg) basaltic stone (EETA 79001; see Fig. 13), which proved to be a shergottite. Sawed surfaces revealed internal pods of dark, shock-produced glass (Fig. 13b), which was full of minute bubbles loaded with trapped gases – argon, krypton, xenon – in relative proportions and isotopic compositions similar to those of the Martian atmosphere (Bogard & Johnson 1983). Furthermore, it proved to be another youthful achondrite: the basalt crystallized only about 175 million years ago –

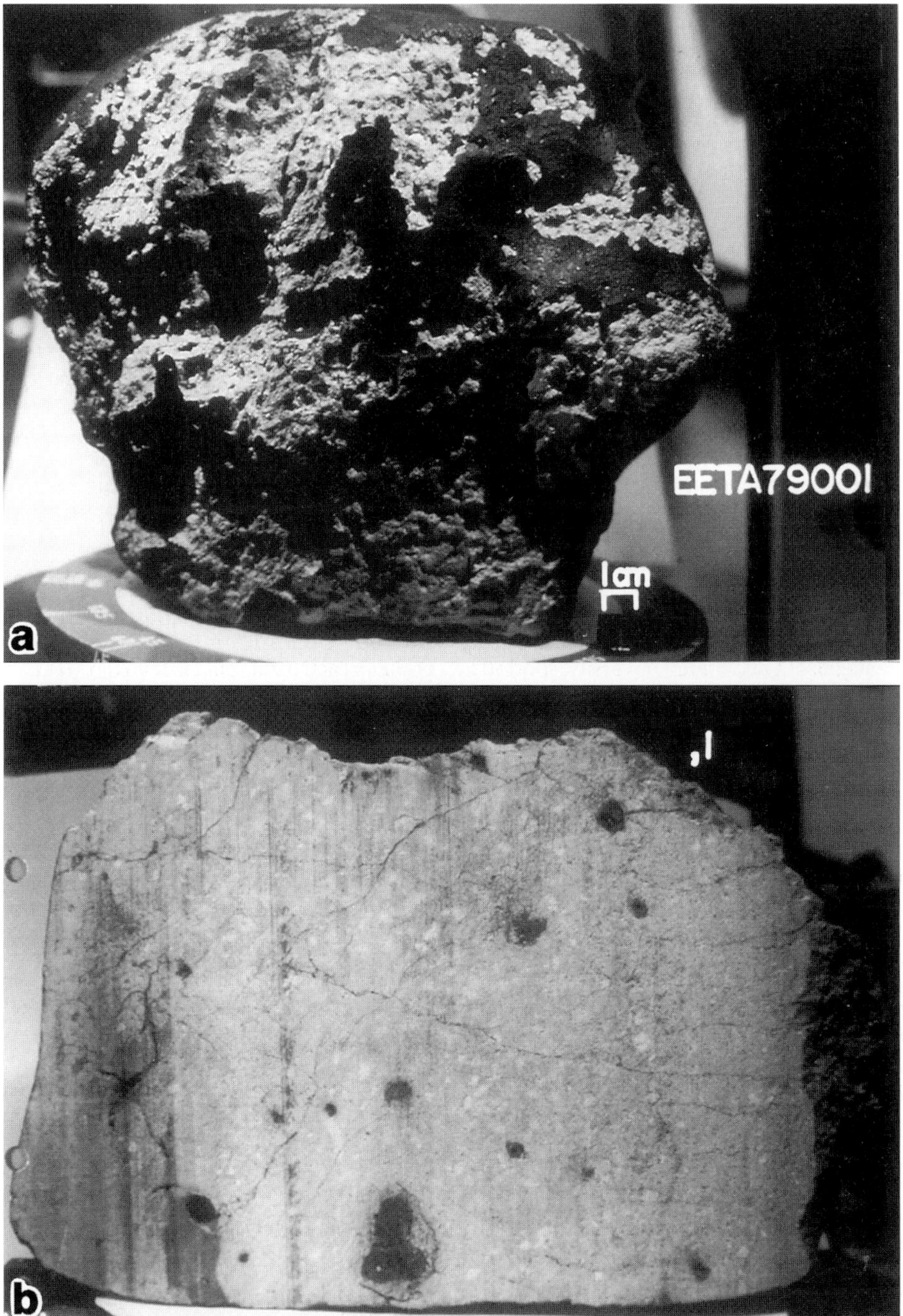

Fig. 13. The Antarctic shergottite, Elephant Moraine 79001. (**a**) The uncut stone with large patches of fusion crust. (**b**) A sawn surface showing pods of dark shock-produced glass which is riddled with minute bubbles containing Martian atmospheric gases (NASA photographs at the Curatorial Facility in Houston; the cube is 1 cm on an edge).

practically yesterday! Scarcely any further proof was needed that this meteorite was a Martian rock, and if the Elephant Moraine shergottite was Martian, were not all the members of the SNC suite Martians? The answer had to be 'Yes'.

How could specimens of Mars get to Earth? Most of us picture giant impacts on Mars, routinely blasting off debris into Earth-crossing orbits. But in the early 1980s specialists in celestial dynamics objected that this could not work, for two reasons: one dynamical, the other statistical. First, computer models showed that an impact of sufficient magnitude to accelerate debris to escape velocity from Mars (5.0 km s^{-1}) would shock, melt or crush the target rock beyond recognition, even if some of it were to fall on Earth. Second, we had found no meteorites from the nearby, much-cratered Moon with an escape velocity of only 2.4 km s^{-1}, so we certainly could not expect to have them from Mars.

Then, *voila*, a meteorite lying on the Antarctic ice sheet solved that problem. On a snowmobile traverse taken on the final day of the US field season of 1981–1982, a visiting glaciologist, Ian Whillans of Ohio State University, spotted a small rock about the size of an apricot. It had a brownish, frothy fusion crust and large white clasts in a brown glassy matrix – totally different from any known meteorite but much like some of the *Apollo 16* lunar highland rocks.

At the curatorial facility in Houston this specimen was labelled ALHA 81005, and a small chip of it was sent to Washington where Dr Brian Mason identified the white clasts as consisting mainly of anorthite (what else?). Samples quickly were sent to 14 research laboratories in several countries and, although members of the meteoritical community rarely agree upon anything, all 14 groups concurred that this was, without any doubt, a meteorite from the Moon. Thus, we call this expedition 'the *Apollo 18* mission': it went to Antarctica and returned with lunar sample ALHA 81005 (Fig. 14), which changed the history of planetary geology.

Now that we had a meteorite from the Moon, dynamicists tested new ways to eject intact rock fragments from planetary surfaces (Melosh 1985). Early results indicated that small unshocked or lightly shocked fragments of lunar surface rocks could be accelerated to escape velocity from the Moon by stress-wave interferences set up on the lunar surface by large impacts. However, this process could not be extended to acceleration of the larger, more deep-seated rocks of the SNC suite to escape velocity from Mars. Then, new tests showed that rocks could escape almost tangentially from Mars-sized bodies if they are entrained in vapour plume jets caused by the force of low-angle impacts (O'Keefe & Ahrens 1986). Some difficult dynamical problems remain unsolved; nevertheless, 'everybody' agrees that the SNC meteorites come from Mars.

Sixteen Martian meteorites are recognized today. All are youthful lavas or cumulate rocks except one: Allan Hills 84001, a 1.9 kg Antarctic pyroxenite that crystallized deep within the Martian crust *c.* 4.5 Ae ago. Pyroxenite is one of the last types of rock in which we would be likely to search for fossils, but it is the one in which a consortium of scientists, led by David McKay at the Johnson Space Center in Houston, announced in 1996 that they had found possible evidence of ancient Martian life. The team offered no proof: they simply offered four lines of evidence, which, taken together, they regarded as good evidence. They invited others to test it (McKay *et al.* 1996).

The news of life on Mars was carried on CNN International where I heard it during the International Geological Congress in Beijing and realized, instantaneously, that I would return home to total uproar. At that time I was chairing the committee that allocates US Antarctic samples for research. The uproar arose as expected. A special committee of scientists who were not, themselves, working on Martian samples evaluated 101 sample requests and assigned a total of 75 mg of samples to 39 laboratories in several countries. This small sample size illustrates the miniscule amounts from which analysts, today, can extract a maximum of crucial chemical, mineralogical, trace element and isotopic data.

Volumes of research results have since been published on this meteorite but, as of this writing, no conclusive proof has been established that it contains fossils. Even members of the original group have conceded that most of their proposed evidence is not uniquely indicative of biological activity. Inasmuch as the burden of scientific proof always rests entirely upon those making a claim, we can only state that as yet we have no positive evidence of life on Mars, past or present.

However, bacterial life on Mars seems to be a perfectly reasonable proposition. Mars clearly has been much wetter in the past than it is today. Beginning in 1972, *Mariner 9* and subsequent missions have imaged long, dendritic valleys, immense canyons, deep channels and enormous deltaic deposits wrought by rampaging waters that have long since vanished. Some of this Martian landscape bears striking similarities to that of the Channeled Scablands that cover

Fig. 14. Lunar meteorite Allan Hills 81005, collected on 'the *Apollo 18* mission', the US expedition of 1981–1982 to Antarctica. (**a**) The stone as it appeared in the snow (photograph by John Schutt). (**b**) A broken surface showing white anorthositic clasts and other rock and mineral fragments embedded in brown glass (NASA photograph taken in the Curatorial Facility at Houston; the cube is 1 cm on an edge).

520 km^2 of eastern Washington State in the USA (Baker 1978, 1982). From 1923 to 1928, the geomorphologist J Harlen Bretz (1882–1981) described the scablands as having been carved out by a cataclysmic debacle he called the Spokane Flood, in which volumes of water suddenly scoured deep channels, huge potholes and high-walled cataracts in basaltic bedrock. Concurrently the waters built gargantuan gravel bars several kilometres long and tens of metres high,

and deposited huge erratic boulders hundreds of kilometres from their sources. Bretz's all too ununiformitarian hypothesis was roundly denounced as 'outrageous', and many of his contemporaries felt that this heresy should be gently but firmly stamped out. Extreme floodwaters were anathema to uniformitarians, whose earliest struggles in the previous century had been waged in opposition to the Noachian deluge. Their successors did not like Bretz's deluge any better.

For the first seven years, Bretz was unable to point to an adequate source of the floodwaters. He could only describe the field evidence and ask his colleagues to consider his hypothesis not by emotion or intuition but by the principles of the scientific method. In a dramatic confrontation in 1927 Bretz presented his evidence before a room packed with his adversaries who clung to their ideas that the scablands were fashioned by glaciers, or by floating icebergs, or by normal but long-continued stream erosion. By 1930, however, Bretz, had found a source for the water. He declared that the floodwaters were released from glacial Lake Missoula, an enormous Pleistocene body of water in southwestern Montana, which had poured across northern Idaho into Washington, through the scablands, and down the Columbia River valley to the sea when an ice dam suddenly failed.

Lake Missoula had been described as early as 1910 by John T. Pardee (1871–1960) of the US Geological Survey. From his letters written in the 1920s it appears that Pardee was considering Lake Missoula as the likely water source that Bretz so urgently needed. However, Pardee had been argued out of this idea (or at least argued out of publishing it) by the chief of the Survey's Hydrology Division, so he remained silent until after Bretz discovered Lake Missoula for himself (Baker 1978). In later years, Bretz found evidence of several episodes of flooding as the ice dam congealed and failed again (Bretz *et al.* 1956). Today, some investigators postulate, from evidence based on ancient shorelines, that up to 100 episodes have occurred.

Even after Bretz pointed to Lake Missoula as the source, many geologists still hesitated to accept the actuality of the Spokane Flood. Few of them visited the remote scablands, and those who did found the immense scale of the landforms almost impossible to comprehend by observers on the ground. In 1956, Bretz returned to the scablands with two young colleagues and re-examined the evidence using all the advantages of aerial photos, new topographic maps, and numerous excavations made during construction of the new Columbia River irrigation system. Their findings confirmed his flood hypothesis in detail (Bretz *et al.* 1956). His theory finally went mainstream in 1965 when participants of a Congress of the International Association for Quarternary Research visited Lake Missoula and the scablands and afterward cabled Bretz: 'Greetings and salutations! . . . We are now all catastrophists' (Bretz 1969).

Seven years later *Mariner 9* sent images to Earth of flood deposits on Mars similar in form but even more catastrophic in scale than those of the scablands (Baker 1982, chapter 7). *Mariner 9* led to the ease of interpretation of the deltaic site strewn with enormous boulders near rounded hills where *Mars Pathfinder* landed in 1997 (Baker 1998, p. 172). J Harlen Bretz lived to see his catastrophic flood theory extended to another planet. In 1980, at the age of 97, he was presented with the Penrose Medal, the highest honour bestowed by the Geological Society of America. Bretz was enormously pleased, although he did lament to his son that he had outlived all of his old adversaries and had no one left to gloat over.

In July 2000, new images from *Mars Global Surveyor* showed fresh-looking gullies in the walls of a large Martian impact crater, doubtless formed by an erosive agent such as water released from permafrost. Perhaps, as was suggested, the gullies formed 'only' a million years ago. Appearances can be deceiving, but the gullies as seen in published pictures actually looked as though they had formed only a few weeks ago. On the whole, it seems very likely that someday in the not too distant future, we will find water on Mars, possibly as springs, fumaroles or seepages of permafrost. And in rock samples from the Martian surface we may detect conclusive evidence of ancient Martian life, and conceivably of present Martian life as well.

If ever we do find living organisms on Mars, they most likely will be simple cellular bacteria or archaea of the types that were the only living things on our own planet during the first 3000 million years or more of its existence. On the Earth such organisms thrive under the most extreme conditions: some of them luxuriate in total darkness around the vents of black smokers along oceanic ridges; others populate minute pore spaces inside rocks subject to the prolonged cold and winter darkness of Antarctica; still others live in minute pore spaces kilometres deep inside thick layers of plateau basalts. At present, the Earth's oldest dated microfossils are found in the *c.* 3465 million year old early Archean Apex Chert of north Western Australia (Schopf 1993). Even earlier evidence

Table 2. *Chronology of Earth's impact scars and the record of life*

Ma BP	Eras of Earth history
65 to present	*Cenozoic Era* 33 Tertiary and 24 Quaternary dated impact features; age of mammals; appearance of *Homo sapiens sapiens* ≥130 000 years ago; of highly technical *Homo* seeking companions in universe *c.* 50 years ago.
250 to 65	*Mesozoic Era* 39 dated impact structures; Chicxulub crater formed 65 Ma coincident with mass extinction at end of Cretaceous Period; flourishing of life; age of reptiles, earliest appearance of mammals.
570 to 250	*Paleozoic Era* 49 dated impact structures, including four large ones nearly coincident with extinction events, but with no links or periodicity established; Cambrian 'explosion' of life *c.* 570 Ma BP followed by proliferation of Palaeozoic marine animals, land plants, insects, reptiles, etc.; mass extinction at end of the Permian Period 250 Ma BP was the most severe on record. *Precambrian Era: 4550 to 570* Ma BP
2500 to 570	*Proterozoic Eon* Vredeford Dome 2200 Ma, ≤300 km diameter; Sudbury Basin *c.* 1850 Ma, 250 km diameter; plus seven dated impact scars >600 Ma, 3 to 30 km diameter. Marine algae, oldest organisms with cell nuclei and chromosomes occur in strata 1400 Ma; oldest multicellular algae in strata *c.* 1200 Ma; organisms capable of locomotion appeared *c.* 1000 Ma.
3800 to 2500	*Archaen Eon* No known impact scars; oldest dated fossils, filamentous microbes *c.* 3465 Ma in Apex Chert, Western Australia; sedimentary rocks showing isotopic signatures of light carbon (^{12}C) indicative of biologic processes at Akilia Island, Greenland, 3820 to 3950 Ma; at Isua, Greenland, 3650 to 3800 Ma; Onverwacht, S. Africa, 3550 to 3590 Ma; at Warrawoona, Australia, 3500 to 3550 Ma.
4550 to 3800	*Hadean Eon* Earth achieved present mass *c.* 4550 Ma; evidence of heavy bombardment and volcanism visible only on Moon; oldest dated minerals, detrital zircons *c.* 4400 Ma old in Western Australia; earliest dated patches of crust, Acasta gneisses, Canada, 3900–4000 Ma. Lunar Period names: e.g., Cryptic, Nectarian, Imbrian, applied to this Eon.

Compiled from Geological Society of Canada crater list, at Crater@gsc.nrcan.gc.ca, Mojzsis & Harrison (2000), and Harland *et al.* (1989).

of biologic activity, based not on fossils but on the presence of light isotopes of carbon, occur in the *c.* 4000 million year old sediments of Akilia Island in southern Greenland (Mojzsis & Harrison 2000).

Glancing back over the history of life on Earth we find that complex cells with nuclei and chromosomes first appeared *c.* 1400 Ma ago, multicellular organisms arose *c.* 1200 Ma ago, and those capable of locomotion first began moving about in the muddy waters of the late Precambrian *c.* 1000 Ma ago. The Earth was already 3550 million years old by then, and almost another 1000 Ma would pass before *Homo sapiens* would evolve *c.* 130 000 years ago (Table 2). Once mankind began to build cities, nearly a dozen civilizations rose and fell before the development, in the twentieth century, of a technical civilization capable of asking if we are alone and sending signals into space in an effort to find out. That we shall receive an answer seems most unlikely. Thousands of millions of years and a long series of contingencies preceded our own existence on the warm, wet, wonderfully habitable Earth. How can we suppose that beings similar to ourselves would be present at this particular moment anywhere in our galaxy, or in distant galaxies. We may well be alone in the universe (Taylor 1998).

The *Apollo* missions and the environmental movement

In 1950, the astronomer Fred Hoyle wrote: 'Once a photograph of the Earth, taken from outside is available . . . once the sheer isolation of the Earth becomes plain to every man whatever his nationality or creed, a new idea as powerful as any in history will be let loose'.

Hoyle surmised that this development in the not-so-distant future might benefit the whole of society by exposing the futility of nationalistic strife (Hoyle 1950, p. 26).

The first images of the whole Earth from space were made by *Lunar Orbiter 1* in August, 1966. They were black and white and some of them showed oblique views of the much-dimpled surface of the Moon in the foreground. The pictures thrilled people interested in the space adventure but fell short of inspiring the public as predicted by Hoyle. Then, at Christmas time in 1968, three astronauts aboard the *Apollo 8* mission rode all the way to the vicinity of the Moon and swung into orbit. Never before had men journeyed so far into space and they gave us our first opportunity to see the whole Earth through their eyes. Their photographs showed us our planet looking like a radiant blue and white sphere isolated in the blackness of space. Did these images let loose a new idea as powerful as any in history? Perhaps not the urgent impulse toward world peace and brotherhood that Hoyle envisioned, but they surely did inspire an enhanced appreciation of our planet.

We can find clear evidence of this by checking through the index volumes of *The New York Times* for the years 1960 to 1970 under the headings: 'Environment' and 'US environmental problems'. There, one can trace a precipitous rise in 1969 in the numbers of items on these subjects (Bethell 1975). From 1960 to 1963, there are zero items under either of these two headings, nor are there any under 'Ecology'. Although, some groups of citizens had been actively working toward bettering the environment throughout the twentieth century, none of their efforts caught the attention of *The New York Times* in the early 1960s.

In 1964 one item was listed on environmental housing design. In 1965, the index included 9 cm of listings; President Lyndon Johnson had begun to take an interest in the environment, and his wife was promoting the beautification of America. In 1966 President Johnson created the Environmental Science Services Administration (ESSA). There were 30 cm of listings that year and 31 cm in 1967.

Then, at Christmas time in 1968, the three astronauts of *Apollo 8* orbited the Moon and snapped pictures of the Earth in space. On the way home Astronaut Lovell welcomed the sight of the approaching Earth 'like a grand oasis in the vastness of space'. Splashdown in the Pacific Ocean took place on 28 December, thus ending the trip called 'the greatest voyage since Columbus' (see cover illustration).

The response was immediate: although the voyage took place in late December, the listings in *The New York Times* index for those last few days caused the total for 1968 to nearly double to 51 cm. On Christmas Day, Archibald MacLeish, Poet Laureate of the USA, wrote of the vision of Earth as seen by the astronauts; on 29 December Dr Barry Commoner warned that man may destroy the environment out of his ignorance; and on 30 December, an editorial called for an all-out effort to halt destruction of the environment.

In 1969, the listings more than tripled to 165 cm. In January alone *The New York Times* printed seven major articles plus numerous editorials and letters on protection of the environment; *Time* magazine hailed *Apollo 8* astronauts William A. Anders, James A. Lovell, Jr, and Frank Borman as Men of the Year; the Sierra Club started a worldwide drive to preserve the environment; UN Secretary-General U-Thant estimated that we had around ten years to prevent irreversible damage to the environment; law schools began offering courses on legal issues relating to the environment and students crowded into them.

On 20 July 1969, the *Apollo 11* Moon landing was televised to half a billion people around the world. The astronauts photographed the Earth rise from the surface of the Moon, emphasizing once again the contrast between our watery cloudy planet, where life abounds, and the desolation of the Moon.

In 1970 the listings in the index nearly tripled again to 559 cm on environmental issues, with much cross-referencing to related items. That year Earth Day was inaugurated with 20 million people responding; in the US alone, the National Environmental Policy Act was enacted into law and the Environmental Protection Agency was formed. One item referred to a plan by the USA and the USSR and other nations to establish 20 stations around the world by 1972 to monitor harmful environmental changes.

A rise from 51 to 559 cm in just two years (1968–1970) qualifies as an explosion! Suddenly public interest was galvanized into preserving our environment – a concern that has expanded since then to a very major issue in our everyday lives. Beyond doubt, the men of *Apollo 8* gave us the concept of 'Spaceship Earth'.

Recapitulation

In this story we have seen that the solitary Earth of classical geology orbits the Sun amid a community of planets that is almost neighbourly. Primitive meteorites have fallen to Earth bearing grains of stardust, and differentiated meteorites have brought us knowledge of several styles of planetary magmatism. The *Apollo* astronauts visited the Moon and brought back samples showing that the Moon is as old as the Earth and has the same oxygen isotopic signature, indicating that the two bodies originated in the same region of the Solar System. Multiple lines of evidence led to the giant impact hypothesis, in which the proto-moon struck the Earth a glancing blow, sending enough vaporized materials from Earth's mantle into orbit around the Earth to ultimately give rise to the Earth–Moon system.

Mapping and sampling of the lunar surface and tying its stratigraphy to dated samples has provided us with our first tangible evidence of events that took place during the earliest, Hadean, period of the Earth's history, of which we have no extant record in terrestrial rocks. Lunar names for the first four epochs of the Precambrian Era have been adopted in geological timescales.

Samples collected from the anorthositic lunar highlands, enabled us to identify an exotic rock found on the Antarctic ice-sheet as a lunar meteorite. Meanwhile, the Viking missions to Mars had beamed back analyses of the Martian soils and atmosphere that subsequently were shown to match the bulk composition and gas content of a remarkably youthful meteorite, a shergottite, found at Elephant Moraine in Antarctica. The discovery of the first lunar meteorite allowed us to seriously consider a Martian origin for that shergottite and for 15 other members of the SNC suite found in Antarctica and on other continents. Today we possess 46 kg of meteorite fragments from the Moon and 66 kg from Mars, with additional samples of both being collected every few years in Antarctica or in the hot deserts of the world. These samples have provided us with crucially important information on the petrologic histories of their parent bodies and contributed significantly to the new discipline of comparative planetology.

All neighbourhoods can be dangerous at times, and that of the Earth is no exception. After viewing the images from spacecraft, Eugene Shoemaker once declared that impact is the most fundamental geological process in the inner Solar System. Every rocky planetary body is marked with impact craters, ranging from micrometre-sized zap-pits in lunar surface rocks to the Moon's multiring basin, South Pole-Aitken, that is 2500 km in diameter. In 1961, multiring basins were discovered in images of the Moon, but they remained unknown on the Earth until the 1980s. Today, at least eight of them, some of which are deeply eroded or buried by younger sediments are well-established, with several likely candidates under investigation.

Besides excavating craters, impacts shock their target rocks to varying degrees that have given rise to the new petrologic discipline of shock metamorphism. Discoveries of shocked quartz grains along with anomalously high values of iridium in samples of the clay layer at the Cretaceous–Tertiary boundary led to the recognition of mega-impacts as possible causes of extinctions and the terminations of geological periods.

Palpable topographic evidence of catastrophic floodwaters in the Channeled Scablands of eastern Washington was described in the 1920s by J Harlen Bretz and roundly rejected as non-uniformitarian. Opposition remained strong even after Bretz identified an adequate source of the water, and full acceptance of his hypothesis awaited the 1960s. In the early 1970s, images from Mars revealed evidence on this small, currently dry planet of erosion and deposition by catastrophic floods that exceeded in magnitude those that formed the Scablands. These Martian images confirmed cataclysmic flooding as a geological process of interplanetary significance.

We have reviewed several examples of the long delays – the stops and starts – that characterize scientific investigations, and, no doubt, those in other disciplines as well. An impact origin for lunar craters was proposed twice: first by G. K. Gilbert in 1891 and again by Ralph B. Baldwin in 1949. Gilbert was the most prominent geologist in America, but he slighted the principle of uniformitarianism by proposing a process that originated outside the Earth and never had been observed in operation. A few years later he damaged his case when his field tests failed to confirm an impact origin for the crater associated with iron meteorites in northern Arizona. Gilbert settled for a volcanic origin, thereby setting back research on impact cratering for half a century. Ralph Baldwin, working outside the scientific community, published compelling evidence for lunar impact craters in 1949 when there was minimal interest in either the Moon or impact processes. Almost by chance, he engaged the interest of a few scientists, particularly Harold C. Urey and Eugene M. Shoemaker, who influenced the US space programme to go to the Moon and carry out geological fieldwork there. The deeply buried, multiringed Chicxulub Crater of Yucatán, which lies at the Cretaceous–Tertiary boundary, was discovered twice: first in 1981 by oil exploration geologists who had located it by its patterns of magnetic and gravity anomalies and studied its drill cores; and a second time in 1991 by scientists at the University of Arizona conducting a literature search. During the ten-year interval, the original report failed to capture the attention of the international community of cratering specialists although they were avidly seeking a crater at the K–T boundary. Today, the Chicxulub Crater is the object of international investigations that have linked it with the mass extinctions that ended the Cretaceous Period and have shown that the crater's outermost ring is the only known example of a fault that cuts through the Earth's crust–mantle boundary.

Hesitant as they were, these are among the significant advances that transformed geology during the twentieth century from a science focused strictly on the Earth to one that extends outward in space to all planetary bodies.

I wish to thank S. R. Taylor for his advice on dating the events in the lunar chronology; W. A. Cassidy for pointing out passages in need of clarification, and C. S. Shoemaker, J. A. Wood, D. E. Brownlee and P. Lafaite of the MNHM, Paris, for graciously allowing me to use their illustrations.

References

ALVAREZ, W. 1997. *T. rex and the Crater of Doom*. Princeton University Press, Princeton, New Jersey.

ALVAREZ, L., ALVAREZ, W., ASARO, F. & MICHEL, H. V. 1980. Extraterrestrial cause for the Cretaceous–Tertiary extinction. *Science*, **208**, 1095–1108.

ALFVÉN, H. 1963. The early history of the Moon and the Earth. *Icarus*, **1**, 357–363.

ANDERS, E. & ZINNER, E. 1993. Interstellar grains in primitive meteorites: diamond, silicon carbide, and graphite. *Meteoritics*, **28**, 490–514.

BAKER, R. V. 1978. The Spokane Flood controversy and the Martian outflow channels. *Science*, **202**, 1249–1256.

BAKER, R. V. 1982. *The Channels of Mars*. University of Texas Press, Austin.

BAKER, R. V. 1998. Catastrophism and uniformitarianism: logical roots and current relevance in geology. *In*: BLUNDELL, D. J. & SCOTT, A. C. (eds) *Lyell: The Past is the Key to the Present*. Geological Society, London, Special Publications, **143**, 171–182.

BALDWIN, R. B. 1949. *The face of the Moon*. The University of Chicago Press, Chicago.

BALDWIN, R. B 1963. *The Measure of the Moon*. The University of Chicago Press, Chicago.

BALDWIN, R. B & WILHELMS, D. E. 1992. Historical review of a long-overlooked paper by R. A. Daly concerning the origin and early history of the Moon. *Journal of Geophysical Research*, **97**(E3), 3837–3843.

BARRINGER, D. M. 1905, Coon Mt. and its crater. *Proceedings of the Academy of Natural Sciences of Philadelphia*, **57**, 861–886.

BETHELL, T. 1975. NASA (that's right, NASA) is a good thing. *The Washington Monthly*, **7**, 5–14.

BINZELL, R. P. & XU SHUI 1993. Chips off of Asteroid Vesta: evidence for the parent body of basaltic achondrite meteorites. *Science*, **260**, 186–191.

BOGARD, D. D. & JOHNSON, P. 1983. Martian gases in an antarctic meteorite? *Science*, **221**, 651–654.

BOISSE, A. 1847. *Recherches sur l'historie, la nature et l'origine des aérolithes*. Cramaux, Mémoires des Société des Lettres, des Sciences et des Arts de l'Avéyron, **7**.

BOWRING, S. A. & HOUSH, T. 1995. The Earth's early evolution. *Science*, **269**, 1535–1540.

BRETT, P. R. 1999. Speech at the unveiling of a plaque in commemoration of Eugene M. Shoemaker at Tswaing Crater, South Africa, during excursion of The Meteoritical Society, 14 July.

BRETZ, J H. 1930. Lake Missoula and the Spokane Flood [abstract]. *Geological Society of America Bulletin*, **41**, 92–93.

BRETZ, J H. 1969. The Lake Missoula floods and the Channeled Scabland. *Journal of Geology,* **77**, 505–543.

BRETZ, J H., SMITH, H. T. U. & NEFF, G. E. 1956. Channeled Scabland of Washington: new data and interpretations. *Bulletin of the Geological Society of America*, **67**, 957–1049.

CAMERON, A. G. W. 2001. From interstellar gas to the Earth–Moon system. *Meteoritics and Planetary Science*, **36**, 9–22.

CAMERON, A. G. & WARD, W. 1976. On the origin of the Moon. *Lunar Science VII*. Lunar and Planetary Institute, Houston, 120–122.

CHAO, E. T. C., SHOEMAKER, E. M. & MADSEN, B. M. 1960. First natural occurrence of coesite. *Science*, **132**, 220–222.

CLAYTON, R. N. & MAYEDA, T. K. 1975. Genetic relations between the Moon and meteorites. *Proceedings of the Lunar Science Conference 6th, Geochimica et Cosmochimica Acta*, Supplement 6, **2**, 1761–1769.

CLAYTON, R.N., GROSSMAN, L. & MAYEDA, T. K. 1973. A component of primitive nuclear composition in carbonaceous meteorites. *Science*, **182**, 485–488.

COHEN, B. A., SWINDLE, T. D. & KRING, D. A. 2000. Support for the lunar cataclysm hypothesis from lunar meteorite impact melt ages. *Science*, **290**, 1754–1756.

DALY, R. A., 1946. Origin of the moon and its topography. *Proceedings of the American Philosophical Society*, **90**, 104–119.

DARWIN, G. H. 1879. The precession of a viscous spheroid and the remote history of the earth. *Philosophical Transactions of the Royal Society of London*, **170**, Part 2, 447–538.

DEUTSCH, A., GRIEVE, R. A. F., AVERMANN, M., *et al.* 1995. The Sudbury structure (Ontario, Canada) a tectonically deformed multiring impact basin. *Geological Rundschau*, **84**, 369–709.

DIETZ, R. S. 1946. The meteorite impact origin of the Moon's surface features. *Journal of Geology*, **54**, 350–375.

DIETZ, R. S. 1959. Shatter cones in cryptoexplosion structures (meteorite impact?). *Journal of Geology*, **67**, 496–505.

DIETZ, R. S. 1961. Continent and ocean basin evolution by spreading of the sea floor. *Nature*, **190**, 854–857.

ELPHIC, R., LAWRENCE, D. J., FELDMAN, W. C., BARRACLOUGH, B. L., MAURICE, S., BINDER, A. B. & LUCEY, P. G. 1998. Lunar Fe and Ti abundances: comparison of *Lunar Prospector* and *Clementine* data. *Science*, **281**, 1493–1496.

FISHER, O. 1881. *Physics of the Earth's Crust*. MacMillan, London.

FRENCH, B. M. & SHORT, N. M. (eds) 1968. *Shock Metamorphism of Natural Materials*. Mono, Baltimore.

GILBERT, G. K. 1893. The Moon's face, a study of the origin of its features. *Bulletin of the Philosophical Society of Washington*, **12**, 241–292.

GILBERT, G. K. 1896. The origin of hypotheses: illustrated by the discussion of a topographic problem. *Science*, New Series, **3**, 1–24.

GRADY, M. M. 2000. *Catalogue of Meteorites* (5th edn). The Natural History Museum, London, Cambridge University Press, Cambridge.

GRIEVE, R. A. F., STÖFFLER, D. & DEUTSCH, A. 1991. The Sudbury structure: controversial or misunderstood? *Journal of Geophysical Research*, **96**, 22,753–22,764.

HAGER, D. 1953. Crater mound (Meteor Crater), Arizona, a geologic feature. *AAPG Bulletin*, **37**, 821–857.

HALLIDAY, I. 1981. New Honorary Member of the Royal Astronomical Society of Canada. *Journal of the Royal Astronomical Society of Canada*, **75**, 152.

HARLAND, W. B., ARMSTRONG, R. L., COX, A. V., CRAIG, L. E., SMITH, A. G. & SMITH, D. G. 1989. *A Geologic Time Scale 1989*. Cambridge University Press, Cambridge.

HARTMANN, W. K. 1981. Discovery of multiring basins: gestalt perception in planetary science. *In*: SCHULTZ, P. M. & MERRILL, P. B. (eds) *Multiring Basins, Proceedings of Lunar and Planetary Science*, **12A**, 79–90; *Geochimica et Cosmochimica Acta*, Supplement **15**.

HARTMANN, W. K. & DAVIS, D. R. 1975. Satellite-sized planetesimals and lunar origin. *Icarus*, **12**, 504–515.

HARTMANN, W. K. & KUIPER, G. P. 1962. Concentric structures surrounding lunar basins. *Communications of the Lunar and Planetary Laboratory*, **1**, 51–66.

HARTMANN, W. K. & WOOD, C. A. 1971. Moon: origin and evolution of multiring basins. *The Moon*, **3**, 3–78.

HEAD, J. W. III. 1999. Surfaces and interiors of the terrestrial planets. *In*: BEATTY, J. K., PETERSEN, C. C. & CHAIKEN, A. (eds) *The New Solar System*. Sky Publishing Company, Cambridge, Mass, 157–174.

HEAD, J. W. III. & WILSON, L. 1979. Alphonsus type dark-halo craters: morphology, morphometry, and eruption conditions. *Proceedings of the Lunar and Planetary Science, 10th Conference*, **3**, 2861–2897.

HILDERBRAND, A. R. & BOYNTON, W. V. 1991. Cretaceous ground zero. *Natural History*, **6**, 47–53.

HILDERBAND, A. R., PENFIELD, G. T., KRING, D. A., PILKINGTON, M., CAMARGO, Z. A., JACOBSEN, S. B. & BOYNTON, W. V. 1991. Chicxulub crater: a possible Cretaceous/Tertiary boundary impact crater on the Yucatán Peninsula, Mexico. *Geology*, **19**, 867–871.

HÖRZ, F., GRIEVE, R., HEIKEN, G., SPUDIS., P. & BINDER, A. 1991. Lunar surface processes. *In*: HEIKEN, H., VANIMAN, D. T. & FRENCH, B. M. (eds) *Lunar Sourcebook: A User's Guide to the Moon*. Cambridge University Press, Cambridge.

HOUGH, R. M., GILMOUR, I., PILLINGER, C. T., ARDEN, J. W., GILKES, K. W. R., YUAN, J. & MILLEDGE, H. J. 1995. Diamond and silicon carbide in impact melt rock from the Ries impact crater. *Nature*, **378**, 41–44.

HOWARD, E. C. 1802. Experiments and observations on certain stony and metalline substances, which at different times are said to have fallen on the Earth; also on various kinds of native iron. *Philosophical Transactions of the Royal Society, London*, **92**, 168–175, 179–180, 186–203, 210–212.

HOYLE, F. 1950. The nature of the universe, Part I: the Earth and nearby space. *Harper's Magazine*, **201**, 23–31.

IVANOV, B. A. & DEUTSCH, A. 1999. Sudbury impact event: cratering mechanics and thermal history. *In*: DRESSLER, B. O. & SHARPTON, V. L. (eds) *Large Meteorite Impacts and Planetary Evolution II*. Geological Society of America Special Paper No. **339**.

KOEBERL, C. 1995. Diamonds everywhere. *Nature*, **378**, 17–18.

LIN, R. P., MITCHELL, D. L., CURTIS, D. W., *et al.* 1998. Lunar surface magnetic fields and their interaction with the solar wind: results from *Lunar Prospector*. *Science*, **281**, 1480–1484.

LIPSCHUTZ, M. E. & ANDERS, E. 1961. The record in the meteorites – IV: origin of diamonds in iron meteorites. *Geochimica et Cosmochimica Acta*, **24**, 83–105.

MACDONALD, G. J. F. 1965. Origin of the Moon: dynamical considerations. *Annals of the New York Academy of Sciences*, **118**, 739–782.

MCKAY, D. S., GIBSON, E. K., JR, THOMAS-KEPRTA, K. L., *et al.* 1996. Search for past life on Mars: possible relic biogenic activity in Martian meteorite ALH84001. *Science*, **273**, 924–930.

MARVIN, U. B. 1986. Meteorites, the Moon and the history of geology. *Journal of Geological Education*, **34**, 140–165.

MARVIN, U. B. 1996. Ernst Florens Friedrich Chladni (1756–1827) and the origins of modern meteorite research. *Meteoritics and Planetary Science*, **31**, 545–588.

MARVIN, U. B 1999. Impacts from space: the implications for uniformitarian geology. *In*: CRAIG, G. Y. & HULL, J. H. (eds) *James Hutton – Present and Future*. The Geological Society, London, Special Publications, **150**, 89–117.

MARVIN, U. B, CAREY, J. W. & LINDSTROM, M. M. 1989. Cordierite–spinel troctolite, a new magnesium-rich lithology from the lunar highlands. *Science*, **243**, 925–928.

MASON, B. & MELSON, W. G., 1970. *The Lunar Rocks*. Wiley–Interscience, New York.

MELOSH, H. J. 1985. Ejection of rock fragments from planetary bodies. *Geology*, **13**, 144–148.

MELOSH, H. J. 1997. Multiringed revelation. *Nature*, **390**, 439–440.

MERRILL, G. P. 1908. The meteor crater of Canyon Diablo, Arizona: its history, origin, and associated meteoritic irons. *Smithsonian Miscellaneous Collections*, **50**, 461–498.

METZGER, A. E., TROMBKA, J. I., REEDY, R. C. & ARNOLD, J. R. 1974. Element concentrations from lunar orbital gamma-ray measurements. *Proceedings of the 5th Lunar Science Conference*, **2**, 1067–1078; *Geochimica et Cosmochimica Acta*, Supplement **5**.

MOJZSIS, S. J. & HARRISON, T. M. 2000. Vestiges of a beginning: clues to the emergent biosphere recorded in the oldest known sedimentary rocks. *GSA Today*, **10**, 1–6.

MORGAN, J. & WARNER, M. 1999. Chicxulub: the third dimension of a multiring impact basin. *Geology*, **27**, 477–410.

MORY, A. J., IASKY, R. P., GLICKSON, A. Y. & PIRAJNO, F. 2000. Woodleigh, Carnarvon Basin, Western Australia: a new 120 km diameter impact structure. *Earth and Planetary Science Letters*, **177**, 119–128.

MULHOLLAND, D. 1980. How high the Moon: a decade of laser ranging. *Sky and Telescope*, **62**, 273–279.

NAKAMURA, Y., LAMMLEIN, G., LATHAM, G., EWING, M., DORMAN, J., PRESS, F. & TOKSÖZ, M. N. 1973. New seismic data on the state of the deep lunar interior. *Science*, **181**, 49–51.

NININGER, H. H. 1956. *Arizona's Meteorite Crater*. American Meteorite Museum, Sedona.

O'KEEFE, J. D. & AHRENS, T. J. 1986. Oblique impact: a process for obtaining meteorite samples from other planets. *Science*, **234**, 346–349.

PARDEE, J. T. 1910. The glacial Lake Missoula. *Journal of Geology*, **18**, 376–386.

PATTERSON, C. 1956. Age of meteorites and the Earth. *Geochimica et Cosmochimica Acta*, **10**, 230–237.

PENFIELD, G. T. 1991. Pre-Alvarez impact. *Natural History*, **6**, 4.

PENFIELD, G. T. & CAMARGO, Z. A. 1981. Definition of a major igneous zone in the central Yucatán platform with aeromagnetics and gravity. *Society of Exploration Geophysicists Technical Program, Abstracts, and Biographies*, **51**, 37.

PIKE, R. J. 1985. Some morphologic systematics of complex impact structures. *Meteoritics*, **20**, 49–68.

POWELL, J. L. 1998. *Night Comes to the Cretaceous: Dinosaur Extinction and the Transformation of Modern Geology*. Freeman, New York.

PYNE, S. J. 1980. *Grove Karl Gilbert, A Great Engine of Research*. University of Texas Press, Austin.

REIMOLD, W. U., KOEBERL, C., BRANDSÄTTER, F., KRUGER, F. J., ARMSTRONG, R. A. & BOOTSMAN, C. 1999. Morokweng impact structure, South Africa: geologic, petrographic, and isotope results, and implications for the size of the structure. *In*: DRESSLER, B. O. & SHARPTON, V. L. (eds) *Large meteorite impacts and planetary evolution II*. Geological Society of America Special Paper **339**, 61–72.

RUBIN, A. E. 1997. A history of the mesosiderite asteroid. *American Scientist*, **85**, 26–35.

SANFORD, S. A. 1996. The inventory of interstellar materials available for the formation of the solar system. *Meteoritics and Planetary Science*, **31**, 449–476.

SATO, M. 1978. Oxygen fugacity of basaltic magmas and the role of gas forming elements. *Geophysical Research Letters*, **5**, 447–449.

SCHOPF, J. W. 1993. Microfossils of the early Archaen Apex chert: new evidence for the antiquity of life. *Science*, **260**, 640–646.

SCHRÖTER, J. H. 1791. *Selenographicshe Fragmente*, **1**. Lillenthal & Helmst, Göttingen.

SCHULTZ, L., FRANCHI, I. A., REID, A. M. & ZOLENSKY, M. E. 2000. *Workshop on Extraterrestrial Materials from Cold and Hot Deserts*. LPI Contribution No. **997**, The Lunar and Planetary Institute, Houston.

SHIMA, M. & SHIMA, M. 1973. Mineralogical and chemical composition of new Antarctic meteorites. *Abstracts of the 36th Annual Meeting of The Meteoritical Society*, Davos, Switzerland, 135–136.

SHOEMAKER, E. M. 1960. Penetration mechanics of high velocity meteorites, illustrated by Meteor Crater, Arizona. *In*: *21st International Geological Congress. Det Berlingske Bogtrykken Copenhagen*, 418–434.

SHOEMAKER, E. M. 1963. Impact mechanics at Meteor Crater, Arizona. *In:* MIDDLEHURST, B. & KUIPER, G. P. (eds) *The Moon, Meteorites, and Comets – The Solar System*. University of Chicago Press, Chicago, **4**, 301–336.

SHOEMAKER, E. M. & CHAO, E. T. C. 1961. New evidence for the impact origin of the Ries Basin, Bavaria, Germany. *Journal of Geophysical Research*, **66**, 3371–3378.

SHOEMAKER, E. M. & HACKMAN, R. J. 1962. Stratigraphic base for a lunar time scale. *In*: KOPAL, Z. & MIKHAILOV, Z. K. (eds), *The Moon*. Academic Press, London & New York, 289–300.

SMITH, J. V., ANDERSON, A. T., NEWTON, R. C., OLSEN, E. J., WYLLIE, P. J., CREWE, A. V., ISAACSON, M. S. & JOHNSON, D. 1970. Petrologic history of the Moon inferred from petrography, mineralogy, and petrogenesis of *Apollo 11* rocks. *Proceedings of the* Apollo 11 *Lunar Science Conference*, **1**, 897–925.

SORBY, H. C. 1864. On the microscopical structure of meteorites. *Philosophical Magazine*, **28**, 157–159.

SPUDIS, P. D. 1993. *The Geology of Multiring Impact Basins*. Cambridge Planetary Science Series, Cambridge University Press, Cambridge.

TAYLOR, S. R. 1998. *Destiny or Chance: Our Solar System and its Place in the Cosmos*. Cambridge University Press, Cambridge.

TERA, F., PAPANASTASSIOU, D. A. & WASSERBURG, G. J. 1974. Isotopic evidence for a terminal lunar cataclysm. *Earth and Planetary Science Letters*, **22**, 1–21.

THERRIAULT, A. M., REID, A. M. & REIMOLD, W. U. 1993. Original size of the Vredefort Structure, South Africa. *Lunar and Planetary Science*, **24**, 1419–1420 (The Lunar and Planetary Science Institute, Houston).

THOMSON, G. 1804. Essai sur le fer malléable trouvé en Sibérie par le Professeur Pallas. *Bibliothèque Britannique*, **27**, 135–154, 209–229.

TILGHMAN, B. C. 1905. Coon Butte Arizona: *Proceedings of the Academy of the Natural Sciences of Philadelphia*, **57**, 887–914.

UREY, H. C. 1952. *The Planets: Their Origin and Development.* Yale University Press, New Haven.

UREY, H. C. 1968. The Moon. *In*: ODISHAW, H. (ed.) *Earth in Space.* Voice of America Forum Lectures, **2**, 151–173.

WALKER, D., STOLPER, E. M. & HAYS, J. F. 1979. Basaltic volcanism: the importance of planet size. *Proceedings of the Lunar and Planetary Science Conference 10th*, **2**, 1995–2105.

WASSON, J. T. 1985. *Meteorites. Their Record of Early Solar-system History*. Freeman, New York.

WASSON, J. T. & WETHERILL, G. W. 1979. Dynamical, chemical, and isotopic evidence regarding the formation locations of asteroids and meteorites. *In*: GEHRELS, T. (ed.) *Asteroids*. University of Arizona Press, Tucson.

WILHELMS, D. E. 1993. *To a Rocky Moon: A Geologist's History of Lunar Exploration*. University of Arizona Press, Tucson.

WOOD, J. A. 1971. Fragments of terra rock in the *Apollo 12* soil samples and a structural model of the Moon. *Icarus*, **16**, 462–501.

WOOD, J. A., DICKEY, J. S. JR, MARVIN, U. B. & POWELL, B. N. 1970. Lunar anorthosites and a geophysical model of the Moon. *Proceedings of the* Apollo 11 *Lunar Science Conference*, 965–988.

WRIGHT, F. E. 1935. The surface features of the Moon. *The Scientific Monthly*, **40**, 101–115.

ZINNER, E. 1998. Stellar nucleosynthesis and the isotopic composition of presolar grains from primitive meteorites. *Annual Reviews of Earth and Planetary Science*, **26**, 147–188.

From graphical display to dynamic model: mathematical geology in the Earth sciences in the nineteenth and twentieth centuries

RICHARD J. HOWARTH

Department of Geological Sciences, University College London, Gower Street, London WC1E 6BT, UK

Abstract: Graphical displays were used early in geophysics and crystallography, mineralogy, petrology and structural geology by the early 1800s, but nineteenth-century geology obstinately remained mainly descriptive. Charles Lyell's quantitative classification of the Tertiary Sub-Era in 1828 was a notable exception. Nevertheless, by 1920 the quantitative approach had become established. W. C. Krumbein, who introduced the computer into geology in 1958, encouraged use of probabilistic sampling and process–response models. Early work focused on databases, statistical data analysis and display. By the 1970s, stochastic simulation, deterministic modelling and spatial 'geostatistics' (pioneered by Matheron and his co-workers), were of growing importance. The introduction of the personal computer and the graphical user interface in the 1980s brought well-proven quantitative methods out of the research environment onto the workbench and into the field. Since the mid-1980s, the analysis, display and modelling of behaviour in three dimensions, underpinned by spatial statistics, computational fluid-flow, visualization technology, etc., has proved of economic benefit to mining, petroleum geology and hydrogeology. Other, computationally intensive, methods likely to be of importance in the Earth sciences are the application of 'robust' statistical methods, increasing use of Bayesian methods in uncertainty (risk) estimation (as a result of a renewed interest in statistical intervals and forecasting), and computational mineralogy.

Although the quantitative display and analysis of Earth science data can be traced back to the measurement of magnetic declination and inclination and, later, gravity (from pendulum observations) in the seventeenth and eighteenth centuries, the concern here is with the history of mathematical geology. This subject includes the application of mathematics, classical statistics and, since the late 1960s, spatial statistics, to assist the interpretation of geological data and the modelling of geological phenomena as an aid both to understanding, and to forecasting, in fields such as resource assessment, petroleum geology and hydrogeology. However, a vital part of all these techniques is the graphical display of either raw (or, in more recent times, perhaps smoothed) data values, to assist interpretation, or the presentation of the results of some modelling process.

The visual language inherent in such displays is now taken for granted to such an extent that it is difficult to imagine that even as late as the 1850s, there was a body of opinion among those in the then newly emerging science of statistics arguing against the 'inexactness' of graphical displays and that it was far preferable to record results solely in the form of tables (Funkhouser 1938, pp. 293, 295). Editors and publishers may have tacitly gone along with this view owing to the high cost of preparing the plates of illustrations.

The appearance of graphical displays of geological data and related thematic maps in publications gradually became more widespread during the last half of the nineteenth century (see Fig. 1), but because the history of this development is little-known today, it is touched on here in order to describe their role at the beginning of the use of quantitative methods in the geological sciences.

The emergence of statistical graphics

Isoline maps

The concept of a line of equal value (isoline or isopleth) as a cartographic tool has a long and complex history. Bathymetric maps of limited parts of rivers and harbours, such as that by the geologist Auguste Bravais (1811–1863) showing the topography of the underwater delta front formed by the river Aar, the third largest river in Switzerland, where it flows into Lake Brienz (in Martins 1844), are known from the sixteenth century onwards (Robinson 1982, p. 211). However, the first published topographic contour map was J. Dupain-Triel's (1722–1805) map of France (Dupain-Triel 1798–1799).

From: OLDROYD, D. R. (ed.) 2002. *The Earth Inside and Out: Some Major Contributions to Geology in the Twentieth Century*. Geological Society, London, Special Publications, **192**, 59–97. 0305-8719/02/$15.00

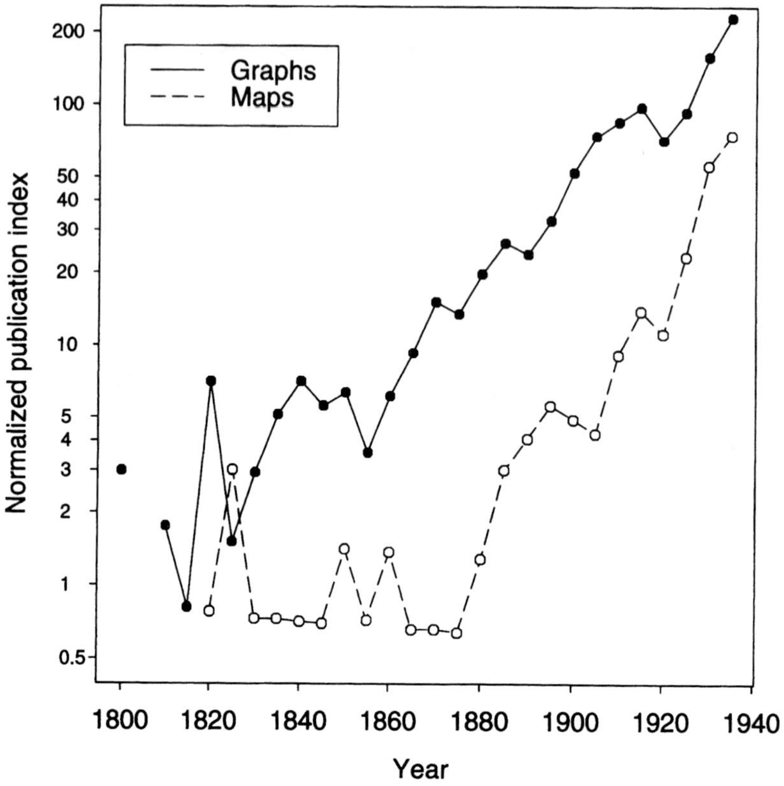

Fig. 1. Count of numbers of statistical graphs and thematic maps per five-year interval (1801–1805, 1806–1810, etc.) in articles on geological subjects in a total of 103 serial publications which existed during the period 1800–1935. Counts (totalling 1942 statistical graphs and 236 thematic maps) have been normalized by dividing through by values of Table 2 (in Appendix). Index is zero where no symbols are shown.

Edmond Halley's (1656–1742) chart of magnetic declination for the North Atlantic (Halley 1701) was the earliest map to show isolines of an observed non-topographic phenomenon, but with the failure of magnetic declination to provide the solution to determining longitude at sea, it had little lasting impact. It was the later publication of a map of temperature distribution for the northern hemisphere by the German explorer, naturalist and geologist Alexander von Humboldt (1769–1859) (1817*a*,*b*) which, on account of his scientific reputation, really excited interest in the thematic mapping of the values of relatively abstract phenomena.

Interest in delineation of the effects of earthquakes prompted some of the earliest isoline maps in the geological sciences. The earliest known map (Fig. 2) was drawn up by two Hungarian academics, Pál Kitaibel (1757–1817) and Ádám Tomtsányi (1755–1831), following an earthquake which occurred on 14 January 1810, whose effects were felt as far away as Prague and Vienna. The map (Kitaibel & Tomtsanyi 1814, pl. ff., p. 110) includes a single, closed, isointensity curve which outlines the area of major damage surrounding the village of Mór, near Székesfehérvár in western Hungary. However, the earliest map to use a definite intensity scale, depicted by multiple isoseismal lines, was drawn by the German geographer and cartographer Auguste Petermann (1822–1878) (in Volger 1856) on the basis of information supplied by Georg H. O. Volger (*fl.* 1822–1897), a student of Swiss earthquakes. By 1862, in a map of the destruction following the Neapolitan earthquake of 16 December 1857, the Irish engineer and seismologist Robert Mallet (1810–1881) included proportional-sized symbols, to indicate intensity at particular localities, as well as isoseismal lines (Mallet 1862, map A, pp. 252–254).

The construction of structure contour maps, which show the subsurface depth to a given

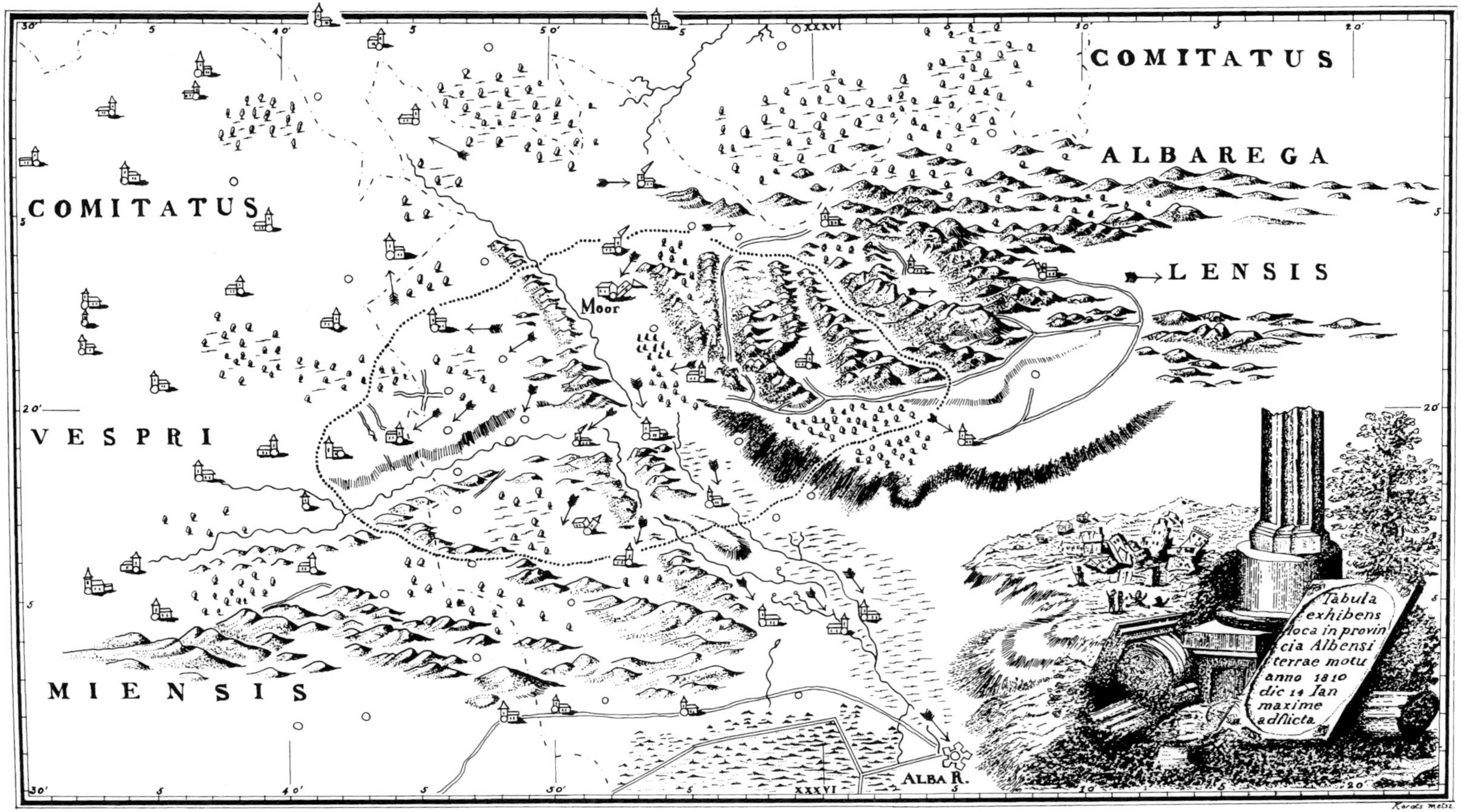

Fig. 2. Map showing earthquake damage of 14 January 1810, surrounding the village of Mór (Moor), near Székesfehérvár in western Hungary. Note dotted isoline delineating the zone of major damage and arrows showing direction of overturning of fallen objects. Original map in Kitaibel & Tomtsanyi (1814, unnumbered plate); this version was redrawn by the author from material supplied by L. Stegena, who included the figure in Stegena (1984, fig. 1). Reproduced with permission of Akadémiai Kiadó Rt., Budapest.

stratigraphic horizon, began with the work of the American geologist Benjamin Lyman (1835–1920), in a report to the Public Works Department of the Indian government on the oil prospects of the Punjab (Lyman 1870). An example of one of his maps (reproduced in Owen 1975, fig. 4–2) shows 100 ft (30.5 m) structure contours on the top of the 'oil bearing bed' for an area of *c*. 230 m^2 of anticlinal folding surrounding an oil seep near Bara Katta, Bannu District, Punjab. Lyman apparently had tried the idea of using 'underground contour lines to give the shape of rock beds' while working on coal, lead and iron deposits in Virginia in 1866 and 1867 (Lyman 1873), but did not publish anything until his report on his work in the Punjab. In Europe, structure contour mapping was used by the geologist Albert De Lapparent (1839–1908) to delineate the top surface of the Gault Clay under the English Channel between Dover and Calais in a report (De Lapparent & Potier 1877) following the second submarine geological survey, conducted in 1875 and 1876, as part of a feasibility study for the first bored Channel Tunnel. In 1885, Gustave F. Dollfus (1850–1931) exhibited structure contour maps for the top of the Gault and Chalk (Fig. 3) in the region around Paris (Dollfus 1888). The earliest structure contour map to be published by the US Geological Survey was a regional map (Orton 1889, pl. LV) showing the top of the Trenton Limestone in western Ohio and eastern Indiana, made by the geologist Edward Orton (1829–1899). Towards the end of the nineteenth century, structure contour maps were regularly appearing in geological reports on coal fields, mineral deposits, and oil and gas fields, and were taken for granted by the time Emmons' (1921) textbook on petroleum geology was published.

By the 1850s, there was growing concern in improving the quality of water supply to cities. In England, the geologist Joseph Lucas (1846–1926) drew the earliest 'hydrogeological' map to show 'contours of the upper surface of water in the Chalk' in a privately published book (Lucas 1874, cited in Mather 1998). This was followed by an isoline map covering an area of *c*. 500 km^2 (Lucas 1877) for an area of the Chalk lying SE of London, and this eventually led to the publication of the earliest true hydrogeological maps in 1878 (Mather 1998). By the turn of the century, such maps were becoming relatively commonplace elsewhere (Linstow 1905; Veatch 1906).

Point-symbol maps

The idea of a map in which a specific symbol indicates the value of a mapped attribute at a given location was current in structural geology by the 1820s, when Carl F. Naumann (1797–1873) used small, locality-specific, strike-bar symbols in his mapping of gneisses in southern Norway (Naumann 1824, vol. 1, plate III, fig. 4). However, maps in which the area of a symbol at a locality is proportional to the value of a quantity only seem to have begun when a British army officer, Henry Harness (1804–1883), used proportional circles in a map of the population of Ireland which he drew up in 1837 (Robinson 1982, p. 205). The idea of proportional symbols was popularized in continental Europe by the work of Petermann and a French engineer Charles J. Minard (1781–1870) in the 1850s (Robinson 1982, p. 205). Minard (1862) also published the first memoir to be devoted to the graphic method. The American geologist James D. Dana (1813–1895) used strike-bar symbols in which the length of the 'stem of the T' was proportional to the amount of dip (Dana 1880), but this idea was largely ignored by others. Other examples of early maps to use point-symbols in the geological literature include mineral production in Russia (Keppen 1894) using proportional squares, and a regional map of heavy-mineral distribution in Quaternary sands of the Netherlands in which the intensity of constant-size circles was made proportional to concentration (Schroeder van der Kolk 1896). The latter style of presentation eventually became widely used in exploration geochemistry following the work of Webb *et al.* (1964).

Proportional line widths were rarely used until they were taken up in the 1950s to show element concentration in segments of stream drainage networks in reconnaissance geochemical-mapping applications (Webb 1958). An exception was a map in an early report on the lignites of Bohemia (Lallemand 1881) showing the volume of lignite transported by water and rail during the year 1879. The style of Lallemand's map was similar to that of maps of traffic flow independently produced by Harness in 1837, and by Minard in 1845 (Funkhouser 1938, p. 301). See Robinson (1982, p. 209) and Tufte (1983, pp. 25, 177) for other examples.

An isolated 'time-lapse' map occurs in Schmidt (1874) and shows the gradually increasing extent of lava flows around the volcanic isle of Santorini (Thíra), in the Aegean, north of Crete, during a series of eruptions which occurred between February 1866 and June 1870. In each of twelve panels, Schmidt showed the area occupied by a new flow (or flows) in red, contrasting with the greenish colour of the rest of the landmass. The series of maps clearly

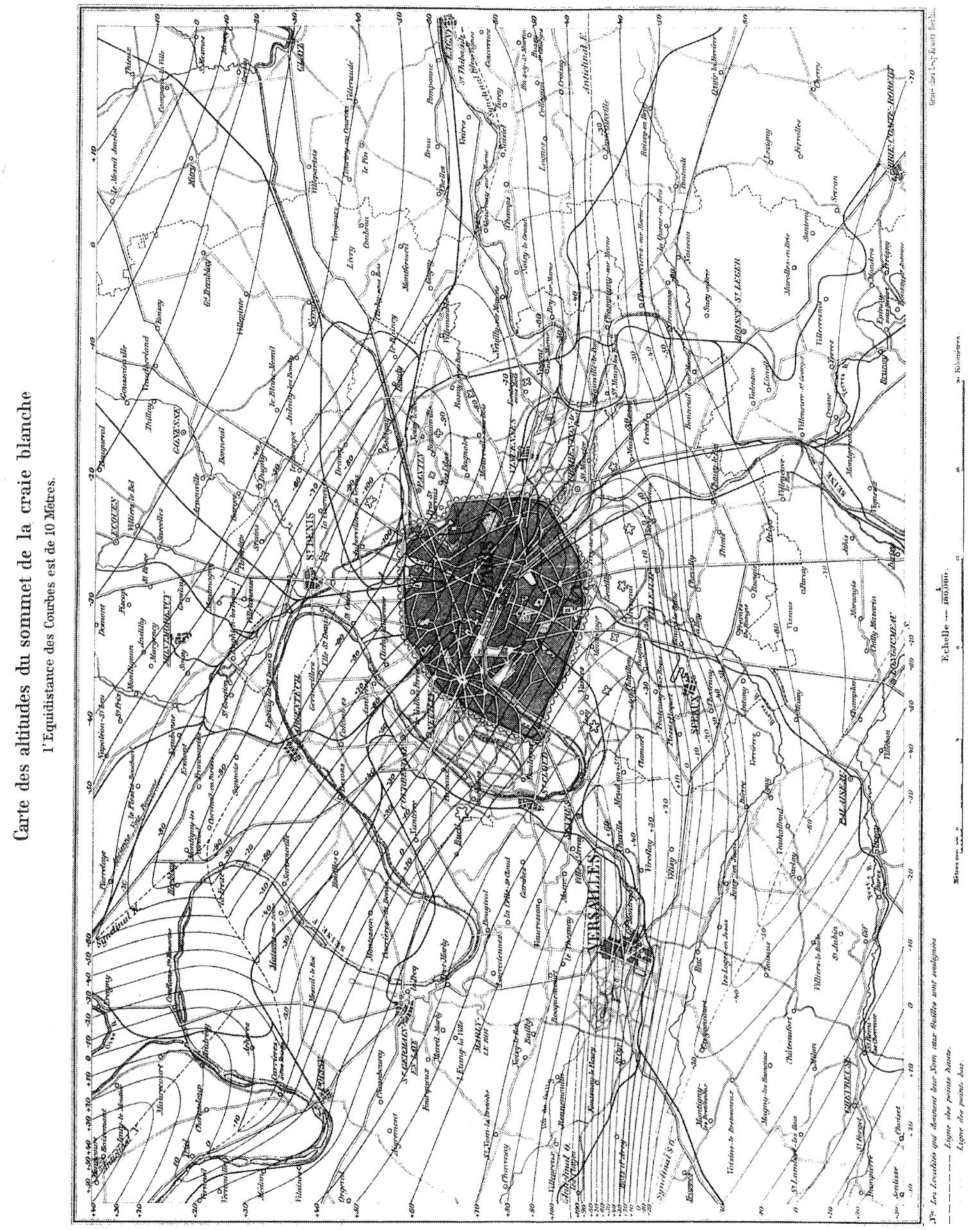

Fig. 3. Structure contours (solid lines, shown red in original) at 10 m intervals on the top of the Chalk in the region of Paris, France. Dashed and pecked lines mark axial lines of anticlines and synclines respectively. Reproduced from Dollfus (1888, unnumbered plate).

shows the emergence and gradual coalescence of flows from two separate vents.

Graphs

The earliest record of the commercial production of printed graph-paper occurs in 1686 (Gunther 1939). It was made by a London scientific instrument-maker, John Warner, in the year following Robert Plot's (1640–1696) ground-breaking publication (Plot 1685) in the *Philosophical Transactions of the Royal Society*, London, of a graph of daily barometric pressure. By the late eighteenth century, graphical displays had begun to come into their own, with the portrayal of experimental and non-meteorological scientific data pioneered by physicists such as Johann H. Lambert (1728–1777) (Lambert 1765); the introduction of line-graphs, bar-graphs and proportional-sized circles to illustrate econometric data by William Playfair (1759–1823) (Playfair 1798, 1801, 1805), younger brother of the Edinburgh mathematician, and Huttonian geologist, John Playfair (1748–1819); and the use of subdivided bar-graphs and proportional squares popularized by the work of Humboldt (1811–1812). Examples of W. Playfair's work are reproduced in Tufte (1983, pp. 32–34, 44, 64–65, 73, 91–92; 1990, p. 107).

Although William Playfair attributed his original interest in graphs to his brother, who had apparently taught him, when they were children, how to make a chart showing daily temperature variations (Funkhouser 1938, p. 289), J. Playfair's own graphical work (e.g. Playfair 1812, plate X, fig. 3) followed firmly in the style current in mathematical publications since the seventeenth century (e.g., Halley 1686) in resembling geometrical figures, without scales or axis annotation. They completely lacked the flair of his younger brother's work. Particularly surprising is the fact that, even in a publication on meteorology, J. Playfair (1805) exhibited his own data in tabular, rather than graphical, form. Funkhouser (1938, p. 289) has suggested that it may have been his younger brother's period of work (*c*. 1780) as an engineering draftsman for James Watt which helped to form his innovative graphic style. Possibly because he became a political exile in France (as a result of debts and his radical political views), W. Playfair's work was ignored in England but it inspired great interest in France and Germany. This eventually led to the appearance of a number of books devoted to the 'graphical method' between the years 1845 and 1854, and by 1890 graphs were regularly appearing in government publications and had become commonplace in most scientific meetings (Funkhouser 1938, pp. 300–330).

Line-graphs

From today's perspective, in which the bivariate scatter-plot (i.e. a graph in which a point corresponding to a particular sample or specimen is plotted on the basis of a grid of $\{x, y\}$ co-ordinates which lie at right angles to each other) is ubiquitous, it may come as a surprise to learn that the scatter-plot *per se* made a relatively late appearance in the geological literature. It was far more usual to have line-graphs, in which the x-axis corresponded to distance or (particularly in economic geology) time, the ordinate y-axis to the attribute in question, and the ordered data points (which usually were not distinguished individually) were joined by a continuous line. An early example (see Fig. 4), drawn by the French engineer and inspector of mines Antoine M. Héron de Villefosse (1774–1852), shows a series of overlaid topographic profiles across the Harz mountain range, south of Brunswick, Germany. The locations and size of a number of principal shafts and adits are also shown (Héron de Villefosse 1808). The British bookseller and geologist William Phillips (1775–1828) published schematized 'scenic views' showing comparative heights of mountains (Phillips 1815, frontispiece), but they became more popular following their use by Humboldt (1825, 1855). This form of illustration became a standard feature of geographical atlases following their incorporation in the magnificent *Physical Atlas of Natural Phenomena* (Johnson & Johnson 1856, plates 2, 9, 11).

Line-graphs also were used by investigators with an interest in hydrology. Examples include a graphical time-series of water levels in the Nile (Girard 1819) drawn by the French engineer Pierre Girard (1765–1836) from observations made when he accompanied Napoleon III's expedition to Egypt. Girard's fellow-countryman, the engineer Benjamin Dausse (1801–1890), drew an elegant composite diagram (Fig. 5) to show the mean height of the River Seine in Paris per month during the period from 1777 until 1836 together with frequency distributions of the water level per month for the same period (reproduced in Élie de Beaumont 1849, plate III, fig. 2).

An early example from mineralogy comes from the work of the British mathematician, physicist and astronomer Sir John F. W. Herschel (1792–1871) who drew a set of graphs to illustrate the absorption of polarized light as a function of wavelength in crystals of the mineral apophyllite (Herschel 1822). Herschel's graphs were drawn in the classical 'mathematical' style and consequently lacked axes, but by the time the French structural geologist, petrologist and mineralogist, Auguste Michel Lévy (1844–1911), published a review paper on the optical determination of minerals in thin section (Michel Lévy 1877), his graphs (Michel Lévy 1877, plates VIII, IX) at least had scaled axes, and their annotation was improved further in Michel Lévy & Lacroix's (1888) textbook.

By the late 1800s, the systematic study of igneous geochemistry was giving rise to the first 'variation diagrams' in which major element oxide compositions (typically Al_2O_3, FeO,

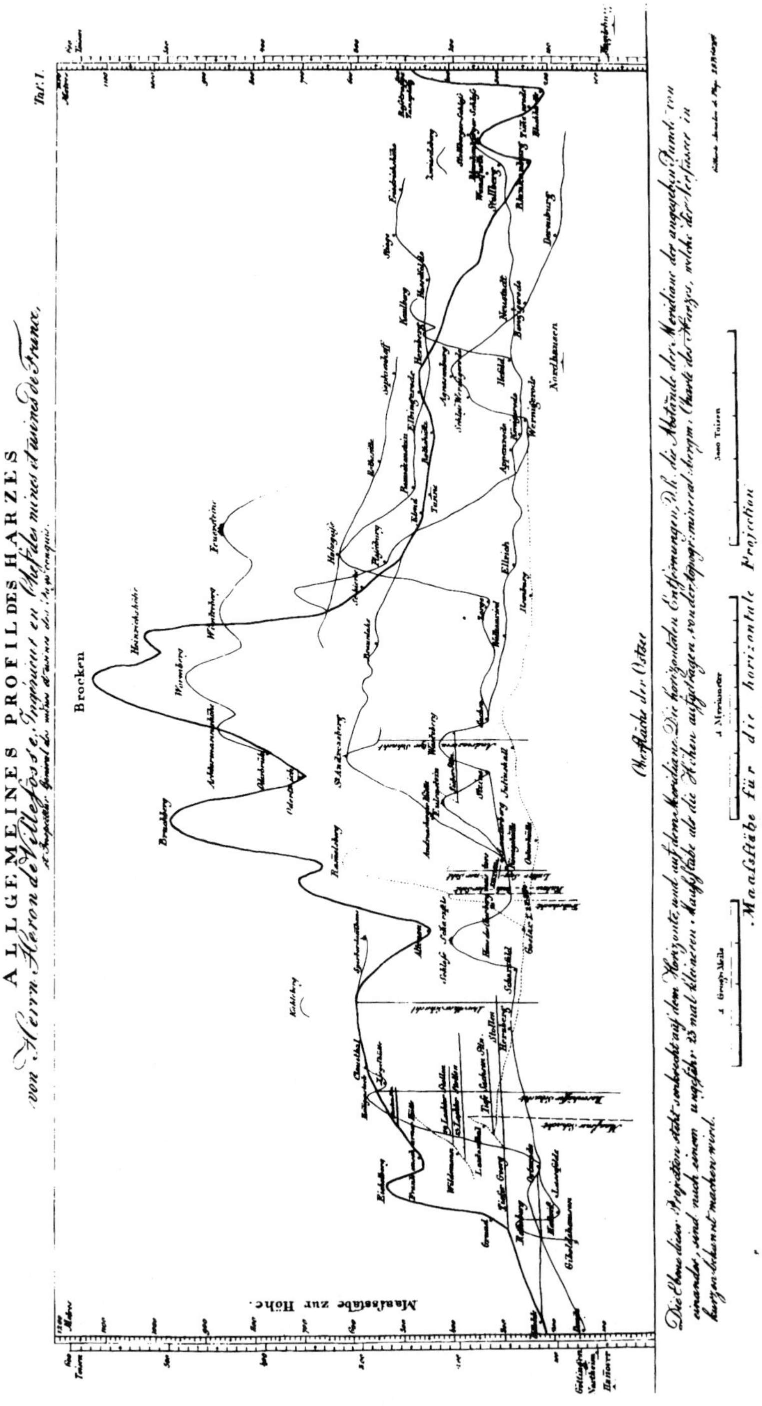

Fig. 4. Superimposed topographic profiles of the Harz mountain range, south of Brunswick, Germany. The locations and size of a number of principal shafts and adits are also shown. Reproduced from plate I of Héron de Villefosse, A. M. 1808. Nivellement des Harzgebirges mit dem barometer. *Gilbert's Annalen der Physik*, **1**, 49–111, with permission of the Science & Society Picture Library. ©Science Museum, London. All rights reserved.

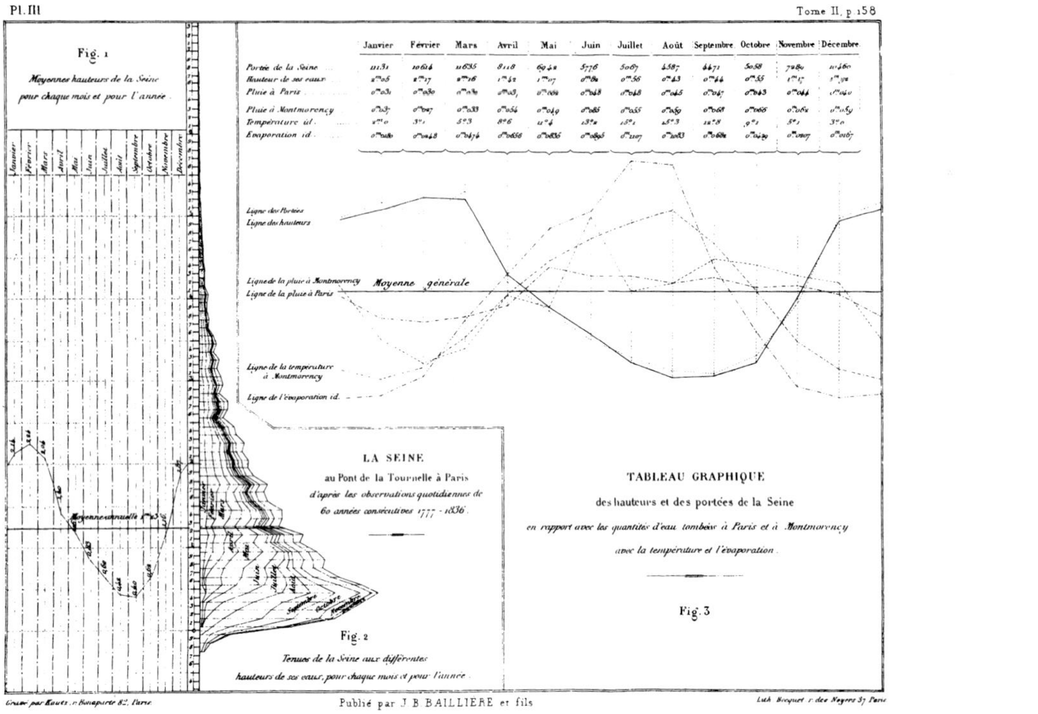

Fig. 5. Line-graphs showing: 'Fig. 1' mean monthly height (m) for the years 1777 to 1836 of the River Seine at the Tournelle Bridge, Paris, France; 'Fig. 2' superimposed frequency distributions for river height (m) per month from 1777 to 1836; 'Fig. 3' a graph and table ('graphic table') of the span and height (m) of water in the Seine, depth of rain fallen in Paris (m) and depth of rain, temperature and evaporation at Montmorency (*c.* 15 km north of the centre of Paris). Reproduced from drawings by Benjamin Dausse in: Élie de Beaumont (1849, plate III).

Fe_2O_3, CaO, MgO, K_2O and Na_2O) were plotted as line-graphs, as a function of increasing SiO_2 concentration, in the same diagram. In early examples of this type of plot, the increasing SiO_2 was shown implicitly (by appropriate ordering of the rock-types), as in the work of the Austrian igneous petrologist Eduard Reyer (1849–1914) (Reyer 1877, plates I, II); or explicitly, by a labelled axis, as was the situation in the work (Iddings 1892) of the American geologist and igneous petrologist Joseph P. Iddings (1857–1920).

The British surgeon and vertebrate paleontologist, George Busk (1807–1886), also used multiple-line plots, which he termed 'odontograms,' to assist the classification of vertebrate remains on the basis of their dentition: the measured characteristics of each tooth were plotted as a function of tooth position in the jaw (Busk 1870). He later used this method to illustrate his identification of teeth found in Brixham Cave, Torquay, Devon, as *Ursus priscus* (in Prestwich *et al.* 1873, pp. 540–548).

Two-dimensional polar co-ordinates were occasionally used in the early literature, e.g. mineralogical applications by the French physicist Jean B. Biot (1774–1862) (Biot 1817) and the British chemist William Ramsay (1852–1916) (Ramsay 1888), but the majority of such plots were related to orientation statistics.

With the opening up of gold mines in the Transvaal, South Africa, by the 1890s, keen attention was being paid to the problems of sampling such a rare and irregularly dispersed ore-mineral and line-graphs began to be used (De Launay 1896; Wybergh 1897) to illustrate the variations in ore-grade along sampling traverses as a guide to improved exploitation.

Rather surprisingly, econometric time-series formed the majority of line-graphs published in the early geological literature, illustrating time-trends of production for the commodity in question. Notable early examples include graphs showing the annual production of copper and tin from mines in Cornwall, England, from 1744 to 1845 (Fig. 6) (Hunt 1846); of silver in the

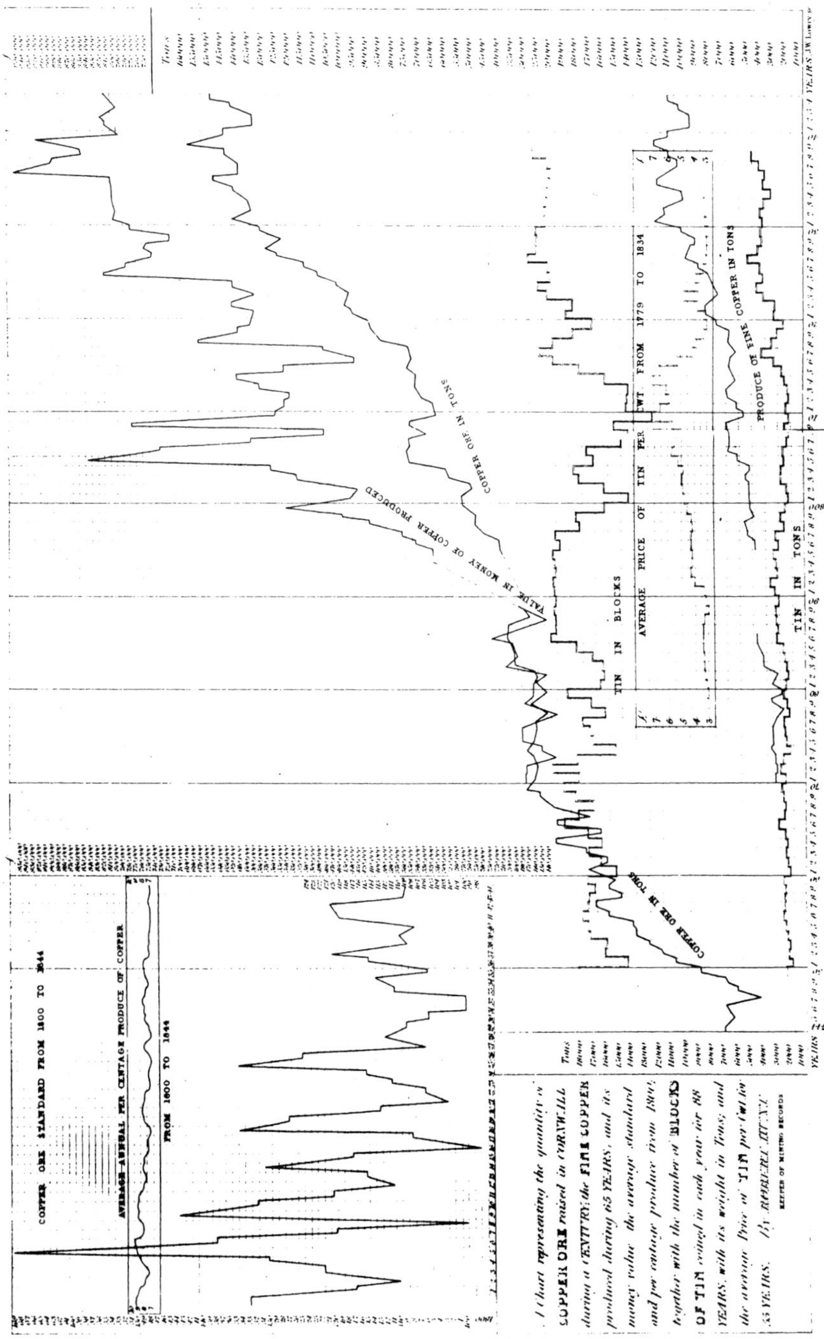

Fig. 6. 'A chart representing the quantity of copper ore raised in Cornwall during a century: the fine copper produced during 65 years; and its money value the average standard and per centage produce from 1800; together with the number of blocks of tin coined in each year for 88 years; with its weight in tons; and the average price of tin per cwt. for 55 years', reproduced from Hunt (1846, plate 9).

Frieberg mining-district, Germany, from 1524 to 1847 (Herder & Gäßfchmann 1849); of coal in northern England from 1821 to 1888, together with the price at Manchester, miners' wages and the number of accidents, and the total number of employees (Knowles 1890); and the production of crude oil from 1859 to 1893 in states of Pennsylvania and New York, and the total United States, together with the price of production (Cadell 1898).

One curious, rather later, example is a paper (Bailly 1905) in which an attempt was made to predict the time-span of future production of sedimentary iron-ore from Luxembourg, France and Germany. In each instance, an exponential growth-curve was fitted to the production rate from 1853 to 1905 in order to predict the year of expected maximum production and, using an estimate of the total reserves, an exponential decline-curve then was applied to obtain the expected production life. Bailly's model predicted that ore supplies would be exhausted in Luxembourg by 1943, Germany by 1953, and France by 2023. His predictions were reprinted in the *Zeitschrift für praktische Geologie* together with an assessment which concluded (Anon. 1910) that the sedimentary iron-ore supplies of Germany, Luxembourg and Italy would be exhausted within 30 years, whereas those of Russia, Sweden and France had expectations of 75, 100 and 700 years, respectively. It would be interesting to know whether this publication helped to influence opinion in pre-war Germany regarding the necessity to obtain additional (external) sources of ore-supply.

Time-series also began to be used in hydrogeology, where plots of groundwater level as a function of time proved to be a useful monitoring tool (Keilhak 1913).

Scatter-plots

Prior to 1900, bivariate 'scatter-plots' (which are sometimes referred to as 'cross-plots') were unusual in Earth science literature. Early examples in non-geophysical publications include a graph by Mallet (1873) showing the increase in volume of slags as a function of temperature. The French mineralogist Julien Thoulet (*fl.* 1843–1922) plotted laboratory experimental data on sedimentation rates (see Fig. 7) (Thoulet 1891, plate I); and both Michel Lévy (1897*a*,*b*) and Iddings (1898*a*,*b*) began to use the scatter-plot as an interpretational tool in igneous geochemistry. Although the scatter-plot was subsequently used particularly widely in geochemical applications, in the 1920s it also began to be taken up (together with line-graphs) by palaeontologists as an aid to morphometric distinction between species (e.g. Swinnerton 1921; Arkell 1926; Davies & Trueman 1927).

Ternary and tetrahedral diagrams

The earliest ternary diagrams (in which composition, in terms of three components, is represented by their relative proportions plotted on the basis of a triangular grid) made their first appearance in papers by Michel Lévy (1897*a*,*b*) and the German mineralogist, Carl Osann (1859–1923) (Osann 1900). The origins and usage of the ternary diagram are more fully discussed by Howarth (1996) and Sabine & Howarth (1998). Tetrahedrons were also widely used in mineralogy and geochemistry to display the relationships between four variables, following their introduction by the German geologist and mineralogist, Hellmut von Philipsborn (1928). A similar idea was later used by the American sedimentologist and mathematical geologist, William ('Bill') C. Krumbein (1902–1979), for the classification of sedimentary lithofacies for mapping purposes, e.g. based on the relative thicknesses of limestone, sandstone, shale and evaporite at different localities in the same formation (Krumbein 1954*a*).

From time-line to bar-chart

Funkhouser (1938, pp. 279–280) attributes W. Playfair's invention of the comparative device of the bar-chart to his awareness of the time-line, a graphical device introduced in 1765 by the British chemist Joseph Priestley (1733–1804), in which the length of an individual's life is shown by a line drawn parallel to the time-axis of a graph, beginning and ending at the years of birth and death of the subject. Exactly the same style applies to the simplest form of a taxonomic range-chart, as used by the British naturalist Samuel Woodward (1821–1865) (Woodward 1854, pp. 414–415) and the geologist Hugh Miller (1802–1856) (Miller 1857, p. 8).

It was the British geologist, Sir Charles Lyell (1797–1875) who, in the third volume of the first edition of his *Principles of Geology* (Lyell 1830–1833) was the earliest to use a 'statistical' count of the relative abundances of extant and extinct species to distinguish between the Eocene, Miocene and Pliocene Series of the Tertiary Sub-Era (see Rudwick (1978) for further discussion of Lyell's approach), but he presented his results simply as tables. The French geologist Joachim Barrande (1799–1883) later used proportional-width lines to indicate the relative abundance of different genera of trilobites in

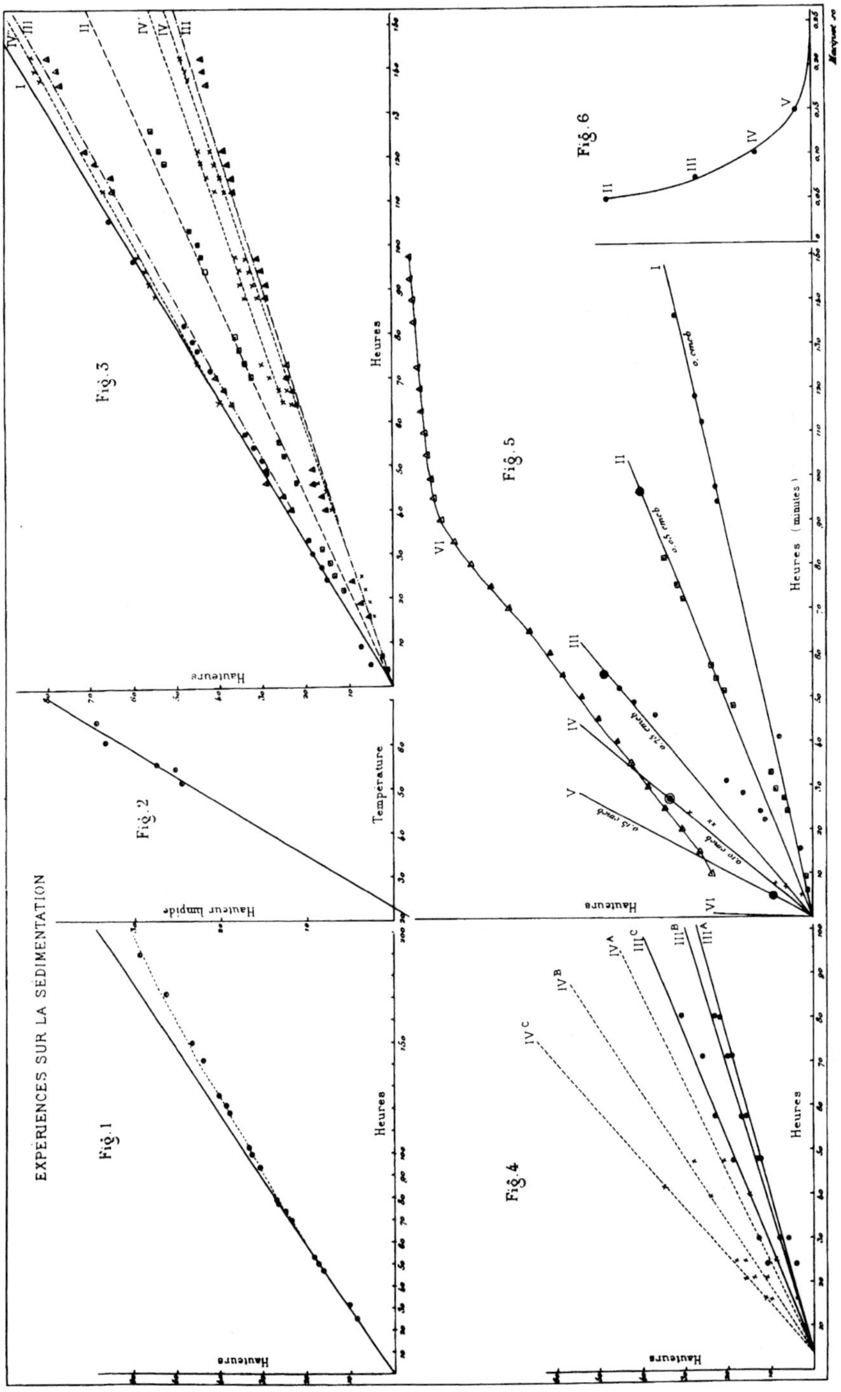

Fig. 7. Six scatter-plots with linear or curved lines fitted by eye illustrating the results of sedimentation experiments. Reproduced from plate I of Thoulet (1891), with permission of the Science & Society Picture Library.

each stratigraphic division of the Silurian Period rocks of Bohemia (Barrande 1852). Many authors subsequently adopted the inclusion of frequency information in taxonomic range charts. By the 1920s, this form of presentation was regularly used to illustrate micropalaeontological or micropalynological results in the form of range-charts for the purposes of biostratigraphic correlation (Goudkoff 1926; Driver 1928; Wray *et al.* 1931). The idea of the time-line also became enshrined in petrology in the form of the mineral paragenesis diagram, first introduced by the Austrian mineralogist Gustav Tschermak (1836–1927) to illustrate the evolution of granites (Tschermak 1863).

In addition to tabular summaries, in his book *Life on the Earth* Phillips (1860, p. 63) used proportional-length bars and proportional-width time-lines (Phillips 1860, p. 80), to illustrate the change in composition of 'marine invertebrata' throughout the 'Lower Palaeozoic' of England and Wales. In the frontispiece to the book, he also showed the relative proportions of eight classes of 'marine invertebral life' in each Period of the Phanerozoic, as constant-length bars subdivided according to the relative proportions of each class (see Fig. 8). A similar presentation was used subsequently by Reyer (1888, p. 215) to compare the major-element oxide compositions of suites of igneous rocks. Proportional-length rectangles (Greenleaf 1896), squares (Ahlburg 1907) and bars (Umpleby 1917) were occasionally used, particularly in publications related to economic geology. In an early paper on stratigraphic correlation using heavy minerals, the German petroleum geologist, Hubert Becker (*b.* 1903) used a range-chart with proportional-length bars to illustrate progressive stratigraphic change in the mineral suite (Becker 1931), but the 'graphic log', based on the proportions of different lithologies in the well-cuttings and drawn as a multiple line-graph, had already been introduced by the American petroleum geologist Earl A. Trager (1920).

Pie diagrams

The division of a circle into proportional-arc sectors to form a 'pie diagram' dates back to the work of W. Playfair (1801) and was used as a cartographic symbol by Minard in 1859 (see Robinson 1982, p. 207). However, apart from occasional applications comparing the composition of fresh with altered rock as a result of mineralization (Lacroix 1899; Leith 1907) or the relative production of metals or coal (Anon. 1907; Butler *et al.* 1920), it was little used by geologists.

Multivariate symbols

Between 1897 and 1909, there was a short-lived enthusiasm for comparison of the major-element composition of igneous rocks using a variety of symbols based mainly on graphic styles which resemble the modern 'star plot' in which the length of each arm is proportional to the amount of each component present in a sample (Fig. 9). The earliest of these was devised by Michel Lévy (1897*a*) but it was Iddings (1903, 1909, pp. 8–22, plates 1, 2) who was a determined advocate for this type of presentation (and for the use of graphical methods in igneous petrology in general). However, the tedium of multivariate symbol construction by hand ultimately prevented the widespread take-up of these methods. For example, although their use was advocated in a 1926 article 'Calculations in petrology: a study for students' by the American geologist Frank F. Grout (*b.* 1880), they were not mentioned in the influential textbook *Petrographic Methods and Calculations* by the British geologist Arthur Holmes (1890–1965), published in 1921 (in which he restricted his discussion to variation and ternary diagrams) Similar multivariate graphical techniques, such as the well-known Stiff (1951) diagram for water composition, were later introduced for comparison of hydrogeochemical data. (For further information, see Howarth (1998) on igneous and metamorphic petrology, and Zaporozec (1972) on hydrogeochemistry.) However, the usage of multivariate symbols did not really revive until it was eased by computer graphics in the 1960s. Figure 10 summarizes the relative frequency of all types of statistical graphs and maps from 1750 to 1935, based on a systematic scan of 116 geological serial publications, plus book collections. Apart from crystallographic applications (which were often undertaken by physicists or other non-geologists), major growth in usage and graphic innovation essentially began in the 1890s.

The rise of statistical thinking

The time-series describing commodity production in economic geology, discussed previously, typify the nineteenth century view of 'statistics' as 'a collection of numerical facts'. Lyell's subdivision of the Tertiary Sub-Era on the basis of faunal counts in 1829 (Lyell 1830–1833) conformed to this somewhat simplistic view, although it is believed that he hoped to verify a general method, a '*statistical* paleontology' (Rudwick 1978, p. 236), which he could apply to earlier parts of the succession. The rapidly

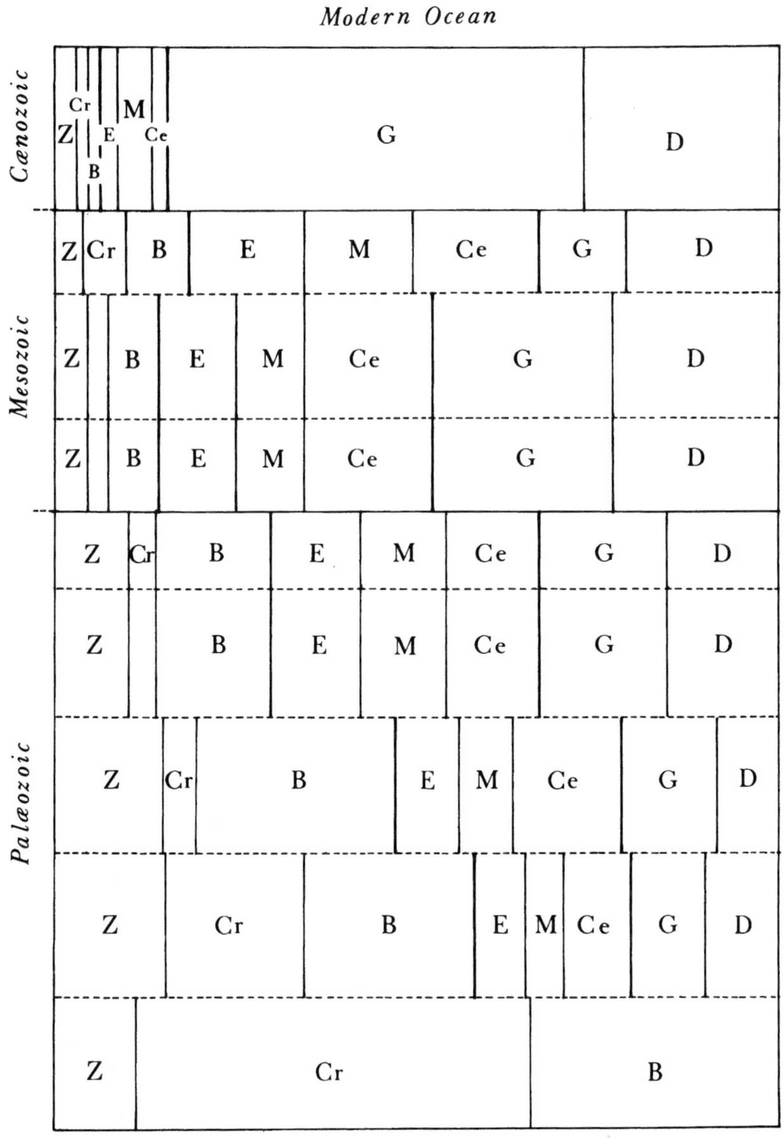

Fig. 8. Divided bar-chart showing 'successive systems of marine invertebral life': Z, Zoophyta; Cr, Crustacea; B, Brachiopoda; E, Echinodermata; M, Monomysaria; Ce, Cephalopoda; G, Gasteropoda; and D, Dimyaria. Redrawn from Phillips (1860, frontispiece).

growing body of mathematical publications on the 'theory of errors' and the method of 'least squares' published in the wake of the pioneering work of the mathematicians Adrien M. Legendre (1752–1833) in 1805 and Carl F. Gauss (1777–1855) in 1809, had little appeal outside the circle of mathematicians and astronomers involved in its development. However, the Belgian astronomer and statistician, Adolphe Quetelet (1796–1874) wrote, in a more approachable manner, on the normal distribution and used statistical maps, in his writings on the 'social statistics' of population, definition of the characteristics of the 'average man,' and

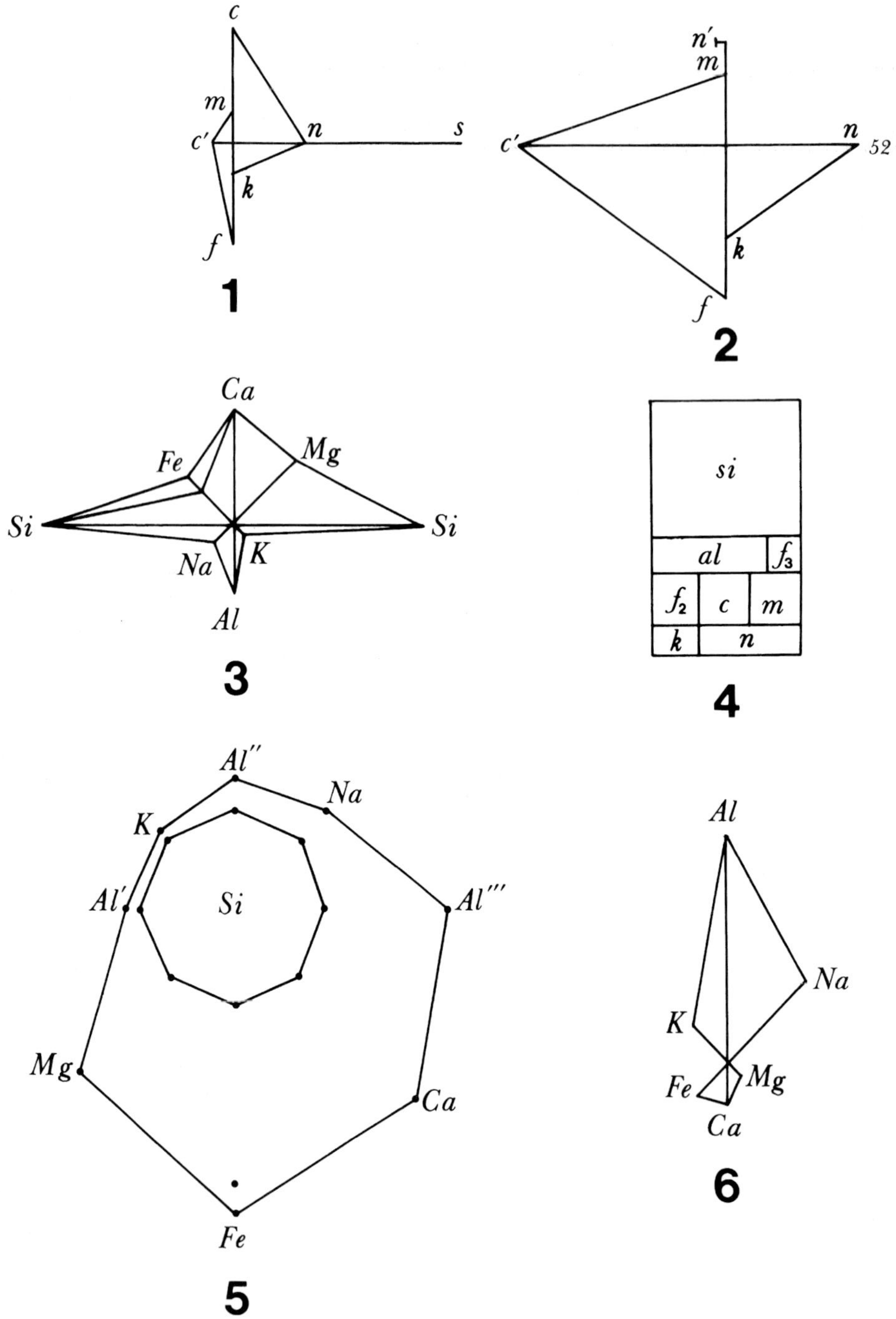

Fig. 9. Different styles of multivariate graphics used to illustrate major element sample composition: 1, Michel Lévy (1897*b*); 2, Michel Lévy (1897*a*); 3, Brøgger (1898); 4, Loewinson-Lessing (1899); 5, Mügge (1900); 6, Iddings (1903). Reproduced from fig. 5 of Howarth, R. J. 1998. Graphical methods in mineralogy and igneous petrology (1800–1935). *In*: Fritscher, B. & Henderson, F. (eds) *Toward a History of Mineralogy, Petrology, and Geochemistry. Proceedings of the International Symposium on the History of Mineralogy, Petrology, and Geochemistry, Munich, March 8–9, 1996*, pp. 281–307, with permission of the Institut für Geschichte der Naturwissenschaften der Universität München. All rights reserved.

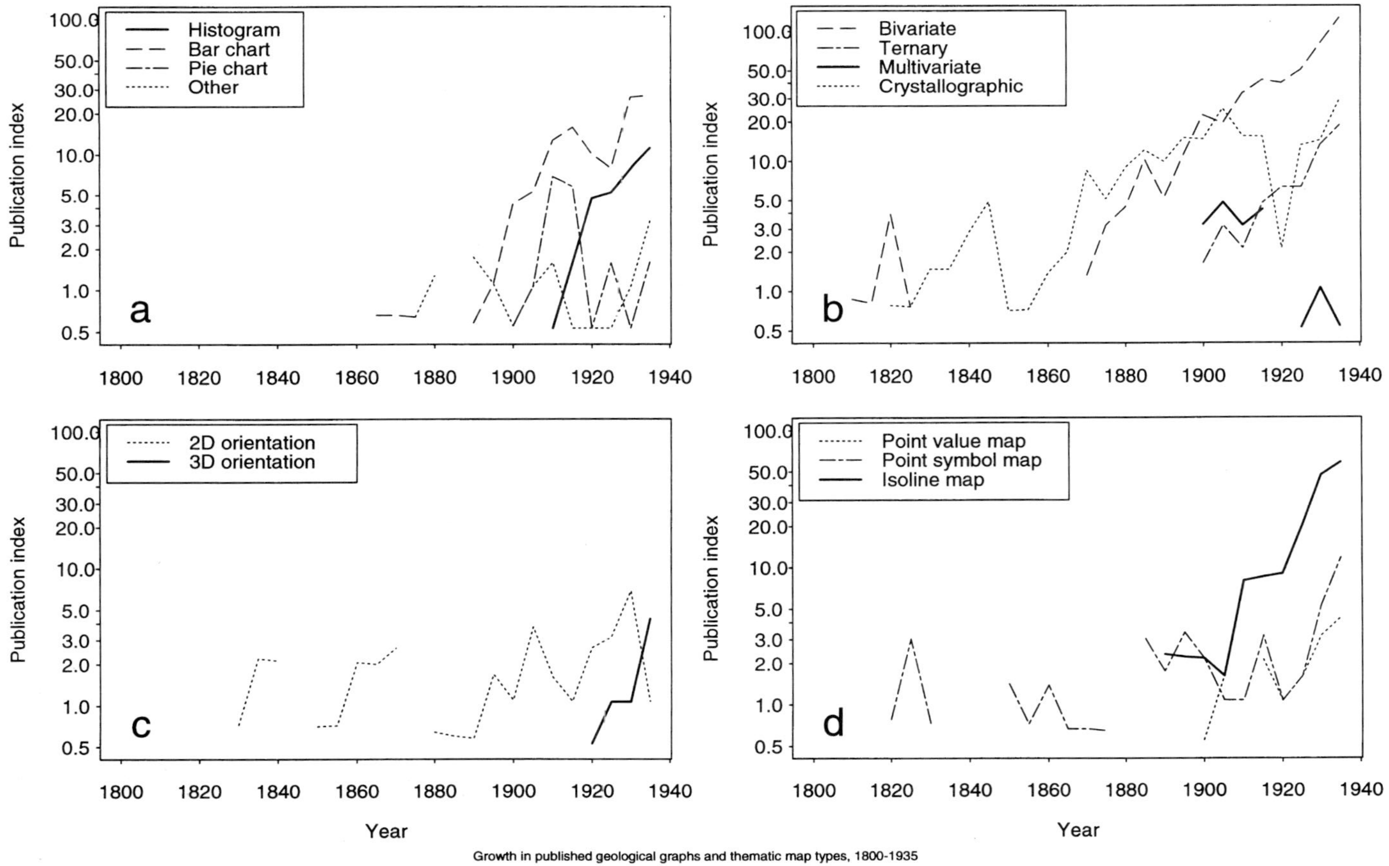

Fig. 10. Normalized publication index for usage of different types of 1942 statistical graphs and 236 thematic maps in systematic scan of more than 100 journals (1800–1935). (**a**) Relative frequency plots: histograms, bar-charts, pie-charts and miscellaneous univariate graphics. (**b**) Bivariate scatter-plots and line diagrams; ternary (triangular) diagrams; multivariate symbols (cf. Fig. 9); and specialized crystallographic and mineralogical diagrams. (**c**) Two-dimensional orientation (rose diagrams, etc.) and three-dimensional orientation (stereographic) plots. (**d**) Point value, point symbol and isoline thematic maps. Counts have been normalized by dividing through by values of Table 2, Appendix. Index is zero where nc symbols are shown.

the statistics of crime (Quetelet 1827, 1836, 1869). As a result, Quetelet's work proved to be enormously influential, and raised widespread interest in the use of both frequency distributions and statistical maps.

In geology, this interest soon manifested itself in the earthquake catalogues of the Belgian scientist Alexis Perrey (1807–1822), who followed Quetelet's advice (Perrey 1845, p. 110) and from 1845 onwards used line-graphs (drawn in exactly the same style as used by Quetelet in his own work) in his earthquake catalogues to illustrate the monthly frequency and direction of earthquake shocks. Other early examples of earthquake frequency polygons occur in Volger (1856). The use of maps showing the frequency of earthquake shocks occurring in a given time-period for different parts of a region was pioneered by the British seismologist John Milne (1850–1913) and his colleagues in Japan (Milne 1882; Sekiya 1887).

In structural geology, attempts to represent two-dimensional directional orientation distributions began in the 1830s, although use of an explicit frequency distribution based on circular co-ordinates only became widespread following the work (Haughton 1864) of the Irish geologist Samuel Haughton (1821–1897). The more specialized study of the three-dimensional orientation distributions did not begin until the 1920s with the work of the Austrian mineralogist Walter Schmidt (1885–1945) and his colleague, the geologist Bruno Sander (1884–1979) who began petrofabric studies of metamorphic rocks. Their work introduced use of the Lambert equal-area projection of the sphere to plot both individual orientation data and isoline plots of point-density. A simpler method of representation, using polar co-ordinate paper, was introduced by Krumbein (1939) to plot the results of three-dimensional fabric analyses of clasts in sedimentary rocks, such as tills. (See Howarth (1999) and Pollard (2000) for further discussion of aspects of the history of structural geology.)

Some early enthusiastic efforts to apply the properties of Quetelet's 'binomial curve' (his approximation of the normal distribution using a large-sample binomial distribution) were misdirected, for example Tylor's (1868, p. 395) attempt to match hill-profiles to its shape. Nevertheless, by the turn of the century, Thomas C. Chamberlin (1843–1928) in America was advocating the use of 'multiple working hypotheses' when attempting to explain complex geological phenomena (Chamberlin 1897) and Henry Sorby (1826–1908) in England was demonstrating the utility of quantitative methods (including model experiments) to gaining a better understanding of sedimentation processes (Sorby 1908).

Nevertheless, statistical applications tended to remain mainly descriptive, characterized by the increasing use of frequency distributions. Examples include morphometric applications in palaeontology (Cumins 1902; Alkins 1920) and igneous petrology (Harker 1909; Robinson 1916; Richardson & Sneesby 1922; Richardson 1923). However, it was the British mineralogist and petrologist William A. Richardson who first made real use of the theoretical properties of the normal distribution. Using the 'method of moments' (Pearson 1893, 1894), which had been developed by the British statistician Karl Pearson (1857–1936), Richardson (1923) successfully resolved the bimodal frequency distribution of SiO_2 wt% in 5159 igneous rocks into two, normally distributed, acid and basic subpopulations and was able to demonstrate their significance in the genesis of igneous rocks.

Another area in which frequency distributions soon grew to play an essential role was in sedimentological applications. Systematic investigation of size-distributions using elutriation and mechanical analysis developed in the second half of the nineteenth century (Krumbein 1932). A grade-scale, based on sieves with mesh sizes increasing in powers of two, was introduced in America by Johan A. Udden (1859–1932) in 1898 (see also Udden 1914; Hansen 1985) and was modified subsequently by Chester K. Wentworth (*b.* 1891) to the size-grade divisions 1/1024, 1/512, 1/256, . . ., 8, 16, 32 mm (Wentworth 1922). Cumulative size-grade curves began to be used in the 1920s (Baker 1920), and both Wentworth and Parker D. Trask (*b.* 1899) tried to use statistical measures, such as quartiles, to describe their attributes (Wentworth 1929, 1931; Trask 1932).

Krumbein had acquired statistical training while gaining his first degree in business management, before turning to geology. This led to his interest in quantifying the degree of uncertainty inherent in sedimentological measurement (Krumbein 1934) and enabled him to demonstrate, using normal probability plots (Krumbein 1938), the broadly lognormal nature of the size distributions and that statistical parameters were therefore best calculated following logtransformation of the sizes. This led to the introduction of the 'phi scale' (given by base-2 logarithms of the size-grades) which eliminated the problems caused by the unequal class intervals in the metric scale. Parameters based on moment measures were eventually augmented by Inman's (1952) introduction of graphical analogues, such as the phi skewness measure.

It soon became apparent that a manual of laboratory methods concerned with all aspects of the size, shape and compositional analysis of sediments was needed. Krumbein collaborated with his former PhD supervisor at the University of Chicago, Francis J. Pettijohn (1904–1999), to produce the *Manual of Sedimentary Petrography* (Krumbein & Pettijohn 1938). In this text, Krumbein described the chi-squared goodness-of-fit test for the similarity of two distributions (Pearson 1900; Fisher 1925), which had been recently introduced into the geological literature (Eisenhart 1935) by the American statistician Churchill Eisenhart (1913–1994). However, although Krumbein discussed the computation of Pearson's (1896) linear correlation coefficient, he rather surprisingly made no mention of fitting even linear functions to data using regression analysis, treating the matter entirely in graphical terms (Krumbein & Pettijohn 1938, pp. 205–211).

The use of bivariate regression analysis in geology began in the 1920s, in palaeontology (Alkins 1920; Stuart 1927; Brinkmann 1929; Waddington 1929), and in geochemistry (Eriksson 1929). The use of other statistical methods was also becoming more widespread, championed, for example, during the 1930s by Krumbein in the United States, and in the 1940s by the British sedimentologist Percival Allen (*b.* 1917), and by Andrei Vistelius (1915–1995) in Russia (Allen 1944; Vistelius 1944; see also selected collected papers (1946–1965) in Vistelius 1967). The foundations of multivariate statistical methods, such as multiple regression analysis and discriminant function analysis (used to assign an unknown specimen on the basis of its measured characteristics to one of two, or more, pre-defined populations), had been laid previously by the British statistician Sir Ronald Aylmer Fisher (1890–1962, Kt., 1952) (Fisher 1922, 1925, 1936). Although these techniques began to make an appearance in geological applications (Leitch 1940; Burma 1949; Vistelius 1950; Emery & Griffiths 1954), with the odd exception – Vistelius apparently carried out a factor analysis by hand in 1948 (Dvali *et al.* 1970, p. 3) – their use was restricted by the tedious nature of the hand-calculations. For example, Vistelius recalls undertaking Monte Carlo (probabilistic) modelling of sulphate deposition in a sedimentary carbonate sequence by hand in 1949, a process (described in Vistelius 1967, p. 78) which 'required several months of tedious work' (Vistelius 1967, p. 34). In the main, geological application of more computationally demanding statistical methods had to await the arrival of the computer.

The roots of mathematical modelling

As Merriam (1981) has noted, mathematicians and physicists have a history of early involvement in the development of theories to explain Earth science phenomena and have underpinned the emergence of geometrical and physical crystallography (Lima-de-Faria 1990). Although in many instances their primary focus was on geophysics, geological phenomena were not excluded from consideration. For example, the Italian mathematician Paolo Frisi (1728–1784) made an early quantitative study of stream transport (Frisi 1762). In the nineteenth century, J. Playfair (1812) applied mathematical modelling to questions such as the thermal regime in the body of the Earth, but he also calculated the vector mean of dip directions measured in the field (Playfair 1802, fn., pp. 236–237); the British mathematician and geologist William Hopkins (1793–1866), who had Stokes, Kelvin, Maxwell, Galton and Todhunter as his Cambridge mathematical tutees, developed mathematical theories to explain the presence and orientation of 'systems of fissures' and ore-veins (Hopkins 1838), glacier motion and the transport of erratic rocks (Hopkins 1845, 1849*a*), the nature of slaty cleavage (Hopkins 1849*b*); and the British geophysicist the Reverend Osmond Fisher (1817–1914) provided mathematical reasoning to explain volcanic phenomena in his textbook *Physics of the Earth's Crust* as well as discussion of the nature of the Earth's interior (Fisher 1881).

As the use of chemical analysis of igneous and metamorphic rocks increased, petrochemical calculations began to be used both to assist the classification of rocks on the basis of their chemical composition and to understand their genesis. This type of study essentially began with the 'CIPW' norm (named after the authors Cross, Iddings, Pirsson and Washington, 1902, 1912) which was used to re-express the chemical composition of an igneous rock in terms of standard 'normative' mineral molecules instead of the major-element oxides.

Another area in which quantitative numerical methods were becoming increasingly important was hydrogeology. Hydrogeological applications in Britain date back to the work of William Smith at the beginning of the nineteenth century (Biswas 1970). Following experiments carried out in 1855 and 1856, the French engineer Henry Darcy (1803–1858) discovered the relationship which now has his name (Darcy 1856, pp. 590–594). He concluded that 'for identical sands, one can assume that the discharge is directly proportional to the [hydraulic] head and inversely proportional to the thickness of the layer traversed' (quoted in Freeze 1994, p. 24). Although Darcy used a physical rather than a mathematical model to determine his law (measuring flow through a sand-filled tube), this can be regarded as the earliest groundwater model study. Thirty years later, Chamberlin (1885) published his classic investigation of artesian flow, which marked the beginning of groundwater hydrology in the United States. The first memoir of the British Geological Survey on underground water supply was published soon afterwards (Whittaker & Reid 1899).

Following the appointment of the American hydrogeologist Oscar E. Meinzer (1876–1948) as chief of the groundwater division of the United States Geological Survey in 1912, quantitative methods to describe the storage and transmission characteristics of aquifers advanced considerably. Meinzer himself laid the foundations with publication of his PhD dissertation as a US Geological Survey water supply paper (Meinzer 1923). Early applications had to make do with steady-state theory for groundwater flow, which only applies after wells have been pumped for a long time. Charles V. Theis (1900–1987) then derived an equation to describe unsteady-state flow conditions (Theis 1935) using an analogy with heat-flow in solids. This enabled the 'formation constants' of an aquifer to be determined from the results of pumping tests. His achievement has been described as 'the greatest single contribution to the science of groundwater hydraulics in this century' (Moore & Hanshaw 1987, pp. 317). Theis (1940) then explained the mechanisms controlling the cone of depression which develops as water is pumped from a well. His work enabled hydrologists to predict well yield and to determine their effects in time and space.

That same year, M. King Hubbert (1903–1989) discussed groundwater flow in the context of petroleum geology (Hubbert 1940). By the 1950s, physical models used a porous medium such as sand (as had Darcy in the 1850s), or stretched membranes, to mimic piezometric surfaces, and analytical solutions were being applied to two-dimensional steady-state flow in a homogeneous flow system However, these analytical methods proved inadequate to solve complex transport problems. The possibility of using electrical analogue models (based on resistor–capacitor networks) in transient-flow problems was investigated first by H. E. Skibitzke and G. M. Robinson at the US Geological Survey in 1954 (Moore & Hanshaw 1987, p. 318). Their work eventually led to the establishment of an analogue-model laboratory at Phoenix, Arizona, in 1960 (Walton & Prickett 1963; Moore & Wood 1965) and more than 100 different models were run by 1975 (Moore & Hanshaw 1987). The use of graphical displays in hydrogeology is discussed in detail in Zaporozec (1972).

The arrival of the digital computer

By the early 1950s, in the United States and Britain, digital computers had begun to emerge from wartime military usage and to be employed in major industries such as petroleum, and in the universities. At first, these computers had to be painstakingly programmed in a low-level machine language. Consequently, it must have come as a considerable relief to users when International Business Machines' Mathematical FORmula TRANslating system (the FORTRAN programming language) was first released in 1957, for the IBM 704 computer (Knuth & Pardo 1980), as FORTRAN had been designed to facilitate programming for scientific applications. Computer facilities did not become available to geologists in Russia until the early 1960s (Vistelius 1967, pp. 29–40), and in China until the 1970s (Liu & Li 1983).

The earliest publication to use results obtained from a digital computer application in the Earth sciences is believed to be Steven Simpson Jr's program for the WHIRLWIND I computer at the Massachusetts Institute of Technology, Cambridge, Massachusetts. His program was essentially a multivariate polynomial regression in which the spatial co-ordinates, and their powers and cross-products, were used as the predictors to fit second- to fourth-order non-orthogonal polynomials to residual gravity data. This type of application later became known as 'trend-surface analysis' (Krumbein 1956; Miller 1956). Simpson presented his results in the form of isoline maps, which had to be contoured by hand on the basis of a 'grid' of values printed out on a large sheet of paper by the computer's Flexowriter (Simpson 1954, fig. 8). However, Simpson also used the computer's oscilloscope display to produce a 'density plot' in which a variable-density dot-matrix provided a grey-scale image showing the topography of the surface formed by the computed regression residuals. This display was then photographed to provide the final 'map' (Simpson 1954, fig. 9).

Nevertheless, it was Krumbein who mainly pioneered the application of the computer in geological applications. Following a short period after World War II working in a research group at the Gulf Oil Company, he developed a strong interest in quantitative lithofacies mapping

(Pettijohn 1984, p. 176), the data being mainly derived from well-logs (Krumbein 1952, 1954*a*, 1956). This interest soon led Krumbein and the stratigrapher Lawrence L. Sloss (1913–1996), based at Northwestern University (Evanston, Illinois), to write a machine-language program for the IBM 650 computer to compute clastic and sand-shale ratios in a succession based on the thicknesses of three or four designated end-members. A flowchart and program listings are given in Krumbein & Sloss (1958, fig. 8, tables 2, 3). The data were both input and output via punched cards, the final ratios being obtained from a listing of the output card deck.

Krumbein was interested in being able to differentiate quantitatively between large-scale systematic regional trends and essentially non-systematic local effects, in order to enhance the rigour of the interpretation of facies, isopachous and structural maps. This led him, in 1957, to write a machine-language program for the IBM 650 to fit trend-surfaces (Whitten *et al.* 1965, iii). It was not long before the release of the FORTRAN II programming language made such tasks easier.

In 1963, two British geologists who had emigrated to the United States, Donald B. McIntyre (*b.* 1923) at Pomona College, Claremont, California, and E. H. Timothy Whitten (*b.* 1927), who was working with Krumbein at Northwestern University, both published trend-surface programs programmed in FORTRAN (Whitten 1963; McIntyre 1963*a*) and in Russia, Vistelius was also using computer-calculated trend-surfaces in a study of the regional distribution of heavy minerals (Vistelius & Yanovskaya 1963; Vistelius & Romanova 1964).

More routine calculations, such as sediment size-grade parameters (Creager *et al.* 1962), geochemical norms (McIntyre 1963*b*) and the statistical calibrations which underpinned the adaptation of new analytical techniques, such as X-ray fluorescence analysis (Leake *et al.* 1970), to geochemical laboratory usage, were all greatly facilitated.

However, it was the rapid development of algorithms enabling the implementation of complex statistical and numerical techniques which perhaps made the most impression on the geological community, as they demonstrated in an unmistakable manner that computers could enable them to apply methods which had hitherto seemed impractical. Examples of early computer-based statistical applications in the west included the following.

(i) The use of stepwise multiple regression (Efroymson 1960) to determine the optimum number of predictors required to form an effective prediction equation (Miesch & Connor 1968).
(ii) The methods of principal components and factor analysis (Spearman 1904; Thurstone 1931; Catell 1952) which were developed to compress the information inherent in a large number of variables into a smaller number which are linear functions of the original set, in order to aid interpretation of the behaviour of the multivariate data and to enable its more efficient representation. The concept was extended, by the American geologist John Imbrie (*b.* 1925), to represent the compositions of a large number of samples in terms of a smaller number of end-members (Imbrie & Purdy 1962; Imbrie 1963; Imbrie & van Andel 1964; McCammon 1966) and proved to be a useful interpretational tool.
(iii) Hierarchical cluster-analysis methods, originally developed to aid numerical taxonomists (Sokal & Sneath 1963), proved extremely helpful in grouping samples on the basis of their petrographical or chemical composition (Bonham-Carter 1965; Valentine & Peddicord 1967).
(iv) Application of the Fast Fourier Transform (FFT; Cooley & Tukey 1965; Gentleman & Sande 1966) to filtering time series and spatial data (Robinson 1969).

Figure 11 shows the approximate time of the earliest publication in the Earth sciences of a wide range of statistical graphics and other statistical methods imported from work outside the Earth sciences (as well as the relatively few examples known to the author in which the geological community seem to have been the first to have developed a method). Note the sharp decrease in the time-lag after the introduction of computers into the universities at the end of World War II, presumably as a result of improved ease of implementation and increasingly rapid information exchange as a result of an exponentially increasing number of serial publications.

In the early years, the dissemination of computer applications in the Earth sciences was immensely helped by the work of the geologist Daniel Merriam (*b.* 1927), at the Kansas Geological Survey, later assisted by John Davis (*b.* 1938), through the dissemination of computer programs and other publications on mathematical geology. These initially appeared as occasional issues of the Special Distribution Publications of the Survey, and then as the Kansas Geological Survey Computer Contributions series, which ran to 50 issues between

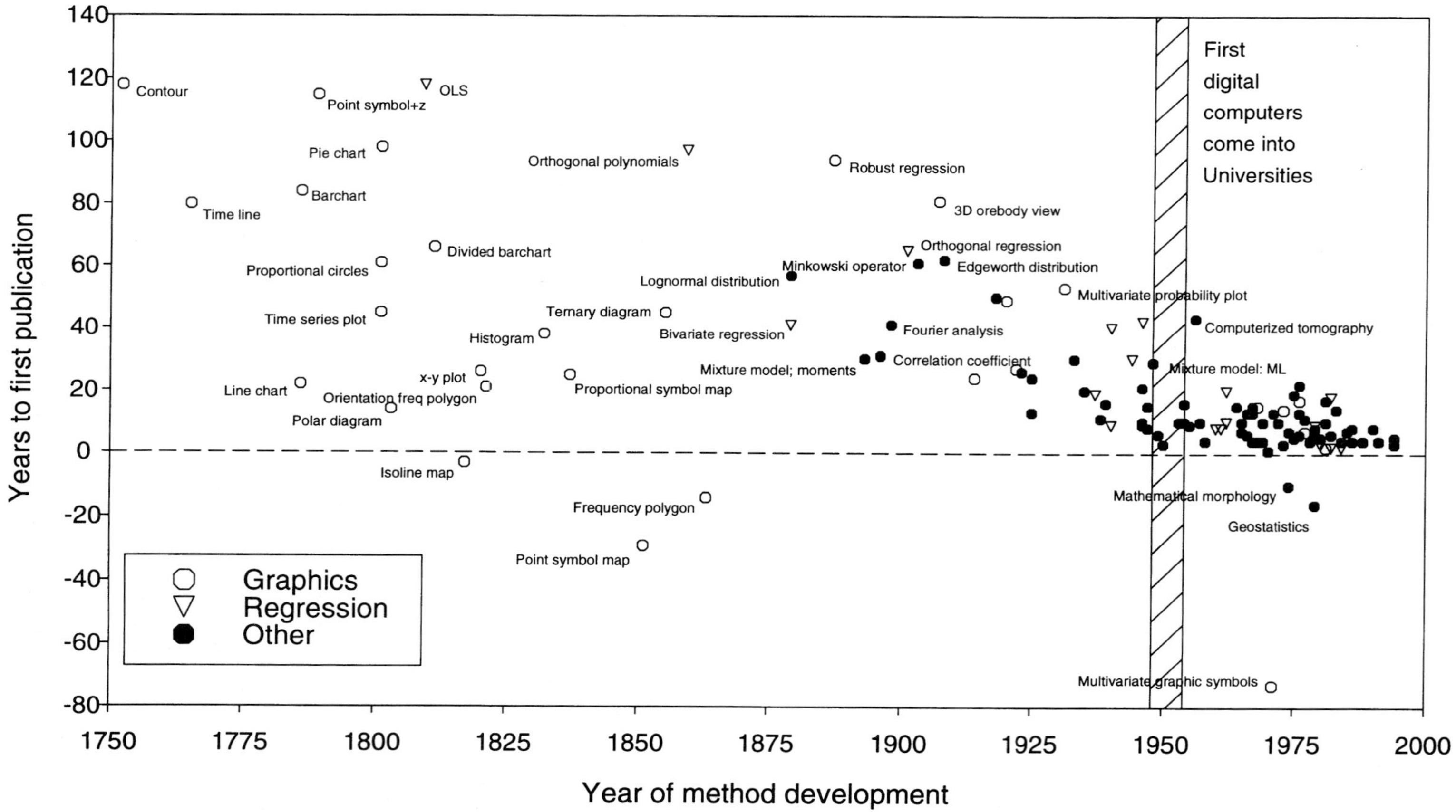

Fig. 11. Time to uptake of 121 statistical methods (graphics or computation) in the Earth sciences from earliest publication in other literature in relation to the years in which the earliest digital computers began to come into the universities following World War II (the few examples in which a method appeared first in the Earth sciences are plotted below the horizontal zero-line).

1966 and 1970. By the end of 1967, Computer Contributions were being distributed, virtually free, to workers across the United States and in 30 foreign countries (Merriam 1999). The Kansas Geological Survey sponsored eight colloquia on mathematical geology between 1966 and 1970.

The International Association for Mathematical Geology (IAMG) was founded in 1968 at the International Geological Congress in Prague, brought to an abrupt end by the chaos of the Warsaw Pact occupation of Czechoslovakia. Syracuse University and the IAMG then sponsored annual meetings ('Geochautauquas') from 1972 to 1997 and Merriam became the first editor-in-chief for the two key journals in the field: *Mathematical Geology*, the official journal of the IAMG (1968–1976 and 1994–1997), and *Computers & Geosciences* (1975–1995).

Sedimentological and stratigraphic applications continued to motivate statistical applications during the 1960s. Krumbein had earlier drawn attention to the importance of experimental design, sampling strategy and of establishing uncertainty ('error') magnitudes (Krumbein & Rasmussen 1941; Krumbein 1953, 1954*b*, 1955; Krumbein & Miller 1953; Krumbein & Tukey 1956); and the work of the émigré British sedimentary petrographer and mathematical geologist John C. Griffiths (1912–1992) reinforced this view (Griffiths 1953, 1962). Following a PhD in petrology from the University of Wales and a PhD in petrography from the University of London, Griffiths worked for an oil company before moving to Pennsylvania State University in 1947, where he remained until his retirement in 1977. An inspirational teacher, administrator and lecturer, he is now perhaps best known for his pioneering studies in the application of search theory (Koopman 1956–1957) to exploration strategies and quantitative mineral- and petroleum-resource assessment (Griffiths 1966*a*,*b*, 1967; Griffiths & Drew 1964, 1973; Griffiths & Singer 1970). The legacy of the work of Griffiths and his students can be seen in the account by Lawrence J. Drew (who was one of them), of the petroleum-resource appraisal studies carried out by the United States Geological Survey (Drew 1990).

Krumbein also introduced the idea of the conceptual process–response model (Krumbein 1963; Krumbein & Sloss 1963, chapter 7) which attempts to express in quantitative terms a set of processes involved in a given geological phenomenon and the responses to that process. Krumbein's earliest example formalized the interaction in a beach environment, showing how factors affecting the beach (energy factors: characteristics of waves, tides, currents, etc.; material factors: sediment-size grades, composition, moisture content, etc.; and shore geometry) were reflected in the response elements (beach geometry, beach materials) and he suggested ways by which such a conceptual model could be translated into a simplified statistically based predictive model (Krumbein 1963). Reflecting Chamberlin's (1897) idea of using multiple working hypotheses in a petrogenetic context, Whitten (1964) suggested that the characteristics of the response model might be used to distinguish between different petrogenetic hypotheses resulting from different conceptual process models. Whitten & Boyer (1964) used this approach in an examination of the petrology of the San Isabel Granite, Colorado, but determined that unequivocal discrimination between the alternative models was more difficult than anticipated.

At this time there was also renewed interest in the statistics of orientation data arising from both sedimentological applications (Agterberg & Briggs 1963; Jones 1968) and petrofabric work in structural geology (see Howarth (1999) and Pollard (2000) for further historical discussion).

The Australian statistician Geoffrey S. Watson (1921–1998), who had emigrated to North America in 1959, published a landmark paper reviewing modern methods for the analysis of two- and three-dimensional orientation data (Watson 1966) in a special supplement of the *Journal of Geology* which was devoted to applications of statistics in geology. This issue of the journal also contained papers in several areas which would assume considerable future importance: the multivariate analysis of major-element compositional data and the apparently intractable problems posed by its inherent percentaged nature (Chayes & Kruskal 1966; Miesch *et al.* 1966), stochastic (probabilistic) simulation (Jizba 1966), and Markov schemes (Agterberg 1966). The American petrologist Felix Chayes (1916–1993) made valiant efforts to solve the statistical problems posed by percentaged data, which also were inherent in petrographic modal analysis, a topic with which he was closely associated for many years (Chayes 1956, 1971; Chayes & Kruskal 1966). A solution was ultimately provided by another British émigré, the statistician John Aitchison (*b.* 1926), then working at the University of Hong Kong, in the form of the 'logratio transformation': $y_i \leftarrow \log(x_i/x_n)$, where the index i refers to each of the first to the $(n{-}1)$th of the n components, while x_n forms the 'basis', e.g. SiO_2 in the case of percentaged major oxide composition (Aitchison 1981, 1982).

A series of observations is said to possess the Markov property if the behaviour of any observation can be predicted solely on the basis of the behaviour of the observations which precede it. Such behaviour may be characterized using a transition probability matrix, which summarizes the probability of any given state switching to another (Allégre 1964). Empirical switching probabilities for the transition from one lithological state to another, e.g. sandstone ⇔ shale ⇔ siltstone ⇔ lignite (data of Wolfgang Scherer, quoted in Krumbein & Dacy 1969), are derived from observations, made at equal intervals along measured stratigraphic sections or well-logs, recording which of a given set of lithologies is present at each position. Although originally pioneered by Vistelius (1949), such applications only came into prominence in the 1960s. This was mainly as a result of renewed interest in cyclic sedimentation, aided by the possibility of using the computer to simulate similar stratigraphic processes (Krumbein 1967). Workers such as Walther Schwarzacher (*b.* 1925), at the University of Belfast (Northern Ireland) and Krumbein concentrated on lithostratigraphic data (Schwarzacher 1967; Krumbein 1968; Krumbein & Dacy 1969). The Dutch mathematical geologist Frederik ('Frits') P. Agterberg (*b.* 1936), who had recently joined the Geological Survey of Canada following a postdoctoral year (1961–1962) at the University of Wisconsin, considered the more general situation of multicomponent geochemical trends (Agterberg 1966). Vistelius undertook a long-term study of the significance of grain-to-grain transition probabilities in the textures of 'ideal' granites and how they change in conditions of metasomatic alteration (Vistelius 1964, revisited in Vistelius *et al.* 1983), although Whitten & Dacey (1975) raised some doubts about the utility of his approach.

The conventional techniques of time-series analysis, as used in geophysics (i.e. power-spectral analysis, enabled by the FFT), also have been applied to sequences of stratigraphic-thickness data as an alternative to the Markov chain approach (Anderson & Koopmans 1963; Schwarzacher 1964; Agterberg & Banerjee 1969). In recent years, increasing interest in the influences of orbital variations on sedimentary processes (on Milankovich cyclicity; see Imbrie & Imbrie 1979, 1980; Schwarzacher & Fischer 1982; Imbrie 1985; and *Terra Nova* 1989, Special Issue **1**, pp. 402–480) has resulted in new techniques being applied to stratigraphic time series analysis, such as the use of Walsh power spectra (Weedon 1989) and wavelet analysis (Prokoph & Barthelmes 1996) which provides not only information regarding the amplitudes (or power) at different frequencies, but also information about their time dependence.

An important application area, in which the role of time is implicit, is that of quantitative biostratigraphy and related methods of stratigraphic correlation. The American palaeontologist Alan B. Shaw first developed the technique of 'graphic correlation', based on correlating the first and last appearances of a series of key taxa in two or more surface- and/or well-sections, while working for the Shell Oil Company in 1958 (Shaw 1995) and, as a result of its simplicity and efficacy, the method is still widely used (Mann & Lane 1995). Quantitative methods for faunal comparison, and seriation of samples based on such information to produce a pseudo-stratigraphy, an approach initially founded on techniques developed in archaeology (Petrie 1899), also began to develop in the 1950s, and the numbers of publications on quantitative stratigraphy increased steadily, until levelling off in the 1980s (Thomas *et al.* 1988; CQS 1988–1997). Since 1972, much of this work has been conducted under the auspices of the International Geological Correlation Programme (IGCP) Project 148 (Evaluation and Development of Quantitative Stratigraphic Correlation Techniques). This was initiated in 1976 as a project on quantitative biostratigraphic correlation under James C. Brower (Syracuse University, New York). Later the same year, its scope was broadened to include equivalent aspects of lithostratigraphic correlation under the leadership of the British geologist John M. Cubitt (at that time also at Syracuse). In 1979 Agterberg took over as project leader and aspects of chronostratigraphic correlation were added in 1981, so that the project then embraced all aspects of quantitative stratigraphic correlation. By the time the project terminated in 1986, some 150 participants in 25 countries had contributed to the research effort. Broadly speaking, the emphasis was on method development to 1981 and applications thereafter. Following cessation of the IGCP project, activities have been co-ordinated by the International Commission of Stratigraphy Committee for Quantitative Stratigraphy, again under the chairmanship of Agterberg. The types of methods and applications covered in the course of this work are discussed in Cubitt (1978), Cubitt & Reyment (1982), Gradstein *et al.* (1985), Agterberg & Gradstein (1988) and Agterberg (1990). See Doveton (1994, chapters 6, 7) for a review of recent lithostratigraphic correlation techniques and the application of artificial intelligence techniques to well-log interpretation.

Computer-based models

Computer simulation has already been mentioned. Early applications were concerned with purely statistical investigations, such as comparison of sampling strategies (Griffiths & Drew 1964; Miesch *et al.* 1964), but computer modelling also afforded an opportunity to gain an improved understanding of a wide variety of natural mechanisms. With the passage of time, and the vast increases in hardware capacity and computational speed, computer-based simulation has become an indispensable tool, underpinning both stochastic methods (Ripley 1987; Efron & Tibshirani 1993) and complex numerical modelling.

Particularly impressive among the early applications were those by the American palaeontologist David M. Raup (*b.* 1933), of mechanisms governing the geometry of shell coiling and the trace-fossil patterns resulting from different foraging behaviours by organisms on the sea floor (Raup 1966; Raup & Seilacher 1969); Louis I. Briggs and H. N. Pollack's (1967) model for evaporite deposition; and the beginning of John W. Harbaugh's (*b.* 1926) long-running investigations of marine sedimentation and basin development (Harbaugh 1966; Harbaugh & Bonham-Carter 1970), which became an integral part of the ongoing geomathematics programme at Stanford University (Harbaugh 1999).

Numerical models have also become crucial in underpinning applications involving fluid-flow, a topic of particular relevance to hydrogeology, petroleum geology and, latterly, nuclear and other contaminant transport problems. The use of analogue models in hydrogeology has already been mentioned. Although effective, they were time-consuming to set up and each hard-wired model was problem-specific. The digital computer provided a more flexible solution. Finite-difference methods (in which the user establishes a regular grid for the model area, subdivides it into a number of subregions and assigns constant system parameters to each cell) were used initially (Ramson *et al.* 1965; Pinder 1968; Pinder & Bredehoeft 1968) but these gradually gave way to the use of finite-element models, in which the flow equations are approximated by integration rather than differentiation, as used in the finite-difference models (see Spitz & Moreno (1996) for a detailed review of these techniques).

Although both types of model can provide similar solutions in terms of their accuracy, finite-element models had the advantage of allowing the use of irregular meshes which could be tailored to any specific application, required a smaller number of nodes and enabled better treatment of boundary conditions and anisotropic media. They were introduced first into groundwater applications by Javandrel & Witherspoon (1969). With increasing interest in problems of environmental contamination, the first chemical-transport model was developed by Anderson (1979). Stochastic (random-walk) 'particle-in-cell' methods were subsequently used to assist visualization of contaminant concentration in flow models: the flow system 'transports' numerical 'particles' throughout the model domain. Plots of the particle locations at successive time-steps gave a good idea of how a concentration field developed (Prickett *et al.* 1981). Spitz & Moreno (1996, table 9.1, pp. 280–294) give a comprehensive summary of recent groundwater flow and transport models.

The use of physical analogues to model rock deformation in structural geology was supplemented in the late 1960s by the introduction of numerical models. Dieterich (1969; Dieterich & Carter 1969) used an approach rather similar to that of the finite-element flow models, discussed previously, to model the development of folds in a single bed (treated as a viscous layer imbedded in a less viscous medium) when subjected to lateral compressive stress. In more recent times, the development of kinematic models has underpinned the application of balanced cross-sections to fold and thrust belt tectonites (Mitra 1992).

Models in which both finite-element and stochastic simulation techniques are applied have become increasingly important. For example, Bitzer & Harbaugh (1987) and Bitzer (1999) have developed realistic basin-simulation models which include processes such as block fault movement, isostatic response, fluid flow, sediment consolidation, compaction, heat flow, and solute transport. Long-term forward-forecasts are required in the consideration of risk which nuclear waste-disposal requires. William Glassley and his colleagues at the Lawrence Livermore National Laboratory, California, are currently trying to develop a reliable model to evaluate the 10 000-year risk of contaminant leakage from the site of the potential Yucca Mountain high-level nuclear waste repository, 160 km NW of Las Vegas, Nevada. This ongoing project uses 1400 microprocessors controlled by a Blue Pacific supercomputer, and the three-dimensional model combines elements of both thermally induced rock deformation and flow modelling (O'Hanlon 2000). In a less computationally demanding groundwater flow problem, Yu (1998) reported significant reductions in processing time for two- and three-dimensional solutions using a Cray Y-MP supercomputer.

The emergence of (Matheronian) 'geostatistics'

Because of their dependence on computer processing, many of the previous applications were first developed in the United States, partly as a product of their relatively easier access to major computing facilities when mainframe machines tended to predominate prior to the mid-1980s. However, what has come to be recognized as one of the most important developments in mathematical geology originated in France. While working with the Algerian Geological Survey in the 1950s, the recently deceased French mining engineer, Georges Matheron (1930–2000), first became aware of publications by the South African mining engineer, Daniel ('Danie') G. Krige (*b.* 1919), who was then working on the problems of evaluation of gold-mining properties (Krige 1975). When Matheron returned to France he continued to work on problems of ore-reserve evaluation. The term *géostatistique* (geostatistics)[1] which Matheron defined as 'the application of the formalism of random functions to the reconnaissance and estimation of natural phenomena' (quoted in Journel & Huijbregts 1978, p. 1) first appeared in his work in 1955 (unpublished material listed in bibliography of Matheron's work; M. Armstrong, pers. comm. 2000). It came to be synonymous with the term *krigeage*, introduced by Matheron in 1960 (M. Armstrong, pers. comm. 2000) in honour of Krige's pioneering work using weighted moving-average surface-fitting (see Krige (1970) for the history of this work), or *kriging* as it has come to be known in the English-language literature. Implicit in all these terms is the analysis of spatially distributed data.

The techniques served two purposes. Firstly, they provided an optimum three-dimensional spatial interpolation method to assist ore-deposit evaluation, with the initial data generally being obtained by grid-drilling the ore-body at the appraisal stage, or through a combination of drilling and chip sampling in an active mine. The key departure from assessment methods used up to that time was Matheron's estimation procedure (Matheron 1957, 1962–1963, 1963, 1965, 1969). Central to this was the idea of fitting a mathematical model which characterized the spatial correlation between ore grades at different locations in the deposit as a function of their distance apart. This function (the experimental variogram) was fitted to the means of the differences in concentration values in all pairs of samples separated by given distance (*d*) taken in a fixed direction (generally defined with regard to the orientation of the deposit as a whole), as a function of *d*. Knowledge of this behaviour then enabled an optimum estimate of the grade at the centre of each ore-block to be made, together with the uncertainty of this estimate (no other spatial interpolation method could provide an uncertainty value). In addition, the directional semivariograms enabled computer simulation techniques to provide models of the ore-deposit which reflected the actual spatial structure of the variation in the ore grades. Based on these simulated realizations, greatly improved estimates of the variation which could be expected in a deposit when mined could be obtained.

Acceptance of this radical new approach to mineral appraisal was not without its difficulties. The work of Matheron and his colleagues at the Centre de Géostatistique (established by Matheron in 1968), Fontainebleau, France, 'encountered no serious problems of acceptance in the Latin-speaking countries of Europe and South America nor in Eastern Europe but at times had stormy receptions from the English-speaking mining countries around the world' (Krige 1977). Such complications gradually eased, following the move to North America of two civil mining engineer graduates of the École des Mines, Nancy: Michel David (1945–2000) went to the École Polytechnique, Montreal, *c.* 1968, and André Journel (*b.* 1944) to Stanford University, California, in 1977. Both had taken Matheron's probability class in 1963, and they persuaded him to start a formal geostatistics programme the following year. Matheron did so, and it was initially taught by Phillipe Formery (A. Journel, pers. comm. 2000). David and Journel soon proved themselves to be able ambassadors for the geostatistical method, both through their English-language publications (David 1977; Journel & Huijbregts 1978), which were more approachable in style for the average geologist than the more formidable mathematical formalism in which Matheron's own work was couched, and through industrial consultancy.

With the passage of time, the geostatistics-based simulation methods originally developed for mine evaluation have come to play an essential role in reservoir characterization in the

[1] Somewhat confusingly, the term 'geostatistics' was independently adopted, particularly in North America, simply to denote the application of statistical methods in geology.

Table 1. *Percentage of papers in Mathematical Geology and Computers & Geosciences by non-exclusive topic*

	M G	*C & G*
Publication time-span	1969–99	1975–99
No. of papers on geology-related topics	1416	1264
Topic		
Statistics	85.4	28.1
Spatial statistics	37.8	8.7
Matheronian geostatistics	26.5	5.7
Mathematical methods		14.7
Petrological and mineralogical calculations		13.3
Data management		12.0
Graphics		11.7
Cartographic methods		10.4
Resource estimation and appraisal	9.9	9.6
Geochemistry	6.1	
Mathematical models	5.6	
Simulation (excluding geostatistics usage)	5.4	8.7
Cluster and principal components analysis, etc.		7.6
Image analysis, image processing		6.2
Orientation statistics		5.5
Laboratory and field instrumentation		5.4

Non-geological papers and topics with under 5% frequency of occurrence are excluded

petroleum industry (Yarus & Chambers 1994) and risking of environmental contamination problems in hydrogeology (Gotway 1994; Fraser & Davis 1998). Furthermore, the practice of geostatistics has attracted the interest and participation of leading statisticians, such as Brian D. Ripley in Britain (Ripley 1981), and Noel A. C. Cressie, formerly in Australia and now in the United States (Cressie 1991). As a result, the use of such methods has now become firmly established as a tool in fields as diverse as climatology, hydrology, environmental monitoring and epidemiology.

Current trends

The spread of geostatistics (in its Matheronian sense), whose development has been driven by mining engineers and statisticians rather than geologists, characterizes a trend evident in the last 30 years from the pages of the leading journals *Mathematical Geology* (which has tended to publish the more theoretical papers) and *Computers & Geosciences*, which took over from the Kansas Geological Survey as major outlets for computer-oriented publications in the field of mathematical geology. Table 1 summarizes the overall most important topics of papers published in the two journals.

A classification of the type of authors contributing papers to these journals (see Fig. 12) shows that from the 1970s until the mid-1980s there was an overall decline in the number of 'geological' authors per publication and, particularly noticeable in *Mathematical Geology*, a corresponding increase in the contributions of mathematicians, statisticians, computer scientists, and mining and other engineers, all of whom will have had a strong mathematical training. This change in authorship should not be too surprising: even in nineteenth century Europe, mining engineers generally had a more rigorous mathematical education than geologists (Smyth 1854).

A literature database search (see Fig. 13) shows that although mathematical and stochastic modelling techniques have played the most important role since the 1960s (particularly in areas such as the characterization of fluid-, heat- and rock-flow, the study of pressure and stress regimes, geochemical modelling of solute transport), the use of physical models has remained relatively constant since the 1980s. It looks as though usage of simulation-based models is beginning to overtake that of purely mathematical models.

These trends reflect a broad change in the interests and requirements of the community engaged in mathematical geology (see Fig. 14). Early topics of interest, such as trend-surface analysis, Markov chains, and the application of multivariate statistics, have given way to geostatistical applications. More recent entrants to the field are fractal and chaotic processes which

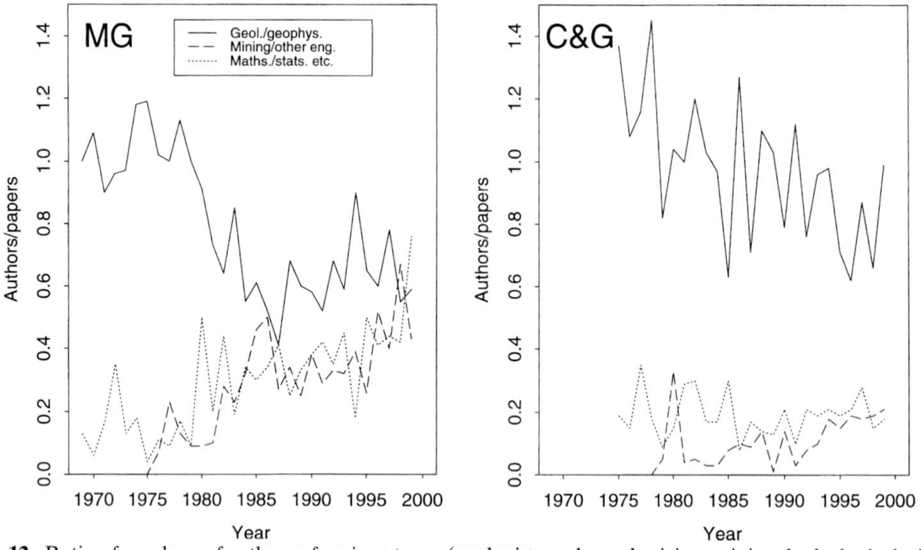

Fig. 12. Ratio of numbers of authors of various types (geologists and geophysicists; mining, hydrological, civil and environmental engineers; mathematicians, statisticians, computer scientists) to number of papers published in *Mathematical Geology* (MG; 1416 non-geophysics articles) and *Computers & Geosciences* (C&G; 1264) from earliest publication to end 1999. Other types of author (e.g. oceanographers, geographers, environmental scientists, etc. not shown).

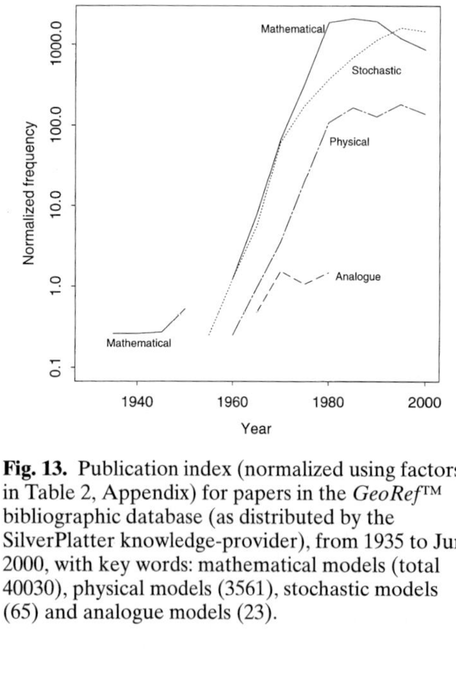

Fig. 13. Publication index (normalized using factors in Table 2, Appendix) for papers in the *GeoRef*™ bibliographic database (as distributed by the SilverPlatter knowledge-provider), from 1935 to June 2000, with key words: mathematical models (total 40030), physical models (3561), stochastic models (65) and analogue models (23).

describe the behaviour of scale-invariant phenomena. Such processes typically describe the size–frequency distributions of phenomena which range in magnitude from the porosity distribution within a rock to the sizes of oil fields (Barton & La Pointe 1995; Tourcotte 1997) and are beginning to be incorporated in geostatistical simulations (Yarus & Chambers 1994). This has happened mainly as a result of the attention gained by the pioneering work of the mathematician Benoit B. Mandelbrot (1962, 1967, 1982).

Image-processing techniques have become increasingly important in the Earth sciences since the late 1960s, driven mainly by the impact of remote-sensing of the Earth and other planetary imagery (Nathan 1966; Rindfleisch *et al.* 1971; Nagy 1972; Viljoen *et al.* 1975), and now are taken for granted, although spatial filtering techniques derived from image-processing have proved useful in other geological contexts, such as geochemical map analysis (Howarth *et al.* 1980). A different image-related area of application has been the development of mathematical morphology by Matheron and his colleague, the civil engineer and philosopher Jean Serra (*b.* 1940). This grew out of petrographic applications of sedimentary iron ores undertaken by Serra in 1964 and 1965 and their applications now underpin the software routinely used in Leitz and other texture-analysis instrumentation (Matheron & Serra 2001). Computer-generated images have also proved invaluable in enabling the visualization of complex three-dimensional, or occasionally higher, relationships which may arise from something as relatively simple as serial-sectioning of a

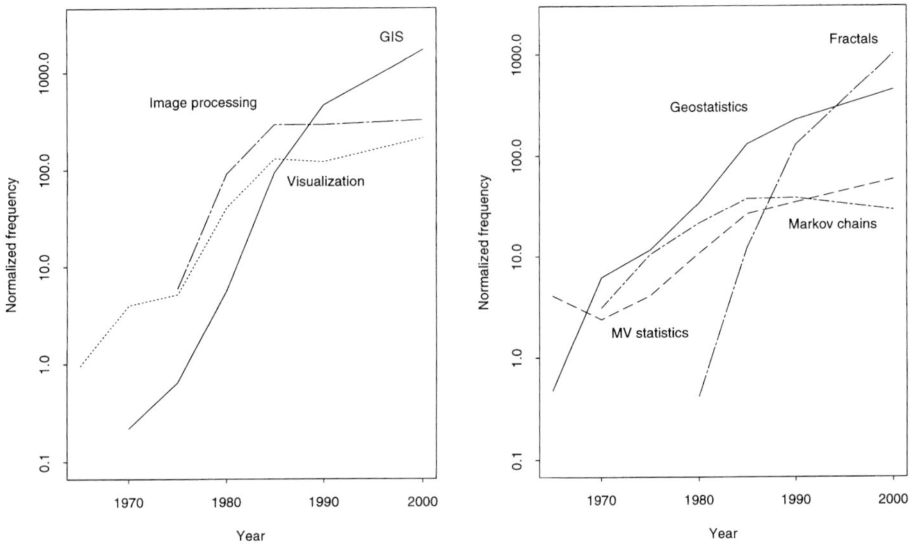

Fig. 14. Publication index (normalized using factors in Table 2, Appendix) for papers in the *GeoRef*™ bibliographic databases, from 1935 to 2000, with the following strings in title or keywords: image processing (total 6094), visualization (2813), geographic information system (GIS; 2692), multivariate (MV) statistics (777), Markov chains (779), geostatistics (4285), and fractals (3437).

fossil-bearing rock (Marschallinger 1998); to fault and other subsurface geometry (Houlding 1994; Renard & Courrioux 1994) and viewing the results of geostatistical simulations (Yarus & Chambers 1994; Fraser & Davis 1998), both of which are crucial in reservoir characterization and mining and environmental geology; or examining the results of integration of topographical, geological, geophysical, and other data by geographical information systems (Bonham-Carter 1994; Maceachren & Kraak 1997; Fuhrmann *et al.* 2000).

The development of computer-intensive methods in statistics, such as the resampling ('bootstrap') techniques of Efron & Tibshirani (1993), for assessing uncertainty in parameter estimates, evidently have considerable potential (Joy & Chatterjee 1998), but may need to be used with care with spatially correlated data (Solow 1985). Similarly, 'robust' methods for parameter estimation and related regression techniques (Huber 1964; Rousseeuw 1983, 1984), which provide the means to obtain reliable regression models even in the presence of outliers in the data, are proving extremely effective (e.g. Cressie & Hawkins 1980; Garrett *et al.* 1982; Powell 1985; Genton 1998).

There is also growing interest in the application of the Bayesian 'degree-of-belief' philosophy as an alternative to the classical 'frequentist' or 'long-run relative frequency' view. In its simplest form, the Bayesian approach could be described as a way of implementing the scientific method in which you state a hypothesis by a prior distribution, collect and summarize relevant data, and then revise your opinion by application of the Bayes rule. This is named for a principle first stated by the British cleric and mathematician Thomas Bayes (*c.* 1701–1761), in a posthumous publication in 1764. It was later discovered independently by the French mathematician Pierre-Simon Laplace (1749–1827) in 1774 (see Stigler (1986) and Hald (1998) for further discussion). Bayes' rule can be expressed as: the probability of a stated hypothesis being true, given the data and prior information, is proportional to the probability of the observed data values occurring given the hypothesis is true and the prior information, multiplied by the probability that the hypothesis is true given only the prior information. In practice, implementation of Bayesian inference is often computer-intensive for reasons which become apparent from the article by Smith & Gelfand (1992). It is true to say that the application of Bayesian statistics is somewhat controversial (see, for example, the arguments advanced for and against the use of Bayesian methods in the 1997 collection of papers in *The American Statistician*, **51**,

241–274). The relatively few geological applications in which Bayesian inference has been used include biostratigraphy (Strauss & Sadler 1989), hydrogeology (Eslinger & Sagar 1989; Freeze *et al.* 1990), resource estimation (Stone 1990), hydrogeochemistry (Crawford *et al.* 1992), geological risk assessment at the Yucca Mountain high-level nuclear waste repository site (Ho 1992), analysis of the time evolution of earthquakes (Peruggia & Santner 1996), and spatial interpolation (Christakos & Li 1998). Bayesian methods are also used in archaeology in connection with radiocarbon dating (Christen & Buck 1998), classification of Neolithic tools (Dellaportas 1998), and archaeological stratigraphic analysis (Allum *et al.* 1999), all of which have obvious geological analogues. There seems to be considerable scope for further use of Bayesian methods in geological applications.

Computational mineralogy is another area which is making rapid strides as a result of advances in processing power. Price & Vocadlo (1996; Vocadlo & Price 1999) believe that before long computational mineralogists will be able to 'simulate entirely from first principles the most complex mineral phases undergoing complicated processes at extreme conditions of pressure and temperature' such as exist within the Earth's deep interior. The results obtained would be used to interpret or extend understanding of laboratory results.

As has been remarked, geostatistical and fluid-transport studies currently are providing some of the most challenging and computationally intensive applications. New techniques being applied include simulated annealing (Deutsch & Journel 1992; Carle 1997), Markov chain Monte Carlo (Oliver *et al.* 1997) and Bayesian maximum entropy (Christakos & Li 1998). Results of recent research are described in Gómez-Hernández & Deutsch (1999).

Conclusion

This account began with the slow growth, during the nineteenth century, of awareness of the utility of hand-drawn graphics as an efficient way to encapsulate information and to convey ideas through the visual medium. The next 50 years saw the beginning of the application of statistical (mainly univariate) and mathematical methods to geological problems. With the spread of computers into civilian use after the end of World War II, the average time-lag of statistical method development (or adaptation) in the geological sciences, compared to its earliest use outside the field, dropped from around 40 years to ten, and since 1985 it has been of the order of one to two years (Fig. 11). Method development time has continued to shorten rapidly as improved computer hardware has become available, both in terms of raw computing power and portability. The increasing dissemination of ideas through journal and book publication and, in the last few years, media such as the Internet, has also improved dramatically the ease of co-working.

The application of computer-intensive methods, coupled with computer-aided visualization, is revolutionizing our capability in fields such as metalliferous mining and reservoir characterization, but the ability to deal effectively with problems involving fluid flow has already had a profound impact in hydrogeological, environmental geology, and environmental contamination applications. The experimental Yucca Mountain nuclear-waste repository study, based as it is on massively parallel processing, is pointing the way towards obtaining significantly improved long-term forecasts of behaviour, as well as better hindcasting. To achieve such goals will, in general, require well-integrated teams of geologists with mathematicians, statisticians and mining engineers. Figure 12 suggests that such team-work is already happening, but the mathematical and statistical skills of many geologists may need to be strengthened if we are to capitalize fully on the opportunity presented by the ongoing technological revolution.

I am grateful to F. Agterberg, G. Bonham-Carter, J. Brodholt, B. Garrett, C. Gotway Crawford, C. Griffiths, E. Grunsky, S. Henley, T. Jones, G. Koch, D. Krige, A. Lord, R. Olea, D. Price, J. Schuenemeyer, S. Treagus and T. Whitten, who all answered my enquiry as to what they thought the five most important innovations in mathematical geology might have been. The resulting diversity was so immense that I have been forced to try to narrow the spectrum to some kind of commonality (or else this article would have grown to book length). In doing so, many interesting ideas have had to fall by the wayside, but nevertheless all their suggestions have been immensely useful. My thanks also go to M. Armstrong and J. Serra for giving me information regarding Georges Matheron's early career, and to G. Bonham-Carter, D. Pollard, D. Price and J. Serra for sending me preprints of papers in press at the time of writing this article. It is some fifteen years since I read Karl Pearson's *History of Statistics in the Seventeenth & Eighteenth Centuries* (ed. E. Pearson 1978). In the Introduction to this text, based on lectures which he gave in the 1920s, Pearson wrote 'I do feel how very wrongful it was to work for so many years at statistics and neglect its history, and that is why I want to interest you in this matter'. This struck a distinct chord, as I was then in exactly the same position, having been teaching statistics and quantitative geology in the Department of Geology at Imperial College, London, for many years. I have been trying to expiate my guilt ever since! I am extremely grateful to the librarians at what was formerly the Department of Geology in the Royal School of Mines (now, sadly, subsumed into the all-embracing Huxley School of Environment, Earth Science and Engineering), Imperial College, The Science Reference Library, the D. M. S. Watson Library, University College London, and The Geological Society, London, throughout the years, without whose assistance in locating dusty volumes from their stack rooms my research would have been impossible to undertake. Photographic work over this time has been carried out by A. Cash and N. Morton (Imperial College), M. Grey (University College), and the Science Museum Library (now the Science Reference Library), and their help is also gratefully acknowledged. I am also grateful to D. Merriam for his referee's comments.

Appendix

An index for the geoscience publication rate from 1700 to 2000 has been derived by comparison of counts of journal holdings in the Geological Society of London with the articles and books recorded in the *GeoRef*™ bibliographic database (as distributed by the SilverPlatter knowledge-provider). Undercount of the latter, pre-1936, has been corrected using robust regression analysis of the *GeoRef*™ counts on the Geological Society journal holdings. Undercount post-1989 has been corrected by extrapolation from the immediately preceding trend for 1982 to 1987. Taking base-10 logarithms of the regression-predicted counts per five-year period yields the final index values of Table 2, which have been used for normalization of Figures 1, 10, 13 and 14.

Table 2. *Publication index: 1700–2000*

Year	1700	1800	1900	2000
0	0.30	1.35	3.86	*5.00*
5	0.30	1.52	3.87	
10	0.30	1.56	3.90	
15	0.30	1.68	3.90	
20	0.41	1.79	3.88	
25	0.55	1.90	3.92	
30	0.70	2.05	3.91	
35	0.70	2.19	3.90	
40	0.70	2.36	3.90	
45	0.70	2.48	3.69	
50	0.70	2.60	3.78	
55	0.70	2.82	3.96	
60	0.76	3.00	4.04	
65	0.78	3.14	4.15	
70	0.83	3.28	4.51	
75	0.90	3.44	4.67	
80	1.03	3.61	4.74	
85	1.12	3.68	4.84	
90	1.21	3.77	*4.87*	
95	1.28	3.83	*4.93*	

Italicized entries based on extrapolated values

References

AGTERBERG, F. P. 1966. The use of multivariate Markov schemes in petrology. *Journal of Geology*, **74**, 764–785.

AGTERBERG, F. P. 1990. *Automated Stratigraphic Correlation*. Elsevier, Amsterdam.

AGTERBERG, F. P. (ed.) 1994. Quantitative Stratigraphy. *Mathematical Geology*, **26**, 757–876.

AGTERBERG, F. P. & BANERJEE, I. 1969. Stochastic model for the deposition of varves in glacial Lake Barlow-Ojibway, Ontario, Canada. *Canadian Journal of Earth Sciences*, **6**, 625–652.

AGTERBERG, F. P. & BRIGGS, G. 1963. Statistical analysis of ripple marks in Atokan and Desmoinesian rocks in the Arkoma Basin of east-central Oklahoma. *Journal of Sedimentary Petrology*, **33**, 393–410.

AGTERBERG, F. P. & GRADSTEIN, F. M. 1988. Recent developments in quantitative biostratigraphy. *Earth-Science Reviews*, **25**, 1–73.

AHLBURG, J. 1907. Die nutzbaren Mineralien Spaniens und Portugals. *Zeitschrift für praktische Geologie*, **15**, 183–210.

AITCHISON, J. 1981. A new approach to null correlation of proportions. *Mathematical Geology*, **13**, 175–189.

AITCHISON, J. 1982. The statistical analysis of compositional data (with discussion). *Journal of the Royal Statistical Society*, Series B, **44**, 139–177.

ALKINS, W. E. 1920. Morphogenesis of brachiopoda. I. Reticularia lineata (Martin), Carboniferous Limestone. *Memoirs and Proceedings of the Manchester Literary and Philosophical Society, London*, **64**, 1–11.

ALLÉGRE, C. 1964. Vers une logique mathématique des series sédimentaires. *Bulletin de la Société Géologique de France*, Series 7, **6**, 214–218.

ALLEN, P. 1944. Statistics in sedimentary petrology. *Nature*, **153**, 71–74.

ALLUM, G. T., AYKROYD, R. G. & HAIGH, J. G. B. 1999. Empirical Bayes estimation for archaeological stratigraphy. *Applied Statistics*, **48**, 1–14.

ANDERSON, M. P. 1979. Using models to simulate the movement of contaminants through groundwater flow systems. In: *Critical Reviews of Environmental Controls* **9**, Chemical Rubber Company Press, Boca Raton, Florida, 97–156.

ANDERSON, R. Y. & KOOPMANS, L. H. 1963. Harmonic analysis of varve time series. *Journal of Geophysical Research*, **68**, 877–893.

ANON. 1907. Eisen und Kohle. *Zeitschrift für praktische Geologie*, **15**, 334–337.

ANON. 1910. Die Eisenerzvorräte der Welt. *Zeitschrift für praktische Geologie*, Supplement (March): *Bergwirtschaffliche Mitteilungen und Anzeigen*, 69–70.

ARKELL, W. J. 1926. Studies in the Corallian lamellibranch fauna of Oxford, Berkshire and Wiltshire. *Geological Magazine*, **63**, 193–210.

BAILLY, L. 1905. Exploitation du minerai de fer oolithique de la Lorraine. *Annales des Mines*, ser 10, **1**, 5–55.

BAKER, H. A. 1920. On the investigation of the mechanical constitution of loose arenaceous sediments by the method of elutriation, with special reference to the Thanet Beds of the southern side of the London Basin. *Geological Magazine*, **57**, 321–332, 363–370, 411–420, 463–467.

BARRANDE, J. 1852. Sur la systèmè silurien de la Bohemie. *Bulletin de la Société Géologique de France*, Serie 2, **10**, 403–424.

BARTON, C. C. & LA POINTE, P. R. (eds) 1995. *Fractals in Petroleum Geology and Earth Processes*. Plenum, New York.

BECKER, H. 1931. A study of the heavy minerals of the Precambrian and Palaeozoic rocks of the Baraboo Range, Wisconsin. *Journal of Sedimentary Petrology*, **1**, 91–95.

BIOT, J. B. 1817. Mémoire sur les rotations que certaines substances impriment aux axes de polarisation des rayons lumineux. *Mémoires de l'Academie royale des Sciences de l'Institut de France, Paris*, **2**, 41–136.

BISWAS, A. K. 1970. *History of Hydrology*. North-Holland, Amsterdam.

BITZER, K. 1999. Two-dimensional simulation of clastic and carbonate sedimentation, consolidation, subsidence, fluid flow, heat flow and solute transport during the formation of sedimentary basins. *Computers & Geosciences*, **25**, 431–447.

BITZER, K. & HARBAUGH, J. W. 1987. DEPOSIM: a Macintosh computer model for two-dimensional simulation of transport, deposition, erosion and compaction of clastic sediments. *Computers & Geosciences*, **13**, 611–637.

BONHAM-CARTER, G. F. 1965. A numerical method of classification using qualitative and semi-quantitative data, as applied to the facies analysis of limestones. *Bulletin of Canadian Petroleum Geology*, **13**, 482–502.

BONHAM-CARTER, G. F. 1994. *Geographic Information Systems for Geoscientists*. Pergamon, Kidlington.

BRIGGS, L. I. & POLLACK, H. N. 1967. Digital model of evaporite sedimentation. *Science*, **155**, 453–456.

BRINKMANN, R. 1929. Statistisch-biostratigraphische Untersuchungen an mitteljurassischen Ammoniten über Artbegriff und Stammesentwicklung. *Abhandlungen der Gesellschaft der Wissenschaften zu Göttingen. Mathematisch-physikalische Klasse*, neue folge, **13**.

BRØGGER, W. C. 1898. Die Eruptivgesteine des Kristianiagebiets. III. Das Ganggefolge des Laurdalits. *Videnskabsselskabets Skrifter*. I. *Mathematisk-naturv Klasse, Christiania*, No. **6**.

BURMA, B. H. 1949. Studies in quantitative paleontology. II: Multivariate analysis – a new analytical tool for paleontology and geology. *Journal of Paleontology*, **23**, 95–103.

BUSK, G. 1870. On a method of graphically representing the dimensions and proportions of the teeth of mammals. *Proceedings of the Royal Society, London*, **18**, 544–546.

BUTLER, B. S., LOUGHLIN, G. F., HEIKES, V. C. 1920. *The ore deposits of Utah*. US Geological Survey Professional Paper No. **111**.

CADELL, H. M. 1898. Petroleum and natural gas: their geological history and production. *Transactions of the Edinburgh Geological Society*, **7**, 51–73.

CARLE, S. F. 1997. Implementation schemes for avoiding artifact discontinuities in simulated annealing. *Mathematical Geology*, **29**, 231–244.

CATELL, R. B. 1952. *Factor Analysis*. Harper, New York.

CHAMBERLIN, T. C. 1885. The requisite and qualifying conditions of artesian wells. *US Geological Survey Annual Report*, **5**, 125–173.

CHAMBERLIN, T. C. 1897. The method of multiple working hypotheses. *Journal of Geology*, **5**, 837–848.

CHAYES, F. 1956. *Petrographic Modal Analysis*. Wiley, New York.

CHAYES, F. 1971. *Ratio Correlation*. University of Chicago Press, Chicago.

CHAYES, F. & KRUSKAL, W. 1966. An approximate statistical test for correlations between propositions. *Journal of Geology*, **74**, 692–702.

CHRISTAKOS, G. & LI, X. 1998. Bayesian maximum entropy analysis and mapping: a farewell to Kriging estimators? *Mathematical Geology*, **30**, 435–462.

CHRISTEN, J. A. & BUCK, C. E. 1998. Sample selection in radiocarbon dating. *Applied Statistics*, **47**, 543–557.

COOLEY, J. W. & TUKEY, J. W. 1965. An algorithm for the machine calculation of complex Fourier series. *Mathematics of Computing*, **19**, 297–301.

CQS, 1988–1997. *Newsletters* 2–8. Committee on Quantitative Stratigraphy under International Commission on Stratigraphy.

CRAWFORD, S. L., DEGROOT, M. H., KADANE, J. B. & SMALL, M. J. 1992. Modeling lake-geochemistry distributions: approximate Bayesian methods for estimating a finite-mixture model. *Technometrics*, **34**, 441–453.

CREAGER, J. S., MCMANUS, D. A. & COLLIAS, E. E. 1962. Electronic data processing in sedimentary size analysis. *Journal of Sedimentary Petrology*, **32**, 833–839.

CRESSIE, N. A. C. 1991. *Statistics for Spatial Data*. Wiley, New York.

CRESSIE, N. A. C. & HAWKINS, D. M. 1980. Robust estimation of the semivariogram. I. *Mathematical Geology*, **12**, 115–125.

CROSS, W., IDDINGS, J. P., PIRSSON, L. V. & WASHINGTON, H. S. 1902. A chemico-mineralogical classification and nomenclature of igneous rocks. *Journal of Geology*, **10**, 555–690.

CROSS, W., IDDINGS, J. P., PIRSSON, L. V. & WASHINGTON, H. S. 1912. Modifications of the quantitative system of classification of igneous rocks. *Journal of Geology*, **20**, 550–561.

CUBITT, J. M. (ed.) 1978. Quantitative stratigraphic correlation. Proceedings of the 6th Geochautauqua held at Syracuse University 28 October, 1977. *Computers & Geosciences*, **4**, 215–318.

CUBITT, J. M. & REYMENT, R. A. (eds) 1982. *Quantitative Stratigraphic Correlation*. Wiley, New York.

CUMINS, E. R. 1902. A quantitative study of variation in the fossil brachiopod Platystrophia lynx. *American Journal of Science*, Series 4, **14**, 9–16.

DANA, J. D. 1880. Geological relations of the limestone belts of Westchester County, New York. *American Journal of Science*, Series 3, **20**, 359–375.

DARCY, H. 1856. *Les Fontaines Publiques de la Ville de Dijon*. Dalmont, Paris.

DAVID, M. 1977. *Geostatistical Ore Reserve Estimation*. Elsevier, Amsterdam.

DAVIES, J. H. & TRUEMAN, A. E. 1927. A revision of the non-marine Lamellibranchs of the Coal Measures and a discussion of their zonal sequence. *Journal of the Geological Society, London*, **83**, 210–259.

DE LAPPARENT, A. & POTIER, A. 1877. Rapport sur l'exploration geologique sous-marin du Pas-de-Calais. *In*: LAVALLEY, A. (ed.) *Chemin de Fer sous-marin entre la France et l'Angleterre. Rapports présentés aux membres de l'Association sur les explorations géologiques faites en 1875 et 1876*. Dupont, Paris, 33–50.

DE LAUNAY, L. 1896. Les mines d'or du Transvaal: districts du Witwatersrand, d'Heidelberg et de Klerksdorp. *Annales des mines*, Series 9, **9**, 5–201.

DELLAPORTAS, P. 1998. Bayesian classification of Neolithic tools. *Applied Statistics*, **47**, 279–297.

DEUTSCH, C. V. & JOURNEL, A. 1992. *GSLIB Geostatistical Software Library and User's Guide*. Oxford University Press, New York.

DIETERICH, J. H. 1969. Origin of cleavage in folded rocks. *American Journal of Science*, **267**, 155–165.

DIETERICH, J. H. & CARTER, N. L. 1969. Stress-history of folding. *American Journal of Science*, **267**, 129–154.

DOLLFUS, G. F. 1888. Notice sur une nouvelle carte géologique des environs de Paris. *Comptes Rendus, 3rd International Geological Congress, Berlin 1885*, 98–220.

DOVETON, J. H. 1994. *Geologic Log Analysis Using Computer Methods*. AAPG, Tulsa, Computer Applications in Geology **2**.

DREW, L. J. 1990. *Oil and Gas Forecasting. Reflections of a Petroleum Geologist*. International Association for Mathematical Geology Studies in Mathematical Geology **2**, Oxford University Press, New York.

DRIVER, H. L. 1928. Foraminiferal section along Adams Canyon, Ventura County, Los Angeles, California. *AAPG Bulletin*, **12**, 753–756.

DUPAIN-TRIEL, J. I. 1798–1799 (*An VII*). Carte de la France où l'on a essayé de donner la configuration de son territoire, par une nouvelle methode de nivellements, Paris (Map, 1 sheet. Reproduced in: ROBINSON, A. H. 1982. *Early Thematic Mapping in the History of Cartography*, Chicago University Press, Chicago, 214).

DVALI, M. F., KORZHINSKII, D. S., LINNIK, Y. V., ROMANOVA, M. A. & SARMARNOV, O. V. 1970. The life and work of Andrei Borisovich Vistelius. *In*: ROMANOVA, M. A. & SARMARNOV, O. V. (eds) *Topics in Mathematical Geology*. Consultants Bureau, New York, 1–12.

EFRON, B. & TIBSHIRANI, R. J. 1993. *An Introduction to the Bootstrap*. Monographs on Statistics and Applied Probability **57**, Chapman and Hall, London.

EFROYMSON, M. A. 1960. Multiple regression analysis. *In*: RALSTON, A. & WILF, H. S. (eds), *Mathematical Methods for Digital Computers*. Wiley, New York, 191–203.

EISENHART, C. 1935. A test for significance of lithological variation. *Journal of Sedimentary Petrology*, **5**, 137–145.

ÉLIE DE BEAUMONT, L. 1849. *Leçons de Géologie Practique*, Vol. 2. Baillière, Paris.

EMERY, J. R. & GRIFFITHS, J. C. 1954. Differentiation of oil-bearing from barren sediments by quantitative petrographic methods. *The Pennsylvania State University Mineral Industries Experimental Station Bulletin*, **64**, 63–68.

EMMONS, W. H. 1921. *Geology of Petroleum*. McGraw-Hill, New York.

ERIKSSON, J. V. 1929. Den kemiska denudationen i Sverige. *Meddelanden Från Statens Meteorologisk-hydrografiska Anstalt*, **5**, 1–96.

ESLINGER, P. W. & SAGAR, B. 1989. Use of Bayesian analysis for incorporating subjective information. *In*: BUXTON, B. E. (ed.) *Geostatistical, Sensitivity, and Uncertainty Methods in Ground-Water Flow and Radionuclide Transport Modelling*. Batelle Press, Columbus, 613–627.

FISHER, O. 1881. *Physics of the Earth's Crust*. Macmillan, London.

FISHER, R. A. 1922. The goodness of fit of regression formulae and the distribution of regression coefficients. *Journal of the Royal Statistical Society*, **85**, 597–612.

FISHER, R. A. 1925. *Statistical Methods for Research Workers*. Oliver & Boyd, Edinburgh.

FISHER, R. A. 1936. The use of multiple measurements in taxonomic problems. *Annals of Eugenics*, **8**, 179–188.

FRASER, G. S. & DAVIS, J. M. (eds) 1998. *Hydrogeologic Models of Sedimentary Aquifers*. SEPM Society for Sedimentary Geology, Tulsa.

FREEZE, R. A. 1994. Henry Darcy and the Fountains of Dijon. *Ground Water*, **32**, 23–30.

FREEZE, R. A., MASSIMAN, J., SMITH, L., SPERLING, T. & JAMES, B. 1990. Hydrogeological decision analysis. I. A framework. *Ground Water*, **28**, 738–766.

FRISI, P. 1762. *A Treatise on Rivers and Torrents; With the Method of Regulating their Course and Channels* (English translation: GARSTIN, J. 1818. Longman, Hurst, Rees, Orme & Brown, London).

FUHRMANN, S., STREIT, U. & KUHN, W. (eds) 2000. Geoscientific visualization special issue. *Computers & Geosciences*, **26**, 1–118.

FUNKHOUSER, H. G. 1938. Historical development of the graphical representation of statistical data. *Osiris*, **3**, 269–404.

GARRETT, R. G., GOSS, T. I. & POIRIER, P. R. 1982. Multivariate outlier detection—an application to robust regression in the Earth sciences [abstract]. *Joint Statistical Meetings of the American Statistical Association, August 16–19, 1982, Cincinnati, Ohio*. American Statistical Association, 101.

GAUSS, C. F. 1809. Determinatio orbitae observationibus quotcunque maxime satisfacientis. *In*: *Theoria Motus Corporum Coelestium in Sectionibus Conicis Solem Ambientium*. Perthes & Besser, Hamburg. 205–224 (English translation, DAVIS, C. H. 1857. Reproduced in: SHAPLEY, H. & HOWARTH, H. E. (eds) 1929. *A Source Book in Astronomy*, McGraw-Hill, New York, 183–195).

GENTLEMAN, W. M. & SANDE, G. 1966. Fast Fourier transforms – for fun and profit. *AFIPS Proceedings of the Fall Joint Computer Conference*, **29**, 563–578.

GENTON, M. G. 1998. Spatial breakdown point of variogram estimators. *Mathematical Geology*, **30**, 853–871.

GIRARD, [P.] 1819. Observations Sur la vallee d'Egypte et sur l'exhaussement seculaire du sol qui la recouvre. *Mémoires de l'Académie royale des Sciences de l'Institut de France, Paris*, **2**, 185–304.

GÓMEZ-HERNÁNDEZ, J. J. & DEUTSCH, C. V. (eds) 1999. Special issue: Modeling subsurface flow. *Mathematical Geology*, **31**, 749–927.

GOTWAY, C. A. 1994. The use of conditional simulation in nuclear-waste-site performance assessment. *Technometrics*, **36**, 129–141.

GOUDKOFF, P. P. 1926. Correlative value of the microlithology and micropalaeontology of the oil-bearing formations in the Sunset-midway and Kearn River oil fields. *AAPG Bulletin*, **10**, 482–494.

GRADSTEIN, F. M., AGTERBERG, F. P., BROWER, J. C. & SCHWARZACHER, W. 1985. *Quantitative Stratigraphy*. Reidel, Dordrecht.

GREENLEAF, J. 1896. The hydrology of the Mississippi. *American Journal of Science*, Series 4, **2,** 29–46.

GRIFFITHS, J. C. 1953. Estimation of error in grain size analysis. *Journal of Sedimentary Petrography*, **23**, 75–84.

GRIFFITHS, J. C. 1962. Statistical methods in the analysis of sediments. *In*: MILNER, H. B. (ed.) *Sedimentary Petrography*. Macmillan, New York, 565–617.

GRIFFITHS, J. C. 1966*a*. Grid spacing and success ratios in exploration for natural resources. *Proceedings of the Symposium and Short Course on Computers and Operations Research in the Mineral Industries*. Pennsylvania State University, Mineral Industries Experimental Station Publication no. 2–65, **1**, Q1–Q24.

GRIFFITHS, J. C. 1966*b*. Exploration for natural resources. *Journal of the Operations Research Society of America*, **14**, 189–209.

GRIFFITHS, J. C. 1967. Mathematical exploration strategy and decision-making. *Proceedings of the 7th World Petroleum Congress, Mexico*, **11**, 87–98.

GRIFFITHS, J. C. & DREW, L. J. 1964. Simulation of exploration programs for natural resources by models. *Colorado School of Mines Quarterly*, **59**, 187–206.

GRIFFITHS, J. C. & DREW, L. J. 1973. The Engel simulator and the search for uranium. *Application of Computer Methods in the Mineral Industry*. South African Institution of Mining and Metallurgy, Johannesburg, 9–16.

GRIFFITHS, J. C. & Singer, D. A. 1970. *Size, shape and arrangement of some uranium ore bodies*. Final Report to US Atomic Energy Commission, Raw Materials Division, Grand Junction, Colorado.

GROUT, F. F. 1926. Calculations in petrology: a study for students. *Journal of Geology*, **34,** 512–558.

GUNTHER, R. T. 1939. *Early Science in Oxford. XII. Dr. Plot and the correspondence of the Philosophical Society of Oxford*. Privately published, Oxford.

HALD, A. 1998. *A History of Mathematical Statistics from 1750 to 1930*. Wiley, New York.

HALLEY, E. 1686. On the height of mercury in the barometer at different elevations above the surface of the earth; and on the rising and falling of the mercury on the change of weather. *Philosophical Transactions of the Royal Society, London*, **16**, 104–116.

HALLEY, E. 1701. *A New and Correct Chart Shewing the Variations of the Compass in the Western & Southern Oceans as Observed in ye Year 1700 by his Ma.ties Command by Edm. Halley*. [map, 1 sheet] Mount & Page, London.

HANSEN, W. B. 1985. Dust in the Wind: J. A. Udden's turn-of-the-century research at Augustana. *In*: DRAKE, E. T. & JORDAN, W. M. (eds) *Geologists and Ideas: A History of North American Geology*. The Geological Society of America, Boulder, 203–214.

HARBAUGH, J. W. 1966. Mathematical simulation of marine sedimentation with IBM 7090/7094 computers. *Kansas Geological Survey Computer Contribution* No. **1**, 1–52.

HARBAUGH, J. W. 1999. Stanford's geomath program. I–II. *Newsletter of the International Association for Mathematical Geology*, **58**, May, 10–11; November, 12–13.

HARBAUGH, J. W. & BONHAM-CARTER, G. 1970. *Computer Simulation in Geology*. Wiley, New York.

HARKER, A. 1909. *The Natural History of Igneous Rocks*. Methuen, London.

HAUGHTON, S. 1864. On the joint-systems of Ireland and Cornwall and their mechanical origin. *Philosophical Transactions of the Royal Society, London*, **154**, 393–411.

HERDER, M. von & GÄßFCHMANN, M. F. 1849. Das Silberausbringen des Freiberger Reviers vom Jahre 1542 an bis mit dem Jahre 1847. *Jahrbuch für den Berg- und hutten-Mann*, 1–19.

HÉRON DE VILLEFOSSE, A. M. 1808. Nivellement des Harzgebirges mit dem barometer. *Gilbert's Annalen der Physik*, **1**, 49–111.

HERSCHEL, J. F. W. 1822. On certain remarkable instances of deviation from newton's scale in the tints developed by crystals, with one axis of double refraction, on exposure to polarized light. *Transactions of the Cambridge Philosophical Society*, **1,** 23–41.

HO, C. H. 1992. Risk assessment for the Yucca Mountain high-level nuclear waste repository site: estimation of volcanic disruption. *Mathematical Geology*, **24**, 347–364.

HOLMES, A. 1921. *Petrographic Methods and Calculations with Some Examples of Results Achieved*. Murby, London.

HOPKINS, W. 1838. Researches in physical geology. *Transactions of the Cambridge Philosophical Society*, **6**, 1–84.

HOPKINS, W. 1845. On the motion of glaciers. *Philosophical Magazine*, Series 3, **26**, 1–16.

HOPKINS, W. 1849*a*. On the transport of erratic blocks. *Transactions of the Cambridge Philosophical Society*, **8**, 220–240.

HOPKINS, W. 1849*b*. On the internal pressure to which rock masses may be subjected, and its possible influence in the production of laminated structure. *Transactions of the Cambridge Philosophical Society*, **8**, 456–470.

HOULDING, S. W. 1994. *3D Geoscience Modelling: Computer Techniques for Geological Characterisation*. Springer, Berlin.

HOWARTH, R. J. 1996. Sources for a history of the ternary diagram. *British Journal for the History of Science*, **29**, 337–356.

HOWARTH, R. J. 1998. Graphical methods in mineralogy and igneous petrology (1800–1935). *In*: FRITSCHER, B. & HENDERSON, F. (eds) *Toward a History of Mineralogy, Petrology, and Geochemistry. Proceedings of the International Symposium on the History of Mineralogy, Petrology, and Geochemistry, Munich, March 8–9, 1996*. Institut für Geschichte der Naturwissenschaften, Munich, 281–307.

HOWARTH, R. J. 1999. Measurement, portrayal and analysis of orientation data and the origins of early modern structural geology (1670–1967). *Proceedings of the Geologists' Association*, **110**, 273–309.

HOWARTH, R. J., KOCH, G. S. Jr, CHORK, C. Y., CARPENTER, R. H. & SCHUENEMEYER, J. H. 1980. Statistical map analysis techniques applied to regional distribution of uranium in stream sediment samples from the southeastern United States for the national uranium resource evaluation program. *Mathematical Geology*, **12**, 339–366.

HUBBERT, M. K. 1940. The theory of ground-water motion. *Journal of Geology*, **48**, 785–944.

HUBER, P. J. 1964. Robust estimation of a location parameter. *Annals of Mathematical Statistics*, **35**, 73–101.

HUMBOLDT, A. VON, 1811–1812. *Essai politique sur la royaume de la Nouvelle-Espagne, avec un atlas physique et géographique, fondé sur des observations astronomiques, des mesures trigonométriques et des nivellements barométriques*. Paris.

HUMBOLDT, A. VON, 1817*a*. Des lignes isothermes et de la distribution de la chaleur sur le globe. *Mémoires de Physique et de Chimie, de la Société D'Arcueil*, **3**, 462–602.

HUMBOLDT, A. VON, 1817*b*. Sur les lignes isothermes (extrait). *Annales de chimie et physique*, **5**, 102–112.

HUMBOLDT, A. VON, 1825. De quelques phénomenes physiques et géologiques qu'offrent les Cordillères des Andes de Quito et la partie occidentale de l'Himalayah. *Annales des sciences naturelles*, **4**, 225–253.

HUMBOLDT, A. VON, 1855. *Schetsen van vulkanen uit de Cordillera's van Quito en Mexiko. Eene Bijtrage tot de Physiognomie der Natuur*. Van den Heuvell, Leiden.

HUNT, R. 1846. *A notice of the copper and tin raised in Cornwall*. Memoirs of the Geological Survey of Great Britain, **1**, Her Majesty's Stationery Office, London.

IDDINGS, J. P. 1892. The eruptive rocks of Electric Peak and Sepulchre Mountain Yellowstone National Park. *Annual Report of the US Geological Survey*, **12**, 577–664.

IDDINGS, J. P. 1898*a*. On rock classification. *Journal of Geology*, **6**, 92–111.

IDDINGS, J. P. 1898*b*. Chemical and mineral relationships in igneous rocks. *Journal of Geology*, **6**, 219–237.

IDDINGS, J. P. 1903. *Chemical composition of igneous rocks expressed by means of diagrams with reference to rock classification on a quantitative chemico-mineralogical basis*. US Geological Survey Professional Paper **18**.

IDDINGS, J. P. 1909. *Igneous Rocks; Composition, Texture and Classification; Description and Occurrence*, Vol. 1. Wiley, New York.

IMBRIE, J. 1963. *Factor and vector analysis program for analyzing geologic data*. Office of Naval Research Geography Branch, Technical Report 6 (ONR Task No. 389–135).

IMBRIE, J. 1985. A theoretical framework for the Pleistocene ice ages: a brief review. *Journal of the Geological Society, London*, **142**, 417–432.

IMBRIE, J. & IMBRIE, K. P. 1979. *Ice Ages: Solving the Mystery*. Macmillan, London.

IMBRIE, J. & IMBRIE, J. Z. 1980. Modelling the climatic response to orbital variations. *Science*, **207**, 943–953.

IMBRIE, J. & PURDY, E. G. 1962. The classification of modern Bahamian carbonate sediments. *In*: HAM, W. E. (ed.) *Classification of Carbonate Rocks: A Symposium*. AAPG, Tulsa, Memoir **1**, 253–272.

IMBRIE, J. & VAN ANDEL, T. H. 1964. Vector analysis of heavy-mineral data. *Bulletin of the Geological Society of America*, **75**, 1131–1156.

INMAN, D. L. 1952. Measures for describing the size distribution of sediments. *Journal of Sedimentary Petrology*, **22**, 125–145.

JAVANDREL, I. & WITHERSPOON, P. A. 1969. A method of analysing transient fluid flow in multilayer aquifers. *Water Resources Research*, **5**, 856–869.

JIZBA, Z. V. 1966. Sand evolution simulation. *Journal of Geology*, **74**, 734–743.

JOHNSON, W. & JOHNSON, A. K. 1856. *The Physical Atlas of Natural Phenomena*. Blackwood, Edinburgh.

JONES, T. A. 1968. Statistical analysis of orientation data. *Journal of Sedimentary Petrology*, **38**, 61–67.

JOURNEL, A. G. & HUIJBREGTS, C. J. 1978. *Mining Geostatistics*. Academic Press, London.

JOY, S. & CHATTERJEE, S. 1998. A bootstrap test using maximum likelihood ratio statistics to check the similarity of two 3-dimensionally orientated data samples. *Mathematical Geology*, **30**, 275–284.

KEILHAK, K. 1913. Grundwasser studien. V. *Zeitschrift für praktische Geologie*, **21**, 29–41.

KEPPEN, A. de, 1894. Aperçu général sur l'industrie minerale de la Russie. *Annales des Mines*, Serie 9, **4**, 180–273, 279–368.

KITAIBEL, P. & TOMTSÁNYI, A. 1814. *Dissertartio de terrae motu in genere, ac in specie Mórensi anno 1810 die 14 januarii orto*. Regiae Universitatis Hungaricae, Budae. (reprinted, with commentary by A. Réthly, 1960, Akédemiai Kaidó, Budapest).

KNOWLES, J. 1890. On the coal trade. *Transactions of the Manchester Geological Society*, **20**, 42–53.

KNUTH, D. E. & PARDO, L. T. 1980. The early development of programming languages. *In*: METROPOLIS,

N., HOWLETT, J. & ROTA, G. C. (eds) *A History of Computing in the Twentieth Century*. Academic Press, New York, 197–273.

KOOPMAN, B. 1956–1957. The theory of search. I–III. *Journal of the Operations Research Society of America*, **4**, 324–346, 503–531; **5**, 613–626.

KRIGE, D. 1970. The role of mathematical statistics in improved ore valuation techniques in South African gold mines. *In*: ROMANOVA, M. A. & SARMARNOV, O. V. (eds) *Topics in Mathematical Geology*. Consultants Bureau, New York, 243–261.

KRIGE, D. 1975. A review of the development of geostatistics in South Africa. *In*: GUARASCIO, DAVID, M. & HUIJBREGTS, C. (eds) *Advanced Geostatistics in the Mining Industry*. Reidel, Dordrecht, 279–293.

KRIGE, D. 1977. Preface. *In*: DAVID, M. *Geostatistical Ore Reserve Estimation*. Elsevier, Amsterdam.

KRUMBEIN, W. C. 1932. A history of the principles and methods of mechanical analysis. *Journal of Sedimentary Petrology*, **2**, 89–124.

KRUMBEIN, W. C. 1934. The probable error of sampling sediment for mechanical analysis. *American Journal of Science*, **27**, 204–214.

KRUMBEIN, W. C. 1938. Size frequency distributions of sediments and the normal phi curve. *Journal of Sedimentary Petrology*, **8**, 84–90.

KRUMBEIN, W. C. 1939. Application of photo-electric cell to the measurement of pebble axes for orientation analysis. *Journal of Sedimentary Petrology*, **9**, 122–130.

KRUMBEIN, W. C. 1952. Principles of facies map interpretation. *Journal of Sedimentary Petrology*, **22**, 200–211.

KRUMBEIN, W. C. 1953. Latin square experiments in sedimentary petrology. *Journal of Sedimentary Petrology*, **23**, 280–283.

KRUMBEIN, W. C. 1954*a*. The tetrahedron as a facies mapping device. *Journal of Sedimentary Petrology*, **24**, 3–19.

KRUMBEIN, W. C. 1954*b*. Applications of statistical methods to sedimentary rocks. *Journal of the American Statistical Association*, **49**, 51–67.

KRUMBEIN, W. C. 1955. Experimental design in the earth sciences. *Transactions of the American Geophysical Union*, **36**, 1–11.

KRUMBEIN, W. C. 1956. Regional and local components in facies maps. *AAPG Bulletin*, **40**, 2163–2194.

KRUMBEIN, W. C. 1963. A geological process-response model for analysis of beach phenomena. *Bulletin of the Beach Erosion Board*, **17**, 1–15.

KRUMBEIN, W. C. 1967. *FORTRAN IV computer programs for Markov chain experiments in geology*. Kansas Geological Survey Computer Contribution, **13**.

KRUMBEIN, W. C. 1968. Computer simulation of transgressive and regressive deposits with a discrete-state, continuous-time Markov model. *In*: MERRIAM, D. F. (ed.) *Computer Applications in the Earth Sciences: Colloquium on Simulation*. Kansas Geological Survey Computer Contribution **22**, 11–18.

KRUMBEIN, W. C. & DACY, M. F. 1969. Markov chains and embedded Markov chains in geology. *Mathematical Geology*, **1**, 79–96.

KRUMBEIN, W. C. & MILLER, R. L. 1953. Design of experiments for statistical analysis of geologic data. *Journal of Geology*, **61**, 510–532.

KRUMBEIN, W. C. & PETTIJOHN, F. J. 1938. *Manual of Sedimentary Petrography. I. Sampling, Preparation for Analysis, Mechanical Analysis, and Statistical Analysis by W. C. Krumbein. II. Shape Analysis, Mineralogical Analysis, Chemical Analysis, and Mass Properties by F. J. Pettijohn*. Appleton-Century, New York.

KRUMBEIN, W. C. & RASMUSSEN, W. C. 1941. The probable error of sampling beach sand for heavy mineral analysis. *Journal of Sedimentary Petrography*, **11**, 10–20.

KRUMBEIN, W. C. & SLOSS, L. L. 1958. High-speed digital computers in stratigraphic and facies analysis. *AAPG Bulletin*, **42**, 2650–2669.

KRUMBEIN, W. C. & SLOSS, L. L. 1963. *Stratigraphy and Sedimentation* (2nd edn). Freeman, San Francisco.

KRUMBEIN, W. C. & TUKEY, J. W. 1956. Multivariate analysis of mineralogic, lithologic, and chemical composition of rock bodies. *Journal of Sedimentary Petrology*, **26**, 322–337.

LACROIX, A. 1899. Le gabbro du Pallet et ses modifications. *Bulletin des Services de la Carte Géologique de la France et des Topographies Souterraines*, **10**, 342–396.

LALLEMAND, C. 1881. Les lignites dans la nord de la Bohême. *Annales des Mines*, Serie 7, **19**, 350–493.

LAMBERT, J. H. 1765. *Beytrage zum Gebrauche der Mathematik und deren Anwendung*, **1**. Buchhandlung der Realschule, Berlin.

LEAKE, B. E., HENDRY, G. L., KEMP, A., *et al.* 1970. The chemical analysis of rock powders by automatic X-ray fluorescence. *Chemical Geology*, **5**, 7–86.

LEGENDRE, A. M. 1805. *Nouvelles méthodes pour la determination des orbites des comètes*. Courcier, Paris (partial translation *in*: SMITH, D. E. (ed.) 1929. *A Source Book in Mathematics*, McGraw-Hill, New York, 576–579; reprinted, Dover, New York, 1959).

LEITCH, D. 1940. A statistical investigation of *Anthracomyas* of the basal *Simulis pulchra* zone of Scotland. *Quarterly Journal of the Geological Society, London*, **96**, 1–38.

LEITH, C. K. 1907. The metamorphic cycle. *Journal of Geology*, **15**, 303–313.

LIMA-DE-FARIA, J. (ed.) 1990. *Historical Atlas of Crystallography*. Kluwer, Dordrecht.

LINSTOW, O. von, 1905. Die Grundwasserverhältnisee zwischen Mulde und Elbe südlich Dessau und die praktische Bedeutung derartiger Unterschungen. *Zeitschrift für praktische Geologie*, **13**, 121–135.

LIU, C. & LI, Y. 1983. Development of mathematical geology in China. *Mathematical Geology*, **15**, 483–492.

LOEWINSON-LESSING, F. 1899. Studien über die eruptivgesteine. *Proceedings of the 7th International Geological Congress 1897, St Petersburg*, 193–464.

LUCAS, J. 1874. *Horizontal Wells, a New Application of Geological Principles to Effect the Solution of the*

Problem of Supplying London with Pure Water. Stanford, London.

LUCAS, J. 1877. The Chalk water system. *Proceedings of the Institution of Civil Engineers*, **47**, 70–167.

LYELL, C. 1830–1833. *Principles of Geology, Being an Attempt to Explain the Former Changes of the Earth's Surface by Reference to Causes Now in Operation* (3 vols). Murray, London.

LYMAN, B. S. 1870. *General Report on the Punjab Oil Lands*. Government of India, Lahore.

LYMAN, B. S. 1873. On the importance of surveying in geology. *American Institute of Mining Engineers Transactions*, **1**, 183–192.

MCCAMMON, R. B. 1966. Principal component analysis and its application in large-scale correlation studies. *Journal of Geology*, **74**, 721–733.

MCINTYRE, D. B. 1963*a*. *Program for Computation of Trend Surfaces and Residuals of Degree 1 Through 8*. Department of Geology, Pomona College, Claremont, California, Technical Report **4**.

MCINTYRE, D. B. 1963*b*. *FORTRAN II Program for Norm and von Wolff Calculations*. Department of Geology, Pomona College, Claremont, California, Technical Report **14**.

MACEACHREN, A. M. & KRAAK, M. J. 1997. Exploratory cartographic visualization. *Computers & Geosciences*, **23**, 335–491.

MALLET, R. 1862. *The First Principles of Observational Seismology as Developed in the Report to the Royal Society of London of the Expedition Made by Command of the Society into the Interior of the Kingdom of Naples to Investigate the Circumstances of the Great Earthquake of December 1857* (Vol. 2). Chapman & Hall, London.

MALLET, R. 1873. Volcanic energy: an attempt to develop its true origin and cosmical relations. *Philosophical Transactions of the Royal Society, London*, **163**, 147–227.

MANDELBROT, B. B. 1962. The statistics of natural resources and the law of Pareto. (Unpublished International Business Machines internal report; reprinted in: BARTON, C. C. & LA POINTE, P. R. (eds) 1995, *Fractals in Petroleum Geology and Earth Processes*, Plenum, New York, 1–12.)

MANDELBROT, B. B. 1967. How long is the coast of Britain? Statistical self-similarity and fractional dimension. *Science*, **156**, 636–638.

MANDELBROT, B. B. 1982. *The Fractal Geometry of Nature*. Freeman, San Francisco.

MANN, K. O. & LANE, H. R. (eds) 1995. *Graphic Correlation*. SEPM Society for Sedimentary Geology, Tulsa, Special Publication **53**.

MARSCHALLINGER, R. 1998. A method for three-dimensional reconstruction of macroscopic features in geological materials. *Computers & Geosciences*, **24**, 875–883.

MARTINS, 1844. Note sur le delta de l'Aar á son embouchère dans le lac de Brienz. *Bulletin de la Société Géologique de France*, Series 2, **2**, 118–124.

MATHER, J. 1998. From William Smith to William Whitaker: the development of British hydrogeology in the nineteenth century. *In*: BLUNDELL, D. J. & SCOTT, A. C. (eds) *Lyell: The Past is the Key to the Present*. The Geological Society, London, Special Publications **143**, 183–196.

MATHERON, G. 1957. Théorie lognormalle de l'échantillonage systematique des gisements. *Annales des Mines*, **9**, 566–584.

MATHERON, G. 1962–1963. *Traité de géostatistique appliquée*. 2 vols, Technip, Paris.

MATHERON, G. 1963. Principles of geostatistics. *Economic Geology*, **58**, 1246–1266.

MATHERON, G. 1965. *Les variables régionalisées et leur estimation*. Masson, Paris.

MATHERON, G. 1969. *Le krigeage universel*. L'Ecole Nationale Superieure des Mines, Paris, Les cahiers du Centre de Morphologie Mathematique de Fontainbleau, **1**.

MATHERON, G. & SERRA, J. 2001. The birth of mathematical morphology, unpublished ms.

MEINZER, O. E. 1923. *Outline of ground-water hydrology, with definitions*. US Geological Survey Water Supply Paper **494**.

MERRIAM, D. F. 1981. Roots of quantitative geology. *In*: MERRIAM, D. F. (ed.) *Down-to-Earth Statistics: Solutions Looking for Geological Problems*. Syracuse University, Syracuse, New York, Geology Contribution **8**.

MERRIAM, D. F. 1999. Reminiscences of the editor of the Kansas Geological Survey Computer Contributions, 1966–70 & a byte. *Computers & Geosciences*, **25**, 321–334.

MICHEL LÉVY, A. 1877. De l'emploi du microscope polarisant a lumière parallele pour la détermination des espèces minérales en plaques minces. *Annales des mines*, Serie 7, **12**, 392–469.

MICHEL LÉVY, A. 1897*a*. Note sur la classification des magmas des roches éruptives. *Bulletin de la Société Géologique de France*, Serie 3, **25**, 326–377.

MICHEL LÉVY, A. 1897*b*. Mémoire sur le porphyre bleu de l'Esterel. *Bulletin des Services de la Carte Géologique de la France et des Topographies Souterraines*, **9**, 1–47.

MICHEL LÉVY, A. & LACROIX, A. 1888. *Les minéraux des roches. 1. Application des méthodes minéralogiques et chimiques a leur etude microscopique par A. Michel Lévy. 2. Données physiques et optiques par A. Michel Lévy et Alf. Lacroix*. Baudry, Paris.

MIESCH, A. T. & CONNOR, J. J. 1968. *Stepwise Regression and Nonpolynomial Models in Trend Analysis*. Kansas Geological Survey Computer Contribution **27**.

MIESCH, A. T., CHAO, E. C. T. & CUTTITTA, F. 1966. Multivariate analysis of geochemical data on tektites. *Journal of Geology*, **74**, 673–691.

MIESCH, A. T., CONNOR, J. J. & EICHER, R. N. 1964. Investigation of geochemical sampling problems by computer simulation. *Quarterly of the Colorado School of Mines*, **59**, 131–148.

MILLER, H. 1857. *The Testimony of the Rocks, Or Geology and its Bearings on the Two Theologies, Natural and Revealed*. Shepard & Elliot, Edinburgh.

MILLER, R. L. 1956. Trend surfaces: their application to analysis and description of environments of sedimentation. I. The relation of sediment-size parameters to current-wave systems and physiography. *Journal of Geology*, **64**, 425–446.

MILNE, J. 1882. The distribution of seismic activity in Japan. *Transactions of the Seismological Society of Japan*, **4**, 1–30.

MINARD, C. J. 1862. *Des tableaux graphiques et des cartes figuratives*. Thunot, Paris.

MITRA, S. 1992. Balanced structural interpretations in fold and thrust belts. *In*: MITRA, S. & FISHER, G. W. (eds) *Structural Geology of Fold and Thrust Belts*. Johns Hopkins Press, Baltimore, 53–77.

MOORE, J. E. & HANSHAW, B. B. 1987. Hydrogeological concepts in the United States: a historical perspective. *Episodes*, **10**, 315–320.

MOORE, J. E. & WOOD, L. A. 1965. Data requirements and preliminary results of an analogue-model evaluation—Arkansas River Valley in Eastern Colorado. *Ground Water*, **5**, 20–23.

MÜGGE, O. 1900. Zur graphischen Darstellung der Zusammensetzung der Gesteine. *Neues Jahrbuch für Mineralogie, Geologie und Palaeontologie, Stuttgart*, **18**, 100–112.

NAGY, G. 1972. Digital image-processing activities in remote sensing for Earth resources. *Proceedings of the Institute of Electrical and Electronics Engineers*, **60**, 1177–1200.

NATHAN, R. 1966. *Digital video-data handling*. Jet Propulsion Laboratory, Pasadena, California, Technical Report **32–877**.

NAUMANN, C. F. 1824. *Beytrage zur Kenntniß Norwegen's* (2 vols). Wienbrad, Leipzig.

O'HANLON, L. 2000. The time travelling mountain. *New Scientist*, **167**, 30–33.

OLIVER, D. S., CUNHA, L. B. & REYNOLDS, A. C. 1997. Markov chain Monte Carlo methods for conditioning a permeability field to pressure data. *Mathematical Geology*, **29**, 61–91.

ORTON, E. 1889. The Trenton Limestone as a source of petroleum and inflammable gas in Ohio and Indiana. *US Geological Survey Annual Report*, **8**, Part 2, 477–662.

OSANN, A. 1900. Versuch einer chemischen Classification der Eruptivgesteine. *Tschermak's Mineralogischen und Petrographischen Mittheilungen*, **19**, 351–469.

OWEN, E. W. 1975. *Trek of the Oil Finders: A History of Exploration for Petroleum*. AAPG, Tulsa.

PEARSON, E. S. (ed.) 1978. *The History of Statistics in the 17th & 18th Centuries Against the Changing Background of Intellectual, Scientific and Religious Thought. Lectures by Karl Pearson Given at University College London during the Academic Sessions 1921–1933*. Griffin, London.

PEARSON, K. 1893. Assymetrical frequency curves. *Nature*, **48**, 615–616.

PEARSON, K. 1894. Contribution to the mathematical theory of evolution. *Philosophical Transactions of the Royal Society, London*, Series A, **185**, 71–110.

PEARSON, K. 1896. Mathematical contributions to the theory of evolution. III. Regression, heredity and panmixia. *Philosophical Transactions of the Royal Society, London*, Series A, **187**, 253–318.

PEARSON, K. 1900. On the criterion that a given system of deviations from the probable in the case of a correlated system of variables is such that it can be reasonably supposed to have arisen from random sampling. *Philosophical Magazine*, Series 5, **50**, 157–175.

PERREY, A. 1845. Memoire sur les tremblements de terreressentis en France, en Belgique et en Hollande, depuis le quatrième siècle de l'ère Crétienne jusqu'a nos jours (1845 inclusiv.). *Mémoires Couronnes et Mémoires des Savants Étrangers, Académie Royale des Sciences et des Belles-Lettres de Bruxelles*, **18**, 1–110.

PERUGGIA, M. & SANTNER, T. 1996. Bayesian analysis of time evolution of earthquakes. *Journal of the American Statistical Association*, **91**, 1209–1218.

PETRIE, W. M. F. 1899. Sequences in prehistoric remains. *Journal of the Anthropological Institute*, **29**, 295–301.

PETTIJOHN, F. J. 1984. *Memoirs of an Unrepentant Field Geologist. A Candid Profile of Some Geologists and their Science*. University of Chicago Press, Chicago.

PHILIPSBORN, H. von, 1928. Zur graphischen Behandlung quaternärer Systeme. *Neues Jahrbuch für Mineralogie, Geologie und Paläontologie*, **57A**, 973–1012.

PHILLIPS, J. 1860. *Life on the Earth; Its Origin and Succession*. Macmillan, Cambridge.

PHILLIPS, W. 1815. *An Outline of Mineralogy and Geology Intended for the Use of Those Who May Desire to Become Acquainted with the Elements of Those Sciences; Especially of Young Persons*. Phillips, London.

PINDER, G. F. 1968. A digital model for aquifer evaluation. *Techniques of Water-Resources Investigations of the United States Geological Survey*, **7**, Chapter C1, 1–18.

PINDER, G. F. & BREDEHOEFT, J. D. 1968. Applications of the digital computer for aquifer evaluation. *Water Resources Research*, **4**, 1,069–1,093.

PLAYFAIR, J. 1802. Illustrations of the Huttonian Theory. *In*: PLAYFAIR, J. G. (ed., 1822) *The Works of John Playfair, Esq*. Constable, Edinburgh, **1**, 1–514.

PLAYFAIR, J. 1805. Meteorological abstract for the years 1797, 1798 and 1799. *Transactions of the Royal Society of Edinburgh*, **5**, 193–202.

PLAYFAIR, J. 1812. Progress of heat in spherical bodies. *Transactions of the Royal Society of Edinburgh*, **6**, 353–370.

PLAYFAIR, W. 1798. *Lineal Arithmetick; Applied to Shew the Progress of the Commerce and Revenue of England during the Present Century; which is Represented and Illustrated by Thirty-three Copper-plate Charts. Being an Useful Companion for the Cabinet and Counting House*. Printed for the Author, Paris.

PLAYFAIR, W. 1801. *The Statistical Breviary; Shewing, on a Principle Entirely New, the Resources of Every State and Kingdom in Europe, Illustrated with Stained Copper-plate Charts, Representing the Physical Powers of Each Distinct Nation with Ease and Perspicacity*. Stockdale, London.

PLAYFAIR, W. 1805. *An Inquiry into the Permanent Causes of the Decline and Fall of Powerfull and Wealthy Nations, Illustrated by Four Engraved Charts. Designed to shew how the Prosperity of the*

British Empire May be Prolonged. Greenland & Norris, London.

PLOT, R. 1685. The history of the weather at Oxford in 1684. *Philosophical Transactions of the Royal Society, London*, **15**, 930–943.

POLLARD, D. D. 2000. Strain and stress: discussion. *Journal of Structural Geology*, **22**, 1359–1367.

POWELL, R. 1985. Regression diagnostics and robust regression in geothermometer/geobarometer calibration; the garnet-clinopyroxene geothermometer revisited. *Journal of Metamorphic Geology*, **3**, 231–244.

PRESTWICH, J., BUSK, G. & EVANS, J. 1873. Report on the exploration of Brixham Cave, conducted by a committee of the Geological Society, and under the superintendence of Wm. Pengelly, Esq., F.R.S., aided by a local committee; with descriptions of the animal remains by George Busk, Esq., F.R.S., and of the flint implements by John Evans, Esq., F.R.S. by Joseph Prestwich, F.R.S., F.G.S., &c., reporter. *Philosophical Transactions of the Royal Society, London*, **163**, 471–572.

PRICE, G. D. & VOCADLO, L. 1996. Computational mineralogy. Comptes Rendus des Seances de l'Académie des Sciences, Paris. Ser II. fasc. A, *Sciences de la Terre et des Planetes*, **323**, 357–371.

PRICKETT, T. A., NAYMIK, T. G. & LONNQUIST, C. G. 1981. A "random walk" solute transport model for selected groundwater quality evaluations. *Illinois State Water Survey Bulletin*, **65**, 1–103.

PROKOPH, A. & BARTHELMES, F. 1996. Detection of nonstationarities in geological time series: wavelet transform of chaotic and cyclic sequences. *Computers & Geosciences*, **22**, 1097–1108.

QUETELET, A. 1827. Recherches sur la population, les naissances, les décès, les prisons, les dépôts de mendicité, etc., dans la royaume des Pays-Bas. *Nouveaux Mémoires de l'Academie Royale des Sciences et des Belles-Lettres de Bruxelles*, **4**, 117–92.

QUETELET, A. 1836. *Sur l'homme et le développement de ses facultés, ou essai de physique sociale* (2 vols). Hauman, Bruxelles.

QUETELET, A. 1869. *Physique Sociale ou essai sur le développement des facultés de l'homme* (2 vols). Muquardt, Brussels.

RAMSAY, W. 1888. Ueber die Absorption des Lichtes im Epidot vom Sulzbachthal. *Zeitschrift fur Kristallographie und Mineralogie, Leipzig*, **13**, 97–134.

RAMSON, I., APPEL, C. A. & WEBSTER, R. A. 1965. Ground-water models solved by digital computer. *Journal of the Hydraulics Division, Proceedings of the American Society of Civil Engineers*, **91**, 133–147.

RAUP, D. M. 1966. Geometric analysis of shell coiling: general problems. *Journal of Paleontology*, **40**, 1178–1190.

RAUP, D. M. & SEILACHER, A. 1969. Fossil foraging behaviour: computer simulation. *Science*, **166**, 994–995.

RENARD, P. & COURRIOUX, G. 1994. Three-dimensional geometric modeling of a faulted domain: The Soultz horst example (Alsace, France). *Computers & Geosciences*, **20**, 1379–1390.

REYER, E. 1877. *Beitrag zur Fysik der Eruptionen und der eruptiv-gesteine*. Hölder, Vienna.

REYER, E. 1888. *Theoretische Geologie*. Schweizerbart, Stuttgart.

RICHARDSON, W. A. 1923. The frequency-distribution of igneous rocks. Part II. The laws of distribution in relation to petrogenetic theories. *Mineralogical Magazine*, **20**, 1–19.

RICHARDSON, W. A. & SNEESBY, G. 1922. The frequency-distribution of igneous rocks. I. Frequency-distribution of the major oxides in analyses of igneous rocks. *Mineralogical Magazine*, **19**, 303–313.

RINDFLEISCH, T. C., DUNNE, J. A., FRIEDEN, H. J., STROMBERG, W. D.& RUIZ, R. M. 1971. Digital processing of Mariner 6 and 7 pictures. *Journal of Geophysical Research*, **76**, 934–947.

RIPLEY, B. D. 1981. *Spatial Statistics*. Wiley, New York.

RIPLEY, B. D. 1987. *Stochastic Simulation*. Wiley, New York.

ROBINSON, A. H. 1982. *Early Thematic Mapping in the History of Cartography*. Chicago University Press, Chicago.

ROBINSON, H. H. 1916. The summation of chemical analyses of igneous rocks. *American Journal of Science*, Series 4, **41**, 257–275.

ROBINSON, J. E. 1969. *Spatial filtering of geological data*. Preprint, IAMG section, 37th Session, International Statistical Institute, London, **22**.

ROUSSEEUW, P. J. 1983. Regression techniques with a high breakdown point. *Bulletin of the Institute of Mathematics and Statistics*, **12**, 155.

ROUSSEEUW, P. J. 1984. Least median of squares regression. *Journal of the American Statistical Association*, **79**, 871–880.

RUDWICK, M. J. S. 1978. Charles Lyell's dream of a statistical palaeontology. *Palaeontology*, **21**, 225–244.

SABINE, P. A. & HOWARTH, R. J. 1998. The role of ternary projections in colour displays for geochemical maps and in economic mineralogy and petrology. *Journal of Geochemical Exploration*, **63**, 123–144.

SCHMIDT, J. F. J. 1874. *Vulkanstudien. Santorin 1866 bis 1872: Vesuv, Bajae, Stromboli, Aetna 1870*. Scholtze, Leipzig.

SCHROEDER VAN DER KOLK, J. L. C. 1896. Beiträge zur Kartirung der quartären Sande. *Zeitschrift der Deutschen Geologischen Gesellschaft*, **48**, 773–807.

SCHWARZACHER, W. 1964. An application of statistical time-series analysis of a limestone-shale sequence. *Journal of Geology*, **72**, 195–213.

SCHWARZACHER, W. 1967. Some experiments to simulate the Pennsylvanian rock sequence of Kansas. *In*: MERRIAM, D. F. (ed.) *Computer Applications in the Earth Sciences*: *Colloquium on Simulation*. Kansas Geological Survey Computer Contribution **18**, 5–14.

SCHWARZACHER, W. 1969. The use of Markov chains in the study of sedimentary cycles. *Mathematical Geology*, **1**, 17–39.

SCHWARZACHER, W. & FISCHER, A. G. 1982. Limestone-shale bedding and perturbations of the Earth's orbit. *In*: ENSELE, G. & SEILACHER, A.

(eds) *Cyclic and Event Stratification*. Springer, Berlin, 72–95.

SEKIYA, S. 1887. Earthquake observations of 1885 in Japan. *Transactions of the Seismological Society of Japan*, **10**, 57–82.

SHAW, A. B. 1995. Early history of graphic correlation. *In*: MANN, K. O. & LANE, H. R. (eds) *Graphic Correlation*. SEPM Society for Sedimentary Geology, Tulsa, Special Publication **53**, 15–19.

SIMPSON, S. M. Jr. 1954. Least squares polynomial fitting to gravitational data and density plotting by digital computers. *Geophysics*, **19**, 255–269, 644.

SMITH, A. F. M. & GELFAND, A. E. 1992. Bayesian statistics without tears: a sampling-resampling perspective. *The American Statistician*, **46**, 84–88.

SMITH, D. G. (ed.) 1989. Milankovitch cyclicity in the pre-Pleistocene stratigraphic record. *Terra Nova*, **1**, 402–480.

SMYTH, W. W. 1854. *Notes on Mining Schools and Mining Academies*. Chapman & Hall, London.

SOKAL, R. R. & SNEATH, P. H. A. 1963. *Principles of Numerical Taxonomy*. Freeman, San Francisco.

SOLOW, A. R. 1985. Bootstrapping correlated data. *Mathematical Geology*, **17**, 769–775.

SORBY, H. C. 1908. On the application of quantitative methods to the study of the structure and history of rocks. *Quarterly Journal of the Geological Society*, London, **64**, 171–233.

SPEARMAN, C. 1904. General intelligence, objectively determined and measured. *American Journal of Psychology*, **15**, 201–293.

SPITZ, K. & MORENO, J. 1996. *A Practical Guide to Groundwater Transport and Solute Transport Modelling*. Wiley, New York.

STEGENA, L. 1984. Hungarian contribution to the geophysical mapping of the world. Proceedings of the Xth INHIGEO Symposium 16–22 August 1982, Budapest, Hungary. *In*: DUDICH, E. (ed.) *Contributions to the History of Geological Mapping*. Akadémiai Kiadó, Budapest, 251–255.

STIFF, H. A. J. 1951. The interpretation of chemical water analysis by means of patterns. *Journal of Petroleum Technology*, **3**, 15–17.

STIGLER, S. M. 1986. *The History of Statistics. The Measurement of Uncertainty before 1900*. Belknap Press, Cambridge, Massachusetts.

STONE, L. D. 1990. Bayesian estimation of undiscovered pool sizes using the discovery record. *Mathematical Geology*, **22**, 309–322.

STRAUSS, D. & SADLER, P. M. 1989. Classical confidence intervals and Bayesian probability estimates for ends of local taxon ranges. *Mathematical Geology*, **21**, 411–427.

STUART, A. 1927. Ontogenetic and other variations in *Volutospina spinosa*. *Geological Magazine*, **64**, 545–557.

SWINNERTON, H. H. 1921. The use of graphs in palaeontology. *Geological Magazine*, **58**, 357–364.

THEIS, C. V. 1935. The relation between the lowering of the piezometric surface and the rate and duration of discharge of a well using ground-water storage. *Transactions of the American Geophysical Union*, **16**, 519–524.

THEIS, C. V. 1940. The source of water derived from wells; essential factors controlling the response of aquifers to development. *Civil Engineering*, **10**, 277–280.

THOMAS, F. C., GRADSTEIN, F. M. & GRIFFITHS, C. M. 1988. *Bibliography and Index of Quantitative Biostratigraphy*. Committee on Quantitative Stratigraphy. Special Publication **1**.

THOULET, J. 1891. Expériences sur la sédimentation. *Annales des Mines*, Series 8., **19**, 5–35.

THURSTONE, L. L. 1931. Multiple factor analysis. *Psychological Review*, **38**, 406–427.

TOURCOTTE, D. L. 1997. *Fractals and Chaos in Geology and Geophysics* (2nd edn). Cambridge University Press, Cambridge.

TRAGER, E. A. 1920. A laboratory method for the examination of well cuttings. *Economic Geology*, **15**, 170–176.

TRASK, P. D. 1932. *Origin and Development of Source Sediments of Petroleum*. Gulf Publishing, Houston, Texas.

TSCHERMAK, G. 1863. Die Entstehungsfolge der Mineralien in einigen Graniten. *Sitzungberichte der mathematischen-naturwissenschaftlichen Klasse der Osterreichischen Akademie der Wissenschaften, Wien*, **47**, 207–220.

TUFTE, E. R. 1983. *The Visual Display of Quantitative Information*. Graphics Press, Cheshire, Connecticut.

TUFTE, E. R. 1990. *Envisioning Information*. Graphics Press, Cheshire, Connecticut.

TYLOR, A. 1868. On the formation of deltas; and on the evidence and cause of great changes in the sea-level during the glacial period with the laws of denudation and river levels. *Geological Magazine*, **9**, 392–500.

UDDEN, J. A. 1898. Mechanical composition of wind deposits. *Augustana Library Publications*, **1**, 1–69.

UDDEN, J. A. 1914. Mechanical composition of clastic sediments. *Bulletin of the Geological Society of America*, **25**, 655–744.

UMPLEBY, J. B. 1917. *Geology and ore deposits of the Mackay region, Idaho*. US Geological Survey Professional Paper **97**.

VALENTINE, J. W. & PEDDICORD, R. G. 1967. Evaluation of fossil assemblages by cluster analysis. *Journal of Palaeontology*, **4**, 502–507.

VEATCH, A. C. 1906. *Geology and underground water resources of northern Louisiana and southern Arkansas*. US Geological Survey Professional Paper **46**.

VILJOEN, R. P., VILJOEN, M. J., GROOTENBOER, J. & LONGSHAW, T. G. 1975. ERTS-1 imagery: an appraisal of applications in geology and mineral exploration. *Minerals Science Engineering*, **7**, 132–168.

VISTELIUS, A. B. 1944. Notes on analytical geology. *Doklady Akademiya Nauk S S S R*, **64**, 27–31 (in Russian).

VISTELIUS, A. B. 1949. On the question of the mechanism of the formation of strata. *Doklady Akademiya Nauk S S S R*, New Series, **65**, 191–194 (in Russian).

VISTELIUS, A. B. 1950. On the mineralogical composition of the heavy fraction of the sand on the lower productive strata of the Volga alluvium. *Doklady Akademiya Nauk S S S R*, **71** (2) (in Russian).

VISTELIUS, A. B. 1964. A stochastic model of crystallization of alaskites and the corresponding transitional probabilities. *Doklady Akademiya Nauk S S S R*, **170**, 653–656.

VISTELIUS, A. B. 1967. *Studies in Mathematical Geology*. Consultants Bureau, New York.

VISTELIUS, A. B. & ROMANOVA, M. A. 1964. Distribution of the heavy mineral fraction in sandy deposits of the Central Kara-Kum. *Doklady Akademiya Nauk S S S R*, **158**, 860–863 (in Russian).

VISTELIUS, A. B. & YANOVSKAYA, T. B. 1963. Programming problems of geology and geochemistry for use with all-purpose electronic computers. *Geologija Rudnykh Mestordzhdenii*, **3**, 34–48 (in Russian).

VISTELIUS, A. B., AGTERBERG, F. P., DIVI, S. R. & HOGARTH, D. D. 1983. *A stochastic model for the crystallization and textural analysis of a fine grained granitic stock near Meech lake, Gatineau Park, Quebec*. Geological Survey of Canada Paper **81–21**.

VOCADLO, L. & PRICE, G. D. 1999. The theory and simulation of the melting of minerals. *In*: WRIGHT, K. & CATLOW, C. R. A. (eds) *Microscopic Properties and Processes in Minerals*, Kluwer, Amsterdam, 561–576.

VOLGER, G. H. O. 1856. Untersuchungen über das letztjährige erdbeben in Central-Europa. *Mittheilungen aus Justus Perthes' Geographischer Anstalt*, **2**, 85–102.

WADDINGTON, C. H. 1929. Notes on graphical methods of recording the dimensions of ammonites. *Geological Magazine*, **66**, 180–186.

WALTON, W. C. & PRICKETT, T. A. 1963. Hydrogeologic electric analog computers. *American Society of Civil Engineers Proceedings, Journal of the Hydraulics Division*, **89**, 67–91.

WATSON, G. S. 1966. The statistics of orientation data. *Journal of Geology*, **74**, 786–797.

WEBB, J. S. 1958. Notes on geochemical prospecting for lead-zinc deposits in the British Isles. *In*: WALTER, A. J. P. (ed.) *A Symposium. The Future of Non-ferrous Mining in Great Britain and Ireland. London, 23–30 September 1958*, Pre-print Paper 19. The Institution of Mining and Metallurgy, London, 23–40.

WEBB, J. S., FORTESCUE, J. & TOOMS, J. S. 1964. *Regional Geochemical maps of the Namwala Concession Areas, Zambia. Based on a Reconnaissance Stream Sediment Survey*. Geological Survey of Zambia, Lusaka.

WEEDON, G. P. 1989. The detection and illustration of regular sedimentary cycles using Walsh power spectra and filtering, with examples from the Lias of Switzerland. *Journal of the Geological Society, London*, **146**, 133–144.

WENTWORTH, C. K. 1922. A scale of grade and class terms for clastic sedimentology. *Journal of Geology*, **30**, 377–392.

WENTWORTH, C. K. 1929. Method of computing mechanical composition types in sediments. *Bulletin of the Geological Society of America*, **40**, 771–790.

WENTWORTH, C. K. 1931. The mechanical determination of sediments in graphic form. *Iowa University Studies in Natural History*, **14**, 1–127.

WHITTAKER, W. & REID, C. 1899. *The Water Supply of Sussex from Underground Sources*. Memoir of the Geological Survey, Her Majesty's Stationery Office, London.

WHITTEN, E. H. T. 1963. *A Surface-fitting Program Suitable for Testing Geological Models which Involve Areally-distributed Data*. Geography Branch, Office of Naval Research, Northwestern University, Evanston, Illinois, Technical Report **2** (ONR Task No. 389–135).

WHITTEN, E. H. T. 1964. Process-response models in geology. *Bulletin of the Geological Society of America*, **75**, 455–464.

WHITTEN, E. H. T. & BOYER, R. E. 1964. Process-response models based on heavy-mineral content of the San Isabel Granite, Colorado. *Bulletin of the Geological Society of America*, **75**, 841–862.

WHITTEN, E. H. T. & DACEY, M. F. 1975. On the significance of certain Markovian features of granite textures. *Journal of Petrology*, **16**, 429–453.

WHITTEN, E. H. T., KRUMBEIN, W. C., WAYE, I. & BECKMAN, W. A., Jr 1965. *A surface-fitting program for areally-distributed data from the earth sciences and remote sensing*. National Aeronautics and Space Administration, Washington, DC, NASA Contractor Report CR 318.

WOODWARD, S. P. 1854. *A Manual of the Mollusca; Or Rudimentary Treatise of Recent and Fossil Shells*, **2**. Weale, London.

WRAY, D. A., SLATER, L. & EDDY, G. E. 1931. The correlation of the Arley Mine of Lancashire with the Better Bed Coal of Yorkshire. *Summary of Progress of the Geological Survey of Great Britain for 1930*, Part 2, 1–17.

WYBERGH, W. 1897. Graphic assay plans. *Transactions of the Institution of Mining and Metallurgy*, **5**, 235–242.

YARUS, J. M. & CHAMBERS, R. L. (eds) 1994. *Stochastic Modeling and Geostatistics. Principles, Methods and Case Studies*. AAPG, Tulsa, Computer Applications in Geology **3**.

YU, Z. 1998. Applications of vector and parallel supercomputers to ground-water flow modelling. *Computers & Geosciences*, **23**, 917–927.

ZAPOROZEC, A. 1972. Graphical Interpretation of water-quality data. *Ground Water*, **10**, 32–43.

Norman Levi Bowen (1887–1956) and igneous rock diversity

DAVIS A. YOUNG

Department of Geology, Geography, and Environmental Studies, Calvin College, Grand Rapids, Michigan, USA

Abstract: By the beginning of the twentieth century, differentiation had emerged as the leading theory to explain the chemical and mineralogical diversity of igneous rocks. Soret diffusion, liquid immiscibility, compositional stratification of magma by gravity (Gouy–Chaperon effect), volatile transport, crystal settling and other processes had been advocated as mechanisms of differentiation, but no consensus was achieved regarding a dominant mechanism.

During a career spent primarily at the Geophysical Laboratory, Washington DC, Norman Levi Bowen (1887–1956) initiated a new approach to petrology. On the basis of experimental studies of rock-forming silicates and on physicochemical principles, Bowen argued against the importance of Soret diffusion, liquid immiscibility, volatile transport and assimilation as major causes of diversity. He formulated a comprehensive theory of differentiation that emphasized the role of separation of crystals from liquid. He reasoned that rocks of the subalkaline igneous rock series, including granite, have been derived from parental basalt by crystal separation, e.g. settling, filter-pressing or armouring of crystals.

Apart from its scientific merits, Bowen's achievement rested on personal, institutional and technical factors that included his determination to dedicate virtually his entire career to solution of the problem of igneous rock diversity; his affiliation with the Geophysical Laboratory; the prior development of the quenching method and the calibration of the temperature scale to very high temperatures; and the influence of Arthur L. Day.

For more than a century, the mineralogical and chemical diversity of igneous rocks has posed one of the perennial problems in igneous petrology. Petrologists continue to acknowledge many complex causes of diversity, and they struggle to assess the relative importance of the mechanisms that have been proposed. Contemporary understanding of the causes of igneous rock diversity, derived from integrated geological, petrographical, experimental, isotopic, chemical and theoretical studies, has been powerfully shaped by the scientific legacy of Norman L. Bowen. During a career that spanned most of the first half of the twentieth century, Bowen conducted a comprehensive programme of experimental studies on the phase equilibria in silicate systems. His work and that of colleagues at the Geophysical Laboratory, Washington DC, established a sound theoretical and experimental basis on which to evaluate proposed mechanisms of diversity. His theory proposing that crystallization–differentiation, or fractional crystallization, is the dominant cause of diversity arguably ranks as the most significant theoretical advance in twentieth-century igneous petrology (Young 1998).

Although petrologists no longer attribute to fractional crystallization the universality accorded to it by Bowen, they still maintain that the process played a major role in the development of large layered igneous intrusions and of many suites of volcanic rocks. In contrast, contemporary petrologists are inclined to account for large volumes of granitoid rocks by anatexis rather than fractional crystallization of basaltic liquid, Bowen's favoured process. They are also more open to a significant role for assimilation, liquid immiscibility, magma mixing, and other processes than was Bowen. They are, moreover, far less enthusiastic than Bowen was about crystal settling as a mechanism of crystal–liquid separation.

Nevertheless, although contemporary petrologists have modified aspects of Bowen's theory, they have constructed their views about diversity on a foundation that he laid. Geschwind (1995) has argued that Bowen's application of the principles of physical chemistry and the data on phase equilibria obtained by controlled experiments to the solution of petrological problems seemed radical to his contemporaries, most of whom were accustomed to thinking that such problems should be solved by geological means. The power of Bowen's methods, however, led igneous petrologists increasingly to adopt his approach to solving problems. The more recent application of isotopic, trace-element and fluid dynamical studies to igneous geology, taken for

From: OLDROYD, D. R. (ed.) 2002. *The Earth Inside and Out: Some Major Contributions to Geology in the Twentieth Century*. Geological Society, London, Special Publications, **192,** 99–111. 0305-8719/02/$15.00

granted by contemporary petrologists, is fundamentally an extension of Bowen's scientific methodology. Apart from the scientific merits of his theory of fractional crystallization and the fruitfulness of his methodology (Yoder 1979; Hargraves 1980), the problem remains as to why Bowen exerted so much influence on igneous petrology. What historical factors shaped Bowen so that he was able to achieve such eminence as a scientist?

The petrological situation prior to Bowen

To suggest an answer to that question, I shall sketch the context for Bowen's work and then review his career, with emphasis on the development of his theory of crystallization–differentiation. About a century ago, petrologists were generalists with broad interests in field work, microscopic petrography, structural aspects of petrology, and the classification of igneous rocks. Many, like Harker (1932) and Teall (1885), were equally at home with metamorphic rocks. For petrologists of that era, the diversity of igneous rocks was one among many interesting challenges. Theoretical reflection about igneous rock diversity was in a state of confusion. Petrologists had discarded the 'two-source hypothesis' of Bunsen (1851) and the hypothesis of Sartorius von Waltershausen (1853) and Durocher (1857) that magmas are successively tapped from increasing depths in concentric acid and basic shells. Thanks to the concepts of the petrographic province (Judd 1886) and consanguinity (Iddings 1892) many petrologists concluded that diversity should be attributed to differentiation. Others such as Loewinson-Lessing (1899) assigned important roles to assimilation and magma mixing.

The developing field of physical chemistry (Servos 1990) provided petrologists with a wealth of ideas about possible mechanisms of differentiation. Lagorio (1887), Teall (1888), Brøgger (1890), Vogt (1891), Iddings (1892) and Judd (1893) invoked the Soret (1879, 1881) effect, in which material was said to diffuse from the hot centre toward the cool walls of a magma chamber. The idea was largely abandoned, however, when Bäckström (1893), Harker (1894) and Becker (1897*a*) assailed it on the grounds that diffusion of heat occurs much more rapidly than diffusion of matter. 'Liquation', or the separation of magma into immiscible fractions of contrasted composition, was suggested by Durocher (1857) and advocated by Rosenbusch (1889) and Bäckström (1893). Vogt, however, claimed that the only evidence of liquid immiscibility that he had seen in his studies of slags was between silicate and sulphide melts. Teall (1888) and Vogt (1903) argued for the importance of eutectics for the understanding of differentiation. Iddings (1892) and Brøgger (1894) toyed with the idea of Gouy & Chaperon (1887) that solutions undergo compositional stratification due to gravity. Becker (1897*b*) originated the application of the process of fractional crystallization to geological situations. He argued that magma compositions would change in response to selective removal of crystals. Harker (1893) suggested that crystallization at the margins of a magma chamber continuously fed by diffusion of matter to growing crystals could produce differentiation, but Becker's point about the slowness of diffusion also felled that idea. Thus there was no consensus about the character of differentiation.

Franz Y. Loewinson-Lessing (1861–1939) asserted that a combination of assimilation, refusion, differentiation and eutectics provided the best basis for a theory of diversity. Loewinson-Lessing (1899, 1911) also insisted on the existence of two fundamental magmas, granitic and gabbroic, independently erupted throughout geological time. Igneous rocks, he said, are derived from these two magmas by differentiation and assimilation. To account for the wide diversity among igneous rocks, Reginald A. Daly (1871–1957) also proposed a combination of differentiation and assimilation. Daly (1914) believed that superheated basaltic magma injected along abyssal magmatic wedges accomplished large-scale magmatic stoping of blocks of country rock, which would sink through the magma to be assimilated near the base of a magma chamber. Both the original basalt magma and the contaminated, syntectic magma, he maintained, would undergo varying degrees of differentiation. Daly rejected molecular diffusion but accepted liquid immiscibility, fractional crystallization and gas transfer as mechanisms of differentiation.

The wide spectrum of opinion surrounding the causes of diversity was symptomatic of the pressing need for reliable experimental data on the thermal and physical properties of minerals and rocks at high temperatures. As a result, the extension of the temperature scale to very high temperatures, the precise measurement of high temperatures by thermoelectric methods, and the determination of phase diagrams for simple oxide systems with synthetic chemically pure starting materials at the Geophysical Laboratory, newly established at Washington in 1905, all held out promise that the problem of igneous rock diversity might be solved.

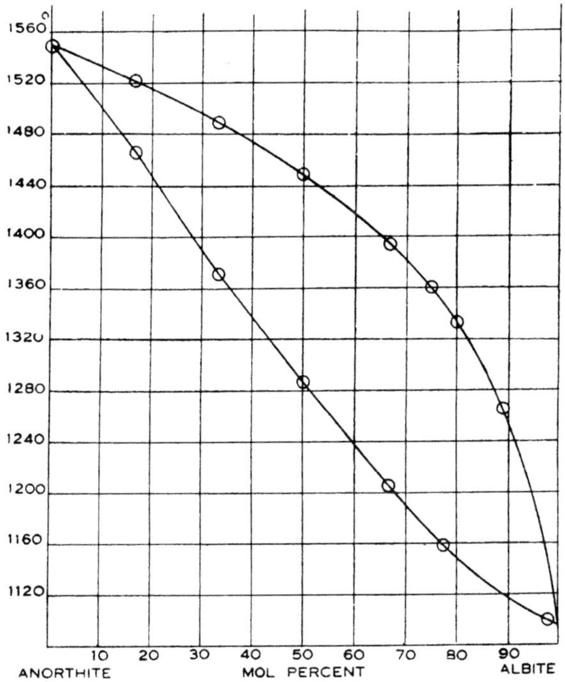

Fig. 1. Phase diagram of plagioclase feldspar at one atmosphere (Bowen 1913). From this diagram Bowen conceived the idea of the continuous reaction series and perceived that fractional crystallization could result in extreme variation in magma composition. Reprinted by permission of *American Journal of Science.*

Bowen's education and early career

Born in Kingston, Ontario, Canada, Norman Levi Bowen (1887–1956) completed his undergraduate studies in his home town at Queen's University in 1909. Bowen wanted to do graduate work with Johan H. L. Vogt (1858–1932) and Waldemar C. Brøgger (1851–1940), but Vogt informed Bowen that the Norwegian language would be of no use after his studies and that Brøgger was occupied with political responsibilities. Bowen therefore enrolled at Massachusetts Institute of Technology (MIT) where he learned igneous petrology from Daly (1914). He also studied with the physical chemists Gilbert N. Lewis (1875–1946) and Arthur A. Noyes (1866–1936), the economic geologist Waldemar Lindgren (1860–1939), and the volcanologist Thomas A. Jaggar (1871–1953), who urged Bowen to spend a year at the Geophysical Laboratory to conduct an experimental study for his doctoral dissertation. The director, Arthur L. Day (1869–1960), welcomed Bowen and suggested that he tackle the system nepheline–anorthite. In determining the phase relations, Bowen quickly realized the advantages of the newly developed quench method of Shepherd *et al.* (1909). Details of Bowen's early years have been reviewed by Yoder (1992).

Upon receipt of his PhD from MIT in 1912, Bowen was besieged with offers of employment at the Geological Survey of Canada, the Hawaii Volcano Observatory, and the US Geological Survey. However, he chose to continue experimental work at the Geophysical Laboratory as assistant petrologist. Day, who had shown that plagioclase might have an ascendant loop phase diagram, but was unable to determine the solidus with the cooling curve method (Day *et al.* 1905), supported Bowen's decision to endeavour to decipher the plagioclase diagram with the recently developed quenching method. Bowen (1913) determined the liquidus and solidus curves for plagioclase, confirmed the ascendant loop phase diagram proposed by Day *et al.* (1905), established the continuous solid solution behaviour of the plagioclase feldspar series suggested by Hunt (1854) and Tschermak (1864), and hinted at the extreme changes in chemical composition possible in magmas as a result of the separation of plagioclase crystals from liquid (see Fig. 1).

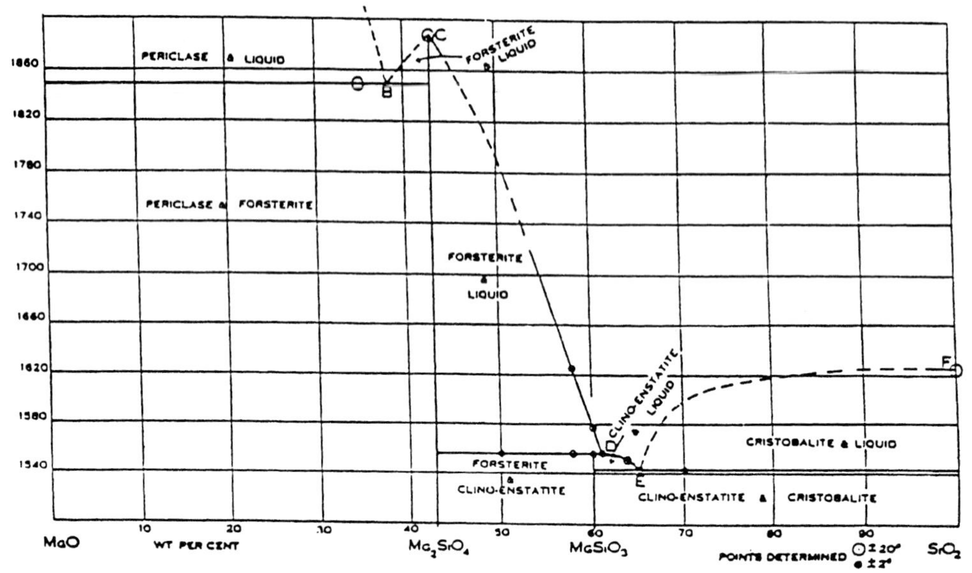

Fig. 2. Phase diagram of $MgO–SiO_2$ at one atmosphere (Bowen & Andersen 1914). From this diagram Bowen conceived the idea of the discontinuous reaction series. Reprinted by permission of *American Journal of Science.*

Bowen next tackled pyroxenes and olivines in the system $MgO–SiO_2$ (Bowen & Andersen 1914). They discovered that forsterite and enstatite bear a reaction relation to one another and maintained that early-formed olivine crystallized from liquids of appropriate composition would react partially or completely with residual liquid upon cooling to precipitate enstatite co-existing with more siliceous liquid (see Fig. 2). These two diagrams displaying continuous and discontinuous reactions contained the germs of virtually the entire theory that Bowen would gradually develop.

In a paper on the system diopside–forsterite–silica, Bowen (1914) demonstrated how paths of liquid composition would be affected during cooling by slow equilibrium crystallization or by fractional crystallization involving sinking of crystals, 'squeezing off' of liquid, or zoning of crystals. He criticized eutectic crystallization as a significant factor in the differentiation of magmas. With eutectic crystallization, Bowen pointed out, cooling liquids finally crystallize completely at the same temperature to the same minerals, no matter what the initial liquid composition. In the system diopside–forsterite–silica, however, the reaction relation of olivine precluded eutectic crystallization. Bowen suggested that fractional crystallization involving the settling of calcic plagioclase and magnesian pyroxene from basic magma was the prime factor in the differentiation of the ordinary sub-alkaline series of igneous rocks.

Bowen (1915*a*) experimentally demonstrated the reality of crystal settling by melting a mixture of diopside and enstatite, then cooling and holding it at 1430°C, so that olivine could crystallize. When the charge was quenched, Bowen found that olivine crystals had concentrated toward the bottom. He also found that pyroxene crystals settled from another melt obtained by heating a mixture of pyroxene and silica and that tridymite crystals concentrated toward the top of a very silica-rich melt. In a final experiment, Bowen allowed a charge to crystallize completely. The lower part consisted of a mixture of olivine and pyroxene and the upper part of pyroxene and silica. Bowen applied his results to the Palisades Sill in New Jersey, a diabase sheet enriched in quartz and alkali feldspar toward the top and characterized by an olivine-rich layer above its base. He claimed that the sill offered a prime example of differentiation by gravitational movements of crystals in magma.

In a study of the system diopside–albite–anorthite, Bowen (1915*b*) pointed out the progressive sodium enrichment of residual liquids, illustrated the effect of equilibrium and fractional crystallization on paths of liquid composition, and attacked the eutectic theory of differentiation. Then, on the basis of this small number of critical experiments, he presented a detailed case for the dominant role of crystallization–differentiation in magmas in a paper entitled 'The later stages of the evolution of the igneous rocks' (Bowen 1915*c*). He disputed the importance of the Soret effect, dismissed the importance of liquid immiscibility, and opposed the idea that magmas undergo gravitative differentiation in the liquid state. Bowen also questioned whether assimilation played the dominant role accorded to it by Loewinson-Lessing and Daly.

Bowen reviewed the available phase diagrams, stressing how different types of crystallization affect paths of liquid composition and the end-points of crystallization. He suggested that crystal settling played an important role in the early stages of differentiation and that squeezing off of residual liquid became important in the final stages of crystallization. He claimed that typical basaltic magma differentiated toward increasingly siliceous compositions and regarded granite as the normal end-product of the fractional crystallization of basaltic liquid. He suggested, however, that parental liquid, normally basaltic, variously differentiated to diorite, granodiorite or granite depending on the rate of cooling. In summary, Bowen attributed the diversity within the common subalkaline igneous rock series, basalt–andesite–rhyolite or gabbro–diorite–granite, to crystallization–differentiation. He intimated that monomineralic rocks like anorthosite and dunite were the products of early settling of plagioclase or olivine, and he envisioned extremely alkaline rocks as differentiates of granitic liquid.

World War I and its aftermath

During World War I, the Geophysical Laboratory, including Bowen, diverted its energies toward putting American glass production on an efficient, scientific basis for the armed forces. After the war, Bowen accepted a teaching position at Queen's University where he remained throughout 1919 and the first half of 1920. His experiments ground to a halt. Lacking the facilities he had enjoyed at the Geophysical Laboratory, he did optical studies of rare minerals from the university collection and published papers on glass. During this period, he also wrote about the origin of anorthosite and responded to criticisms of his theory (Bowen 1917, 1919).

Arthur Day returned as director of the Geophysical Laboratory a year after the conclusion of the war and made it a top priority to attract Bowen back to Washington. Bowen was anxious to get a substantial salary increase to enable him to afford housing in Washington (Young 1998) and told Day that he could do better financially at Queen's. Day called on all his powers of persuasion, including questioning Bowen's commitment to science, to get him back. Day's challenge hit its mark. Back in Washington, Bowen resumed experiments on the melting of orthoclase with George W. Morey (1888–1965), and on the alumina–silica system with Joseph W. Greig (1895–1977). He published results of diffusion experiments in which he placed a layer of diopside beneath a layer of plagioclase, melted the layers, quenched the material after a couple of days, and measured the compositions of the glasses to determine the extent of diffusion from one layer into the next (Bowen 1921). He confirmed Becker's assertion that material diffusion in silicate magmas was negligible in comparison with the rate of diffusion of heat.

Bowen's most significant papers of the 1920s were theoretical extrapolations of his early experiments. Bowen (1922*a*) established the 'reaction principle' and proposed the idea of continuous and discontinuous reaction series. Bowen (1922*b*) showed from thermochemical data that assimilation would typically be an endothermic process that would quickly cease unless a magma possessed substantial superheat. Because of the presence of phenocrysts in chill zones, he argued that magmas do not possess much superheat and denied that assimilation is important for producing diversity. He related assimilation to the reaction principle by showing from the plagioclase diagram that basic magmas could melt acid rock fragments but that acid magmas would simply react with basic rock fragments. Bowen insisted on the inability of granitic magma to melt gabbro, amphibolite or basalt. Greig *et al.* (1931) finally established the melting temperatures of granite and basalt experimentally, showing, contrary to the prevailing opinion, that granite melts at a lower temperature than basalt.

Later, Bowen (1927) argued for the origin of ultramafic rocks by settling and solid-state remobilization of olivine. He spelled out his entire theory in a series of lectures at Princeton University published a few months later as *The Evolution of the Igneous Rocks* (Bowen 1928). Widely hailed for its lucid writing, logic of argument and imaginative insight into the nature of igneous rocks, the book remained a standard text for decades. Although much of the book was a compilation of earlier papers, Bowen used recent experimental studies by his Geophysical Laboratory colleague Joseph Greig (1927) to demonstrate that liquid immiscibility occurs in metal oxide–silica systems only at compositions and temperatures far from those of natural magmas.

Another important aspect of *The Evolution of the Igneous Rocks* was Bowen's response to the claim of another Geophysical Laboratory colleague, Clarence Fenner (1870–1949), that rhyolite magma at Katmai had melted large quantities of basaltic rock (Fenner 1926) and to his use of variation diagrams to support the idea of magma mixing. Bowen repeated verbatim his original article on assimilation (Bowen 1922*b*) but employed variation diagrams for the first time to argue that Fenner had failed to use them properly. He included a chapter on the role of volatiles in magmatism and concluded that, although volatiles were significant in the formation of veins, pegmatites and ore bodies, they played a negligible role during the course of differentiation from basalt to granite.

The goal of crystallization–differentiation

With the publication of *The Evolution of the Igneous Rocks*, Bowen's reputation as the leading thinker among igneous petrologists of his era was sealed. In effect, Bowen had set the terms of debate in igneous petrology for the first half of the twentieth century with his comprehensive theory. While all petrologists were uncomfortable with at least one element or another of Bowen's ambitious theory, Fenner became his most persistent critic. The year after Bowen's book was published, Fenner (1929) issued a paper on the crystallization of basalts. He complained that Bowen paid great attention to the plagioclase solid-solution series as a means for increasing the alkali content of residual liquids but ignored pyroxene and olivine solid-solution series as a means for increasing the iron content of residual liquids. Fenner described many examples of basalt with iron-rich, silica-poor residual groundmasses that consist of magnetite, pyroxene and plagioclase. Whereas Bowen had repeatedly argued that crystallization–differentiation resulted in granitic residual liquids enriched in alkalis and silica, Fenner saw evidence that the course of differentiation led to silica-poor, iron-rich residual liquids.

Another challenge to Bowen's conception of the goal of differentiation came from W. Q. Kennedy (1903–1979). Kennedy (1933) asserted that there are two kinds of primary basalt magma: an olivine-basalt magma type, crystallizing to olivine, titaniferous augite, basic plagioclase and iron ore with alkaline residual material free of quartz; and a more siliceous tholeiitic magma type crystallizing to enstatite-augite (pigeonite), basic plagioclase and iron ore with dominantly quartzofeldspathic residual interstitial material. Kennedy argued that the final differentiate of a basaltic magma depended on the chemical and mineralogical constitution of the parental basalt type. He knew of no instances in which an olivine-basalt parent differentiated to a quartz-bearing rock. Instead, he claimed that late-stage rhyolites typically occur with tholeiitic basalts and that late-stage phonolites generally accompany olivine basalts. Kennedy denied that one primary magma is derived from the other.

Even before these criticisms, Bowen, knowing that his theory was vulnerable because it was based on a limited number of experiments, felt the importance of investigating systems containing iron, potash and water, and at pressures above one atmosphere. During his doctoral work at Yale, J. Frank Schairer (1904–1970) applied to the Geophysical Laboratory for a fellowship to work on his dissertation. Instead, Arthur Day persuaded Schairer to accept a full-time staff position, beginning in the autumn of 1927. Day encouraged him to investigate a system with iron oxides. Schairer teamed up with Bowen in a long series of studies of iron silicate systems. Their first collaborative venture concerned the melting relations of acmite (Bowen & Schairer 1929). Earlier workers such as Doelter (1903) had melted acmite and were in reasonable agreement on a melting-point somewhere between 950°C and 1020°C. Bowen and Schairer, however, discovered that acmite melts incongruently. They also discovered that quartz precipitates directly from melts having very silica-rich compositions near a eutectic point at 850°C. Bowen and Schairer proved experimentally that quartz can crystallize directly from magma.

Bowen *et al.* (1930) studied the system Na_2SiO_3–Fe_2O_3–SiO_2. They pointed out that cooling liquids followed different paths toward several ternary eutectic points, depending on the degree of reaction between early-formed haematite and liquid. They showed that a low degree of fractionation led to an iron-enriched residual liquid and a high degree of fractionation led to a siliceous residual liquid. It was precisely in quickly cooled, little fractionated basalts, they maintained, that late crystallization of iron-rich liquid had been recognized by petrologists like Fenner. In contrast, they said, late-stage siliceous differentiates were typically associated with the more slowly cooled dolerites. Bowen *et al.* (1930) believed that they had demonstrated convincingly that strongly fractionated basaltic liquids proceeded in the direction of silica enrichment. Although Bowen had previously talked as if basaltic liquids always moved toward

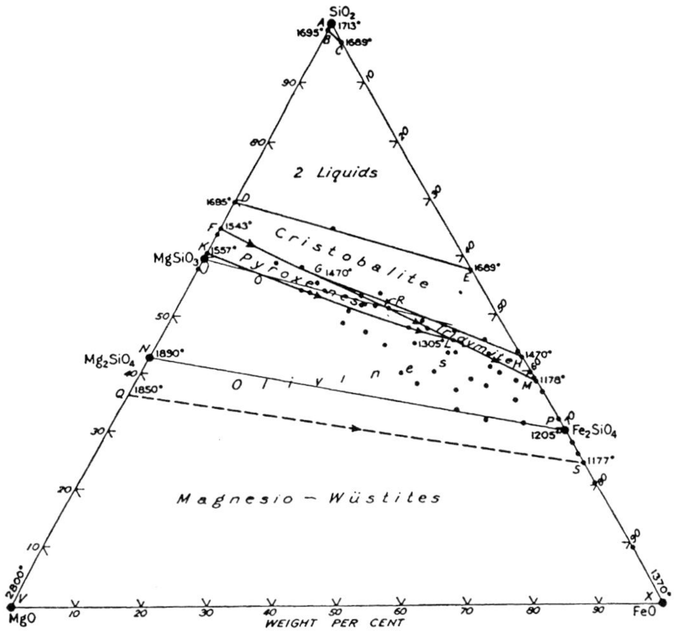

Fig. 3. Phase diagram of the ternary system $MgO–FeO–SiO_2$ at one atmosphere (Bowen & Schairer 1935). In this system, Bowen determined the ascendant loop of olivine and demonstrated that residual liquids could become saturated or undersaturated in silica depending on the degree of fractionation. Reprinted by permission of *American Journal of Science.*

granite, he now conceded that Fenner's iron-enrichment tendency was a possibility. He became an advocate of alternate iron-rich and silica-rich fractionation tracks while trying to make it look as if he had done so all along, but he was not ready to concede that the iron-enrichment trend was equally important as the silica-enrichment trend. Fenner had assumed that the iron enrichment effected by ferromagnesian solid solutions would prevail over the silica and alkali enrichment effected by plagioclase solid solutions. Bowen and his co-authors assumed exactly the opposite.

Bowen and Schairer shifted to the investigation of systems containing ferrous iron. They overcame the technical problem of controlling the oxidation state of iron by passing a slow stream of purified nitrogen through the furnace containing charges in crucibles of electrolytic iron. Consequently, all ferrous silicate systems that they investigated were in equilibrium with metallic iron. Using this method, Bowen and Schairer also found that iron-bearing melts always contained just a little Fe_2O_3 for most compositions. For practical purposes their results could still be portrayed on simplified diagrams.

The investigators began with $FeO–SiO_2$ (Bowen & Schairer 1932), establishing the melting-point of fayalite at 1205°C and demonstrating the pseudo-eutectic crystallization of tridymite and fayalite. There followed three more major studies on the systems $Ca_2SiO_4–Fe_2SiO_4$ (Bowen *et al.* 1933*a*), $CaO–FeO–SiO_2$ (Bowen *et al.* 1933*b*) and $MgO–FeO–SiO_2$ (Bowen & Schairer 1935). In the first two systems, relations among metasilicates ($CaSiO_3–FeSiO_3$), orthosilicates ($Ca_2SiO_4–Fe_2SiO_4$) and hedenbergite ($CaFeSi_2O_6$) were established. Bowen recognized that some of the incongruently melting minerals or solid solutions would melt to a more basic liquid enriched in iron plus tridymite. Rather than providing confirmation of iron enrichment during fractional crystallization, Bowen maintained that this discovery did not weaken the fact that silica was the normal late product of fractional crystallization. It simply emphasized that exceptions should be expected. He regarded the formation of an iron-rich residual liquid as one such exception.

In their report on the system $MgO–FeO–SiO_2$, Bowen & Schairer (1935) presented a liquidus diagram (see Fig. 3) containing fields of cristobalite and tridymite, of two liquids, and of three series of Mg–Fe solid solutions, the magnesiowüstites, the olivines and the pyroxenes. They found no ternary eutectics. They explored the orthosilicate join, Mg_2SiO_4–Fe_2SiO_4, and determined the solidus and liquidus of olivine up to 1500°C. Because iron crucibles melt at 1535°C, determinations at higher temperatures were not possible. From studies of compositions more iron-rich than 40% fayalite, in conjunction with the estimated melting-point of pure forsterite at 1890° ± 20°C, they established the now-familiar solid solution loop of the olivines. They also determined solidus and liquidus points on the metasilicate (pyroxene) join, $MgSiO_3$–$FeSiO_3$, and established the essentially solid solution loop behaviour of the pyroxenes as well as the temperatures at which clinopyroxenes invert to orthopyroxenes.

From the results, Bowen & Schairer (1935) argued that monomineralic pyroxene or olivine rocks were unlikely to be the result of direct crystallization from an ultrabasic magma because of the extremely high magmatic temperatures required and because of the lack of volatiles that might have lowered liquidus temperatures. That left Bowen's favoured notion of olivine or pyroxene accumulation from basaltic liquid with or without solid-state intrusion of the accumulated crystals as the only viable option for the formation of monomineralic ultramafic rocks.

Bowen & Schairer (1935) showed that residual liquids in this system could become undersilicated or oversilicated depending on the degree of fractionation. Hence, they suggested, some basalts might produce undersilicated, alkalic differentiates, although most basalts would produce oversilicated differentiates. Despite efforts to downplay the fact, their experimental studies of iron-bearing systems continued to provide at least some support for the observations of Fenner and Kennedy.

Because they were eager to investigate systems containing both ferrous iron and alkali, Bowen & Schairer (1936) studied the albite–fayalite join, where they encountered a eutectic point at which a large proportion of alkali feldspar crystallized with a small proportion of fayalite. Bowen concluded that fayalite-bearing rhyolite and phonolite from the African rift valleys were the end-product of the alleged iron-enrichment trend.

Because of the liquid cooling paths in virtually all systems that he and Schairer investigated, Bowen (1937) argued that the alkali–alumina silicate system is the 'goal' toward which fractional crystallization moves. Thus, in the classic contribution to emerge from this phase of his work, Bowen (1937) focused on the nature of liquids in 'petrogeny's residual system', $NaAlSiO_4$–$KAlSiO_4$–SiO_2 (see Fig. 4). He pointed out the existence of a thermal trough on the liquidus, and predicted that igneous rocks approaching the system in composition should plot within or very close to the trough. He plotted normative chemical compositions of 40 rhyolites, trachytes and phonolites from the East African rift valley on the diagram. All but three plotted within the thermal trough, and those three were very close. The experimental data, Bowen claimed, showed that rhyolites, trachytes and phonolites formed from residual liquids derived by fractional crystallization of complex magmas.

Work on ferrous silicate systems concluded with a paper on the system $NaAlSiO_4$–FeO–SiO_2 (Bowen & Schairer 1938). They encountered two important ternary eutectics. They showed that compositions in one part of the system would produce relatively silica-poor residual liquids that ultimately crystallized nepheline, and that compositions in another part of the system would produce relatively silica-rich residual liquids that ultimately crystallized tridymite. In both cases, they noted, the eutectic points were relatively enriched in alkali–alumina components and poor in the fayalite component. Bowen & Schairer (1938) believed that this system provided the solution to Fenner's question about the dominance of plagioclase crystallization, leading to enrichment of liquid in albite, versus olivine crystallization, leading to enrichment in fayalite. They concluded that the net result of fractional crystallization in the complex liquid of the system was enrichment in alkali–alumina silicate. Absolute enrichment in iron silicate did not occur, only enrichment in iron silicate relative to magnesium silicate. Bowen was confident that the major goal of crystallization–differentiation had now been experimentally established. Fractional crystallization, he was thoroughly convinced, ordinarily led to the formation of alkali–alumina–silica-enriched, not iron-enriched, liquids. Despite such compelling experimental results, the wind was taken out of Bowen's sails by the fact that the recently discovered coarse-grained, strongly fractionated Skaergaard Intrusion strikingly exhibited Fenner's iron-enrichment trend throughout most of its crystallization history (Wager & Deer 1939). Although the final differentiates of the intrusion were granophyres, their small volume posed a devastating challenge to Bowen's belief that granitic batholiths were derived from basaltic parents.

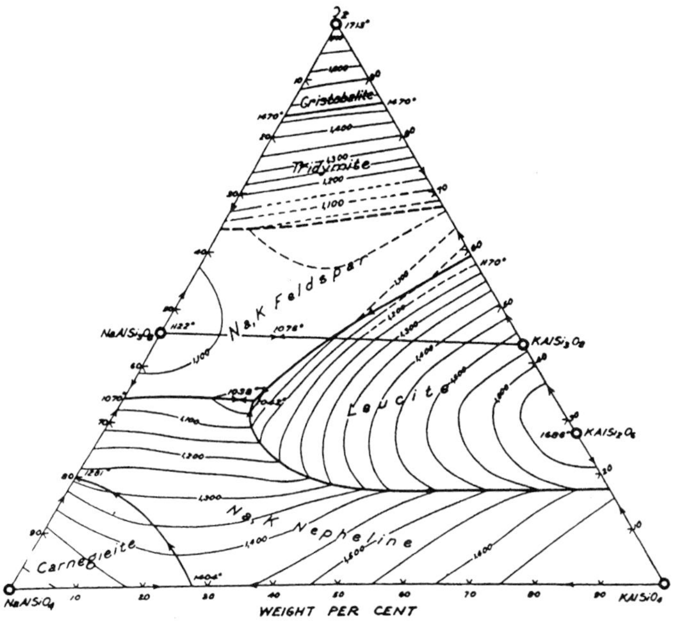

Fig. 4. Phase diagram of 'petrogeny's residual system', $NaAlSiO_4$–$KAlSiO_4$–SiO_2 at one atmosphere (Bowen 1937). Bowen argued that the composition of igneous rocks approaching this system ought to plot near the low-temperature trough on the liquidus surface. Reprinted by permission of *American Journal of Science.*

The final years

Bowen wanted to introduce the methods of experimental petrology into the academic world. After Arthur Day retired in 1936, Bowen left Washington a year later to become the Charles L. Hutchinson Distinguished Service Professor of Petrology at the University of Chicago. During his ten years at Chicago, Bowen established a small experimental laboratory and supervised a handful of doctoral students, but his own experimental research slowed considerably. World War II imposed other obligations on Bowen. Not until his return to the Geophysical Laboratory at the beginning of 1947 was he able to resume substantial experimental work. In the meantime, Bowen not only defended the origin of granite by fractional crystallization but even defended the existence of magmatic granite in opposition to the claims of the granitization school.

In the final phase of his career, Bowen added water to his systems and began to examine the effects of high pressure. He teamed up with Orville Frank Tuttle to study the system MgO–SiO_2–H_2O (Bowen & Tuttle 1949). The high-pressure hydrothermal apparatus designed by Tuttle was particularly valuable for the extensive studies of the granite system to four kilobars water pressure that the two carried out during the late 1940s and early 1950s. The results of their studies vindicated Bowen's contention that granites are magmatic and that granitization processes do not account for the majority of them. Together, Tuttle & Bowen (1958) found that the chemical compositions of granites plot very close to the thermal trough defined by the trend of minimum and eutectic points at low and medium water pressures. They concluded that granites represent the compositions of residual liquids in the granite system. Bowen had always favoured the notion that granite was the end-product of crystallization of basaltic liquid, but at the time of Bowen's death in 1956 Tuttle began to consider that the results of their experiments were compatible with the production of granitic melts by partial melting of deep crustal rocks because water dissolved in silicate melts drastically lowers their liquidus temperatures.

Over a lengthy career, three-quarters of which had been spent at the Geophysical Laboratory, Bowen had produced a staggering body of experimental studies that confirmed the notion that silicate liquid compositions may be changed in various ways by the selective removal of different amounts of different crystals at different rates during the cooling of magmas. For the most part, petrologists were persuaded that fractional crystallization had been established as an important process of differentiation in magmas by Bowen's dazzling array of experiments. Most were also persuaded that his work had demonstrated the magmatic origin of granite. Many were not so sure about granite as the end result of crystallization–differentiation.

Discussion

Bowen's theory left an indelible mark on contemporary igneous petrology (Yoder 1979; Hargraves 1980). His theory of fractional crystallization, of course, gained attention because of its inherent scientific merits and because it was promoted by a scientist of rare intellectual vigour, determination and literary skill. There is more to the story than that, however, because Bowen's influence owed as much to institutional and interpersonal factors and to his approach to science as it did to the content and evidential basis of his theory. The sheer comprehensiveness of the theory, like the natural selection theory of Darwin, who also first described crystal settling (Darwin 1844), guaranteed its great influence. Backed by an unending stream of precise experiments on a wide range of silicate compositions conducted under precisely controlled conditions including high pressure, Bowen's theory accounted for the origin of the large majority of igneous rocks. By varying the initial compositions of magmas, by varying the rates of cooling of those magmas to yield equilibrium or fractional crystallization, and by varying the extent and manner of fractionation of all kinds of crystals, Bowen's theory provided a means for generating almost any kind of silicate liquid. By segregation of crystals, the theory provided an explanation for monomineralic rocks. Because of its breath-taking sweep, igneous petrologists could not ignore the theory. Some were largely convinced, but at some point or other the theory touched on some aspect of magmatism in which someone other than Bowen was an expert. As a result, the theory provided ample opportunity for disagreement with particular features. The theory was so comprehensive that virtually every igneous petrologist had to interact with it in one way or another.

Bowen's ability to construct a theory of such comprehensiveness arose from his almost total focus on the problem of diversity throughout most of his career. Igneous petrology had reached a stage at which a scientist such as Bowen might emerge to specialize on this one problem throughout his career. Unlike most geologists until his time, with the possible exceptions of K. H. F. (Harry) Rosenbusch (1836–1914), Joseph P. Iddings (1857–1920) and Loewinson-Lessing, Bowen was interested almost exclusively in the igneous rocks. Bowen's renowned discourse on the metamorphism of siliceous carbonates was a temporary diversion, necessitated by the fact that he had not yet been able to establish a laboratory at the University of Chicago. Although other geologists had thought much about diversity, they devoted their energies to other concerns as well. Rosenbusch was consumed by descriptive microscopic petrography. Iddings was absorbed by fieldwork, petrography and the classification of the igneous rocks. He mapped igneous rocks in Yellowstone and used the mining districts of Nevada and was a principal architect of the American quantitative (CIPW) classification scheme. Alfred Harker (1859–1939) was interested in petrographic provinces, petrography, field studies of the Hebrides, and the production of textbooks on metamorphism and on petrography for students. Arthur Holmes (1890–1965) was fascinated by radioactivity and geochronology as much as by igneous rocks (see Lewis 2000, 2002). Daly was constantly looking for ways to relate igneous phenomena to tectonics, structure and geophysics, as for example in his contributions to the mechanics of igneous intrusion via magmatic stoping. Brøgger spent much of his time on politics, other facets of geology, and petrography and fieldwork, particularly on the igneous suites of the Oslo district in Norway. Frank F. Grout (1880–1958) was interested in stratiform lopoliths like the Duluth gabbro, the petrology and structure of granitoid batholiths, and the Precambrian geology of Minnesota. Victor M. Goldschmidt (1888–1947) (see Fritscher 2002) was passionately interested in the distribution of chemical elements, X-ray crystallography and crystal chemistry, and laboured incessantly to ascertain the values of ionic radii. These exceptional workers made very important contributions to the theory of diversity, but they were not in a position to propose such a comprehensive theory and back it up with such masses of data. Bowen discovered the seeds of his far-reaching concept in the plagioclase and the $MgO–SiO_2$ systems (Figs 1 and 2) early in his career and doggedly designed virtually all of his future experiments and theoretical arguments around the theme of fractional crystallization. He was so focused on his developing theory that he did not become distracted by mapping, writing textbooks, thinking about classification, or learning much about tectonics or metamorphism.

Bowen's single-minded focus is unthinkable anywhere but at the Geophysical Laboratory, the institution that provided a congenial environment for him to develop and apply his diverse gifts so remarkably. He did his doctoral work and spent more than three-quarters of his professional career at the Laboratory, supplied with the finest equipment and surrounded and assisted by other gifted experimentalists like Olaf Andersen, George Morey, Joseph Greig, Frank Tuttle and, above all, Frank Schairer. At the Geophysical Laboratory, Bowen was freed from the time-consuming preparation and delivery of lectures, the supervision of students, the grading of tests and papers, the drudgery of committee work, the tedium of administrative detail, and other distractions that are the portion of academicians. Moreover, Bowen was the beneficiary, just as he began his professional career, of three recent technical advances: the extension of the available temperature range to around 1550°C, the precise measurement of high temperatures by thermoelectric methods, and the application of the quench method to the determination of silicate phase equilibria. With these achievements in place at the Geophysical Laboratory, Bowen was largely free to determine phase relationships rather than overcome major technical obstacles.

Bowen would undoubtedly have carved out a distinguished scientific career as a professor, but his achievement would have been significantly lessened. While at Queen's University during 1919 and 1920, Bowen found that furnaces were lacking, despite administrative promises to supply him with such facilities. He had to borrow a petrographic microscope from the Geophysical Laboratory. Virtually unable to continue the experimental work he had been doing at the Geophysical Laboratory, Bowen contented himself with a series of optical studies of rare minerals. When Bowen left the Geophysical Laboratory in 1937 for the University of Chicago because of his desire to introduce experimental methods into the academic world, he succeeded in establishing a small laboratory and turning out a handful of PhD students. His own productivity declined, however, because of time consumed in establishing the laboratory, the demands of teaching, the supervision of doctoral students, and two years as chairman of the department. Bowen's experiences at Queen's and Chicago confirm that his productivity as an academician would have been much less than it actually was at the Geophysical Laboratory. In Bowen's case, the institution made the scientist.

Bowen's ties to the Geophysical Laboratory were, of course, the result of various personal influences. In the first place, he might never have gone to the Geophysical Laboratory. After graduating from Queen's Bowen had a great desire to travel to Norway for graduate study with Brøgger and Vogt. The disappointment of Vogt's rejection opened the way for Bowen to attend MIT, where Thomas Jaggar urged Bowen to consider doing experimental work for his doctoral dissertation at the Geophysical Laboratory.

More than any other individual, Arthur Day exercised a profound personal influence, both directly and indirectly, on Bowen. Day influenced Bowen indirectly through his own technical work. Bowen's phase-equilibrium studies would have been far less reliable had not Day spent the years from 1899 to 1911 extending the temperature scale to 1550°C by means of nitrogen gas thermometry and thermoelectric measurement calibrated to the gas thermometer at the Physikalisch-Technische Reichsanstalt in Germany, the United States Geological Survey Laboratory, and the Geophysical Laboratory. When Bowen began his PhD work in the fall of 1910, he was able to take full advantage of Day's technical achievements immediately.

More directly, Day urged Bowen to come to the laboratory for his doctoral work and recommended that he investigate the nepheline–anorthite system. After he finished his degree, Bowen was under pressure to leave the Geophysical Laboratory. Waldemar Lindgren urged Bowen to work for the United States Geological Survey. Jaggar wanted Bowen for the Hawaii Volcano Observatory. Bowen's experimental work, however, had been so productive and enjoyable that he had made a most favourable impression on Day and the rest of the staff of the Geophysical Laboratory. So when Day invited him to accept a staff position, Bowen decided to cast in his lot with the young research institution. The Geophysical Laboratory proved to be a perfect match for Bowen, a rather quiet, retiring person who lacked the charisma requisite for success as a college teacher. Day also provided constant encouragement for Bowen's early career. As soon as Bowen joined the staff of the Geophysical Laboratory, Day supported Bowen's decision to investigate the plagioclase feldspars. Day and Allen had previously undertaken detailed studies of the phase relations of the plagioclase feldspars, work that opened Bowen's eyes to the role of fractional crystallization in differentiation. Day made sure that Bowen was happy at the Geophysical Laboratory, keeping him well paid, often recommending a higher salary for him than for many of his colleagues. When Day returned to the Geophysical Laboratory after World War I, he went to great lengths to persuade Bowen to come back to the Geophysical Laboratory from

Queen's, and after Bowen returned to Washington, Day made sure that Bowen received generous salary increases whenever possible. Because there is little doubt that Bowen's career would have taken a considerably different course had he not spent most of his career at the Geophysical Laboratory, twentieth-century igneous petrology owes an enormous debt to Arthur Day, not only for technical achievements that made it possible for experimentalists like Bowen to obtain such dramatic results, but also for bringing Bowen to the Geophysical Laboratory on three different occasions, for doing all he could to keep him there, and providing him with strong encouragement throughout his career.

Looming over twentieth-century igneous petrology is the shadow of a scientist of single-minded purpose who spent most of his career at an institution that was ideally suited to his talents and temperament and who was guided by an individual of rare ability to judge, develop, and encourage exceptional scientific ability.

Appreciation is due to H. S. Yoder Jr for providing a review of the manuscript. My work on Bowen and the history of igneous petrology has been supported by grants SBR-9601203 and SES-9905627 from the Science and Technology Studies Program of the National Science Foundation.

References

BÄCKSTRÖM, H. 1893. Causes of magmatic differentiation. *Journal of Geology*, **1**, 773–779.

BECKER, G. F. 1897*a*. Some queries on rock differentiation. *American Journal of Science*, Series 4, **3**, 21–40.

BECKER, G. F. 1897*b*. Fractional crystallization of rocks. *American Journal of Science*, **4** Series 4, 257–261.

BOWEN, N. L. 1913. The melting phenomena of the plagioclase feldspars. *American Journal of Science*, Series 4, **35**, 577–599.

BOWEN, N. L. 1914. The ternary system: diopside–forsterite–silica. *American Journal of Science*, Series 4, **38**, 207–264.

BOWEN, N. L. 1915*a*. Crystallization-differentiation in silicate liquids. *American Journal of Science*, Series 4, **39**, 175–191.

BOWEN, N. L. 1915*b*. The crystallization of haplobasaltic, haplodioritic, and related magmas. *American Journal of Science*, Series 4, **40**, 161–185.

BOWEN, N. L. 1915*c*. The later stages of the evolution of the igneous rocks. *Journal of Geology*, **23**, Supplement to No. **8**, 1–91.

BOWEN, N. L. 1917. The problem of the anorthosites. *Journal of Geology*, **25**, 209–243.

BOWEN, N. L. 1919. Crystallization-differentiation in igneous magmas. *Journal of Geology*, **27**, 393–430.

BOWEN, N. L. 1921. Diffusion in silicate melts. *Journal of Geology*, **29**, 295–317.

BOWEN, N. L. 1922*a*. The reaction principle in petrogenesis. *Journal of Geology*, **30**, 177–198.

BOWEN, N. L. 1922*b*. The behavior of inclusions in igneous magmas. *Journal of Geology*, **30**, 513–570.

BOWEN, N. L. 1927. The origin of ultra-basic and related rocks. *American Journal of Science*, Series 5, **8**, 89–108.

BOWEN, N. L. 1928. *The Evolution of the Igneous Rocks*. Princeton University Press, Princeton.

BOWEN, N. L. 1937. Recent high-temperature research on silicates and its significance in igneous geology. *American Journal of Science*, Series 5, **33**, 1–21.

BOWEN, N. L. & ANDERSEN, O. 1914. The binary system $MgO–SiO_2$. *American Journal of Science*, Series 4, **37**, 487–500.

BOWEN, N. L. & SCHAIRER, J. F. 1929. The fusion relations of acmite. *American Journal of Science*, **18** Series 5, 365–374.

BOWEN, N. L. & SCHAIRER, J. F. 1932. The system, $FeO–SiO_2$. *American Journal of Science*, Series 5, **24**, 177–213.

BOWEN, N. L. & SCHAIRER, J. F. 1935. The system, $MgO–FeO–SiO_2$. *American Journal of Science*, Series 5, **29**, 151–217.

BOWEN, N. L. & SCHAIRER, J. F. 1936. The system, albite–fayalite. *Proceedings of the National Academy of Sciences*, **22**, 345–350.

BOWEN, N. L. & SCHAIRER, J. F. 1938. Crystallization equilibrium in nepheline–albite–silica mixtures with fayalite. *Journal of Geology*, **46**, 397–411.

BOWEN, N. L. & TUTTLE, O. F. 1949. The system $MgO–SiO_2–H_2O$. *Bulletin of the Geological Society of America*, **60**, 439–460.

BOWEN, N. L., SCHAIRER, J. F. & POSNJAK, E. 1933*a*. The system, $Ca2SiO_4–Fe_2SiO_4$. *American Journal of Science*, Series 5, **25**, 273–297.

BOWEN, N. L., SCHAIRER, J. F. & POSNJAK, E. 1933*b*. The system, $CaO–FeO–SiO_2$. *American Journal of Science*, Series 5, **26**, 193–284.

BOWEN, N. L., SCHAIRER, J. F. & WILLEMS, H. W. V. 1930. The ternary system: $Na_2SiO_3–Fe_2O_3–SiO_2$. *American Journal of Science*, Series 5, **20**, 405–455.

BRØGGER, W. C. 1890. Die Mineralien der Syenitpegmatitgange der südnorwegischen Augit- und Nephelinsyenite. *Zeitschrift für Krystallographie*, **16**, 1–235.

BRØGGER, W. C. 1894. The basic eruptive rocks of Gran. *Quarterly Journal of the Geological Society of London,* **50**, 15–38.

BUNSEN, R. 1851. Über die Processe der vulkanischen Gesteinsbildung Islands. *Annalen der Physik und Chimie*, **83**, 197–272.

DALY, R. A. 1914. *Igneous Rocks and their Origin*. McGraw–Hill, New York.

DARWIN, C. 1844. *Geological Observations on the Volcanic Islands*. Smith Elder, London.

DAY, A. L., ALLEN, E. T. & IDDINGS, J. P. 1905. *The Isomorphism and Thermal Properties of the Feldspars*. Carnegie Institution of Washington, Washington.

DOELTER, C. 1903. Beziehungen zwischen Schmelzpunkt und chemischer Zusammensetzung der Mineralien. *Tschermaks Mineralogische und Petrographische Mitteilungen*, **22**, 297–321.

DUROCHER, J. 1857. Essai de pétrologie comparée ou recherches sur la composition chimique et minéralogique des roches ignées, sur les phénomènes de leur émission et sur leur classification. *Annales des Mines*, **11**, 217–259.

FENNER, C. N. 1926. The Katmai magmatic province. *Journal of Geology,* **34**, 673–772.

FENNER, C. N. 1929. The crystallization of basalts. *American Journal of Science*, Series 5, **18**, 225–253.

FRITSCHER, B. 2002. Metamorphism and thermodynamics: the formative years. *In*: OLDROYD, D. (ed.) *The Earth Inside and Out: Some Major Contributions to Geology in the Twentieth Century*. Geological Society, London, Special Publications, **192**, 143–166.

GESCHWIND, C.-H. 1995. Becoming interested in experiments: American igneous petrologists and the Geophysical Laboratory, 1905–1965. *Earth Sciences History*, **14**, 47–61.

GOUY, L. G. & CHAPERON, G. 1887. Sur la concentration des dissolutions par la pesanteur. *Annales de Chimie et de Physique*, **12**, 384–393.

GRIEG, J. W. 1927. Immiscibility in silicate melts. *American Journal of Science*, Series 5, **13**, 1–44, 133–154.

GRIEG, J. W., SHEPHERD, E. S. & MERWIN, H. E. 1931. Melting temperatures of granite and basalt. *Carnegie Institution of Washington Year Book*, **30**, 75–78.

HARGRAVES, R. B. (ed.) 1980. *Physics of Magmatic Processes*. Princeton University Press, Princeton.

HARKER, A. 1893. Berthelot's principle applied to magmatic concentration. *Geological Magazine*, **10**, 546–547.

HARKER, A. 1894. Carrock Fell: a study in the variation of igneous rock-masses. Part I. The gabbro. *Quarterly Journal of the Geological Society of London*, **50**, 311–337.

HARKER, A. 1932. *Metamorphism: a Study of the Transformation of Rock Masses*. Methuen, London

HUNT, T. S. 1854. Illustrations of chemical homology. *Proceedings of the American Association for the Advancement of Science*, 237–247.

IDDINGS, J. P. 1892. The origin of igneous rocks. *Bulletin of the Philosophical Society of Washington*, **12**, 89–213.

JUDD, J. W. 1886. On the gabbros, dolerites, and basalts, of Tertiary age, in Scotland and Ireland. *Quarterly Journal of the Geological Society of London*, **42**, 49–97.

JUDD, J. W. 1893. On composite dykes in Arran. *Quarterly Journal of the Geological Society of London*, **49**, 536–564.

KENNEDY, W. Q. 1933. Trends of differentiation in basaltic magmas. *American Journal of Science*, Series 5, **25**, 239–256.

LAGORIO, A. 1887. Über die Natur der Glasbasis, sowie der Krystallisationsvorgänge im eruptiven Magma. *Tschermaks Mineralogische und Petrographische Mitteilungen*, **8**, 421–529.

LEWIS, C. 2000. *The Dating Game: One Man's Search for the Age of the Earth.* Cambridge University Press, Cambridge.

Lewis, C. L. E. 2002. Arthur Holmes' unifying theory: from radioactivity to continental drift. *In*: OLDROYD, D. (ed.) *The Earth Inside and Out: Some Major Contributions to Geology in the Twentieth Century*. Geological Society, London, Special Publications, **192**, 167–184.

LOEWINSON-LESSING, F. Y. 1899. *Studien über die Eruptivgesteine. VII International Geological Congress, 1897*. M. Stassulewitsch, St Petersburg.

LOEWINSON-LESSING, F. Y. 1911. The fundamental problems of petrogenesis, or the origin of the igneous rocks. *Geological Magazine*, **8**, 248–257, 289–297.

ROSENBUSCH, H. 1889. Über die chemische Beziehungen der Eruptivgesteine. *Tschermaks Mineralogische und Petrographische Mitteilungen,* **11**, 144–178.

SARTORIUS VON WALTERSHAUSEN, W. 1853. *Über die Vulkanischen Gesteine in Sicilien und Island and ihre Submarine Umbildung*. Dieterich Buchhandlung, Göttingen.

SERVOS, J. W. 1990. *Physical Chemistry from Ostwald to Pauling: The Making of a Science in America.* Princeton University Press, Princeton.

SHEPHERD, E. S., RANKIN, G. A. & WRIGHT, F. E. 1909. The binary systems of alumina with silica, lime and magnesia. *American Journal of Science*, Series 4, **28**, 293–333.

SORET, C. 1879. Sur l'état d'équilibre que prend au point de vue de sa concentration une dissolution saline primitivement homgène dont deux parties sont portées a des températures différentes. *Archives des Sciences Physiques et Naturelles*, **27**, 48–61.

SORET, C. 1881. Sur l'état d'équililbre que prend au point de vue de sa concentration une dissolution saline primitivement homogène dont deux parties sont portées, a des températures différentes. *Annales de Chimie et de Physique*, **22**, 293–297.

TEALL, J. J. H. 1885. The metamorphosis of dolerite into hornblende-schist. *Quarterly Journal of the Geological Society of London*, **41**, 133–144.

TEALL, J. J. H. 1888. *British Petrography*. Dulau, London.

TSCHERMAK, G. 1864. Chemisch-mineralogische Studie I: Die Feldspatgruppe. *Sitzungsberichte Akademie Wissenschaften Wien*, **50**, 566–613.

TUTTLE, O. F. & BOWEN, N. L. 1958. *Origin of Granite in the Light of Experimental Studies in the System* $NaAlSi_3O_8$–$KAlSi_3O_8$–SiO_2–H_2O. Geological Society of America Memoir **74**.

VOGT, J. H. L. 1891. Om Dannelsen af de vitigste in Norge og Sverige representerede grupper af jemmalmforeskomster. *Geologiska Foreningens i Stockholm Forhandlingar*, **13**, 476–536.

VOGT, J. H. L. 1903. *Die Silikatschmelzlösungen*. Jacob Dybwad, Christiania.

WAGER, L. R. & DEER, W. A. 1939. Geological investigations in East Greenland: Part III. The petrology of the Skaergaard Intrusion, Kangerdluqssuaq, East Greenland. *Meddelelser om Grønland*, **105**, 1–352.

YODER, H. S. Jr (ed.) 1979. *The Evolution of the Igneous Rocks: Fiftieth Anniversary Perspectives*. Princeton University Press, Princeton.

YODER, H. S. Jr 1992. Norman L. Bowen (1887–1956), MIT class of 1912, first predoctoral fellow of the Geophysical Laboratory. *Earth Sciences History*, **11**, 45–55.

YOUNG, D. A. 1998. *N. L. Bowen and Crystallization-Differentiation: the Evolution of a Theory*. Mineralogical Society of America, Washington, DC.

Metamorphism today: new science, old problems

JACQUES L. R. TOURET[1] & TIMO G. NIJLAND[2]

[1]*Department of Petrology, Vrije Universiteit, De Boelelaan 1085, 1081 HV Amsterdam, The Netherlands*

[2]*Rooseveltlaan 964, 3526 BP Utrecht, The Netherlands*

Abstract: A concise history of the discipline of metamorphic petrology is presented, from the eighteenth-century concepts of Werner and Hutton to the end of the twentieth century.

At the beginning of the twenty-first century, can we speak of a crisis in metamorphic petrology? Only a few years ago, it was still considered to be one of the most 'scientific' branches of the Earth sciences, flourishing in all major universities. It was a time when, in a few places, metamorphic petrologists were given official positions in chemistry or physics departments, as the best possible specialists for a discipline like equilibrium thermodynamics, traditionally considered an integral part of chemistry. Currently, the situation is completely different. The irruption of 'exact' sciences in the traditionally 'descriptive' biological and terrestrial disciplines, has been marked by a profusion of new terms such as biogeochemistry and associated 'new' disciplines, all claiming to be drastically different from their predecessors and seeking recognition and independence. Added to a pronounced change in scientific priorities, caused by a growing awareness of the fragility of our environment and the uncertain fate of future generations, the result is an obvious decline in some topic areas, among which is metamorphic petrology. The large population of metamorphic petrologists that was hired during the golden years of university expansion after World War II is now slowly disappearing without being replaced, and public and private funding is redirected to apparently more urgent problems, mostly dealing with the environment.

However, among the three rock types occurring at the Earth's surface or accessible to direct observation in the outer layers of our planet (sedimentary, magmatic, metamorphic), metamorphic rocks are by far the most abundant. Sediments only make up a thin, discontinuous layer at the Earth's surface. Magmas are (partly) formed at depth by partial melting of former metamorphic rocks, but this melting is local, limited in time and space. After crystallization, most volcanic and plutonic rocks are reworked and transformed into metamorphic rocks. The Earth is in constant evolution, characterized by permanent continental masses and temporary oceans, created and collapsing at a timescale of few hundred million years. The oceanic crust, created by magmatic eruptions at mid-ocean ridges, is to a large extent – at least 80% in volume – transformed into metamorphic rocks by sea-floor hydrothermal alteration. So, all together, it is not an exaggeration to claim that most rocks that we can observe are metamorphic. Yet, if the present trend continues, metamorphic petrology will soon join other 'ancient' disciplines, like mineralogy and palaeontology, on the list of endangered species in today's competitive university world.

It is true that metamorphic petrology has always had problems in finding its right place between its neighbours, magmatic and sedimentary petrology, with which it partly overlaps. This is probably one of the reasons why, a century apart, two prominent petrologists have felt the need to make an extensive review of the historical development of their discipline: Gabriel Auguste Daubrée (1857, 1859) and Akiho Miyashiro (1973, 1994), and many others essays can be found (e.g. Hunt 1884; Williams 1890; Yoder 1993). Metamorphic petrologists ourselves, we have drawn on the work of these illustrious predecessors, without attempting to go into the detail of their investigations. To cover everything would require more than one book. We have, however, tried to identify the most important lines of research and thinking, showing that despite considerable developments in methodology, instrumentation and interpretation, some basic questions keep recurring, and probably will do so for years to come.

Metamorphism and magmatism: from the beginning, not easy to define limits and relations

Even now, it is not easy to define metamorphic rocks so as to distinguish them unambiguously

From: OLDROYD, D. R. (ed.) 2002. *The Earth Inside and Out: Some Major Contributions to Geology in the Twentieth Century*. Geological Society, London, Special Publications, **192**, 113–141. 0305-8719/02/$15.00

from sedimentary or magmatic rocks. Metamorphic rocks derive from 'protoliths' (sedimentary, magmatic or metamorphic) formerly exposed at the surface, buried at lesser and greater depths during the subsiding of sedimentary basins or the formation of mountain chains, then brought back to the surface by erosion. Changing pressure and temperature conditions lead to the formation of new minerals, typically formed through (fluid-assisted) solid-state recrystallization. In the early stages, most newly formed minerals are platy (chlorites, micas), and they define a new rock structure/texture: schistosity for low-grade metamorphic rocks (transition from pelite (sediment) to slate, and then to schist); foliation for high-grade rocks (gneiss). But any petrologist knows that structural elements alone cannot give a precise definition, which relies essentially on the presence of characteristic minerals: zeolites at the beginning of metamorphism; and, at highest temperatures, minerals like pyroxene or garnet, which result in rocks devoid of oriented structures. This is the domain of granulites, where metamorphic temperatures can reach 1000°C or more, overlapping the magmatic domain. For these rocks, the distinction between magmatic and metamorphic rocks is by no means clear-cut. Metamorphic rocks, in principle, should not have passed through a melting stage. But partial (or total) melting is common at these high temperatures, resulting in an intricate mixture of both types (migmatites). Moreover, magmatic rocks, once crystallized at depth, may have subsequently been deformed and recrystallized, becoming a new category of metamorphites (orthoderivates). In such cases, the precise characterization of the different rock types requires an advanced knowledge of the conditions of their formation, notably the timing at which the different events have occurred. Is the magmatic rock, with granite as the typical example, the cause of metamorphism, provoking mineral recrystallization at its contact? Or is it its result, the ultimate product of metamorphic transformation? In this respect, metamorphism is closely related to the 'granite problem', a major source of discussion among petrologists for nearly two centuries.

Metamorphism in the period of Neptunism and Plutonism

The Neptunist scheme, proposed by Abraham Gottlob Werner (1774) and developed in the writings of his students such as Jean François d'Aubuisson des Voisins (1819), had little place for what we would call metamorphism. Every rock type was deposited in a stratified form at a given time. Even 'hard rocks' like schist and granite were supposedly deposited from a hypothetical 'primitive' ocean, hotter and more concentrated than the present-day, 'post-Flood' ocean. As observed by Gabriel Gohau (in Bonin *et al.* 1997), this scheme was linear overall, each epoch being characterized by a specific rock type (though Werner did envisage rises and falls of his ocean, and different conditions of storm and calm, to allow for divergences from his general 'directionalist' scheme). The oldest rock was thought to be granite, and the evolution was essentially irreversible: there was only one epoch for the formation of granite, as well as all non-fossiliferous rocks (gneiss, schists), all regarded as 'primitive rocks'.

At the turn of the nineteenth century, Werner's prestige and influence were such that most of continental Europe had accepted his views, despite the fact that students of the French Massif Central, notably Faujas de Saint Fond and Desmarest, had recognized the igneous origin of basalts. But the Scotsman James Hutton went much further. According to his thinking, not only basalt, but even granite, the fundament of the Wernerian system, was an igneous rock, a kind of lava that might be younger than the surrounding rocks. Hutton's *Theory of the Earth*, first published in 1788 and then elaborated in a book of the same title in 1795, corresponded, at least from an early twentieth-century perspective, to the only true 'revolution' that Earth sciences have known (Von Zittel 1899; Geikie 1905). Not only lavas, but also 'plutonic' rocks, notably granite, were supposedly made by fire, at any epoch of the Earth's history, provided that adequate physical conditions (notably temperature) were attained. Note that Hutton remained rather vague about the location and cause of this fire. He simply referred to subterraneous fire or heat and argued that, as the reality of heat was demonstrable by its effects, it was unnecessary to search for its cause. In fact, in this respect Hutton was not a great distance from Werner, who had explained present-day basalt, the only volcanic rock that he recognized, by the underground combustion of coal deposits. For instance, Hutton stated that combustible rocks, issued from the vegetal remnants in sediments, constituted an inexhaustible heat source (Gohau, in Bonin *et al.* 1997).

Yet Hutton's ideas led to the notion of metamorphism. In the Isle of Skye, he had observed that lignite at the contact of basalt was transformed into shiny coal, from which he inferred the igneous origin of basalt. However, he did not use the word 'metamorphism', at least in the sense that it has today (The term 'metamorphosed' is to be found in the *Theory of the Earth* (Hutton 1795, vol. 1, p. 504), but in the context of a long citation (in French) from Jean Philipe Carosi, about the supposed formation of flint ('silex') from a 'calcareous body' under the influence of running water, a notion which Hutton rejected.)

Hutton's ideas were not immediately accepted by the whole scientific community. Several of Werner's students, notably Leopold von Buch, were convinced of the igneous origin of basalt after having seen the active volcanoes in Italy. However, as late as 1863, most popular geology books in France (e.g. Figuier (1863), which was soon translated in neighbouring countries (Beima 1867)), were still much influenced by the Wernerian system. Hutton himself, who had initially studied medicine and then agronomy, before turning to geology, was considered to be an amateur by much of the European establishment. A 'Wernerian Society' was even created in Edinburgh not long after Hutton's death (1803), with the goal of expounding and defending the ideas of the old master of Freiberg. But Hutton found two dedicated disciples, John Playfair and, after his death, Charles Lyell, who proved to be lucid and prolific writers and finally achieved a wide acceptance of his views.

It is remarkable to see how much the dispute relied on theoretical arguments, with only a few people taking a more empirical approach, resorting to the examination of field exposures to decide between both systems. George Bellas Greenough, first president of the Geological Society of London, who travelled through Scotland equipped with Playfair's (1802) exposition of Hutton's work and the Wernerian-inspired *Mineralogy of the Scottish Isles* by Robert Jameson (1800), was a notable exception, but he found the evidence inconclusive (Rudwick 1962). Hutton's friend Sir James Hall (1805, 1812, 1826) sought to carry out experiments to test Hutton's ideas, but without total success.

In the last volume of the first edition of his *Principles of Geology* (Lyell 1833, pp. 374–375), Lyell claimed the paternity of the term 'metamorphism'. Daubrée (1857) gave the year 1825 as the first introduction of the term by Lyell, but despite careful search, Gohau (in Bonin *et al.* 1997) was unable to find the original reference.

A difference of a few years is not really of great importance: the idea was already 'in the air'. Before 1833, the name (often in a slightly different form, '*métamorphose*'), had already been used by a number of other authors, including Ami Boué (1820, 1824) and Léonce Elie de Beaumont (1831) in France. In fact, it seems that the contribution of Ami Boué to the birth of the concept of metamorphism from Hutton's theory is far more important than those of Lyell and Elie de Beaumont, but his writings, still rather difficult to find today, remained relatively 'confidential' (G. Godard pers. comm.). This was not the case with Elie de Beaumont, a powerful and authoritative figure at a time of French economic prosperity, who had been a good field geologist in his younger days, responsible with Dufrénoy for the first edition of the *Carte géologique de la France*. His great idea, developed from the theory of 'central fire' of Fournier (1820, 1837) and Cordier (1828), who themselves developed earlier concepts adumbrated by Descartes, Leibniz and Buffon (Green 1992), was that the Earth had cooled progressively, leading to a thickening of the crust and shrinkage of the outer envelope 'to stay in contact with the molten core' (Elie de Beaumont 1831). In 1833, in his lectures at the Collège de France, he introduced the notion of 'ordinary metamorphism' ('*métamorphisme normal*') 'for the transformations occurring at the bottom of the oceans under the influence of the incandescent core' and 'extraordinary metamorphism' ('*métamorphisme anormal*'), produced by temperature changes at contacts with igneous masses. *Métamorphisme normal* still relied on a vague notion of a Wernerian '*Urozean*', whereas *métamorphisme anormal* was much closer to contact metamorphism as we know it today. The terminology introduced by Elie de Beaumont was soon modified by two French colleagues, leading to the names still used today. Daubrée (1857), who developed the experimental approach initiated by Hall at the time of Hutton, called ordinary metamorphism '*régional*', as opposed to *métamorphisme de juxtaposition* (the *métamorphisme anormal* of Elie de Beaumont) caused by the proximity of eruptive rocks. Daubrée recognized that the latter, soon called 'contact metamorphism' in the international literature, resulted in a loss of pre-existing structure, whereas regional metamorphism led to foliation (*feuilletage*). This regional metamorphism might occur at different times. Thus, in this respect, Daubrée (1857) was close to some views defended by Lyell. However, for pre-Silurian rocks, he still invoked a 'primitive' metamorphism, which was different from any Lyellian or modern concept. In all cases, temperature (only approximately estimated at that time) was not considered to be a dominant factor. Daubrée, with Elie de Beaumont at the Paris Ecole des Mines, then the major geological centre in

France, was impressed by minerals deposited from thermal spas, notably at Plombières in the Vosges (Daubrée 1857). Thus, together with most of his colleagues, he thought that most recrystallizations at depth were induced by circulating solutions. Even granite was thought to be produced by 'aqueous plasticity', not igneous melting (Breislak 1822). Daubrée's ideas were not that different from Werner's conceptions, except that the '*Urozean*' was not thought to be at the Earth's surface, but hidden at depth.

Other scientists were following Hutton more closely regarding the major role of fire and, above all, the uniformity of physical conditions since the beginning of Earth's history. These contrasting views led to controversy, well illustrated by an exchange of notes between Joseph Durocher (1845) with Joseph Fournet (1848), the major defendant of magmatic theories, and Theodor Scheerer (1847), who had joined the Ecole des Mines group from Scandinavia. It would take too long here to report the details of this debate, but essentially it dealt (already!) with the question of the metamorphic or magmatic nature of granite, a recurrent debate which was to rekindle in the twentieth century (see summary by Gohau in Bonin *et al.* (1997, pp. 37–45)). Here, we may only mention that the most extreme 'hydrothermalist' was Achille Delesse, also related to the Ecole des Mines group. His book on metamorphism (Delesse 1857), first printed as a series of papers in the *Annales des Mines*, was later taken as their original reference source by the 'transformist' school. Delesse preferred the name 'general' rather than the normal or regional metamorphism of Daubrée and Elie de Beaumont, and 'special' for contact metamorphism. The first type was characterized by its regional scale, and a usually unseen cause. The second occurred at contacts with volcanic or plutonic rocks. But, in all cases, temperature was not considered to be an important factor. Delesse thought that only effusive lavas were true igneous rocks. But, in most cases, these had little influence on the surrounding rocks. Consequently, igneous rocks were not regarded as a cause of metamorphism; they were not igneous, but, like the surrounding gneiss, were the ultimate product of metamorphism. They could supposedly be formed almost at room temperature under the action of appropriate circulating solutions.

For his demonstration, besides observations which were, indeed, not irrelevant (e.g. the absence of indications of mutual influence between granite and gneiss), Delesse used arguments that may sound surprising today. For instance, granite must soften at the sea shore, as it is easily penetrated by sea-weed! Together with water, under great pressure but at moderate temperature, all rocks which are not clearly volcanic lavas could form from 'a very fluid muddy-paste' ('*une pâte boueuse trés fluide*'), analogous to a cement. Metamorphism occurred during the consolidation of this 'paste' and affected both the surrounding rocks ('*métamorphisme éverse*' or '*exomorphisme*') as well as the plutonic rock itself ('*métamorphisme inverse*' or '*endomorphisme*').

The golden (German) era of descriptive petrography

France was defeated by Prussia in 1870, and French scientists were soon to lose their preeminence on the international scene. Strasbourg, now at the western border of the German nation, became a major university, with a mineralogy chair occupied by Harry Rosenbusch, who together with Ferdinand Zirkel from Leipzig and some others created modern descriptive petrography. The polarizing microscope and techniques of sample preparation (thin sections), elaborated by a small group of British scientists (Davy, Brewster, Nicol and Sorby) during the first half of the century, were by then of high quality, and were to remain largely unchanged for many years. For more than fifty years – the first edition of the *Mikroskopische Physiographie der Mineralien und Gesteine* was published in 1873 and the last in 1929, well after his death – Rosenbusch compiled a descriptive catalogue of all magmatic and metamorphic rock types, worldwide. Discussion of magmatic rocks occupied by far the most important place: more than four-fifths of the *Physiographie*. But he also showed a keen interest in metamorphic rocks, and one of his major Strasbourg achievements was to study the contact aureole of the Andlau granite, in the Vosges (Rosenbusch 1877) (see Fig. 1). Rosenbusch identified several successive zones, based on the rock structure (schists, knotted schists, hornfelses). Contact metamorphism could be clearly related to heating by the intrusive granite. The same process could occur on a larger scale, if caused by a continuous, hidden layer of granite at the base of the continents. This was so evident for Rosenbusch that he did not consider any type other than contact metamorphism for the clay-rich sediments (pelites), which show the most obvious changes during progressive metamorphism. He observed that rocks in the contact aureoles around the Andlau massif did not contain feldspar, and he regarded this an

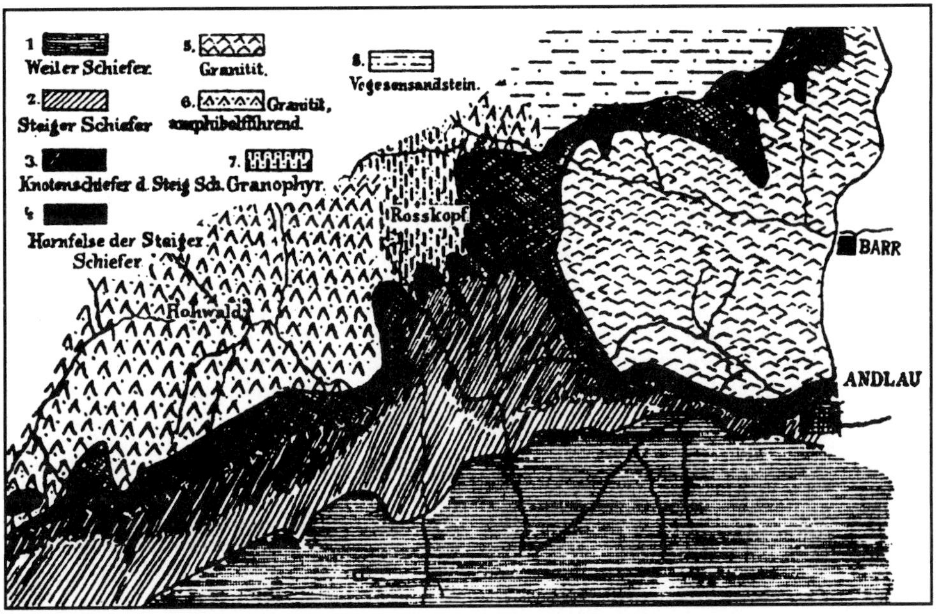

Fig. 1. Contact metamorphism of the Barr-Andlau granite, Vosges (Rosenbusch 1877).

essential feature of contact metamorphism. However, feldspars are major constituents of most rocks occurring in areas of regional metamorphism, which therefore had to be fundamentally different. Rosenbusch ascribed the acquisition of gneissose structure to deformation, mostly of former igneous rocks, and defined the new concept of 'dynamometamorphism'. Both types could be independent, but in general they occurred successively, dynamometamorphism being superimposed on former contact metamorphism to give the typically foliated texture.

It is interesting to note that Rosenbusch's ideas on dynamometamorphism derived directly from some experiments by Daubrée, who showed that deformation could generate heat. However, despite the prominent position of Daubrée in his country's academic system, dynamometamorphism did not become popular in France. The ideas of Rosenbusch were vigorously discussed in France by the followers of Delesse and Elie de Beaumont, notably Alfred Michel-Lévy. Together with Ferdinand Fouqué, who was trained by Rosenbusch himself, Michel-Lévy brought a major contribution to the theory of polarization microscopy. Both authors wrote a book on the determination of the rock-forming minerals – the French equivalent of the *Mikroskopische Physiographie* – which, although it had not the encyclopedic character of the treatise of the master of Heidelberg, attached much greater importance to the determination of feldspars (Fouqué & Michel-Lévy 1878; Michel-Lévy 1888). This had major consequences, not only for igneous rock classification (for the French based on feldspar composition; for the Germans on the colour index), but also for the conception of metamorphism. Michel-Lévy (1887) found feldspar in the contact aureole of the Flamanville granite in Normandy. In consequence, there was, in his view, no fundamental difference between contact and regional metamorphism. He eliminated the old notion of '*terrains primitifs*', a relic from Werner's belief that metamorphism (as we would call it) depended on age and occurred under conditions essentially different from today. Feldspathization could occur at any time, mostly under the influence of '*émanations*' issued from a mysterious source at depth. Deformation was unimportant: '*les actions mécaniques déforment, mais ne transforment pas*' (De Lapparent 1906, p. 1945). This citation is almost literally taken from Pierre Termier (1903: '*les actions dynamiques déforment, mais elles ne transforment point*'), who reached international celebrity with his concept of '*colonnes filtrantes*'. This idea was derived from the observation that, in the Alps, synclinal structures are more strongly metamorphosed and 'feldspathized' than anticlines, supposedly because they were closer to 'vapours emanating from an underlying eruptive centre'.

The first attempts at global interpretation: stress/anti-stress minerals, and depth zones

At the beginning of the twentieth century, descriptive petrography was sufficiently developed to attempt some kind of general interpretation. Rosenbusch had identified successive zones in contact metamorphism, but mainly on structural/textural grounds. The more important observation that regular mineral changes might also occur in regional metamorphism soon followed, albeit hampered by lack of communication between the different schools.

First observations were made by George Barrow (1893) in the Scottish Highlands (Fig. 2). Barrow was a self-taught field geologist employed by the Geological Survey, who had, however, studied science at King's College London and learnt much from George P. Scrope, for whom he acted as an amanuensis. Barrow found a regular sequence of changes in the mineralogy of metamorphic rocks close to a granite intrusion. He defined 'successive areas', based on the occurrence of different aluminium silicates: sillimanite, kyanite, staurolite. This work was politely discussed during its oral presentation at a meeting of the Geological Society – notably by the young Alfred Harker, who was to revisit the issue some years later (Harker 1918) – but it remained more or less unnoticed in the published literature of the time. (Barrow was in dispute with his Survey colleagues about a number of issues, which may account for his ideas being disregarded or discounted for several years.) In 1915, a similar approach was taken by Victor Moritz Goldschmidt in the Trondheim area, Norway (see Fritscher 2002), but without being aware of Barrow's work. So Barrow's ideas were forgotten or ignored for a couple of decades, being eventually resuscitated by Cecil Tilley (1925) and subsequently by Harker himself (Harker 1932). At this time, Barrow's zones were 'completed', with the addition of chlorite, biotite, staurolite and garnet to the index minerals.

It is important to note that the relation to contact metamorphism, which was obvious in the original discovery ('silicates of alumina which are connected to the intrusion'), was then replaced by the notion of regional metamorphism. Harker (1918, 1932), who was extremely influential until the 1930s and 1940s, with his brilliant style and excellent illustrations (Fig. 3), developed the concept of 'stress' versus 'anti-stress' minerals, which to some extent was an elaboration of Rosenbusch's ideas on dynamometamorphism. This was done in response to the ideas of Friedrich Becke (1903) and Ulrich Grubenmann (1904–1906), which he thought too static. According to Harker, stress minerals, characteristic of regional metamorphism, were formed under a strong non-hydrostatic stress regime. The Barrovian region of the Scottish Highlands was taken as the type example of this ('normal') metamorphism. In contact metamorphism, on the other hand, only anti-stress minerals (cordierite, andalusite), stable under a hydrostatic stress regime, were present. By relating the occurrence of metamorphic minerals to deformation, Harker anticipated one of the great developments of structural metamorphic petrology which were to occur after World War II (see below). But his views also had a negative influence. By providing a 'short-cut explanation' (Miyashiro 1973) for the occurrence of metamorphic minerals by an unquantifiable mechanism, they diverted many petrologists' interests towards explanations based on changing physical (pressure or temperature) or chemical (rock and mineral composition) parameters.

Given that the German school had dominated the early stage of descriptive petrography, it should not be a surprise that many followers of Zirkel and Rosenbusch also came from German-speaking countries: Austria and Switzerland. Independently of Barrow, they discovered a regular scheme of mineral evolution during progressive metamorphism, essentially at a regional scale, which they attributed to the depth at which rocks had been transported during orogenic evolution.

Van Hise (1904) proposed four 'depth zones' of metamorphism, against only two for Becke (1903), characterized by the occurrence of a certain number of given minerals, which he called 'typomorphic'. Finally Grubenmann wrote, first alone (1904–1906), then with his successor at Zürich, Paul Niggli (1924), a series of books which remained the basic references in continental Europe in the inter-war period. He defined three depth zones, with names which are still used in some of the geological literature (epizone, mesozone and catazone, in order of increasing depth). Contact metamorphism was assumed to be a local, relatively unimportant phenomenon, which differed only from regional metamorphism by producing different structures. Regional metamorphism was the 'real thing', and all observed metamorphic types were

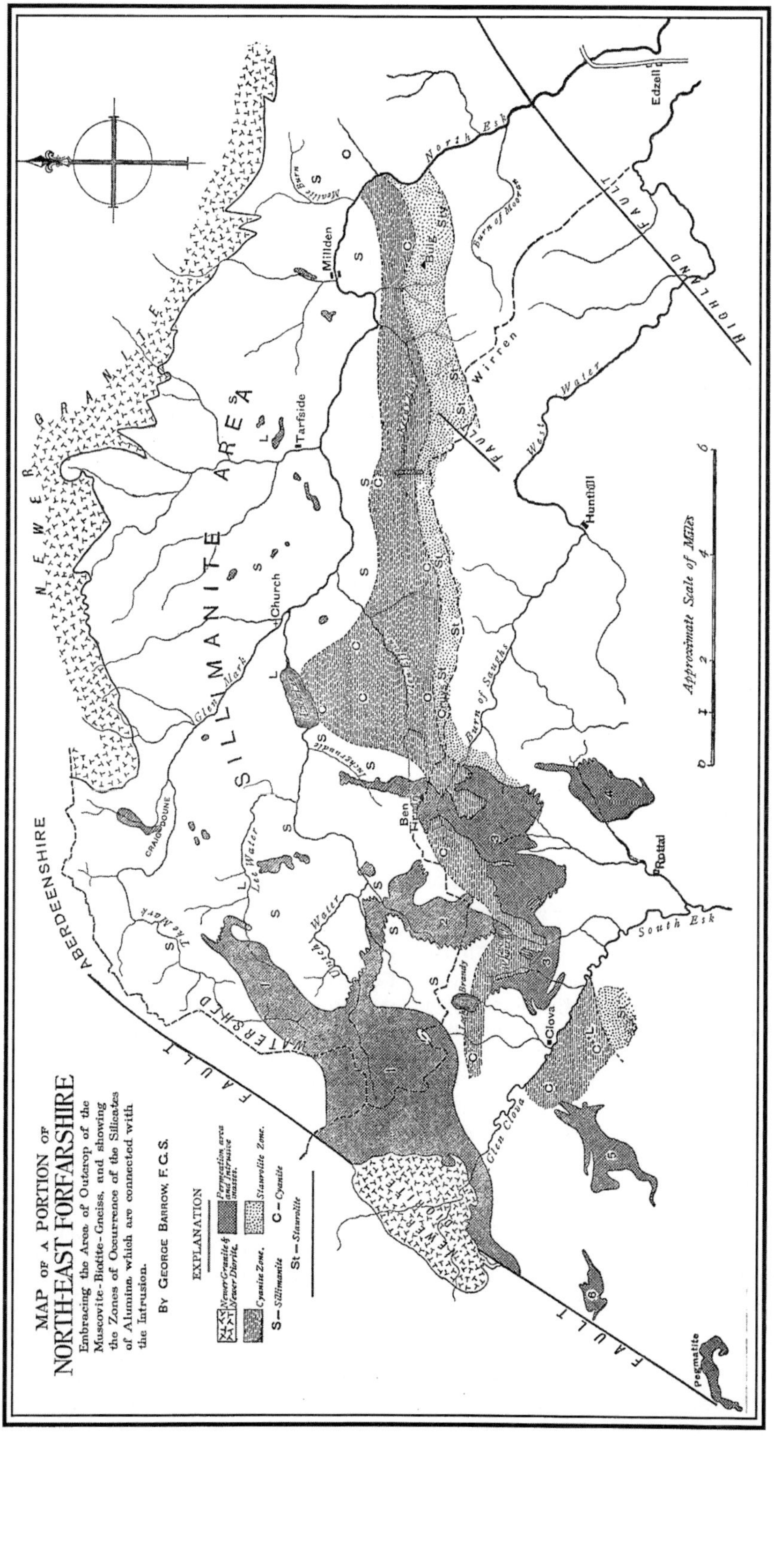

Fig. 2. Original map by George Barrow of progressive metamorphic zones in the Scottish Highlands (later called Barrovian metamorphism). From Barrow 1893, *Quarterly Journal of the Geological Society*.

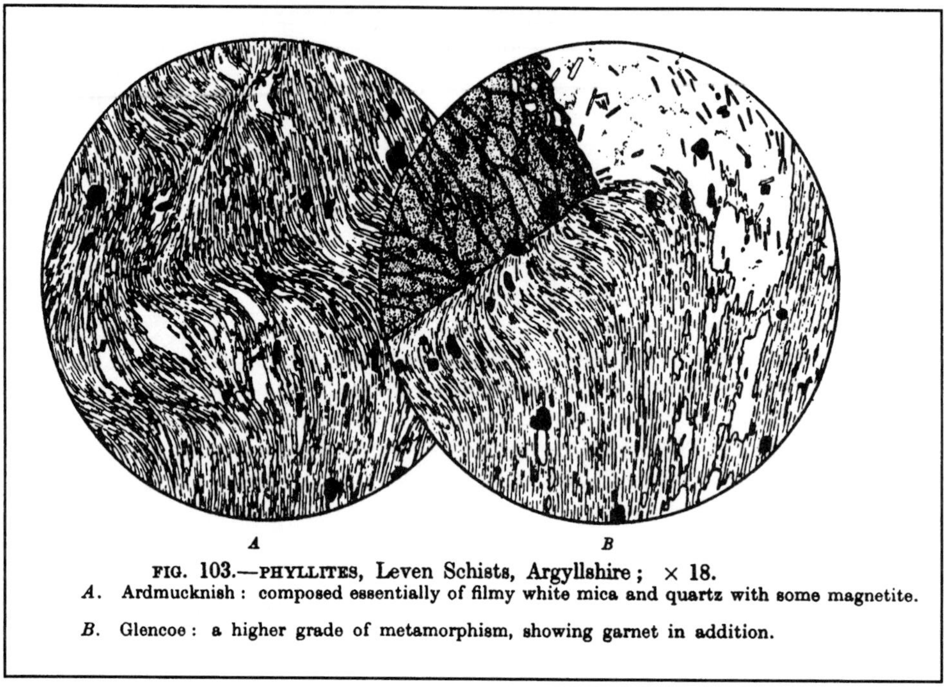
A
B
FIG. 103.—PHYLLITES, Leven Schists, Argyllshire; × 18.
A. Ardmucknish: composed essentially of filmy white mica and quartz with some magnetite.
B. Glencoe: a higher grade of metamorphism, showing garnet in addition.

Fig. 3. Illustrations by Alfred Harker (1932) of metamorphic textures (phyllites from Barrovian metamorphism).

assigned to a given zone on the basis of general impressions of grain sizes (increasing with depth) and mineral compositions. For instance, phyllites, chlorite schists and glaucophane schists were assigned to the epizone; biotite and muscovite-bearing schists and amphibolites to the mesozone; and muscovite-free gneisses, eclogites and granulites to the catazone. Under the influence of Niggli, the cause of metamorphism was regarded as exclusively magmatic: an intrusion at depth, typically a granite, provided the heat source. Mixed rocks (i.e. gneiss and granite), soon to be described from Nordic countries (migmatites), were explained in terms of granite injection, eventually supplemented by later deformation.

The depth-zone system was easily accommodated by the notion of '*métamorphisme géosynclinal*', formulated contemporaneously by the French school, notably Emile Haug (1907–1911). Depth zones correspond to successive layers in geosynclines, closer and closer to the granitic basement (Fig. 4). But contrasting views on the role of granite remained, yielding ongoing discussions between Rosenbusch and Michel-Lévy. Was the magmatic/metamorphic distinction clear-cut, as claimed by Niggli and the upholders of magmatic differentiation, notably Norman Bowen (1928)? Alternatively, were there intermediate rocks, 'feldspathized' gneiss, caused by 'emanations' issued from underlying granite? This view was a kind of tradition in the French school, and was soon to be boosted by a revolution from Scandinavia. The importance of this revolution took a long time to be fully appreciated, but finally it created modern metamorphic petrology.

New light from Scandinavia: migmatites and mineral facies

Petrology (magmatic and metamorphic) has been developed as a real science at a few major European universities (in Germany, Britain and France). At a time when travelling was less easy than today, many interesting field areas were relatively close to the research centres, in a few typical orogenic belts (Caledonian, Variscan, Alpine). But many of these exposures are strongly altered, partly covered by superficial material, or, in the case of the Alps, difficult to reach. Scandinavia provided a very different picture: rocks there have been polished by the

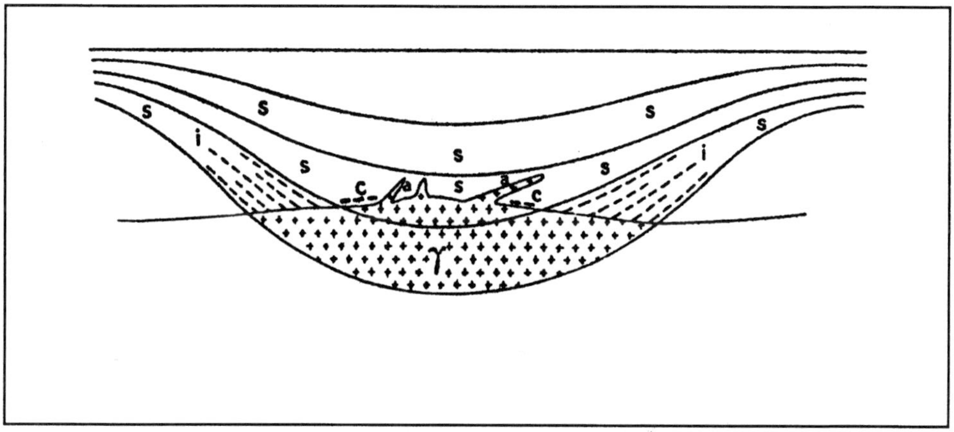

Fig. 4. '*Métamorphisme géosynclinal*', as seen by Haug (1907–1911, fig. 48). Translation of the French caption: 'Schematic section explaining the transformation of a geosyncline bottom, made of schists (s), into granite (γ), with 'lateral impregnation' (i), formation of contact aureoles (c) and apophyses (a) at lower depth'.

recent glaciations, providing excellent exposures. The Norwegian Waldemar Christofer Brøgger, who after his studies of geology in Kristiania went to Germany to study optical mineralogy and microscopic petrography, first with Heinrich Mühl in Kassel, and subsequently under Rosenbusch and Paul von Groth in Strasbourg, brought back these skills, as well as useful contacts, to major universities in Scandinavia (Hestmark 1999). Brøgger became professor, first at Stockholm's Högskole and later at the University of Kristiania (Oslo). Several of his students proved to be notable researchers, able to transpose field observations into an elaborate interpretative system. Notable among these men were Jakob Johannes Sederholm in Finland, and Johan Herman Vogt and Victor Moritz Goldschmidt in Norway. Vogt was to become professor of metallurgy at the Kristiania University and had a profound influence on experimental petrology (Vogt 1903–1904).

Goldschmidt was more a theoretician, who made the breakthrough, essentially by himself, at a very early age (see Fritscher 2002). Sederholm, who started his work earlier (before the end of the nineteenth century), was more field-orientated, also more of a '*chef de file*', who managed to have near him two great scientists, César Eugène Wegmann of Switzerland and Pentti Eskola of Finland, who may be regarded as the real founders of modern metamorphic petrology.

Sederholm and his co-workers on the one hand, and Goldschmidt on the other, operated roughly contemporaneously (during the period 1910–1930). However, they addressed different problems: the transition between gneiss and granite for Sederholm; the relations between rock chemistry and mineral assemblage for Goldschmidt and Eskola. Only after World War II were these approaches more or less integrated.

Sederholm (1907) tried to elucidate the complicated relations between the most common type of high-grade 'crystalline schists', namely gneiss and granite, already an important topic in Nordic geology since the work of Baltazar Keilhau in the early nineteenth century. Both rock types have basically the same mineralogical composition, differing only in structure, a fact that had led Rosenbusch to propose the concept of 'dynamometamorphism'. Sederholm could see that most of the Precambrian Baltic Shield is made of an intricate mixture of granite and gneiss, at all scales, which he named migmatites (see Fig. 5). To explain their formation, he called for a mysterious 'ichor' (literally, the 'blood of a nymph'), which could permeate the rocks, partly dissolving and 'granitizing' them. Migmatites were found to dominate the core of all Precambrian terranes, and were also identified by Sederholm in the Vosges, on the occasion of an excursion to classical 'Rosenbusch' exposures. (In fact, we know now that they constitute the bulk of continental masses, the so-called 'granitic layer' of geophysicists.)

Migmatites have complex textures, for which a profusion of terms was created, mostly by Sederholm himself. He was an excellent linguist, who could write papers in Swedish, Finnish,

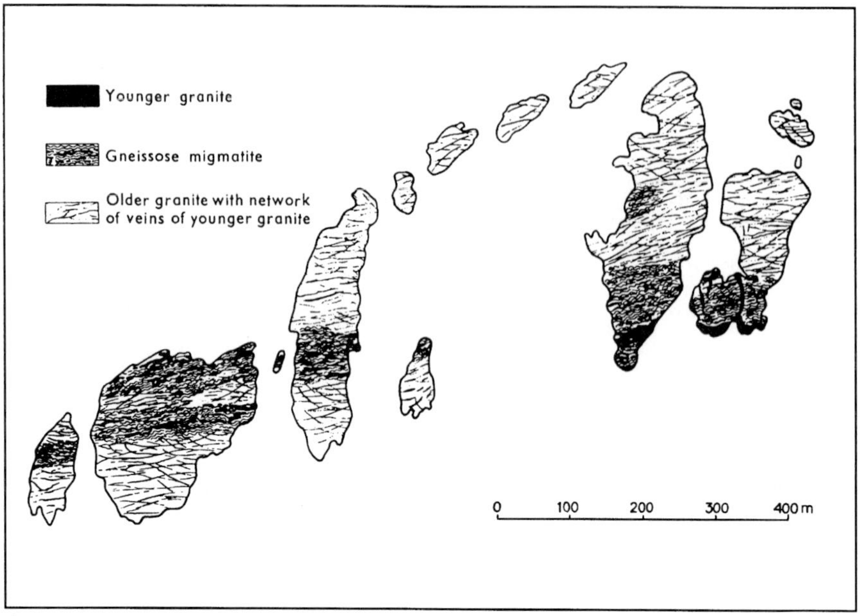

Fig. 5. Map by Sederholm (1907) of granite/gneiss contacts, using the term 'migmatite' for the first time.

German, English and French; he definitely had a flair for terminology. Besides 'migmatites' and 'ichor', he coined names like 'agmatite', 'anatexis', 'deuteric', 'dictyonite', 'homophanous', 'katarchean', 'myrmekite', 'palingenesis', 'palympsest', 'ptygmatic': most terms are derived from Greek and are still found in the petrologic vocabulary. These names did not provide *explanation*, but at least they showed that a simple magmatic explanation, namely the injection of granite dykes into pre-existing gneiss, faced serious difficulties. The 'geometrical' approach received a decisive impulse from the young Wegmann (1929, 1935), who transposed structural techniques elaborated in the Alps to Precambrian areas (see Fig. 6). Development of these techniques would ultimately lead to structural metamorphic petrology, now almost an independent discipline.

The work of Goldschmidt was completely different, but it also started from field observations, this time on metamorphic aureoles around intrusive granite in the Oslo region (Goldschmidt 1911). Having a much broader physicochemical background than most of his contemporaries – except possibly Paul Niggli, who was also an excellent chemist – Goldschmidt discovered systematic relations between rock composition and metamorphic mineral assemblage in hornfels, the highest-grade metamorphic rocks of the contact aureoles. Although it had been vaguely noted, in particular by Barrow, that some minerals preferentially occur in certain rock types, it was more or less tacitly assumed that the role of rock chemistry was not important. The formation of new minerals depended either on changing external conditions (pressure and temperature) or on external introduction of new elements. Goldschmidt demonstrated that rocks are chemical systems, which can be treated according to the laws of physicochemical equilibria, notably the 'phase rule' (sec Fritscher 2002). The importance and pioneering aspect of his work are fully recognized today. Other geologists had, however, already attempted to apply chemical thermodynamics to the study of rocks, notably Becke (1903) who, from the well-known Clausius–Clapeyron equation, had understood that pressure increase should lead to the formation of higher density materials. He applied this 'volume law' to eclogites and, simply by comparing the molar volumes of gabbroic and eclogite mineral assemblages, he concluded that eclogites were high-pressure equivalents of gabbro (Godard 2001).

The work of Goldschmidt on contact aureoles had attracted the attention of Eskola, a student of Sederholm in Helsinki, who had investigated some comparable rocks in the Orijärvi region in southern Finland (Eskola 1915). Eskola came to Oslo, and in 1920, the first comprehensive paper

Fig. 6. Examples of the structural contribution brought by Wegmann (1929) to the study of Precambrian metamorphic complexes. Above: serial profiles, allowing the representation of three-dimensional structures on a plane. Below: block diagram, showing the relation between true (af) and apparent fold axes. The correct axial direction can only be measured along vertical layers.

on the notion of mineral facies was published (see Fig. 7). It is interesting to note that Eskola considered magmatic as well as metamorphic rock types (Eskola 1920):

> A mineral facies comprises all the rocks that have originated under pressure and temperature conditions so similar that a definite chemical composition has resulted in the same set of minerals, quite regardless of their mode of crystallization, whether from magma or aqueous solution or gas, and whether by direct crystallization from solution (primary crystallization) or by gradual change of earlier minerals (metamorphic recrystallization).

But, as the conclusion was rather obvious for magmas, the initial notion of 'igneous facies' was soon replaced by that of high-temperature metamorphic facies (granulite), and only the different metamorphic facies have remained, with the names and broad pressure–temperature (P–T) interpretation that they still have today.

An epoch-making controversy: 'soaks' *contra* 'pontiffs'

Even if the name 'facies' was immediately endorsed by Becke (1921), it was not easy for the Scandinavian newcomers to be recognized by

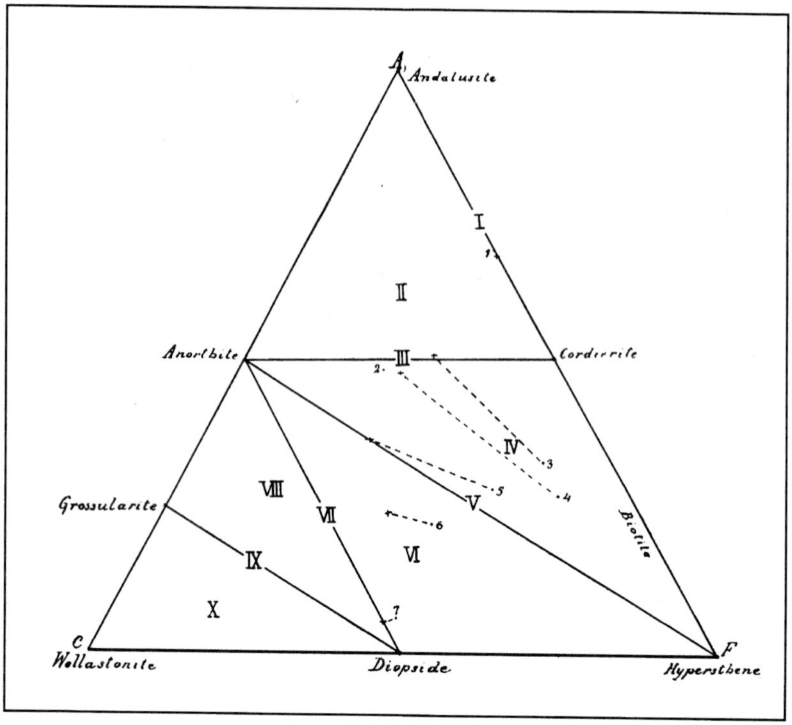

Fig. 7. ACF diagram by Eskola (1920), used for the definition of metamorphic facies. ACF metamorphic parameters: A = aluminium, C = calcium, F = iron + magnesium. I to X: the ten classes of hornfelses observed by V. M. Golsdchmidt (1911), corresponding to bi- or triphase diagnostic metamorphic assemblages. 1 to 7: whole-rock compositions. Dashed lines ending in a cross: correction made by subtracting potassium component, since K-bearing minerals (notably biotite) cannot be adequately represented in the diagram. The corrected compositions give a much better correspondence between chemistry and mineralogy (e.g. 3/III, 5/V, 7/VII).

the international scientific establishment. Mineral facies superficially resembled depth zones, to the point that a number of authors had proposed an equivalence of terminology (e.g. greenschist facies and epizone). At a time when it was not easy to have precise information on the chemistry of mineral and rocks, many petrologists did not see the need to deploy complicated thermodynamic equations. They also failed to see the real novelty of the concept, namely that pressure and temperature do not always show the same relation ('geothermal gradient'), and thus that they could be treated as independent variables. Eskola made repeated attempts to demonstrate the superiority of his facies concept to that of depth zones, but mostly in regional Nordic journals (e.g. *Bulletin de la Commission géologique de Finlande*, 1915; *Norsk Geologisk Tidsskrift*, 1920; *Geologiska Förening i Stockholm Förhandlingar*, 1929), too often considered as subordinate literature. Harker, who saw little room for his stress and anti-stress minerals in chemical thermodynamics, reviewed Goldschmidt's classification of Oslo hornfelses in a rather negative manner, and he almost completely ignored Eskola's work (even though it was clear that the concept of mineral facies would have been the easiest way to explain Barrow's zones). It was significant that when, just before World War II, Eskola finally published the most elaborate version of his work, together with Tom F. W. Barth for the magmatic and Carl W. Correns for the sedimentary rocks (Eskola 1939), he mentioned in his extensive historical introduction all names that counted in the preceding generations, *except* Harker.

However, it is clear that, for thirty years after the introduction of the depth zones or mineral facies concepts, the big question was not the relative merits of the two systems, but the relationships between gneiss and granite (e.g. Raguin 1957): is the granite the cause or the result of metamorphism? Migmatites are at the core of this problem. Sederholm's 'ichor' was supposedly able to transform some pre-existing sediments into homogeneous granite. Wegmann provided a geometrical framework, by defining a 'migmatite front' separating isochemically recrystallized from 'granitized' rocks. These views were enthusiastically endorsed by extreme 'transformists' – Herbert Read and Doris Reynolds in Britain, René Perrin and Marcel Roubault in France – who did not call for fluid media to transport the elements. Granitization supposedly occurred by 'solid-state reaction', by element diffusion through the crystalline structure. This hypothesis was, of course, denied by the magmatists, who relied on experimental evidence. Both camps found vigorous and able defenders, and *The Granite Controversy* by H. H. Read (1957; see Fig. 8) can still be read with pleasure, at least for the quality of the expression (see also Read 1943–1944). Personal attacks were not lacking. Because of the supposedly authoritarian character of the magmatist '*chef de file*', Niggli, they were called 'pontiffs' by the transformists. Bowen replied with the nickname 'soaks', as well as with the devastating appellation of 'Maxwell's Demon' to volatiles in general, which (at the time) could not be demonstrated experimentally. Some authors, notably Jean Jung and Maurice Roques in France, attempted to incorporate the notion of migmatites within the framework of depth zones. Using the example of the French Massif Central, they separated 'ectinites', isochemically recrystallized rocks, from metasomatically transformed migmatites (Jung & Roques 1952). The cause of metamorphism was still believed to be geosynclinal burial (see Fig. 4).The successive ectinite zones, more or less horizontal, apparently corresponded to increasing depth in a geosyncline. Microstructural studies, which had received a great impulse from Bruno Sander in Austria (Sander 1948–1950), soon showed that Jung and Roques' 'zoneography', with its 'migmatite front' cutting obliquely the horizontal ectinite boundaries, could not be reconciled with detailed field observations, even in the supposed type locality (the French Massif Central; Demay 1942; Collomb 1998). However, the apparent simplicity of the system made it attractive to many geologists, who could map rapidly wide areas of poorly exposed, unknown terranes, e.g. in Africa. The problem was that some field geologists, notably in French-speaking countries, failed to represent also the lithologies of the rocks. For instance, limestones, metavolcanics and quartzites could be collectively described as '*micaschistes supérieurs*' or '*gneiss inférieurs*'. Unfortunately, this made their maps almost useless when the concept of mineral facies replaced that of zoneography.

As far as the migmatite problem was concerned, the quarrel between soaks and pontiffs ended in the 1960s with apparent victory for the pontiffs. Experimental petrology showed that solid-state diffusion is very limited, even at high temperatures, and that a rock like a granite can only be formed by crystallization from a melt. Granite magmas can be formed by different processes at different levels, notably in the lower part of the continental crust. This is the domain of the granulites where, as we will see, some of the old questions were to reappear.

The revolution of the 1960s

It is customary in the Earth sciences to envisage a 'revolution' in the 1960s, with the development of plate-tectonic concepts. Plate tectonics, however, was less a drastic change in geological thinking than a consequence of technological progress: the ability, with equipment directly resulting from World War II, to measure remanent magnetism in the lavas emitted at mid-oceanic ridges and ocean-floor mapping (see Barton 2002). The symmetrical magnetic 'zebra' pattern on both sides of the ridges immediately suggested how oceanic crust was created, to disappear by subduction under the continents. But marine geophysics was not the only discipline to be transformed by modern technology. For metamorphic petrology, a number of instruments fundamentally changed the nature and even the scope of the discipline.

Firstly, the electron microprobe (first patented by J. Hillier in the USA in 1947, with the first working instrument being developed by R. Castaing and R. Guinier in 1949, though the instrument did not come into widespread use until the 1960s), allows in situ spot analysis of any mineral phase. Analyses are almost instantaneous, compared to the tedious, time-consuming wet-chemical analysis, especially for silicates. Chemical petrology was reborn, and the importance of this new instrument, now standard in any laboratory, can only be compared to the proliferation of microscope studies during the second half of the nineteenth century.

Modern technology also opened a new field of research for trace-element and isotope geochemistry. Mass spectrometers and other techniques of 'nuclear' mineralogy, at the edge of scientific research before the War, became standard instruments in many geoscience research laboratories. It was now possible to measure, on smaller and smaller samples, the relative proportions of both stable and radioactive isotopes

Fig. 8. Frontispiece of *The Granite Controversy* by H. H. Read (1957) (drawn by D. A. Walton).

in a given rock or mineral. Knowing the decay constants of radioisotopes, the time at which the nuclear reaction started could be calculated. After chemical age determinations, pioneered well before World War II (see Lewis 2002), a new discipline was thus created, geochronology, which in due course went well beyond simple age determination. Radiometric dating is in practice the only way to establish the age of a relatively old rock which does not contain remnants of living organisms (fossils). Since early work by Holmes *et al.* (1957) on the Precambrian of southern Norway and Canada, radiometric investigations have had a major impact on our understanding of the Precambrian, which cannot be dated by fossils, but covers more than four-fifths of Earth history. Not only radiogenic, but also stable isotopes can be used as tracers, for the investigation of most varied processes: mineral crystallization or recrystallization, origin of various rock components, interactions between rocks and fluids, etc. These techniques were developed within the framework of a new discipline, geochemistry, from a name/discipline created by Vernadsky (1924). (The name was first suggested in 1838 by Christian Friedrich Schoenbein, but in a different sense from that which it has today.)

High pressure and temperature experiments have also become an essential part of metamorphic petrology. After the pioneering efforts of Hall, Daubrée and Vogt, a decisive impulse came from the Geophysical Laboratory of the Carnegie Institution, Washington DC, in the United States (Young 1998, 2002) which, through to the present, remains the standard reference for experimental studies. During the first half of the century, most experiments were done on magmatic rocks, which could be treated as dry, fluid-absent systems. Many metamorphic minerals contain volatiles (e.g. CO_2 or H_2O) in their structure, and serious experimentation could only start when volatile-bearing, hydrothermal syntheses could be undertaken at sufficiently high pressures and temperatures. This was achieved around 1950, notably through the work of Hatten Yoder and others (Yoder & Eugster 1954). From this time onwards, a flow of data emerged, not only from the Geophysical Laboratory, but from many places in the world (e.g. Göttingen and Bochum in Germany; Toronto and Ottawa in Canada; the former Soviet Union). The results of this immense research effort drastically changed the perception of the physical conditions (temperature and pressure) at which metamorphic changes occur. In the granite debate, a key argument of the transformist school for the solid-state origin of granite was the supposedly low metamorphic temperatures, well below the melting-point of the water-saturated granite system (about 700°C). Harker (1932, p. 209) thought that muscovite recrystallization could take place in cataclastic (dynamic) metamorphism 'at ordinary temperature' (meaning surface temperature), and that the lowest temperature at which metamorphism could appear (the chlorite zone in Barrovian metamorphism) was also close to this temperature. The decisive factor for the development of metamorphic minerals was not temperature but time and, above all, deformation (or state of stress). Maximum temperatures, corresponding to amphibolite facies, should not exceed 600°C, well below the melting-point of granite.

With regard to pressure, uncertainties were even greater. There was no precise idea about the absolute value of pressure; it could only be roughly estimated from depth of burial (1 km corresponds to roughly 0.3 kbar). The lithostatic pressure was not thought to exceed a maximum of about 3 kbar, corresponding to a depth of 10 km, the supposed thickness of the 'metamorphic layer' in the upper part of the continental crust. In any case, this absolute value was not important, as it was (again) not a controlling factor, in contrast to (in Harkerian thinking) the state of stress (isotropic or anisotropic).

Changing the scope of metamorphism

The flow of experimental data which, after the 1950s, came from many places in the world, completely changed the 'scope' of metamorphism. First of all, the role of pressure became better understood. In 1953, experimental petrologists, notably Lawrence Coes Jr at Norton Company, succeeded in making a pressure vessel able to sustain a pressure of a few tens of kilobars (Coes 1953). Coes synthesized a new, dense modification of silica (later named coesite), and this was followed by the discovery in the Soviet Union of an even denser form (stishovite), stable at the enormous pressure of over 50 kbar (corresponding to a depth of 150 km) (Stishov & Popova 1961). These species were later found near the Earth's surface, mostly at the sites of former meteorite impacts, and for coesite more recently in some terrestrial rocks (eclogites – one of the most frequently discussed metamorphic rock types since its identification by René-Just Haüy in 1822; see Godard 2001). Around 1980, dry experiments could be undertaken at much higher pressures, with the initiation of diamond-anvil techniques. Multianvil, high-pressure apparatus was first developed in Japan (Kawai & Endo 1970), and then diamond-cells by Ho-kwang Mao and Peter Bell at the Geophysical Laboratory (Mao & Bell 1975, 1976). By simply pressing together two opposite diamond-anvils through levers, the investigator could produce extraordinary pressures (more than 1 megabar), corresponding to conditions near the Earth's core/mantle boundary. Thus, mineral-phase transitions could be predicted for depths far beyond any possibility of direct observation.

'Dry' experiments are, however, not directly relevant to metamorphic reactions, which in most cases occur in the presence of a fluid phase. With few exceptions, which incidentally turned out to be among the most difficult (e.g. all discussions with regard to the exact position of the sillimanite–andalusite–kyanite triple-point), metamorphic reactions can be studied only by hydrothermal experiments, notoriously more difficult and more dangerous than dry experiments. For safety and financial reasons, most hydrothermal experiments are limited to pressures of about 10 kbar. These were, however, sufficient to yield a wealth of new data on fluid–mineral interactions at depth and, above all, to show that crustal temperatures, commonly reached at high metamorphic grade, are sufficient to melt many former sediments. Magma, if any, did not have to be introduced from outside, but could be generated in situ by partial (or complete) melting of some metamorphic rocks. The old chicken-and-egg problem of the relationships between metamorphism and magmatism received a new powerful argument: the hen was metamorphic.

Calibrating metamorphic reactions, solving the granite controversy

Most of the hydrothermal experiments that 'flourished' after the 1960s were aimed at calibrating the zones of progressive metamorphism in terms of P and T and solving the granite problem. These types of experiments are rather different, and they were conducted differently in the two places (Washington and Göttingen) which, for many years, were to symbolize these different approaches. The Geophysical Laboratory of Washington was essentially concerned with mineral stability. Species had to be pure (mostly synthetic), in order to determine all the compositional and experimental variables. Data had to be retrieved by computational techniques, and essentially were derived from equilibrium thermodynamics, in order to construct mineral stability fields in the P–T space. At increasing P and T, successive mineral stability fields define a 'petrogenetic grid', a notion first proposed by Bowen in 1940. A number of gifted theoreticians, notably E-An Zen at the Geophysical Laboratory (Zen 1966) and James (Jim) B. Thompson at Harvard (Thompson 1955, 1957), developed this concept along the lines initiated by Goldschmidt (1912) and his student Hans Ramberg (1944, 1949, 1952). They treated metamorphic rocks, as well as mineral assemblages, as chemical multicomponent systems. This approach is now at the core of all modern studies, but its elaboration required one to go back to the literature of the end of the nineteenth or the beginning of the twentieth century (Gibbs, Backhuys Rozenboom, Schreinemakers, Van't Hoff), which had escaped the attention of most petrologists for more than fifty years.

Before coming to a quantitative interpretation of metamorphic assemblages in terms of P and T, later developing into geothermometry and geobarometry, the first results of these experiments showed that metamorphic temperatures were much higher than previously assumed. The first attempt came at the lower metamorphic grade, with the identification of metamorphic zeolites (Coombs 1954), as well as other minerals (prehnite–pumpellyite; Coombs 1960) which defined the lowest-temperature metamorphic facies, immediately following sedimentary diagenesis. These diagnostic minerals were found to occur in a systematic way in volcanic and sedimentary rocks from New Zealand and were later found at the margins of many orogenic belts, notably the western Alps. They were formed well before any platy minerals, like chlorite, capable of giving the rock a typical metamorphic texture (schistosity). Experiments on zeolites showed that the crystallization temperature must be at least 250°C. Therefore, the temperature of the first Barrovian metamorphic zone, defined by chlorite, must be significantly higher (at least 300°C by present-day estimates). At higher temperatures, an important reference would be the Al-silicates triple-point (andalusite–sillimanite–kyanite), not influenced by fluid activity. However, equilibrium conditions were notably difficult to realize. Experiments in the first half of the 1960s suggested possible temperatures as low as 300°C (Miyashiro 1949). But Robert Newton (1966) found 520°C, with even higher temperatures (600°C) obtained by Egon Althaus (1967) and Richardson *et al.* (1969). Now the commonly accepted value is around 500°C, remarkably close to the *c.* 540°C, 3.4 kbar obtained by Olaf Schuiling (1957) in his attempt to calibrate the triple-point from field occurrences. It was soon obvious that many metamorphic reactions should take place at much higher temperatures. Some regional metamorphic assemblages, characterized by the widespread occurrence of volatile-free minerals (orthopyroxene and/or garnet instead of micas or amphibole) are conspicuously similar to contact metamorphic rocks (pyroxene hornfels), obtained close to the hottest intrusions (T at least 900°C). Therefore, the temperature of mineral equilibration must be roughly comparable. We now know that this is the case, and metamorphic minerals typical of ultra-high temperature rocks, like quartz–sapphirine, described from Enderby Land, Antarctica (Ellis *et al.* 1980) and Hoggar, Algeria (Ouzegane & Boumaza 1996), or osumilite, found in regional aureoles around massif anorthosites, e.g. Nain, Canada (Berg & Wheeler 1976) and Rogaland, Norway (Maijer *et al.* 1977), correspond to temperatures in excess of 1000°C (Ellis 1987; Harley 1989). If we remember that most mantle rocks (peridotites) show typical metamorphic textures (equilibrated and deformed in the solid state; see Den Tex 1969), the conclusion is inescapable: the field of metamorphic temperatures extends to more than 1000°C, overlapping the field of magmatic temperatures.

Experiments dealing with the origin of granite were conducted by Jean Wyart and François Sabatier (1958, 1960) in France, partly as a reaction against the most extreme views hold by some 'transformists' (Perrin & Roubault 1939, 1963). But they were developed and systematized by Helmut Winkler and his co-workers at Göttingen, resulting in a book that exerted a strong influence on European petrologists for many years (Winkler 1965). These researchers did not start from pure, synthetic materials, but from natural rocks – a tactic considered almost a crime by purists at the Geophysical Laboratory! Results, however, were spectacular, showing that some metamorphic protoliths, notably metagreywackes, easily melt at temperatures of about 700 to 800°C, well within the range of temperatures reached by many high-grade metamorphic rocks. This provides an easy explanation for the formation of migmatites, which had so puzzled Sederholm and his successors: no need for large-scale element transport or mysterious 'ichor'. Migmatites were simply partly molten rocks, formed near the source at which granite melts are produced. Discussions still continue on the mechanisms by which these partial melts are collected to form massive intrusions, able to rise and cause contact metamorphism in the upper crustal levels. It should be noted that this process is only valid for some granites (S-type granites; White & Chappel 1990), whereas others (I-type granites) have no relation to a metamorphic environment and are best explained by magmatic differentiation. Bowen, who always claimed that magmatic differentiation was the only process by which all magmatic rocks are created, was not wrong, but he had missed the metamorphic counterpart.

From mineral facies to geothermobarometry

The work of generations of experimentalists, field and structural geologists, and geochemists, has had far-reaching implications for all geosciences. We are now able to trace the origin of different rock components (trace-element geochemistry, stable isotopes), date the different stages of the rock evolution (geochronology), and estimate the pressure and temperature at which a given set of coexisting minerals has equilibrated (geothermobarometry). It would be wrong to assume that all these results have been obtained in a harmonious and linear form, without many discussions, controversies, and a number of unsuccessful attempts. Barrow-type zoneography gives only a very approximate estimate of metamorphic pressure and temperature conditions. Most index minerals (chlorite, biotite, garnet, etc.) show considerable solid solutions, resulting in multivariant mineral reactions in P–T space. Corresponding zones are not bounded by a single line, or 'isograd', but by a band, in which the compositions of coexisting minerals progressively adapt to changing pressure and temperature conditions ('sliding' reactions). The identification of these mineral reactions is an essential step in regional analysis, and can be derived relatively easily from the graphical presentation of metamorphic facies (metamorphic parameters; see Spear (1993) for technical details). On a P–T grid, major metamorphic facies appear as fields (greenschist, amphibolite, granulite, etc.) illustrating the successive steps of the metamorphic evolution in a given region (see Fig. 9).

It was soon realized that some facies are more typical for relatively high temperature facies (e.g. granulite facies), and others for high pressures (i.e. eclogite facies). In a given region, the succession of facies can be represented by a line, defining a 'metamorphic gradient'. With the progress of geothermobarometry, it became possible to quantity these gradients, with three major trends: high T–low P (typical examples being found at Buchan in the British Isles, in the Pyrenees, and on the Abukuma plateau in Japan); high P–low T (e.g. western Alps or the Franciscan Range in California); and an intermediate series, sometimes assumed to be the 'normal' type of metamorphism, represented by the Barrovian metamorphism of the Scottish Highlands. It is now clear that this evolutionary trend can be roughly understood in terms of regional pressure and temperature conditions. But attempts to go into more details, as done by

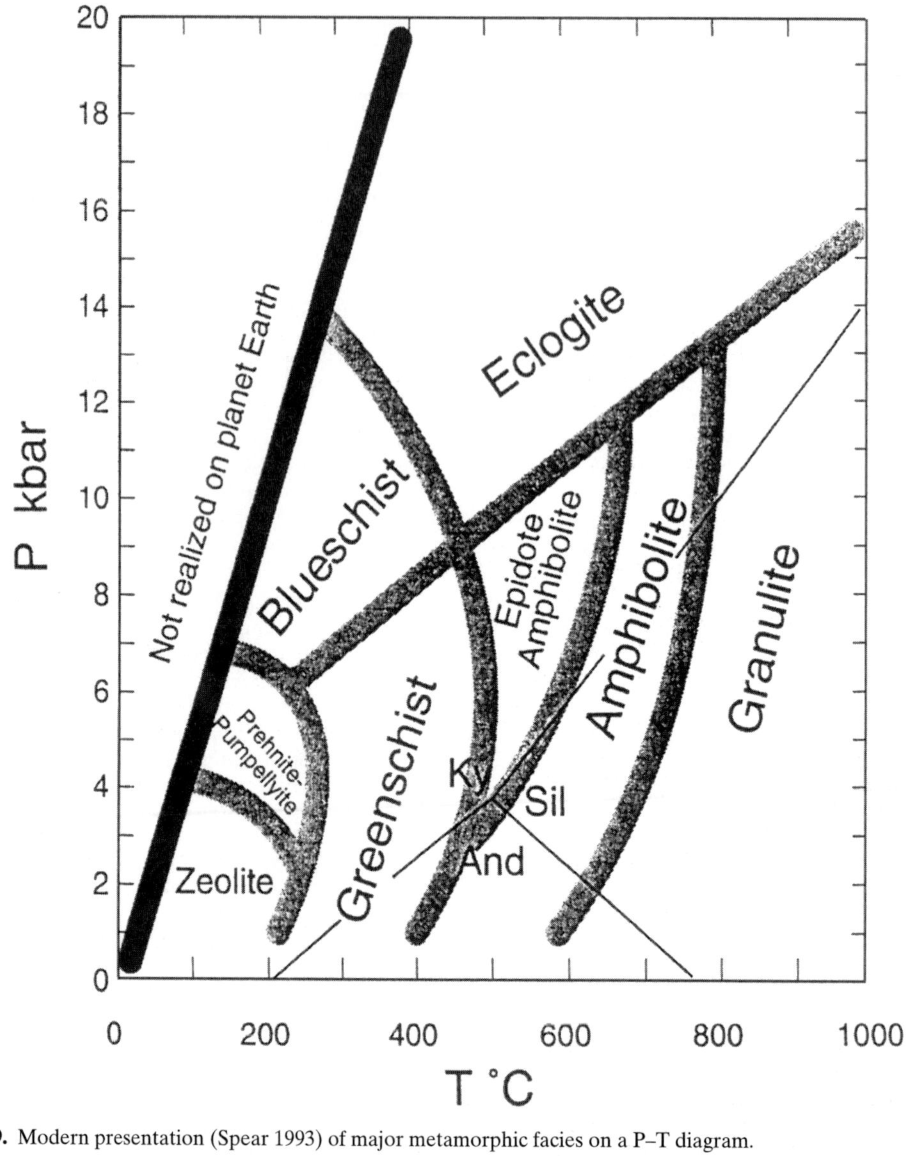

Fig. 9. Modern presentation (Spear 1993) of major metamorphic facies on a P–T diagram.

Helmut Winkler in the third and fourth editions of his *Petrogenesis of Metamorphic Rocks*, in which he increased the number of subfacies to more than 20, have failed (to the point that Winkler himself proposed to abolish even the concept of metamorphic facies in the last edition of his book of 1979). Metamorphic facies (or subfacies) only consider a limited part of the rock system, both for the internal (chemical component) and external (temperature, lithostatic and fluid pressures) conditions. They can only give a rough idea of pressure and temperature conditions, definitely not a quantitative estimate. For this reason, the intermediate trend now tends to be abandoned, leaving only the high-T and high-P metamorphic types, which indeed correspond to well-defined orogenic belts.

Luckily, the progress of geothermobarometry, associated with precise studies of metamorphic textures, has allowed for much better calibration of metamorphic P–T conditions. Single mineral-pair geothermobarometry has benefited from the definition of internally consistent thermodynamic databases, from which reaction curves, mineral stability fields or P–T conditions of mineral equilibration are automatically generated; examples are GEOCALC (Berman & Perkins 1987), THERMOCALC 2.7 (Holland & Powell 1998), TWQ (Berman 1991) and TPF (Fonarev *et al.* 1991). We caution against the dangers of blindly using such computer programs, with insufficient analysis and discussion of their possibilities, as well as a careful microscope study of textures and mineral assemblages. Nevertheless, the experimental thermodynamic approach has had some unexpected results, notably increasing tremendously the range of possible metamorphic conditions. We have already stated that experiments considerably increased the range of metamorphic temperatures. The case of pressure is even more spectacular. For a long time (still in many modern textbooks), the maximum pressures considered are about 10 kbar, in line with average 'high' conditions (intermediate granulites) of, roughly, 800°C and 8 kbar. Data with regard to peridotites and the Earth's mantle had already been compiled by Ringwood (1975), but mantle pressures did not occur in the metamorphic perspective of 'crustal' rocks. Although a separate blueschist facies was introduced by Eskola (1929, 1939), it took considerable time for it to be accepted. Petrologists like Francis Turner and John Verhoogen (1951) considered the formation of glaucophane to be due to solutions derived from (ultra)basic rocks. The work of Wilhelm de Roever (1950, 1955*a*,*b*) on Celebes, as well as discussions of the nature of Franciscan metamorphism (cf. Miyashiro 1994, p. 310), finally led to the full recognition of the role of pressure. In the 1980s, the scale of metamorphic pressure was, suddenly, multiplied by a factor of five. Coesite, which imposes a metamorphic pressure of at least 20 kbar, had been identified in the form of quartz pseudomorphs in eclogites from the southern Urals (Chesnokov & Popov 1965), but this work had been completely ignored. In 1984, Christian Chopin identified and made a correct interpretation of coesite inclusions in pyropes from the Dora Maira massif, western Alps (see also Schertl *et al.* 1991). Simultaneously, David Smith (1984) found coesite inclusions in clinopyroxene in the Norwegian Caledonides. This was soon followed, first in the Kovchetav massif, Siberia, and subsequently elsewhere, by the discovery of a mineral quite unexpected in metamorphic environments, namely diamond (Sobolev & Shatsky 1987, 1990). An astonishing sequence of completely new high-pressure minerals followed (ellenbergerite, Mg–carpholite, etc.). As for ultra-high temperatures, an ultra-high pressure (UHP) metamorphic facies had to be defined, reaching more than 50 kbar (a depth of about 150 km). In some regions (like the Dabie-Shan Mountains, east-central China), UHP rocks cover large areas. Again, experimental petrology, especially the numerous experiments performed by Werner Schreyer and co-workers at Bochum, Germany (Schreyer 1988, 1995), facilitated petrologists' understanding.

The idea that glaucophane schists were (almost entirely) restricted to younger (i.e. Cainozoic and Mesozoic) orogenic belts (e.g. De Roever 1956, 1964; Miyashiro 1973) – a belief reminiscent of Wernerian concepts of there being different types of metamorphism for each epoch – has prevailed for a long time, but was recently shown to be incorrect (Liou *et al.* 1990). The discovery of coesite in the Precambrian of Mali (Caby 1994) has demonstrated the presence of ultra-high pressure metamorphism over large parts of the Earth's history.

The role of fluids

Hydrothermal experiments, as well as the theoretical interpretation of heterogeneous (mineral–fluid) equilibria, have underlined the importance of the fluid phase in almost all metamorphic reactions. The old adage *corpora non agunt nisi fluida*, which had been denied by the transformist school, made a triumphant comeback. If no fluid is present, element transfer is very limited, being insufficient in most cases to form new minerals. Detailed investigations on the mechanisms by which large metamorphic porphyroblasts grow underline the importance of dissolution/precipitation processes, requiring the intervention of a fluid phase. But the existence of this fluid phase is limited. As long as it is present, the composition of some minerals (e.g. garnet) will change in response to ambient P–T conditions. But if, at a certain moment, fluids disappear, then the composition will (in most cases) be fixed. It remains unchanged during subsequent metamorphic evolution, until the time when rocks finally reach the Earth's surface. For rocks recrystallizing at depth, the best way to eliminate pervasive fluids is decompression. Then, fractures will be formed in the rock mass, which drain the fluids. Mineral assemblages in the groundmass will tend to record maximum (peak) P–T conditions. During uplift, further evolution will be restricted to these veins, provided that no new fluids are introduced.

The study of this 'now missing' phase (the expression is from the late Philip Orville) has become one of the most important issues in present-day metamorphic petrology. It can be apprehended either indirectly, from thermodynamic calculations (Eugster 1959; French 1966), or directly from the study of small fluid remnants preserved in some minerals as inclusions (Roedder 1984). Fluid inclusions have long been known. As early as the beginning of the eighteenth century Johann Scheuchzer made drawings of quartz crystals with fluid inclusions, and when Sorby applied the microscope to the study of rocks (his major paper was published in 1858, but his observations started well before; see Judd 1908), among the first objects he saw were fluid (and melt) inclusions, which he investigated with remarkable flair and ingenuity. Fluid inclusions remained a significant part of descriptive petrography in the times of Rosenbusch (1873–1877), Zirkel (1866) and Vogelsang (1867), but they largely disappeared from the literature of petrology during the first half of the twentieth century.

With problems of interpretation, insufficient knowledge of the behaviour of fluid systems at high pressure and temperature and lack of adequate analytical tools, there were many reasons for Bowen's (1928) concern about 'Maxwell's Demon'. In fact, Bowen, who was well informed about fluid inclusions through the thesis work of Tuttle (1949), his most assiduous assistant, ascribed more to fluids in general than to inclusions, but the association between both was soon made. Extreme transformists were no greater supporters, as their motto was precisely the *lack* of any fluid. In most places in the world, the study of fluid inclusions was restricted to their minor role in metallogeny. However, the former Soviet Union was a notable exception and fluid inclusions remained an important topic of study there, with fundamental scientific work done during the darkest years of World War II; and it was from this country that a renewed international interest in fluid inclusions arose after the war.

At the beginning of the 1960s, the situation changed completely. We know much more about the laws governing metasomatism (Korzhinskii 1936, 1959), as well as about the solubilities of different minerals in various fluids (Helgeson 1964; Garrels & Christ 1965; Barnes 1967). Notable progress has been made in the knowledge of fluid systems at high P and T (Kennedy 1950, 1954; Franck & Tötheide 1959) and in the instrumentation (heating/freezing microscopic stage: Roedder 1962–1963). Fluid inclusions are studied not only in ore deposits but in all kinds of sedimentary, magmatic and especially metamorphic rocks: alpine veins and segregation (Poty 1969), high-grade metamorphic rocks (Touret 1971), etc. Complementary to this type of study are investigations on the signatures left in the rock by the passage of fluid flows, by means of stable isotopes or fluid–mineral interactions, which have become essential in modern metamorphic petrology. In principle, fluid inclusions can provide two sets of data: P–T conditions at which inclusions have been formed, giving one point on the metamorphic P–T path; and the chemical composition of this 'now missing' fluid phase. Specific techniques have evolved as a complete subdiscipline, using advanced analytical instruments (Raman spectroscopes, electron and ion microprobes, laser ablation ICP–MS, etc.). Fluid inclusion data play a role in the vast effort now undertaken to model fluid flow through rocks (Spear 1993). For some petrologists, fluids in inclusions may not have completely lost their 'Maxwell Demon' character, but at least, the 'Demon' is now under reasonable control.

Vapour-absent versus fluid-assisted metamorphism: resurgence of old controversies

The importance of fluid phases in medium-grade metamorphic rocks has been demonstrated by several petrologists, notably John Ferry (1976, 1983 1987) in Maine. In higher-grade rocks, fluid inclusion studies have, unexpectedly, revived old controversies. In the 1970s, high-grade metamorphic rocks (granulites) were found to contain large quantities of CO_2-rich fluid inclusions, also occurring in mantle xenoliths brought to the surface by volcanic eruptions. Granulites, like eclogites, were first thought to be relatively rare petrological curiosities, but it was soon realized that they are essential, if not exclusive, constituents of the lower part of the continents. Migmatites occur in this domain and, as temperatures tend to be higher, the degree of partial melting increases: granulites are a major source of granite magmas. Granulites are 'dry rocks', characterized by the widespread occurrence of anhydrous minerals like pyroxene and garnet. Therefore, it was assumed that their formation was only due to an increase of regional metamorphic temperatures. The 'unexpected discovery' (the term is from Winkler) of fluid remnants at this level has fuelled discussions, which recall some aspects of former controversies.

For the advocates of 'vapour-absent' (or fluid-absent) metamorphism, no free-fluid phase exists at the level of granulite formation. All fluids are either dissolved in melts or are bound in mineral structures. Fluid inclusions, if any, are formed at a late stage, unrelated to 'peak' metamorphism. Proponents of this purely magmatic school, which finds strong support in experimental petrology (Thompson 1982; Clemens & Vielzeuf 1987), in equilibrium thermodynamic calculations (Lamb & Valley 1984), or in the stable isotope signature left by circulating fluids (Valley *et al.* 1990), can be seen as the direct successors of Bowen, or even Hutton. The supporters of 'fluid-assisted metamorphism', on the other hand (e.g. Newton *et al.* 1980; Perchuk & Gerya 1993; Touret & Huizenga 1999), admit the existence of a free synmetamorphic fluid phase, in some cases externally derived (i.e. produced from the underlying mantle). These fluids may play a double role: they lower the water activity by dilution, controlling the stability of water-deficient mineral assemblages and rock-melting at a lower and higher temperature, respectively, than for water-saturated systems; and they induce metasomatic transformations that may significantly change rock compositions.

It is clear that the fluid-assisted school reiterates a number of arguments previously advanced by transformists, with, however, a major difference: they do not claim that mineral reactions occur in the absence of fluids but, much to the contrary, that they occur in the presence of fluids of precise composition. Also, they do not deny the possibility of melts, but merely question their ubiquitous importance. In this respect, they are less the direct successors of the extreme transformists, like Read or Roubault, than of the proponents of the 'mineralizers' or 'emanations' of the preceding generations: Termier and '*colonnes filtrantes*', Sederholm and his 'ichor', or, even further back in time, Virlet d'Aoust or Delesse. 'Deep hydrothermalism', which may occur in the form of supercritical fluids or volatile-loaded melts at lower crustal or upper-mantle depth, appears to be a recurrent concept, especially favoured by French or Russian geologists.

The discussions about the 'granulite problem' have not come to an end, and both parties (fluid-absent versus fluid-assisted) have their supporters. A consensus is, however, in sight, with the idea that both processes are complementary, not mutually exclusive, and that they may occur at different scales and places.

Structural petrology and geodynamical interpretation

If we can reconstruct the P–T conditions of metamorphic evolution and if we can add the time factor, then we can propose P–T–t paths (or 'trajectories'), which can be interpreted in terms of plate tectonics. Metamorphic rocks are essential in this respect to model orogenic belts, a most important topic in present-day Earth sciences.

Temporal relationships between the different stages of metamorphic mineral growth may serve this purpose. But it was first necessary to reach the concept of 'plurifacial' metamorphism, namely that a single rock may preserve traces of successive metamorphic episodes. This type of investigation is essentially based on the structural relations between coexisting minerals, carefully appreciated under the microscope. Most metamorphic rocks appear then to be organized in coexisting domains, at a subcrystal size, eventually equilibrated at very different P–T conditions. This fundamental observation gives a new dimension to the concept of structural petrology, the other way of studying metamorphic rocks. A gneiss (metamorphic rock) differs from a granite (magmatic rock) by its structure, but this structure concerns not only the external, macroscopic features (foliation, schistosity), but also the internal organization of the rock-forming minerals. These can be equilibrated or not (mutual replacement), and a precise knowledge of these structural relations is evidently of prime importance for a correct estimation of P–T conditions of rock formation. Structural metamorphic petrology has taken great steps forward in modern times, and from a rather thorough survey of the literature of the 1960s, it is evident that Dutch universities have played an important role in the elaboration of a method of study that has now taken the dimensions of almost a sub-discipline. The name 'structural petrology', in its modern sense, was coined by Emile Den Tex, who took up the chair of mineralogy and petrology at the University of Leiden in 1959 after a long and profitable stay in Australia (Den Tex 1959), succeeding Wilhelm de Roever who had preferred to move to Amsterdam. In Amsterdam, De Roever and his co-workers worked mainly in the Betic cordilleras, Spain, where they identified in some rocks three consecutive stages of metamorphism, each characterized by a different metamorphic facies (glaucophane–schist, greenschist– and almandine–amphibolite facies, respectively). But their work was only published in rather obscure journals (De Roever *et al.* 1961), and remained relatively ignored abroad. This was not the case for the Leiden group, which was extremely active and became internationally known under the guidance of both Den Tex, who applied his structural petrology concepts to the study of solid-state deformation of lower crustal (granulites and eclogites)

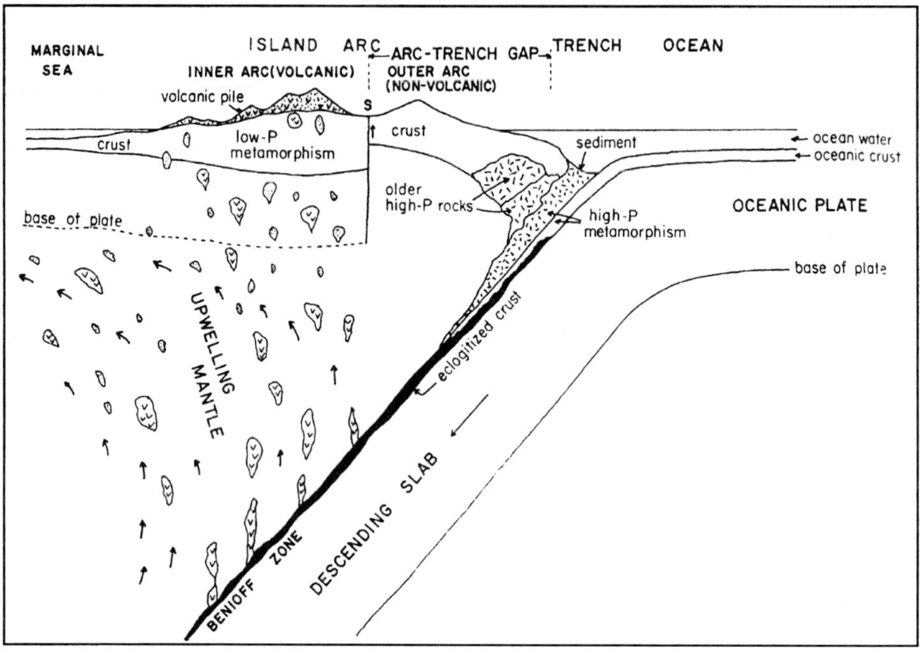

Fig. 10. The model of Miyashiro (1972) illustrating the relations between high-pressure and high-temperature metamorphic types (subduction of oceanic plate below continent).

and mantle rocks, and Henk Zwart, who worked in the Alps and French Pyrenees, notably the Bosost area in the Central Pyrenees (Zwart 1962, 1963). Structural petrology (or mineralogy) is now an extremely active field of research, using instruments or techniques (notably transmission electron microscope, high pressure and temperature experiments, interpretation of dislocations or other defects in mineral structures) that were initially developed in other disciplines, notably metallurgy, material science, or solid-state physics.

Once identified, successive metamorphic facies were found to correspond to different points on a P–T–t 'path' (or trajectory), and nowadays the determination of these is a major objective of metamorphic studies. Their elaboration relies on great instrumental sophistication (high-precision, single-grain analyses for the determination of time, and detailed discussion of a great number of analytical data for pressure and temperature), and many examples are now known with a relatively great accuracy for the most extreme conditions (notably for high-grade metamorphic rocks such as granulites or eclogites). The geodynamic interpretation is facilitated by thermal-model studies, in particular emanating from Oxford University, and culminating in a series of papers by Phil England and co-workers (England & Richardson 1977; England & Thompson 1984; Thompson & England 1984), which outlined the P–T–t paths for specific types of metamorphic rocks.

P–T–t trajectories can then be interpreted in term of plate tectonics. Even though it later proved to rely on the wrong interpretation of field evidence, the concept of paired orogenic belts, proposed by Miyashiro in 1961, played an important role in this respect. In central Japan, two metamorphic belts lie side by side, one corresponding to high-temperature metamorphic type (Abukuma-Ryoke), the other to high pressure (Sanbagawa). This last belt occupies a marginal position in respect to a subduction zone, corresponding to the collision of two oceanic plates. The interpretation by Miyashiro of the relations between the different types of metamorphism is illustrated in Figure 10. The subducted, relatively cold oceanic plate develops high-pressure metamorphism, ultimately leading to the formation of eclogite. During this evolution, it releases fluids (mostly water), which induce melting in the overlying mantle. Alternatively, the plate itself may have melted, on reaching a sufficient depth (of the order of several hundred kilometres), or sunk into the mantle, when the density contrast between mantle rocks and high-pressure eclogites disappear ('slab detachment'). Rising magmas (or fluids, the discussion continues) supposedly induced high-temperature metamorphism in the overriding, collided plate.

In fact, it was found later that the time and space relations that Miyashiro had assumed were mistaken. The two belts are not genetically related: the Sambagawa metamorphism took place far away from the Ryoke Belt (*c.* 2000 km or more away), and later moved to its present position by lateral displacement. However, Miyashiro's model immediately appealed to European geologists, and it was extended to Europe (Caledonian, Variscan and Alpine orogens) by Zwart in 1967. Concurrently, it was recognized, both by Miyashiro (1961, 1967, 1972) and Zwart (1967, 1969), that the type of igneous activity varied with the type of metamorphism. The metamorphic view of orogenesis culminated in Miyashiro's great textbook *Metamorphism and Metamorphic Belts* (1973).

In all cases, however, time and space relations proved to be major problems, when the progress of geochronology and geophysics allowed better understanding of the formation of a mountain chain. Presently, neighbouring orogenic belts were either initially separated by thousands of kilometres, as in Japan, or were made at very different epochs. In central Europe, the Variscan and Caledonian orogens are separated by several hundred million years. In consequence, we have now abandoned the concept of paired belts, but we still see in most single orogens a succession of high-pressure and high-temperature metamorphic stages. For instance, in the western Alps, a Paleocene, high-pressure, eoalpine metamorphism is thought to be followed by a Miocene to Pliocene lepontine high-temperature metamorphic episode. It has also been realized that collision orogens are far more complex than oceanic subduction. Sometimes, in apparent contradiction to basic rheology laws, the oceanic plate appears to have ridden over the continental plate (obduction), leading to the formation of ophiolites, fragments of undisturbed oceanic crust preserved on the continent surface. In other cases, two continental plates may collide (continental subduction). Continent fragments may be carried down to several hundred kilometres depth, and then come back to the surface. The interrelations between the different envelopes of the solid Earth appear to be far more complicated than was formerly assumed. A proper understanding requires advanced studies, in which the three aspects of the modern science of metamorphism – structural, petrological and geochemical – are all performed at the same level of detail and sophistication.

Conclusions

Since its initiation in the first half of the nineteenth century, metamorphic petrology has evolved to a complex discipline, with at least three major trends:

1. mineralogical petrology, mainly concerned with the determination of pressure and temperature conditions at which mineral phases have equilibrated (geothermobarometry);
2. chemical (or geochemical) petrology, which adds the time factor (age determination being made by means of radioactive isotopes), but is primarily concerned with the origin of rock components and the chemical changes that may have happened during metamorphic evolution;
3. structural petrology, which is indispensable for the reconstruction of the kinematics of the rock evolution during an orogenic cycle and its interpretation in terms of global tectonics.

These three approaches are interrelated, with constant feedbacks and 'mutual support'. Thus it is useless to make a complicated solution model for estimating the pressure and temperature conditions at which two neighbouring mineral phases have equilibrated (mineralogical petrology) if these phases have not crystallized at the same time (structural and/or geochemical petrology). All three aspects of metamorphic studies have developed in different directions, to the point that contacts and mutual feedback can be completely lost. Each subdiscipline has developed its own concepts, vocabulary and equipment. Modern analytical instruments 'define' the analysed samples by an extremely narrow beam of light or particles: electrons or ions for analytical probes; lasers for some mass-spectrometric analyses. The volume of analysed material is of the order of a few cubic micrometres or less, so that a single mineral phase is the most common object of study. Field investigations, on the other hand, rely more and more on elaborate geophysical or remote-sensing techniques, which can investigate regions at a subcontinental scale. It has thus become impossible for a single individual to cover all these different aspects at the same level of sophistication. This evolution makes metamorphic petrology an ideal field for well-planned, integrated research, but at the same time it illustrates the risks mentioned at the beginning of this paper. Today, non-specialists may have the impression that mineralogical petrology has reached its ultimate state, with computer programs that everyone can use for estimating pressure and temperature conditions of mineral equilibration. At a time of recession within the university world, it is tempting to favour other disciplines, which may give the impression of more rapid development, thus promising more success in the constant search for grants and other forms of subsidies. The dangers are obvious: science must progress in a coherent way, with consideration for all sectors.

Despite this constant progression, as well as the growing number of concerned disciplines (or subdisciplines), it is remarkable to see how some basic problems, notably the contest between fire (magma) and water (fluid), have remained at the core of research. Although transferring the debate between Werner and Hutton to present-day Earth sciences would be anachronistic, it might well be the case that if these eighteenth-century theorists could come back today, Werner and Hutton would use different instruments and speak another language, but a number of opposing viewpoints in the current debate in metamorphic petrology would still appeal to them.

This paper would never have been started without the friendly request of E. Den Tex to the senior author. It would not have been finished without the help and comments of many colleagues, in the first place D. R. Oldroyd, but also E. Den Tex, H. J. Zwart, T. Andersen, D. Vissers, F. Beunk, G. Gohau and J. Gaudant. Reviews by G. Godard and an anonymous reviewer greatly improved the manuscript.

References

ALTHAUS, E. 1967. The triple point andalusite–sillimanite–kyanite. *Contributions to Mineralogy and Petrology*, **16**, 29–44.

AUBUISSON DES VOISINS, J. F. d' 1819. *Traité de géognosie* (3 vols). F. G. Levrault, Paris (Vol. 3 by A. Burat, 1839).

BARNES, H. L. (ed.) 1967. *Geochemistry of Hydrothermal Ore Deposits*. Holt, Rinehart & Winston, New York.

BARROW, G. 1893. On an intrusion of muscovite–biotite gneiss in the southeastern Highlands of Scotland, and its accompanying metamorphism. *Quarterly Journal of the Geological Society of London*, **49**, 330–356.

BARTON, C. 2002. Marie Tharp, oceanographic cartographer, and her contributions to the revolution in the Earth sciences. *In*: Oldroyd, D. (ed.) *The Earth Inside and Out: Some Major Contributions to Geology in the Twentieth Century*. Geological Society, London, Special Publications, **192**, 215–228.

BECKE, F. 1903. Ueber Mineralbestand und Struktur der kristallinischen Schiefer. *Denkschriften der kaiserliche Akademie der Wissenschaften, Mathematisch-Naturwissenschaftliche Klasse*, **75**, 1–53 (extended German abstract in: *Comptes Rendu IX Session Congress Géologique International*, Part **2**, 553–570).

BECKE, F. 1921. Zur Facies-Klassification der metamorphen Gesteinen. *Mineralogische und Petrographische Mitteilungen*, **35**, 215–230.

BEIMA, E. M. 1867. *De Aarde voor den Zondvloed (bewerkt naar het Fransch en Hoogduitsch van L. Figuier en O. Fraas)*. E. Nijgh en Van den Heuvel & Van Santen, Rotterdam & Leiden.

BERG, J. H. & WHEELER II, E. P. 1976. Osumilite of deep-seated origin in the contact aureole of the anorthositic Nain complex, Labrador, Canada. *American Mineralogist*, **61**, 29–37.

BERMAN, R. G. 1991. Thermobarometry using multi-equilibrium calculations: a new technique, with petrological applications. *Canadian Mineralogist*, **29**, 833–855.

BERMAN, R. G. & PERKINS, E. H. 1987. GEO–CALC: software for calculations and display of pressure–temperature–composition phase diagrams. *American Mineralogist*, **72**, 861–862.

BONIN, B., DUBOIS, R. & GOHAU, G. 1997. *Le métamorphisme et la formation des granites*. Nathan Université, Paris.

BOUÉ, A. 1820. *Essai géologique sur l'Ecosse*. Veuve Courcier, Paris.

BOUÉ, A. 1824. Mémoire géologique sur le sud-ouest de la France, suivi d'observations comparatives sur le nord du même royaume, et en particulier sur les bords du Rhin. *Annales des Sciences Naturelles*, **2**, 387–423; **3**, 55–95, 299–317.

BOWEN, N. L. 1928. *The Evolution of Igneous Rocks*. Princeton University Press, Princeton.

BOWEN, N. L. 1940. Progressive metamorphism of siliceous limestone and dolomite. *Journal of Geology*, **48**, 225–274.

BREISLAK, S. 1822. *Traité sur la structure extérieure du Globe* (3 vols). J. P. Giegler, Milan; Fantin, Paris.

CABY, R. 1994. Precambrian coesite from northern Mali: first record and implications for plate tectonics in the trans-Sahara segment of the Pan-African belt. *European Journal of Mineralogy*, **6**, 235–244.

CHESNOKOV, B. V. & POPOV, V. A. 1965. Increase in the volume of quartz grains in South Urals eclogites (English translation) *Doklady Academy of Sciences S. S. S. R.*, **162**, 176–178.

CHOPIN, C. 1984. Coesite and pure pyrope in high-grade blueschists of the western Alps: a first record and some consequences. *Contributions to Mineralogy and Petrology*, **86**, 107–118.

CLEMENS, J. & VIELZEUF, D. 1987. Constraints on melting and magma production in the crust. *Earth and Planetary Science Letters*, **86**, 207–306.

COES, L. 1953. A new dense crystalline silica. *Science*, **118**, 131–132.

COLLOMB, P. 1998. Deux siècles d'évolution des idées sur le métamorphisme, notamment en France, dans le Massif Central. *Bulletin de la Société Géologique de France*, **169**, 725–736.

COOMBS, D. S. 1954. The nature and alteration of some Triassic sediments from Southland, New Zealand.

Transactions of the Royal Society of New Zealand, **82**, 65–109.

COOMBS, D. S. 1960. Lower grade mineral facies in New Zealand. *Proceedings of the 21st International Geological Congress, Copenhagen,* **13**, 339–351.

CORDIER, L. 1828. Essai sur la température de l'intérieur de la terre. *Bibliothèque universelle des sciences et des arts de Genève*, **1**, 85–118.

DAUBRÉE, A. 1857. Observations sur le métamorphisme et recherches expérimentales sur quelques uns des agents qui ont pu le produire. *Annales des mines*, **5**, 289–326.

DAUBRÉE, A. 1859. Études et expériences synthétiques sur le métamorphisme et sur la formation des roches cristallines. *Annales des mines*, **5**, 155–218, 393–476 (English translation: *Smithsonian Annual Report for 1961*, 228–304).

DE LAPPARENT, A. 1906. *Traité de géologie* (3 vols, 5th edn). Masson, Paris.

DELESSE, A. 1857. Etudes sur le métamorphisme des roches. *Annales des mines*, **5–12**, 89–288, 417–516, 705–772. (See also **5–13**, 321–416 and *Mémoires présentés par divers savants à l'Académie des Sciences*, **17**, 1–97.)

DEMAY, A. 1942. Microtectonique et tectonique profonde: cristallisations et injections magmatiques syntectoniques. *Mémoire du Service de la Carte Géologique de France*. Paris.

DEN TEX, E. 1959. De ontwikkeling van het struktuurbegrip in de Petrologie. *Inaugurele Oratie, Leiden University*. E. J. Brill, Leiden.

DEN TEX, E. 1969. Origin of ultramafic rocks, their tectonic setting and history: a contribution to the discussion of the paper 'The origin of ultramafic and ultrabasic rocks' by P. J. Wyllie. *Tectonophysics,* **7**, 457–488.

DE ROEVER, W. P. 1950. Preliminary notes on glaucophane-bearing and other crystalline schists from south-east Celebes, and the origin of glaucophane-bearing rocks. *Proceedings of the Koninklijke Nederlandse Akademie van Wetenschappen,* **53**, 1455–1465.

DE ROEVER, W. P. 1955*a*. Some remarks concerning the origin of glaucophane in the North Berkeley Hills, California. *American Journal of Science*, **253**, 240–244.

DE ROEVER, W. P. 1955*b*. Genesis of jadeite by low-grade metamorphism. *American Journal of Science,* **253**, 283–298.

DE ROEVER, W. P. 1956. Some differences between post-Paleozoic and older regional metamorphism. *Geologie en Mijnbouw*, **18**, 123–127.

DE ROEVER, W. P. 1964. On the cause of the preferential distribution of certain metamorphic minerals in orogenic belts of different age. *Geologische Rundschau*, **53**, 324–336.

DE ROEVER, W. P., EGELER, C. G. & NIJHUIS, H. J. 1961. Nota preliminar sobre la geologia de la llamada zona mixta tal como se desarolla en el extremo est de la Sierra de los Filabres (SE de España). *Notas y communicationes del Instituto Geologico y Minero de España*, **63**, 223–232.

DUROCHER, J. 1845. Etudes sur le métamorphisme des roches. *Bulletin de la Société Géologique de France*, Series 1, **3,** 546–648.

ELIE DE BEAUMONT, L. 1831. Recherches on some of the revolutions on the surface of the globe; presenting various examples of the coincidence between the elevation of beds in certain systems of mountains, and the sudden changes which have produced the lines of demarcation observable in certain stages of the sedimentary deposits. *Philosophical Magazine*, **10**, 241–264.

ELLIS, D. J. 1987. Origin and evolution of granulites in normal and thickened crust. *Geology*, **15**, 167–170.

ELLIS, D. J., SHERATON, J. W., ENGLAND, R. N. & DALLWITZ, W. B. 1980. Osumilite–sapphirine–quartz granulites from Enderby Land, Antarctica: mineral assemblages and reactions. *Contributions to Mineralogy and Petrology*, **90**, 123–143.

ENGLAND, P. C. & RICHARDSON, S. W. 1977. The influence of erosion upon mineral facies of rocks from different metamorphic environments. *Journal of the Geological Society, London*, **134**, 201–213.

ENGLAND, P. C. & THOMPSON, A. B. 1984. Pressure–temperature–time paths of regional metamorphism. I. *Journal of Petrology*, **25**, 894–928.

ESKOLA, P. 1915. On the relations between the chemical and mineralogical composition in the metamorphic rocks of the Orijärvi region. *Bulletin de la Commission Géologique de Finlande*, **44**, 109–143.

ESKOLA, P. 1920. The mineral facies of rocks. *Norsk Geologisk Tidsskrift*, **6**, 143–194.

ESKOLA, P. 1929. Om mineral facies. *Geologiske Förening i Stockholm Förhandlingar,* **51**, 157–173.

ESKOLA, P. 1939. Die metamorphen Gesteine. *In:* BARTH, T. W., CORRENS, C. W. & ESKOLA, P. (eds) *Die Entstehung der Gesteine*. Springer, Berlin, 263–407.

EUGSTER, H. P. 1959. Reduction and oxidation in metamorphism. *In*: ABELSON, P. H. (ed.) *Researches in Geochemistry.* John Wiley, New York, 397–426.

FERRY, J. M. 1976. P, T, fCO_2, and fH_2O during metamorphism of calcareous sediments in the Waterville–Vassalboro area, south-central Maine. *Contributions to Mineralogy and Petrology*, **57**, 119–143.

FERRY, J. M. 1983. Regional metamorphism of he Vassalboro formation, south-central Maine, U S A: A case study of the role of fluid in metamorphic petrogenesis. *Journal of the Geological Society, London*, **140**, 551–576

FERRY, J. M. 1987. Metamorphic hydrology at 13-km depth and 400–450°C. *American Mineralogist,* **72**, 39–58.

FIGUIER, L. 1863. *La terre avant le déluge*. Libraire Hachette, Paris.

FONAREV, V. I., GRAPHCHIKOV, A. A. & KONILOV, A. N. 1991. A consistent system of geothermometers for metamorphic complexes. *International Geological Review*, **33**, 743–783.

FOUQUÉ, F. & MICHEL-LÉVY, A. 1878. *Minéralogie micrographique*. Librairie Polytechnique Baudry, Paris.

FOURNET, J. 1848. Aperçus sur diverses questions géologiques. *Bulletin de la Société Géologique de France*, Series 2, **6,** 502–518.

FOURNIER, J. 1820. Mémoire sur le refroidissement séculier du globe terrestre. *Société Philomathique, Bulletin,* 81–87, 156–165.

FOURNIER, J. 1837. General remarks on the temperature of the terrestrial globe and planetary spaces. *American Journal of Science,* **32**, 1–20.

FRANCK, E. U. & TÖTHEIDE, K. 1959. Thermische Eigenschaften überkritischer Mischungen von Kohlendioxyd und Wasser bis zu 750°C und 2000 atm. *Zeitschrift für Physikalische Chemie, neue folge,* **22**, 232–245.

FRENCH, B. M. 1966. Some geological implications of equilibrium between graphite and a C–H–O gas phase at high temperature and pressure. *Reviews in Geophysics*, **4**, 223–253.

FRITSCHER, B. 2002. Metamorphism and thermodynamics: the formative years. *In*: Oldroyd, D. (ed.) *The Earth Inside and Out: Some Major Contributions to Geology in the Twentieth Century*. Geological Society, London, Special Publications, **192**, 143–166.

GARRELS, R. M. & CHRIST, C. L. 1965. *Solutions, Minerals and Equilibria.* Harper & Row, New York.

GEIKIE, A. 1905. *The Founders of Geology* (2nd edn). Macmillan, London.

GODARD, G. 2001. Eclogites and their geodynamic interpretation: a history. *Journal of Geodynamics Special Issue on the History of Geodynamics*, **32**, 165–203.

GOLDSCHMIDT, V. M. 1911. Die Kontaktmetamorphose in Kristianiagebiet. *Norsk Videnskapelig Selskap i Oslo Skrifter, Matematisk-Naturvidenskapelig Klasse* **11**, 1–483.

GOLDSCHMIDT, V. M. 1912. Die Gesetze der Gesteinmetamorphose, mit Beispiele aus der Geologie des südlichen Norwegens. *Norsk Videnskapelig Selskap i Oslo Skrifter, Matematisk-Naturvidenskapelig Klasse*, **22**.

GOLDSCHMIDT, V. M. 1915. Geologisch-petrographische Studien im Hochgebirge des südlichen Norwegens. III. Die Kalksilikatgneise und Kalksilikatglimmerschiefer im Trondhjem-Gebiete. *Norsk Videnskapelig Selskap i Oslo Skrifter, Matematisk-Naturvidenskapelig Klasse*, **10**.

GREEN, M. T. 1992. *Geology in the Nineteenth Century*. Cornell University Press, Ithaca.

GRUBENMANN, U. 1904–1906. *Die Kristallinen Schiefer* (2 vols 2nd edn 1910). Gebrueder Bornträger, Berlin.

GRUBENMANN, U. & NIGGLI, P. 1924. *Die Gesteinsmetamorphose.* Gebrueder Bornträger, Berlin.

HALL, J. 1805. Experiments on whinstone and lava (March, June 1798). *Transactions of the Royal Society of Edinburgh*, **5**, 43–75.

HALL, J. 1812. Account of a series of experiments, showing the effect of compression in modifying the action of heat (June 1805). *Transactions of the Royal Society of Edinburgh*, **6**, 71–185.

HALL, J. 1826. On the consolidation of the strata of the Earth (April 1825). *Transactions of the Royal Society of Edinburgh*, **18**, 314–329.

HARKER, A. 1918. The present position and outlook of the study of metamorphism of rock masses. *Quarterly Journal of the Geological Society of London*, **74**, 51–80.

HARKER, A. 1932. *Metamorphism: A Study of the Transformation of Rock Masses* (2nd edn, 1939). Methuen, London.

HARLEY, S. L. 1989. The origin of granulites: a metamorphic perspective. *Geological Magazine*, **126**, 215–247.

HAUG, E. 1907–1911. *Traité de géologie* (3 vols). Armand Colin, Paris.

HELGESON, H.C. 1964. *Complexing and Hydrothermal Ore Deposition*. Pergamon, New York.

HEMLEY, R. J., BELL, P. M. & MAO, H. K. 1987. Laser techniques in high-pressure geophysics. *Science*, **237**, 605–612.

HESTMARK, G. 1999. *Vitenskap of Nasjon: Waldemar Christoffer Brøgger 1851–1905*. Aschehoug, Oslo.

HOLLAND, T. B. J. & POWELL, R. 1998. An internally consistent thermodynamic data set for phases of petrological interest. *Journal of Metamorphic Geology*, **16**, 309–343.

HOLMES, A., SHILLIBEER, H. A. & WILSON, J. T. 1957. Potassium–argon ages of some Lewisian and Fennoscandian pegmatites. *Nature,* **176**, 390–392.

HUNT, T. S. 1884. The origin of crystalline rocks. *Mémoires et comptes rendus de la Société Royale du Canada,* Section **III**, 1–67.

HUTTON, J. 1788. Theory of the Earth; or an investigation of the laws observable in the composition, dissolution and restoration of land upon the globe. *Transactions of the Royal Society of Edinburgh*, **1**, 209–304.

HUTTON, J. 1795. *Theory of the Earth, with Proofs and Illustrations* (2 vols). Cadell, Junior & Davies, London; William Creech, Edinburgh (reprinted 1960, Weldon & Wesley, London).

JAMESON, R. 1800. *Mineralogy of the Scottish Isles* (2 vols). Edinburgh.

JUDD, J. W. 1908. Henry Clifton Sorby, and the birth of microscopical petrology. *Geological Magazine*, Decade 5, **5**, 193–204.

JUNG, J. & ROQUES, M. 1952. Introduction à l'étude zonéographique des formations cristallophylliennes. *Bulletin du Service de Carte Géologique de France*, **50**, 1–61.

KAWAI, N. & ENDO, S. 1970. The generation of ultrahigh hydrostatic pressure by a split sphere apparatus. *Review of Scientific Instruments*, **41**, 1178–1181.

KENNEDY, G. C. 1950. A portion of the system silica–water. *Economic Geology*, **45**, 629–653.

KENNEDY, G. C. 1954. Pressure–temperature–volume relations in CO_2 at elevated pressures and temperatures. *American Journal of Science*, **252**, 225–241.

KORZHINSKII, D. S. 1936. Mobility and inertness of components in metasomatosis. *Doklady Akademii Nauk S. S. S. R., Geological Series*, **1**, 35–60.

KORZHINSKII, D. S. 1959. *Physicochemical basis of the analysis of the paragenesis of minerals*. Consultant Bureau, New York.

LAMB, W. & VALLEY, J. W. 1984. Metamorphism of reduced granulites in low-CO_2 vapour-free environment. *Nature*, **312**, 56–57.

LEWIS, C. L. E. 2002. Arthur Holmes' unifying theory: from radioactivity to continental drift. *In*:

Oldroyd, D. (ed.) *The Earth Inside and Out: Some Major Contributions to Geology in the Twentieth Century*. Geological Society, London, Special Publications, **192**, 167–184.

LIOU, J. C., MARUYAMA, S., WANG, X. & GRAHAM, S. 1990. Precambrian blueschist terranes of the world. *Tectonophysics*, **181**, 97–111.

LYELL, C. 1833. *Principles of Geology, Being an Attempt to Explain the Former Changes of the Earth's Surface, by Reference to Causes Now in Operation* (3 vols, 10th edn, 1867). John Murray, London.

MAIJER, C., JANSEN, J. B. H., WEVERS, J. & POORTER, R. P. E. 1977. Osumilite, a mineral new to Norway. *Norsk Geologisk Tidsskrift,* **57**, 187–188.

MAO, H. K. & BELL, P. M. 1975. Design of a diamond-windowed, high-pressure cell for hydrostatic pressure in the range 1 bar to 0.5 Mbar. *Carnegie Institution of Washington Yearbook 1974*, 402–405.

MAO, H. K. & BELL, P. M. 1976. High-pressure physics: the 1-megabar mark on the ruby R_1 static pressure scale. *Science*, **191**, 851–852.

MICHEL-LÉVY, A. 1887. Sur l'origine des terrains cristallins primitifs. *Bulletin de la Société Géologique de France*, Series 3, **16**, 102–113.

MICHEL-LÉVY, A. 1888. *Les minéraux des roches*. Librairie Polytechnique Baudry, Paris.

MIYASHIRO, A. 1949. The stability relation of kyanite, sillimanite and andalusite, and the physical conditions of the metamorphic processes. *Journal of the Geological Society of Japan*, **55**, 218–223.

MIYASHIRO, A. 1961. Evolution of metamorphic belts. *Journal of Petrology*, **2**, 277–311.

MIYASHIRO, A. 1967. Aspects of metamorphism in the circum-Pacific region. *Tectonophysics*, **4**, 519–521.

MIYASHIRO, A. 1972. Metamorphism and related magmatism in plate tectonics. *American Journal of Science*, **272**, 629–656.

MIYASHIRO, A. 1973. *Metamorphism and Metamorphic Belts*. Allen & Unwin, London.

MIYASHIRO, A. 1994. *Metamorphic petrology*. UCL Press, London.

NEWTON, R. C. 1966. Kyanite–andalusite equilibrium from 700 to 800°C. *Science*, **153**, 170–172.

NEWTON, R. C., SMITH, J. V. & WINDLEY, B. F. 1980. Carbonic metamorphism, granulites and crustal growth. *Nature,* **288**, 45–50.

OUZEGANE, K. & BOUMAZA, S. 1996. An example of ultrahigh-temperature metamorphism: Orthopyroxene–sillimanite–garnet, sapphirine–quartz and spinel–quartz parageneses in Al–Mg granulites from In Hihaou, In Ouzzal, Hoggar. *Journal of Metamorphic Geology*, **14**, 693–708.

PERCHUK, L. L. & GERYA, T. V. 1993. Fluid control of charnockitisation. *Chemical Geology*, **108**, 175–186.

PERRIN, R. & ROUBAULT, M. 1939. Le granite et les réactions à l'état solide. *Bulletin du Service de la Carte Géologique d'Algérie,* **(5) 4**.

PERRIN, R. & ROUBAULT, M. 1963. Remarques sur l'interprétation des données expérimentales: leur liaison avec la genèse des granites stratoïdes. *Science de la terre,* **9**, 73–81.

PLAYFAIR, J. 1802. *Illustrations of the Huttonian Theory of the Earth.* William Creech, Edinburgh.

POTY, B. 1969. La croissance des cristaux de quartz dans les filons sur l'exemple de la Gardette (Bourg d'Oisans) et du Massif du Mont-Blanc. *Science de la terre Mémoire*, **17**, 1–162.

RAGUIN, E. 1957. *Géologie du granite*. Masson, Paris.

RAMBERG, H. 1944. Petrological significance of sub-solidus phase transitions in mixed crystals. *Norsk Geologisk Tidsskrift,* **24**, 42–73.

RAMBERG, H. 1949. The facies classification of rocks: a clue to the origin of quartzo-feldspathic massifs and veins. *Journal of Geology*, **57**, 18–54.

RAMBERG, H. 1952. *The Origin of Metamorphic and Metasomatic rocks*. University of Chicago Press, Chicago.

READ, H. H. 1943–1944. Meditations on granite. *Proceedings of the Geologists' Association*, Part 1, **54**, 64–85, Part 2, **55**, 45–93.

READ, H. H. 1957. *The Granite Controversy.* Thomas Murby, London.

RICHARDSON, S. W., GILBERT, M. C. & BELL, P. M. 1969. Experimental determination of kyanite–andalusite and andalusite–sillimanite equilibria; the aluminum silicate triple point. *American Journal of Science*, **267**, 259–272.

RINGWOOD, A. E. 1975. *Composition and Petrology of the Earth's Mantle.* McGraw-Hill, New York.

ROEDDER, E. 1962–1963. Studies of fluid inclusions. I: Low-temperature application of a dual-purpose freezing and heating stage. *Economic Geology*, **57**, 1045–1061. II: Freezing data and their interpretation. *Economic Geology*, **58**, 167–211.

ROEDDER, E. 1984. Fluid inclusions. *Reviews in Mineralogy*, **12**, 1–646.

ROSENBUSCH, H. 1873 (Band 1)–1877 (Band II). *Mikroskopische Physiographie der Mineralien und Gesteine* (2nd edn, 1885–1887; 4th edn [with A. Wulfing], 1904–1908; 5th edn [by O. A. Mügge], 1924–1927). E. Schweizerbartsche, Stuttgart.

ROSENBUSCH, H. 1877. Die Steiger Schiefer und ihre Kontaktzone an den Graniten von Barr-Andlau und Hohwald. *Abhandlungen zum geologische SpezialKarte Elsass-Lothringen, Strasburg*, **1**, 79–214.

ROSENBUSCH, H. 1898–1910. *Elemente der Gesteinlehre* (2nd edn, 1901; 3rd edn, 1910). E. Schweizerbartsche, Stuttgart.

RUDWICK, M. J. S. 1962. Hutton and Werner compared: George Greenough's geological tour of Scotland in 1805. *British Journal for the History of Science,* **1,** 117–135.

SANDER, B. 1948–1950. *Einführung in die Gefügekunde der Geologischen Körper* (2 vols). Springer, Vienna.

SCHEERER, T. 1847. Discussions sur la nature plutonique du granite et des schistes cristallins qui s'y rallient. *Bulletin de la Société Géologique de France*, Séries 2, **4,** 468–495.

SCHERTL, H. P., SCHREYER, W. & CHOPIN, C. 1991. The pyrope–coesite rocks and their country rocks at Parigi, Dora Maira massif, western Alps: detailed petrography, mineral chemistry and PT–path. *Contributions to Mineralogy and Petrology*, **108**, 1–21.

SCHREYER, W. 1988. Experimental studies on metamorphism of crustal rocks under mantle pressures. *Mineralogical Magazine*, **52**, 1–26.

SCHREYER, W. 1995. Ultradeep metamorphic rocks: the retrospective viewpoint. *Journal of Geophysical Research*, **100 B**, 8353–8366.

SCHUILING, R. D. 1957. A geo-experimental phase-diagram of Al_2SiO_5 (sillimanite, kyanite, andalusite). *Proceedings of the Koninklijke Nederlandse Akademie van Wetenschappen*, Series B, **60**, 220–226.

SEDERHOLM, J. J. 1907. Om granit och gneis. *Bulletin de la Commission Géologique de Finlande*, **23**.

SMITH, D. C. 1984. Coesite in clinopyroxene in the Caledonides and its implications for geodynamics. *Nature*, **310**, 641–644.

SOBOLEV, N. V. & SHATSKY, V. S. 1987. Carbon mineral inclusions in garnets from metamorphic rocks. *Geologia, Geofizica*, **7**, 77–80.

SOBOLEV, N. V. & SHATSKY, V. S. 1990. Diamond inclusions in garnets from metamorphic rocks: a new environment for diamond formation. *Nature*, **343**, 742–746.

SORBY, H. C. 1858. On the microscopic structure of crystals, indicating the origin of minerals and rocks. *Quarterly Journal of the Geological Society, London*, **14**, 453–500.

SPEAR, F. S. 1993. *Metamorphic Phase Equilibria and Pressure–Temperature–Time Paths*. Mineralogical Society of America, Washington.

STISHOV, S. M. & POPOVA, S. V. 1961. A new dense modification of silica. *Geochemistry*, **10**, 923–926.

TERMIER, P. 1903. Les nappes des Alpes orientales et la synthèse des Alpes. *Bulletin de la Société Géologique de France*, Series 4, **3**, 711–766.

THOMPSON, A. B. 1982. Dehydration melting of pelitic rocks and the generation of H_2O-undersaturated granitic liquids. *American Journal of Science*, **282**, 1567–1595.

THOMPSON, A. B. & ENGLAND, P. C. 1984. Pressure–-temperature–time paths of regional metamorphism. II. *Journal of Petrology*, **25**, 929–955.

THOMPSON J. B. JR 1955. The thermodynamic for the mineral facies concept. *American Journal of Science*, **253**, 65–103.

THOMPSON J. B. JR 1957. The graphical analysis of mineral assemblages in pelitic schists. *American Mineralogist*, **42**, 842–858.

TILLEY, C. E. 1925. A preliminary survey of metamorphic zones in the Southern Highlands of Scotland. *Quarterly Journal of the Geological Society of London*, **81**, 100–112.

TOURET, J.[L. R.] 1971. Le faciès granulite en Norvège méridionale. I: Les associations minéralogiques. *Lithos*, **4**, 239–249. II: Les inclusions fluides. *Lithos*, **4**, 423–436.

TOURET, J. L. R. & HUIZENGA, J. M. 1999. Intraplate magmatism at depth: high-temperature lower crustal granulites. *Journal of African Earth Sciences*, **27**, 367–382.

TURNER, F. J. & VERHOOGEN, J. 1951. *Igneous and Metamorphic Petrology* (2nd edn, 1960). McGraw–Hill, New York.

TUTTLE, O. 1949. Structural petrology of planes of liquid inclusions. *Journal of Geology*, **57**, 331–356.

VALLEY, J. W., BOHLEN, S. R., ESSENE, E. J. & LAMB, W. 1990. Metamorphism in the Adirondacks: II. The role of fluids. *Journal of Petrology,* **31**, 555–596.

VAN HISE, C. 1904. *A treatise on metamorphism*. US Geological Survey, Monograph **47**.

VERNADSKY, W. 1924. *La géochimie*. Félix Alcan, Paris.

VOGELSANG, H. 1867. *Philosophie der Geologie und mikroskopische Gesteinstudien*. Bonn.

VOGT, J. H. L. 1903–1904. Die Silikatschmelzlösungen mit besonderer Rücksicht auf die Mineralbildung und die Schmelzpunkt-Erniederung *Norsk Videnskapelig Selskap i Oslo Skrifter, Matematisk-Naturvidenskapelig Klasse,* **1903–1908**, 1–162; **1904–1901**, 1–236.

VON ZITTEL, K.A. 1899. Geschichte der Geologie und Paläontologie bis Ende des 19. Jahrunderts. Drud und Verlag, München und Leipzig (English translation by OGILVIE-GORDON, M. M. 1901. *History of Geology and Paleontology to the End of the Nineteenth Century*. W. Scott, London; reissued, Cramer, 1962).

WEGMANN, C. E. 1929. Beispiele tektonischer Analysen des Grundgebirges in Finland. *Bulletin de la Commission Géologique de Finlande*, **87**, 99–126.

WEGMANN, C. E. 1935. Zur Deutung der Migmatite. *Geologische Rundschau,* **26**, 305–350.

WERNER, A. G. 1774. *Von den äusserlichen Kennzeichen der Fossilien* (English translation by A. V. Carozzi. 1962. *On the External Character of Minerals*. University of Illinois, Urbana).

WHITE, A. J. R. & CHAPPEL, B. W. 1990. Per migma ad magma downunder. *Geological Journal*, **25**, 221–225.

WILLIAMS, G. H. 1890. The Greenstone Schist areas of the Menominee and Marquette regions of Michigan. A contribution to the subject of dynamic metamorphism in eruptive rocks. *US Geological Survey Bulletin,* **62**, 1–241.

WINKLER, H. G. F. 1965. *Petrogenesis of Metamorphic Rocks* (5th edn 1979). Springer Verlag, Berlin & NewYork.

WYART, J. & SABATIER, G. 1958. Mobilité des ions Si et Al dans les cristaux de feldspath. *Bulletin de la Société Française de Minéralogie et Cristallographie*, **81**, 223–226.

WYART, J. & SABATIER, G. 1960. Réaction du granite fondu sous pression d'eau avec la calcite et une argile. Application au métamorphisme de contact. *Bulletin de la Société Française de Minéralogie et Cristallographie*, **83**, 128–133.

YODER, H. S. 1993. Timetable of petrology. *Journal of Geological Education*, **41**, 47–489.

YODER, H. S. & EUGSTER, H. P. 1954. Phlogopite synthesis and stability range. *Geochimica Cosmochimica Acta*, **6**, 157–165.

YOUNG, D. A. 1998. *N. L. Bowen and Crystallization-Differentiation Theory: The Evolution of a Theory.* Mineralogical Society of America, Washington.

YOUNG, D. A. 2002. Norman Levi Bowen (1887–1956) and igneous rock diversity. *In*: Oldroyd, D. (ed.) *The Earth Inside and Out: Some Major Contributions to Geology in the Twentieth Century*. Geological Society, London, Special Publications, **192**, 99–112.

ZEN, E-AN 1966. Construction of pressure–temperature diagrams for multicomponent systems after the method of Schreinemakers–a geometrical approach. *US Geological Survey Bulletin*, **1225**.

ZIRKEL, F. 1866. *Lehrbuch der Petrographie* (2 vols). A. Marcus, Bonn.

ZWART, H. J. 1962. On the determination of polymorphic mineral associations, and its application to the Bosost area (Central Pyrenees). *Geologische Rundschau,* **52**, 38–65.

ZWART, H. J. 1963. Some examples of the relations between deformation and metamorphism from the central Pyrenees. *Geologie en Mijnbouw*, **42**, 143–154.

ZWART, H. J. 1967. The duality of orogenic belts. *Geologie en Mijnbouw*, **46**, 283–309.

ZWART, H. J. 1969. Metamorphic facies series in the European orogenic belts and their bearing on the causes of orogeny. *Geological Association of Canada, Special Paper*, **5**, 7–16.

Metamorphism and thermodynamics: the formative years

BERNHARD FRITSCHER

Ludwig-Maximilians-Universität München, Institut für Geschichte der Naturwissenschaften, Museumsinsel 1, D-80306 Munich, Germany

Abstract: Between 1890 and the 1920s petrologists and mineralogists began to apply concepts of theoretical chemistry – in particular, the concept of chemical equilibrium – to the study of metamorphic rocks. The majority of the petrological community, however, hesitated to apply the new method on a large scale to metamorphism. Focusing on the works of Becke, Goldschmidt and Eskola, some early approaches to a linkage of metamorphic petrology and theoretical chemistry are reviewed. The controversial discussion, particularly of Goldschmidt's classical study of the Christiania area, led Miyashiro to distinguish two paradigms of early twentieth-century metamorphic petrology. With regard to the contemporary discussion, as well as to Miyashiro's interpretation, this paper is concluded by an epilogue on 'image and logic', which is intended to relate the paradigms of early modern metamorphism to different cultures, and 'national styles' of Earth sciences in the nineteenth and early twentieth centuries.

> The most beautiful metamorphosis in the inorganic kingdom occurs when, during formation, formlessness changes to form. Each piece of matter has its own impulse and its own right. The mica schist changes to garnets and makes up rock masses where the mica is nearly completely dissolved, remaining nothing but an inferior binder between those crystals. (Goethe)

The study of the occurrence and formation of metamorphic rocks, as well as the role of metamorphic processes in tectonics and Earth history, is among the most interesting and important features of modern Earth sciences. Starting in the late eighteenth and early nineteenth centuries as a mass of divergent, and sometimes curious, ideas on the transformation of sedimentary and crystalline rocks, metamorphism became a special branch of petrography in the second half of nineteenth century. Around 1900, the first concepts of progressive metamorphism were established and, at the same time, a small group of petrologists and mineralogists began to apply ideas from theoretical chemistry, i.e. the concepts of chemical equilibria, to the study of metamorphic rocks. In the 1960s and early 1970s, Akiho Miyashiro (*b.* 1920) (Miyashiro 1973) succeeded in linking metamorphic petrology firmly to tectonics by his concept of metamorphic belts. And finally, at the end of the century, studies of ultra-high-pressure metamorphism opened up a new and ongoing chapter in Earth science.[1]

For historians of science, metamorphism is also of interest as an indicator of different practices and different cultures of the Earth sciences. Throughout its history, metamorphic geology has been particularly related to the cultural and social context of Earth sciences. Thus, the story of metamorphism is not only the story of the establishment of a new branch of geology and new methods in Earth sciences but also the story of 'geology's road to modernity'. Miyashiro indicated the existence of such a philosophical context by his distinction between two fundamentally different paradigms of metamorphic petrology in relation to the reception of the doctrine of chemical equilibrium in the first half of the twentieth century.

The present paper focuses on the history of the early applications of theoretical chemistry to the study of metamorphism, i.e. on the 'formative years' of 'metamorphism and thermodynamics', covering a period of about three decades from the 1890s to the 1920s. With regard to Miyashiro's 'philosophy' of metamorphism, however, an epilogue is added, discussing some ideas by means of which Miyashiro's paradigms of early metamorphic petrology can be related to different cultures, and 'national styles' of Earth sciences in the nineteenth and early twentieth centuries.

---

[1] In this paper 'metamorphism' is used in a broad sense, i.e. it covers all the ideas on the subject from the middle of the nineteenth century until the end of the twentieth century; for the modern notion of metamorphism see Miyashiro (1973) and Schreyer (1995). 'Metamorphosis' is used to distinguish an overall notion of the 'transformation', or the 'change of form or character' of things (plants, animals or human society). The term does not primarily relate to the transformation of rocks.

From: Oldroyd, D. R. (ed.) 2002. *The Earth Inside and Out: Some Major Contributions to Geology in the Twentieth Century*. Geological Society, London, Special Publications, **192,** 143–165. 0305-8719/02/$15.00

The history of metamorphism is not yet written. Valuable details, however, are given by Winkler (1965), Miyashiro (1973, 1994), Gohau (1997) and Touret & Nijland (2002). For the nineteenth century, Fischer (1961) is recommended, as well as some older textbooks (e.g. Zirkel 1894; Grubenmann & Niggli 1924). Eitel (1925) provides a comprehensive contemporary account of mineralogical and petrological research based on the doctrine of chemical equilibria. Some of the basic papers on metamorphism from the nineteenth and early twentieth centuries were reprinted by Mather & Mason (1939) and Mather (1967). A retrospective view of ultra-high-pressure metamorphism has been given by Schreyer (1995; 1999).

Beginnings: the Kaiser's petrographers, and a natural history of metamorphism

The question of when and where the study of metamorphism began produces different answers depending on the different styles of metamorphism considered, and the different meanings of the term metamorphism. The basic idea of metamorphism – that some rocks, or even classes of rocks, are formed by the transformation of older rocks by means of heat, pressure, water, hot vapours, and so on – was popular during the nineteenth century. This was due not only to specific aspects of early nineteenth-century Earth sciences, in particular, the Vulcanist–Plutonist–Neptunist controversy, but also to some of the cultural and philosophical ideas of the century. In addition to these specific geological and mineralogical aspects, the 'popularity' of metamorphism in the nineteenth century was due to the widespread use of the term 'metamorphosis'. It was used in science, literature and, the arts, denoting an alteration of form, where 'form', according to its classical meaning, meant the sum of the specific (i.e. chemical and physical) 'qualities' of a thing. Goethe's 'Metamorphosis of Plants' provided the most outstanding example of this meaning of metamorphosis. And for most of the German 'nature philosophers' (*Naturphilosophen*) of the early nineteenth century the very formation of the physical world was a kind of process of 'metamorphosis' (Lichtenstern 1990).

Early on, the French geologist Ami Boué (1794–1881) gave, with regard to the 'stratified crystalline rocks', a comprehensive discussion of the transformation of sedimentary to crystalline rocks by heat, infiltrations, volcanic gases and so on (Boué 1820). In his descriptions of such transformations he used the term *métamorphose* from time to time. Commonly, however, he spoke of 'alteration'. Then, in the third volume of his *Principles of Geology* (Lyell 1833, pp. 374–375), Charles Lyell (1797–1875) defined the 'altered stratified crystalline rocks' as a separate class of 'hypogene' or 'nether-formed' (i.e. formed in the Earth's interior) rocks and called them 'metamorphic rocks' (from Greek *meta*, indicating change, and *morphe* = form), the unaltered ones being 'plutonic rocks'.

The beginnings of modern metamorphism are, then, clearly marked. Actually, it was the Franco–Prussian war of 1870–1871, or rather some of its political consequences, that opened up new concepts of metamorphism. France had to cede Alsace Lorraine, which became a crown land (*Reichsland*) of the newly founded German Empire. Two of the measures intended to make the new territory an integral part of the Empire were the foundation of a German University at Strasbourg, and the establishment of the Geological Survey of Alsace Lorraine.

The new Imperial University became the first modern German university, with respect both to its equipment and its staff. One of the first-rate young scientists who were called to Strasbourg was Harry Rosenbusch (1836–1914). He had pioneered the use of the polarizing microscope in petrography, and the first volume of his *Mikroskopische Physiographie der petrographisch wichtigen Mineralien* (1873) had just been completed. He was also called to Strasbourg to contribute to the Alsace Lorraine Survey. Among the localities that Rosenbusch had to survey there was an area of contact metamorphic rocks – he called them *Steiger Schiefer* – in the regions of Barr and Andlau on the eastern side of the Vosges. Rosenbusch described the gradual alterations of the mineral contents of these slates in contact with an intrusion of granite. In this study, he gave the first clear description of progressive metamorphism (Rosenbusch 1875, 1877).

The ideas of metamorphic depth-zones and progressive contact metamorphism were not, however, wholly new in the 1870s. Rosenbusch noted that it was already known that different zones with homogeneous petrographic characteristics and mineralogical composition could be distinguished within areas of contact metamorphism (Rosenbusch 1877). As early as 1848, Wilhelm Haidinger (1795–1871) had defined two depth-zones of metamorphism, which he had called the zones of 'anogen' and 'katogen' reactions. In fact, it was a distinction between the upper regions of the Earth, which are subject to the work of atmospheric agencies (water, air), and its interior parts, where these agencies are

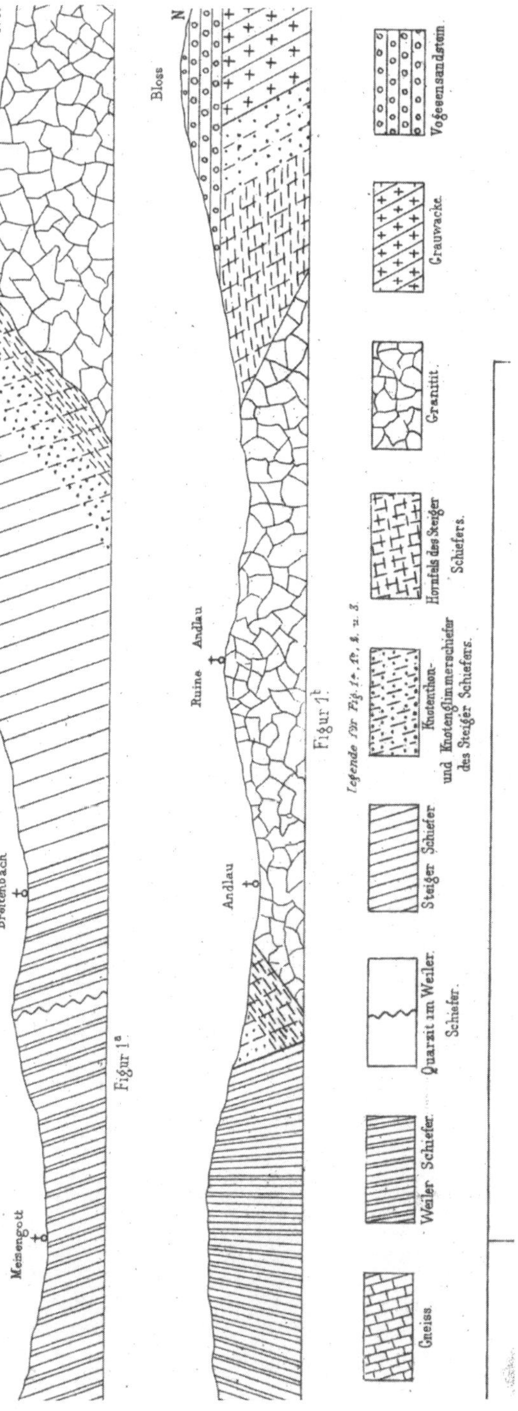

Fig. 1. Two sections of the contact metamorphic rocks of the Steiger Schiefer at Barr-Andlau, the initial site for the modern study of metamorphism (Rosenbusch 1877).

not active. Similar distinctions of metamorphic depth-zones were subsequently made in 1862 by Bernhard von Cotta (1808–1879), and – emphasizing a steady, long-lasting pressure as the essential characteristic of the lower region – by Jakob Johannes Sederholm (1863–1934) in 1891, and Charles Richard van Hise (1857–1918) in 1904. Rosenbusch himself pointed to Joseph Durocher's (1817–1860) statement of a proportionate relation between the degree of (contact) metamorphism and the distance from the magmatic intrusion that had caused the transformation (Durocher 1845–1846).

Notwithstanding these early anticipations, it was not until Rosenbusch's detailed study of the *Steiger Schiefer* that the ideas of metamorphic zones and progressive metamorphism became more widespread among Earth scientists. Rosenbusch's well-known zones of gradually increasing contact metamorphism were (see Fig. 1):

(1) the zone of spotted slates or phyllites (*Knotenthonschiefer*), with occasional contact minerals, mainly chiastolite;
(2) the zone of spotted schists (*Knotenglimmerschiefer*);
(3) and, the zone of 'hornfelses', which represented the highest degree of metamorphism; (in 1875 Rosenbusch had called this the zone of 'andalusite schists', according to its characteristic mineral).

One of Rosenbusch's most significant results was his observation that these zones were made up of just a few minerals, such as quartz, mica, andalusite, chiastolite and staurolite, as well as, though rarely, cordierite, garnet and pyroxene. Quartz was observed to occur in each zone, as well as biotite, whereas feldspars seemed to be completely lacking in these contact metamorphic rocks. Rosenbusch also emphasized the transformation of the calcareous components of the original rocks, i.e. CO_2 was usually replaced by SiO_2.

Rosenbusch's work gave a strong impulse to studies of metamorphism. He was, however, also responsible for some of the later difficulties in establishing theoretical chemistry as a method of the study of metamorphism. In his study of the *Steiger Schiefer*, Rosenbusch showed that in this special case no chemical alterations took place within the metamorphic rocks except for the loss of water. These results were due to some particularities of the Barr–Andlau area. Rosenbusch's followers, however, were often prepared to utilize the idea of contact metamorphism without any essential chemical change (e.g. Sederholm 1891; Kayser 1893; Brauns 1896; Lindgren 1905).

Moreover, the leading concepts of petrography of the time – the concepts of Rosenbusch and Ferdinand Zirkel (1838–1912) at Leipzig – favoured neither a chemical nor an experimental approach to the study of metamorphism. In the 1860s, the polarizing microscope had been introduced to the study of rocks and it promised to be the most effective instrument for a new science of petrography. Hence, the chemical characteristics of rocks became subordinate to the petrographical and stratigraphical ones, and chemical theories of metamorphism – such as, for instance, the theories of Justus Roth (1871) or Carl Gustav Bischof (1847–1855) – lost their influence.

These conceptual features may also have been strengthened by some political 'necessities' of the new science. With the institutionalization of petrography, the workers in the new field needed to demonstrate that it was not just a branch of mineralogy or chemical crystallography. For instance, at Strasbourg Rosenbusch had to share his department with Paul Groth (1843–1927), the leading German mineralogist. Groth never concealed his opinion that mineralogy and chemical crystallography were the essential branches of Earth science (comparable with palaeontology, petrography, etc.). And he argued successfully for this view, in the filling of positions at German universities (Fritscher 1997). Accordingly, the successful use of the petrographical techniques in the study of the stratigraphical characteristics of rocks helped to strengthen the institutional position of petrography, i.e. to prevent it from being subordinated to mineralogy and crystallography.

Rosenbusch's ideas on contact metamorphic zones, and on the occurrence of specific contact metamorphic associations of minerals, became widely known. Nevertheless, they were not genuinely advanced until 1893, when the British surveyor George Barrow (1853–1932) published a paper on contact metamorphic rocks in the Southern Highlands of Scotland. Barrow described metamorphic rocks accompanying an 'intrusion' of 'muscovite–biotite gneiss'. According to the abundance of three minerals, he distinguished three 'zones', i.e. types of metamorphism, within the 'metamorphic area', which he called the 'sillimanite zone' (the 'region of greatest metamorphism'), the 'cyanite zone', and the 'staurolite zone' (Barrow 1893).

Compared with his clear descriptions of the Southern Highland rocks, Barrow's discussion of the causes of metamorphism was relatively brief. He usually spoke of 'thermometamorphism', implying that an elevated temperature was an essential cause of metamorphism.

Pressure was not explicitly mentioned. Barrow, however, emphasized that the special features of metamorphic rocks were due to the depth at which the metamorphism took place, rejecting the hypothesis that the physical conditions of former geological times might have been distinctly different from those now prevailing. Finally, Barrow referred to some regional metamorphic rocks of New Galloway, strengthening the view that the difference between them and the rocks he had examined was 'one of degree, not of kind', i.e. that 'regional metamorphism and contact metamorphism [were] ... much the same thing' (Barrow 1893).

Barrow's study was largely ignored before World War I, with geologists like Ulrich Grubenmann (1850–1924), Friedrich Becke (1855–1931), Victor Moritz Goldschmidt (1888–1947) and Pentti Eskola (1883–1964) apparently being unaware of it before the 1920s. One of the reasons may have been Barrow's interpretation of gneiss as an igneous rock. Moreover, Alfred Harker (1859–1939), who was to become the outstanding figure among British petrologists, had questioned the possibility of distinguishing metamorphic zones at all, only two years before Barrow published his study. Harker's early statements on contact metamorphism ('thermal metamorphism') are to be found in his famous paper on the Shap Granite and its metamorphic aureole in Westmorland (now Cumbria) where metamorphic zones are, as it happens, hardly distinguishable. Referring to Rosenbusch, Harker noted that the zone of metamorphic minerals around the granite 'seem[ed] to be tolerably uniform in different directions', though the changes seemed to increase approaching the granite. Any division of the aureole into distinct rings or zones, however, 'would be arbitrary and artificial, and certainly could not be made to apply alike to the various kinds of rocks metamorphosed' (Harker 1891).

Despite these differences, both Harker's and Barrow's concepts of contact metamorphic areas agreed in one essential characteristic, namely that neither of them allowed much scope for the idea of associations of minerals being in a state of chemical equilibrium. They implicitly assumed that metamorphic processes are such that there are minerals, or associations of minerals, that can be formed only by metamorphic action, and, hence, are 'natural' to metamorphic rocks, just as other minerals may be 'natural' to igneous rocks. Metamorphic actions supposedly required special conditions (namely elevated temperatures and/or higher pressures) for their formation. But investigation of metamorphic rocks did not, however, necessarily demand quantitative knowledge of, or empirical and theoretical investigation of, these conditions. Rather, it would be possible to define 'natural types' of metamorphism by the description and comparison of the individual minerals naturally occurring in rocks.

I call the descriptive approach, represented by Barrow and Harker, the 'natural history of metamorphism'. At the end of this paper I shall return to this approach, and its implications, for it was one of the constituents of the 'chasm' that separated metamorphic petrologists into two parties until the middle of the twentieth century. Now, however, we have to turn to the antithesis of the natural history of metamorphism: what I call the 'science of metamorphism', i.e. the 'construction of metamorphic rocks' according to the principles of theoretical chemistry.

Making space for theoretical chemistry

In 1911, Goldschmidt published his dissertation on the contact metamorphic rocks of the Christiania (Oslo) region in Norway (Goldschmidt 1911*a*). It marked a new epoch in the history of metamorphism since, for the first time, the phase rule was applied to the study of a specific area of metamorphic rocks. Nevertheless, it must be recalled that the reception of theoretical chemistry was well prepared. The essential chemical problems of metamorphic zones and progressive metamorphism – the alteration of minerals by means of heat, solutions, gases and pressure, as well as the occurrence of specific associations of minerals – had been a leading feature of nineteenth-century chemical mineralogy.

The question of chemical equilibria in nature, as well as the common conditions of the formation of peculiar associations of minerals in metamorphic rocks, was anticipated by the doctrine of 'paragenesis'. This idea was formulated by Friedrich Breithaupt (1791–1873), professor of mineralogy at the Freiberg Mining Academy (Breithaupt 1849), and was based in the first instance on observations of ores and their associated minerals. In the last third of the nineteenth century the concept of paragenesis was modified and enlarged – by, amongst others, Goldschmidt's teacher Waldemar Christopher Brøgger (1851–1940) (Brøgger 1890) – and it became a general doctrine of mineral associations.

A second essential problem of progressive metamorphism – the alteration of minerals by heat and pressure – was anticipated by nineteenth-century chemical mineralogy. Understanding the nature of the relations between the crystallographic form and the chemical composition of minerals was one of its most significant problems and was an essential background to Goldschmidt's works. Among the more specific topics of this field of research was the field of mixed crystals (particularly feldspars), which had been discussed throughout the nineteenth century (Schütt 1984), and it was a timely idea at the beginning of the twentieth century (Day & Allen 1905). The specific problem of the alteration of the crystallographic form of minerals by means of high temperatures had also been discussed throughout the nineteenth century. Among the most remarkable examples were Vladimir Vernadsky's (1863–1945) studies on kyanite and sillimanite – an essential stability relation for modern metamorphic petrology (Miyashiro 1949, 1994).

In 1889, while studying with Ferdinand Fouqué (1828–1904) in Paris, Vernadsky had obtained sillimanite by melting siliceous earth with Al_2O_3. Because this result was obtained without any flux (only an excess of SiO_2 seemed to be required), he supposed that sillimanite might be a stable modification of Al_2SiO_5 at higher temperatures. In a letter to Groth, he gave an account of his results concluding that 'at a temperature close to 1400°C ... kyanite is always transformed into another modification (sillimanite?)' (Vernadsky to Groth, 20 June 1889; see also Vernadsky 1889; Bailes 1990). The experiments were carried out at the end of the 1880s, i.e. at the end of the decade of theoretical chemistry in which Jacobus Henricus Van't Hoff (1852–1911) and his students and collaborators at Amsterdam formulated the theory of mobile equilibrium, and the theory of affinity based on free energy. Just a few years later, these theories began to find their way into sedimentology, and also igneous and metamorphic petrology.

The year 1896 may be called the crucial year. That year, Van't Hoff – who had been professor of chemistry at Amsterdam, and also of mineralogy and geology – moved to Berlin. Already at Amsterdam he had considered the possible application of his results on the formation of double salts to the formation of natural salt deposits. While at Berlin, the formation of the famous Stassfurt salt deposit became one of his main fields of research (Van't Hoff & Meyerhoffer 1898–1899; Eugster 1971; Fritscher 1994). Van't Hoff's most important collaborator on these studies was Wilhelm Meyerhoffer (1864–1906) who had written the first book on the phase rule and its applications to chemistry some years earlier at Vienna (Meyerhoffer 1893). Futhermore, in 1896, Reinhard Brauns (1861–1937), professor of mineralogy and geology at the University of Giessen – who was later to become one of Goldschmidt's critics – published his *Chemical Mineralogy*. The treatise included a short account of contact metamorphism and crystalline schists. The account was chiefly based on the textbooks of Rosenbusch (1873–1877) and Zirkel (1894). Brauns made the remarkable statement that the crystalline schists approached a state of chemical equilibrium appropriate to higher pressure and higher temperature within the Earth's interior (Brauns 1896).

Theoretical chemistry was first used in 1896 to characterize metamorphic rocks, i.e. metamorphic minerals. The Austrian petrologist Friedrich Becke, than working at Prague, proposed the so-called 'Becke volume rule' stating that, with increasing pressure (isothermal conditions presumed), the formation of minerals with the smallest molecular volume (i.e. the greatest density) is favoured (Becke 1896). Becke's rule was based on the common opinion that the chemical composition of crystalline schists was analogous to the original igneous rocks (except for a small amount of water), and the observation that the newly formed minerals are of high density (e.g. garnet, muscovite and epidote) which, according to Becke's theory, might be called 'high-pressure minerals'. It will be observed that Becke's rule was obtained by inductive reasoning, not by deduction from chemical principles, and that his principle was a quite judicious one: it is compatible with common-sense reasoning that high pressures must yield minerals of greater density. Actually, the volume rule was already implicitly in use in petrology when Becke published it. Becke himself named Rosenbusch as one of his predecessors with regard to the idea (Becke *et al.* 1903).

Most probably, Becke was also aware of the principle of Henri Louis Le Chatelier (1850–1936), although he did not mention it. In a note, however, Becke remarked that he had emphasized in his lectures the significance of the 'Riecke principle' for the explanation of the textures of crystalline schists since about 1896. The Riecke principle defined a relation between the solubility of a solid and the stress acting on it. Becke applied this principle to explain the phenomenon of mineral alignment in crystalline schists, which, according to his theory of the preferential growth of crystals perpendicular to the direction of the strongest pressure (Becke *et al.* 1903; cf. Durney 1978), was less due to mechanical plasticity than to chemical processes, i.e. dissolution and crystallization.

The essential statement of Becke's theory was the notion of a direct influence of pressure on 'chemical forces'. This idea was established in its definitive form in the 1870s and 1880s by the works of Josiah Willard Gibbs (1839–1903), Van't Hoff and Le Chatelier. It had, however, been discussed on occasions since the early nineteenth century. For example, in the 1820s the Berlin mineralogist Eilhard Mitscherlich (1794–1863) stated that compression could have influenced the chemical and mineralogical composition of igneous rocks evolving from the chemical heterogeneous melts of the primeval Earth. This suggestion would have provided a solution to one of the main problems of Plutonist theory, namely the abundance of compounds of CaO and CO_2 in rocks, while ones composed of CaO and SiO_2 are comparatively rare. Presuming a hot, or even molten primeval Earth, functioning according to the 'normal chemical laws', one would expect compounds of CaO and SiO_2 to predominate, whereas $CaCO_3$ should be relatively rare, since it would have decomposed. Mitscherlich thought 'pressure' could have been the agency that overcame the usual chemical processes, and an appropriate high pressure should have been available during the primeval state of the Earth since a molten Earth would have caused the atmosphere to be filled with hot water vapour (Mitscherlich 1823; see also Fritscher 1991). Here, one may recall the experiments of Sir James Hall (1761–1832) on limestone and marble (Hall 1812; Fritscher 1988). It has to be realized, however, that Hall's experiments related to a single compound, whereas Mitscherlich's theory was concerned with heterogeneous melts.

Better known than Mitscherlich's idea is Henry Clifton Sorby's (1826–1908) postulate of a 'direct correlation of mechanical and chemical forces'. Concerning rock cleavage he stated that pressure could change the chemical affinities since it causes changes in volume (Sorby 1863; see also Durney 1978). To some degree Sorby's postulate may be interpreted as an anticipation of Becke's theory, or even the principle of Le Chatelier, although, in the 1860s, it lacked the necessary theoretical background. Finally, at the end of the century, Van Hise and Sederholm discussed the direct influence of pressure on chemical affinities. The latter, for instance, supposed that pressure might be able to increase the 'chemical energy' of the dissolving capability of water (Sederholm 1891).

The American geologist Van Hise had begun to pave the way for the application of thermodynamics to metamorphism contemporaneously with Becke. In 1898, and in a more comprehensive study in 1904, Van Hise discussed the chemical and physical principles of metamorphism referring, amongst others, to Van't Hoff and Walther Nernst (1864–1941). Van Hise claimed water, accompanied by gases and organic compounds, to be the dominating agency of metamorphism. The essential 'forces' of metamorphism were, according to his thinking, 'dynamic action', 'heat' and 'chemical action' (Van Hise 1898). In 1904, he modified this threefold division of forces to gravity (i.e. mechanical action), heat, light and 'chemical energy' (Van Hise 1904).

Van Hise seems to have been the first to use the term 'energy' in relation to metamorphic processes. Referring to Van't Hoff, he interpreted chemical reactions (caused by heating) as a release and a consumption of energy, respectively, as well as a displacement of the state of equilibrium. Furthermore, he entertained the possibility of solid–solid reactions during metamorphic processes, and he distinguished two 'physicochemical zones' of metamorphism according to the principles of theoretical chemistry: e.g. release and consumption of energy, liberation and absorption of heat, increase and decrease of volume.

His 'modern language' notwithstanding, Van Hise was more a prophet of theoretical chemistry than its pioneer (see Fritscher 1998). His treatise of 1904 was a comprehensive compilation of metamorphic phenomena, whereby metamorphism meant 'any change in the constitution of any kind of rock', including changes due to weathering. The 'physicochemical zones' were, however, similar to earlier distinctions (e.g. by Cotta, Sederholm and Becke; see above); and there was no genuine discussion of metamorphic zones or even of progressive metamorphism.

Concerning progressive metamorphism, the work of Ulrich Grubenmann, the outstanding metamorphic petrologist at the turn of the century, was notable. Together with Becke and Friedrich Berwerth (1850–1918), he had been a member of a group of geologists established by the Viennese Academy of Science to study the crystalline schists of the Eastern Alps. One result of the group's work was Becke's paper of 1903. Another was Grubenmann's well-known classification of metamorphic rocks, as well as his distinction of three metamorphic depth-zones (Grubenmann 1904–1907). Essentially, Grubenmann's classification provided a definition of 'index minerals' for each depth-zone. Grubenmann himself called them 'typomorphic' minerals, according to the suggestion of his friend Becke.

Grubenmann's work was entirely based on observation. He did not undertake experimental work, nor did he discuss phase relations.

Fig. 2. The young Victor Moritz Goldschmidt, in the year of the publication of his classic study on the contact metamorphism of the Christiania region in Norway (1911) (photograph reproduced from Isaksen & Wallem 1911).

Consequently, his first classification was a 'natural history of metamorphism'. Contrary to his British colleagues like Harker, however, Grubenmann implicitly started from the principle that the mineral contents of metamorphic rocks of given chemical compositions are functions of the pressure–temperature (P–T) conditions prevailing at the time of their formation. And, in later editions of his textbook he emphasized that his classification was essentially based on varying P–T conditions, i.e. that the 'typomorphic' minerals were indicators of particular states of chemical equilibrium (Grubenmann 1910; Grubenmann & Niggli 1924).

Constructing metamorphic rocks

The second edition of Grubenmann's classic text on metamorphic rocks was not even a year old when Goldschmidt, then only 23 years old (see Fig. 2), published his doctoral thesis on the contact metamorphism of the Christiania area in Norway (Goldschmidt 1911*a*), ignoring nearly all limitations that might have been set to the application of theoretical chemistry to contact metamorphic petrology.

Goldschmidt demonstrated that the associations of minerals in a natural occurrence of hornfels rocks obeyed the phase rule, which meant that the mineral content of a specific hornfels was completely determined by the components of its original materials, constant pressure and temperature being presumed. Goldschmidt distinguished ten classes of hornfels rocks according to specific mineral associations, which – and this is the crucial point – could be deduced from the range of compositions of the original shales and limestones. Ordered according to increasing calcium content, these associations were (see Fig. 3):

1. andalusite–cordierite
2. plagioclase–andalusite–cordierite
3. plagioclase–cordierite
4. plagioclase–hypersthene–cordierite
5. plagioclase–hypersthene
6. plagioclase–diopside–hypersthene
7. plagioclase–diopside
8. grossular–plagioclase–diopside
9. grossular–diopside
10. vesuvianite–grossular–diopside.

Minerals that occur in all ten of the classes – mainly quartz and biotite – are omitted from this table, and the vesuvianite of the last group is due to the presence of water (there would usually be wollastonite).

Goldschmidt summarized his results in his 'mineralogical phase rule', usually written as $P \leq C$ (Goldschmidt himself gave no mathematical expression for his mineralogical phase rule in his 1911 papers). The rule says that at any pressure and temperature the number of phases (P) cannot be more than the number of components (C), where the phases are the physically different and mechanically separable parts of a system, and components are the minimum number of molecules necessary for the composition of these phases (see Fig. 3).[2]

A year later, Goldschmidt published a second

[2] The 'mineralogical phase-rule' is a reduction of J. W. Gibbs' phase-rule: $P = C + 2 - f$, were f represents the number of degrees of freedom, namely the smallest number of independent variables required to define the state of equilibrium of a system completely. Because petrological processes usually take place at changing PT conditions, there are always two degrees of freedom, i.e. the 'mineralogical phase rule', actually, is $P = C$. Usually (i.e. in natural occurrences of rocks) there are fewer phases than the possible maximum number; thus, the 'mineralogical phase rule' is usually written as $P \leq C$.

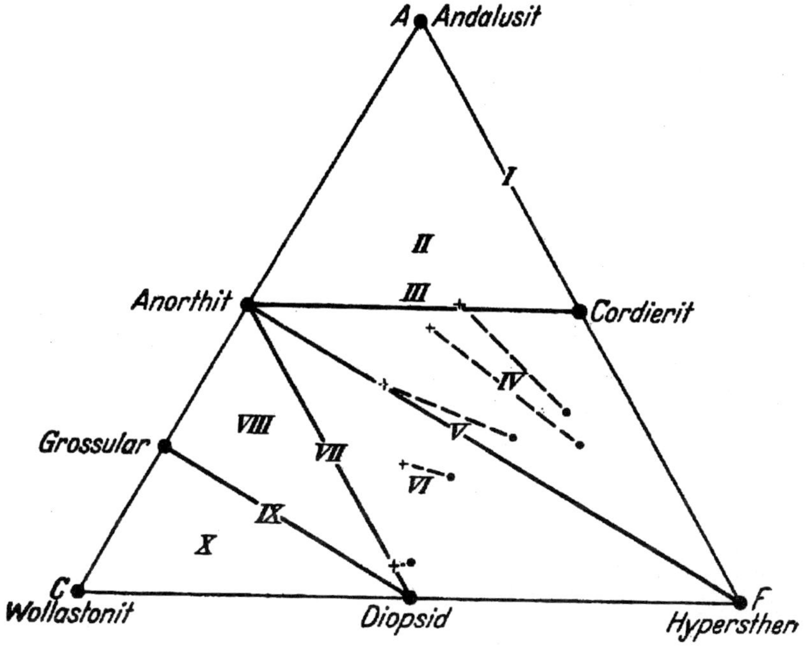

Fig. 3. ACF (i.e. aluminium, calcium, iron) diagram of the hornfels facies by Eskola illustrating Goldschmidt's ten classes of hornfels rocks (from Barth *et al.* 1939; see also Eskola 1920; Mason 1992). The diagram illustrates the mineralogical phase rule stating that in a ternary system a maximum of three minerals can coexist as a stable system. The Roman numerals show the position of the classes according to the results of the chemical analyses. Some hornfels rocks contain biotite, in which the Mg and Fe contents affect their positions in the diagram, i.e. shifting them toward hypersthene. However, since these contents are due to chemical components (K_2O, H_2O) that are not shown in the diagram, Eskola also calculated their ACF values by omitting the oxides of the biotite. He pointed out that the resulting changes are the expected ones, according to the mineralogical phase rule (the shifts of the positions are indicated by broken lines, the new positions by '+').

paper on metamorphism entitled 'The Laws of the Metamorphism of Rocks'. Its concern was Mitscherlich's problem (see above), i.e. the frequent metamorphic reaction: calcite + quartz = wollastonite + CO_2. Considering the curve for the equilibrium partial pressure of CO_2, Goldschmidt determined the temperature/pressure fields for the coexistence of calcite and quartz, i.e. wollastonite (and CO_2), respectively (see Fig. 4), which are meant to indicate different depth zones of the Earth's interior, i.e. of metamorphism. He referred to these theoretical considerations in a further study of the regional metamorphic lime–silica rocks of the Trondheim area, distinguishing various degrees of metamorphism according to the presence of chlorite, biotite and garnet (Goldschmidt 1915).

The significance of Goldschmidt's results for modern Earth sciences is well known (see Mason 1992; Manten 1966; Winkler 1965; Miyashiro 1973), and need not be rehearsed in detail here. Rather, it is more interesting to ask *why* Goldschmidt obviously felt so sure of the applicability of the phase rule to metamorphic petrology. For the majority of his contemporaries, such an application was far less convincing (see below), and for many of them, Goldschmidt's (1912*a*) claim to present the 'laws of metamorphism' might have sounded pretentious. For Goldschmidt himself there was never a shadow of doubt about the soundness of his methods and results. In his inaugural lecture on 'The Problems of Mineralogy', given on 28 September 1914, he claimed that the thermodynamic approach was essential to mineralogy and petrology, whose fundamental questions must be: '[w]hat are the conditions for thermodynamic equilibrium (in geological systems), and why is it that we find some minerals in one occurrence and not in another?' (quoted from Mason 1992).

Goldschmidt himself – notwithstanding the tenor of some of his later critiques (see below) –

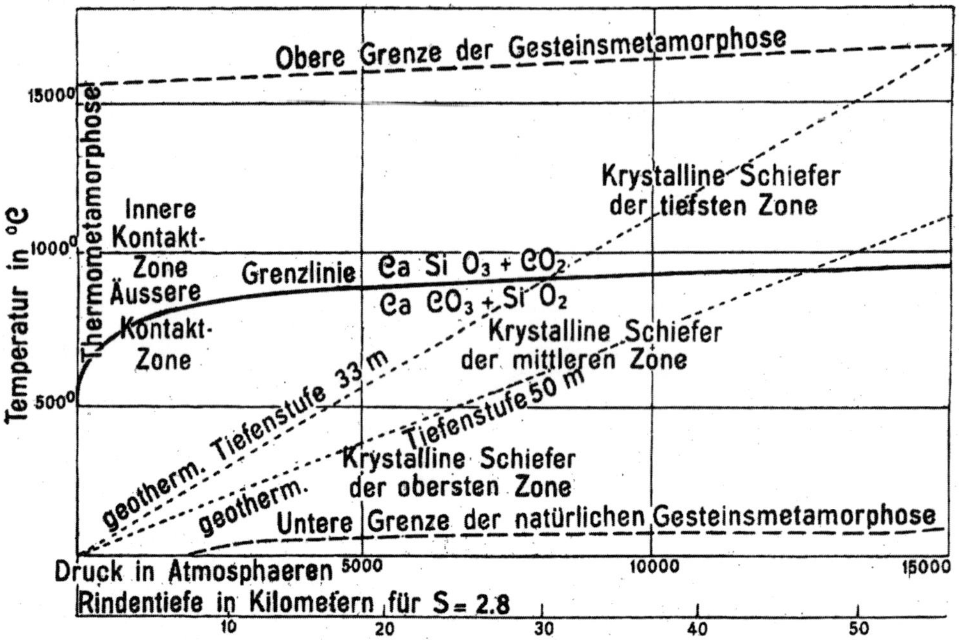

Fig. 4. Temperature–pressure relations in the system $CaCO_3$–$CaSiO_3$–SiO_2 (Goldschmidt 1912*a*; reprinted by Becke 1911–1916). According to the curve for the equilibrium partial pressure of CO_2, Goldschmidt determined the temperature/pressure fields for the coexistence of calcite and quartz (lower part of the diagram), i.e. wollastonite and CO_2 (upper part), respectively, thus indicating different metamorphic depth zones. The upper part of the diagram is thought to represent the P–T conditions of the crystalline schists of the deepest zone, the lower part those of the middle and the uppermost zone. At the left side of the diagram, where conditions of high temperatures and low pressures are represented, Goldschmidt also indicated a similar distinction between an inner contact metamorphic zone (upper part of the diagram) and an outer one (lower part). For an English version of the figure, see Mason (1992).

was well aware of the advantages, as well as the limits of the new thermodynamic approach with regard to the study of metamorphism and metamorphic rocks. His primary aim was a comprehensive and systematic nomenclature of contact metamorphism and meta-sedimentary rocks. Hitherto, the nomenclature had been arbitrary; that is, it reflected a lot of accidental aspects because contact metamorphic phenomena were commonly named according to the features that the observer concerned thought most conspicuous.

In 1898, Wilhelm Salomon (1868–1941) made a first attempt to establish a more systematic nomenclature of contact metamorphic rocks by focusing on their mineral content and chemical composition, whereas characteristics such as grain size or schistosity were used only incidentally (Salomon 1898). Thus Salomon used the characteristic minerals of the rocks, supplemented by local names derived from their natural occurrences. Such a nomenclature, Goldschmidt stated, was sufficient if our knowledge of the mineral content and the composition of rocks was merely empirical. Now, however, this knowledge was much advanced, and we were in a position to discuss the mineral content of the most different contact metamorphic rocks from a common point of view, namely the phase rule, i.e. the doctrine of chemical equilibrium. Moreover, Goldschmidt pointed out that his classes of hornfels rocks were valid only for contact metamorphic rocks of the inner area of clay–slate–limestone series in contact with plutonic rocks. There would be other minerals in contact areas with volcanic rocks, and if the effects of regional metamorphism (i.e. Becke's volume rule) were to be taken into account a different nomenclature would be required.

But it should be realized that Goldschmidt himself – contrary to his later critics – saw no 'artificial characters' in his classification. He rejected purely chemical classifications, such as the CIPW classification, because quantitative classifications, omitting all mineralogical and genetic characteristics, would lead to 'unnatural' ones. A petrographical system that claimed to be a 'natural system' necessarily had to take into account actual mineral compositions. A quantitative chemical system was required, not in place of but in addition to, the mineralogical and genetic classifications. Thus, a mineralogical classification had to be based on those minerals that are characteristic for the rocks – which was obviously the idea of 'typomorphic' minerals of his teacher Becke (Goldschmidt 1911*a*). Accordingly, later on Goldschmidt frequently emphasized that he had found the ten classes of hornfels rocks *before* he realized that these classes were in accordance with the requirements of the phase rule (Goldschmidt 1911*b*).

Following Goldschmidt's arguments, some modern geoscientists may also ask: if the actual mineral composition has to remain the basis of the classification of metamorphic rocks, what is the actual benefit of the application of the phase rule to metamorphism? A simple answer may be that it saved metamorphic petrologists some hundred years of empirical fieldwork, since it represents the 'way of nature' in highly complex processes. The phase rule does not restrict the number of minerals that actually occur, but it states limits to the possible number in a given petrological situation. In this sense Paul Niggli wrote in 1949 that the thermodynamical approach made it possible to establish 'prohibiting signs' whose overall neglect 'by nature' was improbable.

Concerning Goldschmidt's reliance on the applicability of the phase rule to metamorphism, it may be noted that he did not use any new instruments, nor did he undertake any specific experimental work. Rather, his results were obtained by 'descriptive methods', i.e. by conventional methods of the petrography of his day such as chemical analyses, or the study of thin sections and crystallographic properties. In a first preliminary communication of his results Goldschmidt (1909) dealt exclusively with the optical characteristics of the minerals involved. And in his inaugural lecture, mentioned above, he stated that optical characteristics had been one of the essential means for his determination of temperature–pressure ranges.

One reason for Goldschmidt's reliance on the correctness of his method and his results may have been his area of research. Goldschmidt himself pointed out that the Christiania region offers outstanding conditions for the study of contact metamorphism. Contrary to nearly all the contact metamorphic areas in central Europe, the Christiania area has not been subjected to regional metamorphism, i.e. stress need not be taken into account (Goldschmidt 1911*a*). This peculiarity of the Christiania region had been remarked on previously by the Norwegian geologist Baltazar Keilhau (1797–1858) (Keilhau 1840), and by Goldschmidt's teacher Brøgger (1882, 1890; see also Hestmark 1999). Brøgger, moreover, emphasized the regularity of the contact metamorphism of this area, i.e. all the true igneous rocks – notwithstanding their mineralogical composition – have formed a similar series of changes in the adjacent rocks proportional to their masses (Brøgger 1890).

Notwithstanding these regional peculiarities, the essential reason for Goldschmidt's reliance on the applicability of the phase rule to metamorphism was his strong personal background in theoretical chemistry. His father, Heinrich Goldschmidt (1857–1932), had been one of the leading physical chemists of his time. Heinrich Goldschmidt received his doctorate at Prague in 1881, the experimental work for his thesis being undertaken at the newly established chemical laboratory at the University of Graz, which offered one of the best equipped laboratories of the time. And from 1893 to 1896 he was working with Van't Hoff at Amsterdam (Bodenstein 1932). Hence his son was well acquainted with the new theoretical chemistry from his early youth. Among Goldschmidt's later teachers, Becke was an expert in the new theoretical chemistry (see above). Accordingly, the application of theoretical chemistry to metamorphic rocks and other fields of petrology was a matter of course for Goldschmidt, and not, as it was for many of his contemporaries, something strange or obscure.

The younger Goldschmidt's work became widely known and generally acknowledged. Nevertheless, his new methodological approach found no immediate continuation, with the exceptions of Eskola and, in a qualified sense, Paul Niggli (1888–1953). The latter had studied with Grubenmann at Zurich, and in 1912 – the year after Goldschmidt – he received his PhD with a thesis on the chloritoid schists of the St Gotthard area (Niggli 1912*a*; see also Becke 1911–1916). In 1913, Niggli went to the Geophysical Laboratory at Washington where he worked with Norman Levi Bowen (1887–1956) on phase equilibria (see Young 2002). One of the results of these studies was a paper, written with John Johnston (1881–1950), on 'The general principles underlying metamorphic processes' (Johnston & Niggli 1913). Later, Niggli (1938) also wrote a popular account of the application of the phase rule to mineralogy and petrology.

Niggli's early work on phase equilibria was

Fig. 5. Pentti Eskola in 1916, one year after the introduction of his concept of metamorphic facies in his study on the metamorphic rocks of the Orijärvi region (photograph reproduced from Carpelan & Tudeer 1925).

most probably done independently of Goldschmidt; that is, he seems to have been unaware of the parallel work done by his colleague in Norway. Moreover, his early papers on phase equilibria (e.g. Niggli 1912*b*) had a strong theoretical aspect: they did not, like Goldschmidt's studies, relate specifically to metamorphic rocks. Thus, the Finnish petrologist Eskola (Fig. 5) was the only real follower of Goldschmidt. Studying the metamorphic rocks of the Orijärvi region in southwestern Finland (see Fig. 6) Eskola found similar regularities of mineral associations, although there were usually amphiboles instead of Goldschmidt's pyroxenes, which Eskola ascribed to different P–T conditions. Referring to a study of the saturation diagrams by Van't Hoff, and also referring to Goldschmidt (1911*a*) and Johnston & Niggli (1913), Eskola introduced the concept of 'metamorphic facies': a specific metamorphic facies denoted a group of rocks which, at an identical chemical composition, has an identical mineral content, and whose mineral content will change according to definite rules if the chemical composition changes. Eskola emphasized that his new concept did not make any supposition as to the genetic, pre-metamorphic relations of the rocks. In particular, a specific metamorphic facies was not related to any individual occurrence of metamorphic rocks, i.e. it might be found in widely different parts of the world, while in neighbouring localities different facies might occur (Eskola 1915). In the same year, Goldschmidt proposed a similar concept of 'metamorphic facies'. The character of a specific 'metamorphic facies', he stated, was due to its 'geological history'. This meant that the mineral content and texture of a group of metamorphic rocks occurring together are due to their chemical composition and to the variations of temperature, pressure and stress in time. Thus, if there were no such variations, and an identical chemical composition, there would be an identical mineral content (Goldschmidt 1915).

By his definition of metamorphic facies Eskola gave a striking example for what has been said above concerning the essential difference between the 'natural history' and the 'science' of metamorphism. The former pointed to 'ideal types' of metamorphism, realized in specific local occurrences of metamorphic rocks. The latter, by contrast, pointed to the formation or production of metamorphic rocks according to the principles of theoretical chemistry. As Eskola himself pointed out, the definition of a specific metamorphic facies is independent of its actual occurrence in nature.

In 1920, while working with Goldschmidt at Oslo, Eskola recognized that some igneous rocks could be discussed according to the same principles as metamorphic rocks. Therefore, he extended his principle to one of 'mineral facies of rocks' (Eskola 1920). Later, he emphasized that this principle was based on the observation that the mineral associations of metamorphic rocks are, in most cases, in accordance with the principles of chemical equilibrium. The definition itself, however, did not include any assumptions as to an existing state of chemical equilibrium, i.e. it should not include any hypothetical assumption(s). The application of the principle of the 'mineral facies of rocks' simply indicated whether a specific association of minerals represented a state of disequilibrium, or whether it was in accordance with the rules of a specific mineral facies (Barth *et al.* 1939).

Is equilibrium always attained during metamorphism?

As indicated above, Goldschmidt's work and his application of theoretical chemistry to metamorphic rocks became quickly known and widely acknowledged. Nevertheless, as mentioned, his new thermodynamic approach found

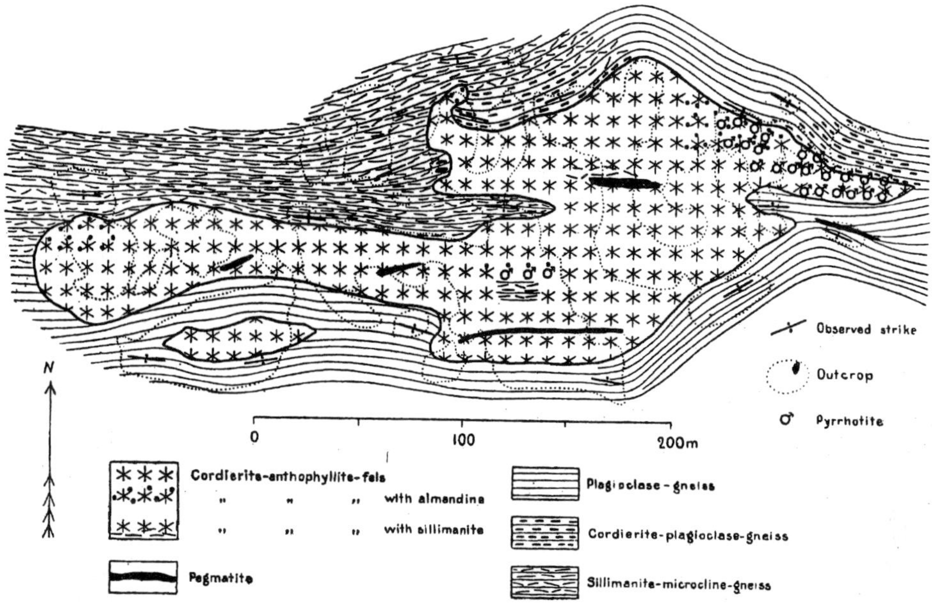

Fig. 6. Occurrence of a homogeneous body of cordierite–anthophyllite rock near Träskböle (Eskola 1914), a metamorphic rock of common occurrence in the Orijärvi area in southwestern Finland. There Eskola observed regularities of mineral associations similar to those observed by Goldschmidt in the Christiania area, which became the starting point of his concept of 'metamorphic facies'.

few immediate followers. Thus, the story of its early reception was not simply one of general agreement or rejection. Some of his contemporary colleagues realized the significance of his study for future research in metamorphism. Becke, for instance, in his reports on the progress of metamorphism of 1911 and 1916, included chapters on contact metamorphism and on the physical–chemical foundations of the doctrine of metamorphism, which were mainly accounts of Goldschmidt's work in the Christiania area (Becke 1911–1916; see also Harker 1918). The major part of the geological community, however, confined its acknowledgment to Goldschmidt's mineralogical results in a narrower sense, discussing them within the traditional concept of paragenesis. His new thermodynamic approach was more or less set aside; at best, it was conceded that it might have been applicable to the Christiania region, with its peculiar geological history. The crucial point for his critics was the question of whether or not chemical equilibrium was always attained during metamorphism, i.e. whether metamorphic rocks could generally be expected to be in a state of chemical equilibrium, or if such a state was exceptional.

An example is provided by Emil Baur's (1873–1944) critique of Goldschmidt's lecture on 'The Application of the Phase Rule to Silicate Rocks', which Goldschmidt gave in 1911 at the Meeting of the German Bunsen Society for Applied Physical Chemistry. Baur, a professor of physical chemistry at Brunswick, objected that in the case of the hornfels rocks of the Christiania area the crystallization would have taken place, at least to some extent, under the action of superheated water. This meant that there should have been supersaturated solutions and, in consequence, a great many different minerals would have been formed contemporaneously. These crystals would not all disappear, even if they were approaching a region of instability. Prior to the application of physical chemistry, i.e. the phase rule, to silicate rocks, a complete compilation of all known paragenetic sequences of igneous rocks, as well as of contact metamorphic rocks, would be required. Only in this way could a truly significant application of physical chemistry to petrology be possible. Goldschmidt, however, simply replied that, if there had originally been more minerals than the phase rule demanded, and if they were therefore remaining, these minerals should be discoverable by thin sections: '[b]ut in these four years I examined nearly 1000 thin sections of the metamorphic rocks of the Christiania area, and there was not one where the requirements of the phase rule were not fulfilled' (Goldschmidt 1911*b*).

Two years later the applicability of the phase rule to metamorphic rocks became the subject of a longer controversy with Johann Koenigsberger (1874–1946) of Freiburg University, who is remembered for his introduction of the notion of 'polymetamorphism' and his discussion of the use of the inversion-points of polymorphic crystals of SiO_2 as geological thermometers (see Fischer 1961). The controversy began with a critical essay by Goldschmidt, John Rekstad and Thorolf Vogt (Goldschmidt *et al.* 1913) on some of Koenigsberger's papers in which he had frequently touched on problems of Norwegian geology. In early 1913, Goldschmidt had already complained about these papers in a letter to Groth: '[h]ere, his [Koenigsberger's] statements on Norwegian geology and mineral occurrences evoked general astonishment. His theory of anatexis [i.e. on the formation of gneiss] is based on three observations along a length of 1200 km. If he had seen more, he would have less said' (Goldschmidt to Groth, 12 January 1913).

In a reply to his critics, Koenigsberger commented on Goldschmidt's application of thermodynamics to petrography. Referring to Brauns (1912), he stated – somewhat misleadingly – that the phase rule was not valid *a priori*, i.e. it should be thought of as a general law of thermodynamics, being inapplicable to unstable compounds, which he thought to be the usual case in metamorphic rocks. Furthermore, Koenigsberger questioned Goldschmidt's priority in applying the phase rule to mineral associations, referring to Emil Baur who had used it in 1903 in his experiments on the system quartz–orthoclase (Koenigsberger 1913; see also Baur 1903). In his reply, Goldschmidt acknowledged Baur's 'excellent description' of a specific system. Baur, however, had said nothing about the general phase rule relations between the number of components and the number of minerals (Goldschmidt *et al.* 1914).

With respect to Koenigsberger's misleading statement on the restricted applicability of the phase rule, Goldschmidt maintained that the rule taught one to distinguish between stable and unstable systems of phases. Then, he recommended Koenigsberger to publish his 'new discovery' on the restrictions of the phase rule in a physics journal (Goldschmidt *et al.* 1914). This ironic statement induced Koenigsberger to write to Goldschmidt's father asking him to try to help settle the controversy with his son. With regard to his own statement on the phase rule, Koenigsberger hastened to say that he only meant that: '[t]he phase rule – as a numerical relation – is only applicable in the case of a complete chemical equilibrium. The first and second law of thermodynamics, however, are valid generally' (Koenigsberger to H. Goldschmidt, 17 April 1914).

The year after this exchange, the Dutch chemist and mineralogist Hendrik Boeke (1881–1918), then working at the University of Halle, cautioned against overestimating the significance of the phase rule for the advancement of natural sciences. He referred particularly to Goldschmidt's application of the phase rule to contact metamorphic rocks as a striking example of such an overestimation. The phase rule, Boeke objected, offered a system of classification that could be misleading in the world of minerals and rocks, i.e. without experimental data on chemical equilibria it could be completely useless (Boeke 1915).

Boeke was a former student of Van't Hoff, and is today considered as a pioneer of the application of physical chemistry to petrography. Goldschmidt himself was well aware that Boeke was his most serious critic. In a letter to Groth he compared him with Niggli (Fig. 7):

> I think that the weak point of them both (in particular, of the second one [Niggli]) is their lack of familiarity with the pure petrographic materials, and methods, and, in consequence, they give a one-sided emphasis to theoretical aspects. Nevertheless, both are to be preferred compared to the majority of their colleagues, in particular, Boeke, who, in my opinion, is better informed with respect to the theoretical aspects than is Niggli (Goldschmidt to Groth, 2 August 1916).

Boeke's critique, indeed, underscored the crucial point concerning the application of thermodynamics, and the doctrine of chemical equilibria, in metamorphic petrology. For Boeke, and many colleagues (e.g. Sederholm, see below), a rock that underwent metamorphism was a highly complex 'system of systems'. Each part by volume made up a system of its own, marked by specific pressure, temperature, components, and a specific solid, liquid or vapour phase. Hence, a state of chemical equilibrium for the entire rock could hardly be attained. Only for crystalline schists could such a state of equilibrium be assumed. In accordance with an early nineteenth-century idea, Boeke acknowledged these rocks to be among the Earth's oldest formations, in which the process of metamorphism had been completed; therefore, they might well have approached a state of chemical

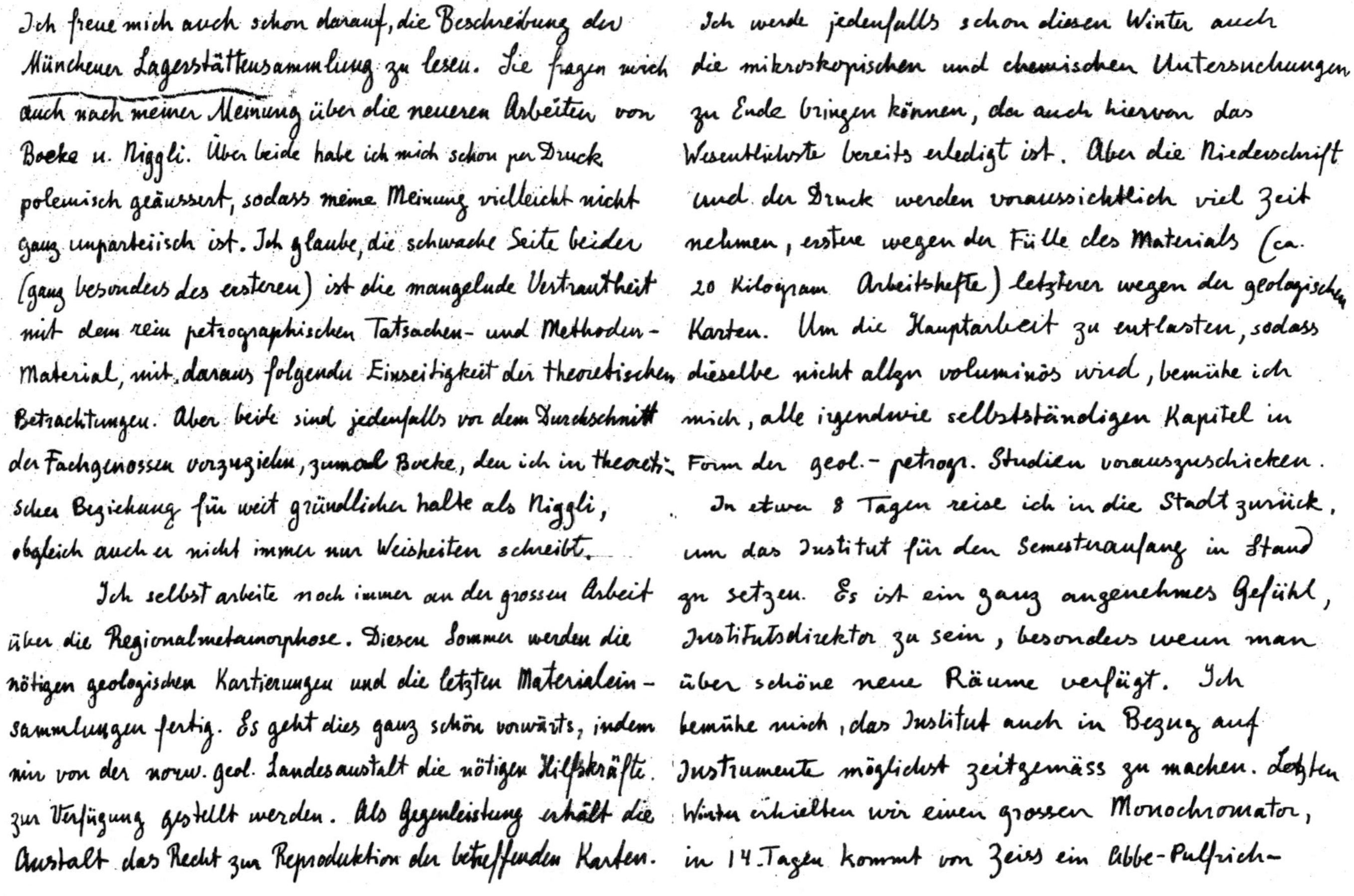

Ich freue mich auch schon darauf, die Beschreibung der Münchener Lagerstättensammlung zu lesen. Sie fragen mich auch nach meiner Meinung über die neueren Arbeiten von Boeke u. Niggli. Über beide habe ich mich schon per Druck polemisch geäussert, sodass meine Meinung vielleicht nicht ganz unparteiisch ist. Ich glaube, die schwache Seite beider (ganz besonders des ersteren) ist die mangelnde Vertrautheit mit dem rein petrographischen Tatsachen- und Methoden-Material, mit daraus folgender Einseitigkeit der theoretischen Betrachtungen. Aber beide sind jedenfalls vor dem Durchschnitt der Fachgenossen vorzuziehen, zumal Boeke, den ich in theoretischer Beziehung für weit gründlicher halte als Niggli, obgleich auch er nicht immer nur Weisheiten schreibt.

Ich selbst arbeite noch immer an der grossen Arbeit über die Regionalmetamorphose. Diesen Sommer werden die nötigen geologischen Kartierungen und die letzten Materialeinsammlungen fertig. Es geht dies ganz schön vorwärts, indem mir von der norw. geol. Landesanstalt die nötigen Hilfskräfte zur Verfügung gestellt werden. Als Gegenleistung erhält die Anstalt das Recht zur Reproduktion der betreffenden Karten.

Ich werde jedenfalls schon diesen Winter auch die mikroskopischen und chemischen Untersuchungen zu Ende bringen können, da auch hiervon das Wesentlichste bereits erledigt ist. Aber die Niederschrift und der Druck werden voraussichtlich viel Zeit nehmen, erstere wegen der Fülle des Materials (ca. 20 Kilogram Arbeitshefte) letzterer wegen der geologischen Karten. Um die Hauptarbeit zu entlasten, sodass dieselbe nicht allzu voluminös wird, bemühe ich mich, alle irgendwie selbstständigen Kapitel in Form der geol.-petrogr. Studien vorauszuschicken.

In etwa 8 Tagen reise ich in die Stadt zurück, um das Institut für den Semesteranfang in Stand zu setzen. Es ist ein ganz angenehmes Gefühl, Institutsdirektor zu sein, besonders wenn man über schöne neue Räume verfügt. Ich bemühe mich, das Institut auch in Bezug auf Instrumente möglichst zeitgemäss zu machen. Letzten Winter erhielten wir einen grossen Monochromator, in 14 Tagen kommt von Zeiss ein Abbe-Pulfrich-

Fig. 7. Letter from Victor Goldschmidt to Paul Groth, 2 August 1916, comparing the petrological work of Boeke and Niggli (by courtesy of the Bavarian States Library, Munich, Manuscript Department).

equilibrium. Only in this case, i.e. for primeval regional metamorphic rocks, the doctrine of chemical equilibrium might be applicable, but hardly ever in the case of contact metamorphism, or dynamometamorphism. In addition, Boeke pointed out that the application of results of chemical equilibria studies to metamorphic rocks had to be done in a different way from that with regard to igneous or sedimentary rocks. Thus the phase rule would be of little help in defining the number of possible minerals within a metamorphic rock (Boeke 1915).

By his comment, Boeke also implicitly indicated that the critics focused on the application of the phase rule to contact metamorphism, whereas the majority of the petrological community conceded that the crystalline schists approached a state of chemical equilibrium. Brauns, for instance, objected to Goldschmidt's application of the phase rule to the hornfels rocks of the Christiania area (Brauns 1912). On the other hand, Brauns himself, more than fifteen years before, had stated that crystalline schists usually approach a state of equilibrium, although the achievement of equilibrium may never be complete (see above; see also Johnston & Niggli 1913; Eitel 1925). Actually, the assumption that crystalline schists represented a state of chemical equilibrium went back to the 1870s. As early as 1874, Gustav Leonhard (1816–1878), professor of mineralogy at Heidelberg (and son of the famous German geologist Karl Caesar Leonhard (1779–1862)) applied the term 'chemical equilibrium' to what he thought were metamorphic rocks. Referring to a contemporary theory of the origin of granite, according to which granite is a 'metasomatic rock' with 'trachytic lava' as its basic material, Gustav Leonhard stated that granite is trachytic matter in a state of 'chemical equilibrium' appropriate to the physical conditions of the Earth's interior (Leonhard 1874). At the turn of the century, Becke even claimed a state of perfect chemical equilibrium as being the essential characteristic of crystalline schists as opposed to igneous rocks. In crystalline schists, Becke maintained, all components are 'mutually harmonizing', and the striking zonal features essential for and characteristic of igneous rocks diminish in crystalline schists (Becke *et al.* 1903; see also Turner 1948).

Concerning these early critiques and the early reception of Goldschmidt's thermodynamic approach, it should be recognized that the critique of the application of the doctrine of phases to petrology already had a kind of tradition when Goldschmidt was young. In the first years of the century, the Austrian mineralogist and petrologist Cornelio Doelter (1850–1930), after a series of experiments on the melting-points of silicate melts, and contrary to his Norwegian colleague Johan Herman Lie Vogt (1858–1932), concluded that the applicability of the doctrine of phases, i.e. Van't Hoff's doctrine of solutions, to silicate solutions was limited, due to the viscosity of silicates (Doelter 1904). Actually, Goldschmidt's mineralogical phase rule was a more exact definition of a result that Vogt had obtained from his studies on slags. In the early 1880s he stated that the formation of minerals in silicate melts, i.e. in igneous rocks (at ordinary pressure and with the absence of volatiles such as water or CO_2 being presumed) depended mainly on the chemical composition of the average mass, i.e. that the minerals were products of the effects of chemical affinity of the main components (or the formation of minerals depends on chemical mass actions). In 1903, Vogt remarked on these early statements that, instead of 'effects of chemical affinity', he would better have said 'states of chemical equilibrium' (Vogt 1903–1904). It was the appeal to those traditional critiques which, in 1915, caused Arthur L. Day (1869–1960), the first director of the Geophysical Laboratory of the Carnegie Institution in Washington, to write to Goldschmidt assuring him of his support in his struggle for a comprehensive application of the phase rule:

> We too have regretted the tendency in certain European literature to deny the application of the phase rule to silicate solutions, and have made an especial effort in our recent papers to meet this opposition. The trouble is due, I think, to technical difficulties and not to questions of principle, and will therefore correct itself with the accumulation of more experience in the study of silicate products. For this reason, we have preferred not to arouse a controversy, but rather to continue our work in the usual way, trusting to the mass of accumulated evidence to overwhelm the opposition (Day to Goldschmidt, 2 March 1915).

Day himself had started his career at the forefront of physical chemistry. Before he joined the US Geological Survey in 1900, he had been, for nearly four years, on the staff of the Physikalisch-Technische Reichsanstalt in Berlin-Charlottenburg, then one of the best-equipped physics laboratories in the world. And, in 1900, he married Helene Kohlrausch, the daughter of Friedrich Kohlrausch (1840–1910), then president of the Reichsanstalt. At Berlin, Day began his investigations on the high-temperature scale, which he continued for about ten years in America. Also at Berlin, he obviously became acquainted with the work of Van't Hoff, who taught physical chemistry at the Berlin University from 1896 (for Day's knowledge of the latest developments in physical chemistry, see, for instance, Day & Shepherd 1905).

In addition to these 'internal' arguments, a more detailed discussion of the early reception of Goldschmidt's thermodynamic approach would have to take into account some external features. One of them would be the philosophical context of the controversy. It is known that the introduction of Gibbs' doctrine of energy was accompanied by an influential, and popular, philosophical movement called 'energetics' (cf. Vernadsky 1908). Its advocates – who called themselves 'the energetics' – claimed 'energy' to be the essential category of science, and society also. The head of the movement was Wilhelm Ostwald (1853–1932), who had introduced Gibbs' doctrine of energy, and his phase rule, to Europe. An analogous philosophy of energy was also influential in the United States around 1900. Van Hise's emphasis on 'energy', for instance, was obviously indebted to it (see Fritscher 1998). Therefore, some early twentieth-century geologists could have regarded Goldschmidt's work as more philosophical than empirical, which could explain the hesitations of many of his colleagues toward his applications of thermodynamics and the doctrine of chemical equilibrium in petrology.

A second 'external' feature could have been Goldschmidt's personality. Goldschmidt was a brilliant scientist and, as indicated above, was well aware of his abilities. He was convinced that he had laid open the 'laws of metamorphism'. Moreover, in his replies to his critics, and his comments on other approaches on the application of the phase rule to geological problems, one can feel his interest in claiming priority in the new field of a metamorphic petrology based on thermodynamics. He was annoyed by references to the speculations of his predecessors in the field (see his replies to Koenigsberger, mentioned above). In a short comment on Niggli's (1912*b*) paper 'On rock series of metamorphic origin', Goldschmidt made some objections to Niggli's theoretical discussion of phase relations of the lime–silica series. But first he hastened to claim that the explanation of Niggli's, and analogous, cases had already been given by himself a year prior to Niggli's 'valuable study' (Goldschmidt 1912*b*; see also Becke 1911–1916). Consequently, Goldschmidt felt affronted when the University of Göttingen, in 1915, announced a prize-competition for a comprehensive and critical essay on contact metamorphism, i.e. on the changes of the chemical and mineralogical composition of contact metamorphic rocks, as well as on the chemical and physical processes caused by metamorphism. In Goldschmidt's opinion, these problems had already been solved by his study of the Christiania area. In a letter to Groth he wrote:

> It has been completely ignored that this problem is already solved. The act of the Göttingen University here is regarded as an insult to the Oslo Academy of Science, which has already awarded my study on the same subject (Goldschmidt to Groth, 28 March 1915).

Epilogue: image and logic

Hitherto, the scope of the discussion has been chiefly limited to a historical description of the formative years of metamorphism and thermodynamics. The historian of science, however, might have chosen a slightly different point of view. Modern Earth scientists are quite right in ascribing the controversial discussion concerning the application of the doctrine of chemical equilibrium to petrology to a lack of adequate experimental methods and instruments (see Yoder 1980; Geschwind 1995). Moreover, it may be recalled that, in the 1950s, the facies concept faced new difficulties. Hatten S. Yoder, for instance, in a study of $MgO–Al_2O_3–SiO_2–H_2O$, found representatives of all the then-defined facies to be stable at the same pressure and temperature, and he also raised the issue of the role of water in metamorphism (Yoder 1989). Furthermore, Miyashiro (1953) showed that the formation of garnet is not, as was commonly thought, necessarily related to high pressures.

As indicated above, however, some features of this discussion, at least in part, were due to its cultural context. Such an 'external dimension' has previously been suggested by Miyashiro's identification of two paradigms in early twentieth-century metamorphism. The first, represented by Grubenmann and Harker, was characterized by the use of the concept of stress minerals, and 'normal regional metamorphism'. The second paradigm (Goldschmidt, Eskola) was characterized by utilization of the concept of a chemical equilibrium, controlled by temperature and pressure, and the recognition of the diversity of regional metamorphism due to pressure (Miyashiro 1994). With respect to the formative years of metamorphism and thermodynamics, a somewhat modified distinction between these two 'styles' of metamorphic petrology has been recommended. The first one, represented by Barrow, Harker, Grubenmann and their followers, has been called the 'natural history of metamorphism'. It was characterized by the description of 'genuinely metamorphic sites', and the distinction of peculiar metamorphic zones according to 'genuinely metamorphic minerals', which were implicitly thought to be the 'embodiment' of the sum of specific metamorphic actions or changes. It should be observed that theoretical chemistry was in no way neglected. On the contrary, it was frequently recommended that it should be held in view. Nevertheless, it did not play an essential role. The second style, represented by Becke, Goldschmidt and Eskola, was characterized by the construction of metamorphic (i.e. mineral) facies according to the principles of theoretical chemistry, whereby the definition of a specific facies does not depend on a specific natural occurrence of metamorphic rocks. In comparison with the descriptive tradition, this style has been dubbed the 'science of metamorphism'.

Definitions of this kind relate mainly to the internal structures of scientific thinking. Thus, it may be of interest to conjecture some of the 'external' features that could have constituted the two styles of early twentieth-century metamorphic petrology, whereby these styles may be related to different practices and different cultures of Earth science in the nineteenth and early twentieth centuries. Here, a comprehensive study by Peter Galison (1997) on the material cultures of modern physics is particularly helpful (see also Jardine 1991; Oreskes 1999). By means of an analysis of the instruments of modern physics, Galison distinguished two competing traditions of experimental practice, which he called the 'homomorphic' and the 'logic tradition'. The first pointed to the 'representation of natural processes in all their fullness and complexity – the production of images of such clarity that a single picture can serve as evidence for a new entity of effect', i.e. the recreation, or visualization, of the 'very form of invisible nature' (Galison 1997). Against this 'homomorphic tradition' Galison juxtaposed the 'logic' one, which used 'counting (rather than picturing) machines' (e.g. electronic counters) 'to aggregate masses of data to make statistical arguments for the existence of a particle or effect'. The logic tradition gave up, or even explicitly rejected, the focus on individual occurrences of the 'homomorphic tradition' (Galison 1997).

Galison's distinction between different experimental practices of modern physics relates to the classical distinction between qualitative and quantitative studies of nature, i.e. between a phenomenological and a 'constructive' approach to experience (Fritscher 1991). It was foreshadowed by Niggli's definition of two essentially different methods of scientific investigation, namely a method of causal explanation and one deploying 'ideal images' (Niggli 1949). In this respect, it can serve as a versatile model for understanding the different aspects of metamorphic petrology. Nevertheless, it has to be realized that the distinction of the basic styles of metamorphism does not concern different experimental practices. Rather, it concerns a different 'handling' of the natural phenomena of metamorphic rocks. The explicit point of the 'logic style' is the construction of metamorphic (mineral) facies according to the principles of theoretical chemistry and experimental results. The natural occurrences of metamorphic rocks, of course, are not omitted. They serve, however, as more or less complete manifestations of those basic principles, not as their model. By contrast, the 'homomorphic style' points toward the reproduction of typical metamorphic zones according to typical sites. The natural process(es) of a specific kind of metamorphism should be represented in all their fullness and complexity. Thus, a single picture serves as evidence for a whole range of metamorphic processes, and this is the essential meaning of what were later called 'Barrovian zones'.

With regard to the possible relations of these styles to different national practices of Earth science, I confine myself to a few observations. The most significant one has to do with the distinction of two lines in the early argumentation concerning the pros and cons of the application of theoretical chemistry to petrology. According to their provenance, the lines may be called the British and the Continental styles, for their differences were due, not least, to the fact that the basic studies on petrology (metamorphism) and thermodynamics were undertaken on the Continent – in the Netherlands, in Vienna and, above all, in Scandinavia. It may be noted that this observation does not neglect Van Hise, or the experimental works of Ernest S. Shepherd, Norman Bowen and others at the Geophysical Laboratory. As indicated above, however, Van Hise was more a 'propagandist' of physical chemistry than an actual practitioner in petrology. And the Geophysical Laboratory of the Carnegie Institution in Washington was not a 'genuine product' of Anglo-Saxon Earth sciences. George F. Becker (1847–1919), for example, the true 'constructor' of the laboratory, noted that his plans, in particular his estimates of the personnel and plant appropriate to a geophysical laboratory, were largely based upon the experience of the Physikalisch–Technische Reichanstalt of Berlin, with modifications appropriate to the American circumstances (Becker *et al.* 1903; see also Cahan 1989). And the early work of A. L. Day was actually a continuation of the high-temperature research that he had started as a member of staff of the German institution (see above). Indeed, in its early decades the Geophysical Laboratory did not fit well in the culture of Anglo-Saxon Earth sciences, in which the chemical and experimental approaches, contrary to the 'Continental style', did not play a leading role. In this connection it may also be recalled that although the Geophysical Laboratory was soon acknowledged worldwide, its actual studies were not widely utilized before World War II (Geschwind 1995; Oreskes 1999).

The characteristic features of the 'British style of metamorphism' were exemplified by Harker's presidential address to the Geological Society of 1918, on the present position of the study of metamorphism, which can be read as the programmatic manifestation of the style. To be sure, Harker did not disregard thermodynamics and the phase rule. On the contrary, he emphasized its outstanding importance: '[t]he Phase Rule . . . means so much for petrology that it must be considered as marking for us a distinct epoch' (Harker 1918). Nevertheless, his paper was a plea for the use of 'ideal images' in the study of metamorphism. Harker pointed to the Scottish Highlands, which might serve as a 'model metamorphic region' (Harker 1918). And, notwithstanding his favourable mention of the phase rule, the main factor in metamorphism should be stress: 'the student of metamorphism must realize how radically some simple physical and chemical principles become modified when applied to bodies in a condition of internal stress; and, moreover, of stress which varies from place to place and from time to time'. Moreover, 'unequal stress' might create 'in some important degree a new chemistry, different from that of the laboratory' (Harker 1918). It may be noted that statements like this – on a 'peculiar geological chemistry', beyond the scope of laboratory facilities – had been important arguments for the constitution of geology as an independent science in the nineteenth century, and hence for the constitution of the culture of British geology (Fritscher 1991).

On the Continent – particularly in Scandinavia and the German-speaking countries – the situation was significantly different. From early modern times, the Earth sciences in these areas were based on mineralogy, crystallography and chemistry. This predominance never actually changed during the nineteenth century, despite the frequent intentions to 'copy' the 'British style' of geology. The incompatibility of the Continental culture of Earth sciences to that of the British one was due to the close relation of British geology to specific features of British society in the nineteenth century (cf. Cannon 1978); such features were missing on the Continent, especially in the German-speaking countries. On the other hand, the German-speaking and the Scandinavian geoscientists were better prepared for the acceptance of the new theoretical chemistry. Therefore, the British and the Continental oppositions to the thermodynamic approach were not necessarily related.

Continental critics also doubted whether theoretical chemistry, and experimental/field work, could be sufficient to cover the whole range of natural processes generating the rocks. Contrary to British critiques – which are relatively easy to understand as a relict of nineteenth-century efforts to 'making space for geology' – the Continental critiques, and particularly the German ones, were more complex. The Continental geoscientists had never cut off their connections with mineralogy and chemistry, as had British geologists. Thus, the background of the Continental (again, particularly German-speaking) critiques was constituted more by peculiar German philosophical ideas than by the 'defence' of the original domain of geology against 'unauthorized claims'.

Goldschmidt's critic Boeke, for instance, was himself a pioneer of the application of theoretical chemistry to geological problems. Nevertheless, Goldschmidt's application of the phase rule to contact metamorphic rocks was an 'exaggeration' so far as Boeke was concerned. A more sophisticated formulation of this specifically German critique, indicating also its philosophical background, was Niggli's philosophical discussion of the doctrine of mineral association. The application of the basic laws of physics to complex natural processes always meant that one would ignore, at least in part, this complexity. Accordingly, Niggli opposed the 'dynamic of formation and changing' to the 'statics of what should be', according to the basic laws of physics (Niggli 1949). This critique related to nineteenth-century German historicism and German idealism, according to which 'simple physical laws' have never been more than an approximation of the full, 'real' nature of things. In this sense, the study of the formation of rocks according to the principles of thermodynamics seems to be a mere theoretical explanation corresponding to 'nature itself', at best only partial. Thus 'nature itself' should be much more complicated than the 'arbitrary constructions' that mathematical physics assumed.

Within the scope of these critiques, we have

also to locate Niggli's teacher Grubenmann, and Sederholm. Sederholm, who had studied with Brøgger at Stockholm, and with Rosenbusch at Heidelberg, discussed 'the nature and causes of metamorphism' in a paper on the eruptive rocks of southern Finland (Sederholm 1891). He noted that many of these rocks are metamorphosed. His concept of metamorphism, however, bore more resemblance to the mid-nineteenth-century ones of Carl Gustav Bischof, and Wilhelm Haidinger than to the 'science of metamorphism'. For Sederholm, metamorphism was not a general change of rocks according to specific laws, but rather the sum of highly complex alterations whereby each mineral is changed individually to some other one by way of pseudomorphism (Sederholm 1891). Each mineral, and each rock, has its own history. Hence Sederholm gave no significant space to the idea of chemical equilibrium or the idea of metamorphic zones being characterized by peculiar associations of minerals.

Goldschmidt was far from such ideas. He was the first petrologist for whom 'nature' was something to be constructed according to simple laws. This attitude, again, might have been due to his father with whom he had done a considerable amount of applied research in his early years. Moreover, in 1917, Goldschmidt became the director of the Norwegian State Commission on Raw Materials. Regarding his new job, he wrote to Groth: '[p]ure science and the lessons at the institute have to be done as an additional job; however, the scientific results of this additional job are quantitatively and qualitatively better than they had been earlier in the main job (Goldschmidt to Groth, 5 December 1918).

The Norwegian Commission on Raw Materials was established for a definite reason, namely Norway's intention to enter World War I, which in fact it never did. But the point brings to our attention one more feature that is normally omitted in discussing the formative years of metamorphism and thermodynamics. The essential discussions on this subject occurred at the eve of World War I and continued during the War. The German-speaking countries suffered greatly from the conflict. Indeed, German science was excluded from international science for about a decade. It is possible, then, that the dominating role of the 'British style' of metamorphic studies up to the 1940s could have had, in part at least, this political cause.

I should like to thank H. Yoder and D. Young for their valuable comments on this paper, and D. Oldroyd for his editorial assistance and patience.

References

BAILES, K. E. 1990. *Science and Russian Culture in an Age of Revolution: V. I. Vernadsky and his Scientific School, 1863–1945*. Indiana University Press, Bloomington.

BARROW, G. 1893. On an intrusion of muscovite–biotite gneiss in the south-eastern Highlands of Scotland, and its accompanying metamorphism. *The Quarterly Journal of the Geological Society, London*, **49**, 330–358.

BARTH, T. F. W., CORRENS, C. W. & ESKOLA, P. 1939. *Die Entstehung der Gesteine*. Springer, Berlin.

BAUR, E. 1903. Über die Bildungsverhältnisse von Orthoklas und Albit. *Zeitschrift für physikalische Chemie*, **42**, 566–576.

BECKE, F. 1896. Ueber Beziehungen zwischen Dynamometamorphose und Molecularvolumen. *Neues Jahrbuch für Mineralogie, Geologie und Petrefaktenkunde*, **1896**, Part 2, 182–183.

BECKE, F. 1911–1916. Fortschritte auf dem Gebiete der Metamorphose. *Fortschritte der Mineralogie, Kristallographie und Petrographie*, **1**, 22–156; **5**, 210–264.

BECKE, F., BERWERTH, F. & GRUBENMANN, U. 1903. Über Mineralbestand und Struktur der krystallinischen Schiefer. *Denkschriften der Kaiserlichen Akademie der Wissenschaften, Mathematisch-naturwissenschaftliche Klasse*, **75**, Part 1 [1913].

BISCHOF, C. G. 1847–1855. *Lehrbuch der chemischen und physikalischen Geologie* (3 vols). A. Marcus, Bonn.

BODENSTEIN, M. 1932. Heinrich Goldschmidt zum fünfundsiebzigsten Geburtstage. *Zeitschrift für Elektrochemie*, **38**, 899–900.

BOEKE, H. E. 1915. *Grundlagen der physikalisch-chemischen Petrographie*. Bornträger, Berlin.

BOUÉ, A. 1820. *Essai géologique sur l'Écosse*. Courcier, Paris.

BRAUNS, R. 1896. *Chemische Mineralogie*. C. H. Tauchnitz, Leipzig.

BRAUNS, R. 1912. Review of V. M. Goldschmidt (1911c). *Neues Jahrbuch für Mineralogie, Geognosie, Geologie und Petrefaktenkunde*, **1912**, 216–217.

BREITHAUPT, J. F. A. 1849. *Die Paragenesis der Mineralien: Mineralogisch, geognostisch und chemisch beleuchtet mit besonderer Rücksicht auf Bergbau*. J. G. Engelhardt, Freiberg.

BRØGGER, W. C. 1882. *Die silurischen Etagen 2 und 3 im Kristianiagebiet und auf Eker, ihre Gliederung, Fossilien, Schichtenstörungen und Contactmetamorphosen*. Brøgger, Kristiania.

BRØGGER, W. C. 1890. Die Mineralien der Syenitpegmatitgänge der südnorwegischen Augit- und Nephelinsyenite. Mit zahlreichen chemisch-analytischen Beiträgen von P. T. Cleve. *Zeitschrift für Krystallographie und Mineralogie*, **16**.

CAHAN, D. 1989. *An Institute for an Empire: The Physikalisch-Technische Reichsanstalt, 1871–1918*. Cambridge University Press, Cambridge.

CANNON, S. F. 1978. *Science in Culture: The Early Victorian Period*. Science History Publications, New York.

CARPELAN, T. & TUDEER, L. O. 1925. *Helsingfors*

Universitet. Lärare och Tjänstemän från År 1828 (2 vols). Söderström, Helsingfors.

COTTA, B. VON 1862. *Die Gesteinslehre.* J. G. Engelhardt, Freiberg.

DAY, A. L. & ALLEN, E. T. 1905. The isomorphism and thermal properties of the feldspars. Part I. Thermal study. *Carnegie Institution of Washington Publications*, **31**, 15–75.

DAY, A. L. & SHEPHERD, E. S. 1905. The phase rule and conceptions of igneous magma. *Economic Geology*, **1**, 286–289.

DOELTER, C. 1904. Die Silikatschmelzen (Erste & zweite Mitteilung). *Sitzungsberichte der Kaiserlichen Akademie der Wissenschaften in Wien, Mathematisch-naturwissenschaftliche Klasse*, **108**, Part 1, No. 4, 177–249, 495–511.

DURNEY, D. W. 1978. Early theories and hypotheses on pressure–solution–redeposition. *Geology*, **6**, 369–372.

DUROCHER, J. 1845–1846. Études sur la métamorphisme. *Bulletin de la Société Géologique de France*, 2nd series, **3**, 546–647.

EITEL, W. 1925. *Physikalisch-chemische Mineralogie und Petrologie: Die Fortschritte in den letzten zehn Jahren.* Wissenschaftliche Forschungsberichte-Naturwissenschaftliche Reihe, **13**, Steinkopff, Dresden & Leipzig.

ESKOLA, P. 1914. *On the Petrology of the Orijärvi Region in Southwestern Finland.* Bulletin de la Commission Géologique de Finlande, **40**.

ESKOLA, P. 1915. *Om sambandet mellan kemisk och mineralogisk sammansättning hos Orijärvitraktens metamorfa bergarter. With an English Summary of the Contents.* Bulletin de la Commission Géologique de Finlande, **44**.

ESKOLA, P. 1920. The mineral facies of rocks. *Norsk Geologisk Tidsskrift*, **6**, 143–194.

EUGSTER, H. P. 1971. The beginnings of experimental petrology. *Science*, **173**, 481–489.

FISCHER, W. 1961. *Gesteins- und Lagerstättenbildung im Wandel der wissenschaftlichen*, Auschauung. Schweizerbart, Stuttgart.

FRITSCHER, B. 1988. Die 'James Hall-Sammlung' in Keyworth. *Berichte zur Wissenschaftsgeschichte*, **11**, 27–34.

FRITSCHER, B. 1991. *Vulkanismusstreit und Geochemie. Die Bedeutung der Chemie und des Experiments in der Vulkanismus-Neptunismus-Kontroverse.* Boethius. Texte und Abhandlungen zur Geschichte der Mathematik und der Naturwissenschaften, **25**. Steiner, Stuttgart.

FRITSCHER, B. 1994. Salzbildung und Kontaktmetamorphose: Zur Aufnahme der physikalischen Chemie in den Geowissenschaften um 1900. *In*: FRITSCHER, B. & BREY, G. (eds) *Cosmographica et Geographica. Festschrift für Heribert M. Nobis zum 70. Geburtstag*, **2**, 281–307. *Algorismus: Studien zur Geschichte der Mathematik und der Naturwissenschaften*, **13**, Institut für Geschichte der Naturwissenschaften, Munich.

FRITSCHER, B. 1997. *Geowissenschaften und Moderne. Studien zur Kulturgeschichte der Mineralogie und Chemischen Geologie (1848–1926).* Habilitationsschrift, University of Munich.

FRITSCHER, B. 1998. The fabrication of rocks: the Geophysical Laboratory and the production of modernity in mineralogy and geochemistry. *In*: FRITSCHER, B. & HENDERSON, F. (eds) *Toward a History of Mineralogy, Petrology and Geochemistry. Proceedings of the Symposium on the History of Mineralogy, Petrology and Geochemistry, 8–9.3.1996, Munich*, 357–379. *Algorismus: Studien zur Geschichte der Mathematik und der Naturwissenschaften*, **23**, Institut für Geschichte der Naturwissenschaften, Munich.

GALSION, P. 1997. *Image and Logic: A Material Culture of Microphysics.* The University of Chicago Press, Chicago and London.

GESCHWIND, C.-H. 1995. Becoming interested in experiments: American igneous petrologists and the Geophysical Laboratory, 1905–1965. *Earth Sciences History*, **14**, 47–61.

GOHAU, G. 1997. Evolution des idées sur le métamorphisme et l'origine des granites. *In*: BONIN, B., DUBOIS, R. & GOHAU, G., *Le métamorphisme et la formation des granites: évolution des idées et concepts actuels.* Nathan, Paris, 11–58.

GOLDSCHMIDT, V. M. 1909. Ueber die Mineralien der Kontaktmetamorphose im Kristiania-Gebiet. *Centralblatt für Mineralogie, Geologie und Paläontologie*, **1909**, 405–410.

GOLDSCHMIDT, V. M. 1911*a*. *Die Kontaktmetamorphose im Kristianiagebiet.* Videnskapsselskapets Skrifter I, Mathematisk-Naturvidenskabelig Klasse, **1**.

GOLDSCHMIDT, V. M. 1911*b*. Anwendung der Phasenregel auf Silikatgesteine. *Zeitschrift für Elektrochemie*, **17**, 686–689.

GOLDSCHMIDT, V. M. 1911*c*. Die Gesetze der Mineralassoziation vom Standpunkt der Phasenregel. *Zeitschrift für anorganische Chemie*, **71**, 313–322.

GOLDSCHMIDT, V. M. 1912*a*. *Die Gesetze der Gesteinsmetamorphose, mit Beispielen aus der Geologie des südlichen Norwegens.* Videnskapsselskapets Skrifter I, Mathematisk-Naturvidenskabelig Klasse, **22**.

GOLDSCHMIDT, V. M. 1912*b*. Zu Herrn P. Nigglis Abhandlung: Über metamorphe Gesteinsserien. *Tschermak's Mineralogische und Petrographische Mitteilungen*, New Series, **31**, 695–696.

GOLDSCHMIDT, V. M. 1915. *Geologisch-petrographische Studien im Hochgebirge des südlichen Norwegens. III. Die Kalksilikatgneise und Kalksilikatglimmerschiefer des Trondhjem-Gebiets.* Videnskapsselskapets Skrifter I, Mathematisk-Naturvidenskabelig Klasse, **10**.

GOLDSCHMIDT, V. M., REKSTAD, J. & VOGT, T. 1913. Zu Herrn Joh. Koenigsberger's geologischen Mitteilungen über Norwegen. *Centralblatt für Mineralogie, Geologie und Paläontologie,* **1913**, 324–328.

GOLDSCHMIDT, V. M., REKSTAD, J. & VOGT, T. 1914. Nochmals Herrn Joh. Koenigsberger's geologische Mitteilungen über Norwegen. *Centralblatt für Mineralogie, Geologie und Paläontologie*, **1914**, 114–118.

GRUBENMANN, U. 1904–1907. *Die kristallinen Schiefer. 1: Allgemeiner Teil. 2: Specieller Teil.* Bornträger, Berlin.

GRUBENMANN, U. 1910. *Die kristallinen Schiefer: Eine Darstellung der Erscheinungen der Gesteinsmetamorphose und ihrer Produkte* (2nd revised edition). Bornträger, Berlin.

GRUBENMANN, U. & NIGGLI, P. 1924. *Die Gesteinsmetamorphose. 1: Allgemeiner Teil.* Bornträger, Berlin.

HAIDINGER, W. 1848. Abriss eines Aufsatzes: Über die Metamorphose der Gebirgsarten. *Sitzungsberichte der königlichen Akademie der Wissenschaften, Mathematisch-naturwissenschaftliche Klasse*, **1**, 51–58.

HALL, J. 1812. Account of a series of experiments, shewing the effects of compression in modifying the action of heat. *Transactions of the Royal Society of Edinburgh*, **6**, 71–185.

HARKER, A. 1891. The Shap Granite, and the associated igneous and metamorphic rocks. *The Quarterly Journal of the Geological Society, London*, **47**, 266–328.

HARKER, A. 1918. Presidential Address [The present position and outlook of the study of metamorphism in rock-masses]. *The Quarterly Journal of the Geological Society, London*, **74**, lxiii–lxxx.

HESTMARK, G. 1999. *Vitenskap og nasjon: Waldemar Christopher Brøgger 1851–1905.* Aschehoug, Oslo.

ISAKSEN, A. & WALLEM, F. B. (eds) 1911. *Norges Universitet. Professorer, Docenter, Amanuenser, Stipendiater samt øvrige lærere og tjenestemænd.* Gyldendal, Christiania.

JARDINE, N. 1991. *The Scenes of Inquiry: On the Reality of Questions in Science.* Clarendon Press, Oxford.

JOHNSTON, J. & NIGGLI, P. 1913. The general principles underlying metamorphic processes. *Journal of Geology*, **21**, 481–516, 588–624.

KAYSER, E. 1893. *Lehrbuch der Geologie für Studierende und zum Selbstunterricht. 1: Allgemeine Geologie.* Ferdinand Enke, Stuttgart.

KEILHAU, B. M. 1840. *Einiges gegen den Vulkanismus (Des Herrn Dr. von Dechen Gutachten über das 1ste Heft der Gaea Norvegica, mit Anmerkungen von B. M. Keilhau).* Keilhau, Christiania.

KOENIGSBERGER, J. 1913. Antwort auf die Bemerkungen der Herren V. M. Goldschmidt, J. Rekstad, T. Vogt. *Centralblatt für Mineralogie, Geologie und Paläontologie*, **1913**, 520–526.

LEONHARD, G. 1874. *Grundzüge der Geognosie und Geologie* (3rd edn, enlarged and revised). C. F. Winter, Leipzig & Heidelberg.

LICHTENSTERN, C. 1990. *Die Wirkungsgeschichte der Metamorphoselehre Goethes von Philipp Otto Runge bis Joseph Beuys.* Acta Humaniora. VCH, Weinheim.

LINDGREN, W. 1905. *The Copper Deposits of the Clifton–Morenci District, Arizona.* US Geological Survey Professional Paper **43**, Washington.

LYELL, C. 1833. *Principles of Geology, being an Attempt to Explain the Former Changes of the Earth's Surface, by Reference to Causes now in Operation* (3 vols). John Murray, London.

MANTEN, A. A. 1966. Historical foundations of chemical geology and geochemistry. *Chemical Geology*, **1**, 5–31.

MASON, B. 1992. *Victor Moritz Goldschmidt. Father of Modern Geochemistry.* Geochemical Society, San Antonio, Special Publication Series **4**.

MATHER, K. F. (ed.) 1967. *Source Book in Geology, 1900–1950.* Harvard University Press, Cambridge (Mass).

MATHER, K. F. & MASON, S. L. (eds) 1939. *A Source Book in Geology.* McGraw Hill, New York.

MEYERHOFFER, W. 1893. *Die Phasenregel und ihre Anwendungen.* F. Deuticke, Leipzig and Vienna.

MITSCHERLICH, E. 1823. Über das Verhältnis der Krystallformen zu den chemischen Proportionen. 3: Über die künstliche Darstellung der Mineralien aus ihren Bestandteilen. *Abhandlungen der königlichen Akademie der Wissenschaften zu Berlin aus den Jahren 1822–23, Physikalische Klasse*, 25–41.

MIYASHIRO, A. 1949. The stability relation of kyanite, sillimanite and andalusite, and the physical conditions of the metamorphic processes. *Journal of the Geological Society of Japan*, **55**, 218–223 (in Japanese).

MIYASHIRO, A. 1953. Calcium-poor garnet in relation to metamorphism. *Geochimica et Cosmochimica Acta*, **4**, 179–208.

MIYASHIRO, A. 1973. *Metamorphism and Metamorphic Belts.* Allen & Unwin, London.

MIYASHIRO, A. 1994. *Metamorphic Petrology.* UCL Press, London.

NIGGLI, P. 1912*a*. *Die Chloritoidschiefer und die sedimentäre Zone am Nordostrande des Gotthardmassivs.* Beiträge zur geologischen Karte der Schweiz, New Series, **36**.

NIGGLI, P. 1912*b*. Über Gesteinsserien metamorphen Ursprungs. *Tschermak's Mineralogische und Petrographische Mitteilungen,* New Series, **31**, 477–494.

NIGGLI, P. 1938. *La loi des phases en minéralogie et pétrographie, 1: Généralités. 2: Applications minéralogiques et pétrographiques de la loi des phases.* Actualités scientifiques et industrielles, **611** and **612**. Hermann, Paris.

NIGGLI, P. 1949. Probleme der Naturwissenschaften erläutert am Begriff der Mineralart. *Wissenschaft und Kultur*, **5**. Birkhäuser, Basel.

ORESKES, N. 1999. *The Rejection of Continental Drift. Theory and Method in American Earth Science.* Oxford University Press, New York.

ROSENBUSCH, H. 1873–1877. *Mikroskopische Physiographie der petrographisch wichtigen Mineralien* (2 vols). Schweizerbart, Stuttgart.

ROSENBUSCH, H. 1875. [Brief an G. Leonhard: Kontaktmetamorphose]. *Neues Jahrbuch für Mineralogie, Geologie und Petrefactenkunde*, **1875**, 849–851.

ROSENBUSCH, H. 1877. Die Steiger Schiefer und ihre Contactzone an den Granititen von Barr-Andlau und Hohwald. *Abhandlungen zur geologischen Specialkarte von Elsass–Lothringen*, **1**, No. 2, 79–393.

ROTH, J. L. A. 1871. Über die Lehre vom Metamorphismus und die Entstehung der krystallinischen Schiefer. *Abhandlungen der physikalischen Klasse der königlichen Akademie der Wissenschaften zu Berlin aus dem Jahr 1871*, 151–232.

SALOMON, W. 1898. Ueber Alter, Lagerungsform und

Entstehungsarten der periadriatischen granitisch-körnigen Massen. *Tschermak's Mineralogische und Petrographische Mitteilungen*, New Series, **17**, 109–283.

SCHREYER, W. 1995. Ultradeep metamorphic rocks: the retrospective viewpoint. *Journal of Geophysical Research*, **100**(B5), 8353–8366.

SCHREYER, W. 1999. High-pressure experiments and the varying depth of rock metamorphism. *In*: CRAIG, G. Y. & HULL, J. H. (eds) *James Hutton – Present and Future*. The Geological Society, London, 59–74.

SCHÜTT, H.-W. 1984. *Die Entdeckung des Isomorphismus. Eine Fallstudie zur Geschichte der Mineralogie und der Chemie*. Arbor Scientiarum. Beiträge zur Wissenschaftsgeschichte, Series A, **9**, Gerstenberg, Hildesheim.

SEDERHOLM, J. J. 1891. Studien über archäische Eruptivgesteine aus dem südwestlichen Finnland. *Tschermak's Mineralogische und Petrographische Mitteilungen*, New Series, **12**, 97–142.

SORBY, H. C. 1863. On the direct correlation of mechanical and chemical forces. *Proceedings of the Royal Society of London*, **12**, 538–550.

TOURET, J. L. R. & NIJLAND, T. G. 2002. Metamorphism today: new science, old problems. *In*: Oldroyd, D. (ed.) *The Earth Inside and Out: Some Major Contributions to Geology in the Twentieth Century*. Geological Society, London, Special Publications, **192**, 113–142.

TURNER, F. J. 1948. *Mineralogical and Structural Evolution of the Metamorphic Rocks*. The Geological Society of America, New York, Memoir **30**.

VAN HISE, C. R. 1898. Metamorphism of rocks and rock flowage. *The American Journal of Science*, 4th series, **6**, 75–91.

VAN HISE, C. R. 1904. *A Treatise on Metamorphism*. Monographs of the United States Geological Survey, **47**, Washington, D. C.

VAN'T HOFF, J. H. & MEYERHOFFER, W. 1898–1899. Über Anwendungen der Gleichgewichtslehre auf die Bildung ozeanischer Salzablagerungen, mit besonderer Berücksichtigung des Stassfurter Salzlagers (2 parts). *Zeitschrift für physikalische Chemie*, **27**, 75–93; **30**, 64–88.

VERNADSKY, V. I. 1889. Note sur l'influence de la haute témperature sur le disthène. *Bulletin de la Société Française de Minéralogie*, **12**, 447–456.

VERNADSKY, V. I. 1908. Beiträge zur Energetik der Krystalle. *Zeitschrift für Krystallographie und Mineralogie*, **45**, 124–142.

VOGT, J. H. L. 1903–1904. *Die Silikatschmelzlösungen mit besonderer Rücksicht auf die Mineralbildung und die Schmelzpunkt-Erniedrigung* (2 vols). Videnskapsselskapets Skrifter I, Mathematisk-Naturvidenskabelig Klasse, 1903, **1** and 1904, **1**, Christiania.

WINKLER, H. G. 1965. *Die Genese der metamorphen Gesteine*. Springer, Berlin.

YODER, H. S. JR. 1980. Experimental mineralogy: achievements and prospects. *Bulletin de Minéralogie*, **103**, 5–26.

YODER, H. S. JR. 1989. Scientific highlights of the Geophysical Laboratory, 1905–1989. *Annual Report of the Director of the Geophysical Laboratory 1988–1989*. Washington, 143–197.

YOUNG, D. A. 2002. Norman Levi Bowen (1887–1956) and igneous rock diversity. *In*: Oldroyd, D. (ed.) *The Earth Inside and Out: Some Major Contributions to Geology in the Twentieth Century*. Geological Society, London, Special Publications, **192**, 99–112.

ZIRKEL, F. 1894. *Lehrbuch der Petrographie*, Vol. 3 (2nd edn, completely revised). Engelmann, Leipzig.

Arthur Holmes' unifying theory: from radioactivity to continental drift

CHERRY L. E. LEWIS

History of Geology Group, 21 Fowler Street, Macclesfield, Cheshire, SK10 2AN, UK (email: clelewis@aol.com)

Abstract: Only ten years after the discovery of radium in 1897, Arthur Holmes (1890–1965) began his studies at the Royal College of Science in London where he completed the very first U/Pb age determination designed specifically for that purpose. His continued interest in radioactivity and its effect on the thermal history of the Earth led to his early recognition that the age of the Earth should be measured in thousands, not hundreds, of millions of years, a subject he pursued for the rest of his career, despite considerable opposition from traditional geologists. Following a short period in Burma, he returned in 1922 to find that not only had attitudes to the age of the Earth changed, but that geologists were embroiled in a new controversy over continental drift. Evidence is put forward that suggests Holmes may have been aware of Wegener's theories virtually from the time they were proposed, and that by 1924 he was already searching for his own theory which would explain all geological processes. His profound understanding of the effects of radioactivity on the internal processes of the Earth, and his advanced knowledge of petrology, placed him in a unique position to develop a mechanism for driving continental plates around the globe. The progression of his ideas for this mechanism – convection currents in the mantle – and the unifying theory that that led to, is traced through his papers and letters to colleagues.

In the last half of the nineteenth century the age of the Earth became a highly contentious issue, due largely to the efforts of the physicist Lord Kelvin (formerly William Thompson), who considered that when calculating that age, 'essential principles of Thermo-dynamics have been overlooked by ... geologists ...' (Thomson 1862). Kelvin's own calculations, based on the thermodynamics of a body cooling from its molten state, originally placed wide limits on the Earth's age that most geologists found acceptable (20–400 million years). But Kelvin remained fascinated by the subject and worked on it for more than 30 years, during which time he reduced his estimate to 100 million years (Ma). When new data on the subject became available in 1893, which reduced the age still further to 24 Ma (King 1893), Kelvin wrote to the physicist, and his one time student, John Perry (Shipley 2001), in the following terms:

> The subject is intensely interesting; in fact, I would rather know the date of the Consistentior Status than of the Norman Conquest; but it can bring no comfort in respect to demand for time in Palæontological Geology. Helmholtz, Newcomb and another, are inexorable in refusing sunlight for more than a score or a very few scores of million years of past time. So far as underground heat alone is concerned you are quite right that my estimate was 100 millions, and please remark that that is all Geikie wants; but I should be exceedingly frightened to meet him now with only 20 million in my mouth (Thompson 1895*a*).

What Kelvin was referring to was the support given by astronomers – who had shown how heat from the sun could not have continued for more than 20 million years – for his own most recent calculations, based on Clarence King's data for the melting temperature of rocks (Thomson 1895*b*). These calculations concurred with King that the age of the Earth could be no more than 20 to 40 million years, with Kelvin's personal preference being for the lower value. The consequent antagonism felt between geologists and physicists was epitomized by Charles Walcott who, on 17 August 1893, delivered his Vice-Presidential address to the American Association for the Advancement of Science:

> Of all subjects of speculative geology, few are more attractive or more uncertain in positive results than geological time. The physicists have drawn the lines closer and closer until the geologist is told that he must bring his estimates of the age of the earth within a limit of from ten to thirty millions of years. The geologist masses his observations and replies that more time is required, and suggests to the physicist that there may be an error somewhere in his data or the method of his treatment (Walcott 1893).

From: OLDROYD, D. R. (ed.) 2002. *The Earth Inside and Out: Some Major Contributions to Geology in the Twentieth Century*. Geological Society, London, Special Publications, **192,** 167–183. 0305-8719/02/$15.00

Walcott continued his address by reviewing at least 15 of the most recent attempts by geologists to calculate the Earth's age, values largely derived by estimating denudation rates of various sediment thicknesses. The ages ranged enormously from Dr Alexander Winchell's 1883 value of three million years 'for the whole incrusted age of the world' to Mr W. J. McGee's 1892 estimate that the mean age of the Earth was 15 billion years, and that seven billion had elapsed since the beginning of Palaeozoic time, although he did subsequently modify these values to 6000 Ma and 2400 Ma, respectively. Walcott surmised:

> From the foregoing estimates of geologic time the only conclusion that can be drawn is that the earth is very old and that man's occupation of it is but a day's span as compared with the eons that have elapsed since the first consolidation of the rocks with which the geologist is acquainted.

Walcott then went on to make his own calculation, as indicated by the sedimentary rocks of North America, and concluded that although estimates could vary by assuming different denudation rates, a result emerged 'that does not pass below 25,000,000 to 30,000,000 years as a minimum, and 60,000,000 to 70,000,000 years as a maximum for post-Archaean Geologic time' (Walcott 1893, p. 675). This upper limit concurred with many other estimates that fell around the 100 million year value, once favoured by Kelvin.

But despite the ever-widening gap between the geologists and physicists during the 1890s, within ten remarkable years of Walcott's speech, discoveries were made by physicists and chemists that would finally facilitate dating of the age of the Earth by methods undreamt of by geologists of that time.

Following the discovery of X-rays by William Röntgen in 1895, in 1896 Henri Becquerel recognized that uranium was also emitting strange rays, which led to Marie Curie's discovery of radium and her coining of the term 'radioactivity' in 1897. In that same year J. J. Thomson identified the electron and realized that the strange rays were in fact streams of electrons. By 1902 the transmutation of radioactive elements had been established by Ernest Rutherford and Frederick Soddy, and the release of helium was recognized as a byproduct of that process in 1903. But perhaps the most crucial discovery for geologists came in 1903 when Pierre Curie demonstrated that during radioactive decay, energy was released in the form of heat. Since the Earth contained significant quantities of radioactive elements, the cooling of the Earth from its molten state would be prolonged for an unimaginable period of time. Thus within only a decade since the discovery of radioactivity, all the elements were in place that would facilitate radiometric dating of the first mineral, a fergusonite.

Using an early form of Geiger counter, the helium atoms being emitted from a very small but very accurately known quantity of radium were directed through a chamber, such that the passage of each particle set up a tiny electric current which gave a 'kick' to the needle of an electrometer. By counting the kicks, the particles themselves could be counted and the production rate of helium measured. Having established the rate of helium production and measured the amount of helium that had accumulated in the mineral, in 1904 Ernest Rutherford obtained an age of 40 Ma for the fergusonite (Rutherford 1905*a*, p. 34), although this was subsequently revised the following year to 500 Ma as the production rate of helium became better quantified (Rutherford 1905*b*, p. 486).

Radioactivity and the age of the Earth

The early life of Arthur Holmes (1890–1965) is now well documented in Lewis (2000, 2001) and it is therefore sufficient to say here that in 1907, only ten years after the discovery of radium, he gained a scholarship to study physics at the Royal College of Science in London. By 1909 he had obtained his BSc in physics and transferred departments to study geology, although he never obtained a degree in that subject since his studies were interrupted by an enforced trip to Mozambique to earn some money, due to his critical financial affairs. In order to gain his Associateship of the Royal College, it was necessary for him to accomplish some individual research in his fourth and final year, and this he commenced in 1910 under the guidance of Robert Strutt (later Lord Rayleigh), then professor of physics. Strutt had worked on dating minerals using Rutherford's helium method, but had also shown that helium escaped from the powdered rock at an alarming rate, thus rendering all helium dates *minimum* values.

Holmes, whose interests spanned both physics and geology, was the ideal candidate to research a new method for dating minerals, and he chose to investigate the uranium–lead decay scheme, although at that time lead had still not been proven to be the final decay product of uranium. Nevertheless, the close association of uranium and lead which were usually found together in rocks, plus the correspondence of increasingly old lead ages with increasingly large Pb/U ratios that had been demonstrated by Bertram Boltwood (Boltwood 1907), clearly indicated a strong correlation between the two elements. However, the Pb/U ages determined by Boltwood had been achieved using Pb and U data obtained from published literature and determined by a variety of analysts, mainly for the purposes of evaluating the chemical composition of the mineral in question. Thus Holmes set out to confirm the relationship between uranium and lead by making the first analysis specifically designed for age dating purposes (Lewis 2001). An age of 370 Ma was obtained from a nepheline syenite from Norway, believed to have been intruded during the Devonian, which concurred well with results found by Boltwood (Holmes 1911, p. 256). In 1911 this first uranium–lead analysis by Arthur Holmes established the decay of uranium to lead as the prime tool for dating rocks until well into the 1960s. It is still widely used today.

In 1913 the discovery of isotopes by Frederick Soddy was to have a major impact on age dating techniques. He had shown that U decayed to ^{206}Pb, and ^{207}Pb was believed to be 'ordinary' lead – lead that had been around since the formation of the Earth and was not derived from radioactive decay. Hitherto, the total amount of uranium and lead in an age sample had been measured, regardless of the provenance of those elements, and while Holmes had recognized that if 'ordinary' lead was present in significant amounts, then the age obtained would be anomalously high, he considered that over geological time the amount of 'ordinary' lead present would become insignificant when compared to that generated by the decay of uranium, and thus it would not affect the determined age.

To complicate the picture still further, Soddy believed that thorium also produced a lead isotope, ^{208}Pb, but early papers by Holmes dismissed this possibility. Holmes considered that because the Th/Pb ratio did not vary consistently with the age of the mineral as determined by the U/Pb method, then Pb derived from Th must be unstable and continue to decay to yet another element, so consequently it would not be stored in the mineral (Holmes & Lawson 1914, 1915). An interesting correspondence between Holmes and Soddy, referred to in the literature (Soddy 1917) but regrettably not located, seemed to solve the problem when Soddy pointed out that in fact only 35% of Th-derived Pb is stable, which accounted for the anomalous results found by Holmes (Holmes 1917). This made Holmes appreciate however, that all ages requiring the determination of lead were 'worthless in the absence of atomic weight determinations' (Holmes 1917, p. 245).

When Soddy first identified the presence of lead isotopes, Holmes initiated a programme of atomic weight determinations in collaboration with his school friend, Bob Lawson, who worked at the Vienna Radium Institute from 1913 and throughout World War I. At that time, the only way physically to distinguish the chemically inseparable isotopes of lead was by determining their atomic weights, but this made even slower and more laborious the already difficult chemical analysis of uranium and lead, resulting in a single age determination taking many months. By the end of 1914 communications with Austria had become very difficult, although Lawson continued to work on the problem during the war, and by 1915 Holmes was able to recalculate the ages of existing dates, following data obtained from the atomic weight determinations. Consequently the age of the oldest mineral hitherto dated was revised from 1640 to 1500 million years, and Holmes reasonably argued that the age of the Earth must therefore be at least 1600 million years (Holmes *et al.* 1915). But to many geologists who had become entrenched in their ideas of an Earth only 100 million years old, all thoughts of such an ancient planet were still totally unacceptable. Fifty years later Holmes recalled one such occasion during a talk being given at the Geological Society of London in March 1915 (From Holmes' unpublished response to his award of the Vetlesen Prize, 14 April 1964. Courtesy of RHUL Archives.):

> I was being violently attacked by the reader of a paper who insisted that the age of the Earth must be less than 100 million years old. In the discussion that followed I had occasion to refer to the isotopes of lead, then newly discovered. But isotopes did not seem to have been heard of in that audience. The reader of the paper insisted that all atoms of lead must have the same atomic weight, and I found myself in an exasperated minority of one.

Confirmation as to how new the idea of isotopes was to geologists at that time is illustrated by the fact that the isotopic numbers of lead are incorrectly reported in the meeting notes as being 106 and 107, instead of 206 and 207.

Despite the fact that many geologists concurred with the view that radioactivity still had to prove itself, for it was suffering from the damage inflicted by the helium method which by now Holmes had shown gave ages almost half those determined by the uranium–lead method, Holmes was driven by a conviction that the ages he had determined were at least of the right order of magnitude. Consequently, in the early war years he continued to pursue with vigour his interest in radioactivity, and in a series of three papers (Holmes 1915*a*,*b*, 1916) he reviewed the thermal history of the Earth and the contribution made by radioactivity to all the major processes within it. However, towards the latter part of the war, Holmes was required by the government to work on more mundane projects for the war effort, such as finding new sources of potash, almost all of which had previously been supplied by Germany, which severely curtailed his time to research radiometric dating. At the end of the war, with Lawson's return from Vienna, access to laboratories capable of verifying age determinations with atomic weight measurements became more difficult and, thus discouraged, most of Holmes's time was spent on his petrographic work from which two important books resulted (Holmes 1920*a*, 1921).

In August 1920 Holmes left Imperial College, where, after nine years he was still only a demonstrator on £200 a year, to work in Burma (Myanmar) as chief geologist to Yomah Oil (1920) Ltd at the vastly improved salary of £1400 a year. But Burma proved a disaster, resulting in financial ruin and personal tragedy for Holmes (Lewis 2000), and within two years he was back in Gateshead, his home town, facing 18 months of unemployment. It was June 1924 before Durham University eventually offered him the opportunity to single-handedly form their new geology department, which he accepted gratefully: 'You will be glad to hear that I have got the Durham post and a Department of my own; and in the first place I want to thank you for your own share in supporting me,' he wrote to his old friend Dr Prior at the Natural History Museum: 'I am glad to feel settled again and am looking forward with great interest to the work of building up from the very start.'[1] Holmes immediately set to work publishing material he must have been preparing while still unemployed, and by the end of the 1920s a prodigious outpouring of papers had earned him a reputation as 'one of the few English geologists with ideas on the grand scale', as the American geologist Reginald Daly described him (Dunham 1966).

While Holmes had been away in Burma he had missed two crucial meetings of the British Association for the Advancement of Science at which he would normally have been present. The first of these, held in Edinburgh in 1921, included a joint discussion between physicists, geologists, astronomers and biologists on the age of the Earth (Rayleigh *et al.* 1921*a*,*b*). Although all the old arguments supporting the traditional methods of dating the Earth were reviewed, the majority of speakers referred to Holmes's work on this subject, and for the first time there seemed to be a general consensus that ages determined by radiometric methods were at least of the right order of magnitude, and that the age of the Earth was around 1600 Ma. As William Sollas, professor of geology at Oxford put it: 'The age of the earth was thus increased from a mere score of millions to a thousand millions and more, and the geologist who had before been bankrupt in time, now finds himself suddenly transformed into a capitalist with more millions in the bank than he knows how to dispose of.' (Rayleigh 1921*b*, p. 282).

Some six months later a similar meeting was held in Philadelphia and again geologists came together with representatives from other scientific disciplines to consider the age of the Earth (Yochelson & Lewis 2001). On his return from Burma Holmes reviewed for *Nature* the publication resulting from this meeting (Chamberlin *et al.* 1922) and noted that 'there is a marked change of opinion in favour of the longer estimates' (Holmes 1923). Thus by 1922, some 18 years after the first rock had been dated by radioactivity, there seemed at last to be a general agreement that the age of the Earth was to be measured in thousands of millions rather than hundreds of millions of years.

The changing mind-set that enabled geologists to accept such an ancient Earth was part of a slow process that was paving the way towards a new thinking about geology that, in Britain at least, seems to have taken a distinct change of direction in the early 1920s. In the vanguard of this new thinking was Arthur Holmes who had not only contributed ground-breaking work to the understanding of radioactivity and its effect on the thermal history of the Earth, but who had also led an almost evangelistic crusade to tell geologists and the world at large about the great antiquity of the Earth. Aged only 22 he had written a short book entitled *The Age of the Earth* (Holmes 1913) in a style readily accessible to the layman, but also of sufficient authority that it was to eventually establish him as the world's expert on the subject. He followed this at regular intervals with both academic and 'popular' articles for magazines (e.g. Holmes 1915*c*, 1920*b*) and newspapers like the *New York Times* (Holmes 1926*a*), and in 1927 an abridged

[1] Holmes to Prior, 4 June, 1924. Courtesy of the Natural History Museum.

and revised edition of his book *The Age of the Earth*, subtitled *An Introduction to Geological Ideas*, was brought out by his publishers in their 'Things to Know' series and subsequently included as a paperback in 'Benn's Sixpenny Library' series (Holmes 1927). Selling for just that, sixpence, Holmes gained an even wider following than before.

It is interesting to note how either Holmes, or his publishers, may have been concerned about the impact his large estimates for the age of the Earth would make on a wider audience, since such an audience would inevitably have contained a number of significant religious figures, some of whom would have still believed in the biblical interpretation of the Earth's creation, in 4004 BC. In my 'sixpenny' copy, for example, a handwritten note on the first page says simply 'See pp 44 and 53'. On turning to those pages one finds, underlined in purple, the words 'geological time must be of the order of 1,500 million years' and, 'the age of the earth may far exceed 1,000 million years', suggesting that the reader is somewhat startled by these values, a reaction that was perhaps anticipated. For in complete contrast to the 1913 edition, where, with remarkable perspicacity for one so young, his closing remarks visualized the contribution that developing a geological timescale would make to the understanding of geology (Lewis 2001), the 1927 edition concludes, surprisingly, with a somewhat maudlin paragraph about 'Our place in the eternal miracle of existence' and our 'adventuring towards the very threshold of Creation', ending with this quote from Alfred Noyes:

> Here, now, the eternal miracle is renewed;
> Now and forever God makes heaven and earth.

It is surprising in as much as this is the only reference to God or religion that can be found in any of Holmes's writings, except for the odd remark in a couple of letters he wrote to Bob Lawson from Mozambique[2] and a popular account where he points out how the age of the Earth 'is more than a million times longer than the whole of the Christian Era – two thousand million years' (Holmes 1939, p. 190). Having discussed Holmes's religious beliefs with many of his colleagues and students, there is a strong consensus, supported by the remarks to Bob Lawson, that he was at the very least 'highly agnostic'.[3] It does seem therefore, that Holmes, or his publishers, may have had an expectation of how the book would be received by the general public and that the last paragraph is a slightly cynical attempt to temper their reactions.

Radioactivity and continental drift

Further evidence for the gradual change in thinking about geology during the early 1920s can be found in reports of the second British Association meeting that Holmes missed while in Burma, which was held in Hull in August 1922 (Evans *et al.* 1923; Wright 1923). A brief notice that 'A discussion on Wegener's hypothesis of continental drift, in which both the geological and astronomical sides will receive attention' was placed in *Nature* the week before (Anon. 1922*a*, p. 263), and many of the speakers who had attended the previous year's meeting on the age of the Earth were present, testifying to interdisciplinary interest in 'the Wegener hypothesis'. This is the first record of the topic being discussed at a British Association meeting and as such it is a milestone in the development of the theory of continental drift in Britain. Significantly, the comments of the majority of speakers whose remarks were recorded, tentatively supported the theory, or at least felt there was a case for further investigation, although Wright's concluding remarks suggest that there was also much opposition.

[2] Holmes to Lawson, 29 June 1911: 'I felt somehow what a fearful meaningless tragedy the whole Universe appeared to be. Have you yet seen Dr. Russell Wallace's newest book 'The World of Life'? It is good while it remains scientific but philosophically and imaginatively it is insanely absurd – the sort of book which ministers will rave over – excusing God's ways to Man and pointing out the contemptibly conceited idea, that the whole purpose of existence is MAN. You might read it during the vac. It is stimulating, but rather disappointing when he harps on the "Purpose of God" '. And again on 11August: 'Christianity makes no progress on the African coast. Mohommedanism holds almost entire sway and I honestly feel it is a better religion altogether for a black. It has the advantage of including Christianity, for Christ is one of the prophets and is considered second only to Mahomet. The boys say Christ was never married and so cannot afford a guide as to how wives and children should be treated – not a bad criticism and very original!' Copies of these letters will form part of the Arthur Holmes Collection being established at the Geological Society in London.

[3] Taped interview with Sir Kingsley Dunham, 6 July 1998, in response to a question about Holmes' religious beliefs: 'Highly agnostic. He said to me one day, "I thought I would give the Cathedral [in Durham] a try, but I simply couldn't take it. I went there thinking I might make some sort of symbolism of it, but I couldn't really make sense of it." So you can take it he was an agnostic and certainly took no interest in the doctrine.'

The British Association discussion is of further interest since Wegener's hypothesis, first published in German in 1912 (Wegener 1912) was not to be translated into English for another two years, although the second edition of his book (Wegener 1920) had just been reviewed in *Nature* in February 1922. The anonymous reviewer, believed to have been the physicist William Bragg (Marvin 1985), noted: 'This book makes an immediate appeal to physicists, but is meeting strong opposition from a good many geologists' (Anon. 1922*b*, p. 202), which suggests that the hypothesis was already being widely discussed, if not in the literature. With his concluding remarks the reviewer accurately predicted the tremendous impact that the theory was to have: 'The revolution in thought, if the theory is substantiated, may be expected to resemble the change in astronomical ideas at the time of Copernicus' (Anon. 1922*b*, p. 203).

It is not clear exactly when Arthur Holmes first started to think specifically about problems relating to continental drift, but throughout his career he had been interested in the thermal history of the Earth and evolution of the crust, which would have led naturally to an interest in continental drift when it became an issue. Alfred Wegener is believed to have first presented his theory of continental displacement to the Geological Association of Frankfurt in January1912, and a week later to the Society for the Advancement of Natural Science, although he claimed to have first had the idea in 1910 when considering a map of the world (Wegener 1967). Holmes is known to have visited Germany during August and September 1912,[4] where he stayed with a friend studying at the Freiburg Institute, and lectured there.[5] Holmes's daughter-in-law recalls Holmes saying that he had met Wegener,[6] but it could not have been during this trip since Wegener was by that time in Greenland.[7] It nevertheless seems quite likely that on this visit to Germany Holmes would have come across discussion of the theory amongst geologists who had heard Wegener lecture earlier that year. Certainly Wegener's ideas caused tremendous discussion right from the start.

In 1914 Holmes became a colleague and close friend of John W. Evans whose discourse on 'the Wegener hypothesis' had opened the 1922 British Association meeting (although Evans was not actually present) and who, two years later, was to write the introduction to the English translation of Wegener's book *The Origin of Continents and Oceans* (Wegener 1924). Evans and Holmes both worked at Imperial College when in 1914 Evans, too old to be called up for active service during World War I, became a demonstrator there at the age of 57. Having been chief geologist to a mining company in India, Evans shared Holmes's interests in mineralogy and petrology, and by 1915 had become a strong advocate of Holmes's arguments in favour of an Earth of great antiquity, based on radiometric dates. They frequently attended Geological Society meetings together so it seems very likely that Holmes accompanied Evans in June 1920 to the Royal Geographical Society, of which Holmes was a member, when G. A. F. Molengraaf, by then a strong advocate of continental drift (Oreskes 1999, p. 92), lectured there on the East Indies archipelago (Molengraaf 1921). Evans was a great linguist (Lewis 2000) and spoke German fluently, and Holmes was certainly able to read it,[8] thus it seems likely that they read Wegener in the original German and discussed his displacement theory together, some time before Holmes went to Burma.

To earn some much-needed money during his period of unemployment on returning from Burma, Holmes wrote reviews of meetings and books, one of which was Charles Schuchert's second edition of his book on historical geology that Holmes reviewed for *Nature* (Holmes 1924). Schuchert, an important palaeontologist of his time, wrote to Holmes thanking him for the 'splendid notice' and commenting on the influence Holmes's first edition of his book *The Age of the Earth* had had on American geologists such as himself and Joseph Barrell. 'Many years ago I read your splendid and ?[9] little book *The Age of the Earth*. I soon directed Barrell to it and I loaned him my copy. So you see you have

---

[4] Holmes to Sederholm, 28 June 1912: 'During August and September I shall be travelling in Germany and Switzerland.' Courtesy of Åbo Akademis Bibliotek, Åbo, Finland.

[5] Holmes had a close friend, Aquila Foster, who in 1910 went to study geology in Freiburg and whose description of his time there made Holmes 'long to visit Germany' (diary entry, 27 February 1911).

[6] Pers. comm., Karla Holmes, November 1998. Unfortunately Karla, herself a geographer who understood the importance of Wegener and his work, was not able to remember the circumstances in which they met. She described it as a 'visit'.

[7] Pers. comm., Mott Greene, e-mail dated 25 September 2000.

[8] Holmes to Sederholm, 24 November 1911: 'I must apologise for writing in my own language but I may say a reply in either English, French or German will be equally acceptable.' Courtesy of Åbo Akademis Bibliotek, Åbo, Finland.

[9] Schuchert's handwriting is, in places, impossible to decipher.

had influence over both of us and now you are to have still more over me through your notice of my *Hist. Geol.* in *Nature*'.[10] This letter from Schuchert marks the start of a five-year correspondence between the two men, through which we gain an important insight into the evolution of Holmes's thinking on continental drift.

Holmes was in fact highly critical of Schuchert's chapter that dealt with the permanency of the continents. Schuchert saw 'Gondwana Land', the missing land bridge that he believed had once joined South America to Africa, as a means of explaining the similarity of rock types and fossils on either side of the Atlantic, but he needed it to disappear during the Cretaceous. Thus Joseph Barrell, an American geologist and close friend of Schuchert, proposed that Gondwana Land had been loaded down by a huge outpouring of magma, which caused the land to sink beneath the waves. Holmes, however, pointed out in his review that unless the magmas were very much more dense than normal, the new surface would be no lower than the original one: 'Pushing the light rocks down to great depths does not solve the problem [of continental drift], which is one of lateral not vertical differences', he explained.

But while many geologists, like Holmes, may have objected to land bridges and accepted that Wegener's theory clarified a number of geological problems, it was the lack of a mechanism to drive continental plates around the globe that was the main stumbling block which caused the majority to dismiss continental drift. However, by the end of 1925, Holmes's understanding of radioactivity and the thermal effects it had on the Earth's interior, coupled with his ability to comprehend the enormous time scales involved, placed him on the brink of formulating just such a mechanism.

Radioactivity and convection currents

Within less than a year of arriving at Durham, Holmes concluded his series of five articles on 'Radioactivity and the Earth's thermal history', started ten years earlier. The first of these two later papers criticized the earlier three in the light of recent developments (Holmes 1925*a*), but the second one, entitled 'The control of geological history by radioactivity' (Holmes 1925*b*), is evidence of the concept forming in Holmes's mind of the great unifying theory he intuitively felt would one day elucidate all geological processes such as the origin of granites, plateau basalts, geosynclines, mountain building, continents and oceans, that were hitherto inexplicable. John Joly, thinking along similar lines, had already proposed a model of 'basaltic cycles' whereby regular worldwide melting of a basaltic layer beneath the continents and oceans facilitated movement of the continents around the globe in a westerly direction, driven by tidal action in the molten layer (July 1923; Wyse Jackson 2001). But Joly's refusal to accept the long timescales proposed by Holmes – 'So far as I know, Joly, among authorities on radioactivity, stands alone in his adherence to such low estimates of geological times' (Holmes 1925*b*, p. 542) – rendered Joly's model, in Holmes's opinion, 'far too simple to match the complex details of geological history. Nevertheless,' Holmes continued in a conciliatory tone, 'Joly has surmised what is so far the only kind of process that even begins to correspond with the dominant facts. And that, without any qualification, is a very great achievement.'

By combining early seismic data with recent density results and the principle of isostasy, Holmes developed a crude model of continental and oceanic substructures in which a granitic continent of varying thickness was underlain by a basaltic layer that turned to peridotite at depth (Holmes 1925*b*, fig. 1). He then calculated how even a sparse distribution of radioactive elements throughout the basaltic layer would, given enough time, promote melting at the base of the crust. As melting occurred, contemporaneous volcanism (flood basalts) would be facilitated by the cracking of the crust under tension, due to the expansion occurring beneath it, the oceanic areas in particular being increased in area by the intrusion of magma. The eventual escape of some of the magma then facilitated cooling and consolidation, but as consolidation began the increased area had to settle down on the contracting substratum (mantle), which led to folding and thrusting (mountain building) in areas of weakness. Many geological anomalies were addressed by this model, but unfortunately, the continued generation of such heat over the whole of geological time (then 1600 million years) led to the embarrassing conclusion that the Earth must be getting hotter, when clearly this was not the case. Consequently the extraneous heat had to be got rid of in some way, and he proposed that 'lateral currents within the earth' would facilitate the escape of this heat from beneath areas where the crust was particularly thick, towards both the oceanic areas where it could escape readily, and to the thinner parts of the crust.

An important feature of this model was the need to have sufficient time for these melting and cooling cycles to occur. Holmes calculated

[10] Schuchert to Holmes, 8 October 1924. Courtesy of Yale University Library.

there had been 20 such cycles since the end of the Precambrian, each of which he correlated with an episode of mountain building that occurred during the cooling and contraction period, and each of which was approximately of 30 million years duration, thus requiring 600 million years in total for the Phanerozoic. This estimate concurred reasonably well with a U/Pb date of 528 Mathat Holmes had for the close of the Precambrian, but even in the late 1920s some orthodox geologists, in addition to Joly, felt uncomfortable with the enormous timescales required. Schuchert for one had such difficulties:

> And the age of the earth since the close of the Proterozoic is of 'the order of, but less than, 600 million years'?[11] Whew! We stratigraphers have yet many new formations to find, and many more breaks – but both are coming right along.[12]

However, Chester Longwell, a colleague of Schuchert's at Yale, appreciated the long-term implications of what Holmes was saying and remarked to Schuchert after reading Holmes's paper (1925*b*): 'Joly and Holmes together have a beautiful theory, and I believe it will be epoch-making.'[13]

Both Holmes and Schuchert were founder members of the Age of the Earth committee that was established in 1926 by the National Research Council in America in an attempt to reconcile the large discrepancies that still existed between radiometric and non-radiometric assessments of the age of the Earth. Despite the general acceptance by geologists that radiometric dates were a more reliable indicator of the Earth's age, as Schuchert's comment above illustrates, they were still presented with a considerable conundrum as to why this great age was apparently not reflected in the rocks. Just where were all the sediments that should have accumulated over such enormous spans of time?

Over the five years that it took to research and publish the committee's findings, Holmes and Schuchert continued their correspondence. Like many Americans, Schuchert was still highly resistant to the concept of continental drift:

> Gondwana is a fact, but I still have finally to get rid of it, or else go into the Wegener following, and Heaven knows I will not trail into that camp!

So he appealed to Holmes:

> Help me along on Gondwana. Give me land to stand on, at least for a time, or cause me to swim among the ammonites with eyes to see how the Tethyian forms got to Peru and Chile without going south into the Atlantic.[14]

But Holmes was not very accommodating, only agreeing that:

> Like you, I have been trying to avoid Wegenerism, for it is a subtle temptation to adopt it and escape all one's difficulties. However, it is necessary to have an adequate force to do the alleged work [of driving continents], and I can find nothing in the fusion of the substratum to produce the desired results.[15]

In other words, at this time, June 1926, he had not yet developed a mechanism for driving continental drift. He went on: 'Gondwana is a brute to get rid of unless one admits a drift from a central region'; and then confessed:

> So I am led to be at least sympathetic to the idea of differential continental drift, merely as a geological deduction. Next year I have to give a course on geophysics and by then, if possible, I must make up my mind.[16]

Which indeed he did, but in the interim a new discovery was to provide him with even more heat for his radioactive engine within the Earth. For some time Holmes had sought an explanation for the extensive intrusions of granite that characterized the early Precambrian. In his investigations with his long-standing friend and collaborator Bob Lawson, they explored the possibility that the radioactivity of potassium might have been considerably greater in the geological past. It was therefore quite a surprise when they realized in the course of their enquiries that there was no need to invoke greater radioactivity of potassium in the past, for even today, because of the large amount of potassium in crustal rocks such as granite, 'potassium as an emitter of radiothermal energy is in the aggregate of the same order of importance as

[11] Schuchert is quoting Holmes (1925b, p. 542).
[12] Schuchert to Holmes 2 June 1926. Courtesy of Yale University Library.
[13] Longwell to Schuchert, 31 January 1926. Courtesy of Yale University Library.
[14] Schuchert to Holmes 2 June 1926. Courtesy of Yale University Library.
[15] Holmes to Schuchert, 16 June 1926. Courtesy of Yale University Library.
[16] A wonderful collection of the lantern slides that Holmes used for teaching, dating from this period, can be found in the archives of Durham University Library, Palace Green, Durham.

uranium or thorium' (Holmes & Lawson 1926). Consequently, a couple of months later in a paper on magmatic cycles (Holmes 1926*b*) Holmes had even more heat within his magmatic layer to dispose of. This time he managed it using 'convection currents aided by tidal disturbances [which] will sooner or later bring to the top an excess of heat' (Holmes 1926*b*, p. 322).

A convecting substratum was by now a clear concept in Holmes's mind, and in August the following year, 1927, he again wrote to Schuchert:

> I have been hoping to find a clear road through all our tectonic difficulties, but every hypothesis raises fresh difficulties of its own. However, on the whole I find a combination of Wegener's ideas with magmatic convection currents inside the earth on a gigantic scale to provide the energy, seems best to fit our needs.[17]

And by November he was quite convinced:

> it is impossible (within the conditions and limits of present day knowledge) to get rid of the lands that formerly occupied the sites of present oceans except by moving them sideways. I can see no alternative at all to continental drift, and I have come to this conclusion from a position of strong prejudice against such processes.[18]

A month later, having given his first public lecture on the subject to the Edinburgh Geological Society (Holmes 1930*a*), Holmes wrote to Schuchert to wish him a happy new year, and confided, 'I am trying to develop the idea of convection current'.[19] Two weeks after that, when he gave his now famous lecture to the Glasgow Geological Society on 12 January 1928, (Holmes 1928*a*, 1931), his early concept of convection currents in the mantle as a mechanism for driving continental plates around the globe had become a profound theory.

Radioactivity and the unifying theory

Although the substratum was generally considered to be solid, Holmes's theory proposed that given enough time, and of all people Holmes knew that there was enough time, it actually behaved like a highly viscous liquid. Taking advantage of the delay between giving his lecture (1928) and it appearing in print (1931), he took time to further evolve ideas about the substratum which he considered was

> probably for the most part in a glassy state; that its temperature is such that although it is rigid, it is devoid of permanent strength except possibly near the top; and that it is and has been the main source of basic and ultrabasic magmas, including the plateau basalts (Holmes 1929).

He required the substratum to be in this state, for

> only currents in a highly viscous glass could get a sufficient 'grip' on the continental undersurfaces to exert the requisite drag upon the overlying material.

Differential heating, generated by the decay of radioactive elements, caused convection cells to form within the substratum, rising beneath continents and descending at their edges. As hot material reached the top of a convecting cell beneath a continent it would travel horizontally for some distance before cooling and descending again (see Fig. 1). As it travelled horizontally it would produce a force that was sufficient to drag the continents sideways, and so they were very slowly pulled apart allowing the substratum to rise up and take their place in the ocean floor. Quasi-sea-floor spreading?

As the continental mass was dragged open and each part moved forward, 'probably overriding the ocean floor along thrust planes' (subduction), mountain building occurred on the continental margins, beneath which the currents descended. Meanwhile, the sites above ascending currents would become disruptive (tensional) basins and the accumulation of excess heat, responsible for the whole process, would be discharged by the development of new ocean floor within these basins. Eventually a new heat distribution pattern would arise which would gradually generate a correspondingly different set of convection currents.

Here indeed was a unifying theory that seemed to explain all the major geological features, but as Holmes appreciated in his conclusions: 'It is not to be expected that the first presentation of a far-reaching hypothesis and its manifold applications can be wholly free from errors'. Nevertheless, he considered that 'its general geological success seems to justify its tentative adoption as a working hypothesis of

---

17 Holmes to Schuchert, 27 August 1927. Courtesy of Yale University Library.
18 Holmes to Schuchert, 12 November 1927. Courtesy of Yale University Library.
19 Holmes to Schuchert, 29 December 1927. Courtesy of Yale University Library.

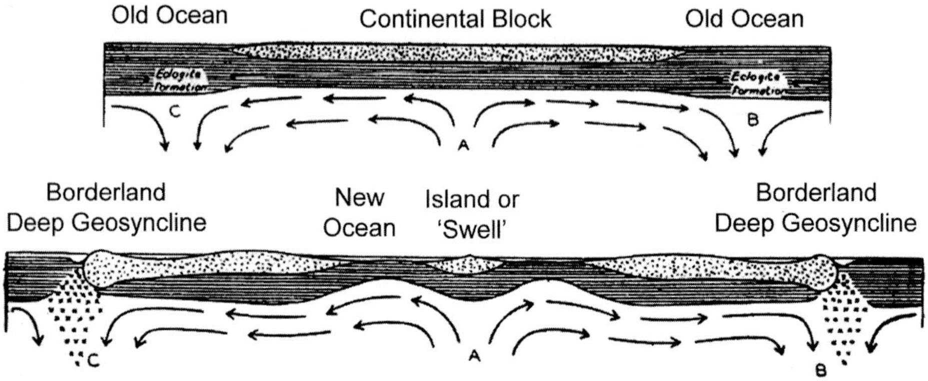

Fig. 1. Continental drift, 1928. Reproduction of figures 2 and 3 from Holmes (1931) illustrating Holmes's 1928 'simplest case' mechanism for driving continental drift. In reality, Holmes recognized, the circulation in the substratum would be much more complex, as ascending currents would not be localized about a single centre. The three most important features of this model are: (1) the generation of new continents caused by the dragging effect of ascending currents travelling horizontally at the top of a convecting cell, which eventually split the old continent apart; (2) that new oceanic crust would fill the tensional basins created where the old continents split; and (3) that old continents would 'founder' on the downward currents, as the high pressures and temperatures generated at continent margins caused recrystallization of the rocks to eclogite facies, the higher density of which would help them to sink.

unusual promise'. But Holmes was ahead of his time.

At this point it is interesting to consider what triggered Holmes's idea that a convecting substratum could provide a force sufficient to move continents and drag them apart. In her book on *The Rejection of Continental Drift*, Naomi Oreskes states that the idea of convective overturn had been around for some time (Oreskes 1999, p. 118), and indeed Holmes's work indicates that it had, for in a paper written in 1929 Holmes revealed that: 'The possibility of convection currents was recognized by A. J. Bull eight years ago in a paper that has not received the attention it deserves', (Holmes 1929, p. 345). Bull was an amateur geologist interested in the tectonics of mountain building, and in the paper Holmes referred to he proposed that convection currents would *act as a drag* on the under-surface of the crust, drawing out the lower portions into schists and gneisses, thereby explaining their presence in mountain chains. In 1921 when Bull's paper was published, Holmes was in Burma, although of course he may well have seen it subsequently, but in 1927 Bull was President of the Geologists' Association. In his presidential address, given in February 1927, he further developed his ideas on mountain building which were then published in the middle of that year (Bull 1927), referencing his earlier work and the dragging effect of convection currents on the base of the crust (Bull 1921, p. 364). A couple of months later we see Holmes, who was an active member of the Geologists' Association, first proposing to Schuchert that convection currents could provide the energy necessary for continental drift. Was it perhaps Bull's paper that had given him the idea that a convecting substratum dragging on the under-surface of the continents, would eventually pull them apart?

Whatever the timing and trigger for Holmes's ideas, he once again found himself at the centre of controversy with few supporters around him, although European geologists seemed to be more in sympathy with his own and Wegener's views than their American counterparts. Schuchert, in particular, was not convinced. Writing to Holmes in November 1928, in response to Holmes's review of the publication of the fourteen papers presented at a symposium in New York in 1926 on the *Theory of Continental Drift as Proposed by Alfred Wegener* (van Waterschoot van der Gracht 1928), Schuchert said rather tersely:

> I have your 'This is what I think about it' and I am glad to get your views. No, no! Gondwana did not move sideways like a crab, but like a continent gone old, it sank out of sight. My dear friend, continents do move, and after you have seen your errors you will come back again into [the] good old fold of orthodox geology. I want you to find the mechanism of why borderlands and continents sink out of

> sight and I have faith that you will find it some day, and then not by the Wegener method. Good cheer to you.[20]

Holmes's review (Holmes 1928*b*) not only raised Schuchert's hackles, but also William Bowie's, who wrote to Schuchert in exasperation:

> Holmes brings out a new thought which is even more impossible than Wegener's hypothesis. That is that the submerged ridge through the Atlantic Ocean is the place at which North and South America separated from Europe and Africa the latter two continents drifting eastward and the Americas drifting westward. I do not see how the same force, operating to send one mass westward, could make another go eastward. I believe that we need to apply elementary physics and mechanics to the continental drift problem in order to show how impossible that drifting would be.[21]

But Holmes showed the courage of his convictions and was not deterred by criticism. Quite the opposite. At a British Association meeting in 1931, held just after Wegener's untimely death, geologists, geographers, mathematicians, meteorologists and physicists came together to discuss 'Problems of the Earth's Crust' in Section E, Geography (Mackinder *et al.* 1931). Of the seven speakers, a mathematician and a geologist were fiercely against drift, Holmes and a meteorologist were strongly in favour, while the remaining three, including Harold Jeffries who had always been a strong opponent on the grounds that no force was sufficient to move continents, were impartial or chose not to show their colours. Holmes reiterated his arguments against a contracting Earth, a theory still favoured by many to explain mountain building, and put forward his ideas on convection currents.

As with his work on the age of the Earth, he had looked at the evidence, devised a model and come up with a theory he knew was rational and logical. Proving it might be harder than with the age of the Earth, for as yet hard data were not forthcoming. In fact, it was another 30 years before the data proved him right. Doris Reynolds, Arthur Holmes's second wife and herself a geologist often at the centre of controversy, summed up the situation when, aged 80, she wrote to John Thackray:

> Geology is always like this, very slow moving. When a new geological discovery or suggestion is made it is quite quick if it is noticed in 20 years, and may take 50 to 100 or more. Then dogmas form obstructions.[22]

The Earth as a radioactive container

Holmes himself was not entirely free from dogma. In 1921 an American astronomer, Henry Russell, calculated the age of the Earth based on the assumption that all lead in igneous rocks was of radioactive origin. He concluded (fortuitously) that the age of the Earth's crust roughly approximated to four billion years, but definitely lay within the range of two to eight billion. On his return to academic life Holmes replied to this interloper by way of a letter to *Nature* (Holmes 1926*c*) where he recalculated Russell's estimate and reduced it to 3200 Ma. He pointed out, however, that only half the amount of lead in igneous rocks was derived from radioactive decay while the other half 'must have originated either in the ancestral sun or during the events that attended the birth of the solar system', in other words, not from radioactive decay subsequent to formation of the crust. He concluded from this 'it is clear that the Earth – as a radioactive container – cannot have existed for so long as 3200 million years' because, he argued, all the ages from the oldest minerals so far dated fell largely in the 1000–1100 million years bracket. At this time the oldest mineral was still only 1525 million years, therefore 'the frequently quoted age of the Earth, 1600 million years, thus appears to be of the right order' (Holmes 1926*c*, p. 482).

Running in parallel with Holmes's work, investigations into the elements and their properties continued at the Cavendish Laboratory in Cambridge. Following Soddy's discovery of isotopes, in 1914 Frederick Aston built the first mass spectrograph, forerunner of the mass spectrometer, and established the phenomenon of isotopy. By 1922 Aston had been awarded the Nobel Prize for Chemistry 'for his discovery, by means of his mass spectrograph, of isotopes in a large number of non-radioactive elements' and by 1927 he had identified the three known isotopes of lead, formerly recognized by their atomic weights, and many other previously unidentified lead isotopes. But even then, ^{207}Pb ('ordinary' lead) was still considered to be a non-radiogenic isotope. Then in 1929 some new

[20] Schuchert to Holmes, 5 November 1928. Courtesy of Yale University Library.
[21] Bowie to Schuchert, 11 October 1928. William Bowie was head of the Geodesy Division at the US Coast and Geodetic Survey. Courtesy of Yale University Library.
[22] Reynolds to Thackray, 17 July 1978. Courtesy of the Natural History Museum.

results indicated that this could not be the case and that ^{207}Pb must result from the decay of a hitherto unknown isotope of uranium. Aston discussed the problem with Rutherford who agreed with his deductions and went on to calculate that the new uranium isotope must have an isotopic number of 235. Indeed, it was ^{235}U which, because it represented less than 1% of total uranium, had so far gone unnoticed. By estimating the rate at which ^{235}U decayed, which was much faster than ^{238}U, and by assuming that equal amounts of each isotope were present when the Earth first formed, Rutherford was able to determine the time it had taken to increase the ratio of the two uranium isotopes from zero to its present-day value. The figure he arrived at, 3400 million years, was the first age of the Earth to be based on isotopic data from a mass spectrograph.

It is therefore difficult to understand why Holmes did not attempt to collaborate with Aston on this work when he certainly knew of this significant new development. In fact, such was Holmes's concern regarding Rutherford's recognition that ^{235}U decayed much faster than ^{238}U, that he wrote a letter to *Nature* putting forward his arguments as to why he thought Rutherford was wrong (Holmes 1930*b*). His concern was, of course, that should Rutherford be proved right, then all pre-existing age determinations would once again have to be recalculated in the light of this new knowledge. Had Holmes at this stage started to collaborate with Aston on age dating, using Aston's mass spectrograph for lead determinations, then progress on the development of radiometric dating techniques might have been advanced by a decade or more. Unfortunately, this development came at a time when Holmes's attention was primarily focused on continental drift. It was also an extremely busy period for him. In the decade after his arrival at Durham he published more than 60 scientific contributions, which he typed himself on an upright typewriter. This work established him as one of the leading authorities on radioactivity and igneous rocks and led to requests for him to lecture around the world. In 1930 he was an exchange professor in Switzerland, in 1931 he toured Finland extensively, and in 1932 was invited to give the prestigious Lowell Lecture series in the United States. In the same period he updated the second editions of his book *The Age of the Earth* and his two books on petrology, as well as fulfilling his normal teaching duties and serving on several committees. At home a young son would have demanded his attention, and the relationship with his then wife-to-be, Doris Reynolds, must also have taken up his time. By any stretch of the imagination he was a busy man and probably just did not have time to develop collaborations with people in other disciplines, such as Aston.

In 1931, the Age of the Earth Committee published its findings (Knopf *et al.* 1931). By that time it had been conclusively shown that the decay rate for ^{235}U was significantly faster than ^{238}U (it is in fact six times faster) and thus it was necessary to recalculate all results based on uranium and lead. In an intensive review of all radiometric data available at that time, Holmes discarded as unreliable all results not based on the U/Pb decay scheme and found no more than thirty U/Pb ages, world-wide, to be acceptable. These he recalculated and deduced from the results that:

> No more definite statement can by made at the present than that the age of the Earth exceeds 1460 million years, is probably not less than 1600 million years, and is probably much less than 3000 million years.

Little progress had been made on dating the age of the Earth since 1915. It was still difficult to find minerals with levels of U and Pb high enough to measure, and methods of analysis were still labour intensive. Holmes had looked at alternative methods, but in the mid-1930s it seemed that potassium was not going to be a suitable element for dating purposes, and by the end of that decade another trial with helium also proved disappointing (Lewis 2001). But during this period, 1936, Alfred Nier commenced work at Harvard University, and a new era in radiometric dating began.

The age of the Earth

Although a physicist by background, Nier was interested in the problems of measuring geological time, and in the late 1930s he dated a suite of 25 rocks and published the results in a series of papers (Nier 1938*a*,*b*, 1939). The oldest sample in this suite, a pegmatite from Manitoba in Canada, gave an age of 2200 million years, which had major implications for the age of the Earth, not least because the age of the whole universe was then considered to be no more than 2000 million years. After publishing the first paper Nier wrote to Holmes whom he considered, 'more than any other was central to ideas ... in which lead isotope studies of ores and rocks played a crucial role, [such as] the construction of an absolute geologic time scale, ages of oldest crust and minerals [and] the age of the Earth' (Nier 1982), but it was to be another six years before their association began in earnest (Lewis 2001).

Following completion of his book *Principles of Physical Geology* in 1944, and his move as Regius Professor of Geology to Edinburgh University, Holmes's publishers requested that he write a fourth edition of his book *The Age of the Earth*, so Holmes wrote to Nier asking for some information. While this fourth edition never came to fruition for some reason, as a result of Holmes's letter a remarkable 20-year correspondence began between the two men, both sides of which are still preserved at the University of Minnesota. Such was Nier's concern about the age of the Manitoba pegmatite, that he analysed another rock from the same area in an attempt to verify his first date. But he obtained an even older result – this rock was 2570 Ma. Holmes was fascinated by the Manitoba results and recalculated them for himself. He found the pegmatite to be 2480 Ma, a value closer to the older result determined by Nier. He wrote to tell Nier in May 1945, and cautiously expressed his views:

> I hope when you get going again [after the Second World War] you will try to clear up the extraordinary discrepancies in the Manitoba results. I feel sure that the true age must be of the order of 2000 m.y. or more, and this is of the greatest interest, not only because the rocks here seem to be the oldest yet found, but also because such a figure shows that current views about the expanding universe need revision – not perhaps to be wondered at![23]

Over the following months Holmes gave the problem of the age of the Earth considerable thought as he devised a new model with which to calculate it from Nier's data (Holmes 1946), and students of the time recall him turning up at class to report that 'Today the Earth is X years old' as it rapidly went up and up. By February 1946 he was again writing to Nier:

> Ever since your isotopic analyses of ore-leads was published I have hoped that it would be possible to calculate from the results the time that has elapsed since the Earth's primeval lead began to be contaminated by radiogenic lead. The acquisition of a calculating machine a few months ago has now made possible the somewhat formidable calculations and I have just completed the work. The age works out at about 3,000 million years by various sets of solutions. . . . This looks like being the first really reliable estimate of the age of the Earth and I should like to salute your work as the means of making it possible.[24]

A few months later Holmes's age of the Earth had gone up another 350 million years (Holmes 1947) as a result of a larger number of solutions.

For the same reasons that Holmes's work on convection currents did not provide all the modern-day answers we now have, Holmes did not get the age of the Earth right because technology could not provide him with the necessary data. He was a man with ideas far ahead of available technology. Nevertheless the model Holmes devised for his calculations is essentially that still in use today for measuring the age of the Earth, as is his model for convection currents in the mantle. In the former case, however, Holmes's contribution to development of the model is acknowledged, albeit shared with other workers who apparently arrived at the same result at much the same time (Gerling 1942; Houtermans 1946), since the model is variously called the 'Holmes–Houtermans model' or the 'Holmes–Gerling–Houtermans' model for dating the Earth. However, given the fact we now recognize that Holmes discovered the principle of initial ratios back in 1932, on which all the models for dating the Earth are based (Lewis 2001), long before Gerling or Houtermans developed their models, there is a strong argument that says it should simply be called the 'Holmes model' for dating the Earth.

After Holmes supplied the model and increased the age of the Earth to 3350 Ma, all that remained was for technology to catch up with his ideas so that the correct age could be determined, and in less than a decade it had done just that. In 1956 Holmes made his final contribution to the age of the Earth debate in a paper entitled 'How Old is the Earth?' (Holmes 1956), read to the Edinburgh Geological Society on his retirement from the department. In it, he hailed Claire Patterson's work on correctly dating the age of the Earth as: 'brilliant joint research [which] owed its almost miraculous success to the development of analytical techniques of the utmost precision and delicacy' (Holmes 1956*b*, p. 319) and records, with what appears to be genuine pleasure and humility, 'my indebtedness to many younger friends who are now boldly accepting the challenge and meeting it with all the resources and superb techniques of the atomic age' (Holmes 1956*b*, p. 331).

[23] Holmes to Nier 21 May 1945. Courtesy of Minnesota University Archives. At that time the age of the Earth was older than the age of the Universe.

[24] Holmes to Nier, 16 February 1946. Courtesy of Minnesota University Archives.

Holmes did argue however, that 'to use the isotopic composition of lead from iron meteorites as part of the basic data for calculating the age of the earth or its crust, is unsound in principle . . . the correct procedure is to use terrestrial materials' (Holmes 1956*b*, p. 323). Accordingly, and now with far more terrestrial data available than he had had access to ten years earlier, he made his own calculations and wrote his closing words on the subject: 'My own attempt to solve the problem from terrestrial evidence alone leads to essentially the same result, which may be expressed as 4,500 ± 100 million years'. His error encompassed Patterson's results of 4550 ± 70 Ma.

In the decade following publication of Holmes's seminal paper on convection currents in the mantle, little progress was made on the theory of continental drift. Nevertheless, in June 1942, during World War II, Holmes started to write his book on the *Principles of Physical Geology* (Holmes 1944) and had to decide whether to include a chapter on continental drift. He wrote to his old friend Reginald Daly, one of the few Americans at that time who believed in the theory, to express a moment of doubt and ask his advice:

> I have been hesitating whether or not to put continental drift into the book. On the whole, I think not. I am still doubtful about it on the Wegener and du Toit scale. We really need a first class palaeontologist to assess the biological evidence at its true value. If you could stimulate someone to do that and then take up the problem in the sane light of your long experience we might expect some real progress. However, while it may be complimentary, it is hardly kind to thrust upon you all the headaches and nightmares that such an enterprise would entail![25]

Sadly we do not have Daly's response, but by now Holmes had been teaching his students about continental drift for nearly 20 years; in fact one recalled that when he went to conferences at other universities he was amazed to discover that what he understood to be the accepted doctrine of continental drift, turned out to be considered heretical and revolutionary, and the old isthmus and land bridge ideas were still advocated.[26]

Eventually Holmes decided he could hardly ignore what he had been preaching for so long and in the last chapter of the book he did indeed include his latest ideas, although couched in 'purely hypothetical' terms. He first reviewed the history of continental drift and pointed out that the idea was far from new, going back much further than Taylor or Wegener, to at least 1858 when Antonio Snider published a map which closely resembled Wegener's 1915 interpretation (Holmes 1944, p. 490). He then put forward all the arguments in favour of continental drift whilst recognizing that the theory could have no scientific value until all the elements

> acquire support from independent evidence. . . . Meanwhile it would be futile to indulge in the early expectation of an all-embracing theory which would satisfactorily correlate all the varied phenomena for which the earth's internal behaviour is responsible (Holmes 1944, pp. 508–509).

His patience was commendable. It was already almost 20 years since he had first proposed his 'all-embracing' theory, but as Daly predicted in his review of Holmes's book, regarding inclusion of a chapter on continental drift, 'for this boldness he will doubtless be chastised' (Reynolds 1968).

Unification

Holmes lived another 21 years after publication of his book containing the chapter on continental drift, just long enough to see the theory of sea-floor spreading proposed, although by that time his contribution to it was almost forgotten. He did, however, allow himself a modest pat on the back in his revised edition of *Principles of Physical Geology*, when he said that 'mantle currents are no longer regarded as inadmissible' (Holmes 1965, p. 1247). In this second edition, fully rewritten in his retirement and published only months before he died, he again reviewed the status of continental drift in the final chapter. Although conceptually not very different from the version he wrote 20 years earlier, the change in tone is very distinct and this time he states with confidence that 'Continental drift, now embracing crustal separation and ocean floor dispersal and renewal, is known to be in operation at the present time' (Homes 1965, p. 1203).

It must have been gratifying for him to realize that it was his pioneering work on radiometric dating that ultimately led to a radiometric timescale for magnetic reversals, which was the key that finally unlocked understanding about sea-floor spreading and all that followed from

[25] Holmes to Daly 23 March 1943. Courtesy of Harvard University Archives.
[26] Pers. comm., Charles Waterston, 8 July 1998

Fig. 2. Arthur Holmes *c.* 1910, 1930 and 1960.

that. Sadly Holmes was not alive when the barcode of magnetic stripes on the sea floor was fully interpreted by applying that timescale. Nevertheless, he clearly appreciated the huge contribution to the understanding of geology that had been achieved by his work and realized that even more problems would be resolved 'as it becomes more generally possible to combine radiometric dating with magnetic measurements' (Holmes 1965, p. 1207).

Today radioactivity is used in an almost limitless variety of ways to tell us about the timing and rates of geological processes and thus it still contributes to our understanding of the evolution of the Earth. It is a crucial tool for geology, and more than any other single individual, Arthur Holmes contributed most to our understanding of that tool. In particular, he appreciated the endless supply of heat generated by radioactivity and the enormous timescales involved. Holmes was a deep thinker and philosopher about the really big geological problems for which he was awarded the Vetlesen Prize in 1964, presented by Columbia University for his 'uniquely distinguished achievement in the sciences resulting in a clearer understanding of the Earth, its history, or its relation to the universe'.[27] The medal is given every two years and it was the founders hope that 'in time this prize will rank in dignity and significance with the Nobel Prizes'.[28]

On the cover of Holmes's second edition of *Principles of Physical Geology* he is described as one of the world's greatest geologists (Fig. 2). This recognition came at the end of a lifetime's contribution to geology, but I suspect he would have been more flattered by the letter he received from Claire Patterson a couple of years earlier, who acknowledged that contribution personally:

> I wish to reiterate my personal indebtedness to your pioneering work in this field. It was outstandingly inspiring and ingenious.[29]

I gratefully acknowledge Research Grant No. 20,235 from the Royal Society that assisted with travel to visit Holmes's family in Vienna, and to research archives in the United States. I should also like to thank H. Torrens, I. Fairchild and other members of the Geology Department at Keele University, for their sustained support of this work.

[27] Grayson Kirk, President Columbia University, to Holmes, 15 January, 1964. Courtesy of the Royal Holloway University of London archives.

[28] Extract is from Kirk's letter above. Since its inception in 1960 there have only been four British recipients of the Vetlesen Prize, and none since 1970. What does this say about the state of British science?

[29] Patterson to Holmes, 12 July 1963. Courtesy of Mrs Lorna Patterson and the California Institute of Technology archives.

References

ANON. 1922*a*. The Hull meeting of the British Association. *Nature*, **110**, 263.

ANON. 1922*b*. Wegener's displacement theory. *Nature*, **109**, 202–203.

BOLTWOOD, B. B. 1907. On the ultimate disintegration products of the radio-active elements. Part II. *The Disintegration Products of Uranium. American Journal of Science*, **23**, 77–88.

BULL, A. J. 1921. A hypothesis of mountain building. *Geological Magazine*, **58**, 364.

BULL, A. J. 1927. Some aspects of the mountain building problem. *Proceedings of the Geologists' Association*, **38**, 145–156.

CHAMBERLIN, T. C., CLARKE, J. M., BROWN, E. W. & DUANE, W. 1922. From the geological view-point; from the palaeontological view-point; from the point of view of astronomy; the radioactive point of view. *Proceedings of the American Philosophical Society*, **61**, 247–288.

DUNHAM, K. C. 1966. Arthur Holmes. *Biographical Memoirs of Fellows of the Royal Society*, **12**, 291–310.

EVANS, J. W., TURNER, H. H., & WRIGHT, W. B. 1923. Discussion on Wegener's hypothesis of continental drift. *Report of the British Association for the Advancement of Science for 1922*, 364–365.

GERLING, E. K. 1942. Age of the Earth according to radioactivity data. *In*: HARPER C. T. (ed.) *Benchmark Papers in Geology, Geochronology: Radiometric Dating of Rocks and Minerals*. Dowden, Hutchinson & Ross, Inc., 121–123 (1973).

HOLMES, A. 1911. The association of lead with uranium in rock-minerals, and its application to the measurement of geological time. *Proceedings of the Royal Society of London*, Series A, **85**, 248–256.

HOLMES, A. 1913. *The Age of the Earth.* Harper, London and New York.

HOLMES, A. 1915*a*. Radioactivity and the Earth's thermal history. Part I: The concentration of the radio-active elements in the Earth's crust. *Geological Magazine*, **6**, 60–71.

HOLMES, A. 1915*b*. Radioactivity and the Earth's thermal history. Part II: Radioactivity and the Earth as a cooling body. *Geological Magazine*, **6**, 102–112.

HOLMES, A. 1915*c*. Radioactivity and the measurement of geological time. *Proceedings of the Geologists' Association*, **26**, 289–309.

HOLMES, A. 1916. Radioactivity and the Earth's thermal history. Part III: Radio-activity and isostasy. *Geological Magazine*, **6**, 265–274.

HOLMES, A. 1917. Comment on 'The stability of lead isotopes from thorium', by F. Soddy. *Nature*, **99**, 244.

HOLMES, A. 1920*a*. *The Nomenclature of Petrology*. Thomas Murby, London.

HOLMES, A. 1920*b*. The measurement of geological time. *Discovery*, **1**, 108–114.

HOLMES, A. 1921. *Petrographic Methods and Calculations*. Thomas Murby, London.

HOLMES, A. 1923. The age of the Earth. *Nature*, **112**, 302–303.

HOLMES, A. 1924. Geology of the United States. *Nature*, **114**, 376–377.

HOLMES, A. 1925*a*. Radioactivity and the Earth's thermal history. Part IV: A criticism of Parts I, II and III. *Geological Magazine*, **62**, 504–515.

HOLMES, A. 1925*b*. Radioactivity and the Earth's thermal history. Part V: The control of geological history by radioactivity. *Geological Magazine*, **62**, 529–544.

HOLMES, A. 1926*a*. Radium uncovers new clues to Earth's age. *In*: SULLIVAN, W. (ed.) *Science in the Twentieth Century*. The New York Times, New York / Arno Press, New York, 175–177 (1976).

HOLMES, A. 1926*b*. Contributions to the theory of magmatic cycles. *Geological Magazine*, **63**, 306–329.

HOLMES, A. 1926*c*. Rock-lead, ore-lead, and the age of the Earth. *Nature*, **117**, 482.

HOLMES, A. 1927. *The Age of the Earth*. (2nd edn.) Harper, London.

HOLMES, A. 1928*a*. Radioactivity and continental drift. *Geological Magazine*, **65**, 236–238.

HOLMES, A. 1928*b*. Continental drift. *Nature*, **122**, 431–433.

HOLMES, A. 1929. A review of the continental drift hypothesis. *The Mining Magazine*, **40**, 205–209, 286–288, 340–347.

HOLMES, A. 1930*a*. Radioactivity and geology. *Transactions of the Edinburgh Geological Society*, **12**, 281–283.

HOLMES, A. 1930*b*. The period of 'actino-uranium' and its bearing on the ages of radioactive minerals. *Nature*, **126**, 348–349.

HOLMES, A. 1931. Radioactivity and Earth movements. *Transactions of the Geological Society of Glasgow for 1928–29*, **18**, 559–606.

HOLMES, A. 1939. The geologist's clock. *In*: COX, I. (ed.) *The Wild Life Around Us, and the Story of the Rocks*. Allen & Unwin, London, 182–190.

HOLMES, A. 1944. *Principles of Physical Geology*. Thomas Nelson, London.

HOLMES, A. 1946. An estimate of the age of the Earth. *Nature*, **157**, 680–684.

HOLMES, A. 1947. A revised estimate of the age of the Earth. *Nature*, **159**, 127–128.

HOLMES, A. 1956. How old is the Earth? *Transactions of the Edinburgh Geological Society*, **16**, 313–333.

HOLMES, A. 1965. *Principles of Physical Geology* (2nd edn). Thomas Nelson, London.

HOLMES, A. & LAWSON, R. W. 1914. Lead and the end product of thorium. Part I. *Philosophical Magazine*, **28**, 823–840.

HOLMES, A. & LAWSON, R. W. 1915. Lead and the end product of thorium. Part II. *Philosophical Magazine*, **29**, 673–688.

HOLMES, A. & LAWSON, R. W. 1926. Potassium and the heat of the Earth. *Nature*, **117**, 620–621.

HOLMES, A., EVANS, J. W., YOUNG, A. P. & SHELTON, H. S. 1915. Discussion on: The radioactive methods of determining geological time. *Abstracts of the Proceedings of the Geological Society of London. Session 1914–1915*, **971**, 63–66.

HOUTERMANS, F. G. 1946. The isotopic abundances in natural lead and the age of uranium, *Naturwissenschaften*, **33**, 185–186, 219.

JOLY, J. 1923. The Movements of the Earth's Surface Crust, *London, Edinburgh and Dublin Philosophical Magazine and Journal of Science*, **45**, Series 6, 1167–1188.

KING, C. 1893. The age of the Earth. *American Journal of Science*, **45**, 1–20.

KNOPF, A., SCHUCHERT, C., KOVARIK, A. F., HOLMES, A. & BROWN, E. W. 1931. Physics of the Earth – IV: The age of the Earth. *Bulletin of the National Research Council of the National Academy of Sciences, Washington, D. C.*, **80**, 440–441.

LEWIS, C. L. E. 2000. *The Dating Game*. Cambridge University Press, Cambridge.

LEWIS, C. L. E. 2001. Arthur Holmes' vision of a geological timescale. *In*: LEWIS, C. L. E. & KNELL, S. (eds) *The Age of the Earth*. The Geological Society, London, Special Publications, **190**, 121–138.

MACKINDER, H., HINKS, A., SIMPSON, G. C., POOLE, J. H. J., GREGORY, J. W., HOLMES, A. & JEFFREYS, H. 1931. Problems of the Earth's crust. *Geographical Journal*, **78**, 433–455.

MARVIN, U. B. 1985. The British reception of Alfred Wegener's continental drift hypothesis. *Earth Sciences History*, **4**, 138–159.

MOLENGRAAF, G. A. F. 1921. Modern deep-sea research in the east Indian archipelago, and discussion. *Geographical Journal*, **57**, 95–120.

NIER, A. O. 1938*a*. Variations in the relative abundances of the isotopes of common lead from various sources. *Journal of the American Chemical Society*, **60**, 1571–1576.

NIER, A. O. 1938*b*. Variations in the relative abundances of the lead isotopes. *Physical Review*, **53**, 680.

NIER, A. O. 1939. The isotopic constitution of radiogenic lead and the measurement of geological time. II. *Physical Review*, **55**, 153–163.

NIER, A. O. 1982. Some reminiscences of isotopes, geochronology and mass spectrometry. *Annual Review of Earth and Planetary Science*, **9**, 1–17.

ORESKES, N. 1999. *The Rejection of Continental Drift*. Oxford University Press, New York.

RAYLEIGH, R. J., SOLLAS, W. J., GREGORY, J. W. & EDDINGTON, A. S. 1921*a*. Joint discussion on the age of the Earth. *Report of the Eighty-Ninth Meeting of the British Association for the Advancement of Science Edinburgh – 1921*. John Murray, London, 413–415.

RAYLEIGH, R. J., SOLLAS, W. J., GREGORY, J. W. & JEFFREYS, H. 1921*b*. The age of the Earth. *Nature*, **108**, 279–284.

REYNOLDS, D. L. 1968. Memorial of Arthur Holmes, January 14, 1890–September 20, 1965. *American Mineralogist*, **53**, 560–566.

RUTHERFORD, E. 1905*a*. Present problems in radioactivity, *Popular Science Monthly*, **67**, 5–34,

RUTHERFORD, E. 1905*b*. *Radioactivity* (2nd edn). Cambridge University Press, Cambridge.

SHIPLEY, B. C. 2001. 'Had Lord Kelvin a right?': John Perry, natural selection, and the age of the Earth, 1894–5. *In*: LEWIS, C. L. E. & KNELL, S. (eds) *The Age of the Earth*. The Geological Society, London, Special Publications, **190**, 91–106.

SNIDER, A. 1858. *La création et ses mystères dévoilés* . . . A. Franck, Paris.

SODDY, F. 1917. The stability of lead isotopes from thorium. *Nature*, **99**, 245.

THOMSON, W. (LORD KELVIN) 1862. On the secular cooling of the Earth. *Transactions of the Royal Society of Edinburgh*, **23**, 157–169.

THOMSON, W. (LORD KELVIN). 1895*a*. Copy of a letter from Lord Kelvin, *Nature*, **51**, 227.

THOMSON, W. (LORD KELVIN). 1895*b*. The age of the Earth, *Nature*, **51**, 438–440.

VAN WATERSCHOOT VAN DER GRACHT, W. A. J. M. (ed.) 1928. *Theory of Continental Drift: A Symposium on the Origin and Movement of Land Masses, both Inter-continental and Intra-continental as Proposed by Alfred Wegener*. American Association of Petroleum Geologists, Tulsa, and Thomas Murby, London.

WALCOTT, C. D. 1893. Geological time as indicated by the sedimentary rocks of North America. *Journal of Geology*, **1**, 639–676.

WEGENER, A. 1912. Die Entstehung der Kontinente. *Petermanns Mitteilungen*, **1912**, 185–195, 253–256, 305–309.

WEGENER, A. 1920. *Die Entstehung der Kontinente und Ozeane*. Friedrich Vieweg, Braunschweig.

WEGENER, A. 1924. *The Origin of Continents and Oceans* (English translation of 3rd German edn). Methuen, London.

WEGENER, A. 1967. *The Origin of Continents and Oceans* (English translation of 4th German edn). Methuen, London.

WRIGHT, W. B. 1923. The Wegener hypothesis. *Nature*. **111**, 30–31.

WYSE JACKSON, P. 2001. John Joly (1857–1933) and his determination of the age of the Earth. *In*: LEWIS, C. L. E. & KNELL, S. (eds) *The Age of the Earth*. The Geological Society, London, Special Publications, **190**, 107–120.

YOCHELSON, E. L. & LEWIS, C. L. E. 2001. The age of the Earth in the United States: 1892–1931. *In*: LEWIS, C. L. E. & KNELL, S. (eds) *The Age of the Earth*. The Geological Society, London, Special Publications, **190**, 139–158.

Russian geology and the plate tectonics revolution

VICTOR E. KHAIN & ANATOLY G. RYABUKHIN

M. Lomonosov Moscow State University, Vorobiovy Gory, Moscow, Russia

Abstract: The suggestion of the concept of 'scientific revolution' by Thomas Kuhn in 1962 was, in itself, a significant event in the history of science, and 'crucial' episodes or 'paradigm shifts' have come to be of special interest in the history of geology (as in other sciences). The appearance of a new paradigm is commonly associated with attempts by the most talented and well-established practitioners to consolidate or sustain the position of the previously prevailing paradigm. For almost 40 years, global theories in geology have been developing under the influence of mobilist ideas. It is no secret that in Russia the mobilist school initially met with serious opposition, and that even up to the present it has had numerous opponents. However, Western, and especially popular, scientific literature usually exaggerates the intensity of the situation and underestimates the contribution of Russian geologists and geophysicists to the development of mobilism and plate tectonics. The present paper describes some of the debates in Russia concerning mobilist doctrines, the work done in that country in the last three decades of the twentieth century from a mobilist perspective, and various theories that had currency in Russia at the end of that century.

In Russia, discussion of the principal factors of tectogenesis has had many vicissitudes in the twentieth century. During the first 70 years of the century, the dominance of vertical, as opposed to horizontal, motion of the Earth's crust was considered self-evident, and the contrary view was regarded as merely the next step in the progress of science. Nevertheless, at present, plate tectonics occupies a defining position in Russian models of tectogenesis – though there are also alternative mobilist concepts that attract support in that country. The aim of this paper is to show the true state of affairs in this field in a retrospective sense, and the conceptual design and principal directions of the ideas that have been developed in Russia in the second half of the twentieth century, and which have adherents there at the end of the century.

The beginnings

The idea of continental drift formulated by Alfred Wegener reached Russia only after World War I, when the Russian version of his famous book *Entstehung der Kontinente und Ozeane* was published first in 1922 in Berlin, then in 1925 in Moscow, and more recently in 1984 in Leningrad. The forewords and commentaries to the second and third editions were written by famous Russian geologists (Professors Georgy Mirchink, Peter Kropotkin and Pavel Voronov). Wegener's publication was received with interest and even sympathy by several eminent Russian Earth scientists including the geologist Aleksey Pavlov, the palaeontologist and stratigrapher Aleksey Borissyak, the leading palaeobotanist African Krishtofovich, and several others. In 1931, Boris Lichkov from Leningrad University even published the title, *Movements of Continents and Climates of the Earth's Past*, based on the notion of continental drift.

Borissyak considered that revising of an actual material within the framework of the hypothesis of continental drift on fold belts and especially the circum-Pacific one represented weighty argument in favour of Wegener's theory. He wrote that: 'it is necessary to recognize, that the little done in this line has already given brilliant results, and that this theory is born powerfully armed' (cited after Borissyak 1922, p. 102). Mobilist reconstructions were used in the lectures on palaeobotany by Krishtofovich to account for plant distributions and the migrations of flora.

Meanwhile, prior to the mid-1930s, a number of the fold-belts in Russia (then USSR) were explored, and the existence of nappe structures was established in the Northern and Central Urals, in the Greater Caucasus and in Transbaikalia. Mobilist works, such as those of Émile Argand and Rudolf Staub, were translated and published in Russia.

But this trend was reversed at the end of the 1930s, mainly under the influence of Michael Tetyayev, an influential and eloquent professor at the Leningrad Mining Institute. He strongly criticized not only continental drift, but also the Suessian contraction hypothesis, and in general the assumption of any major role for horizontal

From: OLDROYD, D. R. (ed.) 2002. *The Earth Inside and Out: Some Major Contributions to Geology in the Twentieth Century*. Geological Society, London, Special Publications, **192**, 185–198. 0305-8719/02/$15.00

movements in the history of the Earth's crust. He considered vertical, oscillatory movements to be the principal type of tectonic movements and horizontal ones as merely subsidiary to, and derivative from, vertical movements. He quickly found a powerful supporter in his disciple Vladimir Beloussov.

But it was not only Tetyayev and Beloussov who criticized the mobilistic theories at that time. The leader of the Moscow school of tectonicians, Nikolay Shatsky, presented a paper to the Geological and Geophysical Sections of the Academy of Sciences of the USSR in 1946 which argued strongly against Wegener's hypothesis. Shatsky's main arguments had to do with what he took to be the contradictions between Wegener's theory and the concept of geosynclines and platforms. He pointed to the existence of deep faults, apparently crossing both the crust and upper mantle and acting over several geological periods, with the consequent inheritance of older structures by younger ones. Specifically, he was concerned that if Wegener's theory were correct the suture between the Andean geosyncline and South American platform would now be in the area of the Atlantic Ocean, so that deep earthquakes would be expected to occur there, contrary to what is known to be the case. The head of the third, Siberian, school of Russian tectonicians, Michael Usov, was also amongst Wegener's opponents. After the basic work by Alexander Peive was published in 1945, the idea of deep-seated faults became popular in Russia (USSR). This concept considered such faults as passing from the crust directly into the mantle, which suggested a close and fixed connection between these two layers such as to exclude any possible 'slippage' or lateral movement of the crust with respect to the mantle. As mentioned, Shatsky's arguments depended on the idea of faults extending from the crust into the mantle.

In consequence, at the beginning of the 1950s practically all the leading geologists and geophysicists in Russia were opposed to continental drift. This position was expressed in a document published in 1951 on behalf of a group of eminent Moscow Earth scientists, which, after discussion, came to the conclusion that the 'fundamental and most universal tectonic movements of the Earth's crust are vertical (oscillatory) movements' and the 'large horizontal displacements of continents suggested in the light of Wegener's ideas definitely have not occurred' (cited after Yury Kossygin 1983, p. 9). It is rather curious that among the proponents of this document were Kropotkin and Peive, who not long after became supporters of mobilism.

In this period of 'fixist reaction', as it has been called by Rudolf Trümpy (1988) who noticed its manifestation also in Western countries, there was a definite tendency to denigrate or deny the existence of nappe structures, previously identified in some of the fold belts of the USSR. Moreover, when Soviet geologists began the exploration of the Ukrainian Carpathians, which became part of the Soviet territory, they reached the conclusion that the nappes suggested earlier by their predecessors from Poland and Czechoslovakia did not exist. Only Professor Oleg Vyalov from Lvov opposed this view. But Beloussov, who obtained permission to visit the Austrian Alps during the Soviet occupation, co-authored a paper with his disciples Michael Gzovsky and Arcady Goriachev in which he rejected the 'nappist' interpretations of the structure of the Alps, declaring that the exposure of rocks in this region was insufficient to allow identification of such complicated structures, owing to the extensive glacial deposits. It was only many years later during an excursion in the Swiss Alps under the leadership of Trümpy – in which Victor Khain (one of the authors of the present paper) participated – that Beloussov accepted the nappe interpretation of the Alpine structures. The tectonists Alexei Bogdanov and Mikhail Muratov, when visiting the Western Carpathians in 1956, arrived at the same conclusion concerning the Carpathian fold system after having previously denied it when working in the Ukrainian Carpathians.

First steps

That was how matters stood by the end of the 1950s. But then the trend of thought changed again, though at first only for a minority of geologists. Russian geology displayed a tendency towards a closer and more accurate observation of phenomena that implied horizontal displacements in the Earth's crust, such as overthrusts (nappes) and large transcurrent faults. The important role of strike-slip faults and overthrusts was stressed by Peive (1960) in his report to the 21st International Geological Congress in Copenhagen. These observations resulted in the publication of a volume entitled *Faults and Horizontal Movements of the Earth's Crust*, edited by Peive (1963), as well as a book by Kropotkin and Kseniya Shahvarostova (1965) entitled *Geological Structure of the Pacific Mobile Belt.*

Still earlier, in 1958, Kropotkin had published a paper with a reviewing palaeomagnetic investigations, noting their importance in evaluating horizontal displacements of the continents. Thus Kropotkin (1958, 1969) was the first Russian (Soviet) scientist to employ palaeomagnetic data as an indication of continental drift and he pointed to their correlation with palaeoclimatic data. Then followed the works by the first Russian (Soviet) explorers of Antarctica (Pavel Voronov 1967, 1968; Sergey Ushakov & Khain 1965), who revived the concept of Gondwana in its mobilistic version.

After visiting the Balkan countries and impressed by the role of ophiolites in their structure, Peive published in 1969 a famous article entitled ‘Oceanic crust of the geological past’. This proved to be a turning point in the study of the structure and evolution of the fold systems of the USSR. Recognition of ophiolites, large overthrusts and nappes followed one after another in the various fold edifices of the vast country, from the Carpathians to Kamchatka and Sakhalin. The best examples of ophiolites were found and described by Andrey Knipper in the Lesser Caucasus (Knipper 1983) and by a group of researchers in the Urals (Savelieva & Saveliev 1977).

In 1967–1968 the neo-mobilistic concept of plate tectonics was definitively formulated in the famous set of papers in the *Journal of Geophysical Research* (translated and published in Russia in 1974) and the no less famous paper on the revolution in Earth sciences by J. Tuzo Wilson (1968), But Beloussov (1970) promptly replied to this paper, strongly opposing the new ideas.

This polemic was discussed by Khain (1970) in the Soviet magazine *Priroda* (*Nature*). Though he had some reservations, Khain shared Wilson' s perspective and in the same year he published the basic postulates of plate tectonics models for the first time in the Soviet literature (Khain 1970).

Meanwhile, two geophysicists, Sergey Ushakov and Oleg Sorokhtin, became the first adherents of the new concept among Russian specialists in this field of research (their activity successfully continues at a very high level even today, see below). Sorokhtin's PhD thesis on the global evolution of the Earth in 1972 was the first of its kind and was published in 1974 (Sorokhtin 1974). The same year saw the publication of Ushakov's first monograph: *Structure and Evolution of the Earth* (Ushakov 1974).

These were the first important works in the Russian literature in which plate tectonics ideas were further developed and connected to those of the global evolution of the Earth. Sorokhtin argued that the tectonic evolution of the Earth, manifested in the lithosphere by plate tectonics, is based on differentiation of the material at the mantle/core boundary, with iron oxide flowing down into the core and silicate melt ascending into the asthenosphere. The layering of the Earth within the mantle and the core was further analysed mathematically by Vladimir Keondjian and Andrey Monin (1976). Sorokhtin also attempted to estimate the duration of a complete convection cycle in the mantle and he identified this cycle with tectonic cycles. This convection was considered as not purely a thermal process but included a chemical-density component. Sorokhtin was also the first to put forward the idea of two types of mantle convection – one-cellular and two-cellular phases – regularly alternating in the course of the Earth's history. The first type of phase was thought to be associated with the formation of the Pangaea super-continent. Subsequently, this idea became widely accepted, both in Russia and in the Western literature (see, for instance, Nance *et al.* 1988).

Among Russian geologists Lev Zonenshain, who was already well known for his work on the tectonics of Siberia and Mongolia, became one of the first and most active proponents of plate tectonics. In the years after he joined the Institute of Oceanology of the Academy of Sciences he assumed a real leadership in this field. In 1976, together with Mikhail Kuzmin and Valery Moralev, he published *Global Tectonics, Magmatism and Metallogeny*; and in 1979 with Leonid Savostin *Geodynamics: An Introduction*, the first detailed exposition of plate tectonic principles in the Russian literature.

Two research groups at the geological faculty of the M. Lomonosov Moscow University were particularly concerned with developing and applying plate tectonics theory. One was organized in the department of geophysics under Vsevolod Fedynsky, and the other in the museum of Earth sciences under the leadership of Sergey Ushakov. The first group concentrated its efforts on developing physical models of the internal development of the Earth, defining the mechanism of motion of lithospheric plates (Fedynsky, Sergey Ushakov, Yury Galushkin, Evgeny Dubinin, Alexandr Shemenda); on global palaeoclimatic reconstructions in the context of plate tectonics, but with special reference to the USSR (Nicolay Yasamanov, Ushakov); and on the development of geodynamic models to account for the distribution of mineral deposits (Alexandr Kovalev, Ushakov, Galushkin).

The second group was organized in the department of dynamic geology under the leadership of Khain. The members of this group chiefly gave their attention to the role and value of plate tectonics in the formulation of a general theory of tectogenesis (Khain, Mikhail Lomize,

Mikhail Volobuev, Nicolay Bozhko), studying the evolution of the main structural elements of the Earth's crust and the regional application of plate tectonics theory (Khain, Lomize, Volobuev, Bozhko, Nicolay Koronovsky, Anatoly Ryabukhin), and also in applying this concept to petroleum geology (Khain, Boris Sokolov).

Resistance to plate tectonics and its reasons

But the expansion of new mobilist ideas in geology met strong opposition in Russia (USSR), mainly from the influential scientists of the older generation – academicians, professors and heads of geological surveys. There were different reasons for such opposition, both objective and subjective. One of them was the popularity of the fixist concept of the evolution of the Earth's crust, elaborated by Vladimir Beloussov, who continued to defend it resolutely and ingeniously until his last days. It is necessary to remark that Beloussov's scientific authority and influence were great not only in Russia. In memoirs about Beloussov, Tuzo Wilson has described him as an inspirational figure: the man 'who at one time headed the Russian scientific collective, who proposed the Upper Mantle Project, who presided at the World Geophysics Congress in 1963 in California, and who ... became one of the most imaginative members of the international community of scientists' (Wilson 1999, p. 192).

Another reason for the success of fixist ideas in Russia was that they could be applied rather successfully to the vast platform regions of that country, where the role of vertical movements was much more evident than that of horizontal movements. Third, the fact that the plate tectonic theory was born in the West and not in the USSR caused some Soviet geologists to be prejudiced against it, since they had been brought up in the conviction that every progressive step in science had first been accomplished in their own country. But the Western origin of plate tectonics was quite natural, for Western scientists were the first to obtain access to new data concerning oceans, whereas Soviet science developed in relative isolation for quite a long period of time. And fourth, the majority of the old generation of the leading Soviet scientists, with their steady fixist mentality, not only never sought to stimulate interest in the new ideas, but actively opposed them.

Even so, vigorous discussions broke out between defenders and opponents of plate tectonics. The first such discussion was organized in 1972 by the department of geology, geophysics and geochemistry of the USSR Academy of Sciences. Kropotkin and Khain spoke in favour of plate tectonics, and Beloussov against it. Other meetings and discussions followed. The number of people adopting plate tectonics steadily grew, but at each annual session of the National Tectonic Committee, plate tectonics was vigorously attacked. Zonenshain organized special conferences, but they only attracted those who were already believers in plate tectonics. The first conference took place in 1987 and five others followed within a two-year interval. In fact, these conferences were quite successful. The number of participants reached 300 and the second and following meetings were attended by several leading figures from the international community.

Yet while the world community of geologists celebrated the 'silver anniversary' of plate tectonics in 1988, a number of papers appeared in our literature which not only posed doubt on the philosophy, but denied the very idea of large horizontal motion of the Earth's crust. The disputes went on at the 'All-Union' tectonic conferences, and at meetings at M. Lomonosov Moscow State University. Within the framework of conferences on the 'Main problems of geology' held at the geological faculty of the M. Lomonosov Moscow University there were lectures by the proponents and opponents of plate tectonics, and theoretical discussions that attracted a large audience from amongst the students. The main theoretical discussion became heated: between Beloussov and his followers, advocates of the orthodox fixist idea, and Khain and his supporters, developing mobilist model of evolution of lithosphere. The debates attracted considerable interest and attention and were not confined to within the walls of the university, being reflected in numerous publications (e.g. Vladimir Smirnov 1989; Evgeny Milanovsky 1984). Vladimir Legler (1989) has made an interesting analysis of the publications in two popular Russian geological journals, *Geotectonics* and the *Bulletin of Moscow Society of Naturalists, Geological Section* for the years 1970–1979. During this period, 443 articles were published about theoretical problems of geotectonics and historical geology in *Geotectonics*, of which 400 (90%) were anti-plate tectonics; while of 154 articles in the *Bulletin*, 148 (97%) were opposed to the theory.

The new 'splash' of discussion was expressed in the publication of a number of critical articles by the professors of leading Russian geological *Hochschulen*. Several professors from the

Moscow Geo-exploration Institute and the M. Lomonosov Moscow University, pointed to difficulties and inconsistencies that were found in the detailed application of the plate tectonics model, casting doubt on the theoretical validity of the concept and the possibility of its application (Vladimir Karaulov 1988; Oleg Mazarovich *et al.* 1988–1989).

Koronovsky (1989) and Khain (1990) from M. Lomonosov Moscow University responded, acknowledging that there were difficulties in the implementation of the model in the investigation of complicated tectonic structures, but pointed to the inconsistencies in the methodical and methodological approaches of their opponents in the solution of the main theoretical problems of geology. The principal value of this discussion, in our view, was that the participants were educating not just one generation of geologists, but were influencing the outlook of the new generation of geologists, which in turn should determine the future progress of geology in Russia.

Plate tectonic reconstructions, global and regional

Despite these not very favourable conditions, mobilism in general, and the plate tectonics concept in particular, kept attracting more and more workers. As soon as Zonenshain joined the Institute of Oceanology, he and his team started working on global and regional palinspastic reconstructions. Global reconstructions for the whole of the Phanerozoic and for the Late Precambrian were published (Zonenshain & Gorodnitsky 1977). A series of reconstructions for the USSR territory was completed and partly published. Zonenshain initiated the work on the Geodynamic Map of the USSR, on the scale of 1: 2 500 000, one of the first of its kind in the world. It was presented at the 28th International Geological Congress in Washington DC in 1989. It was also Zonenshain who published a scheme of the modern plate tectonics of the USSR and adjacent regions, in which a series of small plates and microplates was featured, south and east of the Eurasian plate. A similar pattern is shown in the map of the recent tectonics of China, published by Ma Xingyuan (1988).

The propagation of mobilist views on the structure and evolution of fold belts of the USSR and Eurasia was promoted by a group of tectonicians of the Geological Institute of the USSR Academy of Sciences (Peive, Knipper, Yuri Pushcharovsky, Alexander Mossakovsky, Sergey Samygin, Andrey Perfiliev, Sergey Ruzhentsev, Sergey Sokolov, and others). The same group published the Tectonic Map of Northern Eurasia on a scale of 1: 5 000 000, and the Tectonic Map of the Urals and Central Kazakhstan on a larger scale. At the present time, nappes have been recognized in all fold-belts of Russia (USSR), and even in the platform basement.

Among works worth mentioning there are also regional plate tectonic reconstructions on the Caucasus (Khain, Shota Adamiya, Irakly Gamkrelidze, Manana Lordkipanidze, Lomize, and others), on the Urals (Svyatoslav Ivanov, Victor Puchkov, Zonenshain, and others), and on the NE USSR (Nikita Bogdanov, Solomon Tilman, Leonid Parfenov, and others).

Later Zonenshain, together with Victor Koroteev, organized a collective study of the history of the Urals. It was the world's first palaeo-oceanological expedition on a continent. The results were summarized in Zonenshain *et al.* (1984). An even larger project was realized by a group of Russian and Georgian geologists together with a French team, having as its aim a compilation of a series of palinspastic maps of the Tethys. Leaders of this project were Xavier Le Pichon from the French side, and Zonenshain and Vladimir Kazmin from Russia; the map atlas and the explanatory text were published simultaneously in both countries, and in the international journal *Tectonophysics* (Aubouin *et al.* 1986).

At the same time and subsequently, plate tectonic models were elaborated for other fold systems of the USSR – Tian Shan (Vitaly Burtman *et al.*), Verkhoyansk Chukchi (Leonid Parfenov), Koryak Upland (Sergey Ruzhentsev, Sergey Sokolov), Transbaikalia and Mongolia (Ivan Gordienko), and for the Arctic region as a whole (Zonenshain and Lev Natapov). All these regional works were summarized in a monograph on the plate tectonic synthesis of the territories of the USSR, published simultaneously in our country and in the USA by Zonenshain, Kuzmin, and Natapov (Zonenshain *et al.* 1990).

Alexander Karasik (1980) deciphered the linear magnetic anomalies of the Eurasian Basin of the Arctic Ocean. Palinspastic reconstructions were largely favoured by palaeomagnetic studies made by Alexandr Kravchinsky (1977), Alexey Khramov (1982) and Diamar Pechersky. Using palaeomagnetic data, Mikhail Bazhenov & Burtman (1982; Burtman 1984) demonstrated the secondary nature of the Carpathian and Pamir arcs. Khain (1985) provided evidence to show that the opening of Meso-Cenozoic oceans proceeded not gradually but stepwise, segment by segment, these segments being separated from each other by large transform faults which he called 'magistral'.

Important conclusions were drawn concerning the connection between magmatism and metamorphism and plate tectonics. Contributions include the works by Nikolay Dobretsov (1980), Oleg Bogatikov *et al.* (1987) and Koronovsky & Diomina (1999). Alexander Lisitsin (1988) established general regularities of the sedimentation in oceans, connected with plate tectonic activities, including the avalanche sedimentation of turbidites on continental margins.

Development of the plate tectonic concept

In the 1980s, Russian mobilists started concentrating their efforts on as yet unsolved problems of plate tectonics. One of these was the question of 'plate tectonics manifestations' in the Precambrian, especially in the Early Precambrian. As is well known, opinions on this issue are still divided. While some scientists suggest that plate tectonics phenomena were active already in the Early Precambrian and even in the Archaean, others maintain that its manifestations began only with the Late Precambrian. In the Soviet literature, the first point of view found such advocates as Chermen Borukayev (in his monograph *Precambrian Structures and Plate Tectonics*, 1985) and Andrey Monin (*The Early Geological History of the Earth*, 1987). A somewhat different interpretation is presented in the book by Khain & Bozhko (*Historical Geotectonics: The Precambrian*, 1988). The authors of this latter book point to the evolution of plate tectonics itself during the Precambrian period: from the embryonic stage in the Archaean through a phase of small-plate tectonics in the Early Proterozoic to full-scale plate tectonics in the Late Proterozoic. Recently, the very early stages of the Earth's evolution have been considered in the works of Sorokhtin who, together with Ushakov (1988), has analysed the history of the formation of the World Ocean along with the Earth's crust. According to the calculations by these authors, plate tectonic activity started in the Early Proterozoic. The Archaean was a period of intense spreading, with the piling up of water-rich basalt plates, from which the tonalite–trondhjemite magma fused out to form the cores of Archaean shields, playing the role of subduction.

Mikhail Mints (1999) analysed lithospheric parameters of the Earth and plate tectonics in the Archaean and showed that lithospheric state parameters of the Earth are characterized on the basis of geochronological data. The simatic and sialic segments of the Archaean crust were formed by 3.9–3.8 Ga BP. The Earth's surface physiography was essentially similar to that of the present, but with temperatures several tens of degrees higher than at present. Deep oceanic basins bounded segments of emergent continental areas, with rugged topography. The Early Archaean 'continents' were originally small but rapidly increased in size. Approximately 3.3–3.0 Ga BP, the lithosphere beneath the major cratons (>0.5 × 10^6 km^2) was up to 150–200 km thick. The thickness and temperature distribution within the continental crust and subcontinental lithospheric mantle as well as the temperature of descending mantle flows were close to those at present. At least 3.0 Ga BP, the Archaean continents were characterized by rigidity comparable to that of the present-day continental plates. The mafic–ultramafic composition of the 'oceanic' segments of the lithosphere and the low temperatures of the Earth's surface probably gave rise to a varying buoyancy of the 'oceanic' segments that was necessary for drawing them into mantle convection. By 3.8 Ga BP, the summits of volcanic edifices in the oceans remained below sea level, which accounts for the hydration of rocks in the oceanic lithosphere. These assumptions suggest that plate tectonics had been under way since 3.9–3.8 Ga BP, with the exception of intracontinental processes, which cannot be confidently recognized before 3.1–2.9 Ga BP.

Another issue is intra- and inter-plate tectonics. Khain (1986) showed that the forms in which this tectonic activity (and magmatism) is manifested are various and are not confined to a single mechanism, e.g. the mechanism of mantle plumes and hot spots. In the work by Zonenshain and Kuzmin (1983), the above concept was enlarged to that of 'hot fields'; in an article by Zonenshain (1988), their origin was suggested to be connected to convection in the lower mantle.

In this context, of special interest is the origin of the Central Asian intracontinental mountain belt. Fixist- or 'semi-fixist'-minded geophysicists associate this origin with the ascent of 'anomalous', that is, heated-up and low density, mantle, whereas mobilists interpret this belt as a product of the interaction of the large Eurasian and Indian lithospheric plates with a piling up of intermediate small plates and microplates. A noteworthy contribution has been made by Leopold Lobkovsky (1988) who suggested 'two-layer plate tectonics'. According to this theory, when large plates collide, the material of the lower, viscoplastic part of the crust is forced into the zone of collision, with simultaneous disintegration of the upper, brittle part of the crust into smaller plates, which are thrust over one another.

One of the important phenomena of intraplate tectonics is continental rifting. For the last ten years, its study has become a major geotectonic problem. The most important work on this topic in this country has been accomplished by Milanovsky (1983*a*,*b*, 1987*a*,*b*), Kazmin and Andrey Grachev. The works of Kazmin, who had studied the East African rift system for many years, form one of the most extended studies from the plate tectonics point of view.

Eugeny Mirlin (1985) analysed the whole trend of the evolution of rift zones from narrow downwarps of continental crust to the formation of mature ocean basins with mid-ocean ridges, in connection with the kinematics of lithospheric plates. He stated that the peculiarities of the morphology and deep structure of mid-ocean ridges depend on the uneven rate of the ascent of mantle material during the divergence of plates, which, in turn, depends on the variation of the spreading rate, but this dependence has a non-linear character.

A series of studies by geophysicists from M. Lomonosov Moscow State University has been devoted to the mathematical and physical simulation of zones of divergence and convergence of lithospheric plates. These works concern, in particular, overlapping spreading centres (Shemenda & Grokholsky 1988), transform faults (Dubinin 1987), and intra-plate deformations of the Indian Ocean (Shemenda, 1989). The origin of marginal seas is a special problem that has been speculated upon in a monograph by Nikita Bogdanov (1988), in the works of Sorokhtin, and in some works of the aforementioned physical group in Moscow State University. Opinions on the evolution of marginal seas are divided, just as they are elsewhere in the world. Zonenshain and Leonid Savostin (1979) link the formation of marginal-sea basins to the movement of the overhanging plates above the subduction zones anchored in the mantle. Meanwhile, Anatoly Sharaskin, Zuram Zakariadze and Nikita Bogdanov point to a certain independence in time of the opening of marginal seas and the process of subduction, which should also imply the autonomy of the mechanism of formation of these basins.

In recent years, the attention of researchers has been increasingly focused on problems of deep-Earth dynamics, mainly under the influence of results of seismic tomography. Zonenshain, in a work together with Kuzmin and Natalia Bocharova (1991), examined the problem of hot spots and proposed to distinguish also 'hot fields' using the Pacific Ocean area as an example. He expressed the view that plume tectonics in the context of the whole solar system is more important than plate tectonics, as plume tectonics are manifest in all the planets. Nikolay Dobretsov with Anatoly Kirdyashkin (1994) elaborated a theory of layered mantle convection, supporting it by modelling. Dobretsov also pointed out the periodicity of tectonic and magmatic activity.

Khain has tried to demonstrate the evolution of the plate tectonics concept through the course of its application over a quarter of century (Khain 1988). In another paper (1989) he expressed the view that the time is ripe for the replacement of plate tectonics by a more universal model of global geodynamics, taking into account the processes in the deep interior of our planet and their different manifestations in different Earth layers. A similar opinion was also put forward by Zonenshain and Pushcharovsky. These researchers are convinced that we are on the verge of a new paradigm in the Earth sciences.

On the basis of analysis of global geological processes and interpretation of the results of numerical experiments, Valery Trubitsyn has developed new concepts of global tectonics, updating generally accepted ideas about the neotectonics of oceanic lithospheric plates by attachment of continents. In the modern plate tectonics theory the continents are regarded as passive elements included in oceanic plates, and without an essential influence on global geodynamic processes. But numerical experiments have also shown how the 'floating' continents control global geological processes in forming the 'face of the Earth'. Trubitsyn (1998) analysed this process and compared the Earth to a heat engine, in which the mantle plays the role of the boiler; the oceanic plates have the role of movable parts; and continents act like floating valves regulating heat loss.

Very recently, Mikhail Goncharov (2000) has proposed a 'multi-order level' model for the evolution of the Earth. He distinguishes a hierarchical schema for the convective processes in the mantle. Large-scale convection of the 'first order' occurs within the bulk of the mantle; meso-scale convection of the 'second order' takes place within the upper mantle; while small-scale convection of the 'third rank' takes place within the uppermost mantle. Global ('first order'-convection) is responsible for the movement of continents (with their *c*. 400 km roots) and for the creation and break-up of Pangaea. 'Second order' convection occurs only beneath oceans and is responsible for spreading and subduction. 'Third order' convection takes place as two-stage convection in the asthenosphere + lithosphere, and is held responsible for the generation of systems of transversal rises and depressions in spreading zones – rises being cut by rift valleys and troughs coinciding with transform faults – and of systems of longitudinal rises and depressions in collision zones. In both cases, rises are accompanied by roots, and there are thought to be 'anti-roots' beneath depressions. 'Third order' convection is also held responsible for mantle diapirism beneath back-arc basins and intercontinental ones (Goncharov 2000).

As in other countries, plate tectonics was soon successfully applied in Russia to other branches of the Earth sciences and in particular to petrology and sedimentology. In petrology, the works of Oleg Bogatikov and his team (Bogatikov *et al.* 1987) should be noted, and in sedimentology the fundamental monographs of Alexander Lisitsin (1988) on oceanic sedimentation have been particularly significant.

Plate tectonics applied to mineral deposits

A major connection in the distribution of mineral deposits with plate tectonics has attracted the attention of Soviet and Russian geologists. Alexander Kovalev was a pioneer and active contributor to this problem. His first article on this subject appeared in 1972, and his monograph *Mobilism and Criteria of Geological Prospecting* was published in 1978 (2nd edition, revised and supplemented, 1985). *Global Tectonics, Magmatism, and Metallogeny* by Zonenshain, Kuzmin, and Moralev appeared somewhat earlier in 1976. Andrey Monin and Sorokhtin (1982) described the mechanism of formation of Early Proterozoic iron-ore deposits from the plate tectonics point of view. The same plate tectonics interpretation has been deployed in the work by Sorokhtin (1987), regarding the origin of diamond-bearing kimberlites, as well as alkaline–ultramafic complexes and associated mineral deposits.

It is worth mentioning, however, that the majority of leading Russian metallogenists were for a long time biased against the idea of plate tectonics. Along with the general reasons mentioned above, their attitude towards this theory was much influenced by specific features of regional metallogeny, such as the order of concentration of certain metals in tectonic complexes occurring in certain regions. This sequence was considered to be suggestive of the absence of large horizontal displacements, and the importance of deep faults and block structures in the distribution of deposits was interpreted as evidence of the domination of vertical movements. Actually, neither of these aspects was in contradiction with mobilism, and the manifest zoning in the distribution of certain groups of metals in the Pacific belt, noted by Sergey Smirnov (1955), is well explained from a plate tectonics perspective.

The introduction of new mobilistic ideas has been particularly successful in the field of oil and gas geology. Sorokhtin, Ushakov, and Vsevolod Fedynsky (1974) supported the ideas of Hollis Hedberg about the generation of hydrocarbons in subduction zones. Other studies in this field were focused on the important role of zones of rifting, with their elevated heat and fluid flows. A geodynamic classification of oil and gas basins in general, and of those of the USSR in particular, was proposed by Boris Sokolov & Khain (1982), Evgeny Kucheruk & Elizaveta Alieva (1983) and Kucheruk & Ushakov (1984). The idea of possible oil and gas potential in overthrust zones started to attract adherents with the work of Khain, Konstantin Kleshchev, Sokolov & Vasily Shein (1988).

Starting with the early 1980s, the plate tectonics concept has been progressively applied to the analysis of seismicity in subduction zones. Lobkovsky, Sorokhtin & Shemenda (1980) and Lobkovsky & Boris Baranov (1982, 1984) have studied the seismotectonic phenomena of the inner slopes of deep-sea trenches. These studies have revealed, in particular, possible reasons for tsunamigenic earthquakes. A so-called 'keyboard' (*Klaviatur*) model to account for the most violent earthquakes was put forward by Lobkovsky as a clue to understanding the nature of seismic cycles in subduction zones. He envisaged subduction occurring in front of an island arc, the region between the subducting plate and the islands existing as separate blocks, divided by faults perpendicular to the line of the islands. As subduction proceeded, the blocks act separately from one another, are individually submerged, and sequentially yield to the pressure, each eventually being repulsed from, or springing back from, the island arc. The model was developed in his subsequent works, together with Boris Baranov (1982, 1984: seismotectonic aspects), and Vladimir Kerchman (1986, 1988: mathematical modelling).

Plate tectonics in geological education

For many years, teaching of the geological disciplines in all Russian educational institutions was based on the concept of the geosynclinal evolution of the Earth's crust so that even now mobilist ideas have not found support among the majority of high school teachers of the country. Formerly, the course on geotectonics at the M. Lomonosov Moscow State University was read for the geology students by Beloussov. In his lectures all mobilist ideas were referred to as an amusing historical episode in the development of geology, and plate tectonics theory was just a temporary phenomenon in the evolution of our science. But at the same time and in the same faculty Khain presented mobilist ideas to students of geophysics and geochemistry. The position radically changed after the 27th International Geological Congress in Moscow (1984). A special conference of the geological faculty of the M. Lomonosov Moscow State University revised the curriculums of the fundamental geological disciplines and the programmes of all the fundamental disciplines of the geosciences were reworked. Courses in 'general geology', 'historical geology', 'geotectonics', 'history and methodology of geologic sciences' and others all included plate tectonics. Beloussov refused to read his course according to the new programme; and so Khain began to read the lectures on geotectonics for the geologists instead of him (Ryabukbin 1993).

In 1985, a textbook entitled *General Geotectonics* by Khain and Alexander Mikhaylov was used along with the earlier empirical concepts of evolution of structures of the Earth and expounded the modern mobilist ideas in detail. In subsequent years the new textbooks for the main geological disciplines were published, which are now used in all higher educational institutions of the country: *General Geology* (Alexandra Yakushova, Khain & Vladimir Slavin, 1995); *Geotectonics with Basic Principles of Geodynamics* (Khain & Lomize, 1995); *Historical Geology* (Khain, Nicolay Koronovsky & Nicolay Yasamanovv); *History and Methodology of Geological Sciences* (Khain & Ryabukhin, 1997); *Geology of Mineral Resources* (Victor Starostin & Peter Ignatov, 1997).

Some alternative views

As a result of the growing evidence for, and the rising number of advocates of, plate tectonics, the number of Russian scientists taking the fixist stance has sharply decreased. The most active supporters of fixist ideas are confined to a group of scientists who were former co-workers of Beloussov at the Institute of Physics of the Earth of the USSR Academy of Sciences. This group also includes some university professors and scientists working at research institutes. However, there are now many scientists who recognize the essential role of horizontal movements in the evolution of the Earth's crust, and of oceanic spreading in particular. They are mobilists but do not accept the plate tectonics theory as a whole or accept it only with serious reservations. This group is quite numerous, but their views are diverse.

The fruitful idea of the tectonic delamination of the lithosphere was developed in the 1980s at the Geological Institute of the USSR Academy of Sciences. It was initiated by Peive and developed further by Pushcharovsky, with the active participation of Vladimir Trifonov (1990), Sergey Ruzhentsev, and others. Peive did not oppose the concept of plate tectonics, but considered the idea of tectonic delamination as its useful supplement. Some of his followers attempted to find a contradiction between these two ideas, though without valid arguments. In fact, the concept of tectonic delamination of the lithosphere is gaining more and more support from seismic and magnetotelluric data. At present, this concept, which distinguishes a brittle upper over a ductile lower crust, is developing both abroad and in Russia (the works on two-layer plate tectonics by Lobkovsky and Nikolayevsky).

Another concept set forth as an alternative to plate tectonics was elaborated at the All-Union Geological Institute in St Petersburg by Lev Krasny & Sadovsky (1988): it is the concept of 'geoblocks'. Later it converged with the notion of the fractal structure of the lithosphere, advanced in the Moscow Institute of Physics of the Earth (Mikhail Sadovsky & Valery Pisarenko 1991). The essence of the 'geoblock' theory is very simple. It assumes that the lithosphere is divided into a large number of blocks experiencing both vertical and horizontal movements with respect to each other. The latter assumption refers this theory to the mobilistic trend. It is sufficiently clear that this model is compatible with plate tectonics. The lithospheric plates are, in a way, 'geoblocks', and initially W. J. Morgan called them so. In addition, Krasny singles out a large number of smaller 'geoblocks', many of which are separated by ancient sutures and were independent lithospheric plates in the past, especially those 'geoblocks' that formed part of the basement of old cratons. Subsequently, they could experience differential movement along their borderline sutures. As for oceans, these are taken to be large segments of lithospheric plates, separated by magistral transform faults, which are interpreted as independent 'geoblocks'. So, the question is about the actually observed divisibility of the lithosphere (which is nevertheless subordinate to the principal divisibility into lithospheric plates); or (and) reflecting such a divisibility 'in retrospect'. Also a still smaller-scale divisibility of the brittle upper crust should be considered. These smaller 'geoblocks' are compatible with the terranes and microplates in current Western literature.

In addition to these two concepts, which do not pretend to represent complete global geodynamic models, at least two other attempts have been made in Russia to create such models. Both of them assume the same kinematics of lithospheric plates as does plate tectonics, but they suggest a different interpretation of the geodynamic processes that control these kinematics. One of these models was proposed by Evgeny Artyushkov in his *Geodynamics* (1979) and *Physical Tectonics* (1993), and in a number of later articles. The views of this author demand a special analysis. We consider them disputable and in many respects conjugate with fixism, possessing no advantages over 'classical' plate tectonics. The main features of Artyushkov's model, which distinguish it from 'classical' plate tectonics, are: (1) closed-up convective cells in the mantle are replaced by advective flows ascending from the core surface to the asthenosphere; (2) such flows are presumed to occur not only at mid-oceanic ridges but also within active continental margins and under continents themselves (thus conditions are provided for the subsidence of oceanic lithosphere in seismo-focal zones of active continental margins, and for continental rifting); and (3) lateral displacements of plates, believed to be caused not by the friction at their base by horizontal segments of convective cells in the asthenosphere, but by the gravitational 'disintegration' of the anomalous mantle lens that has accumulated under mid-oceanic ridges owing to an inflow from the lowermost mantle. In addition, Artyushkov denies any substantial extension accompanying the formation of rifts and intracontinental sedimentary basins. He thinks that these processes are mainly determined by eclogitization of the lower crust, induced by the ascent of the anomalous mantle to its base. He also denies the extension, at the initial stage of formation, of passive continental margins. According to his views, oceanic spreading is confined to mid-oceanic ridges and is not supposed to involve abyssal basins for which the same mechanism of eclogitization is evidently inferred. The same explanation is proposed for foredeeps of orogenic belts.

Another different model has been proposed by Kropotkin, the first Russian neomobilist. Kropotkin and his team (1987) considered the plate tectonics model to be imperfect since it does not take into account the large thickness of the lithosphere under the continents (over 400 km); he assumes the absence of a continuous layer of the asthenosphere, and thinks that plate tectonics is unable to explain the prevalence of compression stress over the major part of the Earth's surface and its high absolute value. In this connection, Kropotkin suggests that pulsation of the Earth's volume can be assumed to be the main mechanism of tectogenesis: oceanic spreading occurs during the extension phases, and fold mountain edifices formed during the compression phases; only the 'forced', that is, outward-stimulated mantle convection is admitted. So, mobilism and drifting of lithospheric plates are combined in Kropotkin's model with the notion of pulsation of the Earth's radius, which was at one time suggested as a basis for the so-called pulsation hypothesis of tectogenesis.

It should be said that some advocates in Russia of the latter theory have also attempted to take into account the role of pulsation of the Earth's volume, as distinct from the postulate of the 'classical' plate tectonics about its permanence. Nikita Bogdanov & Dobretsov (1987), and Khain have pointed to a certain periodicity in the formation of ophiolites and glaucophane schists, and the opening of oceans, correlating with the periodicity of fold–nappe deformations and formation of granites, which was ascertained long before. The short-period changes of intensity of volcanism and seismicity in the recent epoch are touched upon in other works (Ellchin Khalilov *et al.* 1987). None of these authors oppose the fact of this periodicity of the endogenic activity of the Earth to the plate tectonic theory, but they do think it necessary to supplement it with the recognition of this phenomenon.

By contrast, Milanovsky (1984, 1987*a*), considering the periodicity of continental rifting in the Earth's history, has favoured a pulsation hypothesis, in combination with the hypothesis of an expanding Earth, as an alternative to plate tectonics. It should be noted in this connection that the present and past dynamics of the Earth are convincing evidence of the simultaneous, and not alternating, manifestations of extension (spreading, continental rifting) and compression (subduction, mountain building) of the lithosphere. And, speaking about the long-term tendency of change of the Earth's volume, there is more evidence for the increase of compression rather than extension (Aslanyan 1982; Kropotkin 1971). However, the hypothesis of an expanding Earth is rather popular among certain Russian geologists.

Conclusions

The suggestion of the concept of 'scientific revolution' by Thomas Kuhn (1962) was, in itself, a significant event in the history of science, and 'crucial' episodes or 'paradigm shifts' have come to be of special interest in the history of geology (as in other sciences). It is generally accepted that the geosciences went through an authentic scientific revolution in the 1960s. This revolution began in the fields of geophysics and geotectonics, and then quickly spread to all other fields of geoscience. As can be seen from the foregoing account, fixist ideas were still dominant in geotectonics at the middle of the century, especially in the USSR, and the new concept encountered strong resistance in Russia from the proponents of geosyncline theory. This was natural. The creation of a new paradigm is not simply an increment of knowledge. It involves a modification of 'world view'; and that, as a rule, does not occur painlessly.

In Western literature the development of geology in the USSR has sometimes been related to the political situation (e.g. Wood 1985). But Beloussov – the principal opponent of plate-tectonics in Russia – never belonged to the political elite. On the contrary, the political elite, knowing his solid, irreconcilable nature, did not want to elect him a member of the Academy of Sciences (the highest level in the Russian scientific hierarchy); and he was refused permission to deliver lectures on tectonics at the M. Lomonosov State Moscow University when he declined to give students the views of his opponents.

As it seems, the example of acceptance of plate tectonics ideas in Russia resembles Kuhn's model of the progress of science. It is difficult to discard ideas to which one has devoted one's creative life. One naturally tries to demonstrate that the old model works. And those who hold the control-levers of the authority may oppose or simply ignore the new ideas. So it was with Wegener's ideas in the United States, and in other countries (Wood 1985; Oreskes 1999). However, the situation in Russia has been rather different from the West, and not entirely as Kuhn's account would lead one to expect, for as we have seen above there are still several, opposed and competing, fundamental geological theories being used and taught in modern Russia. Given this state of affairs, it might seem that, from Kuhn's perspective, geological theory in Russia has not yet fully completed its scientific revolution, for there are still different theories or research programmes being pursued. Nevertheless, plate tectonics presently occupies a dominant place in geological thinking in Russia, as well as in Western countries. This is illustrated by the successful convocation of Zonenshain's conferences on plate tectonics in recent years.

So plate tectonics in Russia has gone through moments of complete denial, doubt and eventually wide acceptance by the majority of geologists. At the beginning of the twenty-first century, most Russian geologists have now adopted plate tectonics, although, as said, opposition has not disappeared completely. Nevertheless, progress has been made, not only in the application of plate tectonics theory to the deciphering of the geological history and structure of the territory of Russia and adjacent seas and oceans, but also in the development of the theory itself. In fact, the alternative geodynamic models can be interpreted as a side-effect of the general revival of studies in the field of theoretical geology, caused by the appearance of plate tectonics.

It must be acknowledged that plate tectonics represents only the tectonics of the upper parts of the solid Earth, and probably is applicable in its classic version only to our planet. The present challenge is to create an authentic global geodynamic model of the Earth, and establish its place in the evolution of the planets.

The authors thank D. Oldroyd (The University of New South Wales) for his interest in our article, and for help in the improvement of the text's English.

References

ALIEVA, E. P. & KUCHERUK, E. V. 1987. Evaluation of oil and gas potential of water areas by geological–geophysical and evolutional–geodynamic methods. Results of science and technology. *Deposits of Fuel Minerals*, **15**. VINITI, Moscow (in Russian).

ARTYUSHKOV, E. V. 1979. *Geodynamics*. Nauka, Moscow (in Russian).

ARTYUSHKOV, E. V. 1993. *Physical Tectonics*. Nauka. Moscow (in Russian).

ASLANYAN, A. T. 1982. Convection and contraction. *Izvestiya Akademii Nauk Armyanskoy S. S. S. R., Nauki o Zemte*, **36**, 3–32 (in Russian).

AUBOUIN, J., LE PICHON, X. & MONIN, A. S. (eds) 1986. The evolution of the Tethys. *Tectonophysics*, **123**, 315.

BAZHENOV, M. & BURTMAN, V. 1982. Kinematics of the Pamir arc. *Geotectonika*, **4**, 54–71 (in Russian).

BELOUSSOV, V. V. 1970. Against the hypothesis of ocean-floor spreading. *Tectonophysics*, **9**, 482–512.

BELOUSSOV, V. V. 1984. Speech in the assembly of the section of geology, geophysics and geochemistry of Academy of sciences of March 13. *Izvestiya Akademii Nauk S. S. S. R., Fizika Zemli*, **12**, 57–58 (in Russian).

BOGATIKOV, O. A., KOVALENKO, V. I., TSVETKOV, A. A., SHARKOV, E. V., YARMOLYUK, V. V. & BUBNOV, S. N. 1987. Series of magmatic rocks: problems and solutions. *Izvestiya Akademii Nauk S. S. S. R. Serya Geologicheskaya*, **6**, 3–12 (in Russian).

BOGDANOV, N. A. 1988. *Tectonics of Deep-sea Basins of Marginal Seas*. Nedra, Moscow (in Russian).

BOGDANOV, N. A. & DOBRETSOV, N. L. 1987. Synchroneity of active tectonic processes in continents and oceans. *Izvestiya Akademii Nauk S. S. S. R. Serya Geologicheskaya*, **1**, 43–52 (in Russian).

BORISSYAK, A. A. 1922. *Course of Historical Geology*. Gosizdat, Petrograd (in Russian).

BORUKAYEV, CH. B. 1985. *Precambrian Structures and Plate Tectonics*. Nauka, Novosibirsk (in Russian).

BURTMAN, V. 1984. Kinematics of the Carpathian structural loop. *Geotectonika*, **3**, 17–31 (in Russian).

DOBRETSOV, N. L. 1980. *Global Petrology*. Nauka, Moscow (in Russian).

DOBRETSOV, N. L. & KIRDYASHKIN, A. G. 1998. *Deep-level Geodynamics*. A. A. Balkema, Rotterdam.

DUBININ, E. P. 1987. *Transform Faults of the Oceanic Lithosphere*. Moscow State University, Moscow (in Russian).

GONCHAROV, M. A. 2000. From plate tectonics to convection of different scales within hierarchically interacting geospheres. *Geophysical Research Abstracts*, **2**, CD-ROM edition.

KARASIK, A. M. 1980. The principal special features of the history and structure of the Arctic Ocean from aerial magnetic data. *In*: VARENTSOV, M. I. (ed.) *Marine Geology, Sedimentation, Lithology and Ocean Geology*. Nedra. Leningrad, 178–193 (in Russian).

KARAULOV, V. B. 1988. Mobilism, fixism and 'concrete' tectonics. *Bulletin of Moscow Society of Naturalists, Geological Section*, **63**, 3–13 (in Russian).

KEONDJIAN, V. P. & MONIN, A. S. 1976. Calculations of the evolution of planets' interiors. *Izvestiya Akademii Nauk S. S. S. R., Fizika Zemli*, **4**, 3–13 (in Russian).

KERCHMAN, V. I. & LOBKOVSKY, L. I. 1986. Simulation of the seismotectonic process in active transitional zones according to the 'key-board' model for strong earthquakes. *Doklady Akademii Nauk S. S. S. R*, **291**, 1086–1091 (in Russian).

KERCHMAN, V. I. & LOBKOVSKY, L. I. 1988. A geomechanical model of tectonic movements of seismogenic blocks in subduction zones with respect to strong earthquakes of thrust and strike-slip type. *Doklady Akademii Nauk S. S. S. R.*, **298**, 1023–1028 (in Russian).

KHAIN, V. E. 1970. Is there a scientific revolution going on in geology? *Priroda*, **1**, 719 (in Russian).

KHAIN, V. E. 1985. Main phases of opening of contemporary oceans in comparison with events on continents. *Vestnik Moskovscogo Universiteta, Geologiya* **4**, 3–11 (in Russian).

KHAIN, V. E. 1986. Problems of intra- and inter-plate tectonics. *In*: YANSHIN, A. L., BEUS, A. A. (eds) *Dynamics and Evolution of the Lithosphere*. Nauka, Moscow, 7–15 (in Russian).

KHAIN, V. E. 1988. Plate tectonics twenty years after (thoughts about past, present, and future developments). *Geotektonika*, **6**, 3–7 (in Russian).

KHAIN, V. E. 1989. Layering of the Earth and multilayer convection as a base of genuine global geodynamic model. *Doklady Akademii Nauk S. S. S. R.*, **308**, 1437–1440 (in Russian).

KHAIN, V. E. 1990. Concerning articles by Mazarovich, O. G, Naydin, D. P. and Zeisler, V. M. *Bulletin of the Moscow Society of Naturalists*, **65**, 7–20 (in Russian).

KHAIN, V. E. 1991. Mobilism and plate tectonics in the USSR *Tectonophysics*, **199**, 137–148.

KHAIN, V. E. & BOZHKO, N. A. 1988. *Historical Geotectonics: The Precambrian*. Nedra, Moscow (in Russian).

KHAIN, V. E. & LOMIZE, M. G. 1995. *Geotectonics with Basic Principles of Geodynamics*. MSU, Moscow (in Russian).

KHAIN, V. E. & RYABUKHIN, A. G. 1997. *History and Methodology of Geological Sciences*. MSU, Moscow (in Russian).

KHAIN, V. E., KLESHEV, K. A., SOKOLOV, B. A. & SHEIN, V. S. 1988. Tectonic and geodynamic setting of oil and gas potential of the territory of the USSR *In*: PUSHCHAROVSKY, YU. M. (ed.) *Current Problems of Tectonics of the USSR* Nauka, Moscow, 46–54 (in Russian).

KHALILOV, E. N., MEKHTIEV, SH. F. & KHAIN, V. E. 1987. On some geophysical data confirming the collision origin of the Greater Caucasus. *Geotectonika*, **2**, 54–60 (in Russian).

KHAIN, V. E., KORONOVSKY, N. V. & YASAMANOV, N. A. 1997. *Historical Geology*. MSU, Moscow (in Russian).

KHRAMOV, A. N. (ed.) 1982. *Palaeomagnetology*. Nedra, Leningrad (in Russian).

KNIPPER, A. L. 1975. *The Oceanic Crust in the Structure of the Alpine Folded Belt*. (South Europe, Western Part of Asia and Cuba). Nauka, Moscow (in Russian).

KORONOVSKY, N. V. 1989. Conceptual alternatives in modern geotectonics (in connection with article by V. Karaulov 'Mobilism, fixism and 'concrete' tectonics'. *Bulletin of the Moscow Society of Naturalists, Geological Section*, **64**, 110–119 (in Russian).

KORONOVSKY, N. V. & DIOMINA, L. I. 1999. The collision stage of the evolution of the Caucasian sector of the Alpine fold-belt: geodynamics and magmatism. *Geotektonika*, **2**, 17–35 (in Russian).

KOSSYGIN, YU. A. 1983. *Tectonics* (2nd edn). Nedra, Moscow (in Russian).

KOVALEV, A. A. 1985. *Mobilism and Criteria for Geological Exploration* (2nd edn). Nedra, Moscow (in Russian).

KRASNY, L. I. & SADOVSKY, M. A. 1988. *The Mosaic Face of the Earth*. Nauka, Moscow (in Russian).

KRAVCHINSKY, A. YA. 1977. *Paleomagnetic and Paleogeographic Reconstructions on Precambrian Platforms*. Nedra, Moscow (in Russian).

KROPOTKIN, P. N. 1958. The importance of palaeomagnetism for stratigraphy and tectonics. *Bulletin of the Moscow Society of Naturalists, Geological Section*, **33**, 57–86 (in Russian).

KROPOTKIN, P. N. 1969. The problem of continental

drift (mobilism). *Izvestiya Akademii Nauk S. S. S. R., Fizika Zemli*, **3**, 3–18 (in Russian).

KROPOTKIN, P. N. 1971. Eurasia as a composite continent. *Tectonophysics*, **12**, 261–266.

KROPOTKIN, P. N. & SHAHVAROSTOVA, K. A. 1965. *Geological Structure of the Pacific Mobile Belt*. Nauka, Moscow (in Russian).

KROPOTKIN, P. N., EFREMOV, V. P. & MAKEEV, V. I., 1987. The stress in the Earth's crust and the geodynamics. *Geotektonika*, **1**, 3–24 (in Russian).

KUCHERUCK, E. V. & ALIEVA, E. R. 1983. *The Present State of Classification of Sedimentary Basins*. VNIIOENG, Moscow (in Russian).

KUCHERUCK, E. & USHAKOV, S. A. 1985. *Plate Tectonics and Oil and Gas Potential (A Geophysical Analysis)*. VINITI, Moscow (in Russian).

KUHN, T. 1962. *The Structure of Scientific Revolution*. University of Chicago Press, Chicago.

LEGLER, V. A. 1989. Plate tectonics as scientific revolution. *In*: ZONENSHAIN, L. P. & PRISTAVAKINA, E. I. (eds) *Geological History and Plate Tectonics of the Territory of the USSR* Nauka, Moscow (in Russian).

LISITSIN, A. P. 1988. *Avalanche Sedimentation and Breaks in Accumulation of Sediments in Seas and Oceans*. Nauka, Moscow (in Russian).

LOBKOVSKY, L. I. 1988. *Geodynamics of Spreading and Subduction Zones in Two-layer Plate Tectonics*. Nauka, Moscow (in Russian).

LOBKOVSKY, L. I. & BARANOV, B. V. 1982. On the question of generation of tsunamis in underthrust zones of lithospheric plates. *In*: SOLOVIEV, S. L. & MASLOV, V. N. (eds) *Processes of Generation and Propagation of Tsunami*. Institute of Oceanology, USSR Academy of Sciences, Moscow, 7–17 (in Russian).

LOBKOVSKY, L. I. & BARANOV, B. V. 1984. A 'keyboard' model of strong earthquakes in island arcs and active continental margins. *Doklady Akademii Nauk S. S. S. R.*, **275**, 843–847 (in Russian).

LOBKOVSKY, L. I. & SOROKHTIN, O. G. 1976*a*. Plastic deformations of the oceanic lithosphere in the zone of underthrust of plates. *In*: SOROKHTIN, O. G. (ed.) *Tectonics of Lithospheric Plates (Dynamics of the Underthrust Zone)*. Institute of Oceanology, USSR Academy of Sciences, Moscow, 22–52 (in Russian).

LOBKOVSKY, L. I. & SOROKHTIN, O. G. 1976*b*. Conditions for absorption of sediments in deep-sea trenches. *In*: SOROKHTIN, O. G. (ed.) *Tectonics of Lithospheric Plates (Dynamics of the Underthrust Zone)*. Institute of Oceanology, USSR Academy of Sciences, Moscow, 84–102 (in Russian).

LOBKOVSKY, L. I., SOROKHTIN, O. G. & SHEMENDA, A. I. 1980. Simulation of island-arc deformations leading to formation of tectonic terraces and to tsunami earthquakes. *Doklady Akademii Nauk S. S. S. R.*, **255**, 74–77.

MA XINGYUAN. 1988. Lithospheric dynamics of China. *Episodes*, **11**, 84–90.

MAZAROVICH, O. A., NAYDIN, D. P & ZEYSLER, V. M. 1988–1989. Palaeomagnetic and historical{o}geological reconstructions. Problems and unsolved questions. Part 1. An occasion for discussion. *Bulletin of the Moscow Society of Naturalists, Geological Section*, **63**, 130–143; Part 2, **64**, 125–147 (in Russian).

MILANOVSKY, E. E. 1983*a*. Major stages of rifting evolution in the Earth's history. *Tectonophysics*, **94**, 599–607.

MILANOVSKY, E. E. 1983*b* *Riftogenesis in Earth History: Riftogenesis on the Ancient Platforms*. Nedra, Moscow (in Russian).

MILANOVSKY, E. E. 1984. The progress and modern situation of the problem of the Earth's expansion and pulsation. *In*: KROPOTKIN, P. N. & MILANOVSKY, E. E. (eds) *Problems of the Earth's Expansion and Pulsation*. Nauka, Moscow, 8–24 (in Russian).

MILANOVSKY, E. E. 1987*a*. Rifting evolution in Earth history. *Tectonophysics*, **143**, 103–118.

MILANOVSKY, E. E. 1987*b*. *Riftogenesis in Earth History: Riftogenesis of the Mobile Belts*. Nedra, Moscow (in Russian).

MINTS, M. V. 1999. Lithospheric state parameters and plate tectonics in the Archaean. *Geotectonika*, **6**, 45–58 (in Russian).

MIRLIN, E. G. 1985. *Divergence of Lithospheric Plates and Riftogenesis*. Nauka, Moscow (in Russian).

MONIN, A. S. 1987. *The Early Geological History of the Earth*. Nedra, Moscow (in Russian).

MONIN, A. S. & SOROKHTIN, O. G. 1982. Evolution of oceans and metallogeny of the Precambrian. *Doklady Akademii Nauk S. S. S. R.*, **264**, 1453–1457 (in Russian).

NANCE, R. E., WORSLEY, T. R. & MOODY, J. B. 1988. Supercontinental cycles. *Science*, **8**, 77–82.

ORESKES, N. 1999. *The Rejection of Continental Drift: Theory and Method in American Earth Science*. Oxford University Press, New York & Oxford.

PEIVE, A. V. 1945. Deep-seated faults in geosynclinal areas. *Izvestia Akademii Nauk S. S. S. R., Seriya Geoloicheskaya*, **5**, 23–46 (in Russian).

PEIVE, A. V. 1960. Faults and their role in the structure of the Earth's crust and deformation of rocks, *Proceedings of the XXI International Geological Congress*, **18**, 280–286.

PEIVE, A. V. (ed.) 1963. *Faults and Horizontal Movements of the Earth's Crust*. Nauka, Moscow (in Russian).

PEIVE, A. V. 1969. Oceanic crust of the geological past. *Geotektonika*, **4**, 5–23 (in Russian).

PUCHKOV, V. N. 1997. Tectonics of the Urals: modern concepts. *Geotektonika*, **4**, 42–61 (in Russian).

PUSHCHAROVSKY, YU. M. 1997. New ideas in tectonics. *Geotektonika*, **4**, 62–68 (in Russian).

PUSHCHAROVSKY, YU. M. & TRIFONOV, V. G. (eds). 1990. *Tectonic Delamination of the Lithosphere and Regional Geological Investigations*. Nauka, Moscow (in Russian).

RYABUKHIN, A. G. 1993. Mobilist ideas in Moscow University. *Vestnik Moscovscogo Universiteta, Geologiya*, Part 1, **3**, 3–13; Part 2, **5**, 39–47 (in Russian).

SADOVSKY, M. A. & PISARENKO, V. F. 1991. *The Seismic Process in the Block Environment*, Nauka, Moscow (in Russian).

SAVELIEVA, G. N. & SAVELIEV, A. A. 1977. Ophiolites of the Voykar–Syntjinsk massif (Polar Urals). *Geotektonika*, **11**, 427–437 (in Russian).

SHATSKY, N. S. 1946. Wegener's hypothesis and geosynclines. *Izvestia Akademii Nauk S. S. S. R., Geologicheskaya*, **4**, 7–12 (in Russian).

SHEMENDA, A. I. 1989. Modelling of intraplate deformations in NE Indian Ocean. *Geotektonika*, **3**, 37–49 (in Russian).

SHEMENDA, A. I. & GROKHOLSKY, A. L. 1988. The mechanism of formation and development of the zone of overlap of spreading axes. *Tikhookeanskaya Geologiya*, **5**, 97–107 (in Russian).

SMIRNOV, S. S. 1955. *Selected Works*. Akademiya Nauk S. S. S. R., Moscow (in Russian).

SMIRNOV, V. I. 1989. *Geology of Mineral Deposits*. Nedra, Moscow (in Russian).

SOKOLOV, B. A. & KHAIN, V. E. 1982. Oil and gas potential of overthrust margins of old mountain edifices. *Sovetskaya Geologiya*, **2**, 53–58 (in Russian).

SOROKHTIN, O. G. 1974. *Global Evolution of the Earth*. Nauka, Moscow (in Russian).

SOROKHTIN, O. G. 1987. Formation of diamond-bearing kimberlites and related rocks from the viewpoint of lithospheric plate tectonics. *In*: MEGELOVSKY, N. V. (ed.) *The Geodynamic Analysis and Regularities of Formation and Distribution of Deposits of Useful Minerals*. Nedra, Moscow, 92–107 (in Russian).

SOROKHTIN, O. G. & USHAKOV, S. A. 1988. Major stages of oceanic evolution. *Doklady Akademii Nauk S. S. S. R.* **302**, 308–312 (in Russian).

SOROKHTIN, O. G. & USHAKOV, S. A. 1991. *Global Evolution of the Earth*. MGU, Moscow (in Russian).

SOROKHTIN, O. G., USHAKOV, S. A. & FEDYNSKY, V. V. 1974. Dynamics of lithospheric plates and origin of oil deposits. *Doklady Akademy Nauk S. S. S. R.*, **214**, 1407–1410 (in Russian).

STAROSTIN, V. I. & IGNATOV, P. 1997. *Geology of Mineral Resources*. MSU, Moscow (in Russian).

TRUBITSYN, V. P. 1998. Role of floating continents in global tectonics of the Earth. *Physics of the Earth*, **1**, 3–10 (in Russian).

TRÜMPY, R. 1988. Cent ans de la tectonique de nappes dans les Alpes, *Compte Rendu de l'Académie des Sciences, Paris*, Section 2, **302**, 1–13.

USHAKOV, S. A. 1966. Earth's crust dynamics in transitional zones from continents to oceans of Atlantic type. *Doklady Akademii Nauk S. S. S. R.*, **171**, 315–317 (in Russian).

USHAKOV, S. A. 1974. *Structure and Evolution of the Earth. Physics of the Earth*. VINITI, Moscow (in Russian).

USHAKOV, S. A. & KHAIN, V. E. 1965. Structure of the Antarctic based on geological-geophysical data. *Vestnik Moskovscogo Universiteta, Geologiya*, **3**, 23–31 (in Russian).

USHAKOV, S. A. & YASAMANOV, N. A. 1984. *Continental Drift and Climates of the Earth*. Mysl, Moscow (in Russian).

VORONOV, P. S. 1967. The Antarctic and problems of Gondwana break-up. *Bulletin of the Soviet Antarctic Expedition*, **65**, 44–57 (in Russian).

VORONOV, P. S. 1968. *Continental Drift: Pro and Contra*. Geograficheskoye obshestvo S. S. S. R., Leningrad (in Russian).

WILSON, J. T. 1968. A revolution in Earth sciences. *Geotimes*, **13**, 10–16.

WILSON, J. T. 1999. Vladimir Vladimirovich Beloussov – an inspirational personality. *In*: SHOLPO, V. N. (ed.) *Vladimir Vladimirovich Beloussov*. UIPHE, Moscow, 189–192 (in Russian).

WOOD, R. M. 1985. *The Dark Side of the Earth*. Allen & Unwin, London.

YACUSHOVA, A. F., KHAIN, V. E. & SLAVIN, V. I. 1995. *General Geology*. MSU, Moscow (in Russian).

ZONENSHAIN, L. P. 1988. Problems and ways of evolution of plate tectonics. *Sovetskaya Geologiya*, **12**, 106–115.

ZONENSHAIN, L. P. & GORODNITSKY, A. M. 1977. Palaeozoic and Mesozoic reconstructions of continents and oceans. *Geotektonika*, **2**, 3–23; **3**, 3–24 (in Russian).

ZONENSHAIN, L. P. & KUZMIN, M. I. 1983. Intra-plate volcanism and its importance for understanding processes in the Earth's mantle. *Geotektonika*, **1**, 28–45 (in Russian).

ZONENSHAIN, L. P. & KUZMIN, M. I. 1997. *Palaeogeodyamics the Plate Tectonic Evolution of the Earth*. American Geophysical Union, Washington DC.

ZONENSHAIN, L. P. & SAVOSTIN, L. A. 1979. *Geodynamics: An Introduction*. Nedra, Moscow (in Russian).

ZONENSHAIN, L. P., KUZMIN, M. I. & MORALEV, V. M. 1976. *Global Tectonics, Magmatism and Metallogeny*. Nedra, Moscow (in Russian).

ZONENSHAIN, L. P. KORINEVSKY, V. G., KAZMIN, V. G., PECHERSKY, D. M., KHAIN, V. V. & MATVEENKOV, V. V. 1984. Plate tectonic model of the South Urals development. *Tectonophysics*, **109**, 95–135.

ZONENSHAIN, L. P., KUZMIN, M. I. & KONONOV, M. V. 1985. Absolute reconstructions of the Paleozoic oceans. *Earth and Planetary Science Letters*, **74**, 103–116.

ZONENSHAIN, L. P., KUZMIN, M. I. & NATAPOV, L. M. 1990. *Geology of the USSR: A Plate-tectonics Synthesis*. American Geophysical Union, Washington, Geodynamic Series, **21**.

ZONENSHAIN, L. P., KUZMIN, M. I. & BOCHAROVA, N. Yu. 1991. Hot-field tectonics. *Tectonophysics*, **197**, 215–250.

Plate tectonics, terranes and continental geology

HOMER E. LE GRAND

Faculty of Arts, Monash University, Clayton 3800, Victoria, Australia

Abstract: The 'modern revolution' in the Earth sciences is associated with the emergence of plate tectonics in the late 1960s. The assumption that the crust of the Earth was composed of a small number of rigid, non-deformable, mobile plates enabled a quantitative, kinematic description of current geological processes and reconstructions of past plate interactions. The simple model of plate theory *c.* 1970, for example its depiction of a subduction zone, has since undergone considerable refinement. However, some geologists, especially those concerned with questions of continental tectonics, contend that plate theory in its current form is of limited value in addressing questions of continental tectonics, and prefer to employ the concept of allochthonous terranes in characterizing, describing and interpreting regional geology. These geologists may understandably take the view that plate tectonics is a kinematic grand generalization but thus far not particularly useful in making sense of the rocks at the local level.

The 'modern revolution' in the Earth sciences is associated with the emergence of plate tectonics in the late 1960s.[1] This had two major phases. In the first, which could be called the sea-floor spreading phase, the concept of sea-floor spreading provided not only a plausible mechanism for the horizontal displacement of continents but also explanations for such recently discovered phenomena as magnetic striping of the sea floor, relatively high heat flow over the oceanic ridges, the distribution of deep- and shallow-focus earthquakes, and the age profile of different parts of the sea floor. This 'dynamic' and empirically based phase was followed by a phase marked by the emergence of more idealized, kinematic models of plate interactions in which blocks of crust were treated as idealized crustal units rotating around Euler poles constrained by correlations between oceanic and continental rock ages based on the rapidly developing magnetic reversal timescale (see Glen 1982).[2] Plate theory marked the culmination of a half-century of debates over crustal mobilism and, in a grand synthesis, drew together developments in many branches of the Earth sciences. The rapid and widespread acceptance of plate theory, J. Tuzo Wilson forcefully argued (1976, p. vii), 'has transformed the earth sciences from a group of rather unimaginative studies based upon pedestrian interpretations of natural phenomena into a unified science that holds the promise of great intellectual and practical advances'. Over the past 30 years, it could be argued with some justice that a major feature of this revolution has been a gradual shift toward a more physical and quantitative geology. However, some geologists judge plate theory in its current form to be of limited value in addressing questions of continental structures and tectonics. One response to the perceived shortcomings of plate tectonics, especially with respect to problems of regional geology, is the employment of the concept of what have been variously denominated accreted, exotic, suspect, allochthonous or tectonostratigraphic terranes.

Well into the early 1960s, there was little reason for geologists not to assume that explanatory frameworks based on the study of the continents over two centuries could be readily applied to processes and structures beneath the seas. For most geologists in the English-speaking world, the crust of their Earth was relatively stable. North and South America, for example, were each thought to be composed of an ancient core to which mountain belts had accumulated through the geosynclinal cycle. Mountain chains might be elevated or eroded, the continents might grow slowly on their margins but, broadly speaking, the continents and ocean basins had been essentially permanent features of the surface of the globe since their formation. Few took seriously the notion of continental drift. Extrapolating from the relatively well-known continents to the little-known ocean basins, it seemed obvious that granitic continents could

[1] This story has been told in varying ways by Marvin (1973), Hallam (1973), Menard (1986), Le Grand (1988), Stewart (1990) and Oreskes (1999).
[2] This point has been emphasized to me by H. J. Harrington (pers. comm.).

From: OLDROYD, D. R. (ed.) 2002. *The Earth Inside and Out: Some Major Contributions to Geology in the Twentieth Century*. Geological Society, London, Special Publications, **192,** 199–213. 0305-8719/02/$15.00

not possibly move through the unyielding basaltic oceans. Plate tectonics strikingly reversed this situation. By the early 1970s, the revolution was essentially over. A vast amount of new and unexpected data had been harvested from the ocean floors. Their history, structure, volcanicity, magnetization, heat flow and other characteristics were different from anything known about the continents. Along with this new data had come both new theories of, and new evidence for, great lateral motion of the continents. Plate tectonics, the triumphant version of continental drift, was developed largely by geophysicists and geologists to make sense of this deluge of novel geophysical and geological data from the sea floor, much of which had no continental counterparts.

By the mid-1970s plate tectonics formed for most Earth scientists the theoretical framework for understanding and describing the workings of the Earth's outer shell or, as one influential textbook (Wyllie 1976) was titled, *The Way the Earth Works*.[3] From one perspective, it constituted only the culmination of a series of theories of global mobilism originating with Alfred Wegener in the early years of the twentieth century. Wegener, Alexander du Toit and others over a period of half a century had gathered and marshalled palaeontological, stratigraphic and geophysical evidence, taken mostly from the continents, to support the view that the continents had once been joined together but had been broken apart and moved to their current locations. Ironically, for many early plate theorists the continents were merely uninteresting excrescences on a fascinating sea floor. Could scientists apply this new framework to the continents to solve or resolve problems that had bedevilled generations of land-based researchers?

For plate theorists, their rigid, non-deformable plates and plate boundaries were almost geometrical entities which one could use to calculate the kinematics of plate motions and interactions over time. Dan McKenzie's Earth, for example, was a geometrical construct on which transform faults were arcs of circles defined by the poles of rotation of idealized plates; ridges and trenches were merely 'lines along which crust is produced and destroyed' (McKenzie & Parker 1967, p. 1276). Jason Morgan (1968, p. 1959) offered his version as 'a geometrical framework with which to describe present-day continental drift'. The third member of the early plate triumvirate, Xavier Le Pichon, put forward 'a geometrical model of the surface of the earth' (1968, p. 3661) which, though necessarily involving 'great simplifications and generalizations' (p. 3679) enabled him to give 'a mathematical solution which can be considered a first-approximation solution to the actual problem of earth's surface displacements' (p. 3674).

Field geologists concerned with the problems of continental geology and its history could not easily begin to use ocean-derived plate theory to solve them. They were in possession of an enormous store of knowledge and detailed maps of the geology of the continents, but knowledge of ocean-floor geology consisted of a rapidly growing fund of widely scattered data. Only very coarse models were available to indicate how oceanic crust might be connected to and interact with continental crust.[4] Once one moved away from the geometrician's globe, the local expressions of past and present plate movements were conjectural, diverse and different from place to place. How might one infer from this large-scale, general and coarsely grained idealized model solutions for the finer grained, specific problems of local geology that, though well known and well mapped, had seemed heretofore intractable? If this could be done, then the resistance of land-locked Earth scientists to the new sea-born theory could be overcome. From a cognitive perspective, the challenge lay in the necessarily uncertain and speculative extrapolation from processes thought to occur in the relatively youthful sea floor to explain the much more ancient structures of the continents. But, I suggest, that challenge lay too in the scientific and social interests of most land-based geologists. They rapidly gave assent – or at least lip service – to plate theory at a general level but controversies abounded over its applications to regional and local geological problems.[5]

There was considerable initial resistance to the very idea of global mobilism, especially from more senior Earth scientists who had invested their careers in a fixist approach to continental

[3] Cf. Glen (1975).

[4] As early as the 1930s, the National Research Council included, among the several major geophysical research problems for the community to address, that of the nature of the 'join' between the oceanic and continental crusts, particularly at what we now know to be a passive margin, e.g. the Atlantic Ocean floor joining the North American and South American continents. In spite of enormous advances in several branches of geophysics that could be brought to bear on this problem, a detailed cross-section of that join is still extremely tenuous more than a half-century later.

problems. The social interests of those who had achieved positions of authority through their work on continental features might well lead them to oppose extrapolations from the sea floor to the continents. These land-based Earth scientists had invested years in meticulous local mapping, fieldwork and analysis, and ever more refined synthesizing and theorizing. They had thereby achieved positions of authority in their chosen period or region or technique or structure.[6] 'Teddy' Bullard incisively commented (1975, p. 5): 'Such a group has a considerable investment in orthodoxy.... To think the whole subject through again when one is no longer young is not easy and involves admitting a partially misspent youth'. Mason Hill, for example, the architect of the previously accepted view of the San Andreas system, 'used to shake with rage when somebody would get up and talk about the San Andreas transform'.[7]

'The new global tectonics' was the agenda for the Penrose Conference, organized by Bill Dickinson, held on 15–20 December 1969 at the Asilomar Conference Grounds in Pacific Grove, California. It marked a major turning point in attempts to apply plate tectonics to the continents.[8] Among the participants were John Dewey, Jack Bird, Seiya Uyeda, Clark Burchfiel, Clark Blake, Greg Davis, Tanya Atwater, Peter Coney and Warren Hamilton. Dewey and Hamilton were already formulating approaches to continental geology based upon plate tectonics and their first papers bracketed the conference. Atwater gave a presentation that inspired many of the participants to try their own hand at plate tectonics-based interpretations. Several of those present were also to take part in the later development of, and debates about, the terrane approach to regional geology.

Dewey, soon after the adumbration of plate tectonics, and just before Asilomar, had begun to apply the new tectonics to construct an overview of orogeny on convergent Atlantic-type plate margins. Pursuing a suggestion of Tuzo Wilson (1966), he proposed that the Appalachians and other mountains bordering the Atlantic had been pushed up through collisions resulting from the openings and closings of the Atlantic, e.g. a second convergence of the Atlantic and African plates had thrust up mountains in Virginia and Pennsylvania (Dewey 1969*b*). Subsequent to the conference, Dewey and Bird extended this view to other convergent plate margins including the North American Cordillera, the Andes and the Himalayas in a broad-brush paper that was to be quite influential. They believed, contrary to some at the time, that 'plate tectonics is too powerful and viable a mechanism in explaining modern mountain belts to be disregarded in favour of *ad hoc*, actualistic models for ancient mountain belts', and that understanding of all mountain belts could come only from the new global tectonics (Dewey & Bird 1970, p. 2626). Their presentation included many sketches of cross-sections of crust presenting in simplified form their ideas.

Warren Hamilton was one of the very few North American geologists in the early 1960s to advocate large-scale crustal mobility as a solution to regional geological problems. In 1961, for example, he proposed as a 'speculation' and a 'radical explanation' for geological correlations that Baja California had once been part of that mainland but had been both shifted 100 miles to the west and transported northward some 250 miles along the San Andreas Fault to its present location (Hamilton 1961, p. 1307). By late 1967 Hamilton was 'aware of this great surge in plate tectonics, but didn't comprehend it . . .'.[9] In the fall of 1968, he visited the Scripps Oceanographic Institution where he encountered a group of graduate students including Tanya Atwater and Jean Francheteau who were 'totally up to speed on plate geometry'. In his words he was 'led by the hand' by them through plate tectonics. The new global tectonics 'meshed beautifully with my . . . background in

[5] As is common with novel, over-arching, conceptual frameworks, plate theory was assimilated at different rates and to different depths in different specialties and in different regions. At a functional level it could be said that different groups of geoscientists were operating with different versions of plate theory depending on their backgrounds and the problems that they were trying to solve (Glen, unpublished data; Le Grand 1988, pp. 75, 80–99, 163–164).

[6] For the roles of technical and social interests in scientific controversies see, *inter alia*, Bourdieu (1975), Latour (1987), Le Grand (1988, pp. 80–99), McAllister (1992).

[7] Interview with D. L. Jones taped by H. E. Le Grand on 18 January 1990, Berkeley.

[8] One rule for the Penrose Conference is that proceedings are not published as such nor are formal minutes kept; the emphasis is upon frank and free-ranging discussion initiated by a few speculative, provocative presentations. Dickinson (1970*a*) did, however, publish a report and overview of the Conference and has kindly supplied considerable additional information.

[9] Interview with W. B. Hamilton taped by H. E. Le Grand on 22 January 1990, USGS, Denver.

descriptive global geology. All of sudden here was a framework for it'. He set out to write a synthetic paper on the geology of California. 'Mesozoic California and the underflow of Pacific mantle' (Hamilton 1969) appeared the same month as the Asilomar Conference. In it he proposed that much of California was made up of island arcs, oceanic crust, abyssal hills and other sea-floor materials that had been scraped off more than 2000 kilometres of Pacific floor that had been subducted beneath the North American plate. Dewey recalls Hamilton saying at Asilomar, 'My God, we must be able to explain things like the Franciscan and the Coast Ranges, all those things, in terms of plate tectonics'.[10] Hamilton's interpretation was certainly a radical one at the time. As he later remarked, 'it was totally contrary to the way practically every Californian geologist looked at it . . . and there was quite a bit of resentment among the natives'.[11]

The most notable event at Asilomar was the presentation by Tanya Atwater. She proposed an elegant solution of the San Andreas Fault in terms of plate kinematics guided by sea-floor magnetics as an age control. She drew together both continental geology and oceanography in a quantitative way and provided refinements in the geometrical kinematics of plate motions. She thought that plate movement could be related to many of the features of continental geology (Atwater 1970, p. 3513) and provided models which were designed to provide 'testable predictions for the distribution of igneous rocks' and also the timing and amount of deformation. Although she treated only schematically configured, not geologically specific, crustal units, her approach made an immediate and profound impression on many of the participants.[12] Davy Jones, who was to be an architect of the terrane programme, describes his reaction when he learned of her work as follows: 'That was the first application of plate tectonics to a real setting and she was able to show people who had been fussing with the San Andreas Fault all of their lives that they were completely missing the story. It was a marvellous paper and that's what convinced us that plate tectonics was the way to go'.[13]

For a few years after Asilomar, there seems to have been an almost euphoric belief, or at least an incautious optimism, that problems of continental geology would quickly yield to the new global tectonics. The initial successes of Atwater, Dewey, Bird, Hamilton, Dickinson (1970*b*) and others seemed to show the way forward. In the first few years of the 1970s, there was a revolutionary fervour: many geologists rushed into print with redescriptions of their patches of ground with reference to so-called plate tectonics corollaries; those who forbore such descriptions were regarded as troglodytes. More than one Earth scientist refers to that era as being cluttered with premature, simplistic, cursory or naive interpretations. Plate tectonics was not to be confused with continental tectonics. The marine magnetic record that had proved so critical for Atwater represented only a small fraction of the geological timescale. Dewey and Bird's sketches were only that, and drawn to a very large scale. Hamilton's syntheses, though highly suggestive, were grand generalizations. Indeed, Hamilton (1995, p. 3) himself recently commented that the complex nature of plate interactions and their boundaries 'invalidates many of the tectonic and magmatic models which clutter the literature' and even now 'few of the geologists and petrologists who work with the structures . . . produced by convergent-plate interactions, and few of the geophysicists who model subduction, have familiarized themselves with the characteristics of actual plate systems'. How helpful would this global theory be in explaining this outcrop or that group of hills? For a field geologist to apply plate tectonics to his 'patch' is not unlike trying to explain the flight of a cricket ball using general relativity theory. One has to make certain simplifying assumptions. For example, may one properly treat the plates as absolutely rigid, knowing full well that the continents, which presumably record previous plate movements, also record considerable deformation?

---

[10] Interview with J. F. Dewey taped by H. E. Le Grand on 21 December 1988 at Department of Geology, Oxford University.

[11] Interview with W. B. Hamilton taped by H. E. Le Grand on 22 January 1990, USGS, Denver.

[12] Interview with B. C. Burchfiel taped by H. E. Le Grand on 26 April 1990 at Earth and Planetary Sciences, Massachusetts Institute of Technology, Cambridge; interview with P. J. Coney taped by H. E. Le Grand on 15 February 1990 at Department of Geosciences, University of Arizona, Tucson; interview with G. A. Davis conducted at the University of Southern California, Department of Geological Sciences by telephone by H. E. Le Grand on 14 May 1990; interview with J. F. Dewey taped by H. E. Le Grand on 21 December 1988 at Department of Geology, Oxford University; interview with W. B. Hamilton taped by H. E. Le Grand on 22 January 1990, USGS, Denver.

[13] Interview with D. L. Jones taped by H. E. Le Grand on 15 May 1990, Berkeley.

Dan McKenzie, one of the pioneer plate theorists, himself sought to 'modify plate tectonics to describe continental, as well as oceanic, tectonics'. But, as he has recently remarked (McKenzie, pers. comm., 2000), '[T]his problem'is much harder than plate tectonics. . . . Plate tectonics was clearly defined as a kinematic theory: one that is concerned with geometry. It is not a dynamic theory. . . . I myself do not describe continental tectonics as plate tectonics, because continental deformation occurs in wide zones where the idea of rigidity is of limited use'. Peter Molnar, a noted tectonician, takes a similar view. For him, the major importance of plate tectonics for most geologists in the 1970s was that it convinced them that continental drift had occurred. However, though it was a useful framework on a global scale, in terms of unravelling problems of continental tectonics, 'plate tectonics is a poor approximation for the tectonics of many continental regions' (Molnar 1988, pp. 131–133). Plate tectonics may account well for the behaviour of oceanic lithosphere but continental lithosphere differs in buoyancy, thickness and rheology. In particular Molnar holds (p. 133) that 'The broad, diffuse deformation of the western United States . . . is much more complex than the rigid-body displacements of a small number of large plates, and finding a simple and accurate way to represent the deformation of continents remains a major task'. Molnar bluntly concludes (p. 137), 'The tectonics of continents has found plate tectonics an inadequate paradigm'. He muses more recently (P. Molnar, pers. comm., 1999) that it is 'no wonder field geologists had not discovered plate tectonics, for diffuse deformation and widespread strain makes recognising rigid plates within continents difficult'.

Dewey's work in the mid-1970s similarly brought home to him the complexities of applying plate tectonics to smaller-scale geological problems at plate boundaries:

> What I found out from this work was that not only is plate tectonics too simple to understand the geology of rock masses at plate margins, in fact it's bound to lead to such bloody enormous complexity that we may never work it out. . . . Plate tectonics is a simple concept but the kinematics tends to some immense complications. You can build wonderfully complicated models as I did in that paper . . . but taking the results [of fieldwork] and working backwards to a model, a unique model, whew! Very hard! The value of models is not that they give us solutions but give us an idea of how to proceed.[14]

In this respect, Tanya Atwater's model remains a 'unique masterpiece'. In a similar vein McKenzie (pers. comm., 2000) comments that 'Geologists such as John Dewey and Jack Bird recognized that the geological continental record contains structures and stratigraphy produced by plate boundaries, and have sketched plate geometries that could generate the features concerned. But, they are unable to show that the motions involved were those of rigid plates, and in many cases I suspect, but cannot yet prove, that they were not'.

It is in this context that controversy over what Earth scientists call variously accreted, suspect, exotic, allochthonous or tectonostratigraphic terranes erupted in the 1970s and continues today. Geologists – not rocks or other forms of evidence – open, sustain and close geological controversies. Geologists make extensive use of field observations and other data, preferred techniques, and information presented in journals, books, reports, maps and so forth. However, neither the rocks nor other facts speak for themselves: it is the geologist who makes 'the mute stones speak' for one or another side in a controversy. It is only after a controversy is settled that the 'facts of the matter' are agreed. What is at the centre of arguments over the terrane concept is not a clash between rival theories but rather preferred means of extending the global theory of plate tectonics to address problems of local geology. There is a perhaps inevitable tension in this respect between divergent and convergent thinkers. Convergent thinkers, that is, Earth scientists aiming at a global or regional synthesis, especially an explanatory one, often must make various potentially treacherous simplifying assumptions, generalize from myriad particulars of varying quality and enter less familiar subjects that strain the synthesizer's understanding. Divergent thinkers, that is, Earth scientists who are intimately familiar with a wide range of the particulars of a patch of ground, and indeed may have spent much labour collecting them, may well mount heated objections to the effect that various details have been ignored, distorted, misunderstood or misinterpreted. Such, for example, was part of the negative reaction of

[14] Interview with J. F. Dewey taped by H. E. Le Grand on 21 December 1988 at Department of Geology, Oxford University.

specialists to Wegener's promulgation of continental drift.[15]

The terrane concept[16] and research programme were initially developed in the 1970s to account for several puzzling features of western North America. These included the presence of an apparently 'Asian' assemblage of fossils in the Cache Creek group in Canada, the highly complex Franciscan Formation in the San Francisco Bay area, and the structure of the Klamath Mountains in California. Most of the loose-knit group who were involved in the early days of the terrane programme had for some years prior to the advent of plate tectonics conducted fieldwork in one or more of these locations and were associated with either the United States or Canadian Surveys.

In 1950 geologists found some unusual marine microfossils while mapping in the Canadian Cordillera near Cache Creek in British Columbia. These assemblages of fusilinids dating from the Permian period (290–200 Ma) were very different from those typical of nearby areas and of the southern and southwestern interior of North America. Instead, they seemed to be identical with those common in rocks in Asia. The presence of this 'Asian' fauna in Canada was explained in terms of a Tethyan 'seaway' connecting Asia with North America (Thompson *et al.* 1950). Wilbert R. Danner (1965, p. 120) commented that the juxtaposition of this Permian so-called 'Tethyan fauna' with the distinctive North American Permian fauna had from that time been 'commonly believed' to be due to the deposition of Tethyan (i.e. Asian) fauna 'in a Permian Tethyan seaway extending from the Mediterranean region to New Zealand and Japan and apparently extending across the Pacific Ocean to the Yukon, British Columbia, Washington and Oregon'. In other words, if one assumed that the continents did not move laterally but might be uplifted or flooded, apparent palaeontological anomalies could be explained by adducing 'sea bridges' analogous to the 'land bridges' used by others to explain similarities between groups of land animals on continents now separated by oceans. Danner himself, however, had reservations about this (Danner 1965 p. 120): 'The difference between the Tethyan faunas and other Permian faunas of North America, however, may be more that of an environmental facies than that of an isolated seaway'. He reaffirmed this in a 1966 paper that had a wider circulation (Johnson & Danner 1966). Danner's own scepticism concerning the existence of a Tethyan seaway seems to have made little impact, though the identification by him and others of an 'Asian fauna' in northwestern North America soon generated an abortive attempt at explanation in terms of plate tectonics.

Tuzo Wilson (1966) had met quick success with his proposal that many of the major features of eastern North America were due to collisional tectonics arising from the opening, closing, and reopening of the Atlantic Ocean. But, as West Coast geologists are fond of saying, the Pacific is different. Wilson sought to apply his model to the Pacific. He speculated by analogy with the finding of 'European' deposits in eastern North America that the presence of Asian fusilinids in northwestern North America meant that Asia had collided with North America through plate action and he suggested that Alaska was a part of ancient Asia left behind when the Pacific had reopened (Wilson 1968). As one geologist (no attribution by request) active at the time puts it now: 'Wilson made some great calls but this was not one of them!' Geologists familiar with the details of the geology agreed that the Asian fusilinids presented problems but there were several lines of evidence to suggest that the West Coast had faced an ocean for at least 600 Ma: there was no analogy with the Atlantic. Danner himself (1970) was highly critical of Wilson's suggestion of large-scale crustal mobility and he preferred an explanation in terms of facies changes due to environmental differences that was consonant with crustal fixity. Wilson's suggestion was ignored or dismissed as an interesting but ill-founded speculation, no doubt confirming for the moment the fears of many analysers and fieldworkers who were concerned about the use or, worse, misuse of their hard-won data by plate enthusiasts.

James W. H. Monger made more effective use of the new global tectonics in trying to account for the Permian 'Tethyan fauna' near Cache Creek described by Danner and others. Danner supervised Monger's thesis at the University of British Columbia on the stratigraphy and structure of a complicated package of rocks in the Cascade Mountains. In 1965 he joined the Canadian Geological Survey and there met John Wheeler, who had worked with Danner on the

[15] See, for example: Frankel (1976), Le Grand (1988, pp. 55–99).

[16] Terranists prefer the term 'concept' to 'theory' or 'hypothesis', in part because they regard it as subsumable within plate tectonics and in part because they consider it to be an empirical generalization.

Cache Creek, and Hubert Gabrielse. They were engaged in applying the geosyncline concept to explain the existence in the Cordillera of what appeared to be a number of parallel, tectonic 'belts' of rocks which 'were very strange' in that they differed considerably from one another in their composition and fossil content. Monger's first field assignment was to 'go and look at a group called the Cache Creek Group which runs right down the middle of British Columbia'.[17] For three years Monger worked on the stratigraphy. To assist him with the fossils, especially the fusilinids, he enlisted Charles Ross, who was already establishing a considerable reputation as a palaeontologist with special expertise in fusilinids and other foraminfera. For well-studied index fossils it is possible for geologists to consult standard monographs in order to classify and date their own specimens. Nonetheless, recourse will often be had to a recognized expert even for 'routine' fossils, because there is often much tacit knowledge and much else that does not find its way into the literature.

Monger & Ross (1971) concluded that the western part of the Cordillera could be divided into three separate, parallel belts. The eastern and western belts were very similar. They contained fusilinids mostly of the family Schwagerenidae along with other marine fossils forming the kind of non-Tethyan Permian assemblage commonly found elsewhere in North America. However, between these two belts was a very different one. It was a typically 'Asian' or Tethyan assemblage, consisting of fusulinids mostly of the family Verbeekinidae together with remnants of crinoids and algae (Monger & Ross 1971, p. 261). Ross's expertise led them to conclude that although there were marked differences in the groups of fusilinids, they were contemporaneous, at least in part. Besides differences in the fossils, the rocks themselves were so different as to suggest differences in the environment of their formation, e.g. the central belt contained extensive, thick deposits of nearly pure limestone whereas the other two contained only scattered, thin deposits mixed with other material (Monger & Ross 1971, p. 270). The central belt also contained an abundance of ribbon cherts, characteristic of a clear, deep-water marine environment.

How could one make sense of this puzzle? Monger recalls that the geosyncline concept was of little use: 'We had the "eugeosyncline" which really means nothing but some marine volcanics and sedimentary rocks and this was the model: you had the eastern Mallard Belt, the western Fraser Belt, the eugeosyncline and the miogeosyncline, which didn't work for anybody. There was no other model and it didn't tell you anything other than you called this a eugeosyncline and that a miogeosyncline. It was more labelling than explanation'.[18] Monger & Ross (1971) put forward two alternatives. The first was that the difference in fusilinid assemblages was due mostly to local environmental differences, but of course this did not address the underlying issue of how the 'Asian' fusilinids had got to Canada. The second alternative, and the one that they favoured, though cautiously, involved large crustal mobility: 'the possibility exists that faunas living in widely separated regions [with different environments] subsequently may have been transported bodily for considerable distances and brought into contact with one another' (Monger & Ross 1971, p. 273). They put forward with some diffidence several plate tectonics-based models of how this might occur. Monger stresses that at the time 'The physical basis of plate tectonics really didn't concern us at all. We just tended to accept it and, said, "look, we have these different things side by side and here's a way things could come together"'.[19] Their preferred model was one in which the eastern and western belts had once formed a single, coherent island arc which had been emplaced first, followed by a piece of oceanic crust. Then, transcurrent faulting had slid part of the inner, island arc belt to the 'outside' of the oceanic belt thus enclosing it (Monger & Ross 1971, p. 276). Their preferred model had one significant implication that they did not immediately pursue. Suppose this part of the Cordillera had indeed not formed in place, but had been added to the North American continent from parts unknown. What would this mean for the five hundred or so miles of Canada that lay to the west of these belts?

The Francisan Formation constituted a second problem. It literally surrounds Berkeley and is on the doorstep of Stanford University and the Menlo Park branch of the US Geological Survey. P. B. King, in his influential

[17] Interview with J. W. H. Monger conducted by H. E. Le Grand on 12 February 1990 at Canadian Geological Survey, Vancouver.
[18] Interview with J. W. H. Monger conducted by H. E. Le Grand on 12 February 1990 at Canadian Geological Survey, Vancouver.
[19] Interview with J. W. H. Monger conducted by H. E. Le Grand on 12 February 1990 at Canadian Geological Survey, Vancouver.

overview of 1959, described it as 'an odd-looking, much deformed, thoroughly indurated series of greywacke, shales, bedded cherts, limestone lenses, and interbedded basaltic lavas cut by many ultramafic serpentine intrusions. . . . This assemblage is typically eugeosynclinal. A curious feature of the Franciscan is that its base is nowhere visible'. Edgar (Ed.) Bailey of the USGS and head of the 'Franciscan Friars', a group of USGS geologists that included among others Clark Blake, Porter Irwin and D. L. (Davy) Jones, had made a systematic effort over the years to unravel the complexities of the Franciscan but to little avail. Their summary in 1964 of all that was then known about the Franciscan acknowledged that 'both major and minor structures in nearly all areas of the Franciscan are inadequately understood, in spite of the mapping that has been done by many competent geologists over a period of more than half a century' (Bailey *et al.* 1964, p. 148). In the new paradigm of plate tectonics, the Franciscan was interpreted as a subduction complex, but little detailed guidance could be gleaned from the new global theory in terms of the origins of the pieces, their interactions over time, or their relationships.

For Porter Irwin, the Klamath Mountains, extending from northwestern California into southwestern Oregon, were his back garden. He began work on them in 1953 and from 1957 was USGS project chief for their systematic mapping. For nearly three decades he spent his field seasons there, walking over the rugged landscape and mapping its geology. In the 1950s he divided the Klamaths into four belts: Eastern Jurassic, Western Palaeozoic and Triassic, Central Metamorphic, and Eastern Klamath. Each had distinctive rocks and fossils and, as one moved to the west, the belts appeared to be progressively younger. Irwin recollects that it seemed clear the belts 'hadn't grown together there, they were pieces that had formed somewhere else and been brought together, I didn't know whether they were formed ten miles or a thousand miles away'.[20] To try to express this sense of three-dimensional juxtaposed blocks of crust he decided: 'Well, rather than calling them "belts" I'd call them "terranes" '. Irwin's (1972) definition was a descriptive one with no reference to fault-boundedness: 'The term 'terrane' as used herein refers to an association of geologic features, such as stratigraphic formations, intrusive rocks, mineral deposits, and tectonic history, some or all of which lend a distinguishing character to a particular tract of rocks and which differ from those of an adjacent terrane'.

The original terrane group, dubbed the 'Menlo Park Mafia' by some, began to coalesce around the USGS in Menlo Park, California. Davy Jones assumed the role of Godfather: the foremost, and most ardent, spokesperson for the terrane concept. A senior palaeontologist at Menlo Park, in the early years of the terrane programme he provided not only much of the drive but also considerable cognitive and social resources. Alaskan geology was particularly fertile ground for demonstrating the potential of what was emerging as the terrane programme. Put simply, Alaska was a mess: a seemingly senseless jumble of rocks once characterized by Hamilton as the 'garbage dump of the Pacific'. Jones had from the late 1950s spent many field seasons in Alaska. One of the major difficulties he and others faced in trying to unravel this jumble was establishing the stratigraphic correlation or lack thereof of adjacent packages of rocks. But, many of the packages contained igneous and heavily metamorphosed rocks, for which the techniques of palaeontologists were worthless, and radiolarian cherts that, though sedimentary and common in the Cordillera, could prior to 1972 be dated only coarsely. The importance of such palaeontological controls were forcefully brought home to Jones in 1972 when he claimed – largely on the basis of lithology and indirect fossil evidence – that a large Palaeozoic section of SE Alaska was a terrane which had been moved from northern California. It was rejected by *Science* but appeared as a USGS publication (Jones *et al.* 1972) that, according to Jones, was still in the mid-1970s 'being laughed at'.[21] Jones was not, however, deterred by this unfavourable reception from pursuing the use of terranes as a means of mapping and investigating relationships among packages of rocks in the Cordillera.

The catalyst for formal co-operation in understanding the Cordillera in terms of the new tectonics seems to have been a discussion among Monger, Greg Davis, Jones and others at the Annual Meeting of the Geological Society of America in Washington, DC, over 1–3 November 1971. On 1 November, a symposium session, 'Plate tectonics in geologic history', was convened by Jack Bird, Clark Burchfiel and Gary Ernst. Papers included one by Dickinson on 'Evidence for plate tectonic regimes in the past',

[20] Interview with P. Irwin taped by H. E. Le Grand on 19 January 1989, Menlo Park.
[21] Interview with D. L. Jones taped by H. E. Le Grand on 15 May 1990, Berkeley.

Hamilton on Indonesia, Monger and others on 'Plate tectonic evolution of the Canadian Cordillera', and Peter Coney on 'Cordilleran tectonic transitions and North American plate motion' and Dewey on 'Plate models for the evolution of the Alpine Fold Belt'. On the following day, Burchfiel gave his and Davis's paper 'Nature of Paleozoic and Mesozoic thrust faulting in the Great Basin area of Nevada, Utah and southeastern California' (previously selected as one of the two 'outstanding papers' given earlier at that year's Cordilleran Section meeting). During the meeting, Monger, Davis and others also talked about the Franciscan. Jones offered Menlo Park as a meeting place to discuss unravelling the whole North American Cordillera. When this meeting on 'Cordilleran Tectonics' was held, Monger recollects: 'twenty or so of us got in a room for a weekend and just talked about whether or not correlations could be drawn between parts of the Cordillera in terms of tectonic entities and each person was given an area to describe ... the USGS had Alaska, the CGS had British Columbia ... and everybody was feeding off everybody else'.[22] The result over the next few years was a large number of papers, including reviews and overviews. Jones himself had a more particular agenda.

Within six years not only was there a dating scale based on Mesozoic radiolarian cherts, but Jones had established the Rad[iolarian] Lab at Menlo Park and was in the process of building a dating scale for Palaeozoic cherts.[23] As Jones later put it, the ability to date and correlate by age radiolarian cherts, 'just blew the whole Cordillera apart'.[24] For Jones, plate tectonics provided the mechanism for chopping up, combining and moving around pieces of crust through which he could make sense of Alaska. The pieces themselves could be characterized in terms of Irwin's definition of 'terranes'. Moreover, Jones succeeded in gaining very substantial funding from George Gryc, an old Alaska hand and collaborator, who in 1976 was appointed head of the newly established Office for National Petroleum Reserves in Alaska (ONPRA). This not only aided the rapid development of the Rad Lab, but also enabled Jones to take teams from Menlo Park to Alaska to apply the terrane concept first-hand. Peter Coney, who was a frequent visitor to Menlo Park and coined the term 'suspect terrane' (Coney *et al.* 1980), received funding from the USGS to support summer research in Alaska while on leave from his university position. He recalls: 'one year when we were working in Alaska Davy put together a team – two palaeomagicians, geochemists, biostratigraphers, a structural geologist, a sandstone petrologist – all working together and the interdisciplinary character of it was very exciting because you realized there was no way you were going to solve it by yourself'.[25]

Jones's core set included Irwin and Clark Blake and others at Menlo Park, Monger at the Canadian Survey, Peter Coney, and several scientists at Stanford. The most notable of the latter was Alan Cox, who as a former member of the USGS at Menlo had been a central figure in the development of the reversal magnetostratigraphy so important to the final acceptance of sea-floor spreading (Glen 1982) and was then dean of science at Stanford. By the mid-1970s they aimed to develop the terrane concept further and to apply it to remap western North America in terms of terranes. Their approach was underpinned by the conviction that most of western North America is made up of a chaotic assortment of rock packages that have come from elsewhere, sometimes from thousands of miles to the south and west, and then have been plastered onto the older North American craton. To underpin this approach, they used not only published and unpublished information on stratigraphy, palaeontology, palaeomagnetism, structural geography, petrology and geochemistry gathered by others, but also conducted their own fieldwork.

The value of such collaborative endeavours is well illustrated by the first major success of the terrane programme: the identification in 1977 (by Jones, Silberling & Hillhouse) of the *Wrangellia* terrane, that earlier Monger & Ross (1971) had described as a fragment of oceanic crust, not a continental fragment. Jones and his collaborators (Jones *et al.* 1977, p. 2565) now

[23] For a study of Jones's construction and use of the Rad Lab as a 'choke-hold' and 'stronghold', see: Le Grand & Glen (1993).

[24] Interview twith D. L. Jones taped by H. E. Le Grand on 18 January 1990, Berkeley.

[25] Interview with P. J. Coney taped by H. E. Le Grand on 15 February 1990 at Department of Geosciences, University of Arizona, Tucson. The range and sophistication of methods has grown steadily over the past two decades. For example, a recent reassessment of terranes in Scotland and northern England (Stone *et al.* 1999) is based upon the analysis and computer mapping of a large database of systematic geochemical analyses of stream sediments, allowing the different terranes of Scotland to be mapped and revealed according to their different subsidiary chemical contents.

proposed that this large crustal unit dubbed *Wrangellia*, was a large 'coherent' terrane extending along the Pacific margin of North America from the Wrangell Mountains in Alaska to Vancouver Island and that it had come from far to the south of its present location. Their claim was reinforced by the combination of data from the three specialties of the authors. Jones was a palaeontologist and able to draw upon data from the Rad Lab. Silberling was a stratigrapher and also a palaeontologist. Hillhouse was a palaeomagnetist. The combination of palaeomagnetic with more traditional data was of key importance. One might argue endlessly about whether or not fossils indicate distant origin or compressed facies change, but the consilience of tropical fossils with a tropical palaeolatitude proved to be very persuasive. The 'hard' empirical evidence was telling: even geologists who were in general opposed to the claims of the terranists accepted that *Wrangellia* had come from afar.

In 1978, the year following the identification of *Wrangellia*, the Pacific section of the Society of Economic Palaeontologists and Mineralogists organized a symposium at Sacramento, California, on the Mesozoic palaeogeography of the western United States (Howell & McDougall 1978). It provided an opportunity for the terranists to present their case. David Howell, who began at Menlo Park in 1974, soon joined the Mafia as a result of his work as an editor. His aim was to produce a palaeogeographic volume 'based upon a palinspastic reconstruction we all agreed upon' but in California, as he relates, 'everything fell apart, because all of the Mesozoic bodies of rock were being hotly contested and you couldn't relate one to the other in any agreeable fashion'.[26] This gave him contact with both the terrane concept and key people, including Jones, in the Mafia. Beginning in the mid-1980s Howell codified the terrane concept through his textbooks (1989, 1995) and spearheaded the extension of the programme worldwide through joint projects and conferences (Howell 1985; Howell *et al.* 1988).

The tone was set by the editors (Howell & McDougall 1978, p. viii) who stated that the border of the western United States was a passive margin in the Palaeozoic (and subject to tensional strain). In the Mesozoic an episode of mountain-building occurred as the result of plate convergence (compression). They claim this marked 'the beginning of a protracted series of continental rifting, accretion, and island arc construction that persisted through the Mesozoic'. They noted the difficulty of making a palaeogeographic reconstruction for the Cordilleran region because of the allochthonous nature of many terranes. They also stated: 'In most instances the allochthonous aspect of specific terranes is documented, but conclusions regarding the original setting remain unknown or equivocal'. The symposium was also significant for the change of rhetoric vis-a-vis allochthonous terranes. There is little evidence of the tentative nature of earlier papers; it seemed, at least for most of the people involved in the symposium, that there was no argument about whether there were such things as allochthonous terranes. They were an accepted geologic fact. By 1987 the programme launched at the 'secret meeting' at Menlo Park, to remap the Cordillera in terms of terranes, had been completed in the form of four maps covering the Cordillera from Alaska to Mexico (Silberling & Jones 1987). Perhaps equally significant was the official adoption by the USGS of rules for the nomenclature of terranes.

From 1983 terranists, including many new recruits, had begun to extend this approach to other regions of North America, e.g. the Appalachians and even to the North American craton itself. Indeed, for North American terranists virtually all of the vast lands west of longitude 111° from the imposing Brooks Range in Alaska south down the Cordillera through California and then almost all of Mexico are 'suspect' as to their birthplace and history. Conferences in other countries further promoted the terrane approach (e.g. Hashimoto & Uyeda 1983; Howell *et al.* 1984; Howell & Wiley 1988). Collaborative projects were undertaken to construct world-wide maps of terranes (e.g. Howell 1985), and the terrane programme begun by the Menlo Park Mafia is actively pursued in other regions around the Pacific Rim, including Japan, China, Australia, South America and New Zealand as well as in Europe. The aims of this ambitious undertaking are four-fold according to the 'manifesto', as the leading terranists describe it (Jones *et al.* 1983 p. 32): '(1) to identify, characterize, and portray on terrane maps all major allochthonous terranes; (2) to relate their faunal and floral characteristics through time to major palaeobiogeographic provinces; (3) to establish palaeolatitudes through time; (4) and finally, in the case of the Cordillera and Pacific region, to attempt palaeogeographic reconstructions of the palaeo-Pacific Ocean (Panthalassa) and surrounding

[26] Interview with D. G. Howell taped by H. E. Le Grand on 19 January 1989, Palo Alto.

cratonal regions'. More informally, such activities had proselytizing as a fifth aim: '[O]ur purpose was to go to a variety of places and in part use the conference as a vehicle to explain to the indigenous community where we were with the whole [terrane] concept and in part at gaining familiarity with new areas to which the terrane approach could be applied'.[27]

As was the case in North America, there were arguments over the novelty and utility of the terrane concept as well as skirmishes between the more enthusiastic converts to terranes and both plate theorists and committed field geologists. For terranists, this expansion in domain, if successful, would generate additional cognitive and social power. Clearly, the terrane programme and its supporters would become more significant if the programme could be shown to apply to regions beyond the Cordillera. To date, over four thousand publications, maps and abstracts which invoke or critique terranes have appeared since that term took on the specific connotation of 'fault-bounded geologic entities of regional extent, each characterized by a geologic history that is different from the histories of contiguous terranes' (Jones *et al.* 1983, p. 22).

Almost all Earth scientists today acknowledge that some accretion has occurred. To many, it seems that the terrane concept is a natural outgrowth and extension of plate tectonics theory; indeed, that it is merely an anticipatable corollary of plate tectonics. Yet, from the outset, considerable controversy has surrounded this approach. The controversy does not involve a clash between fundamentally incompatible theories; rather, more typical of much everyday science, it turns on and throws into sharp relief differing judgements of, and disagreements over, matters of approach, style, preferred methods and techniques, and aims within an shared, overarching theoretical framework. Conflict over facts and their interpretation is conjoined with the conflicting specialist and social interests of the geologists themselves.[28] One might explicate the terrane controversy in many ways. As noted earlier, I emphasize the tension between analysers or divergent thinkers who seek increasingly fine-grained information about, and an explanatory account of, a piece of the crust, and synthesizers or convergent thinkers who seek large-scale patterns and overarching conceptual schemes. For the former, maps may often be ends in themselves; for the latter, means to ends. This tension is certainly evident in the relationships between terranists and plate theorists and, perhaps ironically, between terranists and those specializing in the detailed study of smaller crustal units.

Terranists have been criticized by some plate theorists aiming at large-scale regional or global syntheses. Many proponents of terranes believe, not altogether without reason, that their work highlights the inadequacies of plate tectonics as applied to smaller-scale regional and local problems (e.g. Howell 1989, p. 51) and serves as a useful, empirically based corrective. Jones and others refer disparagingly to 'naive plate tectonics models' and 'Deweygrams'.[29] Rather than providing hasty generalizations, terranists argue, a more prudent course of action is to identify and characterize all the pieces of the puzzle, the terranes, before trying to interpret their emplacement and subsequent history in terms of plate tectonics.

This is reflected in the very terminology of terranes, which by USGS convention are named simply after some geographical feature associated with the terrane, thus avoiding any implied theoretical context or interpretation and any suggestion of genetic relationships with other terranes. Hamilton, like some other convergent thinkers, takes strong exception to this practice. He argues first that once a rock package is labelled as 'suspect', others in the field may well assume this without further analysis. Second, he contends that geographic names rather than

---

[27] Interview with D. G. Howell taped by H. E. Le Grand on 19 January 1989, Palo Alto. This is a twist on the usual 'centre–periphery' relationship, in which data are normally gathered in the 'provinces' and sent to the 'centre' for processing, e.g. Jones's Rad Lab. Howell, Jones, and others at the 'centre' travelled to the 'periphery' to gain support and acquire data.

[28] Glen (1994, pp. 39–91) shows the influence of disciplinary specialization on theory choice and the selection and application of differing standards of appraisal in a continuing scientific debate.

[29] The term refers to idealized representations of plate interactions, together with cross-sections of crust such as those pictured in Dewey & Bird (1970). David Howell (1989), a member of the 'Menlo Park Mafia' and systematizer of the terrane programme, coined the phrase. He recalls (1989b, Tape 1, Side 1): 'I, in a public meeting, once referred to those as "Deweygrams" and I didn't mean to be insulting; I meant that they were very lucid and simplistic renderings of a perception of how, for instance, . . . an island arc accretes. But when you then go to an area you don't find just a big hunk of island arc with a knife-sharp boundary which is the suture zone and then the adjoining terrane or whatever it was. It clearly was much more complicated than that' interview with D. G. Howell taped by H. E. Le Grand on 19 January 1989, Palo Alto.

descriptive ones or ones which suggest genetic relationships in effect 'fence out anyone but the local aficionados because the outsider can keep track of lithologic packages or genetic names such as ophiolite or island arc or back arc basin ... whereas if you hit him with thirty different geographic names in a long article he simply can't'.[30] Terranists understandably argue from their perspective that such a nominalist approach avoids premature interpretation and synthesizing.

Those critical of the terrane approach such as A. M. C. Sengor (1990*a*,*b*) deny significant value or originality to the terrane concept and relegate what some of them term 'terrane theory' – although few, if any, terranists themselves have claimed theoretical status for their conceptual structure – to the waste-bin of rehashed, conventional, descriptive geological methods and techniques.[31] These critics often regard it as just a corollary or conceptual adjunct of plate tectonics and decry what they perceive to be a 'terrane bandwagon' while denying to terranes any novelty or heuristic power. Critics of the terrane programme argue that terranists themselves have simplistic and outmoded understandings of plate theory; that is, they are not conversant with contemporary plate theory as opposed to the cartoons of the 1970s. This view is forcefully expressed by Hamilton:

> [T]he California group in general have a two-dimensional view of plate tectonics. They think of motion as either more or less perpendicular subduction or more or less strike-slip. ... [S]o most of the Californians are still thinking of California as being assembled strictly by Andean-type, steady-state subduction, ... of subduction as a one-sided process ... picturing subduction as the rolling over a [fixed] hinge of a subducting plate, whereas in fact the hinge always migrates back into the subducting plate with extension occurring in the over-riding plate. They just haven't learned how modern systems work, so I think much of their work is going to have to be done again.[32]

There is of course, exaggeration on both sides of the fact–theory divide. Were one to take literally the first of the 'four steps' of the terranist manifesto, i.e. 'to identify, characterize, and portray on terrane maps all major allochthonous terranes', that itself appears to be a never-ending task which would mean that the next three steps would never be undertaken. First, as a methodological point, one would have to describe in some considerable detail without any preconceptions every conceivable crustal unit in non-theory-laden language. This is in accord with the formal presumption that every crustal unit is 'suspect' until proven 'innocent'; that is, that the crustal unit should be presumed to be allochthonous until proved otherwise and, moreover, that it is legitimate to infer fault boundaries even if they have not been directly observed. Moreover, there are social and cognitive mechanisms for generating more and more and smaller and smaller terranes. Each time one remaps an area, each time one uses new techniques, one tends to produce a finer and finer grained study and therefore more and more detail and to find new, previously unmapped features. In a very real sense there is an unspoken mandate to produce a more finely divided taxonomy than previous workers, thereby implying a more profound understanding. To remap an area and find nothing new could well be construed as a failure. Thus, every time an area is mapped, finer and finer subdivisions and greater and greater specificity of components and structures are delineated. Only partly tongue in cheek, one might predict that the number of terranes is inversely proportional to the distance from major centres of geological study and fieldwork and directly proportional to the number of geologists who have worked in the region.

Terranists have also been criticized for overemphasizing the far-travelled nature of terranes. There is a subtext here, particularly with reference to the early battleground of western North America. Through the late 1970s and into the early 1990s terranists claimed south to north displacements of thousands of kilometres for some terranes, largely on the basis of palaeomagnetic data. This seemed to be discordant with other geological data which suggested much smaller displacements and thus to pose a potential threat to the conventional plate tectonic analysis in terms of transport of crustal blocks northwards along the San Andreas transform system. Were processes at work that perhaps were not accounted for in terms of more

[30] Interview with W. B. Hamilton taped by H. E. Le Grand on 22 January 1990, USGS, Denver.

[31] Peter Coney, among others, explicitly makes the point that the terrane approach 'was never a theory. It never was. It was a method of analysis ... to know what you've got, so you can talk about what's there'. Interview with P. J. Coney taped by H. E. Le Grand on 15 February 1990 at Department of Geosciences, University of Arizona, Tucson.

[32] Interview with W. B. Hamilton taped by H. E. Le Grand on 22 January 1990, USGS, Denver.

orthodox plate interactions? The mammoth, influential *Geology of North America* opted for large displacements (Oldow *et al.* 1989).

Alternatively, was there a flaw in the generally accepted apparent polar wander path for North America that served as the reference frame the palaeomechanisms involved?[33] Robert Butler (Butler *et al.* 1984) argued that data from plutons consistent with a larger-than-expected transport along the San Andreas might be explained by the tilting of those plutons that could lead to erroneous apparent polar wandering paths and thus the calculation of a greater latitudinal motion than had actually occurred. Support for this view came from the work of others. European palaeomagneticists (e.g. Courtillot *et al.* 1994) have pointed out that the apparent motion of the magnetic poles constructed from data taken from the southwestern United States conflict with that based on data taken from the northeastern United States, the former entailing far less transport along the San Andreas than the latter. Data taken from sites in Europe and Africa yield polar positions consonant with that from the southwestern United States. Since the Earth could at any one time have only one set of magnetic poles, they suggest – agreeing with Butler – that the apparent polar wandering paths which formed the basis for postulating larger-than-expected displacements on the San Andreas should be questioned. Recently, a further major study by Dickinson & Butler (1998, p. 1268) has reaffirmed this conclusion and presents a comprehensive argument that 'the hypothesis of major tectonic transport in excess of distances inherent in current models is unnecessary and in conflict with geologic observations'. Instead, they contend that any discrepancies are removed through better data sets and making proper allowance for the tilting of formations after magnetization and the effects of sedimentary compaction. Not all are convinced. Ted Irving (pers. comm., 2001) stands by his original determinations implying very large-scale transport and finds it implausible that compaction and tilting should exactly account for what are otherwise calculated to be significant displacements. Even if Butler's argument be accepted, however, it would seem to strike only at terranists' claims of large and rapid transport of crustal blocks, not at their key point that one should not assume that just because two blocks are now adjacent they were formed in place. Moreover, it might contribute to a rapprochement between terranists who are overtly critical of 'naive' plate tectonics and plate tectonicists who may regard such criticisms as touching upon the adequacy of accepted plate mechanisms.

Terranists have been criticized not only by those pursuing a more convergent approach. On venturing new interpretations of crustal units they have been harshly criticized by some field geologists for trespassing on the pieces of crust to which they have devoted their careers. Howell observes that most geologists work in only one or two areas for good reason: 'It's a lot of work to learn all the stratigraphies, the names, the places and you are never satisfied and there are no artificial boundaries to a given area – you can expand it as much as you can'.[34] Many were expert in, and had in fact mastered, the finely detailed geology of the crustal units and domains that the terranists sought to subsume in their newly defined taxonomy. Such a conceptual reordering, reinterpretation, redefinition and reclassification of what was regarded as well-established and defined geology appeared to turn bodies of knowledge won through strenuous efforts of mapping and study by teams of distinguished geologists into mere grist for the terranists' mill. Howell freely acknowledges this source of criticism from some old Alaskan hands: 'guys who have worked an awful lot, sweated blood, witnessed deaths, been threatened themselves to extract this information . . . take a certain amount of umbrage at some dandy coming in and spending a summer reinterpreting all the rocks. . . . We were bold and at various times may have given the impression of arrogance; frankly, we were just enthusiastic'.[35]

Although the terrane programme's detractors may be correct in their claim that the terrane programme offers nothing theoretically or methodologically new, the terranists' claim of new solutions has mobilized efforts at reassessment of longstanding, intractable problems and led to the generation of new, refined data and their reinterpretation. A recent, particularly elegant example of this is a new approach to terrane identification and analysis using geochemical data (Stone *et al.* 1999). This uses a large data base of chemical analyses of stream sediments in Scotland: the sediments serve as a proxy for the composition of the underlying

---

[33] Hagstrum (1993), though a proponent of the Baja-British Columbia interpretation of large displacements, provides a balanced review of this controversy up to that time.
[34] Interview with D. G. Howell taped by H. E. Le Grand on 19 January 1989, Palo Alto.
[35] Interview with D. G. Howell taped by H. E. Le Grand on 19 January 1989, Palo Alto

bedrock (Stone *et al.* 1999, p. 146) and thus as an indicator of variations in the chemistry among, and within, possible terranes. The relative abundance of selected elements that are reflective of the mineralogy can then be mapped by computer to yield 'pictures' of terranes.

The terrane controversy has also helped draw researchers from the ether of plate theory abstraction down to the empirical, fine-scale reassessment of previous and long-established maps, data, fossils, monographs and interpretations in terms of the standards that define the terrane programme.[36] It may also have provided impetus to its detractors further to refine alternative explanations couched in plate tectonics terms. The terrane programme may be a bandwagon, but as Coney refreshingly put it: 'it was a bandwagon – and in part deliberately orchestrated, but it's been a great ride and a hell of a good way to do regional geology'.[37] John Dewey has occasionally put spokes in the wheels of that bandwagon but he too asserts the fundamental importance of field geology and mapping for, as he (pers. comm.) puts it: 'Rocks, fossils and minerals are immensely complicated systems but they – with geophysical and geochemical data and the ideas that stem from them – are the substance of our science'.

Many of the ideas presented in this paper were developed in lengthy dicussions with W. Glen (but I take responsibility for their form as presented here). I am also grateful to L. Harrington for his comments.

References

ATWATER, T. 1970. Implications of plate tectonics for the Cenozoic tectonic evolution of western North America. *Geological Society of America Bulletin*, **81**, 3513–3536.

BAILEY, E. H., IRWIN, W. P & JONES, D. L. 1964. *Franciscan and related rocks, and their significance in the geology of western California*. California Division of Mines and Geology, Bulletin **183**.

BOURDIEU, P. 1975. The specificity of the scientific field and the social conditions of the progress of reason. *Social Science Information*, **14**, 19–47.

BULLARD, E. C. 1975. The emergence of plate tectonics: a personal view. *Annual Review of Earth and Planetary Science*, **3**, 1–30.

BUTLER, R. F., GEHRELS, G. E., MCCLELLAND, W. C., MAY, S. R. & KLEPACKI, D. 1989. Discordant palaeomagnetic poles from the Canadian Coast Plutonic Complex: regional tilt rather than large-scale displacement? *Geology*, **17**, 691–694.

CONEY, P. J., JONES, D. L. & MONGER, J. W. H. 1980. Cordilleran suspect terranes. *Nature*, **288**, 329–333.

COURTILLOT, V., BESSE, J. & THÉVENIAUT, H. 1994. North American Jurassic apparent polar wander: the answer from other continents? *Physics of the Earth and Planetary Interiors*, **82**, 87–104.

DANNER, W. R. 1965. Limestone of the western cordilleran eugeosyncline of southwestern British Columbia, western Washington and northern Oregon. *In*: *Dr. D. N. Wadia Commemorative Volume*. Mining, Geological & Metallurgical Institute of India, Calcutta, 113–125.

DANNER, W. R. 1970. Paleontologic and stratigraphic evidence for and against sea floor spreading and opening and closing oceans in the Pacific northwest. *Geological Society of America Abstracts with Programs*, **2**, 84–85.

DEWEY, J. F. 1969*a*. Structure and sequence in paratectonic British Caledonides. *In*: KAY, M. (ed.) *North Atlantic – Geology and Continental Drift*. AAPG, Tulsa, Memoir **12**, 309–335.

DEWEY, J. F. 1969*b*. Evolution of the Appalachian/Caledonian orogen. *Nature*, **222**, 124–129.

DEWEY, J. F. & BIRD, J. M. 1970. Mountain belts and the new global tectonics. *Journal of Geophysical Research*, **75**, 2,625–2,647.

DICKINSON, W. R. 1970*a*. Second Penrose conference: the new global tectonics. *Geotimes*, **15**(4), 18–20, 22.

DICKINSON, W. R. 1970*b*. Relations of andesites, granites, and derivative sandstones to arc-trench tectonics. *Reviews of Geophysics and Space Physics*, **8**, 813–860.

DICKINSON, W. R. & BUTLER, R. F. 1998. Coastal and Baja California paleomagnetism reconsidered. *Geological Society of America Bulletin*, **110**, 1268–1280.

FRANKEL, H. 1976. Alfred Wegener and the specialists. *Centaurus*, **20**, 305–324

GLEN, W. 1975. *Continental Drift and Plate Tectonics*. Charles E. Merrill, Columbus.

GLEN, W. 1982. *The Road to Jaramillo: Critical Years of the Revolution in Earth Science*. Stanford University Press, Stanford.

GLEN, W. 1994. *The Mass-Extinction Debates: How Science Works in a Crisis*. Stanford University Press, Stanford.

HAGSTRUM, J. T. 1993. North American apw: the current dilemma. *EOS: Transactions of the American Geophysical Union*, **74**, 65, 68–69.

HALLAM, A. 1973. *A Revolution in the Earth Sciences*. Clarendon Press, Oxford.

HAMILTON, W. B. 1961. Origin of the Gulf of California. *Geological Society of America Bulletin*, **72**, 1307–1318.

HAMILTON, W. B. 1969. Mesozoic California and the underflow of the Pacific mantle. *Geological Society of America Bulletin*, **80**, 2409–2430.

HAMILTON, W. B. 1995. Subduction systems and magmatism. *In*: SMELLIE, J. L. (ed.) *Volcanism Associated with Extension at Consuming Plate Margins*. Geological Society, London, Special Publications, **81**, 3–28.

[36] These are captured in what is sometimes referred to as the 'Manifesto' (e.g. Jones *et al.* 1983; Howell 1995).
[37] Interview with P. J. Coney taped by H. E. Le Grand on 15 February 1990 at Department of Geosciences, University of Arizona, Tucson.

HASHIMOTO, M. & UYEDA, S. (eds) 1983. *Accretion Terranes in the Circum-Pacific Region: Proceedings of the Oji International Seminar on Accretion Tectonics*. Terra Scientific Publishing Company, Tokyo.

HOWELL, D. G. 1985. *Tectonostratigraphic Terranes of the Circum-Pacific Region*. Circum-Pacific Council for Energy and Mineral Resources, Houston.

HOWELL, D. G. 1989. *Tectonics of Suspect Terranes: Mountain Building and Continental Growth*. Chapman & Hall, London.

HOWELL, D. G. 1995. *Principles of Terrane Analysis: New Applications for Global Tectonics* (2nd edn of HOWELL 1989). Chapman & Hall, London.

HOWELL, D. G. & MCDOUGALL, K. A. (eds) 1978. *Mesozoic Paleogeography of the Western United States: Pacific Coast Paleogeography Symposium 2. Pacific Section*. Society of Economic Paleontologists and Mineralogists, Los Angeles.

HOWELL, D. G. & WILEY, T. J. (eds) 1988. *Proceedings of the 4th International Tectonostratigraphic Terrane Conference (Nanjing, China, 1988)*. Nanjing University, Nanjing.

HOWELL, D. G., JONES, D. L., COX, A. & NUR, A. (eds) 1984. *Proceedings of the Circum-Pacific Terrane Conference (Stanford, 1983)*. Stanford University, Stanford.

IRWIN, P. 1972. Terranes of the western Paleozoic and Triassic belt in the southern Klamath Mountains, California. USGS Professional Paper 800-C, C103–C111.

JOHNSON, J. H. & DANNER, W. R. 1966. Permian calcareous algae from northwestern Washington and southwestern British Columbia, *Journal of Paleontology*, **40**, 424–432.

JONES, D. L. IRWIN, W. P. & OVENSHINE, A. T. 1972. Southeastern Alaska – a displaced continental fragment? USGS Professional Paper 800-B, B211–B217.

JONES, D. L., SILBERLING, N. J. & HILLHOUSE, J. 1977. Wrangellia – a displaced terrane in northwestern North America. *Canadian Journal of Earth Sciences*, **14**, 2565–2577.

JONES, D. L., HOWELL, D. G., CONEY, P. J. & MONGER, J. W. H. 1983. Recognition, character, and analysis of tectonostratigraphic terranes in western North America. *In*: HASHIMOTO, M. & UYEDA, S. (eds) Accretion Terranes in the Circum-Pacific Region: Proceedings of the Oji International Seminar on Accretion Tectonics. Terra Scientific Publishing Company, Tokyo, 21–35.

KING, P. B. 1959. *The Evolution of North America*. Princeton University Press, Princeton.

LATOUR, B. 1987. *Science in Action*. Harvard University Press, Cambridge.

LE GRAND, H. E. 1988. *Drifting Continents and Shifting Theories*. Cambridge University Press, Cambridge.

LE GRAND, H. E. & GLEN, W. 1993. Choke-holds, radiolarian cherts, and Davy Jones's locker. *Perspectives on Science*, **1**, 24–67.

LE PICHON, X. 1968. Sea-floor spreading and continental drift. *Journal of Geophysical Research*, **73**, 3661–3697.

MCALLISTER, J. W. 1992. Competition among scientific disciplines in cold nuclear fusion research. *Science in Context*, **5**, 17–49.

MCKENZIE, D. P. & PARKER, R. L. 1967. The north Pacific: an example of tectonics on a sphere. *Nature*, **216**, 1276–1280.

MARVIN, U. B. 1973. *Continental Drift: The Evolution of a Concept*. Smithsonian Institution Press, Washington.

MENARD, H. W. 1986. *The Ocean of Truth*. Princeton University Press, Princeton.

MOLNAR, P. 1988. Continental tectonics in the aftermath of plate tectonics. *Nature*, **335**, 131–137

MONGER, J. W. H. & ROSS, C. A. 1971. Distribution of fusilinaceans in the western Canadian cordillera. *Canadian Journal of Earth Sciences*, **8**, 259–278.

MORGAN, J. 1968. Rises, trenches, great faults and crustal blocks. *Journal of Geophysical Research*, **73**, 1959–1982.

OLDOW, J. S., BALLY, A. W., LALLEMANT, A. G. & LEEMAN, W. P. 1989. Phanerozoic evolution of the North American cordillera, United States and Canada. *In*: BALLY, A. W. & PALMER, A. R. (eds) *Geology of North America – An Overview*. Geological Society of America, Boulder, 139–232.

ORESKES, N. 1999. *The Rejection of Continental Drift: Theory and Method in American Earth Science*. Oxford University Press, New York.

SENGÖR, A. M. C. 1990*a*. Lithotectonic terranes and the plate tectonic theory of orogeny: a critique of the principles of terrane analysis. *In*: WILEY, T. J, HOWELL, D. G. & WONG, F. L. (eds) *Terrane Analysis of China and the Pacific Rim*. Circum-Pacific Council for Energy and Mineral Resources, Houston, Earth Science Series, **13**, 9–44.

SENGÖR, A. M. C. 1990*b*. Plate tectonics and orogenic research after 25 years: a Tethyan perspective. *Earth-Science Reviews*, **27**, 1–201.

SILBERLING, N. J. & JONES, D. L. 1987. *Folio of the Lithotectonic Terrane Maps of the North American Cordillera: Miscellaneous Field Studies Maps MF-1874-A, B, C, D*. USGS, Menlo Park.

STEWART, J. S. 1990. *Drifting Continents and Colliding Paradigms*. Indiana University Press, Bloomington.

STONE, P., PLANT, J. A., MENDUM, J. R. & GREEN, P. M. 1999. A regional geochemical assessment of some terrane relationships in the British Caledonides, *Scottish Journal of Geology*, **35**, 145–156.

THOMPSON, M. L., WHEELER, H. E. & DANNER, W. R. 1950. Middle and upper Permian fusilinids of Washington and British Columbia. *Contributions of the Cushman Foundation for Foraminfera Research*, **1**(3–4), 46–63.

WILSON, J. T. 1966. Did the Atlantic close and then reopen? *Nature*, **211**, 676–681.

WILSON, J. T. 1968. Static or mobile Earth: the current scientific revolution. *Proceedings of the American Philosophical Society*, **112**, 309–319.

WILSON, J. T. 1976. *Continents Adrift and Continents Aground: Readings from Scientific American*. W. H. Freeman, San Francisco.

WYLLIE, P. J. 1976. *The Way the Earth Works*. John Wiley & Sons, New York.

Marie Tharp, oceanographic cartographer, and her contributions to the revolution in the Earth sciences

CATHY BARTON

Department of History, University of Maryland Baltimore County, 1000 Hilltop Circle, Baltimore, MD 21250, USA

Abstract: In the early 1950s, two American geologists, Bruce Charles Heezen (1924–1977) and Marie Tharp, began mapping the sea floor to improve understanding of ocean-basin geology and to connect the oceans to the continents theoretically. Both were researchers at the Lamont Geological Observatory of Columbia University, now Lamont–Doherty Earth Observatory. Heezen and Tharp used the 'physiographic mapping' technique, which makes it possible to relate topographic features to underlying geology. The diagrams mostly utilized light and texture, rather than colour, and were sketched using a hachuring technique. Heezen collected data for research purposes and Tharp used his information to compile their physiographic diagrams. During this process, she confirmed previous predictions when she made an important discovery: a rift on the Mid-Atlantic Ridge. Tharp's visual interpretations of the sea-floor data contributed to the reintroduction of continental drift theory and the 1960s geological revolution. At a time when most women were excluded from scientific careers, Tharp, initially a research assistant, succeeded in this competitive arena. Working with Heezen as a geologist and cartographer, she had an unusual opportunity to participate in the era's exciting discoveries; and her contributions were acknowledged. While their data-gathering activities and analyses stimulated change and contributed to the revolution in the Earth sciences, Heezen and Tharp were not directly involved in the plate-tectonics revolution, but favoured expanding-Earth theory.

This paper examines the empirical and cartographic work relating to the ocean floors, and the ideas of Marie Tharp (*b*. 1920) and Bruce Heezen (1924–1977), with particular emphasis on the work of the former. It details their contributions to the Earth sciences revolution of the 1960s, through the provision of fresh empirical information and its presentation in a form that, in itself, led to new ways of thinking about the Earth.

The collaborators: Tharp and Heezen

Marie Tharp, one of the first women employed by the Lamont Geological Observatory, arrived at Columbia in 1948 to do drafting and computing for Maurice Ewing's graduate students. One of these, Bruce Heezen, had begun collecting extensive amounts of Atlantic ocean-floor data in 1947. As he went to sea for long periods of time, the task of organizing his data quickly became Tharp's responsibility and shortly thereafter she began working exclusively on Heezen's research projects. She collated, organized and eventually mapped the sounding data, and the two developed a professional collaboration that continued until Heezen's death in 1977 (Fig. 1). Cartography documented Heezen's data on Atlantic crossings, as the publication of papers was frequently delayed, and Tharp's interpretations of the raw bathymetric data were important contributions. As ocean-floor contour maps were classified as confidential from 1952 until 1961, the collaborators utilized the so-called 'physiographic method', developed by the Columbia University geomorphologist Armin K. Lobeck (1886–1958), in order to publish their discoveries related to sea-floor bathymetry and topography (Tharp 1999). When the security classification changed, the collaborators continued using this method, though they also produced a variety of raised-relief globes, table-top relief models, and contour maps.

Heezen and Tharp were geologists as well as map-makers and their work contributed to the reintroduction of mobilism or drift theory, which had been largely rejected by the American geological community for four decades (Oreskes 1999). Their research on fracture zones was also significant for the introduction of plate tectonic theory. Heezen was one of the greatest gatherers of data from the ocean basins and he made several important discoveries about sea-floor currents and sediments, which conflicted with the old 'permanentist' geological paradigm. His initial research concerned Atlantic ocean-floor canyons produced by continental run-off and earthquake-induced turbidity currents (Erickson

From: OLDROYD, D. R. (ed.) 2002. *The Earth Inside and Out: Some Major Contributions to Geology in the Twentieth Century*. Geological Society, London, Special Publications, **192,** 215–228. 0305-8719/02/$15.00

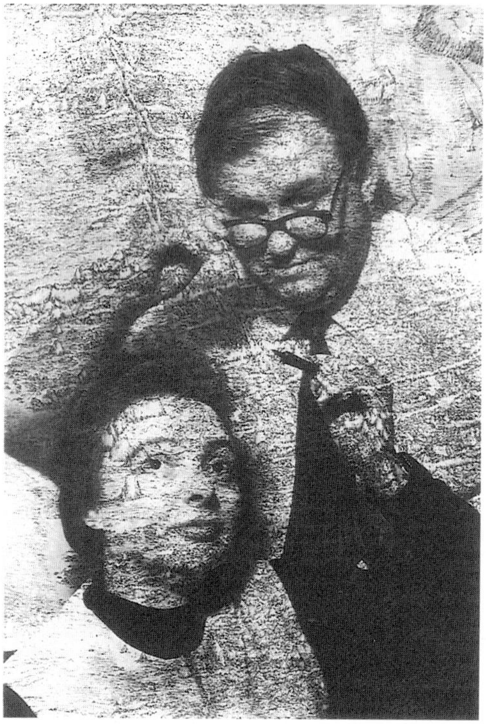

Fig. 1. Heezen and Tharp are perusing a film transparency of their diagram, photographer Robert Brunke, Date 1968 (Copy of promotional photograph from the Heezen Collection, Library of Congress.).

et al. 1952). Heezen pursued a research programme that included the collection and analysis of many different kinds of data, so that it took approximately 30 years for the collaborators to map all the ocean basins. The culmination of their mapping efforts, the 'World Ocean Floor Panorama', was completed in 1977 (Heezen & Tharp 1977).

Tharp's early life and education

Marie Tharp's mother was 40, and her father 50, when she was born at Ypsilanti, Michigan, in 1920.[1] Her only sibling was a half-brother, 17 years older, and she was reared as if she were an only child. Tharp's mother was a school teacher who retired upon marrying. Her father was a soil surveyor for the United States Department of Agriculture and the family moved some 30 times before Tharp was 15.[2] She often accompanied her father on surveying expeditions. By her own account she 'was tall and skinny and wore glasses and had flat feet'. So she 'wasn't part of the cliques that were in high school'.[3] An avid reader, Tharp decided to study literature. Her first choice was St John's College in Annapolis, but this institution did not admit women until many years later. She could have attended The University of Chicago, but in the event she chose Ohio University. It was, like St John's, a small secular school. At Ohio, she attempted to create her own version of the well-known University of Chicago literature classics course. She received a BA in English in 1943.[4]

The bombing of Pearl Harbour and the reality of World War II changed Tharp's life and redirected her education. As in previous eras, the needs of nations at war increased educational and employment opportunities for women. After fortuitously taking a course in geology at Ohio, she was recruited into the University of Michigan at Ann Arbor's petroleum geology programme. The curriculum was designed to attract women into the geological sciences by guaranteeing them jobs in the petroleum industry. The women in the Michigan programme were called the 'PG' or petroleum geology girls. Tharp arrived at Ann Arbor in 1943 having received her BA from Ohio in January, rather than June as was usual, and she was the only one in her Michigan class who finished with a Master's degree. She took courses in petroleum geology, mineralogy, petrography, invertebrate palaeontology and structural geology. Kenneth K. Landes, the department chair, taught petroleum geology, and she studied 'architectonics' with the structural geologist Armand J. Eardley.[5] Micropalaeontology was not offered; it was taught on the job at oil companies.[6] There were approximately ten women in Tharp's class. They spent the summer semester doing geological fieldwork at Camp Davis in Wyoming.[7]

In 1944, Tharp worked as a junior geologist for the US Geological Survey and the following year she received her MS in geology. While some

[1] Archival source (see Appendix): Tharp 1999*b* 5.
[2] Archival source: Tharp 1999*b*, 1.
[3] Archival source: Tharp 1999*b*, 42.
[4] Archival source: Tharp 1999*b*, 13.
[5] Archival source: Tharp 1999*b*, 52.
[6] Marie Tharp, pers. comm.
[7] Archival source: Tharp 1997, 3–4.

of Tharp's colleagues became employed in micropalaeontology, which frequently attracted women, she was not interested in this field as she was looking for more challenging opportunities. After graduating from Michigan, Tharp was offered a job at Standard Oil and Gas in Tulsa. She moved to Oklahoma, but found working in the office of an oil company unrewarding and there were no prospects for advancement. So Tharp decided to study mathematics at Tulsa University. In 1948, she received a BS in mathematics and relocated to New York State where she found a job as a research assistant at the Lamont Laboratory, Columbia University.

Collaboration with Bruce Heezen

Tharp found her early years at Lamont exciting, for many discoveries of importance were being made and she had the opportunity to participate in them. Her employment situation was amicable and the atmosphere was fairly relaxed when it came to job titles and seniority.[8] Initially, there were not many people competing against one another and researchers had considerable freedom in their choice of projects. During the 1950s, Heezen and Tharp planned Lamont cruises to include areas where sea-floor data had not been collected previously, but as time passed more graduate students and scientists vied for Lamont's limited ship-time and the collaborators lost those privileges. Although the positions that Tharp held were funded by external grants and government money, unemployment was not a concern as funds for defence-related projects were plentiful.[9] Her responsibilities were numerous and Tharp became essential to the new practice of oceanographic cartography at Lamont. Because making maps of ocean-basin topography depended upon knowledge of both geological and geographical processes, and on mathematical skills, Tharp's education matched her tasks perfectly. Working as a 'human computer', she converted the bulk of the raw bathymetric data, gathered in numerical form, into graphs, profiles and maps (Tharp & Frankel 1986, p. 52). Later, she had assistants who did the calculations and by the mid-1960s machine computers eliminated the tedious processes. When Heezen and Tharp began their systematic map-work in 1952, Tharp was promoted from research assistant to research geologist, a position that she held until 1961 when she became a research scientist. In 1963, her title became research associate, a position she held until 1968. According to Tharp, after she and Heezen began making their physiographic diagrams, she never thought seriously about getting a PhD as she did not have time to commute to night school.

Female research assistants, in the 1950s and 1960s, and through to the present, rarely received recognition such as Tharp was given. Although women were encouraged to enter careers in science after World War II, and doctoral-granting universities had greater financial resources, not many women were employed as tenured faculty or in higher administrative positions. Certain jobs were reserved for women only and their status became marginal (Rossiter 1995). Women working at the Scripps Institution of Oceanography during the mid-twentieth century were not allowed to undertake research at sea, largely because the navy funded most projects, and naval regulations applied. As a type of fieldwork, going to sea was considered exciting, and the less prestigious on-shore tasks were left for women to complete. Since large amounts of oceanographic data were needed to obtain relatively small amounts of information, ambitious scientists chose research that produced results quickly, while women were given work that their male counterparts rejected (Oreskes 2000). From Tharp's experience, this seems to have been the case at Lamont as well as other institutions.

As a research assistant and computer, Tharp's situation at the Lamont Geological Observatory was comparable to that of other women employed by large universities and research institutions. She too was excluded from research at sea until the mid-1960s and the positions that she held were untenured. However, while Heezen's scientific reputation towered over Tharp's, her contributions did not become entirely anonymous, as did those of some other women who worked in the sciences (Oreskes 1996). She co-authored a book (Heezen *et al.* 1959) and influential papers (Heezen *et al.* 1964*a*,*b*). Beginning in the early 1960s, they published papers with graphs, charts, contour diagrams and profiles that related to their research on the fracture zones intersecting the Mid-Atlantic Ridge. Looking back, Tharp believes that she was rewarded by the excitement of the early years, the variety of projects that she worked on, and the recognition she received during a period of intensive and fruitful scientific discovery.

---

[8] Archival source: Tharp 1997, 8.
[9] Archival source: Tharp 1997.

The collaboration of Heezen and Tharp was mutually beneficial, but for different reasons. As Heezen spent long periods of time at sea and also taught geology at Columbia, he did not have time to organize his data. Tharp did the calculations that were essential to creating the physiographic diagrams and, as she was the first person to interpret the depth-sounding data in visual form and speculate on trends that they suggested, she made significant contributions.[10] As with other couples working in the sciences, the collaborators' friendship, mutual interests and shared enthusiasms enhanced their careers and, working together, they accomplished more than they could have done alone or separately (Pychior *et al.* 1996). Tharp gained recognition, and working with Heezen advanced her career, even though her status became dependent upon his.

During the 1950s, Heezen was one of Ewing's favourite students, but Heezen eventually fell out with Ewing, a geophysicist, who tended to be impatient with traditional geological methods (Le Grand 1988). In the mid-1960s Ewing tried to remove Heezen and suspended his Lamont ship-time.[11] In 1965, lack of space at Lamont and the previously mentioned problems prompted Tharp to work from her home, which was close at hand. In 1968, her Lamont position was abruptly terminated, and Heezen was obliged to find funds for the preparation of diagrams and related maps from the Office of Naval Research (ONR). Tharp continued working at home as an ONR-funded oceanographer.[12] Her lack of tenure made her vulnerable to dismissal and her position was adversely affected by Heezen's changed circumstances, but she was employed continuously; and in the mid-1960s, Tharp was able to go to sea. However, Heezen lost his Lamont ship privileges at that time and the collaborators began accompanying cruises on other institutions' vessels.

In 1965, Tharp was a researcher aboard a cruise of the *Eastward*, a Duke University vessel designed with the help of advice from Heezen and Robert S. Menzies. The conditions on board were sometimes arduous. Tharp recalls that as the boat rocked wildly and waves broke over the deck, volunteers retrieved basketball-sized manganese nodules obtained from the sea floor before they washed overboard.[13] She was primarily concerned with paperwork, however, and brought the previously obtained profiler records along, so that the ship did not collect superfluous data. On another *Eastward* cruise the corer was lost and many sea-floor photographs had to be taken, in order to compensate. The ship followed the Western Boundary Current in the North Atlantic near the Blake Plateau and the Blake–Bahama Basin. Heezen and his student Charles Hollister found evidence for deep geostrophic contour-currents as agents shaping the continental rise and other sediment bodies. The sounding data revealed inconsistencies that were later confirmed as being due to 'drifts' (which are formed by the action of deep currents and appear similar in form to drifted snow), through the analysis of current activity, sediment cores, and bottom photographs (Heezen *et al.* 1966). In time, they discovered sediment drifts in the North Atlantic, the Indian Ocean, and around Antarctica.[14] In 1968, Tharp was an ONR specialist on the *USNS Kane*. Heezen was chief scientist on this Global Ocean Floor Analysis Research (GOFAR) expedition, the objective of which was to determine the trends of specific fracture zones intersecting the Mid-Atlantic Ridge. Confirming proposed trends, the survey followed the sharply defined Scarp D, the most southerly offset in the North Atlantic, discovered by Tharp on a previous *Eastward* expedition (Heezen 1968).

After Heezen died in 1977, Tharp's association with Lamont ended abruptly. Still working at home, she took the title of oceanographic cartographer and finished their final project. Once they began mapping, Tharp recalls, she never considered anything else.[15] While Tharp's accomplishments were exceptional because she overcame educational and employment barriers that limited opportunities for women of her generation, she did not identify with the era's revolutionary geological theories.

Physiographic mapping

The Heezen and Tharp physiographic diagrams were part of a larger genre of geological illustrations and representations, originally produced to accompany written research. As such, they

---

[10] In 1952, Tharp discovered the rift on the Mid-Atlantic Ridge 'in' the new data. In 1964, she identified Scarp D, a fracture-zone intersecting the Mid-Atlantic Ridge.
[11] Archival source: Tharp 1999*b*, 23.
[12] Archival source: Tharp 1997, 20.
[13] Marie Tharp, pers. comm., October 1999.
[14] Archival source: Tharp 1999*a*, Tanya Levin interview, 24 May 1997, 13–95.
[15] Archival source: Tharp 1997, 21–22.

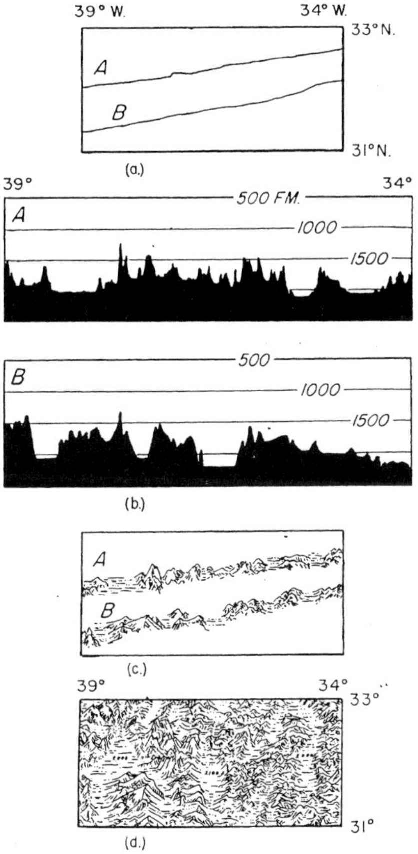

Fig. 2. Method for preparing a physiographic diagram from the *Floors of the Ocean*, authors Bruce C. Heezen, Marie Tharp, and Maurice Ewing, Date 1959 (Heezen *et al.* 1959, p. 4) (a) Positions of sounding lines (A, B) are plotted on chart. (b) Soundings are plotted as profiles at 40:1 vertical exaggeration. (c) Features shown on profiles are sketched on chart along tracks. (d) After all available sounding profiles are sketched the remaining unsounded areas are filled in by extrapolating and interpolating trends observed in a succession of profiles.

are elements of the 'visual language of geology' (Rudwick 1976, 1992), essential to the communication of new ideas. 'Transforming' the ocean basins into 'observed objects' increased their conceptual accessibility and led to theory change, though the collaborators received criticism as well as public acclaim for their physiographic diagrams (Wood 1985; Mukerji 1990).

In a physiographic diagram, hypothetical extrapolation may be used to fill in 'the unknowns'; but this was an invaluable feature since, despite the masses of information collected, there were few sea-floor data for some areas, when a representation of the ocean floors of the whole globe was attempted. When terrestrial regions are mapped, the shape and size of the landforms are known. By comparison, the preparation of a marine physiographic diagram requires one to postulate patterns and relief trends on the basis of cross-sections compiled from selected data soundings (Tharp 1999). As the diagrams were compiled using relatively small amounts of data, when compared to the immensity of the ocean basins, objections were made by proponents of different philosophical perspectives on the nature of cartographic representation. But Tharp and Heezen were undeterred and generally ignored the objections. Figure 2 illustrates their method of preparing a diagram.

Lobeck (1939) had called his products 'diagrams' rather than maps because the view of the terrain that they presented was not in fact consistent with the map-base. Sketched in black ink, using the hachuring method, features were drawn in oblique perspective. Not every feature was illustrated. It was impossible to do so because there could never be enough space, but the method gave the viewer a vivid idea of how the ocean basins would look if drained of water (Thrower 1996). The first Heezen and Tharp physiographic diagrams of the North Atlantic Ocean floor were primarily compiled from bathymetric soundings. Heezen collected much of these data, as well as sediment core samples, seismic refraction and reflection data, and magnetic and gravity readings. Lobeck, and later Heezen and Tharp, used the concept of 'physiographic provinces' to simplify the study of geomorphology and to illustrate the systematic development of features, rather than formation by chance. (Physiographic provinces are regional units divided by relief features created by forces within the Earth's crust; Lobeck 1939.) The collaborators assumed that the floor of the Atlantic was composed of three major types of topography: (1) continental margins; (2) smooth abyssal plains; and (3) the rugged Mid-Atlantic Ridge. When diagramming began in the Atlantic they had to ascertain the location of the boundaries separating these provinces. Large numbers of soundings were used to delineate the Mid-Atlantic Ridge but since the number needed was a function of the frequency and amplitude of the peaks, far fewer were needed to define the featureless abyssal plains.

Tharp produced the first two-dimensional profiles after rows of soundings had been plotted in the form of graphs, whose co-ordinates were depth and distance covered. Adjacent profile sheets were taped together in long strips. Heezen preferred to see the data in profile form, which was easier to evaluate, and he developed a method to simplify the plotting of profiles. His technique required fewer mathematical conversions.[16] Significant peaks, valleys, and changes of slope were selected from hundreds of ship-crossing profiles. In this era Heezen and Tharp worked with data collected by ships as they crossed the Atlantic Ocean on a roughly east or west course. Soundings were plotted so as to yield cross-section profiles.[17] Tharp calculated significant changes of slope, and made the majority of selections from data that indicated topographical trends from samples that she took at different latitudes. On the initial west–east profiles of the North Atlantic, the ratio of the vertical to horizontal co-ordinates was 40: 1. The vertical exaggeration was essential to emphasize differences in height that would otherwise have seemed insignificant or imperceptible to the viewer (Heezen 1960).

When Heezen and Tharp began mapping the South Atlantic, they were able to utilize data obtained during the cruise of the German vessel *Meteor* in 1925–1927, plotting the information with a vertical exaggeration of 100: 1. Lamont obtained the *Vema* as a research vessel in the 1950s and data from this ship were also utilized, again with an exaggeration of 100: 1. Thus time and paper were saved.[18] The ship tracks were traced on the map according to latitude and longitude with spot-depths indicated along the track. (A spot-depth represented one sounding, written on the diagram in numerical form.) After linking up the two-dimensional profiles, Tharp used an oblique perspective to transpose three-dimensional topographic information onto a larger chart. In the next step, several three-dimensional profiles were lined up and studied. According to Tharp, she lined them up and studied them at a scale of 1: 10 000 000. If enough data were collected at some later time, profiles could be worked up at 1: 1 000 000. Each new profile was compared with the earlier data, always aiming to obtain the correct trends of the sea-floor terrain. Then the adjoining sheets were connected and adjusted, keeping the original data in mind.[19] After all the available profiles were drawn at the correct locations and after the trends had been estimated, the blank areas of the physiographic diagram were sketched in 'hypothetically'. Eventually, regional soundings were added as numbers; these were representative of large areas such as abyssal plains.[20] The result was a three-dimensional map. Tharp made most of the decisions in this whole process and Heezen performed the final task of editing.

Heezen and Tharp used the US Navy master sheets and the International Hydrographic Bureau's 'General Bathymetric Chart of the Oceans' (GEBCO) sheets for data and as base maps. Their first physiographic diagram of the North Atlantic (Elmendorf & Heezen 1957) was drawn at a scale of 1: 5 000 000, which was the one used in the 6050 Series of classified US Navy maps that the collaborators used as their base map. Their next mapping project, the South Atlantic, used the GEBCO maps as a base, as they were compiled on a 1: 10 000 000 scale. At that time there were insufficient South Atlantic data to enable them to use a 1: 5 000 000 scale.[21] Many worksheets were produced, in pencil on tracing paper for first drafts and later on transparent frosted acetate. When deemed satisfactory, the physiographic diagrams were reproduced in ink, sometimes of varying shades, on sheets of heavy paper, frosted acetate, or what was then called linen, depending on what the diagrams were to be used for. Tharp experimented with a variety of materials.

Completing the North Atlantic took 70 working sheets, each covering a 10° × 10° area. Initially, the information on each sheet was transferred by hand, with pen and ink, to the master sheet, which was then photographed using a 17-inch process camera. The photograph was then placed in a mosaic composite of the North Atlantic and, when completed, the whole was rephotographed. All information was enlarged or reduced to conform to scale. In the mid-1960s, after fracture zones had been confirmed in the Atlantic, Tharp used colour-coded province maps as one of the preliminary steps in compiling the GEBCO maps. Province maps delineated physiographic provinces but not topographic features. According to her recollection, their data indicated that the intersection of the fracture zones with the Mid-Atlantic Ridge was at an angle and she preserved the angular

[16] Archival source: Tharp 1997, 14; and Marie Tharp, pers. comm.
[17] Archival source: Tharp 1997, 29.
[18] Archival source: Tharp 1997, 15.
[19] Archival source: Tharp 1997, 14.
[20] Archival source: Tharp 1999*b*, 33.
[21] Archival source: Tharp 1999*b*., 29–30.

relationship, working outwards from the ridge towards land.[22] She used the colour-coded province maps to ensure that the offset trends intersecting the Mid-Atlantic Ridge were correct. Unlike the physiographic diagrams, these maps allowed the representation of various depths by means of colour. Figure 3 illustrates a small section of a colour-coded province map for part of the equatorial region of the Atlantic at a scale of 1: 1 000 000. The West African coast is visible on the right. To verify fracture-zone trends in later diagrams, Heezen and Tharp cruised out to specific locations and took soundings. They wanted empirical confirmation of a trend, in order to be as accurate in their findings as possible.[23] Later physiographic diagrams benefited from data gathered during the International Geophysical Year (IGY) of 1957–1959 and the International Indian Ocean Expedition of 1964. Many other nations and oceanographic institutions contributed information, including the Russians.[24]

Beginning in 1950, Heezen contracted to produce the Atlantic Ocean section of the GEBCO map series. Nineteen countries were involved in this project and data were acquired from all of them.[25] Heezen and Tharp eventually extended their mapping efforts to all the ocean floors, and by the mid-1960s their physiographic diagrams, in the form of painted maps, reached a wide audience in the *National Geographic Magazine* (Heezen & Tharp 1967, 1968, 1969, 1971). In 1964, the National Geographic Society became interested in printing a physiographic diagram of the Indian Ocean to illustrate discoveries made during the International Indian Ocean Expedition. The Society published physiographic diagrams of the North and South Atlantic in 1968, following one for the Indian Ocean in 1967, and later published maps of the Pacific (West Central Pacific) (1969) and the Arctic (1971). The Society hired the Austrian artist, Heinrich Berann, to paint in the collaborators' black-and-white physiographic diagrams. They worked closely with Berann on the 'map' of each ocean. Heezen believed that the oceans were a single unit, and thus their last project was the depiction of the world's oceans on a single map-sheet, 'The World Ocean Floor Panorama', funded by the ONR. It took from 1973 to1977 to complete, as new sections of the map frequently involved changing old areas. While work sketches had been drawn at 1: 5 000 000, 1: 10 000 000, and 1: 1 000 000, the overall panorama was compiled on a smaller scale, 1: 23 001 000 (see Fig. 4).[26]

Tharp's (re)discovery, and the importance of the earthquake data

During the mapping process Tharp confirmed an earlier prediction that was of great importance to the return of mobilism and to later scientific theory: the *rift* on the Mid-Atlantic Ridge. In 1952, she detected the rift, a deep V-shaped structure near the crest of the ridge, after comparing six sheets of cruise profiles (Tharp & Frankel 1986). Sewall & Wiseman (1938) had predicted the existence of this rift from their early surveys in the Indian Ocean and the Arabian Sea, but Heezen and Tharp were the first to gather the new Atlantic data and translate them into a convincing visual format. Heezen initially hesitated to accept the existence of the rift, as it hinted at continental drift, which he considered too radical a theory. Many American geologists, including Heezen and Tharp, opposed the idea of moving continents, and mobile ocean basins had not been taken seriously in the USA. However, making analogical comparisons, Heezen compared the rift on the Mid-Atlantic Ridge with continental features, such as the Rift Valley in East Africa. By 1953, he was convinced that the ridge was rifted, but he collected additional evidence before making a public announcement. The collaborators looked for confirmation of a rift valley in the shallow-focus earthquake data from the ridge zones by superimposing a transparency map of earthquake epicentres over the rift locations on the Mid-Atlantic Ridge map (Oreskes 1999). Heezen proposed that the rift extended around the world after Tharp suggested that the rifts be compared with the belts of earthquakes, also observed extending around the Earth. This was an extrapolation using analogy. In retrospect, Tharp emphasizes that the earthquake data were critical to the general discovery of the rifts, as the collaborators lacked extensive sounding data from oceans other than the Atlantic.

The concept of the general existence of rifts on mid-oceanic ridges led to other theories and discoveries. In 1962, Heezen proposed the

---

[22] Archival source: Tharp 1997, 16.
[23] Archival source: Tharp 1997, 17.
[24] Archival source: Tharp 1999*a*, Tanya Levin interview.
[25] Archival source: Tharp 1999*a*, Ronald Doel interview.
[26] Archival source: Tharp 1999*b*, 33.

Fig. 3. Section of a colour-coded province map of the Mid-Atlantic Ridge in the equatorial Atlantic. The red, yellow and green areas show highest elevation. Reproduced from a map worksheet in the Heezen Collection, Library of Congress (photographer Gary North, authors Bruce C. Heezen, Marie Tharp, Date 1960).
Fig. 4. The World Ocean Floor Panorama, authors Bruce C. Heezen and Marie Tharp, Date 1977 and copyright by Marie Tharp 1977. Reproduced by permission of Marie Tharp, 1 Washington Ave., South Nyack, NY 10960.

sequential or 'genetic' development of oceans from mid-ocean ridges. The Indian and Atlantic Oceans supposedly began as continental rifts and slowly grew, and the rifts in East Africa, the Red Sea and the Gulf of Aden were comparable features at different stages of development (Heezen 1962).

Physiographic diagrams: a reflection of changing scientific attitudes

Many scientific disciplines, not least geology, frequently proceed by the use of visual thinking and aesthetic considerations rather than deductively through logic or inductively from empirical data (Miller 1981). This 'aesthetic' method contributed significantly to the reintroduction of the notion of continental drift. The rifted Mid-Atlantic Ridge suggested that the Earth's crust had moved laterally and the diagrams contributed to the demise of geology's old permanence theory. But Heezen and Tharp did not propose that continental drift *caused* the rift. Rather, the diagrams attracted the attention of other geoscientists who made an acceptable case for continental drift and later for plate tectonic theory, which incorporated 'drift' (Le Grand 1988). In 1958, Heezen acknowledged that palaeomagnetic studies and other new data implied lateral continental motion, but he advocated cxpansionism, not drift. Expansion had been proposed previously in the twentieth century.[27]

In 1960, Heezen and Tharp promoted expansion while Harry Hess (1906–1969) proposed what became, after modifications, an acceptable model for continental drift. By that time, tectonics had begun to play a key role in the collaborators' research programme and Heezen used maps of the Earth's major tectonic features to bolster his argument for expansion. According to Heezen, the most important tectonic factors influencing sea-floor topography were crustal extension, strike-slip faulting, normal faulting, and subsidence (Heezen 1962). The work of the Australian geologist S. Warren Carey of the University of Tasmania, a staunch expansionist, influenced Heezen. In 1956, Carey organized a major conference on continental drift and Heezen participated (Carey 1956). He and Carey advocated a relatively rapid rate of expansion (Le Grand 1988). The Columbia structural geologist Walter Bucher (1889–1965), who in 1933 had proposed that the Earth underwent alternating periods of expansion and contraction, also advised Ewing and his students at Lamont.[28] The collaborators believed that mantle material welled up between the separating continents, which were pushed aside as the Earth expanded, and produced the mid-oceanic ridges by a form of sea-floor spreading. A prominent factor in Heezen's advocacy of expansion was that he, and most of his Lamont colleagues, believed that oceanic rifts and trenches were similar crustal features produced by tension. In addition, he did not accept that excess crust could be subducted back into the Earth's interior at the oceanic trenches. Heezen rejected subduction as he could not visualize the geometry of convection cells or currents as being such as to cause compression at the trenches (Heezen 1962). The early physiographic diagrams reflected his tensional hypothesis: the ocean basins were supposedly stretched apart as the Earth expanded, with rifts opening in all directions (Heezen *et al.* 1959; Menard 1986). The collaborators believed that an ocean basin was structurally one unit, with all major features, including the mid-ocean ridge system and the continental margins, being minor splinters and fissures in the floor of one 'grand crack': the ocean basin.

According to Felix Vening-Meinesz (1887–1966), another pioneer of sea-floor research, the presence of negative gravity anomalies over deep-sea troughs, accompanied by seismic activity in these regions, indicated down-buckling of the Earth's crust. Vening-Meinesz, Harry Hess and Ewing had accompanied the 1936–1937 cruise of the submarine *Barracuda* to the Caribbean, with Ewing collecting trench gravity data (Bowin 1972). After this expedition, he concentrated on equipment and the technical aspects of data collection rather than the development of theories. Hess, however, immediately began to develop Vening-Meinesz's ideas on down-buckling and convection currents (Le Grand 1988). He called these regions of crustal compression 'tectogenes' (Oreskes 1999). Lamont scientists, however, especially Ewing, did not accept crustal compression at the trenches, as they believed that the gravity data collected over trenches were inconclusive (Heezen 1962). The 'confusion' at Lamont continued until the evidence for

[27] Heezen was familiar with the literature on expansion. He cited Taylor (1910) and Eyged (1957) in his most descriptive work on the topic, the paper 'The deep-sea floor' (Heezen 1962).
[28] Marie Tharp, pers. comm., October 1999.

Fig. 5. Physiographic diagram of the South Atlantic, authors Bruce C. Heezen and Marie Tharp, Date 1961 and copyright by Marie Tharp 1961. Reproduced by permission of Marie Tharp, 1 Washington Ave., South Nyack, NY 10960.

continental drift became overwhelming in the mid-1960s.

The Heezen and Tharp physiographic diagrams, especially of the Atlantic, reflected the rapidly changing comprehension of processes shaping the ocean basins, during a brief period of revolutionary scientific activity as defined by Thomas Kuhn (Kuhn 1962). In 1961, the diagram of the South Atlantic, with the exception of a few equatorial fracture zones, illustrated few departures from the theories that prevailed during the 1950s (see Fig. 5). The early physiographic diagrams had land-like features, while later versions pictured a different world. Late in the 1960s, continental drift was becoming acceptable and plate tectonic theory was quickly developing as a viable alternative to the old paradigm. The appearance of the diagrams changed drastically in the 1968 Geological Society of America's edition of the North Atlantic after decisive geomagnetic core studies had convinced the majority at Lamont of continental drift, although Heezen and Tharp did not completely abandon expansion (Heezen & Hollister 1971). The most significant change in the appearance of their diagrams was the style of topographic symbolism. The edges of many features, especially mountain peaks, 'became' jagged and sharp (see Fig. 6). The pronounced angularity, emphasized by thicker black lines, and the ordered definition of the later editions contrasts with the more random, 'softer' edges of peaks in the early maps. On the rifted Mid-Atlantic Ridge, the incorporation of many regularly spaced offsets gave this great feature a new and forbidding appearance. Offsets along the ridge were exaggerated to emphasize the extent of displacement.[29] Eventually, the Mid-Atlantic Ridge region became filled with sharp, closely spaced peaks.

The appearance of light and shadow in the diagrams also changed as the new paradigm was adopted. In the early diagrams, the shadowing was not as strong as in later editions and there were no sharply defined sources of light. In the 1968 diagram, the floor of the central Atlantic Ocean was represented as if it were brightly illuminated from a point to the south, with light impinging on the face of the peaks, and highlighting the appearance of the fracture zones. The sketched lines and shading of these features was dark, accentuating their angularity and depth. These changes reflected the significance of these features in the development of plate tectonics theory and the understanding of the Earth's behaviour, even though the collaborators were not themselves supporters of plate tectonics.

The collaborators' years of research and analysis of offset fracture zones intersecting the Mid-Atlantic Ridge inspired others, such as J. Tuzo Wilson, to consider their origin and the direction of crustal motion at these features. While fracture zones had been discovered on the floor of the eastern Pacific by H. W. Menard of the Scripps Institution, in California, Heezen and Tharp established their existence in the Atlantic. During the drafting process, Tharp noticed trends and the collaborators looked for additional fracture zones using the earthquake data (Heezen *et al.* 1964*b*). They discovered irregular patterns along the central rift valley, which led them to believe that offsets on the ridge occurred at angular breaks of between 80 and 100° (Heezen & Tharp 1965). Tuzo Wilson cited this and other data, when he proposed that these features were not ordinary offset faults, but a new class of faults that occurred on mid-ocean ridges and are locally transformed into zones of crustal movement. According to Wilson, the motion along these faults was opposite to that of the usual strike-slip faults. The 'new' type of fault did not extend across the ridge, but joined the next segment of the rifted ridge. He named these features 'transform faults' (Wilson 1965) and they were soon incorporated into the emerging new theory of global tectonics.[30]

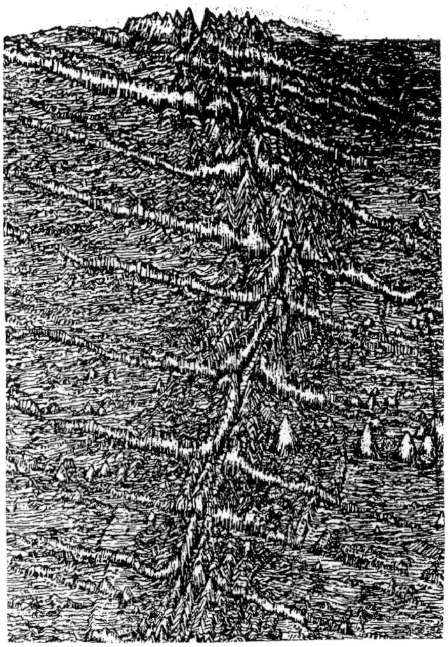

Fig. 6. Section of a physiographic diagram of the North Atlantic ocean floor, showing the jagged nature of the offset fracture zones on the Mid-Atlantic Ridge, *c.* 1968. Reproduced from the Heezen Collection, Library of Congress (photographed by Gary North, authors Bruce C. Heezen and Marie Tharp).

[29] Archival source: Tharp 1999*a*, Tanya Levin interview, 24 May 1997, 142.

[30] In this important paper Wilson cited Bucher (1933), Carey (1956), Heezen (1962) and several other works that laid the foundation for the concept of transform faults.

The collaborators' analysis of the data collected during the International Indian Ocean Expedition illustrates that scientific observations are theory-laden (cf. Hanson 1961); theoretical assumptions can dictate what is discerned in or inferred from data. While the relative symmetry of the Mid-Atlantic Ridge had inspired Heezen and Tharp to consider expansion, the complex and asymmetric nature of the Indian Ocean topography failed to change their expansion model. Rather, the new data strengthened their belief that continental drift, utilizing a simple pattern of convection currents inside the Earth, was not a feasible option and they continued to advocate expansion (Heezen & Tharp 1965). According to Tharp, Heezen was essentially a uniformitarian, who believed that observable processes could be used to explain the geological record (Tharp 1982*a*). Many scientific disciplines employ analogy (Hesse 1981) and as geologists, Heezen and Tharp gathered data and often used analogies to help in their analysis. However, to comprehend all the forces shaping ocean-basin topography, these tools were not sufficient as: (1) subduction does not occur on the continents; (2) the movement of the offset fracture zones intersecting the Mid-Atlantic Ridge differed from existing examples of fault systems on land; and (3) it was necessary to consider physical laws and go beyond geological fieldwork at the surface in order to understand how the Earth's crust behaves, as Wilson (1951) had suggested. In addition, researchers who were not deeply involved in data collection were able to distance themselves from specific research problems and propose broad explanatory theories (Menard 1986).

Conclusions

While Heezen and Tharp helped revive Wegener's mobilism, and their physiographic diagrams reflected the latest findings by leaders in the field of oceanographic research, visual representation and analogy could take the collaborators only so far and they were not involved in establishing plate tectonics *per se* (Le Grand 1988). The collaborators' cartographic endeavours stimulated scientific change by revealing critical elements of sea-floor topography and behaviour. These included: (1) the rifted Mid-Atlantic Ridge; (2) the extension of the mid-oceanic ridges around the planet; (3) the idea of the sequential or genetic development of oceans from continental rifts; and (4) the angled nature of the faults intersecting the Mid-Atlantic Ridge. But this newly documented knowledge, no matter how essential, could not in itself propel ideas beyond a certain point and transcend assumptions that were firmly established in the collaborators' minds. Nevertheless, Heezen and Tharp successfully continued their mapping and data gathering into the 1970s. Their efforts were vital to scientific change, even though, after the plate tectonics revolution, their method remained the same and their theoretical ideas did not change radically. Even if the collaborators are not usually acknowledged for substantial theoretical contributions to the revolution in the Earth sciences, their physiographic diagrams, globes and related artifacts may well be considered milestones in the history of cartography, and their work undoubtedly contributed to the eventual grand change in geological theory that occurred in the 1960s.

Perhaps one might say that Heezen and Tharp were (together) the Tycho Brahe of the Earth sciences revolution, providing essential empirical information but not able to break free of older ways of thinking. Or insofar as they did so, they pursued an idea that (so far as most geologists are concerned) led to a dead-end.

I wish to thank S. Herbert for her invaluable assistance; G. Fitzpatrick, at the Library of Congress, for encouraging my interest in the physiographic diagrams; G. North, at the Library of Congress, for taking the photographs of the diagrams in the Heezen Collection; and Marie Tharp for her encouragement and giving me the opportunity to interview her.

Appendix

Archival sources

THARP, M. 1997. Interview conducted by Gary North, 21 November. One session, one video-cassette; preliminary transcript. The Heezen Collection, The Library of Congress, Washington DC.

THARP, M. 1999*a*. Reminiscences of Marie Tharp: interviews conducted in four sessions by Ronald Doel on 14 December 1995 and 18 December 1996, and by Tanya Levin on 24 May 1997 and 28 June 1997. Preliminary transcript. (These are part of the Lamont–Doherty Earth Observatory Oral History Project. Oral History Research Office, Columbia University. On file at the American Institute of Physics Neils Bohr Library, College Park, MD.)

THARP, M. 1999*b*. Interviews conducted by the author on 25–26 October. Three sessions, three audiotapes; preliminary transcript. The Heezen Collection, The Library of Congress, Washington DC.

References

BOWIN, C. 1972. Puerto Rico trench negative anomaly belt. *In*: SHAGAM, R., HARGRAVES, R. B., MORAN, W. J., VAN HOUTEN, F. B., BURK, C. A., HOLLAND,

H. D. &. HOLLISTER, L. C. (eds) *Studies in Earth and Space Sciences: A Memoir in Honor of Harry Hammond Hess*. The Geological Society of America, Memoir **132**, 339–362.

BUCHER, W. H. 1933. *The Deformation of the Earth's Crust: An Inductive Approach to the Problems of Diastrophism*. Princeton University Press, Princeton.

CAREY, S. W. (ed.) 1956 (reprinted 1959). *Continental Drift: A Symposium being a Symposium on the Present State of the Continental Drift Hypothesis, held in the Geology Department of the University of Tasmania, March 1956*. Geology Department, The University of Tasmania, Hobart.

ELMENDORF, C. H. & HEEZEN, B. C. 1957. Oceanographic information for engineering submarine cable systems. *Bell System Technical Journal*, **36**, 1047–1093.

ERICSON, D. B., EWING, M. & HEEZEN, B. C. 1952. Turbidity currents and sediments in the north Atlantic. *AAPG Bulletin*, **36**, 489–511.

EYGED, L. 1957. A new dynamic conception of the internal constitution of the Earth. *Geologische Rundschau*, **46**, 101–121.

HANSON, N. R. 1961. *Patterns of Discovery: An Inquiry into the Conceptual Foundations of Science*. Cambridge University Press, Cambridge.

HEEZEN, B. C. 1960. The rift in the ocean floor. *Scientific American*, **203**, 98–110.

HEEZEN, B. C. 1962. The deep-sea floor. *In*: RUNCORN, S. K. (ed.), *Continental Drift*. Academic Press, New York, 235–288.

HEEZEN, B. C. 1968. 200,000,000 years under the sea: the voyage of the *U. S. N. S. Kane*, *Saturday Review*, 7 September, 63.

HEEZEN, B. C. 1969. The world rift system. *Tectonophysics*, Special Issuc **8**, 269–279.

HEEZEN, B. C. & EWING, M. 1952. Turbidity currents and submarine slumps and the 1929 Grand Banks earthquake. *American Journal of Science*, **250**, 849–873.

HEEZEN, B. C. & HOLLISTER, C. D. 1971. *The Face of the Deep*. Oxford University Press, New York.

HEEZEN, B. C. & THARP, M. 1954. Physiographic diagram of the western North Atlantic. *Bulletin of the Geological Society of America*, **65**, 1261.

HEEZEN, B. C. 1956. Physiographic diagram of the North Atlantic. *Bulletin of the Geological Society of America*, **67**, 1704.

HEEZEN, B. C. & THARP, M. 1961. *Physiographic Diagram of the South Atlantic*. Geological Society of America of America.

HEEZEN, B. C. & THARP, M. 1963. Oceanic ridges, transcurrent faults, and continental displacements. *Geological Society of America*, Special Papers **76** and **78**.

HEEZEN, B. C. & THARP, M. 1965. Tectonic fabric of the Atlantic and Indian Oceans and continental drift. *In*: BLACKETT. P. M. S., BULLARD, E. & RUNCORN, S. K. (eds) *A Symposium on Continental Drift*. The Royal Society, London, 90–106.

HEEZEN, B. C. & THARP, M. 1967. Indian Ocean floor. Painted by Heinrich C. Berann. *National Geographic Magazine*. October, Special map supplement.

HEEZEN, B. C. & THARP, M. 1968. Atlantic Ocean floor. Painted by Heinrich C. Berann. *National Geographic Magazine*. June, Special map supplement.

HEEZEN, B. C. & THARP, M. 1969. Pacific Ocean floor. Painted by Heinrich C. Berann. *National Geographic Magazine*. October, Special Map Supplement.

HEEZEN, B. C. & THARP, M. 1971. Arctic Ocean floor. Painted by Heinrich C. Berann. *National Geographic Magazine*. October, Special map supplement.

HEEZEN, B. C. & THARP, M. 1973. *USNS Eltanin* cruise 55. *Antarctic Journal*, **8**, 137–141.

HEEZEN, B. C. & THARP, M. 1977. *World ocean floor panorama (map)*. Mercator projection.

HEEZEN, B. C., EWING, M. & THARP, M. 1959. *The Floors of the Oceans: Part I. The North Atlantic*. Geological Society of America, Special Paper **65**.

HEEZEN, B. C., BUNCE, E. T., HERSHEY, J. B. & THARP, M. 1964*a*. Chain and Romanche fracture zones. *Deep-Sea Research*, **11**, 11–33.

HEEZEN, B. C., GERARD, R. D. & THARP, M. 1964*b*. *Vema* Fracture Zone in the Equatorial Atlantic. *Journal of Geophysical Research*, **69**, 733–739.

HEEZEN, B. C., HOLLISTER, C. D. & RUDDIMAN, W. F. 1966. Shaping of the continental rise by deep geostrophic contour currents. *Science*. **152**, 502–508.

HESSE, M. B. 1981. The function of analogies in science. *In*: TWENEY, R. D., DOHERTY, M. E. & MYNATT, C. R. (eds) *On Scientific Thinking*. Columbia University Press, New York, 345–348.

KUHN, T. S. 1962. *The Structure of Scientific Revolutions*. The University of Chicago Press, Chicago.

LE GRAND, H. E. 1988. *Drifting Continents and Shifting Theories*. Cambridge University Press, New York.

LOBECK, A. K. 1939. *Geomorphology: An Introduction to the Study of Landscapes*. McGraw–Hill, New York.

MENARD, H. W. 1986. *The Ocean of Truth: A Personal History of Global Tectonics*. Princeton University Press, Princeton.

MILLER, A. I. 1981. Visualizability as a criterion for scientific acceptability. *In*: TWENEY, R. D., DOHERTY, M. E. & MYNATT, C. R. (eds) *On Scientific Thinking*. Columbia University, New York.

MUKERJI, C. 1990. *A Fragile Power: Scientists and the State*. Princeton University Press, Princeton.

ORESKES, N. 1996. Objectivity or heroism? On the invisibility of women in science. *Osiris*, **11**, 87–113.

ORESKES, N. 1999. *The Rejection of Continental Drift: Theory and Method in American Earth Science*. Oxford University Press, New York.

ORESKES, N. 2000. *Laissez-tomber*: military patronage and women's work in mid-twentieth century oceanography. *Historical Studies in the Physical and Biological Sciences*, **30**, 373–392.

PYCHIOR, H. M., SLACK, N. M. & ABIR-AM, P. G. (eds) 1996. *Creative Couples in the Sciences*. Rutgers University Press, New Brunswick.

ROSSITER, M. W. 1995. *Women Scientists in America*

Before Affirmative Action: 1940–1972. Johns Hopkins University Press, Baltimore.

RUDWICK, M. J. S. 1976. The emergence of a visual language for geological science 1760–1840. *History of Science*, **14**, 149–195.

RUDWICK, M. J. S. 1992. *Scenes from Deep Time: Early Pictorial Representations of the Prehistoric World*. The University of Chicago Press, Chicago & London.

SEWALL, R. B. & WISEMAN, J. D. H. 1938. The relief of the ocean floor in the southern hemisphere. *Compte rendu du Congrès International de Géographie (Amsterdam)*, **2**, 135–140.

TAYLOR, F. B. 1910. Bearing of the Tertiary mountain-belt on the origin of the Earth's plan. *Bulletin of the Geological Society of America*, **21**, 179–226.

THARP, M. 1982*a*. Mapping the ocean floor – 1947 to 1977. *In*: SCRUTTON, R. A. & TALWANI, M. (eds) *The Ocean Floor: Bruce Heezen Commemorative Volume*. John Wiley, New York, 19–31.

THARP, M. 1982*b*. The complete bibliography of Dr. Bruce C. Heezen. *In*: SCRUTTON, R. A. & TALWANI, M. (eds), *The Ocean Floor: Bruce Heezen Commemorative Volume*. John Wiley, New York, 3–17.

THARP, M. 1999. Connect the dots: mapping the sea floor and discovering the Mid-Ocean Ridge. *In*: LIPPSETT, L. (ed.) *Lamont–Doherty Earth Observatory: Twelve Perspectives on the First Fifty Years 1949–1999*. Lamont–Doherty Earth Observatory, Palisades, 31–37.

THARP, M. & FRANKEL, H. 1986. Mappers of the deep. *Natural History*, **10**, 49–62.

THROWER, N. J. 1996. *Maps and Civilization: Cartography in Culture and Society*. The University of Chicago Press, Chicago & London.

WILSON, J. T. 1951. On the growth of continents. *Proceeding of the Royal Society of Tasmania for 1950*, 85–11.

WILSON, J. T. 1965. A new class of faults and their bearing on continental drift. *Nature*, **207**, 343–347.

WOOD, R. M. 1985. *The Dark Side of the Earth*. George Allen & Unwin, Boston.

From terrestrial magnetism to geomagnetism: disciplinary transformation in the twentieth century

GREGORY A. GOOD

History Department, West Virginia University, Morgantown, WV 26506–6303, USA

Abstract: In 1900, researchers interested in Earth's magnetism generally proclaimed all facets of magnetic phenomena to be within their purview. Most researchers in this field referred to themselves as 'magneticians' first and physicists or geologists second. After World War II, specialization increased. A number of distinct research areas appeared over several decades: the geodynamo theory and the study of the core–mantle boundary; palaeomagnetism and its growing connection to geology; the production of induced fields in Earth's crust; and, among others, the electromagnetic phenomena of the upper atmosphere and near space. The former unity dissolved and the field fragmented. One result of fragmentation has been a loss of memory and a consequent misinterpretation of an important part of the history of geoscience. This paper relates the challenges of recovering a history obscured by later events.

When most geologists think of studies of Earth's magnetism in the twentieth century, they think of palaeomagnetism, and with good reason. The investigation of, for example, reversals of direction of Earth's magnetism played a critical role in the acceptance of continental drift and plate tectonics, one of the central developments in geology during the century (Le Grand 1998; Frankel 1998). It's a dramatic story, and magnetic reversals themselves, seeming simultaneously unexpected and unsettling, have caught the imagination of a broader public. The many facets of the story of continental drift and plate tectonics, including the role of palaeomagnetism, are thoroughly analysed in Naomi Oreskes' *The Rejection of Continental Drift* (1999, pp. 263–267).

One must remember, however, that there is much more to Earth's magnetism than palaeomagnetism's importance in plate tectonics. When most *geophysicists* think of this broad phenomenon, they think toward one of two extremes of Earth's environment: the depths of the core–mantle boundary where the main geomagnetic field is produced, or the heights of the magnetosphere where the planet's magnetic field interacts with the solar wind and begins the chains of events that lead to magnetic disturbances and the *aurora polaris*. Investigations of the phenomena of these two realms relate to two other critically important stories of twentieth-century geoscience.

Interestingly, all the major streams of geomagnetic research explored in this paper – palaeomagnetism, the origin of Earth's main magnetic field, fields induced in Earth's crust and mantle, and ionospheric–magnetospheric phenomena – witnessed their great periods of dramatic success simultaneously in the mid-twentieth century. From roughly the end of World War II until the landing on the Moon in 1969, one dramatic discovery followed another. In palaeomagnetism, the work in the 1950s and 1960s of Allan Cox, Richard Doel, Brent Dalrymple, Donald H. Tarling and Ian McDougall, among others, established a timescale for reversals in the main geomagnetic field. This ultimately supported the famous Vine–Matthews–Morley hypothesis, which linked palaeomagnetism firmly to sea-floor spreading and plate tectonics. Concerning the origin of the main geomagnetic field, important developments in this story occurred in the 1940s and 1950s, with the first theories of a self-sustaining dynamo, proposed by Walter Elsasser and Edward Bullard, and the rotational theory of P. M. S. Blackett. These theories, while not immediately successful, started a new direction in geomagnetic research. The third major stream, of fields induced in Earth's crust and of conductivity, included both global and local investigations (Parkinson 1998). The fourth stream, the study of near-Earth space, included the investigation of the interaction of Earth's magnetic field in that region with the solar wind by Eugene Parker, the discovery of polar substorms and *aurora polaris*, of whistlers, and more (see, for example, Akasofu 1996; Hufbauer 1998; Stern 1989, 1996; Van Allen 1983; Cliver 1998).

All these areas of magnetic research, however, are much larger than these descriptions imply. Scientists and historians alike tend to be blinded by the bright lights of successful research. The successes of mid-twentieth century geomagnetic research helped drive the

From: OLDROYD, D. R. (ed.) 2002. *The Earth Inside and Out: Some Major Contributions to Geology in the Twentieth Century*. Geological Society, London, Special Publications, **192,** 229–239. 0305-8719/02/$15.00

historical process of specialization. Palaeomagnetic researchers, main field theorists, conductivity/induced field researchers, and space scientists began moving in progressively more independent directions. Despite some continuing overlap in instrumentation and/or theory, the rigours of their respective fields demanded ever more concentration. As this specialization has continued, people looking back have had a hard time seeing past the bright lights of the mid-century to a time before, when researchers in geomagnetism conducted their research for reasons unconnected with plate tectonics, the geodynamo, crustal conductivity or magnetospheric interactions. This does not imply that the different specializations were or are mutually irrelevant, since indeed palaeomagneticians (to borrow a word from W. D. Parkinson), for example, have placed significant constraints on viable theories of the origin of the main field. And while a palaeomagnetician quite likely would not understand the calculations of flow patterns in the outer core, the results would still be of interest. Specialisation has been partial and primarily methodological and institutional. This paper tells a straightforward story of these events and places the drama of mid-twentieth-century geomagnetism in the context of the longer story of generations of successful and interesting research (cf. Parkinson 1998).

The story told, moreover, concentrates on research questions and methods, leaving aside crucial social, institutional and cultural issues related to the development of geophysics. Undeniably, industrial/economic interests, the military use and support of geophysics, and the politics of the World Wars and the Cold War all played important roles in this history. The International Geophysical Year, the establishment of World Data Centres and space exploration also influenced the development of geophysics in many ways. Doel (1997), among others, has begun investigation of the history of these matters, which require much more vigorous pursuit.

Terrestrial magnetism

In 1900, investigations of Earth's magnetism flourished as never before. All the major European powers and their colonial empires, along with the United States, Japan, and several other nations, established magnetic observatories and sent out teams of researchers to map magnetic declination and other variables (Merlin & Somville 1910; Chapman & Bartels 1940, pp. 955–957). In 1904, Louis Agricola Bauer established the Department of International Research in Terrestrial Magnetism, better known as the DTM, at the Carnegie Institution of Washington, to fill in the gaps in these surveys around the world. The DTM also brought regular observations of the changing magnetic field to places that were previously bereft of observatories (Good 1994; Bauer 1912–1927).

The explicit goal of this frenetic global activity was to understand Earth's magnetism in its entirety. Many geomagnetic scientists at the beginning of the twentieth century were inspired by the two nineteenth-century giants, Alexander von Humboldt (Rupke 1997) and Carl Friedrich Gauss (Dunnington 1955). Humboldt had attempted the impractical: to grasp the dynamic phenomena of the Earth and the Cosmos in one mind, and to reveal and revel in their interconnections. This included both magnetic and electric phenomena. From his work sprang the institutionalization of 'terrestrial physics' and 'cosmic physics', which continued as well recognized branches of physics into the early twentieth century (Walker 1866; Conrad 1938). Magnetic researchers in 1900 saw their chosen phenomena in this context and directed their research in such a way as to endeavour to fulfil Humboldt's vision.

Researchers in 1900 had a critically important practical advantage over Humboldt. Geomagnetic research, with its requirements of observatories, instruments and international activity, was expensive. During the half century since Humboldt's last magnetic researches, the fiscal and organizational vitality of many nations had increased significantly. They could now afford not only to survey their home territories, but their extensive colonial empires. Germany, France, the Netherlands, Britain and the United States, in particular, did this (Pyenson 1985, 1989). A few nations and ambitious individuals like Roald Amundsen and Robert F. Scott, in the rush for the polar regions, likewise equipped their expeditions for magnetic research (e.g. Chree 1903; Good 1991). The DTM took advantage of the largess of Andrew Carnegie's private fortune to launch the most far-ranging and systematic of these global enterprises.

Magnetic researchers in the early twentieth century tended to be interested in all aspects of Earth's magnetic phenomena. Consider two of the more important theorists: Adolf Schmidt and Arthur Schuster. Schmidt, who directed the Prussian magnetic observatory in Potsdam, followed in the footsteps of Edward Sabine by publishing an extensive compendium of magnetic data. Whereas Sabine's numerous 'Contributions to Terrestrial Magnetism' assembled data from magnetic surveys of many countries and individuals, Schmidt collected tables of data from many observatories for the systematic study of time variations (Schmidt 1903–1926). Schmidt assembled these and other data to answer diverse theoretical questions. What were the causes of magnetic storms (Schmidt 1899)? Did electric currents flow through the surface of the Earth (Schmidt 1939)? What caused secular variation (Schmidt 1932)?

Schuster, trained as a physicist by Helmholtz and Maxwell, published his first important magnetic work in 1889: an application of Gauss's spherical harmonics to the problem of the diurnal variation of Earth's magnetism (Schuster 1889). This work provided the basis for future studies of induced electromagnetic fields in the crust and in the upper atmosphere. This gave rise to two apparently independent, yet closely related, areas of research: electrical conductivity of the crust and electrical currents in the ionosphere and beyond. Schuster also published on the causes of magnetic storms (Schuster 1911) and on the causes of Earth's main magnetic field (Schuster 1912). This inclusivity was common to leading researchers around 1900.

Sydney Chapman presented a most useful guide to geomagnetic research in the early twentieth century in his acceptance speech for the first Chree Medal in 1941. Charles Chree, who died in 1928, had been Director of the Kew (meteorological and magnetic) Observatory from 1893 to 1925. In his Chree Address, Chapman related the research careers of Chree, Schmidt and Bauer, saying that these three – plus the Dutch Willem van Bemmelen, the Indian N. A. F. Moos, and Edward Walter Maunder and Arthur Schuster in England – 'epitomise the progress of earth magnetic science during nearly half a century' (Chapman 1941, p. 630). He characterized the different 'gifts' that each researcher brought to the science: Moos and Chree's mathematical ability and indefatigable treatment of data; Bauer's 'fiery enthusiasm' and 'wide views'; Maunder's familiarity with events on the surface of the Sun, and Schuster's 'brilliant sorties' and 'striking theoretical conclusions' (Chapman 1941, pp. 632–633). Chapman divided the rest of his discussion into the consideration of time relationships and distribution of geomagnetism over space, a traditional division that closely parallels the studies of solar–terrestrial relationships and deep-Earth magnetic phenomena today. These traditions of terrestrial and cosmic physics relate intimately to the multifaceted development of geophysics and space physics. This, however, is not the place to pursue the story of the physical study of the Earth *in extenso* (Doel 1997; Good 2000).

As the cases of Schmidt and Schuster indicate, geomagnetic research in 1900 was not merely 'Baconian' or inductive, as, indeed, it was not in earlier centuries either. That is, scientists were not aimlessly collecting reams of data. (This popular characterization of 'Baconianism' does not do justice to Francis Bacon, but this issue need not be entered into here.) Their data collection was directed by theory. Even the activity of the Carnegie's DTM – with its dozens of technician–expeditionaries off around the world, with its magnetic survey vessels cruising the oceans, and with its observatories in Peru and Australia automatically generating extensive data relating to numerous types of phenomena – was undertaken to answer questions. Bauer had written his dissertation at Berlin on the analysis of the main magnetic field and secular variation (Good 1994; Bauer 1895). The data available, he lamented, were inadequate to the theoretical studies that needed to be undertaken. In order to explain the production of the main field, the cause of secular variation, the diurnal variations and magnetic storms, data collection guided by theory was required.

Primarily, the theories of Carl Friedrich Gauss and James C. Maxwell (Garland 1979; Harman 1998; Hunt 1991) provided that guidance. Geomagnetic research from the 1890s onwards was in the hands of investigators trained in physics. They exploited the data obtained during the 'Magnetic Crusade' (Morrell & Thackray 1981) and the first International Polar Year (Millbrooke 1998). They applied Gauss's spherical harmonic analysis with ever-greater sophistication. Schmidt, Bauer, Schuster and others firmly entrenched the habit of treating geomagnetism and geoelectricity exclusively in terms of field theory; and they made it clear that the future of explaining the main field and disturbance fields lay in this direction.

Geomagnetism

We no longer remember, and it seems unlikely today, but in 1938 'geomagnetism' was a new word in English. Germans had written of *Erdmagnetismus* for nearly a century, but to anglophone and francophone researchers the subject had long been 'terrestrial magnetism' and '*magnétisme terrestre*'. Sydney Chapman suggested the change. Although his reasons were linguistic and pragmatic, numerous changes were sweeping through this research community, which made the change more than a matter of linguistic convenience.

A publishing event in 1940 marked a critical period in the history of geomagnetic research: after a decade of collaboration, Sydney Chapman (then of Imperial College London) and Julius Bartels (then the Director of the Geophysical Institute, Potsdam) published their monumental treatise with the simple title, using the new word: *Geomagnetism* (Chapman & Bartels 1940). The authors noted in the preface (Vol. 1, p. vii) that no general treatments had been published on the subject since Edward Walker (1866) and Éleuthère Mascart (1900) and that these works answered 'few of the questions which most interest modern workers on geomagnetism'. Researchers were now investigating solar and cosmic ray physics, geophysical prospecting and radio communication. Perhaps more importantly, these researchers employed new methods of physical and mathematical analysis.

Geomagnetism, according to Chapman & Bartels (1940, Vol. 1, pp. vii–viii), stood 'between solar physics and the mainly more local terrestrial science of meteorology, on the one hand, and on the other, the universal science of physics'. Indeed, it encompassed parts of each of the neighbouring fields. The topics covered in the book reflected this, including for example: Earth's main field; secular variation; magnetic anomalies and geological prospecting; periodic variations due to the Sun and Moon; magnetic disturbances; solar–terrestrial connections; earth currents; *aurora polaris*; atmospheric conductivity and the ionosphere; statistical and harmonic analysis of periodic phenomena and the main field; physical theories of the main field; electromagnetic induction within the Earth; and much more. Nothing, it seems, was omitted.

As with all watershed works, however, Chapman & Bartels's *Geomagnetism* did turn its back on part of the history of its subject. It began a movement in new directions. Their compendium not only incorporated the accomplishments of Gauss and Maxwell and the data of the expeditions and observatories of the nineteenth and early twentieth centuries, it also incorporated elements of the 'new physics'. Many theories based in older natural philosophy did not merit discussion even in the final historical chapter of the book (Chapman & Bartels 1940, Vol. 2, pp. 898–937). The authors faced the future and their book provided the platform for launching the next generation of researchers. These new researchers carried their investigations along diverging trajectories: palaeomagnetism; theories of the main geomagnetic field; investigations of induced fields and currents (and conductivity); and studies of the upper atmosphere and near space.

This paper recounts the 'stories' of palaeomagnetism, the origin of the main magnetic field, and of induced fields/crustal conductivity in the twentieth century. (The history of investigations of near-space researches will mostly be reserved for another publication.) These stories are worth recalling because the development of plate tectonics and of dynamo theories completely changed the reasons for investigating the Earth's magnetism. The contexts of magnetic research before 1950 seldom even merit mention in histories written in recent years, because these histories have often focused explicitly on how plate tectonics came to be accepted or because they have seen the pre-dynamo days as non-theoretical and essentially uninteresting. Hence, although earlier contexts of investigations of rock magnetism have been seen – legitimately – as irrelevant to plate tectonics and so have largely been omitted or forgotten, they do in fact embody a significant part of the history of geomagnetic research in the twentieth century. Likewise, although interest in the causes of the main geomagnetic field motivated generations of researchers before 1950, that early work has been seemingly eclipsed by the development of dynamo theories.

Rock magnetism

The study of 'rock magnetism' is larger than the study of palaeomagnetism. That is, the subject is not just about what the magnetism of rocks can tell us about the past condition of the Earth. This is certainly part of the story, but so are two other main topics: the connection of remanent magnetism to local anomalies and the study of rock magnetism as a subject in its own right. Nevertheless, consider the pre-1950s history of palaeomagnetism first.

The utility of rock magnetism for revealing the history of the main field and secular variation far pre-dates its connection to the research questions of polar wander and plate tectonics. Since the discovery of secular variation in 1634 by Henry Gellibrand, researchers had been trying to explain the slow variation of declination and had placed it in the context of numerous research agendas. Edmond Halley famously sought to explain secular variation by hypothesizing the existence of a magnetic shell surrounding a magnetic 'nucleus' inside the Earth, the two revolving at slightly different speeds. The 'four-pole theory' of Halley was revived in the early nineteenth century by Christopher Hansteen and encouraged consideration of a secular periodicity even into the early twentieth century. In the late nineteenth century, investigators were critically aware that little was known of the history of Earth's magnetism. If one were generous, good data then extended back perhaps three hundred years (now four hundred) to the 1580s (with the work of Robert Norman), and that only for declination and dip (Jackson *et al.* 2000). Good data for magnetic intensity were available only since the work of Gauss and Wilhelm Weber in the 1830s, with less useful data going back to Humboldt and Jean Charles Borda in the 1790s. While such a short reach might have seemed acceptable when the planet was thought to be only a few thousand years old, by 1900 Earth's history was generally accepted to be much longer. While a few nineteenth-century scientists contemplated an Earth as young as a few tens of millions of years, most thought in terms of hundreds of millions, from the 1820s until the discovery of radioactivity (Thomas 1998, pp. 13–16).

The most general motivator, then, for palaeomagnetic research in the early twentieth century was to provide better data for the explanation of secular variation. It was also thought that these data might help to explain the production of the main field. Bauer, certainly, saw the need to extend the palaeomagnetic data-set in these terms. He combined physical palaeomagnetic research with what might be called 'archival palaeomagnetism'. That is, he searched old publications, sea captains' logs, etc., trying to wring the best information possible from a 'barely damp rag' (Jackson & Barraclough 1998; Good 1994; Bauer 1908). From the 1930s onwards, investigators at the DTM began systematic research on remanent magnetism largely to fulfil the Department's original remit, related to the main field and secular variation (McNish 1937; Johnson & McNish 1938; Graham 1949; Le Grand 1994, 1998). Another important question that motivated investigation of palaeomagnetism around 1950 was whether remanent magnetism really reflects the field when a rock formed (Le Grand 1994, 1998).

Rock magnetism transformed

A comparison of how rock magnetism and its history were treated in three landmark books spanning 1940 to 1964 will give some idea of how they were transformed during these 25 years. (Other publications were no less important – notably T. Nagata's *Rock Magnetism* (1953) – but these will suffice to make a few important points.)

The three books here compared are Chapman & Bartels's (1940) *Geomagnetism*, P. M. S. Blackett's (1956) *Lectures on Rock Magnetism*, and Edward (Ted) Irving's (1964) *Paleomagnetism and its Application to Geological and Geophysical Problems*. The points these three examples make are deceptively simple. First, the context of a research problem area changed over time. Second, the changes in that context affect the history that we select to write.

Consider Chapman & Bartels's chapter 'Magnetism and geology: magnetic prospecting' (Chapman & Bartels 1940, Vol. 1, pp. 137–158). In all of their massive two-volume study, this short chapter is the only one to discuss rock magnetism. Granted, both Chapman and Bartels faced more toward the cosmos than toward the solid Earth. Even so, most of this chapter discussed mapping of magnetic anomalies and related it to the mapping of gravitational anomalies and the locating of ore bodies. Chapman and Bartels described Schmidt's field balance, the use of local variometers, and the reduction of observations. They did not ask what rock magnetism had to say about any large theoretical matter – not the history of the magnetic field; not drift or polar wander; not even theories of magnetization in general. The one question they did ask was: Can magnetic anomalies tell us how these crustal rocks were magnetized? Were 'highly susceptible rocks' magnetized by induction by the present geomagnetic field (Chapman & Bartels 1940, Vol. 1, pp. 145–146)? They concluded that the existence of strong negative anomalies indicates 'that magnetic rocks may be *permanently* magnetized in directions differing from that of the present field'. They explicitly avoided deciding between 'whether this permanent magnetization was produced by the general field of the earth at the time of congelation or metamorphosis, or whether other causes must be considered' (Chapman & Bartels 1940, Vol. 1, p. 146).

In a section titled 'Induced or remanent rock magnetism' Chapman & Bartels (Vol. 1, pp. 154–157) noted that local anomalies with reversed polarization exist, only to recite the possible ways this phenomenon could be produced other than by a reversed general magnetic field – induction from a larger local body, overturning of strata, etc. And while they closed with the massive Pilansberg, South Africa dyke system and its reversed polarization, the authors merely said that it 'seems to defy any attempt at explanation other than that of a complete reversal of the magnetic field *in that region* [emphasis added] at the time of the formation' (p. 156). Chapman & Bartels wrote later in the book (p. 701) that, nevertheless, the outright rejection of evidence of geomagnetic reversals as unreliable was 'perhaps too dogmatic'.

By the time Blackett wrote his book, much had changed. At the DTM, Johnson, Murphy & Torreson (1948) had published their 'Pre-history of the Earth's magnetic field'. John Graham had completed much of his research on rock magnetism (e.g. Graham 1949, 1955). Most importantly, Blackett himself had developed and ultimately rejected a 'fundamental' theory that attributed magnetism generally to massive rotating bodies, which will be discussed later in this paper. Also in the late 1940s, the first rough attempts at dynamo theories were made by Elsasser and Bullard (Nye 1999). Blackett came at rock magnetism with the history of the Earth's field prominently in mind. He reviewed all the useful literature back to the 1890s.

> Such information [about past geomagnetic conditions] would be of great importance for its own sake but would be of immense value in an attempt to understand the physical mechanism giving rise to the field. . . . Without the study of rock magnetism we had no possibility of knowing whether the field might not have been vastly different in the distant past, perhaps a thousand or more times greater or smaller (Blackett 1956, p. 5).

While Blackett studied magnetization itself as a phenomenon, he was more interested in the 1950s to connect rock magnetism to global problems, especially to the cause of the main field (Blackett 1956, p. 7).

Blackett raised another critical perspective in which rock magnetism had great importance: the possibility that the Earth's magnetic dipole had reversed itself suddenly and repeatedly (Blackett 1956, pp. 6–8). At first, this larger implication was, for Blackett, mainly part of discovering the history of Earth's main magnetic field. But he conducted this research during the early 1950s, when Keith Runcorn and Kenneth Creer were proposing polar drift as a way to explain 'odd' palaeomagnetic readings. Arthur Holmes was still discussing continental drift and the work of Wegener in his *Physical Geology* (Holmes 1944), which Blackett read while working on rock magnetism in the early 1950s. Blackett saw possible reversals, and the mapping of a drift, as an indication, if not a proof, that his own theory connecting geomagnetism and rotation was wrong. If Earth's magnetism were tied fundamentally to its rotation, this magnetism could not reverse unless Earth's rotation did – a wholly unlikely scenario.

Blackett's experimental investigation contributed to the rapid specialization of magnetometers in the 1950s, some more accurate and precise, others more portable. Although his theory failed, his instrument design helped ultimately to validate plate tectonics theory (Nye 1999). Likewise, the work of Packard & Varian (1954) on the proton-precession magnetometer proved effective in the mapping of magnetically striped oceanic crust. Fluxgate magnetometers (useful in aeromagnetic surveys), Zeeman-effect magnetometers (extremely sensitive and useful in space probes) and cryogenic magnetometers (useful in palaeomagnetic work), taken altogether, represented the broad-ranging effects of new technologies, based on electronics and physical research, on geophysics in the mid-twentieth century (Parkinson 1983, pp. 44–59). They also demonstrate that specialization has been embodied in instrumentation.

Edward Irving's book presented a transformed view of rock magnetism. His shift to the term palaeomagnetism was purposeful, indicating that he was mainly interested in rock magnetism as evidence of past conditions. He particularly highlighted 'the hypothesis of continental drift' and pointed out that palaeomagnetic measurements provided 'numerical tests' that could refute it, with evidence of a type different from the data originally used by Alfred Wegener (Irving 1964, pp. vi–vii). Although Irving was certainly aware of the connection of palaeomagnetic data to theories of the origin of the main field, this had dropped out of his story (Irving 1964, p. 4). His historical section discussed measurements of reversed magnetization as early as Alexander von Humboldt. Curiously, Irving went on from Humboldt to state that a number of other intensely anomalous rock outcroppings had been found in the nineteenth century and that these were termed *punti distinti* or *points isolés*. His history had the same character, with the clear application of a selection criterion. Irving isolated these scientific acts from their historical contexts and saw them in the light of his own concerns (Irving 1964, pp. 6–8).

This selective approach to history is not unusual and I do not mean to criticize Irving. Indeed, this kind of selectivity is inherent in all history – not just the history of science written by scientists. One is easily drawn to tracing out a 'family tree' when looking backward from a strong preoccupation, such as one's current research. Even reflecting back on one's own life, it is common to forget or gloss over the confusion, or the dead-end project, or the one simply left behind. Oral histories and memoirs substantiate this repeatedly.

In the case of palaeomagnetism, its history

was unselfconsciously rewritten with succeeding generations, as the focus of palaeomagnetic research itself shifted. In the end, even historians looking back at the history of palaeomagnetism have written mainly about its importance in the plate tectonics story. But in the 1940s and earlier, that was not generally the context of research in rock magnetism. The utility of palaeomagnetic data in evaluating the hypothesis of continental drift was a connection that few imagined before the 1950s, although Paul Mercanton apparently first suggested this connection in 1926 (Mercanton 1926).

The main field and the geodynamo

The drive to explain the origin of the main field (and secondarily of secular variation) necessarily intersected with interest in palaeomagnetism, as noted above. One of the main reasons for this was that researchers felt that information provided by historical observations did not cover a long enough stretch of Earth's record. Much research on the origin of the main magnetic field, however, had nothing to do with rock magnetism. Indeed, most developments in this line were driven by the capabilities of physics and mathematics.

Arthur Schuster outlined the available explanations in his 'Critical examination of the possible causes of terrestrial magnetism' in 1912 (Good 1998, pp. 355–356). He thought it was premature to rule out permanent magnetization since the effect of very high pressures on magnetization were not understood. The possibility of an inductive effect from electrical currents inside the Earth was, he thought, overrated. The first of these explanations had roots in William Gilbert and Edmond Halley's ideas, and the second in those of André-Marie Ampère. Schuster rejected another idea, popular in the nineteenth century, that Earth's main field was induced by external, cosmic causes, and concluded that one of the most promising ideas connected Earth's magnetism to its rotation. If molecules were magnetic or if they carried an electric charge, rotation could produce the main field. But Schuster drew no firm conclusion.

Interest in this possible explanation persisted throughout the 1920s, 1930s and 1940s, when it was picked up again by Blackett. As outlined by Mary Jo Nye (1999, pp. 74–76), S. J. Barnett, Albert Einstein, Johannes de Haas, H. A. Wilson, W. F. G. Swann and A. Longacre all explored the issue. Barnett, Swann and Longacre attempted to measure the magnetism of rotating bodies. Einstein and de Haas considered the matter theoretically. When Blackett conceived the idea himself in the late 1940s, he quickly tracked down this literature through Chapman & Bartels (1940, Vol. 2, p. 705). In 1947, Blackett 'splashed' this revived theory across the world's headlines and began a multifaceted effort to test the idea once and for all. Objections against it arose from diverse quarters: evidence of magnetic reversals in stars was discouraging, as were arguments from quantum electrodynamics raised by Wolfgang Pauli and others. Most telling, however, were the results of geophysical tests. Following a suggestion of Edward Bullard, Keith Runcorn measured Earth's magnetic field deep in mine shafts. Bullard had noted that if magnetism were due to a distributed cause, such as Blackett's, then the field should be less within Earth's surface. Results, initially equivocal, ultimately tended against Blackett. Blackett's own elaborate laboratory experiments went against a fundamental relation between rotation and magnetism and in 1952 he published his negative results (Nye 1999, pp. 78–87). He shifted his energy to applying his sensitive magnetometers to palaeomagnetism and to testing continental drift.

A second and quite different type of theory emerged alongside Blackett's. In 1919, Joseph Larmor published a short article that discussed the possibility that dynamo action inside the Sun produces its magnetic field. Despite Thomas Cowling's argument in 1934 that an axially symmetrical field cannot be maintained by a dynamo, Walter Elsasser began exploring this type of theory in 1939, in 'On the origin of the Earth's magnetic field' (Brush & Bannerjee 1996, pp. 223–224). Elsasser wrote:

> The terrestrial field is traced here to the existence of thermoelectric currents in the metallic interior of the earth. The currents owe their existence to inhomogeneities continually created by turbulent convective motions (Elsasser 1939, p. 489).

As Elsasser later related, this was not strictly a dynamo theory. Nevertheless, it is where dynamo theory began. According to Stephen G. Brush and S. K. Banerjee, Elsasser's contribution to dynamo theory laid 'the foundation of the modern theory of terrestrial magnetism' (Brush & Banerjee 1996, p. 224). After the interruption due to World War II, Elsasser returned to this problem in 1946 with 'Induction effects in terrestrial magnetism'. His recognition that toroidal fields can exist in Earth's core provided the basis for a self-exciting dynamo, namely one in which induction effects reinforce the existing magnetic field (Brush & Banerjee 1996, p. 225). Bullard stepped into this picture in 1948 and

1954, supplying a more detailed theory of how a self-exciting dynamo might work.

Although the general idea of a self-sustaining dynamo in the outer core quickly gained general (but not universal) acceptance, some aspects remained controversial for years. G. E. Backus and others pointed out defects in the earlier theories and developed improved dynamo models in the late 1950s. Others involved in this included Stephen Childress and Glynn Roberts, A. Herzenberg and E. N. Parker. In the 1960s, Raymond Hide developed the 'magnetohydrodynamic wave hypothesis' as an alternative to motion of the outer core relative to the mantle. In this hypothesis, waves oscillating through the liquid outer core caused secular variation. P. H. Roberts and S. Scott, meanwhile, worked on the idea of 'frozen flux', in which the magnetic field moves with the fluid core material. This critically important area of geomagnetic research continued through the rest of the twentieth century, with important work being undertaken by David Gubbins, Jeremy Bloxham, and others (Brush & Banerjee 1996, pp. 227–231).

Geologists might be forgiven if they wonder what this had to do with their work. There was, however, one main area of intersection. The magnetic polarity reversals on which so much geochronology and so much research in plate tectonics now depend are in principle explainable only with a geodynamo (Gubbins *et al.* 2000). That the details of the physical and mathematical analysis of that dynamo might be beyond the reach of many geologists shows how far the partitioning of geomagnetic research developed during the twentieth century.

From crustal conductivity to the equatorial electrojet

Two lines of research could scarcely appear less related than variations in electrical conductivity of the crust and mantle and the existence of electrical currents in the ionosphere or beyond. However, in 1889 Schuster demonstrated that electrical currents in the upper atmosphere cause the daily variations in geomagnetic measurements (Schuster 1889). Schuster discovered that external currents produce internal fields and currents, and that these can be used to study conductivity of the crust. Sydney Chapman proposed a simple model of global conductivity dependent only on depth in 1919, which was further developed by Albert Price and B. N. Lahiri (1939) and by Chapman & Bartels (1940) (Parkinson 1998, p. 362). The broad outlines of global crustal conductivity established, R. Banks pushed the idea to greater depths around 1970.

Studies of local variation in crustal conductivity also took off. The Carnegie Institution's magnetic observatories began systematic measurements of Earth currents and conductivity in the 1920s. Around 1950, Andrei Nikolaivich Tikhonov and Louis Cagniard developed the methods of 'magnetotellurics', in which measurements of potential differences between probes are combined with readings of appropriate magnetic variations to study conductivity at various depths (Parkinson 1998, p. 363). In the 1960s, Albert Price extended this method to include horizontal variation. The direction of geomagnetic changes can indicate the local gradient of conductivity. Walter Jones, Ulrich Schmucker, Peter Weidelt, John Weaver, Ian Gough, and others, extended this work to the detailed study of bodies of various types and the development of new instruments. Robert Parker applied inversion methods to magnetotellurics in 1970, as this study was pushed deeper into the Earth. Studies of conductivity variation on the ocean floor and of electromagnetic induction in the oceans was pursued extensively from the 1970s into the 1990s (Parkinson 1998, pp. 363–364). These research topics represent one of the most significant interaction zones between geology and geophysics in the late twentieth century.

It should also be remembered, though, that studies of currents and fields in the ionosphere and near space followed paths that were largely ignored by geologists, and rightly so. The investigations included studies of the *aurora polaris* (Silverman & Egeland 1998), magnetic substorms (Akasofu 1998), the interaction of the solar wind and the magnetosphere (Akasofu 1983), the electrical ring-current in the ionosphere called the equatorial electrojet (Chapman 1951), and other important phenomena of the upper atmosphere and near space. The picture of the history of geomagnetic research in the twentieth century must ultimately encompass all of these various investigations and the communities of scientists involved, but that is a task far beyond the scope of this paper.

The transformation of disciplines

This paper has emphasized the fragmentation of terrestrial magnetism into several diverging specializations: palaeomagnetism, work on the geodynamo, crustal conductivity, and ionospheric and magnetospheric research. As J. A. Jacobs writes in his recent text *Geomagnetism* (1987): Chapman & Bartels surveyed about 100 000 pages of literature to write their compendium. When Matsushita and Campbell wrote theirs in 1967, they faced a much more daunting prospect. By 1987, Jacobs felt compelled to call on a long series of experts to each write about a single specialization (Jacobs 1987, Vol. 1, p. vii).

There is an element of truth to this story, but it is important not to carry this thought too far. While certainly geomagnetic researchers in the late twentieth century have tended to specialize more than their colleagues did a century earlier, the process has been necessary. There are, despite the tight focus on separated problem areas, significant continuities across the whole of geomagnetic research. Frequently, the instrumentation is the same. All draw on similar physical theory and mathematical techniques. There are connections among phenomena. Indeed, there are many examples of individuals who continue to work in more than one of the problem areas. Nevertheless, career imperatives and institutions enforce specialization. And, when looking back, writers seldom seem aware of the joint kinship shared by researchers in palaeomagnetism, the main field, and space physics – let alone geophysical prospecting, radio physics and cosmic ray studies. Historical investigations must at least acknowledge how specialization has affected our ability to see this past, or the writing of this history will be 'presentist' in the worst sense. Major developments in science sometimes induce a sort of amnesia, which we must constantly fight against if we are to write histories faithful to the contexts of their times.

Special thanks are due to all of the participants in the Rio sessions for discussions that broadened and deepened this investigation, and to R. E. Doel, W. D. Parkinson, and A. Jonkers for their critiques of the draft manuscript. I didn't address all of their recommendations in the final revision, but I have taken them to heart and will be considering their other suggestions in future writing. I also thank S. Solomon, director of the Department of Terrestrial Magnetism, Carnegie Institution of Washington, for graciously hosting and supporting my historical research during a sabbatical year in 1998–1999.

References

AKASOFU, S.-I. 1983. Evolution of ideas in solar–terrestrial physics. *Geophysical Journal of the Royal Astronomical Society*, **74**, 257–299.

AKASOFU, S.-I. 1996. Search for the 'unknown' quantity in the solar wind: a personal account. *Journal of Geophysical Research*, **101** (A5), 10531–10540.

AKASOFU, S.-I. 1998. The rise and fall of paradigms and some longstanding unsolved problems in solar–terrestrial physics. *In*: KOKUMUN, S. & KAMIDE, Y. (eds), *Substorms-4. International Conference on Substorms*. Terra Scientific, Tokyo; Kluwer Academic, Dordrecht & Boston, 21–25.

BAUER, L. A. 1895. *Beiträge zur Kenntnis des Wesens der Säkularvariation des Erdmagnetismus*. Dissertation, University of Berlin.

BAUER, L. A. 1908. The earliest values of the magnetic declination. *Terrestrial Magnetism and Atmospheric Electricity*, **13**, 97–104.

BAUER, L. A. 1912–1927. *Researches of the Department of Terrestrial Magnetism* (6 vols). Carnegie Institution of Washington, Publication **175**. Washington, DC.

BLACKETT, P. M. S. 1956. *Lectures on Rock Magnetism*. The Weizmann Science Press of Israel, Jerusalem.

BRUSH, S. G. & BANERJEE, S. K. 1996. Geomagnetic secular variation. *In*: BRUSH, S. G. *Nebulous Earth: the Origin of the Solar System and the Core of the Earth from Laplace to Jeffreys*. Cambridge University Press, Cambridge, 220–232.

BULLARD, E. C. 1948. The secular change in the Earth's magnetic field. *Monthly Notices of the Royal Astronomical Society, Geophysical Supplement*, **5**, 248–257.

BULLARD, E. C. 1954. Homogeneous dynamos and terrestrial magnetism. *Philosophical Transactions of the Royal Society of London,* Series A, **247**, 213–278.

CHAPMAN, S. 1938. Geomagnetism or terrestrial magnetism? *Terrestrial Magnetism and Atmospheric Electricity*, **43**, 321.

CHAPMAN, S. 1941. Charles Chree and his work on geomagnetism. *The Proceedings of the Physical Society,* **53**, 629–634.

CHAPMAN, S. 1951. The equatorial electrojet as detected from the abnormal electric current distribution above Huancayo, Peru, and elsewhere. *Archiv für Meteorologie, Geophysik, und Bioklimatologie*, Series A, **4**, 368–390.

CHAPMAN, S. & BARTELS, J. 1940. *Geomagnetism* (2 vols). Clarendon Press, Oxford.

CHREE, C. 1903. *Magnetic Observations Made at the 'Southern Cross' Antarctic Expedition, 1899–1900, at Cape Adare*. The Royal Society, London.

CLIVER, E. 1998. Solar–terrestrial relations. *In*: GOOD, G. A. (ed.) *Sciences of the Earth: An Encyclopedia of Events, People, and Phenomena*. Garland Publishing, New York & London, **2**, 776–787.

CONRAD, V. (ed.). 1938. *Physik der Atmosphäre: Ergebnisse der kosmischen Physik* (3rd supplemental volume to *Gerlands Beiträge zur Geophysik*) Akademische Verlagsgesellschaft, Leipzig.

DOEL, R. E. 1997. The earth sciences and geophysics. *In*: KRIGE, J. & PESTRE, D. (eds) *Science in the 20th Century*. Harwood Academic, Amsterdam, 391–416.

DUNNINGTON, G. W. 1955. *Carl Friedrich Gauss, Titan of Science: A Study of his Life and Work*. Hafner, New York.

ELSASSER, W. M. 1939. On the origin of the Earth's magnetic field. *Physical Review*, **55**, 489–498.

ELSASSER, W. M. 1946. Induction effects in terrestrial magnetism. *Physical Review*, **69**, 106–116; **70**, 202–212.

FRANKEL, H. 1998. Continental drift and plate tectonics. *In*: GOOD, G. A. (ed.) *Sciences of the Earth: An Encyclopedia of Events, People, and Phenomena.* Garland Publishing, New York & London, Vol. 1, 118–136.

GARLAND, G. D. 1979. The contributions of Carl Friedrich Gauss to geomagnetism. *Historia Mathematica*, **6**, 5–29.

GOOD, G. A. 1991. Follow the needle: seeking the magnetic poles. *Earth Sciences History*, **10**, 154–167.

GOOD, G. A. 1994 Vision of a global physics: the Carnegie Institution and the first world magnetic survey. *History of Geophysics*, **5**, 29–36.

GOOD, G. A. 1998. Geomagnetism, theories between 1800 and 1900. *In*: GOOD, G. A. (ed.) *Sciences of the Earth: An Encyclopedia of Events, People, and Phenomena.* Garland Publishing, New York & London, Vol. 1, 350–357.

GOOD, G. A. 2000. The assembly of geophysics: scientific disciplines as frameworks of consensus. *Studies in the History and Philosophy of Modern Physics*, **31**, 259–292.

GRAHAM, J. W. 1949. The stability and significance of magnetism in sedimentary rocks. *Journal of Geophysical Research*, **54**, 131–167.

GRAHAM, J. W. 1955. Evidence of polar shift since Triassic time. *Journal of Geophysical Research*, **60**, 329–347.

GUBBINS, D., KENT, D. V. & LAJ, C. (eds) 2000. Geomagnetic polarity reversals and long-term secular variation. *Philosophical Transactions of the Royal Society of London*, **A358**, 869–1223.

HARMAN, P. M. 1998. *The Natural Philosophy of James Clerk Maxwell.* Cambridge University Press, Cambridge.

HOLMES, A. 1944. *Principles of Physical Geology.* Thomas Nelson, London.

HUFBAUER, K. 1998. Solar wind. *In*: GOOD, G. A. (ed.) *Sciences of the Earth: An Encyclopedia of Events, People, and Phenomena.* Garland Publishing, New York & London, Vol. 2, 774–776.

HUNT, B. J. 1991. *The Maxwellians*. Cornell University Press, Ithaca.

IRVING, E. 1964. *Paleomagnetism and its Application to Geological and Geophysical Problems.* John Wiley & Sons, New York.

JACKSON, A. & BARRACLOUGH, D. 1998. Contemporary use of historical data. *In*: GOOD, G. A. (ed.) *Sciences of the Earth: An Encyclopedia of Events, People, and Phenomena.* Garland Publishing, New York & London, Vol. 1, 115–118.

JACKSON, A., JONKERS, A. R. T. & WALKER, M. R. 2000. Four centuries of geomagnetic secular variation from historical records. *Philosophical Transactions of the Royal Society of London*, Series A, **358**, 957–990.

JACOBS, J. A. 1987. *Geomagnetism* (3 vols). Academic Press, London.

JOHNSON, E. A. & MCNISH, A. G. 1938. An alternating-current apparatus for measuring small magnetic moments. *Terrestrial Magnetism and Atmospheric Electricity*, **53**, 393–399.

JOHNSON, E. A., MURPHY, T. & TORRESON, O. W. 1948. Prehistory of the Earth's magnetic field. *Journal of Geophysical Research*, **43**, 349–372.

LAHIRI, B. N. & PRICE, A. 1939. Electromagnetic induction in non-uniform conductors, and the determination of the conductivity of the Earth from terrestrial magnetic variations. *Philosophical Transactions of the Royal Society of London*, Series A, **237**, 509–540.

LE GRAND, H. 1994. Chopping and changing at the DTM 1946–1958: M. A. Tuve, rock magnetism, and isotope dating. *History of Geophysics*, **5**, 173–184.

LE GRAND, H. 1998. Paleomagnetism. *In*: GOOD, G. A. (ed.) *Sciences of the Earth: An Encyclopedia of Events, People, and Phenomena.* Garland Publishing, New York & London, Vol. 2, 651–655.

MCNISH, A. G. 1937. Electromagnetic methods for testing rock-samples. *Terrestrial Magnetism and Atmospheric Electricity*, **42**, 283–284.

MASCART, É. 1900. *Traité de magnétisme terrestre.* Paris, Gauthier–Villars.

MATSUSHITA, S. & CAMPBELL, W. H. (eds) 1967. *Physics of Geomagnetic Phenomena.* Academic Press, New York.

MERCANTON, P. L. 1926. Inversion de l'inclinaison magnétique terrestre aux âges géologiques. *Terrestrial Magnetism and Atmospheric Electricity*, **31**, 187–190.

MERLIN, E. & SOMVILLE, O. 1910. *Liste des Observatoires Magnétiques et des Observatoires Séismologues.* Observatoire Royal de Belgique, Brussels.

MILLBROOKE, A. 1998. International Polar Years. *In*: GOOD, G. A. (ed.) *Sciences of the Earth: An Encyclopedia of Events, People, and Phenomena.* Garland Publishing, New York & London, Vol. 2, 484–487.

MORELL, J. & THACKRAY, A. 1981. *Gentlemen of Science: Early years of the British Association for the Advancement of Science.* Clarendon Press, Oxford; Oxford University Press, New York.

NAGATA, T. 1953. *Rock Magnetism.* Maruzen, Tokyo.

NYE, M. J. 1999. Temptations of theory, strategies of evidence: P. M. S. Blackett and the Earth's magnetism, 1947–1952. *British Journal for the History of Science*, **32**, 69–92.

ORESKES, N. 1999. *The Rejection of Continental Drift: Theory and Method in American Earth Science.* Oxford University Press, New York.

PACKARD, M. E. & VARIAN, R. 1954. Free nuclear induction in the Earth's magnetic field (abstract). *Physical Review*, **93**, 941.

PARKINSON, W. D. 1983. *Introduction to Geomagnetism.* Scottish Academic Press, Edinburgh.

PARKINSON, W. D. 1998. Geomagnetism, theories since 1900. *In*: GOOD, G. A. (ed.) *Sciences of the Earth: An Encyclopedia of Events, People, and Phenomena.* Garland Publishing, New York & London, Vol. 1, 357–365.

PYENSON, L. 1985. *Cultural Imperialism and Exact Sciences: German Expansion Overseas 1900–1930.* Peter Lang, New York.

PYENSON, L. 1989. *Empire of Reason: Exact Sciences in Indonesia 1840–1940.* E. J. Brill, New York.

RUPKE, N. A. 1997. Introduction: the liberal standard

of science literacy of the mid-nineteenth century. *In*: HUMBOLDT, A. 1997. *Cosmos: A Sketch of the Physical Description of the* Universe (translated by E. C. Otté). Johns Hopkins University Press, Baltimore, Vol. 1, vii–xlii. (Otté's translation was first published 1858.)

SCHMIDT, A. 1899. Über die Ursache der magnetischen Stürme. *Meteorologische Zeitschrift*, **9**, 385–397.

SCHMIDT, A. 1903–1926. *Archiv des Erdmagnetismus* (7 vols). Königlich Preussischen Akademie der Wissenschaften, Potsdam.

SCHMIDT, A. 1932. Das Rätsel der erdmagnetischen Säkularvariation. *Terrestrial Magnetism and Atmospheric Electricity*, **37**, 225–230.

SCHMIDT, A. 1939. Zur Frage der hypothetischen die Erdoberfl˙che durchdringeneden Ströme, mit einem Zusatz vom J. Bartels. *Gerlands Beiträge zur Geophysik*, **55**, 292–302.

SCHUSTER, A. 1889. The diurnal variation of terrestrial magnetism, with an appendix by H. Lamb. On the currents induced in a spherical conductor by variation of an external magnetic potential. *Philosophical Transactions of the Royal Society of London*, Series A, **180**, 467–518.

SCHUSTER, A. 1911. On the origin of magnetic storms. *Proceedings of the Royal Society of London*, **85**, 44–50.

SCHUSTER, A. 1912. Critical examination of the possible causes of terrestrial magnetism. *Proceedings of the Physical Society of London*, **24**, 121–137.

SILVERMAN, S. & EGELAND, A. 1998. Auroras since the International Geophysical Year. *In*: GOOD, G. A. (ed.) *Sciences of the Earth: An Encyclopedia of Events, People, and Phenomena*. Garland Publishing, New York, Vol. 1, 66–70.

STERN, D. P. 1989. A brief history of magnetospheric physics before the spaceflight era. *Reviews of Geophysics*, **27**, 103–114.

STERN, D. P. 1996. A brief history of magnetospheric physics during the space age. *Reviews of Geophysics*, **34**, 1–31.

THOMAS, R. D. K. 1998. Age of the Earth, since 1800. *In*: GOOD, G. A. (ed.) *Sciences of the Earth: An Encyclopedia of Events, People, and Phenomena*. Garland Publishing, New York, Vol. 1, 19–23.

VAN ALLEN, J. 1983. *Origins of Magnetospheric Physics*. Smithsonian Institution Press, Washington, DC.

WALKER, E. 1866. *Terrestrial and Cosmical Magnetism (Adams Prize Essay for 1865)*. Deighton, Bell, Cambridge.

Sedimentology: from single grains to recent and past environments: some trends in sedimentology in the twentieth century

EUGEN & ILSE SEIBOLD

Geological Institute, Freiburg University, Richard Wagner Strasse 56, D-79104 Freiburg, Germany

Abstract: Before 1900, Henry Clifton Sorby, the 'founder of sedimentary petrography', covered many aspects of the study of sedimentary rocks. During the first half of the twentieth century, there followed numerous detailed descriptions of such rocks and actualistic ideas were increasingly introduced. Specialization began between the two World Wars. The use of single grains and their statistical evaluation, especially of heavy minerals, and the investigation of clay minerals, were stimulated by the needs of the oil industry, together with regional descriptions, including facies studies on land and on the sea bottom. Specialization further increased between 1945 and 1968, with an explosion of publications. Ongoing field and laboratory studies, and new concepts such as the origin of turbidites, or diagenesis – especially in carbonate rocks – were treated in much greater detail. Again the oil industry was one of the major driving forces. Since 1968, global aspects gained greater attention, as for example with the Deep-sea Drilling Project. Geophysics contributed to facies and basin analysis. Extraterrestrial factors such as variation in Earth's orbit or bolide impacts, and their indications in sediments, came to be considered important for understanding world climates, and also evolution. Cross-disciplinary and international approaches have become, and continue to be, of growing importance.

Up to the late nineteenth century sedimentary rocks were chiefly studied because they contain fossils, better and more useful for stratigraphy than the sequences of Steno's '*sedimenti*' from 1669. But in addition, since the middle of the nineteenth century their compositions have attracted increasing interest, especially after the introduction of thin sections and their microscopical investigation. However, at the beginning an interest in igneous rocks prevailed, even though Henry Clifton Sorby (1826–1908), the 'founder of sedimentary petrography' (Folk 1965), had published relevant papers with sedimentological aspects from as early as 1850. Up to 1908 he added many important and even quantitative ideas (see reprints with commentaries in Carozzi 1975).

Detailed descriptions of sedimentary rocks followed, as in the famous monographs of Lucien Cayeux (1864–1944), from an 'introduction' in 1916 to ferruginous rocks (1909, 1922), siliceous rocks (1929), carbonates (1935) to phosphates (1935, 1939–1950). Assar Hadding (1886–1962) added many genetic features (as in 1929).

Nevertheless the 'echo' in geology remained modest. Gradually it became stronger, not least since it came to include the deployment of actualistic concepts by Johannes Walther (1860–1937) in 1893–1894, further propagated by Amadeus W. Grabau (1870–1946), who dedicated his classical *Principles of Stratigraphy* (1913) to Walther.

Sedimentology can therefore be seen as a fusion of sedimentary petrology and stratigraphy (Doeglas 1951) or defined simply as the science of sedimentary deposits (Wadell 1932*a*). Historical remarks can be found in textbooks like those of Pettijohn (1949), Milner (1922, 1962) or in Middleton (1978). New concepts are treated in Ginsburg (1973) or Dott (1988).

During World War I the progress in sedimentology was delayed because of the emphasis on applied geology and the interruption of international connections. Afterwards it expanded gradually, but in so many directions that we can select only a few trends, mostly from our own experience. Even important geochemical, pedological, geophysical, biological or mathematical aspects, and applied sedimentology, can be treated only marginally here. Our language barrier, especially where Russian is concerned, is a further restriction. Some relevant information was given by Lew B. Ruchin in 1958 or by Nicolai Mikhailovich Strakhov (1900–1978) in 1970.

The main trends in sedimentological research are indicated below by the time intervals in which they appeared.

From: OLDROYD, D. R. (ed.) 2002. *The Earth Inside and Out: Some Major Contributions to Geology in the Twentieth Century*. Geological Society, London, Special Publications, **192,** 241–250. 0305-8719/02/$15.00

1918–1945: specialization

With the worldwide expansion of hydrocarbon exploration after World War I clastic sedimentary rocks received special attention. *Heavy minerals* were used for stratigraphical correlation and as indicators for provenance. Early pioneers are mentioned in a summary by Boswell (1933). However, with better logging methods and progress in micropalaeontology these interests declined. Nevertheless, to give an example from the period after World War II, one may mention a classic investigation of the molasse north of the Alps, largely based on heavy minerals (Lemcke *et al.* 1953).

Other *single-grain approaches* were devoted to the shape of cobbles (Chester Keller Wentworth (1891–1969 in 1919), or sands (e.g. Hakon Wadell (1895–1962) (1932b, 1935), or to their surfaces (André Cailleux (1907–1986) in 1952).

Statistical methods became fashionable around 1930, starting with the measurement of different parameters and ending with attempts to decipher environmental conditions from grain-size distribution curves through sorting, skewness, etc. (e.g. William Christian Krumbein (1902–1979) in 1932 and 1934). This approach never became really successful, not only because of the influence of diagenesis in fossil rocks but also because it neglected *fractionating* and *qualifying* these grains. Carl Wilhelm Correns (1833–1980) introduced this method with material from the *Meteor* Cruise 1925/27 to the South Atlantic (Correns 1937); and, to look ahead, Francis P. Shepard (1897–1985) used it for the study of sediments in the Gulf of Mexico (Shepard *et al.* 1960) and an outstanding application was published in 1971 by Michael Sarnthein from the Persian/Arabian Gulf.

Up to now the study of *fabrics*, too, has been and still is useful in sediments and sedimentary rocks, especially for transport problems. Many examples are treated in Potter & Pettijohn (1963).

Progress in the investigation of finer particles has depended on new technical equipment. For decades *silt* belonged to a *terra incognita*. Only with the use of stereoscan microscopy since 1965 could nannoplankton and other silt particles be investigated in detail, as a crucial tool for deep-sea stratigraphy. *Clay minerals* became accessible after the introduction of X-ray powder diffraction techniques in the 1930s, based on the work of William Lawrence Bragg (1890–1971). Therefore argillaceous sediments were not treated in the voluminous publications of Lucien Cayeux. Clay minerals were used as indicators for climate and provenance and as factors in compaction. Authorities in clay mineralogy included Georges Millot (1917–1991) in France and Ralph E. Grim (1902–1989) in the United States, who gave summaries of much previous work (Millot 1964; Grim 1953, 1962).

In many of the contexts mentioned above, *organisms* can be important, too. Of course these are also excellent environmental indicators. Their hard parts act as clastic particles and are extremely sensitive to sorting. They may form rocks with specific fabrics as in coral reefs or stromatolites. Therefore biology and palaeontology are involved. (See the classic and monumental two volumes, edited in 1957 by Joel W. Hedgepeth and Harry S. Ladd, with numerous stimulating recent and fossil examples.) With their soft parts they contribute organic matter to sediments, culminating in the formation of black shales, altogether a realm of geochemistry, as, for example, treated by Degens (1965).

Many of these approaches have improved regional studies of recent marine sediment distribution and sedimentation processes. Pioneers for the Baltic were Stina Gripenberg (1934) and Otto Pratje (1948) (the latter for the southern North Sea too; Pratje, 1931), and Andrei Dmitrievich Arkhangelsky (1927) for the Black Sea. Fundamental general conclusions considering all aspects of sedimentation are due to William Henry Twenhofel (1875–1957) in 1932 and 1950. Following its foundation in 1931, he was, from 1933 to 1946, associate editor of the *Journal of Sedimentary Petrology*. Another authority was Parker Davies Trask (1899–1961), who became editor of *Marine Sediments* in 1939 (see Trask 1939).

It took some time until it was recognized that the present is not always a reliable key to the past, because our short 'Holocene' is such a special case in geological history. But Lucien Cayeux had recommended a differentiated approach to actualism as early as 1941.

Single-grain research marked the beginning of *micropalaeontology*. Before 1920 this was 'little more than a subject of cloistered research by a few specialists' (Owen 1975 p. 522). Because the oil industry needed better tools for stratigraphic correlations, rapid progress began. It was no coincidence that the *Journal of Paleontology* was founded in 1928 by the Society of Economic Paleontologists and Mineralogists (SEPM) in the same year as the publication of Joseph Augustine Cushman's (1881–1949) famous textbook on foraminifera (1948). More and more microfossils were used in environmental reconstructions, too.

Horizontal correlation has to be combined with vertical aspects, the first task when analysing drilling results. A breakthrough was the invention of electric logging by the Schlumbergers after the primitive beginning in Pechelbronn within the Rhine graben in 1927. Gamma ray logging has become available commercially since 1940, and neutron logging since 1941. Today a combination of up to a dozen logging tools is sent down into a drill hole. Needless to say, *stratification* – its types and origins – has been a fundamental problem in geology since its beginning. Thanks to 'Walther s law of correlation of facies' (Middleton 1973) one can combine vertical profiles with lateral features. Walther had been inspired by Amanz Gressly (1814–1865) (1838). A special case in cyclicity was the well-known contribution by James Marvin Weller (1899–1970) (1930) on cyclothems in coal-bearing Palaeozoic rocks. It stimulated the discussion of possible external factors versus sedimentary ones. But up to now the explanation of cyclic sedimentation has remained a problem without a generally agreed solution (see, Duff *et al.* 1967; Einsele & Seilacher 1982).

1945–1968: explosion of publications

Most of the foregoing methods and results served as the basis for the further evolution of sedimentology. But some trends in particular became intensified.

Explosion of publications

Ongoing specialization on the one hand and the growing need for syntheses on the other have produced an explosion of publications. New journals like *Sedimentology* (since 1962), *Sedimentary Geology* (since 1967), or special new series such as *Developments in Sedimentology* (since 1964) were initiated by a single publishing house. Special publications were issued by SEPM and other societies or institutions, and many textbooks, including the classic *Sedimentary Rocks* by Francis John Pettijohn (1904–1999), appeared in 1949. In the United States, the work of Robert Folk (1968), Gerald Friedman and John E. Sanders (1978; Friedman *et al.* 1992) should be mentioned; in England, Richard C. Selley (1976); in Germany, Wolf von Engelhardt, Hans Füchtbauer and German Müller (1964–1988); and in Russia, L. B. Ruchin (1958). In general, second editions grew substantially. But how can we digest the wealth of information in 19 chapters on 433 pages of a publication devoted exclusively to *Current Ripples* (Allen 1968)? Or a volume about *Glacigenic Sediments* with more than 2200 references (Brodzikowski & van Loon 1991)?

Processes of sedimentation

Sediments and sedimentation processes can be studied in the field or by experiments. Marine fieldwork continued in areas of shallow water: the Barents Sea (Klenova 1960); the Bering Sea (Lisitsyn & Jerusalem 1968); offshore California; the NE coast of the USA with Kenneth O. Emery (1914–1998) (Emery *et al.* 1954); L. V. Illing (1954) on the Bahamas; Henry Stetson in the Gulf of Mexico (API Project; Shepard *et al.* 1960); and the Persian/Arabian Gulf by the Kiel and SHELL Groups (Purser 1973). Monumental Soviet atlases covered sedimentological aspects of whole oceans, beginning with the Indian and Atlantic Oceans (G. B. Udintsev, 1975, 1989–1990).

Experiments in sedimentology have a long tradition, in the twentieth century beginning with Jacobus Henricus van't Hoff's (1852–1911) laboratory studies of marine evaporites (1912). In 1901, he became the first winner of the Nobel Prize in chemistry. As early as 1907 Grove Karl Gilbert (1843–1918) used flume experiments to study the behaviour of different materials during erosion, transport and sedimentation, and not only for engineering purposes (Gibbert 1914). The 'Hjulstrom curve' (1935) and its refinement by Akae Sundborg (1956) became famous in this regard. Ralph Alger Bagnold is well known for his research on wind action (Bagnold 1941). Since the 1960s, John Robert Lowren Allen has combined field observations with experiments (1970, 1985).

The greatest impact, however, was due to Philip Henry Kuenen (1902–1976). From 1937, following a hint from Reginald Aldworth Daly (1871–1957) in 1936, Kuenen explained the origin of submarine canyons by turbidity current erosion. Fieldwork and experiments explained typical sedimentary sequences, with graded bedding etc. which are found as 'flysch' in orogens all over the world (Kuenen & Migliorini 1950; Bouma 1962). Such ideas spread like an epidemic, being used to explain offshore cable breaks, the formation of deep-sea fans with deep-sea sands, and abyssal plains. For years they became an outstanding part of the sedimentological '*Zeitgeist*', like the subsequent ideas about carbonate platforms, or methane gas hydrates today.

Facies and palaeoenvironment reconstruction

These efforts were the prerequisite for a better understanding of facies and their use to reconstruct palaeoenvironments. After World War II most initiatives were taken by oil geologists looking for reservoir and source rocks, for stratigraphic or tectonic traps. *Deltaic environments* (e.g. Morgan & Shaver 1970) have been studied for centuries, but their sedimentological details have only become known since the work of W. A. Johnston (1921) on the Fraser River near Vancouver. The oil industry, however, pushed further investigations systematically: since 1950 at the deltas of the Rhine, Rhone and Orinoco (Tjeerd H. van Andel, Cornelis K. Kruit, Hendrick Postma), the Mississippi (e.g. H. N. Fisk) and the Niger (John Robert Lowren Allen). Detailed bibliographies can be found in Moore (1966). *Coastal environments* became economically important with their sands in barrier island complexes (Davies *et al.* 1971; Schwartz 1973). Tidal flats remained classic areas of investigation since Rudolf Richter's (1887–1957) activities in the research institute Senckenberg am Meer in Wilhelmshaven, which he had founded in 1929, followed by Hans Erich Reineck (1918–1999) (1972) (see also L. M. J. U. van Straaten (1954) for the Netherlands, and, in general, Robert N. Ginsburg (1975)). Coral reefs became especially interesting to oil companies after the post-war discovery of the Leduc field in Alberta. Their patchiness, lagoons and basements were intensively studied during and after the Pacific War (Preston Cloud *et al.* 1956; Emery *et al.* 1954). Norman D. Newell *et al.*'s (1953) publication on the Permian reef complex in southern USA remains a monumental facies study.

In general, investigations of *carbonate formation* were greatly intensified in the 1950s. Here, too, the trend was from description, especially of thin sections, to facies interpretations (see Friedman 1998). Some examples are given by A. V. Carozzi (1960, 1989), E. Flügel (1982), W. E. Ham (1962), G. V. Chilingar *et al.* (1967) and J. L. Wilson (1975). Numerous publications were devoted to dolomites (e.g. Braithwaite 1991).

But *continental environments* became attractive too. Research – chiefly by SHELL – in the deserts of the Near East, with their dunes and wadis, to assist understanding of the Dutch and North Sea gas fields with their Permian reservoir rocks, has added up to a fascinating story, a summary of which was given by Kenneth W. Glennie (1970). It gives pleasure to rediscover in these respects the work of many former pioneers, such as Walther (1900), Ralph A. Bagnold (1941) and Edwin D. McKee (1979), and to learn again that carefully documented observations keep their significance longer than many transiently fashionable ideas.

Later, sedimentary environments were treated by Selley (1970), Rigby & Hamblin (1972) and Reineck & Singh (1973).

Diagenesis

Sediments can be seen as 'collecting boxes of unbalanced conditions' (Hans Füchtbauer, lecture at Tübingen, 10 February 1978: '*Sammeltöpfe des Ungleichgewichts*'). Therefore diagenesis is active from the beginning of sedimentation – a genuine field for geochemistry (and increasingly for microbiological geochemistry, even in very deep layers (e.g. Charles Curtis 1987)). These fields cannot be treated in detail here. A history of the concept of diagenesis was written by G. Dunoyer de Segonzac (1968), however, and there are general description of diagenesis by Larsen & Chilingar (1967) and Füchtbauer (1988) in W. von Engelhardt *et al.* 1964–1988.

Again, the basis of these investigations is the study of single grains, including the cements between them, i.e. of the microfacies. To reconstruct fossil environments one has to have in mind diagenetic alterations by dissolution, replacement of primary minerals by secondary ones, recrystallizations, overgrowths of detrital grains, etc. The whole complicated fate of a sediment from its deposition, burial and subsidence, to its uplift(s), has to be approached to give information about porosities and permeabilities, so important for the petroleum industry.

Focal topics were 'chemical' sediments like limestones (e.g. Bathurst 1958, 1971), dolomites, cherts (for discussions, see Carozzi 1975) and evaporites (see Kirkland & Evans 1973). Phosphates were treated by V. E. McKelvey (1967). The boron content of clay minerals was used as indicator for palaeosalinities (see Couch 1971); their spatial arrangement as an indicator for compaction (see Moon 1972); coalification and kerogen alteration as quantitative proxy data of diagenesis (Tissot & Welte 1989); illite crystallinity as an indicator for anchimetamorphosis, i.e. a zone between diagenesis and metamorphism (e.g. Kübler 1968). The interrelations between compaction and tectonic subsidence and rise together with the influence of fluids have become increasingly sophisticated.

Progress has depended strongly on new techniques: those giving access to increasingly small concentrations of elements and their isotopes from spectroscopical methods since the middle of the 1920s up to the development of mass spectrometry, microprobe, atomic absorption spectrometry and other techniques; and those which investigated mineralogy or fabric in originally porous media (details in Tucker 1988). Of course chemical and microfacies analysis belong together.

1968–2000: deep-sea drilling, more geophysics and extraterrestrial factors

All these activities in sedimentology were continuing, but some especially important breakthroughs were: (1) the application of continuous reflection seismics during the last decades; (2) the Deep-Sea Drilling Programme since 1968; and (3) the importance of extraterrestrial factors.

Continuous reflection seismics

Continuous reflection seismics became the tool for *sequence stratigraphy*, propagated since 1975 mainly by Peter R. Vail at EXXON (Vail 1987). Soon it was applied globally. However, Augustin Lombard (1905–1997) preformulated this concept to some degree in his *Géologie sédimentaire* in 1956, and Laurence Sloss (1913–1996) goes back even further (see Sloss 1988). Transgressions and regressions are key events. Therefore sea-level changes became an important issue in sedimentology. A classic paper was that of Haq *et al.* (1987). To the even geologically dramatic sea-level variations typical of an icehouse world, more and more aspects of slow variations during greenhouse periods had to be added.

Carbonate platforms like those of the Bahamas – so sensitive because of dissolution or precipitation processes during low or high sea-level stands – were 'experimental areas' since the work of Leslie V. Illing (1954) (see Kendall & Schlager 1988). The effects of sea-level changes are not restricted to shallow waters. Down the continental slope to the abyssal realm mass-flows and turbidity currents – and even methane gas hydrates – are influenced by sea-level variations. Oil geologists were especially interested in deep-sea sands occurring in fans, such as the ones off the Amazon river or in the Tertiary beneath the North Sea (e.g. Walker 1978; Shanmugam & Moiola 1988).

Deep-Sea Drilling Programme

The Deep-Sea Drilling Programme opened many new opportunities for sedimentology, too. The knowledge of different sediment types with their microfossils increased dramatically as illustrated by a comparison of the classic summaries by Gustav Arrhenius from 1963 with that of Cesare Emiliani from 1981. Maximal core lengths of 23 m recovered by Börje Kullenberg in 1944/45 were surpassed by orders of magnitude and their evaluation founded a new branch of science, *palaeoceanography*. Palaeoclimatology also received new and decisive stimuli.

Stratigraphic resolution was refined in a fascinating way, mainly by the use of oxygen isotopes and the approach by Cesare Emiliani (1922–1995) since 1955. Additionally, cyclic variations of the Earth's orbit were used (see Hays *et al.* 1976). For sedimentologists, many new data on accumulation and erosion rates was produced. What progress since the first few data from Wolfgang Schott (1905–1989) in 1935!

Sequence analysis of different facies types and the wealth of new quantitative data resulted in a better understanding of *basin evolution* and its modelling during long-lasting subsidence (see Einsele 1992). To summarize: basin analysis can stand as a multidisciplinary and increasingly quantitative method in Earth sciences, leading from the study of single grains and observations to integrative models. Due to the many factors involved, it seems at present that 'the process of model making is far more important than the product' (Dott 1988, p. 360), which conversely has to be checked finally in sediment samples, cores or outcrops. Most of these efforts are continuing vigorously.

Extraterrestrial factors

Extraterrestrial factors influencing sedimentation as cyclic variations of the Earth's orbit were mentioned above. Based on ideas of James Croll (1821–1890) (1875), Milutin Milankovitch's (1879–1985) astronomical analysis is now widely used to explain climatic variations, especially in the Pleistocene (see Hays *et al.* 1976).

Other well-known cyclic features are of course the different types of annual layers caused by seasonal variations, as of glacial meltwaters or organic production. Even smaller periods can be studied in tidal deposits back to the Precambrian (G. E. Williams 1989).

Finally, many crater-like morphological features, such as Meteor Crater in Arizona, could be explained by impacts of meteorites or asteroids (see Marvin 2002). Until the 1930s, most astronomers believed the Moon's craters were giant extinct volcanoes. In the 1970s, the *Apollo* results confirmed impact origins. But a decade before, in the 1960s, the impact nature of the south German Ries Kessel could be demonstrated by the occurrence of shocked metamorphic minerals such as coesite and stishovite, shatter cones, planar elements in quartz, plagioclase or olivine, strange breccias and even melted rocks (tectites) as part of a new sediment type: *impactites*. Impacts into the oceans can produce tsunami deposits, as around the Gulf of Mexico with the Chicxulub Crater (see Barnes & Barnes 1973; Smit & Romein 1985; Marvin 1990; Melosh 1992; Montanari & Koeberl 2000).

Last but not least, an iridium enrichment in sediments became celebrated by the classic paper of Louis and Walter Alvarez (Alvarez *et al.* 1980). With its theory to explain the Cretaceous/Tertiary boundary extinctions, especially of the dinosaurs, it produced an impact in the geological and palaeontological literature comparable with Kuenen's explanation of turbidity currents and their sedimentary types.

Conclusions

During the second half of the last century, on the one hand *specialization* and the use of new techniques in sedimentology increased significantly. More specialists produced more data and the number of publications in new journals or series with overwhelmingly detailed bibliographies has exploded. (Hopefully, new communication technologies will prevent us from being drowned in this flood.)

On the other hand, and as a consequence of our growing responsibility to help solve general global problems, *cross-disciplinary* and *international* approaches have had to be promoted – in education, textbooks or projects and programmes as in the deep-sea drilling activities. One of the unifying concepts has been that of actualism, according to which one tries to explain past by recent conditions. A special trend has been, finally, the progression to basin analysis, i.e. from single grains to environments; and also from individuals to teams. The challenge to include global aspects like impacts from outer space, the effects of cyclic variations of our Earth's orbit, or the quantification of 'geofluxes' of gases, fluids, organic and inorganic materials and their possible consequences for our climates, has been growing more and more.

Of course, internationalization saves us from the fate of some of our great forerunners who attacked global issues with only regional experiences. But 'ground truth' remains essential.

Meanwhile, daily life in sedimentology continues. The show has to go on. Personnel, research vessels, laboratories with more and more expensive equipment are at work, and we can predict many trends for the near future just by extrapolation.

But, as in the past, hopefully some unexpected driving forces for further progress will emerge: new discoveries, such as the deep-sea hot vents first indicated by metalliferous sediments in the Red Sea (Degens & Ross 1969); cold seeps with their precipitates and specialized organisms; methane gas hydrates in or even on the sea bottom; or meteorite impact indicators.

As a form of the '*Zeitgeist*' and many new ideas 'in the air', new concepts have always been developed by researchers, called 'pioneers', if the time was ripe for them and consequently they became successful: Kuenen's turbidity currents were a direct example in sedimentology; Hess's and Dietz's sea-floor spreading an indirect one.

Each theory generates a 'fashion' (Carozzi 1975) like 'turbidites', 'tidalites', 'clathrates', or the application of Milankovich cycles to the study of deep-sea sediments. The fashions contribute to progress because they produce new observations and data, which sometimes do not mesh with existing concepts.

Finally we should repeat that progress in sedimentology has been substantially improved by two facts in particular: the contributions by engineers and technicians with their new equipment, procedures and techniques; and the influence of the oil industry. In its companies – but also within intelligent large-scale programmes like deep-sea drilling – cross-disciplinary approaches are a necessity, and this remains an important message for everybody.

We are grateful for many discussions and comments which helped us to concentrate the material, especially to H. Füchtbauer, D. Horn and M. Lutz for emphasizing the input of oil geology, and to D. Oldroyd for his editorial assistance. We appreciate the critical reviews of F. van Veen and A. V. Carozzi.

References

ALLEN, J. R. L. 1968. *Current Ripples*. North Holland, Amsterdam.

ALLEN, J. R. L. 1970. *Physical Processes in Sedimentation*. Allen & Unwin, London.

ALLEN, J. R. L. 1985. *Experiments in Physical Sedimentology*. Allen & Unwin, London.

ALVAREZ, L. W., ALVAREZ, W., ASARO, E. & MICHEL, H. V. 1980. Extraterrestrial cause for the Cretaceous/Tertiary extinction. *Science*, **208**, 1095–1108.

ARKHANGELSKI, J. A. D. 1927. On the Black Sea sediments and their importance for the study of sedimentary rocks' origin. *Bulletin of the Moscow Society of Natural History, Geology Section*, New Series, **5**, 199–298 (in Russian).

ARRHENIUS, G. 1963. Pelagic sediments. *In*: HILL, M. N. (ed.) *The Sea*. Interscience, New York, Vol. 3, 655–727.

BAGNOLD, R. A. 1941. *The Physics of Blown Sand and Desert Dunes* (2nd edn, 1954). Methuen, London.

BARNES, V. E. & BARNES, M. A. (eds) 1973. *Tektites*. Benchmark Papers in Geology, Dowden, Hutchinson & Ross, Stroudsburg.

BATHURST, R. G. C. 1958. Diagenetic fabrics in some British Dinantian limestones. *Liverpool–Manchester Geological Journal*, **2**, 11–36.

BATHURST, R. G. C. 1971. *Carbonate Sediments and their Diagenesis* (2nd edn, 1975). Developments in Sedimentology **12**, Elsevier, Amsterdam.

BOSWELL, P. G. H. 1933. *On the Mineralogy of the Sedimentary Rocks*. Murby, London.

BOUMA, A. H. 1962. *Sedimentology of some Flysch Deposits*. Elsevier, Amsterdam.

BRAITHWAITE, C. J. R. 1991. Dolomites, a review of origins, geometry and textures. *Transactions of the Royal Society of Edinburgh*, **82**, 99–112.

BRODZIKOWSKI, K. & VAN LOON, A. J. 1991. *Glacigenic Sediments*. Developments in Sedimentology, **49**, Elsevier, Amsterdam.

CAILLEUX, A. 1952. Morphoskopische Analyse der Geschiebe und Sandkorner und ihre Bedeutung für die Palaoklimatologie. *Geologische Rundschau*, **40**, 11–19.

CAROZZI, A. V. 1960. *Microscopic Sedimentary Petrography*. Wiley, New York.

CAROZZI, A. V. 1975. *Sedimentary Rocks: Concepts and History*. Benchmark Papers in Geology, **15**, Dowden, Hutchinson & Ross, Stroudsburg.

CAROZZI, A. V. 1989. *Carbonate Rock Depositional Models: A Microfacies Approach*. Prentice Hall, Englewood Cliffs.

CAYEUX, L. 1909, 1922. *Les minerais de fer oolithiques de France. Étude des gîtes minéraux de la France* (*Minerais de fer primaires*, 1909. *Minerais de fer secondaires*, 1922). Service Carte Géologique de France, Imprimerie nationale, Paris.

CAYEUX, L. 1916. *Introduction à l'étude pétrographique des roches sédimentaires. Mémoires pour servir à l'explication de la carte géologique detaillée de la France*. Imprimerie nationale, Paris.

CAYEUX, L. 1929. *Les roches sédimentaires de France: roches siliceuses. Mémoires pour servir à l'explication de la Carte géologique de la France*, **23**, Imprimerie nationale, Paris.

CAYEUX, L. 1935. *Les roches sédimentaires de France: roches carbonatées*. Masson, Paris (English translation: CAROZZI, A. V. 1970, Hafner, Davien, Connecticut).

CAYEUX, L. 1939–1950. *Les phosphates de chaux sédimentaires de France* 1970 (3 vols). Service de la Carte géologique de la France, Imprimerie Nationale, Paris.

CAYEUX, L. 1941. *Causes anciennes et causes actuelles en géologie*. Masson, Paris (English translation: CAROZZI, A. V. 1971, Hafner, New York).

CHILINGAR, G. V., BISSELL, H. J. & FAIRBRIDGE, R. W. 1967. *Carbonate Rocks: Origin, Occurrence and Classification*. Developments in Sedimentology, **9A**, Elsevier, Amsterdam.

CLOUD, P. E., SCHMIDT, R. G. & BURKE, H. W. 1956. *Geology of Saipan, Mariana Islands. General geology*. US Geological Survey, Professional Paper **280A**.

CORRENS, C. W. 1937. Die Sedimente des äquatorialen Atlantischen Ozeans. *Deutsche Atlantische Expedition Meteor 1925–1927, Wissenschaftliche Ergebnisse*, **3E**.

COUCH, E. L. 1971. Calculations of paleosalinities from boron and clay mineral data. *Bulletin American Association Petroleum Geologists*, **55**, 1829–1837.

CROLL, J. 1875. *Climate and Time in their Geological Relations: A Theory of Secular Changes of the Earth's Climate*. Daldy, London.

CURTIS, C. 1987. Mineralogical Consequences of organic matter degradation in sediments, Inorganic/organic diagenesis. *In*: LEGGETT, J. K. & ZUFFA, G. G. (eds) *Marine clastic sedimentology*, Graham & Trotman, 108–123.

CUSHMAN, J. A. 1948. *Foraminifera: Their Classification and Economic Use* (4th edn). Harvard University Press, Cambridge, Mass.

DALY, R. A. 1936. Origin of submarine canyons, *American Journal of Science*, Series 5, **31**, 401–420.

DAVIES, D. K., ETHRIDGE, F. G. & BERG, R. R. 1971. Recognition of barrier environments. *AAPG Bulletin* **55**, 550–565.

DEGENS, E. T. 1965. *Geochemistry of Sediments*. Prentice Hall, Englewood Cliffs.

DEGENS, E. T. & ROSS, D. A. (eds) 1969. *Hot Brines and Recent Heavy Metal Deposits in the Red Sea*. Springer, Berlin.

DOEGLAS, D. J. 1951. From sedimentary petrology to sedimentology. *Proceedings of the 3rd International Congress of Sedimentology*, 15–22.

DOTT, R. H. Jr. 1988. Perspectives: something old, something new, something borrowed, something blue – A hindsight and foresight of sedimentary geology. *Journal of Sedimentary Petrology*, **58**, 358–364.

DUFF, P. MCL. D., HALLAM, A. & WALTON, E. K. 1967. *Cyclic Sedimentation*. Elsevier, Amsterdam.

DUNOYER DE SEGONZAC, G. 1968. The birth and development of the concept of diagenesis (1866–1966). *Earth Science Reviews*, **4**, 153–201.

EINSELE, G. 1992. *Sedimentary Basins – Evolution, Facies, and Sediment Budget*. Springer, Berlin.

EINSELE, G. & SEILACHER, A. (eds) 1982. *Cyclic and Event Stratification* (2nd edn, 2000). Springer, Berlin.

EMERY, K. O., TRACEY, J. L. & LADD, H. S. 1954. *Geology of Bikini and nearby atolls*. US Geological Survey, Professional Paper **260A**.

EMILIANI, C. 1955. Pleistocene temperatures. *Journal of Geology*, **63**, 538–578.

EMILIANI, C. 1981. The oceanic lithosphere. *In*: EMILIANI, C. (ed.), *The Sea*, Vol. 3. Wiley Interscience, New York.

ENGELHARDT, W. VON, FÜCHTBAUER, H. & MÜLLER, G. 1964–1988. *Sediment-petrologie*. Schweizerbart, Stuttgart.

FLÜGEL, E. 1982. *Microfacies Analysis of Limestones*. Springer, Berlin.

FOLK, R. L. 1965. Henry Clifton Sorby (1826–1908),

the founder of petrography. *Journal of Geological Education*, **13**, 43–47.

FOLK, R. L. 1968. *Petrology of Sedimentary Rocks*. Hemphills, Austin.

FRIEDMAN, G. M. 1968. Carbonates. *In*: GOOD, G. A. (ed.) *Sciences of the Earth – An Encyclopedia of Events, People, and Phenomena* (2 vols). Garland Publishing, New York & London, Vol. 1, 77–81.

FRIEDMAN, G. M. & SANDERS, J. E. 1978. *Principles of Sedimentology*. Wiley, New York.

FRIEDMAN, G. M., SANDERS, J. E. & KOPASKA-MERKEL, D. C. 1992. *Principles of Sedimentary Deposits – Stratigraphy and Sedimentology*. MacMillan, New York.

GILBERT, G. K. 1914. *The transportation of debris by running water*. US Geological Survey, Professional Paper **86**.

GINSBURG, R. N. (ed.) 1973. *Evolving Concepts in Sedimentology*. Johns Hopkins, Baltimore.

GINSBURG, R. N. 1975. *Tidal Deposits*. Springer, Berlin.

GLENNIE, K. W. 1970. *Desert Sedimentary Environments*. Developments in Sedimentology, **14**, Elsevier, Amsterdam.

GRABAU, A. W. 1913. *Principles of Stratigraphy* (2 vols). Dover, New York. (2nd edn, 1924, Seiler, New York).

GRESSLY, A. 1838. *Observations géologiques sur le Jura soleurois. Nouveaux Mémoires.*

GRIM, R. E. 1953. *Clay Mineralogy* (2nd edn, 1968). MacGraw-Hill, New York.

GRIM, R. E. 1962. *Applied Clay Mineralogy*. MacGraw-Hill, New York.

GRIPENBERG, S. 1934. A study of the sediments of the North Baltic and adjoining seas. *Fennia*, **60**, 1–231.

HADDING, A. 1929. The prequaternary sedimentary rocks of Sweden. III.: Sandstones. *Lunds Universitet Arscrift*, New Series, **25**, 1–287.

HAM, E. (ed.) 1962. *Classification of Carbonate Rocks – A Symposium.* AAPG, Memoirs, **1**.

HAQ, B. U., HARDENBOL, J. & VAIL, P. R. 1987. Chronology of fluctuating sea level since the Triassic. *Science*, **235**, 1156–1167.

HAYS, J., IMBRIE, J. & SHACKLETON, N. J. 1976. Variations in the Earth's orbit: pace maker of the Ice Age. *Science*, **194**, 1121–1132.

HEDGPETH, J. W. & LADD, H. S. (eds) 1957. *Treatise on Marine Ecology and Paleoecology* (2 vols). Geological Society of America, Memoir **67**.

HJULSTRÖM, F. 1935. Studies of the morphological activity of rivers as illustrated by the River Fyris. *Bulletin of the Geological Institute, Uppsala*, **25**, 221–527.

ILLING, L. V. 1954. Bahaman calcareous sands. *AAPG Bulletin*, **38**, 1–95.

JOHNSTON, W. A. 1921. *Sedimentation of the Fraser river debris*. Geological Survey of Canada, Memoir **125**.

KENDALL, C. G. S. C. & SCHLAGER, W. 1981. Carbonates and relative changes in sea level. *Marine Geology*, **44**, 181–212.

KIRKLAND, D. W. & EVANS, R. 1973. *Marine Evaporites: Origin, Diagenesis, and Geochemistry.* Benchmark Papers in Geology. Dowden, Hutchinson & Ross, Stroudsburg.

KLENOVA, M. V. 1960. *The Barents Sea Geology*. Moscow Izdenije Akademii Nauk S. S. S. R., Moscow (in Russian).

KRUMBEIN, W. C. 1932. A history of the principles and methods of mechanical analysis. *Journal of Sedimentary Petrology*, **2**, 89–124.

KRUMBEIN, W. C. 1934. Size frequency distributions of sediments. *Journal of Sedimentary Petrology*, **4**, 65–77.

KÜBLER, B. 1968. Evaluation quantitative du métamorphisme par la cristallinité de l'illite. *Bulletin Centre Recherches, Société National des Pétroles d'Aquitaine*, **2**, 385–397.

KUENEN, P. H. 1937. Experiments in connection with Daly's hypothesis on the formation of submarine canyons. *Leidse Geologische Mededelingen*, **8**, 327–335.

KUENEN, P. H. & CAROZZI, A. V. 1953. Turbidity currents and sliding in geosynclinal basins of the Alps. *Journal of Geology*, **61**, 363–373.

KUENEN, P. H. & MIGLIORINI, C. I. 1950. Turbidity currents as a case of graded bedding. *Journal of Geology*, **58**, 91–127.

LARSEN, G. & CHILINGAR, G. V. (eds) 1967. *Diagenesis in Sediments.* Developments in Sedimentology, **8**, Elsevier, Amsterdam.

LEMCKE, K., ENGELHARDT, W. VON & FÜCHTBAUER, H. 1953. Geologische und sedimentpetrographische Untersuchungen im Westteil der ungefalteten Molasse des suddeutschen Alpenvorlandes. *Geologisches Jahrbuch, Beiheft* **11**, Hannover.

LISITSYN, A. P. & JERUSALEM, I. P. S. T. 1968. *Recent Sedimentation in the Bering Sea.*

LOMBARD, A. 1956. *Géologie sédimentaire: les séries marines*, Masson, Paris.

MCKEE, E. D. (ed.) 1979. *A study of global sand seas.* US Geological Survey, Professional Paper **1052**.

MCKELVEY, V. E. 1967. *Phosphate deposits.* US Geological Survey Bulletin, **1252D**.

MARVIN, U. B. 1990. Impact and its revolutionary implications for geology. Geological Society of America, Special Paper **247**, 147–154.

MARVIN, U. 2002. Geology: from an Earth to a planetary science in the twentieth century. *In*: OLDROYD, D. (ed.) *The Earth Inside and Out: Some Major Contributions to Geology in the Twentieth Century.* Geological Society, London, Special Publications, **192**, 17–58.

MELOSH, H. J. 1992. Impact crater geology. *In*: NIERENBERG, W. A. (ed.) *Encyclopedia of Earth System Science*. Academic Press, San Diego, Vol. 2 591–605.

MIDDLETON, G. V. 1973. Johannes Walther's law of correlation of facies. *Bulletin of the Geological Society America*, **84**, 979–988.

MIDDLETON, G. V. 1978. Sedimentology – history. *In*: FAIRBRIDGE, R. W. & BOURGEOIS, J. (eds) *The Encyclopedia of Sedimentology*. Dowden, Hutchinson & Ross, Stroudsburg, 707–712.

MILLOT, G. 1964. *Géologie des argiles*. Masson, Paris (English translation: 1970, *Geology of Clays*, Springer, Berlin).

MILNER, H. B. 1922. *An Introduction to Sedimentary Petrography*. Murby, London.

MILNER, H. B. 1962. *Sedimentary Petrography* (2 vols). Allen & Unwin, London.

MONTANARI, A. B. & KOEBERL, C. 2000. *Impact Stratigraphy*. Lecture Notes in Earth Sciences, **93**, Springer, Berlin.
MOON, C. F. 1972. The microstructure of clay minerals. *Earth Science Reviews*, **8**, 303–321.
MOORE, D. 1966. Deltaic sedimentation, *Earth Science Reviews*, **1**, 87–104.
MORGAN, J. P. & SHAVER, R. H. (eds) 1970. Deltaic sedimentation: modern and ancient. Society of Economic Paleontologists and Mineralogists, Special Publications, **15**.
NEWELL, N. D., RIGBY, J. K., FISHER, A. G., WHITEMAN, A. J., HICHOX, J. L. & BRADBURY, J. S. 1953. *The Permian Reef Complex of the Guadaloupe Mountains Region, Texas and New Mexico*. Freeman, San Francisco.
OWEN, E. W. 1975. *Trek of the oil finders: a history of exploration for petroleum. AAPG*, Memoir **6**.
PETTIJOHN, F. J. 1949. *Sedimentary Rocks*. Harper, New York (2nd edn, 1957; 3rd edn, 1975).
POTTER, P. E. & PETTIJOHN, F. J. 1963. *Paleocurrents and Basin Analysis* (2nd edn, 1977). Springer, Berlin.
PRATJE, O. 1931. Die Sedimente der Deutschen Bucht. *Wissenschaftliche Untersuchungen*, Abteilung Helgoland, **18** (6).
PRATJE, O. 1948. Die Bodenbedeckung der sudlichen und mittleren Ostsee und ihre Bedeutung für die Ausdeutung fossiler Sedimente. *Deutsche Hydrographische Zeitschrift*, **1**, 45–61.
PURSER, B. H. (ed.) 1973. *The Persian Gulf – Holocene Carbonate Sedimentation and Diagenesis in a Shallow Epicontinental Sea*. Springer, Berlin.
REINECK, H. E. 1972. Tidal flats. *In*: RIGBY, J. K. & HAMBLIN, W. K. (eds) *Recognition of Ancient Sedimentary Environments*. Society of Economical Paleontologists and Mineralogists, Special Publications, **17**, 146–159.
REINECK, H. E. & SINGH, I. B. 1973. *Depositional Sedimentary Environments*. Springer, Berlin.
RIGBY, J. K. & HAMBLIN, W. K. (eds) 1972. *Recognition of Ancient Sedimentary Environments*. Society of Economical Paleontologists and Mineralogists, Special Publications, **16**.
RUCHIN, L. B. 1958. *Grundzuge der Lithologie* (translated from the Russian). Akademie Verlag, Berlin.
SARNTHEIN, M. 1971. *Oberflächensedimente im Persischen Golf und Golf von Oman II: Quantitative Komponentenanalyse der Grobfraktion. Meteor Forschungsergebnisse*, **C5**, Borntraeger, Berlin & Stuttgart.
SCHOTT, W. 1935. Die Foraminiferen in dem aequatorialen Teil des Atlantischen Ozeans. *Deutsche Atlantische Expedition Meteor 1925–1927, Wissenschaftliche Ergebnisse*, **3**, 43–134.
SCHWARTZ, M. L. 1973. *Barrier Islands*. Benchmark Papers in Geology, **9**, Dowden, Hutchinson & Ross, Stroudsburg.
SELLEY, R. C. 1970. *Ancient Sedimentary Environments*. Chapman & Hall, London.
SELLEY, R. C. 1976. *An Introduction to Sedimentology*. Academic Press, London.
SHANMUGAM, G. & MOIOLA, R. J. 1988. Submarine fans: characteristics, models, classification, and reservoir potential. *Earth Science Reviews*, **24**, 383–428.
SHEPARD, F. P., PHLEGER, F. B. & VAN ANDEL, TJ. H. (eds) 1960. *Recent Sediments, Northwest Gulf of Mexico*. AAPG, Tulsa.
SLOSS, L. L. 1988. Forty years of sequence stratigraphy. *Bulletin of the Geological Society* America, **100**, 1661–1665.
SMIT, J. & ROMEIN, A. J. T. 1985. A sequence of events across the Cretaceous/Tertiary boundary. *Earth and Planetary Science Letters*, **74**, 155–170.
STRAATEN, L. M. J. U. 1954. Composition and structure of recent marine sediments in the Netherlands. *Leidse Geologische Mededelingen*, **19**, 1–110.
STRAKHOV, N. M. 1970. Evolution of concepts of lithogenesis in the Russian geology (1870–1970). *Lithology and Mineral Resources*, **2**, 157–177 (in Russian).
SUNDBORG, A. 1956. The River Klaralven, a study of fluvial processes. *Geografiska Annaler*, **38**, 125–316.
TISSOT, B. P. & WELTE, D. H. 1989. *Petroleum Formation and Occurrence* (2nd edn). Springer, Berlin.
TRASK, P. D. (ed.) 1939. *Recent Marine Sediments*. AAPG, Tulsa.
TUCKER, M. 1988. *Techniques in Sedimentology*. Blackwell, Oxford.
TWENHOFEL, W. H. 1932. *Treatise of Sedimentation*. Williams Wilkins, Baltimore (2nd edn, 1936, Dover, New York).
TWENHOFEL, W. H. 1950. *Principles of Sedimentation* (2nd edn). McGraw-Hill, New York.
UDINTSEV, G. B. (ed.) 1975. *Geological–Geophysical Atlas of the Indian Ocean*. USSR Academy of Science, Moscow (in Russian and English).
UDINTSEV, G. B. 1989–1990. *International Geological–Geophysical Atlas of the Atlantic Ocean*. Ministry of Geology, USSR Academy of Science, Moscow (in Russian and English).
VAIL, P. R. 1987. *Sequence Stratigraphy. Workbook. Fundamentals of Sequence Stratigraphy*. Annual Convention Short Course Notes, AAPG, Tulsa.
VAN'T HOFF, J. H. 1912. *Untersuchungen über die Bildungsverhältnisse der ozeanischen Salzablagerungen*. Akademische Verlagsgesellschaft, Leipzig.
WADELL, H. 1932*a*. Sedimentation and sedimentology, *Science*, **75**, 20.
WADELL, H. 1932*b*. Volume, shape, and roundness of rock particles. *Journal of Geology*, **40**, 443–451.
WADELL, H. 1935. Volume, shape, and roundness of quartz particles. *Journal of Geology*, **43**, 250–280.
WALKER, R. G. 1978. Deep-water sandstone facies and ancient submarine fans: models for exploration for stratigraphic traps. *AAPG Bulletin*, **62**, 932–966.
WALTHER, J. 1893–1894. *Einleitung in die Geologie als Historische Wissenschaft. I: Bionomie des Meeres, II: Die Lebensweise der Meeresthiere, III: Lithogenesis der Gegenwart*. Fischer, Jena.
WALTHER, J. 1900. *Das Gesetz der Wüstenbildung in Gegenwart und Vorzeit*. Reimer, Berlin.
WELLER, J. M. 1930. Cyclical sedimentation of the Pennsylvanian Period and its significance. *Journal of Geology*, **38**, 97–135.

WENTWORTH, C. K. 1919. A laboratory and field study of cobble abrasion. *Journal of Geology*, **27**, 507–521.

WILLIAMS, G. E. 1989. Precambrian tidal sedimentary cycles and Earth's paleorotation. *EOS*, **70**, 33–41.

WILSON, J. L. 1975. *Carbonate Facies in Geologic History*, Springer, Berlin.

Some personal thoughts on stratigraphic precision in the twentieth century

HUGH S. TORRENS,
Formerly School of Earth Sciences, Keele University, ST5 5BG, UK

Abstract: Surprisingly, most of the major elements of today's stratigraphic column were in place by 1850. By then, the ideas that stratigraphy concerned geological time relations, and that a palaeontological identity of 'best' fossils (like ammonites) was an indication of time-equivalence, were starting to be accepted. By 1900, thanks to the work of people like Henry Shaler Williams (USA) and Sydney Savory Buckman (UK) stratigraphy was starting to concern itself with the precision with which biochronological time-scales could be created, especially in the Jurassic. By then, Buckman had demonstrated the great extent to which particular lithologies could cross time-lines and equally how well such rapidly evolving fossils as ammonites could be used to discriminate time. But from 1960, facing new demands for energy, and the growth of new 'earth science', focussing on numerical methods in geophysics and geochemistry using computers, such field-based 'historical geology' was progressively perceived as boring, out-dated, and expensive. Many new techniques, which ignored, or worse, assumed time-equivalence, now evolved. Fossils by their unique nature had given unique signatures to discriminate time. But some of the new methods relied on binary repetitions, not unique to time, and may suggest a false precision. This paper attempts a, now near-impossible, investigation of the temporal precisions that stratigraphic methods, both old and new, might attain. It concludes that we need to pay greater attention to the incompleteness of the stratigraphic record and to the chronological precision with which we can investigate that record. It now seems almost axiomatic that the harder you look at a rock the more incomplete the record of its stratigraphy appears to become.

Introduction

The science of geology is all about *time*. As Derek Ager said, our science's 'outstanding characteristic is its appreciation of time' (Ager 1987, p. 116). So stratigraphy must first and foremost concern questions of time. It is the only area of geology that is truly unique, other branches of geology are too often borrowed bits of physics, chemistry or biology. But stratigraphy has been, and still often is, hindered by the words it uses. Stratigraphers write endlessly about effecting 'correlations', between different rocks, cores, outcrops, etc. This word has been used in English in many ways. The *Oxford English Dictionary* (1989, 2nd edn) gives five different uses of it, only two of which are supported by geological examples: 'The condition of [rocks] being correlated', by Murchison in 1849; and 'the action of correlating or bringing [rocks] into mutual relation', by Jukes-Browne in 1886. Neither of these yet carried any time connotation.

In the century since, the word has acquired highly confused uses in geology (Rodgers 1959). Rodgers pointed out that the word acquired a specialized stratigraphic meaning, from about 1890 in the United States, to demonstrate synchrony between different rocks. The claim that fossils could achieve this can be traced much further back in England. William Smith and John Phillips had observed that the similarity of fossil occurrences between Yorkshire and the south of England made it axiomatic that 'deposits of equal antiquity enclose analogous fossils' (Smith & Phillips 1825). Joseph Prestwich was soon writing of synchrony and correlation between rocks as being synonymous (Prestwich 1847, pp. 355, 364 & 375).

But the word correlation still too often retained its old usage to imply merely the demonstration between rocks of 'correspondence in character and in stratigraphic position' (Hedberg 1976, p. 14). This usage was carefully preserved in the most recent *International Stratigraphic Guide* (Salvador 1994, p. 15). Others instead demanded that the word be used only to 'identify rocks of the same age in different places' (Donovan 1966, p. 27); or to 'establish the time equivalents of two spatially separate stratigraphic units' (Raup & Stanley 1978, p. 207); or, more honestly, to note that 'its most usual application is in stratigraphy where, unless otherwise stated, it means equivalence in time' (Challinor 1978, p. 70). The problem was, and too often still is, that the single word is still being

From: OLDROYD, D. R. (ed.) 2002. *The Earth Inside and Out: Some Major Contributions to Geology in the Twentieth Century*. Geological Society, London, Special Publications, **192,** 251–272. 0305-8719/02/$15.00

too often used in its old meaning, to imply only correspondence in character and stratigraphic position (as by Penn *et al.* 1979, pp. 28, 33 & 37 or Hopson *et al.* 2001, p. 193), with no time element invoked. In my opinion, the word, when used in geology, must only be in relation to statements about time equivalence or synchrony (Cope *et al.* 1980, p. 4).

Whatever the semantic problems, by the beginning of the twentieth century, stratigraphy had made great strides from its hesitant beginnings around the start of the nineteenth century. It had advanced to such an extent by mid century (1850) that the great majority of the Prehistoric Time-Scale in use today had been established. Vallance (1968) has recorded the date of introduction of each of the current terms for geological Systems and Eras. The great majority (save those Precambrian) predate 1841. The growth of this Time-Scale, based largely on fossil distributions, has been the subject of a useful study by Berry (1987).

The progress made in refining stratigraphy, and understanding its chronological complexity, at the beginning of the twentieth century, can be demonstrated in the work of two particular contributors: Sydney Savory Buckman (1860–1929) in Britain, and Henry Shaler Williams (1847–1918) in the United States. Buckman had, in 1889, elegantly demonstrated the extent, both geographically and chronologically, to which rocks of a particular, identical lithology could be diachronous. This was in his classic study of Jurassic sands in the south of England (Buckman 1889) (still for this reason separately termed Bridport, Yeovil, and Midford Sands today; see Fig. 1).

In 1893 Buckman next demonstrated with what detail the stratigraphic record could be read, when the best guide fossils, in that case ammonites, were used, and how episodic and incomplete that record proved to be when one did such an analysis (Buckman 1893). This paper has been described as 'one of the all-time classical landmarks of stratigraphy' (Callomon 1995, p. 134; see also Dietze & Chandler 1997). In 1901, Buckman further explained how there was:

> no local limit to time, and there can be no local limit to a time-table. Whether the records of the rocks in distant localities may be sufficiently perfect to enable their dates to be stated with as great exactitude as in my time-table is another matter . . . [My] time-table is a means whereby Jurassic events over a large part of Europe can be exactly dated now; and there is good reason to think the same may be said of a far wider field in the future (Buckman 1901).

Unlike Buckman, who remained an amateur for most of his geological career, Williams was an academic, based at Cornell and Yale Universities (Brice 1989). In an 1893 paper Williams (1893) urged that:

> chronological time periods in geology are not only recognized by means of the fossil remains preserved in the strata, but it is to them chiefly that we must look for the determination and classification on a time basis.

By 1895, when he published his remarkable book *Geological Biology: An Introduction to the Geological History of Organisms*, Williams had, by studying fossil diversities through time, been able to estimate the durations of each 'biological life-period'. He claimed, with remarkable insight, that, whatever the age of the Earth was then thought to be (just before the discovery of radioactivity), Palaeozoic time had made up 65% of the time elapsed since the earliest Cambrian, Mesozoic time contributed 20%, and Cainozoic time only 15% (Williams 1895, p. 64). These figures are close to today's estimates, based on 'black box science' since provided from physics and chemistry (now 54%, 34% & 12%; see Remane 2000*a*). Both Williams and Buckman were pioneers in forcing stratigraphic attention towards the subtle, but critically important, questions of time, the unique feature of geology.

But Robert Muir Wood asserted in 1985 that 'stratigraphy and fossil correlation [were] the backbone of [old] Geology', whereas the new 'Earth Science is revealed by geophysics and geochemistry' (Muir Wood 1985, p. 190). As Robert Dietz noted (1994, p. 2): 'Geology had evolved from an observational and field science into largely a laboratory science with instrumental capabilities that have improved data collection and data processing by orders of magnitude'. But not all are agreed that the move from field to laboratory has been has such an advance, since, as Dewey has rightly noted (1999, p. 3), 'core, field-based geology is [still] the most important, challenging and demanding part of the science'. Largely as a result of contemporary changes in attitude toward old and new, the teaching of stratigraphy has declined to a surprising extent. It is a complex and 'difficult' subject, which became no longer seen as central. Francis Pettijohn had warned of this in 1984 in America, likening the rush into the exotic peripheries around what he saw as the true core of geology – with stratigraphy at its centre – to 'a doughnut with nothing in the middle' (Pettijohn 1984, p. 203). For the more recent situation in the United States see Salvador (1992). Jan

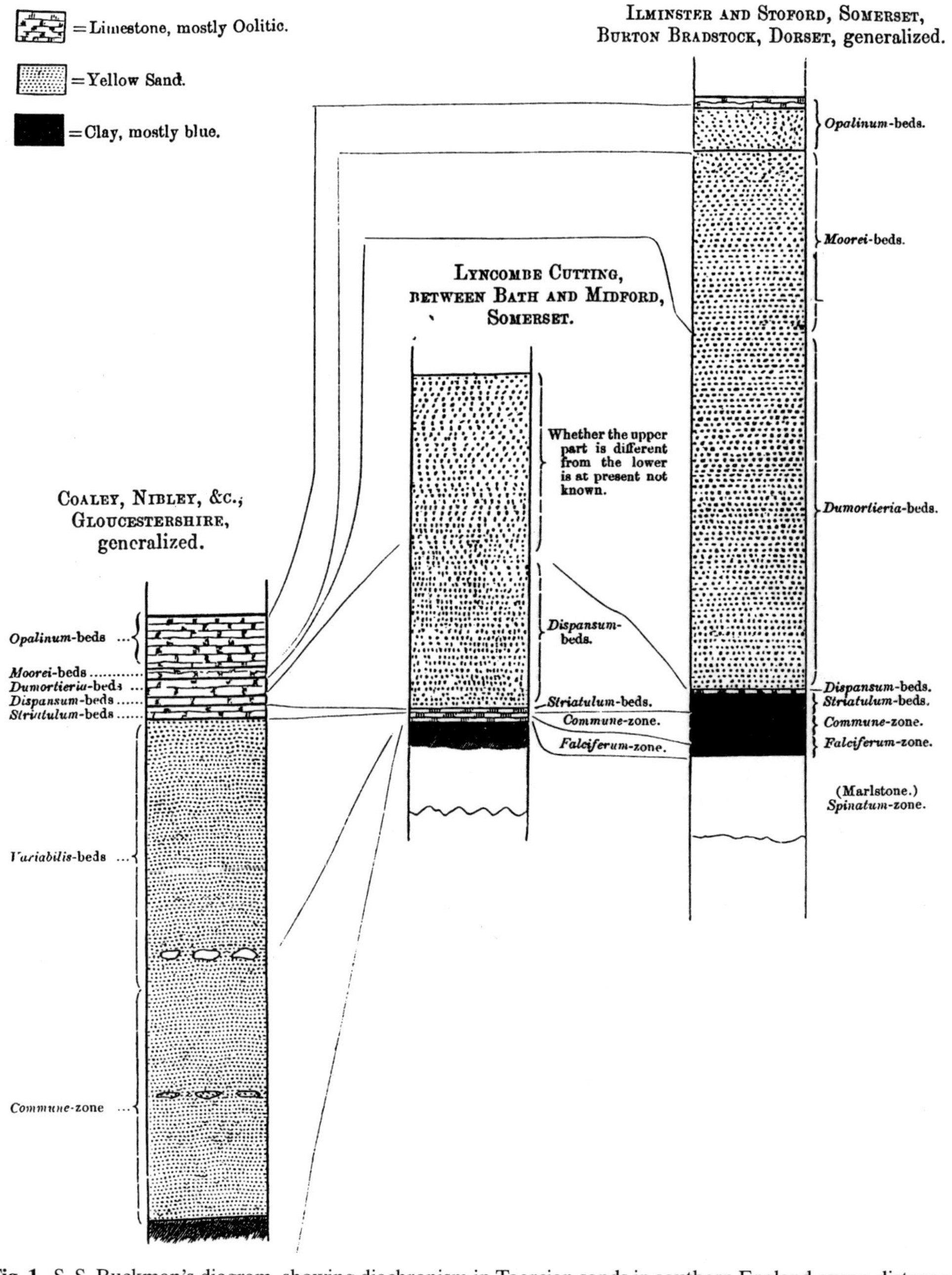

Fig. 1. S. S. Buckman's diagram, showing diachronism in Toarcian sands in southern England over a distance of about 150 km. Lines drawn between columns are effectively time-lines. Note the occurrence of the Striatulum-beds in limestone in left and centre, but in clay on right. These sands, younging to the south, are variously younger to the right and centre and older to the left (Buckman 1889, Fig. 1 – the distance between Cols 1 and 2 is about 50 km while the maximum between Cols 2 and 3 is about 90 km).

Houghton Brunn has informed me (*in lit.* July, 2000) that in France 'the teaching of Historical Geology is practically abandoned except for the efforts of some lecturers. I think it's a great pity as time is essential for the understanding of geology'. But such changes allow some humour, as revealed by Robin Brett, past President of the International Union of Geological Sciences: 'ask a geologist what is 3 + 2, he says the answer is about 5; ask a geochemist and he says 5, plus or minus 2; ask a geophysicist and he takes you to a corner and says: what answer do you need' (Brett 2000).

The introduction of numbers reminds us that throughout the twentieth century the question of the numerical, 'exact', Age of the Earth has been much debated, from Lord Kelvin onwards (Thomson 1894, Burchfield 1975, Lewis 2000). It was during a lecture to engineers on 3 May 1883 that Thomson memorably remarked:

> I often say that when you can measure what you are speaking about, and express it in numbers, you know something about it; but when you cannot measure it, when you cannot express it in numbers, your knowledge is of a meagre and unsatisfactory kind (Mackay 1977, p. 148).

History has proved the converse is often true in stratigraphy. The history of stratigraphy also reveals how, now that we can measure so many things so precisely, it is important that we understand exactly *what* our measurements are attempting to reveal and how precise they are. The *Oxford English Dictionary* (1989, 2nd edn) quotes of that vital word, 'precision', a 1974 source which notes how 'radioactive isotope dates invariably include their precision, that is the repeatability, yet most earth scientists still take these figures as measures of accuracy'.

Biological stratigraphy

Biostratigraphy, a term introduced at the beginning of the twentieth century in 1904 (Sylvester-Bradley 1979, p. 94), involving the use of fossils in stratigraphy, continued its rise until three-quarters of the way through the twentieth century. It then met two problems to obstruct its progress, at least in Europe. The intractable first was that the discovery of good guide fossils with precise chronostratigraphic attributes, became too expensive or, worse, impossible in oil exploration. The second problem, in the interpretation of rocks, was that the expertise needed to identify such guide-fossils was not a skill that was easily taught. It is experience-based and gaining such takes time and money. In today's cost-centred, computer-modelled geology, expertise in, for example, the lore of ammonites, is nearing the extinction that physics and chemistry now proclaim of the dinosaurs. A former colleague of mine, Brian Holdsworth, who made notable advances in both the Deep Sea Drilling Project and on previously undatable (Devonian to Permian) rocks in Nevada and Alaska using fossil radiolaria even wrote (*in lit.* November, 1999) 'recent advances in sequence stratigraphy and seismology have sadly rendered biostratigraphy redundant'. Beris Cox, a stratigrapher equally early-retired from the British Geological Survey, wrote how:

> important it is to ensure that there are always personnel who are sufficiently trained and expert to sustain and maintain the ammonite-based chronostratigraphic standard . . . which underpins so called event stratigraphy and sequence stratigraphy, which are prevalent approaches to chronostratigraphy in the oil industry today (Cox 1990, pp. 175 & 183).

The powerfully different utilities of fossils, whether in groups or often even as single specimens, in the precise 'dating' of rocks was, to me, best demonstrated during international centenary celebrations for the Hungarian Geological Institute in September, 1969, when a colloquium on Mediterranean Jurassic geology was held. An excursion went to Villany, southern Hungary, whose quarries had yielded fine ammonite faunas before the First World War (Loczy 1915). These had always been dated as Callovian but included highly anomalous material, like the eponymous ammonite *Villania* (Till 1909). This, if Callovian, remained so stubbornly unique that the author of the ammonite *Treatise* (Arkell 1957, L196) had to create a whole new subfamily for the single specimen on which it had been based. The solution to, and the stratigraphic range of, the anomalous faunas found here was revealed during this excursion, when it was found that not all the old ammonite material here had come from the single, famous, Cephalopod bed (No. 5). Beds unconformably below this (numbered 1 to 3) also yielded fossils, but these had been previously mis-identified as of assumed Callovian age. These included:

(1) nautiloids, reported by a French delegate to confirm the existing, but wrong, stratigraphic assignment of these beds to the Middle Jurassic;
(2) brachiopods, which Derek Ager realized were in fact older (Pliensbachian, Lower Jurassic); and
(3) a single ammonite, *Apoderoceras*, which not only confirmed the brachiopod evidence, but demonstrated that the precise age of the bed that here lay immediately below the Cephalopod bed was Taylori Subzone (Jamesoni Zone, Pliensbachian, Lower Jurassic).

An enormous, previously unsuspected, time gap, with a duration of at least 25 Ma, or five

Jurassic Stages, stood revealed between Beds 3 and 5. These conclusions were abundantly supported by later investigations of additional museum material (Ager & Callomon 1971). The three different fossil groups found had given three quite different degrees of precision to the dating of these beds: the nautiloids to Series (i.e. Middle Jurassic); brachiopods to Stage (i.e. Pliensbachian); and ammonites to Subzone (i.e. Taylori) level.

Biostratigraphic precision

The rapidity of ammonite evolution, with the often wide extent of ammonite migration, means they can yield closely controlled time-signatures at their lowest – Horizon (i.e. below Subzone) – level. What is more important, biological data from the fossil record carries unique time-signatures as a result of the uni-directional nature of evolution. Much geophysical or geochemical data is not unique in the same way, but is merely repetitious, often only 'binary'.

Biochonologically precise fossils like ammonites allow uniquely precise dating of many rocks and of some 'events', such as the start of opening in the South Atlantic Ocean (Reyment & Tait 1972). This was achieved by study of both non-marine and marine fossils, the latter including ammonites. There are, however, several problems with ammonites, no matter how great the chronostratigraphic precision they can give, such as homoeomorphy (Kennedy & Cobban 1976). As one example, the Cretaceous Vascoceratids used by Reyment & Tait (1972, plate 5) are homoeomorphic with Jurassic Tulitids, although this example has not yet misled stratigraphers. Another real problem with ammonites is their often strong provinciality, which especially continues to bedevil international correlations, as for example of beds across the Jurassic–Cretaceous boundary; 'the most problematical of all system/period boundaries' (Remane 2000a, p. 4). A third problem, which also still continues, is the low quality of basic data recorded for too many fossils. Alan Shaw, pioneer of Graphic Correlation, noted in 1971 how 60% of a literature-sample he then took lacked *any* statement of a measured stratigraphic position from which these fossils had come (Shaw 1971). Just to show that nothing bio-stratigraphic is ever simple, there has also been deceit and fraud, as in the salting of many non-Himalayan fossils into the cacophony of over 450 publications on that highly complex area by V. J. Gupta (born 1942), unknown to many of his 128 co-authors, who became unwittingly involved (Talent 1995). No fossils could be less precise in occurrence here, whether geographic or stratigraphic.

In an important paper, Callomon (1984, p. 83) introduced the concepts of stratigraphic 'time-resolutions <delta t>' and 'secular resolving-powers $<R_t>$' for different fossil groups throughout the Phanerozoic. The numbers deriving from these concepts can be calculated from the number of separate biochronological divisions (i.e. resolutions) that can be discriminated within a particular time period or interval divided into the 'best estimate' radiometric age duration in years of that same time interval.

Such 'secular resolving-powers $<R_t>$' between the different fossil groups vary enormously, just as the resolution of an electronic scanning microscope is so much greater than that of a hand-lens. Any value of $<R_t>$ above 1000 indicates a high resolving-power and one less than 100 a low power. Callomon showed that ammonites had high resolving-powers, often over 1000, so that these fossils can, at best, allow regular, and precise, time resolutions down to the order of 100 000 years for each ammonite horizon. The same sort of resolution seems possible in the Cenozoic record (Miller 1990). In a later paper Callomon analyzed the secular resolving-powers of a whole range of Phanerozoic fossils (Callomon 1995, table 4). These ranged from Cambrian trilobites, with $<R_t>$s as high as 970, through Silurian graptolites (425), Devonian ammonoids (880), Carboniferous goniatites (975), to Jurassic (at best 1260) and Cretaceous ammonites (at best 360). The figures derived from these concepts deserve to be more used.

Such analyses mean that in biostratigraphy, regular resolution below the 100 000-year per horizon level, will prove difficult. It means that, in effect, biochronology cannot normally discriminate between 100 000 years and 'instantaneous' in the geological record. Such problems encouraged stratigraphers to investigate 'events' as a new means of correlation and brought about – often inconclusive – discussions of what words like 'rapid', 'sudden', and 'abrupt' might mean in stratigraphy and sedimentology (Dott 1983, Van Loon 1999).

Yet many unfounded claims continue to be made of 'high' biochronological resolutions for all sorts of fossils by all sorts of geologists. One of the most remarkable is that made for dinosaur bones by Lucas (1991). But, with such low secular resolving-powers ($<R_t>$s here under 10), such fossils must only be used when all else fails. (I note that Lucas's claim was appropriately made in a volume that imprecisely commemorated the '150th anniversary of the invention of dinosaurs' a year too early (Torrens 1993, pp. 274–275).)

Very often the best, or indeed any, guide-fossils simply do not occur in rocks when sampled. This is especially the case with subsurface explorations, where recovery of any large fossils is difficult or impossible. Then one is constrained to use other microfossils, with lower secular resolving-powers, or other methods. Such microfossils are normally not as effective as macrofossils in providing time contours through rocks. Wignall (1990), comparing Kimmeridgian ostracods with foraminifera, confirmed that the ostracods which had been used here as zonal indices, were 'among *the* most facies-controlled fossils' to occur in those rocks. Their supposed precision in indicating the passage of time must be questioned. Cope *et al.* (1980, pp. 6–7) discussed the utility of various other Jurassic fossil groups as time indices and noted that brachiopods could, on occasion, invert their sequential orders, in an even more blatant example of fossil distributions controlled by facies, rather than by time. But whatever the fossils used, we should recall Rudolf Trümpy's comment on stratigraphy in mountain belts. Here 'even a bad fossil is more valuable than a good working hypothesis' (Trümpy 1971, p. 295). Bob Dott reminds me that it may sometimes be crucial simply to know that a particular rock is as imprecisely dated as 'Mesozoic'.

The International Commission on Stratigraphy and the precision of 'golden spikes'

In the absence of fossils, or because many fossils have proved to be provincial in their geographic distributions when new technologies became available to help the stratigraphic analysis of rocks, many other methods of time correlation have been attempted. To help better integrate these new techniques with traditional biostratigraphy, an important breakthrough in international stratigraphic thinking occurred in the early 1960s led by British geologists, although whether these were Jurassic, or Silurian specialists (Ager 1973, chapter 7; Lawson 1974) remains unclear. This was the principle of choosing to define the bases of units at particular horizons, or time-lines, at 'Global Stratigraphic Sections and Points' (GSSPs). This replaced earlier attempts, which instead tried to define both the top and bottom of an inclusive 'Stratotype' section, based on original usage. The definition of GSSPs has since evolved to allow conceptual, and sometimes actual, marker points (so-called 'Golden Spikes') where a marker is hammered into, or painted on, rock at a particular locality. These have served to give ultimate definitions to horizons being considered (see review by Holland 1986). They allowed the use of all available means in attempting correlations with those, now defined, 'points in time'. An example is the potential GSSP for the Pliensbachian, based on both ammonite faunas and isotope stratigraphy (Hesselbo *et al.* 2000*a*). But many problems have been created in the recent drive to define more GSSPs, as the Cenozoic case studies, outlined by Aubry *et al.* (1999, 2000), for proposed Eocene–Oligocene and Palaeocene–Eocene GSSPs, reveal.

All this was achieved by the gradual evolution of the International Commission on Stratigraphy which has been responsible for issuing stratigraphic guidance up to their most recent, abridged, version of the *International Stratigraphic Guide* (Murphy & Salvador 1999). The Commission, through its, currently, seventeen Subcommissions, has also been active in promoting the complex decision-making processes which have made possible the definition of a large number of GSSPs, all over the world and throughout the Phanerozoic stratigraphic column. This has given much new precision to what exactly is to be correlated on a global scale. All GSSPs arbitrated up to 2000 are recorded on the recent *International Stratigraphic Chart* (Remane 2000*a*). For another example of an important stratotype decision, that defining the base of the Cambrian, see Brasier *et al.* (1994).

Some remarkable facts are also revealed in this new *International Stratigraphic Chart*. For example, none of the Cretaceous Stage names, long in use from the previous nineteenth century, have yet been formalized. Of the two Systems (Jurassic and Silurian), whose students gave early leads in striving for basal boundary stratotype definitions, the Silurian has complete coverage, with eight defined GSSPs at Stage/Age level, while the Jurassic has only two GSSPs, out of a potential eleven, revealing the strangely slow progress now being made in that System, once so much in advance of others in stratigraphic methods. This may be no bad thing in view of the great complexities involved in such global correlation work. Barry Webby reminds me that some Silurian GSSPs are still 'not entirely settled and the Silurian subcommission is now refocusing on some levels, which suggests they may not have been originally so well researched' (*in lit.* 31 July, 2001). This despite the fact that all Silurian Stages have been allocated their GSSPs, as indicated in Remane (2000*a*).

Non-biostratigraphical methods of correlation

Some of these methods have been reviewed in a fascinating volume edited by Dunay & Hailwood (1995). They were inspired by the real problems facing those trying to correlate unfossiliferous rocks. The eleven papers in this volume fall into five categories: mineralogical, chemical, isotopic, luminescence, and cyclo-stratigraphic. As can be seen from this array, there are now many new forces at work across the enormous field of stratigraphy, whether inspired by recent progress involving space exploration and Moon landings, the Deep Sea Drilling Project or, crucially, the need to find oil. All these have changed attitudes both within stratigraphy and to stratigraphy.

One of the most important aspects of this change in attitude has been reviewed by Gould (1989) and Dott (1998). This is the newly encouraged separation, and consequent drift apart, of the analytical and synthetic aspects of geology. These, in stratigraphy, are on the one hand those causal, often timeless, processes involving 'hard science' against, on the other, those historical or time-bound aspects of 'what actually happened', or 'soft science'. Gould, regretting the new status-ordering between the two, offered a passionate plea for a higher status for the latter, 'soft', or historical sciences. Dott urged that geologists should continue to be schizoid, and support both.

Some of these new non-biostratigraphic techniques, such as cyclo-stratigraphy and magneto-stratigraphy, discussed below, use basically repetitious, often binary, data preserved in rocks. Here the problem of what may be called the 'bar-code effect' comes into operation, as in (a) cyclo-stratigraphy or (b) magneto-stratigraphy. Bar-codes allow parallel lines of differing thickness to be read by laser or light guns (Hedley 1987). The problem is that, if one bar-line is missed or remains unread, the bar-code becomes that, not of the next object, but that of a quite different object. The proximity of the next object, becomes no proximity at all. Such problems of collocation were discussed by Baker when he revealed the many problems created by the headlong rush to move from old card-indexing systems to new on-line catalogues in libraries, thereby creating 'a kind of self-inflicted on-line hell . . . [whereas] any drawer of the out-of-date paper catalogue [had] represent[ed] the equivalent of a filtered-component search of a very sophisticated sort' (Baker 1997, pp. 150–156). In a similar way, the problems of repetitious 'bar-coded' data in stratigraphy may not be as simple as we hope.

Cyclo-stratigraphy

The ideas of both James Croll (1821–1890) from Scotland and Milutin Milankovich (1879–1958) from Serbia were destined to be ignored until they were capable of better demonstration. Their ideas were, in both cases, too multidisciplinarian for ready assimilation by geologists at those times. When they found support from the study of ocean-cores from 1976 (Milankovich 1995, p. 158) it was realized what a potentially powerful tool cyclo-stratigraphy might offer to stratigraphers.

A. G. Fischer in the United States pioneered this work there (Fischer 1991). House, in two important papers, has urged the potential of this completely new approach to an absolute timescale, by measuring orbital cycles and sedimentary microrhythms in rocks (House 1985, 1986). The whole subject of cyclicity, and of cyclic events in stratigraphy, has been the subject of major volumes published in 1991 (Einsele *et al.* 1991) and 1995 (House & Gale 1995). The problem once again, as with so many major new advances in stratigraphy, is that there is, as yet, no independent method of resolving time at and below the levels at which these cycles may be discriminated in the geological record. As Dott has pertinently noted: 'the Milankovich theory is very accommodating, for it provides a period to suit nearly every purpose – 19 000, 23 000, 41 000, 100 000 and 400 000 years' (Dott 1992, p. 13). A fascinating study by Algeo & Wilkinson (1988) confirms that much caution is needed before assigning Phanerozoic sedimentary cycles to any particular Milankovich cycle origin. Another problem is that at least some of these cycles in the stratigraphic record may not be due to original depositional features, but be diagenetic in origin. There is always the unresolved complication of potential gaps in recording of such cycles (Hallam 1986, Weedon & Hallam 1987). But the potential of this method in the future clearly remains enormous, as work on the most recent part of the geological record shows. Here, in the Plio–Pleistocene, use of the astronomical polarity timescale, quite independent of radio-isotopic dates, has provided a detailed timescale for the last 5.3 Ma and attempts are now under way to extend this back into the Miocene (Hilgen *et al.* 1997).

Magneto-stratigraphy

Magnetic events were revealed by the remarkable revolution which the Deep Sea Drilling Project effected in magneto-stratigraphy from the 1960s. This project has been the subject of three fine reviews: an outsider's view given by Glen (1982), with Menard (1986) and Hsü (1992) providing participants' views. Magneto-stratigraphy has proved particularly effective in Cenozoic rocks because of their more recent date and their better record on the world's present ocean-floors. A fine example of the international collaboration now possible in these rocks, which are paradoxically often not so amenable to normal biostratigraphic methods, is given by Aubry *et al.* (1986). This study integrates biostratigraphy and magneto-stratigraphy. The last technique has revolutionized this part of the stratigraphical column, but the precision thus acquired seems still uncertain. Its early potential in trying to solve a long-standing stratigraphic problem at the Jurassic–Cretaceous boundary was revealed by Ogg & Lowrie (1986).

Chemical stratigraphy

Isotopic age determinations using radioactivity are the chemically-derived age determinations most used in stratigraphy (Callomon 1984), even if the relation between these and biostratigraphy has never been clear or obvious. Thus Thirlwall (1983) pointed to a large discrepancy between such radiometric isotopic age dates, based on chemical analysis, and palaeontological data, based on fossil fish. He suggested this could be explained by particular (swimming) Silurian fish having had dispersal rates extending over ten million years between only Scotland and England. The earlier demonstration of the potential speed of dispersal of a recent gastropod (walking on foot) round the world in only 3000 years (Ager 1963, p. 167) suggested this was not going to be a solution to appeal to many palaeontologists.

Chemo-stratigraphy is another, more recent, method of enormous future potential in stratigraphy. This uses the changing chemistry in rocks to effect correlations. These can be either isotopic or elemental. Strontium-isotope stratigraphy was first proposed in 1948 and was sufficiently refined by the 1980s to be useful (McArthur 1991). A recent review gives fuller details of the important applications of this method in stratigraphy (McArthur 1994). Another study raises more important questions about the precision of biostratigraphic correlations over long distances. This claims, using $^{87}Sr/^{86}Sr$ data, that many Stage boundaries in the late Cretaceous of Antarctica have been misdated (McArthur *et al.* 2000*b*). Chemo-stratigraphy, in the case of carbon isotopes, can also add an important global dimension to potential correlations, once a basal stratotype (GSSP) has been chosen, as the study of the Dob's Linn basal Silurian GSSP in Scotland demonstrates (Underwood *et al.* 1997).

Event stratigraphy

Event stratigraphy attempts to effect time correlations on the basis of 'events' supposedly recorded in rocks, such as, for example, storms, hurricanes, volcanic eruptions, earth impacts by meteorites etc. or sea-level changes. A fine survey of such geological events is provided by Einsele *et al.* (1991). Proving (or assuming) that such 'events' are the same as others recorded in other rocks allows time correlations between the two. That infamous anti-continental-drifter, Harold Jeffreys (1891–1989), was, however, right to point out, in the closely related field of archaeology, that 'the problem [with events] is rather for the archaeologist to prove to the seismologist the contemporaneity of the observed phenomena, than for the archaeologist to assume their contemporaneity and date his sections from this assumption' (Daniel 1950, p. 255). The term 'event stratigraphy' was first used as such in 1973 by Ager in a stimulating volume (Ager 1973, p. 63; 1993, p. 99), although such 'events' have been sought, and explained, in geology for very much longer. Ager – surprisingly for such a 'father of event stratigraphy' – had noted that:

> we are always on dangerous grounds if we accept [event stratigraphy] as anything other than a last resort in the absence of really adequate evidence of evolving lineages [of fossils] (Ager 1973, p. 62).

Many examples of this 'new' stratigraphy could be cited, as one example see Hesselbo *et al.* (2000*b*)'s careful integration of chemical, physical and biological evidence in recognition, and explanation, of an early Toarcian Oceanic Anoxic Event (OAE). But, as with words like 'rapid' and 'abrupt', the question of the rarity of such 'events' must also be considered (Gretener 1967). The best reviews remain those by Dott (1983, 1996). Dott (1996, p. 245) tellingly asks that we distinguish geological events that provide positive deviations in normal energy and sediment accumulation (storms, eruptions, tsunamis, asteroid impacts, etc.) from those that provide negative intensities, which instead produce condensed sequences, hardgrounds, and, all too often, generate hidden breaks in the stratigraphic record.

The influence of impact events, post-1980, inspired a remarkable attempt by Hsü (1989) to show how 'inevitable the improbable' was in geology. Warnings were soon issued that the Earth is overdue for an asteroid to strike that could kill 10% of the Earth's population. McCall felt he now had to point out that 'no single person has succumbed (so far as we know) to asteroid impact during the last two millennia' (McCall 2001). So the more important consideration, if only to those seeking greater precision in stratigraphy, might be to wonder how best we might define 'a geological event' and its duration.

The scale of the problem of what constitutes an 'event' becomes clear in the case of the above-mentioned Early Toarcian OAE. Chemostratigraphy can give important insights into such OAEs. Recent strontium-isotope profiling of this part of the Jurassic column has suggested that, in Yorkshire, the relative durations of ammonite sub-zones can vary by factors of up to thirty times. It is immediately clear that we can no longer simply assume that all such, as here ammonite, sub-zonal or horizonal durations are equal. The same investigation also suggested that 'the early Toarcian OAE persisted for about 0.52 Ma' (McArthur *et al.* 2000*a*). Is an 'event' that lasted over half a million years an event? How long can such an event last and still be regarded as a single event? The lesson from this important study is how much is to be gained from integrated teams, with different sorts of stratigraphers, working together.

Some geological events of potential use in time correlation

Bentonites

Volcanic bentonite (ash-fall) bands, from their nature, are liable to prove some of the best of stratigraphic events. Their use was pioneered before the Second World War in the USA (Moore 1939, p. 27). Such bentonite bands are amenable to K/Ar and Ar/Ar radiometric dating and when, as pioneered in the Cretaceous of the western interior of North America, they can be well integrated with biostratigraphic data, they produced highly accurate timescale data (Obradovich & Cobban 1975). An important series of papers has been published by Messrs Bergstrom, Huff and Kolata on Ordovician and Silurian ash-layers in North America and Europe. These include some of the largest fall-out deposits in the whole Phanerozoic and 'because each ash bed was deposited in a matter of weeks, the base of a K-bentonite . . . is an isochronous surface with time resolution unattainable by other means' (Kolata *et al.* 1996, p. 66). Despite this, attempts to correlate one of these bentonites across the present Atlantic Ocean are still debated (Huff *et al.* 1996, p. 287). Such Middle Ordovician K-bentonites in New York State and Quebec, when integrated with the graptolite biostratigraphy, have demonstrated that an 'acute conflict' currently exists between the two. The alternative model proposed by Goldman *et al.* (1994) remains to be independently tested.

'Storm' deposits

Apart from the question of their durations, other problems of proving geological 'events' are clearly revealed in a study of such deposits by Ager, who in 1986 published his re-interpretation of the basal, 'Littoral Lias', facies in South Wales. These are marginal facies, comprising conglomerates, coquinoid limestones, cross-bedded skeletal grainstones, and oolites. Ager argued that the Sutton Stone here should no longer be regarded as a littoral deposit, laid down by a transgressing sea over a period of early Jurassic time but instead 'as a mass flow deposit laid down very rapidly, probably by a storm . . . one Tuesday afternoon' (Ager 1986). Others, who had geologically mapped the area, soon disagreed that these rocks could ever have represented any such single event. They preferred to see the deposit as the result, not of one depositional event, but those of 'many Tuesday afternoons' (Fletcher *et al.* 1986). Fletcher (1988) later gave a more detailed analysis, which apparently confirmed that these deposits were the result of a complex, long history, not of any single event. Palaeontological work since has supported the view that these deposits 'cannot be attributed to a single storm event' (Johnson & McKerrow 1995). Of course, as soon as it is recognized that there is no evidence that such an event actually occurred, any stratigraphic precision it might have provided disappears.

Sequence stratigraphic events

As a result of the increasing demands for oil and gas after the Second World War, with the development of the Deep Sea Drilling Project, other remarkable changes in attitudes to, and methodologies in, stratigraphy were forced upon it. This was nowhere more so than in the case of seismic stratigraphy, from which emerged sequence chronostratigraphy, a procedure that has attracted numerous advocates and critics since the 1970s. Seismological investigation, directed at the superficial layers of the Earth's crust, reveals numerous reflector horizons, which may represent bedding planes, unconformities, faults, or other significant lithological changes. Each unconformity-bounded 'package' of rock is called a 'seismic sequence'. Sequence stratigraphy ultimately relies on the recognition of 'events', in this case supposedly generated by worldwide changes in sea-level, as revealed by such reflector horizons. Succinct introductions to the topic are provided by Prothero (1990, pp. 258–265) and Leeder (1999, pp. 258–266). Dott (1996, p. 244) has noted that this 'presently seems to be the dominant paradigm in sedimentary geology'.

Larry Sloss was the pioneering figure here (Sloss 1963), which gives his current (dissenting) opinion – that such sequence boundaries have only local origins (Sloss 1991) – all the more credibility. If penetration of this technique into oil companies' research is taken to have occurred in 1975, we have the personal view of the chief protagonist (Peter Vail) as to how and when this revolution happened (Vail 1992). In Vail's opinion, the resulting 'renaissance of stratigraphy ranks in importance with the [other] plate tectonic revolution', which started at the same time, in the 1960s (Dott 1992, p. 13). Vail noted that the 1975 AAPG conference (Payton 1977) had been critical in advancing the speed of take-up of this new technique. There is now an enormous literature, involving both seismic, off-shore and non-seismic, land- or core-based, data, which it would be hard for one so ignorant as this author to review properly. However the real problem remains, as with impact as a cause of mass extinctions, that there is no consensus on the reliability and precision of sequence stratigraphy as a means of effecting time correlations. This much becomes clear from the writings of Sloss (1991), Miall (1997), and Wilson (1998).

Miall has been particularly incisive in his discussions of the limitations of sequence stratigraphy and in a series of papers has questioned much of the methodology used, especially the relationship of these sequences to time (see Miall 1992, 1994, 1995). In particular Miall (1992, p. 789) demonstrated a minimum 77% successful correlation with the standard, Exxon chart using four columns of geological data. But these did not record actual geological data but pseudo-sections which had been *randomly* generated (see Fig. 2).

Miall also pointed out that the claimed chronological precision of much of sequence stratigraphy is again greater than that of any available alternative and so is effectively untestable. While some sea-level changes clearly 'peaked simultaneously' across the (then

Ties with Exxon chart in columns 1-4:
—— Events correlated to within ±0.5 m.y.
- - - Events differing by >0.5 m.y., <1 m.y.
······ Events differing by >1 m.y.

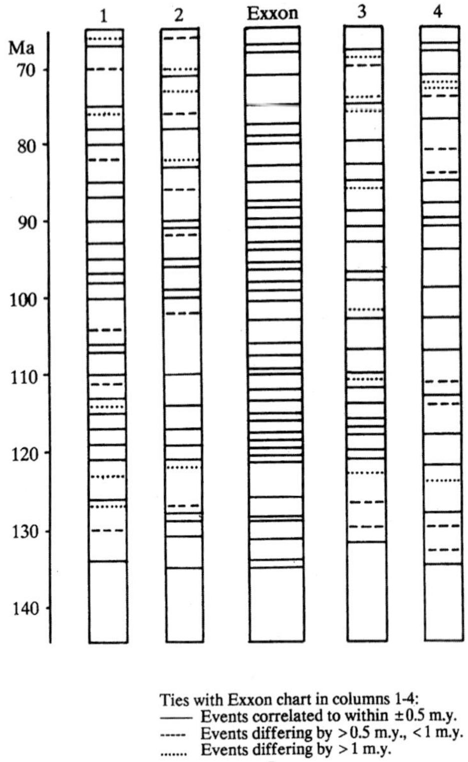

TABLE 1. RESULTS OF CORRELATION EXPERIMENT

Section	No. of events	±0.5 m.y. ties No.	%Fit	±1 m.y. ties No.	Total ties No.	%Fit	Mismatches
1	32	22	69	5	27	84	5
2	28	18	64	6	24	86	4
3	31	21	68	3	24	77	7
4	27	17	63	7	24	89	3

Fig. 2. Miall's Correlation 'experiment' showing the 40 Cretaceous sequence boundaries (Fig. 2, centre column) of the 1988 Exxon global-cycle chart. These were compared with other event boundaries in four other 'sections' (Fig. 2, Nos 1–4). Table 1 (right) shows 'the high degree of correlation of all four sections with this Exxon chart', the lowest correlation success being with No. 3, at 77% fit. 'The catch is that all four of these test sections were constructed by random-number generation' (Miall 1992, p. 789)!

smaller) Atlantic Ocean during the Cretaceous (Hancock 1993), it is notable that in the third edition of Miall's *Principles of Sedimentary Basin Analysis* (Miall 2000) the author plays down any supposedly worldwide eustatic control on such sequences, in favour of more local tectonic causes. Exactly this question – are such sequence boundaries tectonic or eustatic in origin? – was being asked in 1991 (Aubry 1991). No consensus on the origins of sequence boundaries, and thus the precision of their stratigraphic potential, has yet been reached. A fascinating discussion of the evolution of sequence-stratigraphic ideas has recently been published (Miall & Miall 2001), and should be required reading for all who study, or teach, stratigraphy.

Impact: the ultimate event

In May, 1979, the famous Alvarez extraterrestrial Cretaceous–Tertiary (K–T) impact theory was proposed (Alvarez 1979*a*; Alvarez *et al.* 1979). This was at first based only on a 20–25-fold increase in the abundance of iridium found in limestones in northern Italy. It was initially proposed with the expectation that this anomaly would prove to have been due to a supernova explosion, although the expected plutonium 244, osmium, and platinum increases had 'not yet been detected'. Soon afterwards, in September, 1979, the Alvarez team reported that this anomaly could not have been due to a supernova, but that 'the 25 fold increase in iridium . . ., which they found difficult to explain as an aspect of the sedimentary record at Gubbio, suggested that the Ir came from a solar system source, not a supernova' (Alvarez 1979*b*). Thus the evidence at first advanced in support of the K–T impact theory was entirely chemo-stratigraphic.

In June, 1980, it was announced that the K–T iridium anomaly had now proved to be more widespread and was due to an asteroid impact (Alvarez *et al.* 1980, Alvarez 1983). 'Impactology' was born. Its influence throughout the whole of geology has since been incredible. One historian has written that impact carries 'genuinely revolutionary implications that are fatal to the uniformitarian principle itself' (Marvin 1990, p. 147). The most impressive aspect, from a historical viewpoint, is the interdisciplinary nature of much, but not all, of the enormous amount of research which impact studies have inspired (Alvarez 1990). But it is notable that impactology was at first supported by chemical evidence, rather than the physical evidence that can best support it.

Conway Morris urged more recently that ecological evidence must also be much more involved in such investigations, saying of the mass-extinctions of life at the K–T boundary, that 'at one level we can just as easily substitute the trigger for these extinctions being Martians waving laser-cannons rather than asteroids or a comet' (Conway Morris 1995, p. 292). In an incisive early review of the whole impact revolution, Van Valen rightly criticized the Alvarez's claims that their own evidence was experimental (i.e. 'hard') as 'misleading propaganda' (Van Valen 1984, p. 122).

We must be concerned here only with the 'fallout' of impactology on stratigraphy. After the claim that a K–T impact event had been recognized, the search began to find the impact site. Two such craters have special interest for the imprecision with which they were first dated. One was the Duolun impact crater in China, reported in *New Scientist* (Fifield 1987). This briefly then became a candidate for a dino-extinguishing event at the K–T boundary, if only in a English newspaper. But this impact-object, when dated, proved to have struck eighty million years too early (Ager 1993, p. 179)! This was not precise stratigraphy. The other candidate proved a more serious one. This was the Manson crater in northwestern Iowa, the largest – 35 km – crater then recognized in the United States. This was proposed as the K–T boundary candidate on the basis of $^{40}Ar/^{39}Ar$ dating of shocked microcline from the resulting structure (Kunk *et al.* 1989).

Physical evidence

The clearest evidence by which to confirm, and date, impact comes when not only the impact crater is preserved, or can be revealed by seismic and then borehole evidence (as in the case of Chicxulub, Mexico), but can also be partially dated by examining what it struck and whether the physical fallout from the impact can be documented in the surrounding rocks, as in the case of the Manson microcline. Such physical evidence has been the subject of a fine review by Koeberl (1996), but which significantly ignored the many, often subtle, biochronological and extinction questions raised by such impact studies. When such physical evidence was properly investigated for the Manson crater, it emerged that it could not have been the K–T 'killer crater'. A sanidine clast from the melt-matrix breccia of this impact gave a new date of *c.* 73.8 Ma. This was consistent with the biostratigraphic level into which diagnostically shocked, metamorphosed mineral grains had been found ejected in the stratigraphic record nearby, at a lower level in the Pierre Shale, of South Dakota (Izett *et al.* 1993). The Manson crater, like the Duolun Crater, proved to predate the features it was hoped it would explain – but here by 'only' 9 Ma. This again was imprecise stratigraphically, and only demonstrated how important 'wishful thinking' could become in impact stratigraphy.

The best documented example of the precise dating of a crater by its physical ejecta seems to be provided by Australia's – 160 km – Neoproterozoic Acraman crater in South Australia (Gostin *et al.* 1986; Williams 1986). It has the best documented crater-*cum*-ejecta impact on record, although one too old to have had much perceivable biological effect. However Frankel, an enthusiast for impact as the causal agent behind most of the major geological extinctions, and hence of most System-level stratigraphic boundaries, notes that

> the possibility that [this] major impact wiped clean the biological slate and allowed new life-forms (e.g. the Ediacara fossil assemblage) to evolve must be seriously considered (Frankel 1999, p. 146).

When this ejecta-recognizing approach was taken to the now celebrated K–T candidate, Chicxulub crater in Mexico, using diagnostic physical evidence, good evidence for the date and potential scale of a terminal Cretaceous impact there was uncovered. A marker-bed of large microtektites and the thickest ejecta layer known from this impact were found in several places nearby, like southern Haiti (Maurrasse & Sen 1991) in support of a major 'event' nearby.

The potential stratigraphic scale of such impact events is indicated by the title of the International Geological Correlation Project (IGCP) No. 384. The first results of this project were published in 1998 under the title *Impact and Extraterrestrial Spherules: New Tools for Global Correlation* (Detre & Tooth 1998). The same project also started a new international journal in 1997, called *Sphaerula*. Impacts, if proven to be global in effect, must have real stratigraphic potential.

The separate stratigraphic problem of distinguishing multiple impacts often closely coupled in time has also emerged in the late Eocene record. Here two impacts have been documented which are variously calculated to have been separated by anything between only 2 Ka (Glass 2000) to between 10–20 Ka (Vonhof *et al.* 2000). But at most sites where records of these two should be expected, either 'one of the ejecta layers is missing, or the two ejecta layers are indistinguishable' (Vonhof *et al.* 2000). This demonstrates the problems that the available stratigraphic record produces, even when, as here, there is great expectation of what is likely to be present.

Any consensus on the extent, and biological effects, of the K–T boundary event remains obstinately polarized amongst geologists. Some prefer to see the cause of the extinction at this boundary as partly or wholly due to volcanic events over a much longer period of time than the short-lived event implied by impact. This volcanic scenario has a prehistory as well as a history. The history can be said to have started in 1985, with the paper by Officer and Drake (1985). The prehistory need only be taken back as far as Vogt 1972 (Courtillot 1999, p. 58). Such volcanism is now being proposed as an explanation for other second-order mass extinctions, like the Karoo–Ferrar flood basalt volcanism to explain an early Jurassic extinction (Palfy & Smith 2000).

Work using physical evidence of impact is in stark contrast to some of the earlier evidence proposed to explain the first, merely chemical, discoveries of K–T iridium anomalies, with associated concentrations of phosphatic fossils, in the 'fish clays' of Denmark. These were immediately used to prove the impact must have occurred near Denmark. The most extraordinarily subtle ocean currents had then to be invoked to explain the more fishy aspects of the evidence found here (Allaby & Lovelock 1983, pp. 95–99). The paper by Rocchia *et al.* (1990) was crucial in indicating that the iridium anomaly at the original, Gubbio, locality in Italy was much more extensive stratigraphically (and thus must have lasted 'longer') than had previously been realized (see Fig. 3).

The problem of anomalous iridium concentrations must depend on how complete the stratigraphic record can be shown to be at the different localities that show such anomalies. This must now be our final consideration.

How complete is the stratigraphic record?

Nearly a century ago Buckman reminded us of the vital importance of separating sedimentary from chronological records in stratigraphy: 'the amount of deposit can be no indication of the amount of time, ... the deposits of one place correspond to the gaps of another' (Buckman 1910, p. 90). On the related question of the adequacy of the sedimentary rock record, Buckman noted earlier how fossil:

> species may occur [together] in the rocks, but such occurrence is no proof that they were contemporaneous . . . their joint occurrence in the same bed [may] only show that the deposit in which they are embedded accumulated very slowly (Buckman 1893, p. 518).

The basic truth of these statements is still often ignored. The abundance of any particular material, element, mineral, chemical or fossil, in the stratigraphical record need *not* prove either

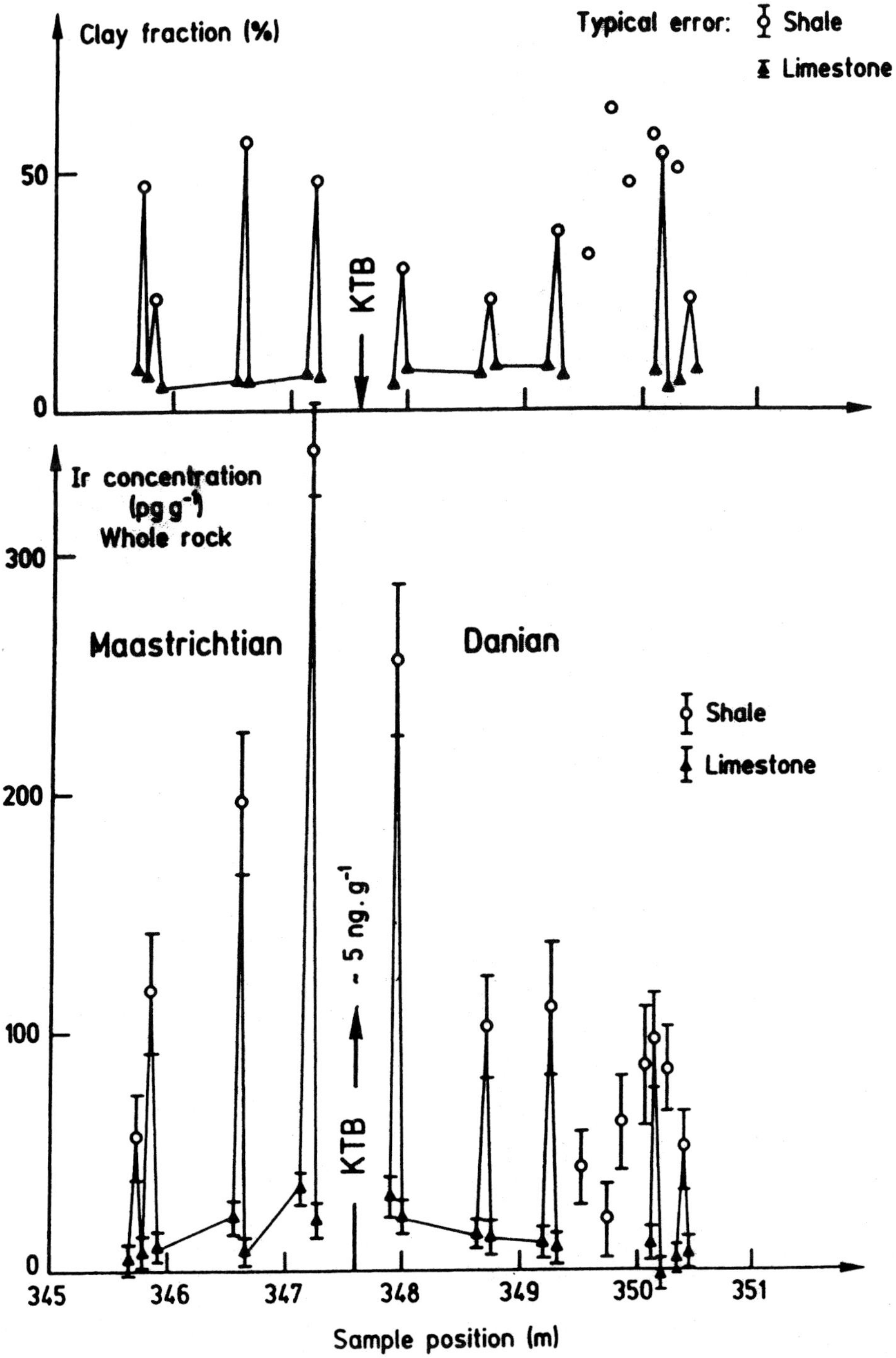

Fig. 3. Whole-rock iridium concentrations across six metres of rock straddling the K–T boundary at Gubbio, Italy, with the K–T boundary (KTB) marked. Concentrations of Ir in limestones are much lower than Ir concentrations in shales which stand out as maxima. 'The existence of such Ir spikes in shales is not due to the occurrence of isolated 'Ir events', but to post-depositional enhancements related to dissolution of carbonates' (Rocchia *et al.* 1990, pp. 214–215).

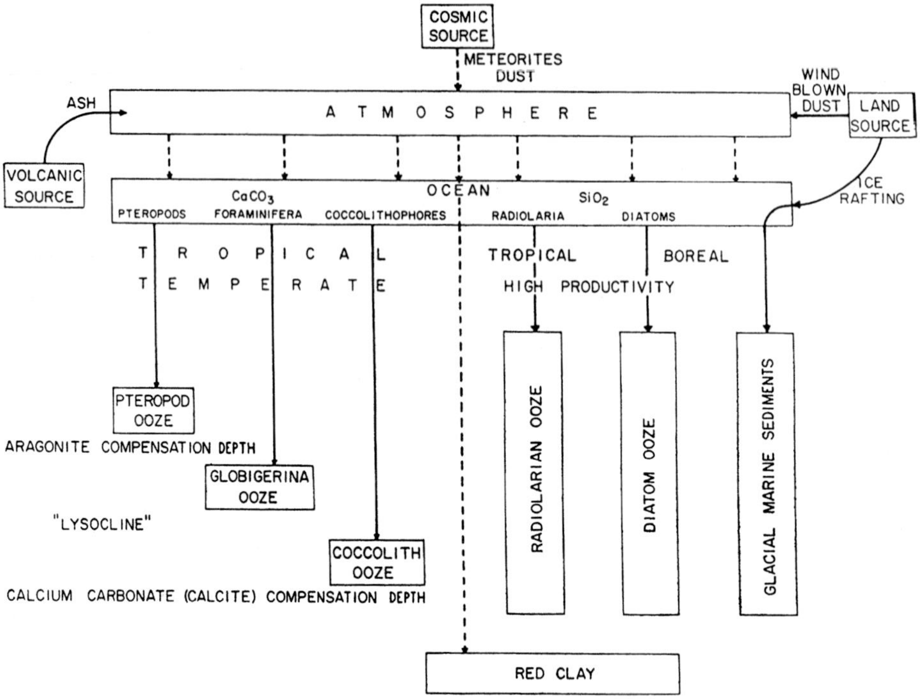

Fig. 4. A cumulative diagram demonstrating 'pelagic sedimentation in the ocean', from Hay (1974, Fig. 2).

the origin, or the contemporaneity, of that material. Attempts to assess the 'stratigraphic completeness of the stratigraphic record' by using timescales based on sedimentation rates as proposed by Schindel (1982) or Sadler & Strauss (1990) prove inappropriate because they take no account of the many gaps, erosion surfaces and all the other complexities of what has been called litho-chronology by Callomon (1995, p. 140).

Similarly doomed are some of the attempts to assess the origins of some concentrations of fossils, whether of Palaeozoic nautiloid cephalopods (Holland *et al.* 1994) as the remains of fossils that lived together 'in schools' and then 'suffered mass mortality', or the geologically later 'belemnite battlefields' (Doyle & Macdonald 1993). These latter may be post-mortal accumulations of a nearly original ecological assemblage, as proposed, but they may as well be entirely condensed and accumulated over much longer periods of time, and concentrated together only because of the lack of any sedimentary dilutant, as in the fossil 'cemeteries' that Buckman worked on. The presence of a 'cemetery deposit' of fossils can never prove those fossils suffered a catastrophic death.

Similar considerations apply to the Danish K–T boundary 'fish clay' or the 'fish mortality horizon' which was claimed 'may represent the first documented, direct evidence of a mass kill event associated with the bolide impact' at the K–T boundary on Seymour Island. These may equally have had secondary, condensed, and thus residual, origins, rather than a primary origin as a 'mass kill associated with an impact event' proposed for Seymour Island. The first, condensed origin, was rejected as an explanation here only because of the fish horizon's 'inescapable' relationship with an iridium anomaly below it (Zinsmeister 1998).

'Anomalous' abundances of iridium also need not have impactal origins. Some can have been derived through condensation, as Rampino originally noted (1982), and as Hallam (1984) and Ager (1993) have more recently supported. One only has to follow the diagram showing the processes involved in getting such normal, but still cosmic, iridium deposited in pelagic sedimentations on the ocean floor given by Hay (1974, p. 3) to realize how such an insoluble material as cosmically derived iridium-rich dust might end up, condensed and isolated, on ocean floors. Most other potential dilutants would simply have been removed by chemical solution on their way down to the sea-floor (see Fig. 4).

The surprises in this field might be first, how different the past might prove from the present, in matters involving compensation depths and solubilities of organic materials; and second, how very condensed and incomplete pelagic deposits can prove to be. We need careful stratigraphic studies of abyssal clays with overall low accumulation rates, such as Kyte & Wasson's (1986) study of a thickness of only 24 metres ranging over more than 70 Ma from the central North Pacific. This gave confirmation of a major impact event having been recorded here, by showing that in this condensed abyssal sequence there was a significant, and surely here primary, increase in Ir concentration at the K–T boundary.

At more distant sections in rocks of shallower water origin (such as Stevns Klint, Denmark), analysis showed how:

> a pulse of calcite dissolution in shallow water coincided precisely with the era [K–T] boundary, and [that] this event played a major role in the formation of the Fish clay in eastern Denmark, which is a condensed series of smectitic clay-rich layers from which much calcite has dissolved. [Such evidence suggested that] no single catastrophe can account for the major biotic extinctions which occurred at the end of the Cretaceous period [here] (Ekdale & Bromley 1984).

In other words there are anomalies and anomalies, which need to be carefully and separately analyzed. It was at a Danish locality that the 160-fold increase of 'anomalous iridium', the highest recorded in the original research, suggested it had to have had a sudden, extra-terrestrial origin (Alvarez *et al.* 1980, p. 1100; Frankel 1999, pp. 19–21). Its extra-terrestrial origins need not be in dispute, but stratigraphers need to ask if all such extra-terrestrial material had to have arrived suddenly, through impact, or could have arrived by more slowly accumulated concentration.

The same problem emerged at El Kef, in Tunisia, chosen in 1989 as the Global Stratotype Section and Point (GSSP) for the base of the Danian, and thus the Cenozoic (Cowie *et al.* 1989, p. 82). The question was again: how complete is the critical K–T section at this boundary here? Its great incompleteness has been confirmed by MacLeod & Keller (1991), and in a more recent paper by Donze *et al.* (1996). None the less, this region is still regarded as 'unique in its documentation of one of the most critical intervals of Earth history. The most complete succession [here] is however that of El Kef [GSSP for the Danian]' (Remane 2000*b*). The same situation re-emerged at the first K–T iridium anomaly locality, Gubbio in Italy, when a more extended vertical extent of 'the iridium anomaly' was investigated. Here Ir associations with clay minerals were thought due 'to post-depositional enhancements related to dissolution of carbonates ... in a sequence characterized by a low sedimentation rate' (Rocchia *et al.* 1990; see Fig. 3). The same problem faces the claim that the iridium anomaly detected in the English Ludlow Bone Bed, Upper Silurian, had a single primary, impactal origin. This occurrence again demonstrates a secondary, condensed, origin (Schmitz 1992; Smith & Robinson 1993), like some of the 'anomalous' sequences known at the K–T boundary.

The real problem, as with sequence stratigraphy, is the difficulty of achieving accurate calibrations of rates and durations of many of these geological processes and, or, events, as Dingis (1984) has pointed out. Indeed, the initial idea of using iridium concentrations, to the single-minded extent that was first proposed by the Alvarezes, as a sedimentary rate-metre (Frankel 1999, p. 19), has now been re-invented as a means of measuring rates of sedimentation, and to prove the completeness of sequences containing iridium 'anomalies' (Bruns *et al.* 1996, 1997). This marks a return to the original, pre-impactal, intentions of the Alvarez team before their work revealed 'over anomalous' amounts of iridium. One man's anomaly has become another's normality. Wallace (1991) and Sawlowicz (1993) have discussed different ways in which iridium can become 'anomalously' abundant in sediments.

Another real problem when discussing stratigraphic precision is again conceptual. A recent paper on dinosaur abundances near their critical terminations in Montana and North Dakota was highlighted on the cover of *Science*. It supposedly proved, of dinosaur remains found here close to the terminal Cretaceous boundary, that 'Dinosaurs were going strong till the last minute [of the Cretaceous]' (Sheehan *et al.* 2000). But space and time are not the same, even in a science as unscientific as geology! Buckman had noted in 1893 how fossil 'species may occur together in the rocks [e.g. in space], yet such occurrence is no proof that they were contemporaneous [e.g. in time]' (Buckman 1893, p. 518). Others have added to this confusion. Gould (1992), when discussing extinctions at the K–T boundary at Zumaya, Spain used an ammonite found spatially 'within inches' of that boundary to prove these ammonites had become extinct at the time of that boundary. Hudson

(1998, p. 414), noting two occurrences a short distance (less than 1 metre), whether below the boundary clay in Montana or the Raton Formation, asked 'can either distance be regarded as "well below" the boundary?'. The answer to this rhetorical question depends on precise separation of those quite different entities; space and time.

This problem has now also reached the museum. A recent acquisition on display at the Manchester Museum, England, excavated from underground caves at Guelhemmerberg, near Maastricht in November 1999, claims to 'record the exact point in time of the end Cretaceous extinction . . . when many animals, including the dinosaurs, became extinct' (Anon. 2000, pp. 15–16). If only the stratigraphic record *could* so precisely record such matters!

Conclusion

The *Quo Vadis* conference of 1982 urged on participants the need for:

> a better understanding of the degree of accuracy and precision that can be reached in regional and global correlations, and more insight into the nature and interrelation of physical, chemical and biological processes in space and time (Seibold & Meulenkamp 1984, pp. 65–66).

While discussing the problems of using eustatic events in stratigraphy Dott pointed out in 1992 that:

> one of the consequences of the renaissance of stratigraphy during the past two decades [using such a wide range of techniques] has been the rekindling of enthusiasm for eustasy and for cycles of several kinds. This has even resulted in a fervent new orthodoxy, which Sloss (1991) has appropriately dubbed '*neoneptunism*' (Dott 1992, p. 13).

The general incompleteness of the stratigraphic record in the Eocene was specifically commented upon by Aubry (1995) who, in a later important abstract, also reminded us of the vital consequences for both sequence stratigraphy and geochronology of the stratigraphic record being, as it is so often shown to be throughout the geological record, incomplete. She noted that

> the challenge for the next decade was to document further the architecture of the stratigraphic record using the temporal component as an essential component, a fact that sequence stratigraphy has somehow failed to recognize (Aubry 1996).

Van Andel (1981) and Bailey (1998) have equally urged a reappraisal of those features of the rock record such as 'perceived cycles and sequences', because of the sheer complexity of that record which often embraces gaps and in which record there may often be 'more gap than record'. Zeller (1964) in a fascinating paper has equally shown how easy it is, through human nature, to discern cycles in stratigraphy.

The critical point is that, amid all the wars of words about 'hard' and 'soft' science, or whether 'all science is either physics or stamp collecting' (as Ernest Rutherford memorably said (Birks 1962, p. 108)), no consensus on either the cause, the extent, or precise timing of the extinctions, even at the K–T boundary, has yet emerged (Glen 1994, Courtillot 1999, Frankel 1999). There is a near consensus that there was a large impact at or near the K–T Boundary in Mexico. But its effect on terminal Cretaceous life around the world is much less clear and perhaps must remain so. The lack of consensus becomes clear by comparing the detailed biostratigraphic data assembled by MacLeod *et al.* (1997), with the response from Hudson (1998).

The authors of a recent paper (Albertao *et al.* 2000) were duly forced to draw the K–T boundary at two quite different stratigraphic horizons in NE Brazil when trying to define this boundary there, depending on whether biological data or physical evidence were invoked. This was another site which provided 'no direct evidence for an impact origin'. One gets a clear view of the lack of consensus by comparing the American view of the debate given by two of its main American protagonists (Alvarez 1997; Frankel 1999) with that of a French competitor (Courtillot 1999).

The need to return to more careful assessment of *all* temporal components in stratigraphy is the most important lesson from all the new stratigraphies, in which the last fifty years have been so prolific. One cause for some future optimism is the way in which graphic correlation (Shaw 1995), which uses statistical analysis of first and last appearances in ranges of fossil taxa, has been demonstrated as a means of investigating the degree of completeness of incomplete sequences (Macleod 1995*a*, *b*). Another is the potential of the methods used by McArthur *et al.* (2000*a*) in integrating strontium isotope profiles to document durations of geological events, with the ammonite biozones used in biochronology. The future lies, not in complaining about 'the current imprecisions of biostratigraphical correlation' (Jeppsson & Aldridge 2000, p. 1,137) but, in integrating stratigraphical studies in the way McArthur *et al.* (2000*a*) have demonstrated.

Localities (Fig. 5)	1 BB	2 Ch	3 WH	4 HP	5 Be-CF	6 Se	7 LH/HH	8 BA	9 SL	10 Cl	11 Ob	12 Br-L	13 Du	Average
Resolution: Stages										†	†			
scope*	3	3	3	3	3	3	3	3	3	1	2	3	3	
number	3	3	3	3	3	3	3	3	3	1	2	3	3	
% completeness	100	100	100	100	100	100	100	100	100	100	100	100	100	100
Resolution: Zones					†	†			†					
scope	14	14	14	14	11	9	14	14	10	8	7	14	14	
number	11	8	11	9	6	8	9	9	8	8	7	9	11	
% completeness	78	57	78	64	43	89	64	64	80	100	100	64	78	74
Resolution: faunal horizons														
scope	56	56	54	56	45	37	56	56	42	32	29	56	56	
number	20	18	21	23	14	10	21	22	20	22	20	22	29	
% completeness	36	32	39	41	31	27	38	39	48	69	69	39	52	43

* Only the Lower Bathonian is represented in the Inferior Oolite. But even at Substage level (Lower and Upper Aalenian, Lower and Upper Bajocian, Lower Bathonian), at which the maximum scope would be 5, the representation would be everywhere 100% complete.
† These sections have exposed only parts of the Inferior Oolite, either cut off at the tops by erosion or covered at the base.

Fig. 5. The three differing 'completenesses' of the geological record, in percentages, as revealed using three different levels of resolution, based on ammonite biochronologies, in the Inferior Oolite of southern England. At Stage level (e.g. Aalenian, Bajocian, Bathonian, etc.) all thirteen sections show complete records where rocks of these ages are exposed (average 100%). At the next lowest, Zonal, level of resolution, completeness varies from 100% to 43% (average 74%); while at the lowest available, Faunal Horizon, level, completeness varies from 69% to 27% (average 43%) (Callomon 1995, p. 147).

Only when such integrated studies are properly attempted may we be able to start to investigate the biological consequences of some of the more extraordinary events to which the Earth has been subjected over its long history. Until then stratigraphy will indeed remain a 'science in a crisis' (Glen 1994). For as Buckman (1921, p. 2) so presciently recorded long ago: 'additions to fauna *decrease* the imperfection of the zoological, but *increase* that of any local geological record: the gaps caused by destruction stand revealed more plainly'. Buckmans's claim has been entirely confirmed by Callomon (1995; see Fig. 5).

It does indeed seem that the harder you look at rocks the less complete their record of the passage of time becomes. Van Andel has said the same. To him, it:

> appears that the geological record is exceedingly incomplete and that the incompleteness is greater the shorter the time-span at which we look. [He too urges] 'the need for a vastly increased care in stratigraphy and chronology' (Van Andel 1981, p. 397).

I thank the editor, D. Oldroyd (Sydney), for his attempts to guide this difficult paper through the editorial process. As Dietz (1994, p. 8) has noted: 'scientists now know more and more about less and less'. This is particularly true in stratigraphy. The same move (also confirmed by Dietz), which has taken geology from the field into the laboratory, has had a similarly negative effect on academic library provision, which has caused new difficulties. In the face of so much 'information', more and more literature gets locked or thrown away. The Senate's reaffirmation of library disposal policy at my former university in November, 1999, makes chilling reading to all of us who care about even recent history. It read: 'old and superseded texts can be misleading or worthless and unsought material can obstruct the search for relevant items'.

My attempt to combat such attitudes in this paper has also had to be biased towards those parts of the stratigraphic column with which I have experience. It has been equally influenced by a lifetime spent attempting to teach the central importance of stratigraphy to declining numbers of students of geology. I thus hope this paper will provoke as much as it informs. I have also tried to repay a long-held debt to J. Callomon (London), who first showed me how subtle and complex the stratigraphic record so often is. For specific help I thank W. Cawthorne (London), C. Lewis (Macclesfield) and G. Papp (Budapest). I am most grateful to A. Rushton (Keyworth), R. Dott (Wisconsin) and B. Webby (North Ryde, New South Wales) who were all sufficiently provoked to make many comments on earlier versions, which has improved it.

References

AGER, D. V. 1963. *Principles of Paleoecology*. McGraw–Hill, New York.

AGER, D. V. 1973. *The Nature of the Stratigraphic Record*. Macmillan, London (3rd edn, John Wiley, Chichester).

AGER, D. V. 1986. A reinterpretation of the basal 'Littoral Lias' of the Vale of Glamorgan, *Proceedings of the Geologists' Association*, **97**, 29–35.

AGER, D. V. 1987. The excitement of traditional stratigraphy. *Geology Today*, July–August, 116–117.

AGER, D. V. 1993. *The New Catastrophism*. Cambridge University Press, Cambridge.

AGER, D. V. & CALLOMON, J. H. 1971. On the Liassic age of the "Bathonian" of Villany (Baranya). *Annales Universitatis Scientiarum Budapestinenis. . . . Sectio Geologica*, **14**, 5–16.

ALBERTAO, G. A., MARINI, F., OLIVEIRA, A. D., DELICIO, M. P. & MARTINS JR, P. P. 2000. *Peculiarities concerning the K/T Boundary in N E Brazil*. Paper/poster presented to the IGC at Rio de Janeiro, Session 25–6.

ALGEO, T. J. & WILKINSON, B. H. 1988. Periodicity of Mesoscale Phanerozoic sedimentary cycles and the role of Milankovich orbital modulation. *Journal of Geology*, **96**, 313–322.

ALLABY, M. & LOVELOCK, J. 1983. *The Great Extinction*. Secker & Warburg, London.

ALVAREZ, L. W. 1983. Experimental evidence that an asteroid impact led to the extinction of many species 65 million years ago. *Proceedings of the National Academy of Sciences, U. S. A.*, **80**, 627–642.

ALVAREZ, L. W., ALVAREZ, W., ASARO, F. & MICHEL, H. V. 1980. Extraterrestrial cause for the Cretaceous–Tertiary extinction. *Science*, **208**, 1095–1108.

ALVAREZ, W. 1979. Dinosaur extinction possibly linked to extra-terrestrial cause, *Episodes*, **2**, July 1979, 30.

ALVAREZ, W. 1979*b*. Anomalous iridium levels at the Cretaceous/Tertiary boundary at Gubbio, Italy. *In*: CHRISTENSEN, W. K. & BIRKELUND, T. (eds) *Proceedings of the Cretaceous–Tertiary Boundary Events Symposium*, **2**. Copenhagen University, Copenhagen, 69.

ALVAREZ, W. 1990. Interdisciplinary aspects of research on impacts and mass extinctions: a personal view. *Geological Society of America Special Paper* **247**, 93–97.

ALVAREZ, W. 1997. *T. rex and the Crater of Doom*. Princeton University Press, Princeton.

ALVAREZ, W., ALVAREZ, L. W., ASARO, F. & MICHEL, H. V. 1979. Experimental evidence in support of an extra-terrestrial trigger for the Cretaceous–Tertiary extinctions, *EOS*, **60**, 734.

ANON. 2000. *Manchester Museum, Annual Report*, August 1999–31 July 2000. Manchester.

ARKELL, W. J. 1957. *Treatise on Invertebrate Paleontology. Part L, Mollusca 4, Cephalopoda Ammonoidea*, University of Kansas and Geological Society of America, Kansas & New York.

AUBRY, M.-P. 1991. Sequence stratigraphy: eustasy or tectonic imprint? *Journal of Geophysical Research*, **96**, 6641–6679.

AUBRY, M.-P. 1995. From chronology to stratigraphy. *In*: BERGGREN W. A., KENT, D. V., AUBRY, M.-P. & HARDENBOL, J. (eds) *Geochronology, Time Scales and Global Stratigraphic Correlation*, Society for Sedimentary Geology, Special Publications **54**, 213–274.

AUBRY, M.-P. 1996. On the incompleteness of the stratigraphic record: implications for sequence stratigraphy and geochronology, *Abstracts of the 30th International Geological Congress, Beijing*, **2**, 10.

AUBRY, M.-P., BERGGREN, W. A., VAN COUVERING, J. A. & STEININGER, F. 1999. Problems in chronostratigraphy, *Earth-Science Reviews*, **46**, 99–148.

AUBRY, M.-P., VAN COUVERING, J. A., BERGGREN, W. A. & STEININGER, F. 2000. Should the golden spike glitter? With comments and a response. *Episodes*, **23**, 203–214.

AUBRY, M.-P., HAILWOOD, E. A. & TOWNSEND, H. A. 1986. Magnetic and calcareous-nannofossil stratigraphy of the lower Palaeogene formations of the Hampshire and London basins, *Journal of the Geological Society, London*, **143**, 729–735.

BAILEY, R. J. 1998. Stratigraphy, meta-stratigraphy and chaos. *Terra Nova*, **10**, 222–230.

BAKER, N. 1997. *The Size of Thoughts*. Vintage, London.

BERRY, W. B. N. 1987. *Growth of a Prehistoric Time Scale*. Blackwell, Palo Alto.

BIRKS, J. B. (ed.) 1962. *Rutherford at Manchester*. Haywood, London.

BRASIER, M., COWIE, J. & TAYLOR, M. 1994. Decision on the Precambrian–Cambrian boundary stratotype. *Episodes*, **17**, 3–8.

BRETT, R. 2000. Frontiers of life, *Brazil 2000 IGC News*, 1–3.

BRICE, W. R., 1989, *Cornell Geology Though the Years*. Cornell University Press, Ithaca.

BRUNS, P., RAKOCZY, H., PERNICKA, E. & DULLO, W.-C. 1997. Slow sedimentation and Ir anomalies at the Cretaceous/Tertiary boundary, *Geologische Rundschau*, **86**, 168–177.

BRUNS, P., DULLO, W.-C., HAY, W. W., WOLD, C. N. & PERNICKA, E. 1996. Iridium concentration as an estimator of instantaneous sediment accumulation rates. *Journal of Sedimentary Research*, **66**, 608–612.

BUCKMAN, S. S. 1889. On the Cotteswold, Midford and Yeovil Sands. *Quarterly Journal of the Geological Society, London*, **45**, 440–474.

BUCKMAN, S. S. 1893. The Bajocian of the Sherborne district, *Quarterly Journal of the Geological Society, London*, **49**, 479–522.

BUCKMAN, S. S. 1901. Jurassic brachiopoda. *Geological Magazine*, Decade 4, **8**, 478.

BUCKMAN, S. S. 1910. Certain Jurassic ('Inferior Oolite') species of ammonites and brachiopoda. *Quarterly Journal of the Geological Society, London*, **66**, 90–110.

BUCKMAN, S. S. 1921 *Type Ammonites*, **3** (Part 30), The Author, Thame.

BURCHFIELD, J. D. 1975. *Lord Kelvin and the Age of the Earth*. Macmillan, London.

CALLOMON, J. H. 1984. The measurement of geological time. *Proceedings of the Royal Institution of Great Britain*, **56**, 65–99.

CALLOMON, J. H. 1995. Time from fossils: S. S. Buckman and Jurassic high-resolution geochronology. *In*: LE BAS, M. J. (ed.) *Milestones in Geology*, Geological Society, Memoir No. **16**, 127–150.

CHALLINOR, J. 1978. *A Dictionary of Geology*. University of Wales Press, Cardiff.

CONWAY MORRIS, S. 1995. Ecology in deep time. *Trends in Ecology and Evolution*, **10**, 290–294.

COPE, J. C. W., GATTY, T. A., HOWARTH, M. K.,

MORTON, N. & TORRENS, H. S. 1980. *A correlation of Jurassic rocks in the British Isles, Part One*. Geological Society of London Special Report, **14**, 1–73.

COURTILLOT, V. 1999. *Evolutionary Catastrophes: The Science of Mass Extinction*. Cambridge University Press, Cambridge.

COWIE, J. W., ZIEGLER, W. & REMANE, J. 1989. Stratigraphic Commission accelerates progress 1984 to 1989. *Episodes*, **12**, 79–83.

COX, B. M., 1990. A review of Jurassic chronostratigraphy and age indicators for the UK. *In*: HARDMAN, R. F. P. & BROOKS, J. (eds) *Tectonic Events Responsible for Britain's Oil and Gas Reserves*. Special Publications **55**, The Geological Society, London, 169–190.

DANIEL, G. E. 1950. *A Hundred Years of Archaeology*, Duckworth, London.

DETRE, C. H. & TOOTH, I. (eds). 1998. *Impact and Extraterrestrial Spherules: New Tools for Global Correlation*, Papers Presented to the 1998 Annual Meeting of IGCP 384, Hungarian Geological Institute, Budapest.

DEWEY, J. F. 1999. Reply when awarded the Wollaston Medal. *Geological Society Awards 1999*. London.

DIETZ, R. 1994. Earth, sea and sky: life and times of a journeyman geologist. *Annual Reviews of Earth and Planetary Science*, **22**, 1–32.

DIETZE, V. & CHANDLER, R. B. 1997. S. S. Buckman und der Inferior Oolite, *Fossilien*, **4**, 207–213.

DINGIS, L. 1984. Effects of stratigraphic completeness on interpretations of extinction rates across the Cretaceous–Tertiary boundary. *Paleobiology*, **10**, 420–438.

DONOVAN, D. T. 1966. *Stratigraphy: An Introduction to Principles*. Thomas Murby, London.

DONZE, P., BEN ABDELKADER, O., BEN SALEM, H., MAAMOURI, A.-L., MEON, H. *et al.* 1996. At K–T boundary, the stratotypical section (El Kef, NW Tunisia) shows a concomitance of three different events. *Abstracts of the 30th International Geological Congress, Beijing*, **2**, 111.

DOTT, R. H. Jr 1983. Episodic sedimentation. How normal is average? How rare is rare? Does it matter? *Journal of Sedimentary Petrology*, **53**, 5–23.

DOTT, R. H. Jr 1992 *An introduction to the ups and downs of eustasy*. Geological Society of America Memoir, **180**, 1–16.

DOTT, R. H. Jr 1996. Episodic event deposits versus stratigraphic sequences – shall the twain never meet? *Sedimentary Geology*, **104**, 243–247.

DOTT, R. H. Jr 1998. What is unique about geological reasoning?, *GSA Today*, October, 15–18.

DOYLE, P. & MACDONALD, D. I. M. 1993. Belemnite battlefields. *Lethaia*, **26**, 65–80.

DUNAY, R. E. & HAILWOOD, E. A. (eds). 1995. *Non-biostratigraphical Methods of Dating and Correlation*. Special Publications, **89**, The Geological Society, London.

EKDALE, A. A. & BROMLEY, R. G. 1984. Sedimentology and ichnology of the Cretaceous–Tertiary boundary in Denmark: implications for the causes of the terminal Cretaceous extinction. *Journal of Sedimentary Petrology*, **54**, 681–703.

EINSELE, G., RICKEN, W. & SEILACHER, A. (eds) 1991. *Cycles and Events in Stratigraphy*. Springer, Berlin.

FIFIELD, R. 1987. Chinese find giant crater. *New Scientist*, **113**, No. 1543, 19.

FISCHER, A. G. 1991. Orbital cyclicity in Mesozoic strata. *In*: EINSELE, G., RICKEN, W. & SEILACHER, A. (eds), *Cycles and Events in Stratigraphy*. Springer, Berlin. 48–62.

FLETCHER, C. J. N., DAVIES, J. R., WILSON, D. & SMITH, M. 1988. Tidal erosion, solution cavities and exhalative mineralization associated with the Jurassic unconformity at Ogmore, South Glamorgan. *Proceedings of the Geologists' Association*, **99**, 1–14.

FLETCHER, C. J. N. *et al.* 1986. The depositional environment of the basal 'Littoral Lias' in the Vale of Glamorgan–a discussion of the reinterpretation by Ager (1986), *Proceedings of the Geologists' Association*, **97**, 383–384.

FRANKEL, C. 1999. *The End of the Dinosaurs: Chicxulub Crater and Mass Extinctions*. Cambridge University Press, Cambridge.

GLASS, B. P. 2000. *Upper Eocene impact/spherule layers: a status report*. Paper presented to the I. G. C. at Rio de Janeiro, Session 25–6.

GLEN, W. 1982. *The Road to Jaramillo*. Stanford University Press, Stanford.

GLEN, W. 1994. *The Mass-Extinction Debates: How Science Works in a Crisis*. Stanford University Press, Stanford.

GOLDMAN, D., MITCHELL, C. E., BERGSTROEM, S. M., DELANO, J. W. & TICE, S. 1994. K–bentonites and Graptolite Biostratigraphy in the Middle Ordovician of New York State and Quebec. *Palaios*, **9**, 124–143.

GOSTIN, V. A., HAINES, P. W., JENKINS, R. J. F., COMPSTON, W. & WILLIAMS, I. S. 1986. Impact ejecta horizon within late Precambrian shales, *Science*, **233**, 198–200.

GOULD, S. J. 1989. *Wonderful Life*. Penguin Books, London.

GOULD, S. J. 1992. Dinosaurs in the haystack, *Natural History*, March, 2–13.

GRETENER, P. E. 1967. Significance of the rare event in geology. *Bulletin of the American Association of Petroleum Geologists*, **51**, 2197–2206.

HALLAM, A. 1984. Asteroids and extinction – no cause for concern. *New Scientist*, **104**, No. 1429, 30–32.

HALLAM, A. 1986. Origin of minor limestone–shale cycles: climatically induced or diagenetic? *Geology*, **14**, 609–612.

HANCOCK, J. M. 1993. Transatlantic correlations in the Campanian–Maastrichtian stages by eustatic changes of sea-level. *In*: HAILWOOD, E. A. & KIDD, R. B. (eds) *High Resolution Stratigraphy*, Special Publications **70**, The Geological Society, London, 241–256.

HAY, W. W. (ed.) 1974. *Studies in Paleo-Oceanography*. Society of Economic Paleontologists and Mineralogists, Special Publications **20**, Tulsa.

HEDBERG, H. D (ed.) 1976, *International Stratigraphic Guide*. John Wiley, New York.

HEDLEY, D. 1987. Barcodes – selling by numbers. *Esso Magazine*, **144**, 18–21.

HESSELBO, S. P., MEISTER, C. & GROECKE, D. R. 2000*a*. A potential global stratotype for the Simemurian–Pliensbachian boundary (Lower Jurassic). *Geological Magazine*, **137**, 601–607.

HESSELBO, S. P., GROECKE, D. R., JENKYNS, H. C., BJERRUM, C. J., FARRIMOND, P. *et al.* 2000*b*. Massive dissociation of gas hydrate during a Jurassic oceanic anoxic event. *Nature*, **406**, 392–395.

HILGEN, F. J., KRIJGSMAN, W., LANGEREIS, C. G. & LOURENS, L. J. 1997. Breakthrough made in dating of the geological record. *EOS*, **78**, 285–289.

HOLLAND, C. H. 1986. Does the golden spike still glitter? *Journal of the Geological Society, London*, **143**, 3–21.

HOLLAND, C. H., GNOLI, M. & HISTON, K. 1994. Concentrations of Palaeozoic nautiloid cephalopods. *Bollettino della Societa Paleontologica Italiana*, **33**, 83–99.

HOPSON, P. M., FARRANT, A. R. & BOOTH, K. A. 2001. Lithostratigraphy and regional correlation of the basal Chalk. *Proceedings of the Geologists' Association*, **112**, 193–210.

HOUSE, M. R. 1985. A new approach to an absolute timescale from measurements of orbital cycles and sedimentary microrhythms. *Nature*, **315**, 721–725.

HOUSE, M. R. 1986. Towards more precise time-scales for geological events. *In*: NESBITT, R. W. & NICHOL, I. (eds) *Geology in the Real World – The Kingsley Dunham Volume*. Institution of Mining and Metallurgy, London, 197–206.

HOUSE, M. R. & GALE, A. S. (eds), 1995 *Orbital Forcing: Timescales and Cyclostratigraphy*, Special Publications **85**, The Geological Society, London.

HSÜ, K. J. 1989. Catastrophic extinctions and the inevitability of the improbable. *Journal of the Geological Society, London*, **146**, 749–754.

HSÜ, K. J. 1992. *Challenger at Sea: A Ship that Revolutionised Earth Science*. Princeton University Press, Princeton.

HUDSON, J. D. 1998. Discussion on the Cretaceous–Tertiary biotic transition. *Journal of the Geological Society of London*, **155**, 413–419.

HUFF, W. D., KOLATA, D. R., BERGSTROEM, S. M. & ZHANG, Y-S. 1996. Large-magnitude Middle Ordovician volcanic ash falls in North America and Europe. *Journal of Volcanology and Geothermal Research*, **73**, 285–301.

IZETT, G. A., COBBAN, W. A., OBRADOVICH, J. D. & KUNK, M. J. 1993. The Manson impact structure: ^{40}Ar/^{39}Ar Age and its distal impact ejecta in the Pierre Shale in Southeastern South Dakota, *Science*, **262**, 729–732.

JEPPSSON, L. & ALDRIDGE, R. J. 2000. Ludlow (late Silurian) oceanic episodes and events. *Journal of the Geological Society, London*, **157**, 1137–1148.

JOHNSON, M. E. & MCKERROW, W. S. 1995. The Sutton Stone: an early Jurassic rocky shore deposit in South Wales. *Palaeontology*, **38**, 529–541.

KENNEDY, W. J. & COBBAN, W. A. 1976. *Aspects of ammonite biology*. Special Papers in Palaeontology, **17**.

KOEBERL, C. 1996. Chicxulub – the K–T boundary impact crater: a review of the evidence, and an introduction to impact crater studies. *Abhandlungen der Geologischen Bundesanstalt*, **53**, 23–50.

KOLATA, D. R., HUFF, W. D. & BERGSTROEM, S. M. 1996. *Ordovician K–bentonites of Eastern North America*. Geological Society of America Special Paper, **313**.

KUNK, M. J., IZETT, G. A., HAUGERUD, R. A. & SUTTER, J. F. 1989. ^{40}Ar–^{39}Ar dating of the Manson impact structure: a Cretaceous–Tertiary boundary crater candidate. *Science*, **244**, 1565–1568.

KYTE, F. T. & WASSON, J. T. 1986. Accretion rate of extraterrestrial matter: iridium deposited 33 to 67 million years ago. *Science*, **232**, 1225–1229.

LAWSON, J. D. 1974. Review of: AGER (1973), *Palaeontological Association Circular*, **75**, 11–12.

LEEDER, M. 1999. *Sedimentology and Sedimentary Basins*. Blackwell Science, London.

LEWIS, C. 2000. *The Dating Game: One Man's Search for the Age of the Earth*. Cambridge University Press, Cambridge.

LOCZY, L. Von 1915. Monographie der Villanyer Callovien–Ammoniten. *Geologica Hungarica*, **1**, 255–507.

LUCAS, S. G. 1991. Dinosaurs and Mesozoic biochronology, *Modern Geology*, **16**, 127–137.

MCARTHUR J. M. 1991. Strontium-isotope stratigraphy. *Geology Today*, **7/6**, 5i–5iv.

MCARTHUR J. M. 1994. Recent trends in strontium isotope stratigraphy. *Terra Nova*, **6**, 331–358.

MCARTHUR J. M., DONOVAN, D. T., THIRLWALL, M. F., FOUKE, B. W. & MATTEY, D. 2000*a*. Strontium isotope profile of the early Toarcian (Jurassic) oceanic anoxic event, the duration of ammonite biozones and belemnite palaeotemperatures, *Earth and Planetary Science Letters*, **179**, 269–285.

MCARTHUR J. M., CRAME, J. A. & THIRLWALL, M. F. 2000*b*. Definition of late Cretaceous stage boundaries in Antarctica using strontium isotope stratigraphy. *Journal of Geology*, **108**, 623–640.

MCCALL, J. 2001. Keep watching the skies–but not in fear. *Geoscientist*, **11** (3), 12–17.

MACKAY, A. L. 1977. *The Harvest of a Quiet Eye*. Institute of Physics, Bristol.

MACLEOD, N. 1995*a*. Graphic correlation of new Cretaceous/ Tertiary (K/T) boundary successions from Denmark, Alabama, Mexico and the southern Indian Ocean. *In*: MANN, K. O. & LANE, H. R. (eds) *Graphic Correlation*, SEPM Special Publications **53**, 215–233.

MACLEOD, N. 1995*b*. Graphic correlation of high-latitude Cretaceous–Tertiary (K/T) boundary sequences from Denmark, the Weddell Sea and Kerguelen Plateau: comparison with the El Kef (Tunisia) boundary stratotype. *Modern Geology*, **20**, 109–147.

MACLEOD, N. & KELLER, G. 1991. How complete are Cretaceous/Tertiary boundary sections? A chronostratigraphic estimate based on graphic correlation. *Geological Society of America Bulletin*, **103**, 1439–1457.

MACLEOD, N., RAWSON, P. F., FOREY, P. L., BANNER, F. T., BOUDAGHER-FADEL, M. K. *et al.* 1997. The

Cretaceous–Tertiary biotic transition. *Journal of the Geological Society, London*, **154**, 265–292.

MARVIN, U. B. 1990. Impact and its revolutionary implications for geology. *Geological Society of America Special paper*, **247**, 147–154.

MAURRASSE, F. J.-M. R. & SEN, G. 1991. Impacts, tsunami and the Haitian Cretaceous–Tertiary boundary layer. *Science*, **252**, 1690–1693.

MENARD, H. W. 1986. *The Ocean of Truth*. Princeton University Press, Princeton.

MIALL, A. D. 1992. Exxon global cycle chart: an event for every occasion? *Geology*, **20**, 787–790.

MIALL, A. D. 1994. Sequence stratigraphy and chronostratigraphy. *Geoscience Canada*, **21**, 1–26.

MIALL, A. D. 1995. Whither stratigraphy? *Sedimentary Geology*, **100**, 5–20.

MIALL, A. D. 1997. *The Geology of Stratigraphic Sequences*. Springer, Berlin.

MIALL, A. D. 2000. *Principles of Sedimentary Basin Analysis*, 3rd edn. Springer, Berlin.

MIALL, A. D. & MIALL, C. E. 2001. Sequence stratigraphy as a scientific enterprise: the evolution and persistence of conflicting paradigms. *Earth-Science Reviews*, **54**, 321–348.

MILANKOVICH, V. 1995. *Milutin Milankovich 1879–1958*. European Geophysical Society, Kaltenburg-Lindau.

MILLER, K. G. 1990. Recent advances in Cenozoic marine stratigraphic resolution. *Palaios*, **5**, 301–302.

MOORE, R. C. 1939. *Meaning of facies*. Geological Society of America Memoirs, **39**, 1–34.

MUIR WOOD, R. 1985. *The Dark Side of the Earth*. George Allen & Unwin, London.

MURPHY, M. A. & SALVADOR, A. 1999. International stratigraphic guide–an abridged version. *Episodes*, **22**, 255 271.

OBRADOVICH, J. D. & COBBAN, W. A. 1975. A time-scale for the late Cretaceous of the western interior of North America. *Geological Association of Canada Special Paper*, **13**, 32–54.

OFFICER, C. B. & DRAKE, C. L. 1985. Terminal Cretaceous environmental events, *Science*, **227**, 1161–1167.

OGG, J. G. & LOWRIE, W. 1986 Magnetostratigraphy of the Jurassic/Cretaceous boundary. *Geology*, **14**, 547–550.

PALFY, J. & SMITH, P. L. 2000. Synchrony between early Jurassic extinction, oceanic anoxic event, and the Karoo–Ferrar flood basalt volcanism. *Geology*, **28**, 747–750.

PAYTON, C. E. (ed.) 1977. Seismic stratigraphy–applications to hydrocarbon exploration. *Memoirs of the American Association of Petroleum Geologists*, **26**, 1–516.

PENN, I. E., MERRIMAN, R. J. & WYATT, R. J. 1979. *The Bathonian Strata of the Bath–Frome Area*. Institute of Geological Sciences, Report **78/12**.

PETTIJOHN, F. J. 1984. *Memoirs of an Unrepentant Field Geologist*. University of Chicago Press, Chicago & London.

PRESTWICH, J. 1847. On the probable age of the London Clay. *Quarterly Journal of the Geological Society, London*, **3**, 354–377.

PROTHERO, D. R. 1990. *Interpreting the Stratigraphic Record*. W. H. Freeman, New York.

RAMPINO, M. R. 1982. A non-catastrophist explanation for the iridium anomaly at the Cretaceous/Tertiary boundary. *Geological Society of America Special Paper*, **190**, 455–460.

RAUP, D. M. & STANLEY, S. M. 1978. *Principles of Paleontology*. W. H. Freeman & Co., San Francisco.

REMANE, J. 2000*a*. *Explanatory Note and International Stratigraphical Chart*. UNESCO, Division of Earth Sciences, Paris.

REMANE, J. 2000*b*. *4 Year Report of the International Commission on Stratigraphy for the Period 1996–2000*, presented to the IUGS, Rio de Janeiro, August 2000.

REYMENT, R. A. & TAIT, E. A. 1972. Biostratigraphical dating of the early history of the South Atlantic Ocean, *Philosophical Transactions of the Royal Society of London*, Series B, **264**, 55–95.

ROCCHIA, R., BOCLET, D., BONTÉ, P., JÉHANNO, C., CHEN, Y. *et al.* 1990. The Cretaceous–Tertiary boundary at Gubbio revisited. *Earth and Planetary Science Letters*, **99**, 206–219.

RODGERS, J. 1959 The meaning of correlation. *American Journal of Science*, **257**, 684–691.

SADLER, P. M. & STRAUSS, D. J. 1990. Estimation of completeness of stratigraphical sections using empirical data and theoretical models. *Journal of the Geological Society, London*, **147**, 471–485.

SALVADOR, A. 1992. The teaching of stratigraphy: replies to a questionnaire. *GSA Today*, **2**, 142–143.

SALVADOR, A. 1994. *International Stratigraphic Guide (2nd edition)*. IUGS and Geological Society of America, Trondheim and Boulder, Co.

SAWLOWICZ, Z. 1993. Iridium and other platinum-group elements as geochemical markers in sedimentary environments. *Palaeogeography, Palaeoclimatology, Palaeoecology*, **104**, 253–270.

SCHINDEL, D. E. 1982. Resolution analysis: a new approach to the gap in the fossil record. *Paleobiology*, **8**, 340–353.

SCHMITZ, B. 1992. An iridium anomaly in the Ludlow Bone Bed from the Upper Silurian, England. *Geological Magazine*, **129**, 359–362.

SEIBOLD, E. & MEULENKAMP, J. D. 1984. *Stratigraphy–Quo Vadis*. Studies in Geology, **16**. American Association of Petroleum Geologists, Tulsa.

SHAW, A. B. 1971. The butterfingered handmaiden. *Journal of Paleontology*, **45**, 1–5.

SHAW, A. B. 1995. Early history of graphic correlation. *In*: MANN, K. O. & LANE, H. R. (eds) *Graphic Correlation*, SEPM Special Publications **53**, 1519.

SHEEHAN, P. M., FASTOVSKY, D. E., BARRETO, C. & HOFFMANN, R. G. 2000. Dinosaur abundance was not declining in a '3 m gap' at the top of the Hell Creek Formation, Montana and North Dakota. *Geology*, **28**, 523–526.

SLOSS, L. L. 1963. Sequences in the cratonic interior of North America, *Geological Society of America Bulletin*, **74**, 93–114.

SLOSS, L. L. 1991. The tectonic factor in sea level change: a countervailing view. *Journal of Geophysical Research*, **96**, 6609–6617.

SMITH, R. D. A. & ROBINSON, R. B. 1993. Discussion

on an iridium anomaly in the Ludlow Bone Bed. *Geological Magazine*, **130**, 855–856.

SMITH, W. & PHILLIPS, J. 1825. Investigations of the geological structure of the north eastern portion of Yorkshire. Paper read to the Yorkshire Philosophical Society, 2 February 1825. *In*: *MSS Scientific Communications to General Meetings*, **1** (Yorkshire Museum archives).

SYLVESTER-BRADLEY, P. C. 1979. Biostratigraphy. *In*: FAIRBRIDGE, R. W. & JABLONSKI, D. (eds) *Encyclopaedia of Paleontology*. Dowden, Hutchinson & Ross, Stroudsburg, 94–99.

TALENT, J. A. 1995. Chaos with conodonts and other fossil biota. *Courier Forschungsinstitut Senckenberg*, **182**, 523–551.

THIRLWALL, M. F. 1983. Discussion on implications for Caledonian plate tectonic models of chemical data . . ., *Journal of the Geological Society, London*, **140**, 315–318.

THOMSON, W. 1894. *Popular Lectures and Addresses: Geology and General Physics*. Macmillan, London, **2**.

TILL, A. 1909. Neues Material zur Ammonitenfauna des Kelloway von Villany (Ungarn). *Verhandlungen der k.-k. Geologischen Reichsanstalt (Wien)*, **1909**, 191–195.

TORRENS, H. S. 1993. The dinosaurs and dinomania over 150 years. *Modern Geology*, **18**, 257–286.

TRÜMPY, R. 1971. Stratigraphy in mountain belts. *Quarterly Journal of the Geological Society, London*, **126**, 293–318.

UNDERWOOD, C. J., CROWLEY, S. F., MARSHALL, J. D. & BRENCHLEY, P. J. 1997. High-resolution carbon isotope stratigraphy of the basal Silurian stratotype. *Journal of the Geological Society, London*, **154**, 709–718.

VAIL, P. R. 1992. The evolution of seismic stratigraphy and the global sea-level curve. *Geological Society of America Memoir*, **180**, 83–91.

VALLANCE, T. G. 1968. The beginning of geological system. *Scan*, November 1968, 28–34.

VAN ANDEL, T. H. 1981. Consider the incompleteness of the geological record. *Nature*, **294**, 397–398.

VAN LOON, A. J. 1999. The meaning of 'abruptness' in the geological past. *Earth-Science Reviews*, **45**, 209–214.

VAN VALEN, L. M. 1984. *Review of*: SILVER, L. T. & SCHULTZ, P. H. (eds), Geological Implications of Impacts of Large Asteroids and Comets on the Earth. *Paleobiology*, **10**, 121–137.

VONHOF, H. B., SMIT, J., BRINKHUIS, H., MONTANARI, A. & NEDERBRAGT, A. J. 2000. Global cooling accelerated by early late Eocene impacts? *Geology*, **28**, 687–690.

WALLACE, M. W., KEAYS, R. R. & GOSTIN, V. A. 1991. Stromatolitic iron oxides: evidence that sea-level changes can cause sedimentary iridium anomalies. *Geology*, **19**, 551–554.

WEEDON, G. P. & HALLAM, A. 1987. Comment and reply on 'Origin of minor limestone–shale cycles'. *Geology*, **15**, 92–94.

WIGNALL, P. B. 1990. Ostracod and foraminifera micropaleoecology and its bearing on biostratigraphy. *Palaios*, **5**, 219–226.

WILLIAMS, G. E. 1986. The Acraman impact structure. *Science*, **233**, 200–203.

WILLIAMS, H. S. 1893. The making of the geological time scale. *Journal of Geology*, **1**, 180–197.

WILLIAMS, H. S. 1895. *Geological Biology: An Introduction to the Geological History of Organisms*. Henry Holt & Co., New York.

WILSON, R. C. L. 1998. Sequence stratigraphy: a revolution without a cause? *In*: BLUNDELL, D. J. & SCOTT, A. C. (eds) *Lyell: The Past is the Key to the Present*, Special Publications **143**, 303–314, The Geological Society, London.

ZELLER, E. J. 1964. Cycles and Psychology. *Kansas Geological Survey Bulletin*, **169**, 631–636.

ZINSMEISTER, W. J. 1998. Discovery of fish mortality horizon at the K–T boundary on Seymour Island. *Journal of Paleontology*, **72**, 556–571.

'As chimney-sweepers, come to dust': a history of palynology to 1970

WILLIAM A. S. SARJEANT
Department of Geological Sciences, University of Saskatchewan, 114 Science Place, Saskatoon, Saskatchewan, S7N 5E2, Canada

Abstract: A brief overview is given of the various fields of palynology, their practical applications being stressed. Particular attention is thereafter paid to the history of palaeopalynology, here considered as the study of pre-Quaternary palynomorphs. This is presented as three stages: the period of pioneer discoveries (to 1918); years of slow progress (1919–1945); and a post-World War II period of accelerating discoveries (1946–1970). Developments concerning the different groups of palynomorphs during these periods are successively presented, under six headings: spores and pollen; dinoflagellates (and acritarchs); prasinophytes; scolecodonts; chitinozoans; and other palynomorphs. The changes brought about in palynology by improving preparation techniques and microscopical equipment are stressed. A brief overview is attempted concerning the developments since 1970, consequent upon ever-expanding research, new preparation techniques and new technology. As conclusion, an overview is presented of the history of palynology and likely future developments are discussed.

'Golden lads and girls all must,
As chimney-sweepers, come to dust'.
(Shakespeare, *Cymbeline*, **IV**. ii. 258)

Palynology is indeed the examination of dust, contemporary or ancient; though it is concerned with the organic particles in particular, the other components of dust need to be dealt with, if only to eliminate them. It is a subdiscipline overlapping the fields of botany, zoology and palaeontology. Originally included within the fields of microscopy and micropalaeontology, it was given separate identity by the coining of that term by H. A. Hyde and D. A. Williams (1944), who derived the name from the Greek *palunein* (*παλυνειν*): to strew or sprinkle, flour or dust'. Originally it comprised only the study of spores and pollen, but its compass has enlarged over the years. J. W. Funkhouser (1959) included also a wide array of other groups of small microfossils: coccolithophorids, dinoflagellates, diatoms, desmids, fungal elements, fragments of higher plants, microforaminifera and even radiolaria. This broadening to include microfossils with walls of $CaCO_3$ or SiO_2 proved unacceptable; a reasonable present-day definition might be as follows:

> Palynology is the study of microscopic objects of macromolecular organic composition (i.e. compounds of carbon, hydrogen, nitrogen and oxygen), not capable of dissolution in hydrochloric or hydrofluoric acids.

Essentially, then, as Jansonius and McGregor noted (1996, p. 1), its compass is circumscribed more by the techniques required to produce palynological assemblages than by any biological unity in the material studied. The frequently made claim that 'micropalaeontology deals with large microfossils; palynology, with small microfossils' cannot be sustained, since coccoliths (formed from $CaCO_3$) and archaeomonads (formed from SiO_2) are smaller than most palynomorphs, while the largest spores and acritarchs are readily visible to the unaided eye. It is also inappropriate to designate palynomorphs as 'acid-insoluble microfossils', since they are readily destroyed by sulphuric, nitric or other acids.

Early studies of palynology: its applications

Though the development of palynology was subsequently to depend so much upon the use of the microscope, the earliest observations of pollen preceded the development of that instrument. The recognition of sexuality in plants occurred still earlier – perhaps as early as the time of the Assyrians (see Wodehouse 1935, pp. 23–26). The earliest recorded observations of pollen took place, however, in the seventeenth century. The English botanist Nehemiah Grew (Fig. 1; miscited as 'N. Green' by Jansonius & McGregor, 1996, p. 1) made the first detailed description of the structure of flowers, noting that the anthers served as 'the Theca or Case of a great many

From: OLDROYD, D. R. (ed.) 2002. *The Earth Inside and Out: Some Major Contributions to Geology in the Twentieth Century*. Geological Society, London, Special Publications, **192**, 273–327. 0305-8719/02/$15.00

Fig. 1. Nehemiah Grew (1641–1712); from the portrait by R. White.

Fig. 3. Rudolph Jakob Camerer (1665–1721); from a portrait by an unknown artist.

Fig. 2. Marcello Malpighi (1628–1694); from the portrait by Tabor.

extreme small Particles either globules or otherwise convex' which, when seen under a magnifying glass, differed in size, colour and shape in different plants (Grew 1682). Almost at the same time, the Italian physician Marcello Malpighi (1687; Fig. 2) made similar observations.

It is not clear, however, that either of these naturalists perceived the sexual function of pollen. That discovery is credited instead to a German botanist, Rudolph Jakob Camerer (or Camerarius, 1692; Fig. 3), who observed that the stamens were the male sexual organs and that, unless fertilized by those small particles, the ovules could not develop into seeds (see discussion in Wodehouse 1935, pp. 18–23).

With the construction of the first microscopes by Robert Hooke, Antoni van Leeuwenhoek and others, the study of pollen and spores was greatly facilitated; however, the history of the development of microscopes is told by Bradbury (1967) and does not require repetition here. A major contributor to the understanding of flower and pollen morphology, and of the processes of pollen dispersal and pollination, was a Dutch clergyman, Johannes Florentinus Martinet. In course of discussing these topics in the fourth volume of his widely translated *Katechismus der*

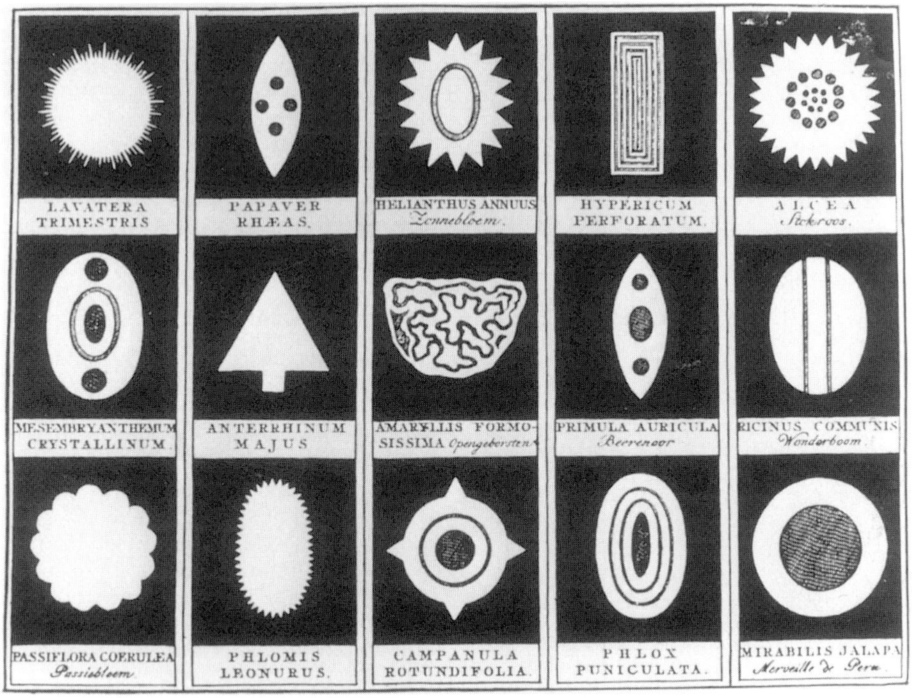

Fig. 4. The earliest illustrations of pollen grains, by J. F. Martinet (1779), reproduced from Jonker (1967, Plate 1).

Natuur (1779), he presented what Jonker (1967) has called 'very primitive illustrations of fifteen pollen grains' (see Fig. 4).

The recognition of the reproductive function of pollen and spores was by that time affecting approaches to plant classification, but further advances in knowledge came slowly. Manten (1967, p. 12) cites in particular the work of three German scientists:

> The first one is H. von Mohl, who published, in 1834, the first detailed descriptive classification of pollen forms. The second is C. J. Fritzsche, who lived around the middle of the nineteenth century and did most of his work in Russia. He observed and pictured the first fine structure of the pollen wall very accurately. The third is C. A. H. Fischer, who lived about half a century later. Only his late nineteenth-century work dealt with pollen. He studied thoroughly the pollen of about 2,200 plant species, a much more complex study than any which had been made before.

Prior to their work, the phenomenon of hay fever had been recognized by an English physician, John Bostock (1819, 1828), who gave a lengthy and precise account of the symptoms of what he termed 'catarrhus aestivus' or 'summer catarrh'. This evoked some controversy, since it was not understood why only certain persons were afflicted (see discussion in Manten 1967, pp. 13–14). Only with the work of the Germans J. W. Weichardt (1905) and A. Wolff-Eisner (1906) was it recognized that hay fever is an allergic reaction excited by a specific antigen to which the individual is sensitized and not till 1911 did an Englishman, Leonard Noon, succeed in treating what was by then called pollinosis with pollen extracts.

An immense expansion of studies in medical palynology has ensued, the story of which is told by Coca *et al.* (1931) and Durham (1936, 1948); O'Rourke (1996) provides an up-to-date review. Subsequently, it was recognized, by Lord Energlyn of Caerphilly in the 1970s, that the concentration of fossil spores in mine dusts correlated with increased incidence of pneumokoniosis; this showed that even ancient spores might have medically adverse effects.

Another practical aspect of pollen study was opened up by R. Pfister (1895), who showed it was possible to demonstrate the geographical and botanical origin of honey by its pollen content. Subsequent researches confirmed his conclusions and led to the use of what has come to be alternatively called *melittopalynology* or

Fig. 5. A youthful Christian Gottfried Ehrenberg (1795–1876); from a portrait by an unknown artist.

melissopalynology, as a means for enforcing the standards of purity and proper description of foods by commercial companies. G. B. Jones and Vaughan M. Bryant Jr (1992) give a good account of the development of this study, in particular in the United States, and have later (1996) presented a modern overview.

A third area in which palynology has proved of importance is in law enforcement – the discipline of *forensic palynology*. This commenced late, with the successful use of pollen content in muds as evidence during the prosecution of a murderer in Australia in 1959. As Bryant *et al.* (1990) have demonstrated, it remains a line of investigation still very much underemployed in criminal investigations (see also Bryant 1996).

A fourth specialized area of palynological study is *copropalynology*, the analysis of the pollen/spore content of recent and fossil excreta to determine the dietary preferences and environmental circumstances of animals and humans formerly living (see Sobolick 1996). *Entomopalynology* is devoted to the study of pollen grains adhering to the bodies of insects, as a means for determining the symbiotic relations between insects and plants and for plotting insect migrations (see Pendleton *et al.* 1996).

All these fields form a part of a larger subdiscipline, called by the Germans *aktuopalynology* and by English-speaking palynologists *actuopalynology* – the study of present-day palynomorphs. The study of pre-Holocene palynology is distinguished as *palaeopalynology*. The two fields overlap in the Quaternary but, since most of the early Quaternary pollen and spores are of types still being produced by living plants, their study is usually considered a component of actuopalynology.

Quaternary studies

The first observation of fossil Quaternary pollen was by the great German microscopist Christian Gottfried Ehrenberg (1795–1876, Fig. 5) who reported *Pinus* pollen in sediments from northern Sweden (1837*a*). A Swiss naturalist, J. Früh (1885), succeeded in enumerating most of the common tree pollen. The Swedish geologist Filip Trybom (1888), having noted the resistant character of pine and spruce pollen during studies of lake sediments, percipiently pointed out how useful these microfossils might be for stratigraphical palaeontology. Shortly afterward, another Swiss geologist, F. E. Geinitz (1887), drawing upon Früh's studies, showed how pollen in peats could be used to elucidate their origin and botanical composition.

The earliest quantitative presentation of pollen-analytical data was by a German plant physiologist, Carl A. Weber (1893, 1896); but Weber avoided making interpretations from those data. The Danish geologist G. F. L. Sarauw (1897) presented quantitative information on pollen distribution, but did not meaningfully compare percentage compositions. Other Scandinavian investigators were soon following up their work. The recognition that the percentage compositions of pollen assemblages could differ in successive peat layers came almost simultaneously in Finland and Sweden, from investigations by Harald Lindberg (1905) and Gustaf Lagerheim (1895; Lagerheim in Witte 1905). Later that year, Lagerheim presented a detailed analysis of pollen observed in samples from the Kallsjö swamp in Skurup (Scania, Sweden), showing an upward decrease of pine, birch, alder and elm pollen, whereas ash, oak and lime pollen were increasing. His findings were included in a paper by N. O. Holst (1908), who recognized that the careful study of successive layers would give key information on plant migrations and their proportions in Quaternary floras – evidence which would reveal the climatic changes that were taking place.

However, it was left to another Swedish geologist, Ernst Jakob Lennart von Post

Fig. 6. Ernst Jakob Lennart von Post (1884–1951); uncredited photographer, reproduced from Traverse (1988, fig. 1.4b).

(1884–1951; Fig. 6) to take up this study. Von Post developed techniques of plotting, in diagrammatic form, the fluctuations in pollen numbers through successive layers of Quaternary deposits (1916, 1918, 1927; see also Selling 1951; Manten 1967). His work transformed pollen analysis into a major tool for dating Quaternary sediments and interpreting past environments. Within Sweden, it inspired the studies of Gunnar Erdtman (1897–1973; Fig. 7), who built up a palynological laboratory in Solna, near Stockholm, and whose work was to become so influential at the international level that he was nicknamed 'the pope of palynology' (Erdtman 1967; Sarjeant 1973). The work of that laboratory has been ably continued, since Erdtman's death, by the US microscopist John R. Rowley (Fig. 8).

The application of these techniques to human prehistory was developed in Denmark by Johs. Iversen (e.g. 1941). Among other discoveries, it was perceived that the spread of weed pollen enabled dating of the inauguration of grain farming in different countries. (The pollen of wheat and other grains is morphologically indistinguishable from grass pollen and could not, therefore, be recognized.)

Palynological techniques are now being used worldwide by geochronologists, prehistorians

Fig. 7. Otto Gunnar Elias Erdtman (1897–1973) on left, with William S. Hoffmeister (1901–1980); uncredited, reproduced from Traverse (1988).

Fig. 8. John Rowley and Eszther Nagy at the International Palynological Congress in Brisbane, Queensland (photograph by the author, 1 September 1988).

and archaeologists, wherever environmental conditions permit. They have contributed immensely to our understanding of human history and its relation to the changing climates of the Pleistocene and Holocene (see Bryant & Holloway 1996 for a succinct account of the use of palynology in archaeology).

Palaeopalynology: the earliest discoveries (to 1918)

Pollen and spores

The earliest pre-Quaternary report was by the German geologist Heinrich R. Göppert, who reported pollen from the Miocene brown coals of Salzhausen, Hessen (1836). Shortly afterwards, Ehrenberg observed *Pinus*-like pollen in Late Cretaceous flint flakes (1837*a*) and in other German Tertiary lignites (1838). In 1848, Göppert took a technological stride forward when he used dilute hydrochloric acid to extract pollen grains of the Pinaceae from Tertiary limestones of Radoboj, near Varazdin, Croatia. During the later nineteenth century, further scattered reports were published of fossil pollen in sediments of Cretaceous to Late Tertiary age, e.g. by H. von Duisburg (1860) and Georg Fresenius (1860).

Palaeozoic spores were first observed in petrological thin sections of coals from Lancashire, England, by H. Witham of Lartington (1833), who misinterpreted them as vessels within the stems of monocotyledonous plants. Subsequently John Morris (1840) observed the macrospores of *Lepidodendron (Lycopodites) longibractus*, but interpreted them merely as 'thecae' or 'capsules' of organic matter. In 1848, Göppert likewise observed macrospores but again misinterpreted them, designating them as *Carpolithes coniformis*; this name was to be used as late as 1881 by Otto Feistmantel, even though he recognized the bodies to be macrospores.

The fact that the macrospores commonly occurred within a mass of microspores was first noted by the eminent botanist Joseph D. Hooker (1848), who observed them in situ in thin sections of sporangia of *Lepidostrobus*. Friedrich Goldenberg (1855), studying disjunct material, noted that macrospores of similar type occurred in both *Lepidodendron* and *Sigillaria*, an observation confirmed by the French palaeobotanist René Zeiller (1884). William Carruthers (1865) described a *Lepidostrobus* cone in which he believed that the macrospores were distributed one per scale, mistaking them to be sporangia since he did not observe the actual sporangium walls. Philipp W. Schimper (1870) and Edward W. Binney (1871) described cones with macro- and microspores in place, while William C. Williamson, in a series of papers (1871, 1872, and others), reported both dispersed and in situ microspores and macrospores.

A major technological advance in palynology was presaged when Franz Schulze (1855) developed a reagent – a mixture of potassium chlorate and nitric acid – that could be used to macerate coal without destroying the contained microfossils. This technique was employed, along with methods using potassium hydroxide and hydrofluoric acid, by Paulus F. Reinsch during studies of Carboniferous, Permian and Triassic coals from Germany and Russia (1881, 1884). He found that the volume of spores was sometimes immense, comprising 80% of some coals. Utilizing modern analogues, he calculated the rate of spore production per plant. Over 600 species of microspores and megaspores were distinguished by him and assigned to different plant groups (cryptogams, *Lepidodendron* and Filicales); however, Reinsch did not name them, merely giving them numbers. The presence of parasitic growths on some larger megaspores was reported for the first time.

A second major advance came with the work of Robert Kidston (in Bennie & Kidston 1886), who not only noted that megaspores were regularly found attached in tetrads but also differentiated the two layers in the megaspore wall. He presented clear morphological descriptions, but did not employ binomial nomenclature, instead identifying the morphotypes by Roman numerals (as *Triletes* I, etc.). He and his colleague, J. Bennie, demonstrated that these spores could be used to characterize and correlate individual coal seams in the Scottish Carboniferous. However, Kidston's work was not swiftly followed up, either in Great Britain or elsewhere.

The first Triassic microspores (or miospores, as they came to be called) were reported from England. William J. Sollas (1901) isolated them from the Rhaetian (Late Triassic) bryophyte *Naiadita lanceolata*, and Leonard J. Wills (1910) reported Middle Triassic forms from Bromsgrove, Worcestershire. Permian miospores were first recorded from Stassfurt, Germany, by H. Lück (1913); but Lück's work, presented in a doctoral thesis, was never published and both the Permian and Triassic studies were likewise long in being followed up.

Dinoflagellates

The first person to recognize fossil dinoflagellates was Ehrenberg, who reported his discovery in a paper presented to the Berlin Academy of Sciences in July 1836. He had observed clearly tabulate dinoflagellates in thin flakes of Cretaceous flint and considered those dinoflagellates to have been silicified. Along with them, and of comparable size, were spheroidal to ovoidal bodies bearing an array of spines or tubes of variable character. Ehrenberg interpreted these as being originally siliceous and thought them to be desmids (freshwater conjugating algae), placing them within his own Recent desmid genus *Xanthidium*.

Though summaries of Ehrenberg's work appeared earlier, it was not published in full until 1837 or 1838; the date is uncertain (see Sarjeant 1970*a*). In the meantime, some of his slides had been displayed to the Academy of Sciences in Paris. One of its members, C. R. Turpin, disagreed with Ehrenberg's findings, instead considering the spiny objects to be reproductive bodies of the freshwater bryozoan *Cristatella*; his critique (1837) and Ehrenberg's response (1837*b*) were published quickly, both apparently preceding in publication date the work itself. It was Ehrenberg's opinion that gained general acceptance; the spiny bodies were to continue to be called 'xanthidia' until well into the twentieth century.

Ehrenberg visited England in the summer of 1838, a visit which encouraged a group of English

Fig. 9. Gideon Algernon Mantell (1790–1852).

microscopists to study the spiny bodies in the Late Cretaceous flint flakes (Sarjeant 1978*a*, 1982, 1991*b*). Among those interested was the palaeontologist and stratigrapher Gideon Algernon Mantell (1790–1852; Fig. 9). Observing that the spiny microfossils frequently showed distortions, he concluded that they could not be composed of silica. This observation was confirmed when, upon heating the flints, he found that the microfossils blackened. He concluded that they must be of organic composition (1845) and later proposed the name *Spiniferites* for them (1850). Unfortunately, this was done so obscurely that the name he had proposed did not come to the attention of other scientists for more than a century (see Sarjeant 1967*a*, 1970*a*, 1992*a*).

In 1843, Ehrenberg reported the first Jurassic 'xanthidia' from Poland. Both dinoflagellates and 'xanthidia' were illustrated in his massive *Mikrogeologie* (1854), still the largest single volume ever to be published on microfossils. In the ensuing years, there were no further published accounts of undoubted dinoflagellates. However, the 'xanthidia' gained intermittent mention, being reported from Early Palaeozoic strata of New York State by M. C. White (1862) and from English Eocene strata by E. W. Wetherell (1892) – who, most unusually, was aware of Mantell's work. In addition, another

Fig. 10. Alfred Gabriel Nathorst (1850–1921), from portrait by an unknown artist.

American, J. A. Merrill (1895), reported them from the Early Cretaceous of Texas, but compounded Ehrenberg's error by not only considering them siliceous but also to be sponge spicules (see Sarjeant 1966*a*).

Among the plankton collected by a German expedition studying the Humboldt Current were spinose micro-organisms very comparable to the 'xanthidia'. H. Lohmann, who published this record (1904), called them 'ova hispida' and re-attributed Ehrenberg's fossil species, giving them such names as *Ovum hispidum furcatum* – a procedure unacceptable under the rules of biological nomenclature, since a trinomen can be applied only to intraspecific taxa, not to species. Lohmann's compatriot Theodor Fuchs (1905) decided that the 'planktonic eggs' were those of copepods. The third German microscopist, Reinsch, was more percipient, suggesting instead that the spiny bodies were cysts of dinoflagellates and naming them 'palinospheres' (1905). Unfortunately his illustration showed, not a dinoflagellate cyst, but a tasmanitid.

Prasinophytes

In 1852, Hooker recorded 'spheroidal bodies' of microscopic size from the English Silurian, considering them to be 'Lycopod seed-cases'. Subsequently, the Canadian geologist J. William Dawson reported similar bodies from the Devonian sediments of Ontario, believing them to be spore-cases and naming them *Sporangites huronensis* (1871*a*,*b*). Dawson later reported similar bodies from Devonian sediments of the United States, Brazil and Bolivia (see Muir & Sarjeant 1971). Startlingly to the citizens of Chicago, these spheroidal bodies even turned up in that city's water supply; however, these were twice-reworked Devonian forms from nearby boulder clays (Johnson & Thomas 1884).

In Tasmania, similar microfossils were concentrated in such abundance in Permian sediments as to form a combustible deposit variously named 'dysodil', 'tasmanite' or 'white coal'. In a paper published only in summary, T. S. Ralph (1865) reported them and interpreted them as algae. No name was given until 1875, when Edwin T. Newton described them in detail, noting the numerous small pits in the wall, and named them *Tasmanites punctatus*. This irritated Dawson, who insisted that his own name *Sporangites* had priority (1886); however, since it was later to be shown that Dawson's genus brought together several unrelated types of microfossils, the spheroidal microfossils came to be styled 'tasmanitids'. They were reported by E. Wethered (1886) from the Carboniferous of Monmouthshire, Wales and, as noted, the modern form illustrated by Reinsch (1905) was unquestionably a tasmanitid.

Small, round, black objects had been found some years earlier in the Late Precambrian (Eocambrian) Visingsö Formation of Sweden by G. Linnarsson (1880). These proved a focus for argument. Alfred G. Nathorst (1886; Fig. 10) thought they might be small branchiopods (*Estheria*); Gerhard Holm (1887) disagreed, suggesting they might either form a part of the shell of the inarticulate brachiopod *Discina* or might be of plant affinity. The latter view was favoured by Carl Wiman (1895), in a comprehensive review of these 'problematica'. Though there is as yet no definite conclusion on them, it seems likely, as suggested by Muir and Sarjeant (1971), that these are early tasmanitids.

Scolecodonts

Segments of the fossil jaw of a polychaete worm were first reported, from Silurian strata on the Estonian island of Saaremaa, by Eduard Eichwald (1854), but they were misinterpreted as fish teeth. A year later, impressions of whole polychaete worms with poorly preserved jaws were described from Italian Tertiary deposits by

Abramo Massalongo (1855). Subsequently E. Ehlers, a specialist on recent polychaetes, recorded them from the Jurassic Solenhofen Stone of Bavaria, Germany, demonstrating their affinity and proposing the generic names *Eunicites* and *Lumbriconereites* (1868*a*,*b*). Extensive studies by George J. Hinde of material from England, Wales, Canada and Sweden (1879, 1880, 1882, 1896) established a basis for the nomenclature of what he regarded as being isolated components of annelid jaws; but study of them lapsed thereafter for almost 50 years.

Other palynomorphs

In a review of the algae published in 1849, F. T. Kützing named a living freshwater colonial form as *Botryococcus braunii*. Though masses of 'yellow bodies' were found by P. Bertrand and B. Renault (in Renault 1889–1900) to be present in boghead coals, it was not considered that they were factors in the genesis of those deposits; indeed, E. C. Jeffrey (1910) dismissed them as being spores. The Russian biologist M. Zalessky (1914) appears to have been the first to recognize that *Botryococcus* was an oil-producer, but his opinion remained to be confirmed.

The yellow-green alga *Vaucheria* was reported by Rudolph Ludwig (1857) from German Miocene brown coals, but has not been recorded subsequently in the fossil state.

The acid-resistant inner shell linings of foraminifera were first illustrated, from Late Cretaceous chalk flints, by Henry Deane (1849, figs 17–18), who styled them 'Polythalamia'. They were not to be reported again for more than a century (see Stancliffe 1996).

Palaeopalynology: inching along (1919–1945)

Spores and pollen

The foundations laid before World War I were slow in being built upon. The concept of introducing a formal nomenclature for dispersed spores and pollen was criticized by Reinhardt Thiessen (1920), who thought this should only be done after their affinity to a plant genus had been demonstrated. The US palaeobotanist H. H. Bartlett disagreed; instead, in a critical review of Reinsch's work, he validated the single generic name (*Triletes*) which Reinsch had tentatively introduced (see discussion in Jansonius & McGregor, 1996, p. 2). Bartlett regarded a separate nomenclature as a prerequisite to meaningful work on the correlation of coal seams. J. Zerndt inaugurated, in 1930, a series of studies of Polish megaspores and there was a scatter of other taxonomic publications on them.

The pioneer work of Kidston on Carboniferous coal seam correlation was not to be followed up in Great Britain for almost 50 years. Only in 1930 did the value of using spores for this purpose come again to be recognized (Slater *et al.* 1930). However, despite further demonstrations of their value by Arthur Raistrick (Raistrick & Simpson 1933; Raistrick 1935), such studies lapsed once again.

The major breakthrough in palynological research came from work on German Tertiary lignites, initiated by Robert Potonié (1889–1974; Fig. 11). In a series of papers (1932, 1934, and others) he demonstrated the value of pollen in Tertiary correlation, adopting the concept of a nomenclature independent from that applied to the whole plant. His work, and that of his student A. C. Ibrahim (1933), laid the foundation for all subsequent studies of fossil pollen and spores.

In the United States, Carboniferous spores were coming to be employed in correlation, at least by the US and Illinois Geological Surveys, through the studies of James M. Schopf (1938), Robert Kosanke (1943) and others; a useful compilation of then-current knowledge was presented by Schopf, L. Richard Wilson and R. Bentall (1944). R. P. Wodehouse, a major worker on living pollen grains, wrote an important account comparing the grains in the Eocene Green River oil shales with modern forms (1933). In Russia, researches on Palaeozoic spores were begun by Sofia N. Naumova (1939), but lapsed during World War II.

Studies of Permian miospores were begun anew in India by C. Virkki (1937) and in the Soviet Union by A. A. Liuber (1938; Liuber & L. E. Val'ts, 1941). Triassic terrestrial microfloras were reported by H. Hamshaw Thomas (1933) in South Africa and, rather incidentally, by the palaeobotanist Thomas M. Harris in England (1938), while L. H. Daugherty (1941) described Late Triassic palynomorphs from Arizona, USA. However, palynological studies of both systems again lapsed thereafter.

Dinoflagellates

Though they had gained brief attention in a work by his unrelated namesake Walter Wetzel (1922; Fig. 12), it was only through the work of Otto Wetzel (Fig. 13) that serious studies of fossil dinoflagellates were resumed, after a hiatus of almost 80 years. Wetzel, studying

Fig. 11. Robert Potonié (1889–1974) with stratigrapher Suzanne Durand (France) at the International Palynological Congress, Utrecht, The Netherlands (photograph by the author, 3 September 1966).

Fig. 12. Konrad Alois Siegmund Karl Walter Wetzel (1887–1978); in a field near Kiel, Germany (by courtesy of Dr Werner Prange).

Fig. 13. Otto Christian August Wetzel (1891–1971); photo *c.* 1955 (by courtesy of Dr Werner Prange).

Fig. 14. Georges Victor Deflandre (1897–1973) with Mrs Sahni, Director of the British Sahni Institute of Palaeobotany, in the Laboratoire de Micropaléontologie, Ecole Pratique des Hautes Études, Paris (photograph by Georges Deflandre, 14 May 1965).

Fig. 15. Maria Lejeune-Carpentier (1910–1995) in her laboratory at the University of Liège (photograph by the author, 8 November 1979).

Cretaceous flints from the Baltic region (1932, 1933), correctly identified many of his fossils as dinoflagellates and, unaware of Mantell's work, recognized independently that the so-called 'xanthidia' were of organic, not siliceous, composition. He erected a new genus, *Hystrichosphaera*, to accommodate them, considering it to be of uncertain systematic position. The new name 'hystrichospheres' swiftly supplanted the older name 'xanthidia'; it was to remain in currency for almost 30 years.

Three other microscopists undertook significant systematic researches within the ensuing decade. Georges Deflandre (1897–1973; Fig. 14) made extensive studies of the microfossils of French flints (1935, 1936, 1937); he noted that specimens of *Hystrichosphaera* exhibited a transverse girdle and a pattern of lines suggesting the dinoflagellate plate arrangement, but since he considered that the position of the spines meant that the girdle could not have contained a flagellum, he did not believe they were dinoflagellates. Maria Lejeune, later Lejeune-Carpentier (1910–1995; Fig. 15) of Liège, Belgium, published careful restudies of Ehrenberg's types and gave descriptions of additional species (e.g. 1936), observing regularly shaped openings in 'hystrichosphere' walls without perceiving their implication (see Sarjeant & Vanguestaine 1999).

Alfred Eisenack (1891–1982; Fig. 16) not only described (1935, 1936*a*, *b*) the first Jurassic assemblages to be reported since the first brief mention by Ehrenberg, but also described assemblages from the Oligocene amber-bearing sediments of East Prussia, now Kaliningrad, Russia (1938*a*), and reported what he considered to be 'hystrichospheres' from German Silurian deposits (1931, 1938*b*). The Welsh palaeontologist Herbert P. Lewis carried their record even further back, when he observed them in the Ordovician of Montgomeryshire, Wales (1940).

However, though Eisenack had employed acetic acid to extract microfossils from some Silurian limestones, this was still the 'stone age' of dinoflagellate study. Almost all of the work of those three scientists and their contemporaries was done through examining microfossils enclosed in thin chert flakes, with consequent problems in resolution of fine detail. Their use in biostratigraphical correlation had not even begun.

The recognition of dinoflagellates with motile cells containing siliceous supporting structures dates back to Ehrenberg (1838, 1840, 1854). However, the discovery of forms apparently having a siliceous motile wall, and exhibiting a plate tabulation, came only through the work of M. Lefèvre (1932, 1933) on Tertiary sediments from Barbados and Deflandre's work on material from the New Zealand Tertiary (1933). It is usually assumed that Lefèvre's generic name for them, *Peridinites*, had narrow priority over Deflandre's name, *Lithoperidinium*; but this remains a matter for question. Certainly their discoveries were almost simultaneous.

Fig. 16. Alfred Eisenack (1891–1982) on his seventy-fifth birthday (photograph by Werner Wetzel, Tübingen).

Prasinophytes

This period saw little work on the tasmanitids. David White (1929) treated them as spores, as did Thiessen (1925*a*) and F. Thiergart (1944). Eisenack (e.g. 1932), describing forms from the Lower Palaeozoic of the Baltic region, at first placed them into the existing algal genus *Bion*, but later he utilized his species *B. solidum* as type for a new genus, *Leiosphaera* (1935). This grew to include not only the thick-walled, porate forms, but also thinner-walled forms without mural pores; through being contrasted with the spinose 'hystrichospheres', these thinner-walled types came to be called 'leiospheres'.

A genus described by Otto Wetzel (1933) from the Late Cretaceous of Germany (*Pleurozonaria*) was shown subsequently to be a tasmanitid.

In 1941, the German palaeobotanist Richard Kräusel reviewed the constituent species of Dawson's genus *Sporangites*, transferring the species *huronensis* to *Leiosphaera*. Unaware of his work, the three US microscopists Schopf, Wilson and Bentall (1944) reconsidered both *Sporangites* and *Tasmanites*, rejecting the former name and transferring all its algal species to *Tasmanites*. However, even though tasmanitids had been shown to be important components of oil shales and other deposits of economic importance, they remained a group almost unknown to geologists at large.

Scolecodonts

After a hiatus of over 50 years, work on these palynomorphs was begun anew by two US palaeontologists, Carey Croneis and Harold W. Scott (1933). Noting the similarity in general form, if not in composition, to the already well-known conodonts, they proposed the name 'scolecodonts' for the disjunct components of polychaete jaw apparatuses. This name came to be widely used by other micropalaeontologists. In the USA, Clinton R. Stauffer published two papers, respectively on Ordovician and Devonian forms (1933, 1939) and E. R. Eller (1933) began a series of studies of Palaeozoic forms that was to be continued intermittently for 36 years. Karel Žebera (1935) reported them from the Silurian of Czechoslovakia and Eisenack (1939) described them from both the Silurian and Jurassic of the Baltic regions, the latter record constituting the first report of Mesozoic scolecodonts.

Chitinozoans

It was Eisenack who discovered, in those same Baltic Silurian sediments, the first of the usually flask-shaped microfossils which he styled 'chitinozoans' (1931). The name, based on a misunderstanding of their chemical constitution – yes, a resistant organic substance, but not chitin – implied immediate recognition of their affinity to the animal, not the plant, kingdom. Eisenack considered initially that they were protozoans, perhaps related to the thecamoebians (Rhizopoda). Later (1932) he suggested an affinity to the Euglenoidea, a group of freshwater flagellated algae; while Charles L. Cooper (1942), reporting the first North American forms, proposed instead an affinity with the Hydrozoa on the basis of their similarity to gonothecae, the structures enclosing hydrozoan reproductive bodies.

Other palynomorphs

Of the wide range of other palynomorphs nowadays recognized, only a few attracted attention during this period. Further work by Zalessky (1926), Thiessen (1925*b*) and, in particular, Kathleen Blackburn (1936) confirmed the significance of *Botryococcus* as a source of oil in boghead coals.

A group of small, hollow ellipsoidal bodies with long, thread-like appendages was reported from Baltic Cretaceous flints by Otto Wetzel (1933) under the generic name *Ophiobolus*. Deflandre, describing a second genus (*Dimastigobolus*) from the Late Cretaceous of France (1935, 1937), considered they were fossil flagellates and erected a family *incertae sedis*, the Ophiobolidae, to accommodate them. Though the generic name *Ophiobolus* was later shown by A. R. Loeblich III (1967) to be preoccupied and the new name *Scuticabolus* proposed, they continue to be styled 'ophiobolids'.

It was again Eisenack who first observed, in his Baltic Silurian sediments, black rodlets of brittle organic material, swelling at each end into bulbs or tripods (1932). In a more extended study (1942), he coined the name 'melanosclerites' for them. He speculated that they were supporting structures from the lower trunks of ancestral Ctenophorans, the comb-jellies, a group otherwise unknown as fossils, but placed them provisionally into an order Melanoskleritoitidea of uncertain affinity.

Years of expansion (1945–1970)

The acceleration in spore-pollen studies in the years following World War II was reported very fully by Alfred Traverse (1974; Fig. 17). As he noted, it was at first slow. The utility of miospores for coal-seam correlation continued to be perceived; Potonié and his associates (in particular Gerhard O. W. Kremp) transferred their attention from Tertiary to Carboniferous coals and further developed their classification of dispersed spores (Kremp 1952; Potonié 1956, 1962, 1967, 1970; Potonié & Kremp 1954, 1955, 1956*a*, *b*). Researches in The Netherlands by Sijbren J. Dijkstra (e.g. 1946) and in Belgium by Pierre Piérart (e.g. 1955, 1958) on Carboniferous megaspores, demonstrated afresh their potential in coal seam correlation. Accounts of Carboniferous microfloras of Turkey were published by Bülent Ağrali (e.g. 1963) and by K. Yahşiman (1964).

In Great Britain, studies of Carboniferous miospores were begun anew, following the creation of the National Coal Board, by A. H. V. Smith, Mavis Butterworth and others (Smith 1962; Butterworth 1966; Smith & Butterworth 1967). Another major factor in British research was Leslie R. Moore's recognition of their importance in stratigraphical correlation (1946); when Moore was appointed to the chair of the University of Sheffield's Department of Geology, he made palaeopalynology a principal focus for that department's research. The work of such persons as Leonard G. Love (e.g. 1960), Herbert J. Sullivan (e.g. 1958, 1962; Fig. 17), Roger Neves (1958; Fig. 18), Edward G. Spinner (e.g. 1965) and Bernard Owens (Owens & Burgess 1965; Owens 1970) greatly expanded knowledge of British Carboniferous spores, while John B. Richardson began what was to prove a lifetime study of Devonian spores (1962; Fig. 19) and George F. Hart (1960; Fig. 20) commenced work on Permian assemblages from eastern and southern Africa.

By this time, Devonian palynology was also attracting the attention of Maurice Streel in Belgium (e.g. 1964); he was to establish an important school of palynology at the University of Liège. Arlette Moreau-Benoît (1966, 1967) studied the Devonian palynomorphs of Anjou, France. Devonian spores were also reported from Lithuania by A. I. Venozhinskene (1964)

Fig. 17. Herbert J. Sullivan (centre), with Alfred Traverse (left) and Jan Jansonius, at the International Palynological Congress, Cambridge, England (photograph by the author, 3 July 1980).

Fig. 18. Roger Neves (right) talking with dinoflagellate and acritarch palynologist S. J. 'Jack' Morbey, at the International Palynological Congress (photograph by the author, 3 July 1980).

Fig. 19. John Brian Richardson on field work near Fountainstown in southern Ireland (photograph by the author, 16 September 1982).

and from Svalbard (Spitzbergen) by J. O. Vigran (1964). In Canada, D. Colin McGregor was beginning a lifelong study of Devonian palynofloras, with work on assemblages from Melville Island and eastern Canada (1960, 1961 and later papers; Fig. 21). His colleagues at the Geological Survey of Canada, P. A. Hacquebard and M. Sedley Barss, published studies of Carboniferous and Permian microfloras (Hacquebard 1957, 1961; Hacquebard & Barss 1957; Barss 1967).

In Australia, work on Permian spores had begun already, with studies by J. A. Dulhunty in New South Wales and Tasmania (1946, 1947; Dulhunty & R. Dulhunty 1949) and by Basil E. Balme in Western Australia (1952 and later papers; Fig. 22). Balme's work was later extended to the crucial Permian–Triassic boundary section in Pakistan (1970) and supplemented, in Western Australia, by that of Kenneth Segroves (e.g. 1969). Gondwana megaspores were studied by Dijkstra (1953) in Brazil, while the microspores were the subject of a major study by Hart (1965) in South Africa and were studied in India by D. C. Bharadwaj (1962 and later papers). G. Leschik, who had worked earlier on Zechstein (Permian) microfloras of Germany (1956), subsequently studied those of Namibia (1959). An account of British Permian saccate and monosaccate miospores was given

Fig. 20. George Frederick Hart (right) expounds his ideas to Alfred Traverse during the International Palynological Congress in Brisbane, Queensland (photograph by the author, 1 September 1988).

by Robin F. A. Clarke (1965), but his work was long in being followed up.

In the Soviet Union, palynology in the post-war years had been given massive governmental support; indeed, Jansonius & McGregor (1996, p. 3) pointed out that:

> A 1961 address book by Kuprianova and Aleshina contains nearly 500 names of the 'most active' [palynologists] and two decades later that number had grown to about 800. . . . Kuprianova (1960) had estimated that the number of palynologists . . . increased from 350 to 1000 from 1953 to 1959 alone. Their influence on the progress of palynology outside the Soviet Union, however, while significant, was far from proportional to their numbers, as a consequence of linguistic and political barriers.

The stratigraphical span of these studies was considerable. Monographs were published on Oligocene microfloras by I. M. Pokrovskaya (1956; Fig. 23); on Late-Cretaceous to Palaeogene assemblages by E. D. Zaklinskaya (1963; Fig. 23); on Late Cretaceous forms by A. F. Khlonova (1962); on Jurassic to Palaeocene and on Permian sporomorphs by S. R. Samoilovich (1953, 1961); on Jurassic to Cretaceous and on Triassic microfloras by V. S. Maliavkina (1949, 1964); and on Permian to Triassic assemblages

Fig. 21. D. Colin McGregor on field work in southern Ireland (photograph by the author, 14 September 1982).

Fig. 22. The Danish palynologist Kaj Raunsgaard Petersen on left, and Australian palynologist Basil E. Balme at the International Polynological Congress in Brisbane, Queensland (photograph by the author, 1 September 1988).

Fig. 23. Elena Dmitrievna Zaklinskaya (1910–1989), on left, with I. M. Pokrovskaya at the International Palynological Congress, Utrecht, The Netherlands (photograph by the author, 2 September 1966).

by L. M. Variukhina (1971), while Naumova continued to publish on Palaeozoic assemblages (e.g. 1953). G. M. Bratsheva published a palynological study of the Late Cretaceous and Palaeogene of the eastern USSR (1969) and E. V. Semenova (1970) described a Jurassic microflora from the Donbassa region of Ukraine. In addition, important taxonomic studies were published by N. A. Bolkhovitina (1961, 1968) and a massive three-volume overview of palynology by Pokrovskaya (1966).

It is noteworthy that virtually all the Soviet palynologists were women. That was also the case in certain countries of eastern Europe, and for the same basic reason – the slaughter of males during the wars that had afflicted those lands. Notable work was done on Carboniferous megaspores by M. Brzozowska (with Žoldani, 1958) and J. Karczewska (1967) in Poland and by M. Kalibová-Kaiserová (1951 and later papers) and Blanca Pacltová (1966) in Czechoslovakia. Eszther Nagy (1968*a*, *b*, and later papers; Fig. 8) worked on Hungarian Tertiary microfloras. M. Rogalská (1954, 1956) studied the spores and pollen of Polish Jurassic lignites and T. Orłowska-Żwolińska (1971 and later papers) examined Poland's Triassic microfloras.

However, even in those countries, there were also a few male palynologists. Researches on Polish Miocene pollen, for example, were undertaken by Stefan Mackó (1957). In Hungary, Ferenc Góczán (1956) wrote on the pollen of Jurassic coals and Miklós Kedves was beginning to publish a long series of studies of the palynofloras of Hungary (1960, 1961 and later papers). In Romania, Nikolae Balteş was studying Cretaceous palynofloras (1965).

In the German Democratic Republic, the most important figures were Wilfried Krutzsch (1957, 1959; Fig. 24) and Eberhard Schulz (1962, 1965; Fig. 24). Unfortunately, increasing bureaucratic restrictions upon publication – and indeed, upon all communication with colleagues in other lands – soon stopped the flow of palynological publications from that unhappy country.

In China there was virtually no work in palynology during this period, the single exception being a series of papers by Ouyang Shu (e.g. 1962) on Permian assemblages from Chekiang, Shansi and Yunnan.

In France, studies of Carboniferous miospores were undertaken by Boris Alpern and Jean-Jacques Liabeuf in the Lorraine and Saar coalfields (1966 and later papers). Accounts of Triassic microfloras were published by W. Klaus in Austria (1960), Karl Mädler in the Federal Republic of Germany (1964),

Fig. 24. Jan Jansonius (left) talking to Eberhard Schulz (centre) and Wilfried Krutzsch during the International Palynological Congress in Utrecht, The Netherlands (photograph by the author, September 1966).

Fig. 25. US Geological Survey palynologists Robert Haydn Tschudy (1908–1986) and Glenn R. Scott in the Tepee Butte country near Pueblo, Colorado (photograph by the author, 2 September 1961).

Bernhard W. Scheuring in Switzerland (1970) and Henk Visscher in The Netherlands (1966).

A first report of pre-Devonian spores came from the Early Silurian of Libya, by William S. Hoffmeister (1959). Ten years later, John Richardson and T. Richard Lister reported Late Silurian to Early Devonian spores from the Welsh Borderland (1969); but Silurian studies proved slow in developing, in part perhaps because of the monotonous morphology of Silurian spores.

The United States saw continuing work on Carboniferous microfloras; its history is recounted in detail by Aureal T. Cross & Kosanke (1995). This was mostly undertaken under the auspices of state geological surveys; the work by Kosanke himself (e.g. 1950) and by Russell W. Peppers and H. W. Pfefferkorn (1970) for the Illinois Geological Survey well exemplifies the high standards attained. Marcia Winslow's study of Carboniferous palynomorphs from Ohio (1962) was undertaken for the US Geological Survey, as were the studies by Robert H. Tschudy which demonstrated how palynomorphs could be used in facies interpretation in the Late Cretaceous and Early Tertiary (1961; Fig. 25). Other researches on Carboniferous spores were conducted in Mississippi, Alabama and Tennessee by F. W. Cropp (1960, 1963), in Oklahoma by Charles J. Felix and Patricia P. Burbridge (1967) and Charles F. Upshaw and Richard W. Hedlund (1967), and in West Virginia and Pennsylvania by John A. Clendening (1962 and later papers) and Daniel Habib (1968). Cross and Mart P. Schemel (Cross 1947; Cross & Schemel 1952) began studies of microfloras from the Appalachian basins, Cross's work on these continuing for many fruitful years.

Two other major figures in US palynology during this time were William S. Hoffmeister (Fig. 8), already mentioned above, and L. Richard Wilson. Hoffmeister strove unsuccessfully to patent the method of using microfossils in prospecting for petroleum (1954) and undertook a wide-ranging joint study of Palaeozoic spore distribution in North America (Hoffmeister, Frank L. Staplin and R. E. Malloy 1955). Wilson's researches had begun on Quaternary pollen, but his work was later to span much of the Middle to Late Palaeozoic, including an early account of Permian microfloras (1962) and

Fig. 26. G. K. 'Joe' Guennel on Boreas Pass, Colorado (photograph by the author, 1 September 1969).

Fig. 27. Glenn Rouse at the International Palynological Congress, Utrecht, The Netherlands (photograph by the author, 2 September 1966).

a pioneer study of the problems caused to palynologists by recycling, stratigraphic leakage and faulty techniques (1964). These were problems still relatively unfamiliar to palynologists, though G. K. 'Joe' Guennel had already reported the finding of Devonian spores in a Middle Silurian reef (1963; Fig. 26) and Traverse and two colleagues had discovered contemporary pollen in drilling-mud thinners (Traverse *et al.* 1961). The problem was soon to be stressed again when Upshaw and W. B. Creath reported some 70 Late Carboniferous spores from a cavern fill in Late Devonian limestone (1965). The occurrence of reworked palynomorphs in contemporary sediments offshore was reported by Edward A. Stanley (1966).

Though his own studies were important, Wilson's major contribution was indirect, through the researches of his numerous students at the University of Oklahoma. The other principal US centres for academic research in palaeopalynology were at Pennsylvania State University, under Traverse, and at Michigan State University under Cross. (A good summary of work on Palaeozoic megaspores generally is given by Scott and Hemsley (1996)).

Post-Palaeozoic studies included work by J. A. Doyle (1969) who considered the major floristic changes in the late Early Cretaceous (Aptian–Albian) – the changes that marked the rapid expansion of the angiosperms. Two remarkable genera prominent in the Late Cretaceous to Palaeocene, *Wodehousia* and *Aquilapollenites*, were considered at length by Glenn E. Rouse (1957; Fig. 27) and Stanley (1961*a*, *b*; Fig. 28). Stanley also studied Cretaceous and Palaeocene plant microfloras from South Dakota and Alaska (1965, 1967*a*), while Warren S. Drugg (1967) wrote generally on palynomorphs from these levels in California. Researches specifically on Palaeocene microfloras were published by William C. Elsik (1968), Harry A. Leffingwell (1970; Fig. 29) and Douglas J. Nichols & Traverse (1971). Traverse's (1955*a*) study of the Brandon Lignite (Oligocene) of Vermont was especially important, since it included also a combined morphographic/natural system for classifying Tertiary pollen. A paper by Elsik (1966), considering biological degradation of fossil pollen grains and spores, is noteworthy; it was to serve as a prelude to his important later studies of fungal spores.

Fig. 28. Edward Stanley at the International Palynological Congress, Utrecht, The Netherlands (photograph by the author, 3 September 1966).

Researches on Jurassic to Early Cretaceous microfloras in France were undertaken by J. Levet-Carette (1963, 1966) and in Sweden by Hans Tralau (1967 and later papers) and Dorothy Guy, later Guy-Ohlson (1971 and later papers). Carla Gruas-Cavagnetto began a life-time study of the Palaeogene microfloras of France (1967 and later papers; Chateauneuf & Gruas-Cavagnetto 1968). Dionijs (Dennis) Burger (1966) initially studied microfloras from The Netherlands, subsequently settling in Australia and undertaking major studies of its Mesozoic biostratigraphy. Noel J. de Jersey inaugurated a long series of palynological studies for the Government of Queensland with reports of Triassic and Jurassic microfloras (1959, 1962). The Australian palaeobotanist Isabel C. Cookson (Fig. 30), after writing an account of Quaternary palynomorphs from Kerguelen Archipelago in the Southern Ocean (1947), progressively expanded her interests downward into Tertiary and Mesozoic palynofloras (1950 and later papers; see also Baker 1973).

In 1955 R. D. Woods was excitedly promoting 'Spores and pollen as a new stratigraphical tool for the oil industry'. In truth, their potential in subsurface correlation had been under investigation by the Royal Dutch/Shell Group very much earlier, in the mid-1930s (Hopping 1967).

Fig. 29. Harry Leffingwell (left) and Carboniferous palynologist Geoffrey Clayton, during the meeting of the American Association of Stratigraphic Palynologists in Dublin, Ireland (photograph by the author, 14 September 1982).

Fig. 30. Isabel Clifton Cookson (1883–1973); uncredited photograph, reproduced in Baker (1973).

However, though the decision to develop palynological research had been taken at that time, its implementation was delayed by World War II, so that Wilson's much earlier paper on the correlation of sedimentary rocks by fossil spores and pollen (1946) was, in contrast, quite timely. Even so, it was not until 1962 that the maturing of industrial attitudes to palynology was to be marked by the holding, by the Society of Economic Paleontologists and Mineralogists, of a symposium on 'Palynology in oil exploration' in San Francisco, California (Cross 1964).

A 'decimal code system' for the characterization of pollen grains and spores was developed and applied during Royal Dutch/Shell's investigations in Venezuela (see Kuyl *et al.* 1955; Hopping 1967). Among the results of that company's work, the pioneer study by Jan Muller of the environmental distribution of palynomorphs in the sediments of the Orinoco delta (1959) was to prove of pivotal importance to future palaeoecological studies, while the long series of researches by Thomas van der Hammen (e.g. 1954) not only established a meaningful subdivision of the Tertiary sediments of Venezuela, Colombia and Guyana, but also furnished a fresh nomenclatural approach, capable of ready application in other countries (1956).

Spores and pollen subsequently enabled a comparable breakthrough to be attained in understanding the geology of the Late Cretaceous to Eocene sequence in Nigeria (P. M. J. van Hoeken-Klinkenberg 1964, 1966). Work by French companies thereafter extended knowledge of Cretaceous geology to much of west Africa – not only Nigeria, but Senegal, Gambia, Portuguese Guinea and the Ivory Coast (Serge Jardiné 1967). Potonié himself was involved in a study of the Late Cretaceous palynofloras of Gabon (Belsky, Boltenhagen & Potonié 1965).

Researches under the auspices of other oil companies were by then taking place elsewhere in the world. N. Goubin, of the Institut Français du Pétrôle, described the Permian to Jurassic microfloras of Madagascar (1965); Roberto Daemon, L. P. Quadros and L. C. da Silva (1967) described spores from the Devonian of the Paraná Basin, Brazil (at that time considered a potentially very important source of petroleum); and there were joint studies, such as that by R. K. Kar, G. Kieser and K. P. Jain (1972) on Permo–Triassic assemblages of Libya – work sponsored by the Compagnie Française des Pétrôles, but undertaken in co-operation with the Birbal Sahni Institute of Palaeobotany in Lucknow, India. As the years passed, even though many companies (including Royal Dutch/Shell) became increasingly reluctant to see their palynological results published and available to competitors, palynologists from petroleum companies have regularly participated in national and international meetings; their work has continued to contribute massively to our knowledge of past palynofloras and environments.

Significant spin-offs from oil company researches have been the development of two new techniques for palynological investigation. C. C. M. Gutjahr (1966) demonstrated how carbonization measurements could aid in determining the petroleum potential of sediments that might have undergone metamorphism, while Pieter van Gijzel (1967) demonstrated how the intensity of autofluorescence correlated with the rank of coals and enabled the ready distinguishing of reworked palynomorphs in mixed assemblages.

Certain academic research centres figured prominently in palynological research during these years. The numerous contributions from the Birbal Sahni Institute, though devoted primarily to Indian palynomorphs (e.g. C. P. Varma and M. S. Rawat 1963; C. F. K. Ramanujam, 1966 on Tertiary pollen) have included several studies

Fig. 31. B. Srinivasin Venkatachala at the International Palynological Congress, Utrecht, The Netherlands (photograph by the author, September 1966).

of assemblages from other countries, such as those by B. S. Venkatachala and Kar (1967; Fig. 31) on the Permian of Pakistan and that by Venkatachala, Kar and S. Raza (1968) on the Carboniferous miospores of Romania.

In England, the University of Cambridge had not only become an important focus for Quaternary studies, but had also developed a school of palynological research, under Norman F. Hughes (Fig. 32). Hughes not only undertook, personally or with associates, studies of particular spore-pollen genera (e.g. 1961) and of Cretaceous assemblages (e.g. Hughes & J. Moody-Stuart, 1966), but also strove to establish international publication standards and policies (1970, 1975) – not always with success. He and his students Geoffrey Playford and Mary Dettman – both later to be prominent in Australian palynology – wrote on the Carboniferous palynofloras of Svalbard (Hughes & Playford 1961; Dettman & Playford 1962) while the New Zealand palynologist R. Ashley Couper, who had already described the Mesozoic microfloras of his homeland (1953), made a wide-ranging study of British Mesozoic palynomorphs during his stay at Cambridge (1958).

Earlier works on New Zealand assemblages had been done by M .T. Te Punga (1948, 1949); they were to be continued by David J. McIntyre (1962 and later papers). In Argentina, noteworthy studies have included those of Carlos A. Menéndez and Elba D. Pothe de Baldis, on Permian to Tertiary palynomorphs from that country and Paraguay (Menéndez 1951 (and later papers); Menéndez & Pothe de Baldis 1967) and the work on Mesozoic and Tertiary microfloras by Wolfgang Volkheimer (e.g. 1968). Another palynologist meriting mention is Kiyoshi Takahashi, whose numerous papers on Japanese and other east Asian palynomorphs (1961 and later papers) have set a high standard for palynological researches in those countries.

Dinoflagellates and acritarchs

In the first five years after World War II, the few studies of fossil dinoflagellates that were made were by persons already engaged in this field of research. Despite wartime problems, both Lejeune-Carpentier and Deflandre had continued their investigations, with Lejeune-Carpentier describing species from Late Cretaceous flints (e.g. 1946, 1951) and Deflandre enlarging his studies of Jurassic and Cretaceous assemblages (1947*a*, *b*). Deflandre also recorded 'hystrichospheres' from Silurian and Carboniferous strata (1942, 1945, 1946). In addition, he reported the first fossil dinoflagellates with calcareous shells and proposed a classification for them (1947*c*, 1948). In 1948, the Belgian palynologist André Pastiels, noted for his studies of Carboniferous microfloras, made a single contribution to this field when he published the first extended description of an Eocene assemblage (see also Sarjeant 1986).

It was Deflandre who gave the first extended consideration to '*le problème des hystrichosphères*' (1947*d*). After assessing other possibilities, he concluded that they were certainly marine planktonic organisms, but not necessarily attributable to a single group; consequently he considered correctly that the 'Order Hystrichosphaeridia' was a polyphyletic assemblage. Thirteen years later, when I tackled the same question, I could do little more than echo Deflandre's findings (Sarjeant 1960*a*, 1961*a*).

The earlier methods of study initially continued with little change. When Deflandre's student Lionel Valensi made an extended examination of Middle Jurassic assemblages, he based it very largely upon microfossils – mostly 'micrhystridia' – that were contained in chert flakes (1948, 1953). Valensi's later studies demonstrated to prehistorians how the sources

Fig. 32. Norman F. Hughes (1918–1994) on left, talking with the Norwegian palynologist Svein Manum during the International Palynological Congress in Brisbane, Queensland. (Photograph by the author, 1 September 1988).

of the flint weapons of early man might be determined from their microfossil content (1955, 1960). When Eisenack returned to Germany from imprisonment in Siberia and set up a laboratory in Tübingen (see Gocht & Sarjeant 1983; Sarjeant 1985), he used pre-War methods of extraction of palynomorphs from limestones. Eisenack's work extended from the Silurian down into the Ordovician (1948) and Cambrian (1951). Though studies of Jurassic and Oligocene palynomorphs were resumed by Eisenack (1954, 1957) and work on Aptian (late Lower Cretaceous) forms initiated (1958a), his principal work thereafter was on the Palaeozoic.

Studies of Cretaceous palynomorphs in flints by Otto Wetzel were continuing (e.g. 1951). After a 30 year hiatus, his namesake Walter Wetzel also re-entered the field, with papers on palynomorphs from Danian (Palaeocene) and from Early Jurassic strata (1952, 1955). The problems presented by his work have been discussed by Sarjeant (1984*a*) and the researches of the two Wetzels are considered at length by Linda F. Dietz, Sarjeant and Trent A. Mitchell (2000).

Isabel Cookson travelled to Europe in the early 1950s to initiate studies of dinoflagellates from Australia and Papua New Guinea, working at first with Deflandre (Deflandre & Cookson 1954, 1955) and briefly with Hughes (Cookson & Hughes 1964). Her most successful collaboration, though, was with Eisenack; its consequence was a long series of papers begun in 1958 and only concluded with the posthumous publication by Hans Gocht of their last results.

In all this work, the interest continued to be in the microfossils as objects in themselves, as evidence for the rich microscopic life in past periods. Though the geological circumstances of the finds were noted – Eisenack, in particular, used his microfossils to give broad dates to rocks included in Baltic boulder clays – their relative age and exact stratigraphic position were matters of less significance. (Deflandre indeed based most of his work on Cretaceous microfossils upon flints collected from the paths of the Jardin des Plantes, where his laboratory was situated.)

During the late 1950s, however, the question of the utility of dinoflagellates and 'hystrichospheres' as a means for biostratigraphical correlation came at last to be investigated. In Germany, three of Eisenack's students were involved in this. Two of them – Hans Gocht and Gerhard Alberti – were refugees from the Communist regime of east Germany. Gocht had already undertaken studies of Oligocene assemblages (1952 and later papers); he enlarged his

Fig. 33. William R. Evitt at the International Congress of Palynology, Utrecht, The Netherlands (photograph by the author, September 1966).

work by beginning studies of Early Cretaceous and Jurassic assemblages (e.g. 1957, 1970). Alberti, by heart a trilobite worker, had been directed by the East German government to work instead on marine palynomorphs, with a view to aiding subsurface exploration for petroleum. Upon arrival in Tübingen, he published his results rather reluctantly (1959*a*, *b*, 1961) and then departed to resume work on trilobites at another university.

In Glasgow, Charles Downie had been studying, for his doctoral thesis, the Late Jurassic (Kimmeridgian) stratigraphy of Dorset, England. When he was given an appointment at the University of Sheffield, he was persuaded by Moore to look at the palynomorph content of his materials. His discoveries led to a first paper (Downie 1957) which indicated their biostratigraphical potential and caused him to give me, his first graduate student, the task of investigating their utility in Jurassic correlation. I succeeded in demonstrating this, in a series of papers (Sarjeant 1959 and later papers) that included two proclamations of their stratigraphical value (1963*a*, 1965) and the first attempt at an international correlation chart relating dinoflagellate distribution to Jurassic ammonite zones (1964). I was also to produce, with Downie, the first bibliography and index of marine palynomorphs (Downie & Sarjeant 1965) and to collaborate with him not only in taxonomic studies (Downie & Sarjeant 1963), but also in attempts to formulate standard taxonomic, terminological and descriptive approaches for dinoflagellates (Downie *et al.* 1961; Downie & Sarjeant 1966). I was privileged to report the first discovery of Triassic dinoflagellates (Sarjeant 1963*b*) and, with David M. Churchill, the first undoubted fossil non-marine dinoflagellates, from lakes and swamps in southwestern Australia (Churchill 1960; Churchill & Sarjeant 1962, 1963).

Almost coincidentally with my own studies, three other persons were investigating the stratigraphical potential of Late Mesozoic marine palynomorphs. The third of Eisenack's students of that time, Karl W. Klement, was engaged in describing assemblages from the Malm (Late Jurassic) of Germany (1957, 1960); he was later to emigrate to the United States and, before his early death, to become a specialist on calcareous algae (see Hanson 1982). Among a group of palynologists working for Imperial Oil in western Canada, Stanley A. J. Pocock was investigating Jurassic assemblages (1962). Thus it came about that the three of us and Alberti were to describe closely similar genera of complex chorate 'hystrichospheres' within less than eighteen months, under the four generic names *Systematophora, Polystephanosphaera, Hystrichosphaerina* and *Hystrichosphaeridium* (Klement 1960; Sarjeant 1961*b*; Alberti 1961; Pocock 1962) – this, at a time when the number of persons worldwide who were studying fossil dinoflagellates scarcely attained double figures. Yes, it was a remarkable coincidence, but it was also an indication of the growing importance of studies of fossil marine palynomorphs.

The fourth person set the task of investigating their use in stratigraphy was William R. Evitt (1964; Fig. 33). Like Alberti, he had published earlier studies of trilobites before being required by the Jersey Production Research Company to assess the use in stratigraphical correlation of dinoflagellates and 'hystrichospheres'. His demonstration, on the basis of detailed morphology, process position and style of shell openings, that the fossil dinoflagellates were cysts and not motile forms (Evitt 1961) was unexpected; his further demonstration that, on the same basis, the great majority of the so-called 'hystrichospheres' were likewise dinoflagellate cysts (Evitt 1963*a*) came almost as a bombshell. Downie and I, having been initially unconvinced by Evitt's ideas, were forced by his evidence to accept them. When Evitt had christened as 'acritarchs' the former hystrichospheres without demonstrable dinoflagellate affinity, we joined him in proposing that they be treated as a group *incertae sedis* and divided into subgroups showing a measure of morphological unity (Downie *et al.* 1963). These subgroups were given the suffix 'morphitae' and did not have specified type genera, since this would enable the continued use of subgroup names even after the affinity of a particular constituent genus had been determined.

Our approach proved initially controversial. Eisenack, in particular, long resisted it, proclaiming *'die Einheitlichkeit der Hystrichosphären'* – the unity of the hystrichospheres (1963*a*, *b*). Only in 1969 did he eventually concede that he had been wrong and adopted the name 'acritarch'. Mädler (1963) erected a Class Hystrichophyta, embracing two orders, Hystrichosphaerales (hystrichospheres) and Leiosphaeridiales (leiospheres); though he continued to advocate these suprafamilial taxa as late as 1967, they were never to gain currency. In contrast, the Russian palynologist Boris V. Timofeyev, who had written extensively on Palaeozoic 'hystrichospheres', essentially purloined the D–E–S (Downie–Evitt–Sarjeant) classification, as it had come to be called; he proposed virtually identical groupings, again without specified types, but with the altered suffix 'morphyda' (1965, 1967). The Timofeyev classification, like that of Mädler, gained no adherents.

A different problem arose when Pocock and his two Imperial Oil colleagues, Frank L. Staplin and Jan Jansonius (Figs 17, 24), proposed that the subgroups of what they styled 'acritarchous hystrichospheres' should have defined types, in addition formulating two new groupings – 'Leiosphaeritae' and 'Baltisphaeritae' – with a shortened suffix (Staplin *et al.* 1965). However, the D–E–S classification prevailed for more than 30 years and is only now being abandoned, in consequence of new phytochemical evidences concerning acritarch affinity.

While Evitt's work was in progress, research on fossil dinoflagellates had already been expanding. At the University of Kiel, Walter Wetzel supervised the studies of three graduate students. Dorothea Maier (1958, 1959) and Barbara Klumpp (1953) undertook studies of north German Eocene to Miocene assemblages, their work being of sharply contrasting quality (see Sarjeant, 1981, 1983). Peter Morgenroth wrote on palynomorphs from the Palaeocene to Oligocene and the Early Jurassic of north Germany (1966*a*, *b*, 1968, 1970). In Tübingen, Eisenack oversaw the studies of Ellen Gerlach (1961) on north German Oligocene and Miocene assemblages and, later, those of Johannes Agelopoulos (1967) on north German Eocene microfossils.

At Cambridge, Norman Hughes oversaw researches by Robin F. A. Clarke, who later published, jointly with Jean-Pierre Verdier, a major study of Late Cretaceous dinoflagellate assemblages (1967). Another of Hughes's students was Geoffrey Norris, who was to make many distinguished contributions to palynology and to collaborate with me on a descriptive index of fossil dinoflagellate and acritarch genera (Norris & Sarjeant 1965). In Paris, Deflandre's student Martine Rossignol was beginning studies of cysts in Pleistocene and Recent sediments of Israel and the eastern Mediterranean (1962, 1963); these would go far towards demonstrating the validity of Evitt's concepts. Evitt himself was by then setting up a research school at Stanford University, California, his collaboration with Susan Davidson on theca/cyst relations of dinoflagellates being an early product (Evitt & Davidson 1964) and his own work on cyst openings (Evitt 1967) among other contributions of lasting value.

In Belgium the study of Tertiary assemblages, inaugurated by Pastiels, was at length resumed by Jan de Coninck (1965, 1969 and later papers). In eastern Europe, work on Mesozoic dinoflagellates was beginning – on Jurassic assemblages by Lilia Dodekova in Bulgaria (1967, 1969) and Dan Beju in Romania (1969) and on Cretaceous forms by Hanna Górka in Poland (1963) and by Nikolae Balteş in Romania (1963, 1964). Far away in Novosibirsk, Tamara F. Vozzhennikova was beginning lifelong studies of Siberian Mesozoic and Tertiary dinoflagellate assemblages (1960 and later papers) – the only significant work on the group in the whole USSR.

The major centre for studies of dinoflagellates and acritarchs was, however, in Sheffield, where Charles Downie supervised studies by a succession of students, who were to contribute massively to the better comprehension of these microfossils (his work is discussed by Sarjeant 1984*b*, 1999, 2000; Owens & Sarjeant 1999). David Wall, having initially studied English Lias (Early Jurassic) assemblages (1965*a*), emigrated to the USA and initiated wide-ranging studies of dinoflagellate cyst distribution in Pleistocene and modern sediments, in part in association with the former Sheffield technician Barrie Dale (e.g. Wall 1965*b*; Wall & Dale 1967); Dale was later to settle in Norway and, personally or through his associates, to contribute greatly to the knowledge of the distribution of dinoflagellates in Recent sediments and waters. Graham L. Williams initially studied the London Clay (Williams & Downie 1966*a*, *b*, *c*); he was later to become a pre-eminent figure in the research team of the Geological Survey of Canada.

By that time, I was not only enlarging my studies to include French Jurassic assemblages, but also examining material from the Speeton Clay (early Cretaceous) of Yorkshire (1966*a*, *b*, *c*, *d*, 1968; Neale & Sarjeant 1962) – studies abruptly terminated when my collections were destroyed in a fire at Nottingham University, where I was establishing my own research school. The first of my research students was David B. Williams, who studied the distribution of dinoflagellate cysts in the North Atlantic Ocean as indicators not only of water depth and proximity to shorelines, but also of oceanic circulation (Williams & Sarjeant 1967; Williams 1968). Roger J. Davey studied Chalk (Late Cretaceous) assemblages (1969, 1970); he and I were involved in collaborative taxonomic researches with Downie and Graham Williams (Davey *et al.* 1966; Davey & Williams 1966*a*, *b*). When, towards the end of this period, I presented further papers at international meetings on dinoflagellate cysts as biostratigraphical indices (1967*b*, 1970*c*), I had a much greater fund of information to draw upon.

Though Downie continued to make personal contributions, his prime concern had shifted to the acritarchs (see Sarjeant 1999). The earliest of his Palaeozoic researches was on a Tremadocian (earliest Ordovician) assemblage (1958) from Shropshire, England, but soon they expanded to comprise Silurian microfloras from that county (1959, 1963), Early Cambrian and late Precambrian acritarchs from Scotland and the Grand Canyon, Colorado (1962, 1969), and joint work with Wall on Permian acritarchs (Wall & Downie 1963). Downie's student J. Richard Lister completed a major study of Silurian acritarchs, presenting evidence for the dinoflagellate affinity of some of them (1970*a*, *b*); unfortunately, Lister's studies were destined never to be fully published.

The most extensive work on Palaeozoic acritarchs during this period was in Russia. Naumova, in Moscow, and Timofeyev, in Leningrad, were the prime figures. Unfortunately both of them initially misinterpreted these marine planktonic organisms as being spores of terrestrial or marine plants. Naumova, in five papers covering Riphean (Late Proterozoic) to Silurian assemblages (starting in 1949), never relinquished that concept. Timofeyev, perhaps the only distinguished male palynologist of the USSR, consistently ignored Naumova's work and was overoptimistic in his reports of trilete marks on Early Palaeozoic specimens (Timofeyev 1955). Ultimately, however, he accepted that many of his forms were of planktonic character, initially naming them 'hystrichospheres' (Timofeyev 1956) and then adopting his own classification (see p. 297). His work was of variable quality but great importance; it is discussed in detail by Jankauskas & Sarjeant (2001).

Other major contributors to acritarch research during this period included François Stockmans and his wife, Yvonne Willière, who described Belgian Silurian to Carboniferous assemblages in a series of papers (e.g. 1960), early accepting the acritarch concept (1963). Their work was to be followed up by further studies of Silurian assemblages by Francine Martin (1966, 1967), who also extended her research to Belgian Ordovician acritarchs (1969*a*, *b*), and by Michel Vanguestaine, who initiated the study of Cambrian assemblages (1967, 1968). Jean Deunff published many accounts of Ordovician to Devonian assemblages, in particular from Brittany, France (1951, 1954*a*, and later papers) but also from the Devonian of Canada (1954*b*, 1957, 1961*a*), from the Saharan region of Algeria (1961*b*) and from Tunisia (1966). Fritz H. Cramer reported acritarchs widely from Spain and Canada (1964 and later papers; Cramer & Díez 1968, 1970), also using them to try to reconstruct the motions of continental plates in the Silurian (1970). Middle Triassic assemblages from Switzerland were described by Marita Brosius and Peter Bitterli (1961), while Permian and Triassic assemblages were reported from western Canada by Jansonius (1962) and from the Permian of Pakistan by myself (Sarjeant 1970*b*).

Paul Tasch's report of Permian hystrichosphaerids from Kansas (1963) was always viewed dubiously and was ultimately discounted by Evitt (1985). In contrast, François Calandra's description (1964) of a tabulate dinoflagellate from the Late Silurian of Tunisia, *Arpylorus antiquus*, was considered sound and was long to remain the earliest undoubted record of that group (see Sarjeant 1978*b*).

An overall review of developments since 1970 is presented by Robert A. Fensome, James B. Riding and F. J. R. 'Max' Taylor (1996).

Prasinophytes

The taxonomic history of this group in the early post-War period was bound up with that of the 'hystrichospheres'. Up to 1952 only two genera,

Tasmanites and *Pleurozonaria*, had been described. Within the ensuing 15 years, however, 14 additional genera were named. Three of these were described from the Silurian to Devonian of Brazil (Brito & Santos 1965; Brito 1965; F. W. Sommer & Norma M. van Boekel 1963) and another from the Devonian of Oklahoma (Wilson & Urban 1963). Further new genera were reported from the Carboniferous of Saudi Arabia (Hemer & Nygreen 1967); the Permian of Western Australia (Segroves 1967); the Early Jurassic of Germany (Mädler 1963); and the Cretaceous to Tertiary of Western Australia, New Guinea and Svalbard (Cookson & Manum 1960). Two genera were named from the Palaeogene of Hungary (Kedves 1962, 1963; Kriván-Hutter 1963); one from the Tertiary of California (Norem 1955); another from the Miocene of Hungary (Hajos 1964); and the latest from the Neogene of Hungary (Nagy 1965). Of these, two (*Pseudolunulidia* and *Quisquilites*) are bean-shaped; the former is probably a synonym of the latter (Wilson, quoted in Muir & Sarjeant 1971). The others are all spheroidal, though often compressed to a disc shape, with walls variously porate: most appear likely to be junior taxonomic synonyms of *Tasmanites*, but this remains to be demonstrated.

The number of species of *Tasmanites* itself likewise increased greatly during this period, in particular through work on the Devonian of the Amazon basin of Brazil by Sommer and van Boekel (Sommer 1953 and later papers; Sommer & van Boekel 1963; van Boekel 1963) and by Eisenack in Germany (1958*b*, 1963*d*). Again, it is likely that these names include many taxonomic synonyms.

Eisenack (1958*b*) accepted that his own *Leiosphaera solida* was a taxonomic synonym of *Tasmanites punctatus*, rendering the generic name *Leiosphaera* redundant. He placed the thin-walled leiospheres instead into a new genus, *Leiosphaeridia*.

Five years later, Mädler set up the Order Tasmanales within his new Class Hystrichophyta (1963), incorporating both these morphotypes (see p. 297). Mädler's proposal was at the outset redundant since, a year earlier, Wall (1962) had demonstrated the close similarity of *Tasmanites* to the reproductive bodies of the living algal genera *Pachysphaera* and *Halosphaera*. These two genera were placed by Wall into the Class Chlorophyceae. However, almost at the same time, the algologist T. Christiansen (1962) was subdividing that class, on the basis of differing life cycles; he erected the new Class Prasinophyceae, which included both those modern genera. In recognition of Wall's work, Downie (1967) reallocated *Tasmanites* to that class. Electron-microscope studies by Ulrich Jux (1968, 1969) subsequently confirmed Wall's work and Downie's action; in contrast, his joint suggestion that Norem's genus *Tytthodiscus* was a thecamoebian (Jux & Moericke 1965) has found few adherents.

The characteristic porate walls of these prasinophytes find no parallels in the leiospheres. These remain an *incertae sedis* group, being most often placed into the acritarch subgroup Sphaeromorphitae.

Two other genera, nowadays considered to be prasinophytes, were erected during this period. *Cymatiosphaera*, a spheroidal form patterned with polygonal meshes of variable height, had been named by Otto Wetzel (1933) but was only validated many years later by Deflandre (1954). The Danian (early Palaeocene) genus *Pterospermopsis* W. Wetzel (1952) was so named since, from the outset, its close resemblance to the living alga *Pterosperma* was perceived. Eisenack (1972) was to claim, quite without justification, that the type of Wetzel's genus was unstudyable; he erected his own genus *Pterospermella* as substitute. Though I have demonstrated the invalidity of Eisenack's premise (Sarjeant 1984a), the later, quite superfluous name continues in use. Both genera were treated as acritarchs during this period, being placed respectively into the acritarch subgroups Herkomorphitae and Pteromorphitae. (For an excellent summary of later work on prasinophytes, see Guy-Ohlson 1996.)

Scolecodonts

During the petroleum exploration in the Devonian of Brazil, Frederico W. Lange (Fig. 34) reported articulated, as well as dispersed, scolecodonts on shale surfaces (1947, 1949*a*). Roman Kozlowski (1956), employing chemical extraction techniques, obtained further well-preserved jaw apparatuses from the Polish Ordovician. His method was used on a larger scale by Zofia Kielan Jaworowska (1961, 1966); she concurred with her predecessors in considering dispersed scolecodonts to be normally incapable of precise systematic assignation. In her *magnum opus* on this group, Kielan-Jaworowska (1966) described many jaw apparatuses in detail and presented a preliminary phylogeny of certain groups. The microstructure of living and fossil scolecodonts was described by K. W. Schwab (1966).

Their stratigraphical range was expanded by reports of Permian scolecodonts from Germany by H. Kozur (1967) and from Poland by H. Szaniawski (1968); the latter author also reported further finds in the Polish Ordovician

Fig. 34. Frederico Waldemar Lange (1911–1988) at a meeting of the Commission International sur le microflore du Paléozoique (acritarches), Bordeaux, France (photograph by the author, 25 November 1964).

Fig. 35. Charles Collinson (photograph by the author, 10 September 1969).

and Silurian (1970). Philippe Taugourdeau described Siluro-Devonian and Carboniferous forms from boreholes in the Algerian Sahara (1968).

In 1970, Kozur attempted to integrate the two taxonomies – that for jaw apparatuses and that for individual scolecodonts. Jansonius and J. H. Craig (1971) considered his approach to be premature and it has dropped from use. However, the recognition of scolecodonts as being components of the proboscidal armatures ('jaws') of polychaete worms, and not of annelids, was by then universal. (For a useful summary of present knowledge, see H. Szaniawski 1996.)

Chitinozoans

It was their recognition as biostratigraphical tools by the French petroleum industry, and that industry's concern with discovering oil concentrations in Palaeozoic strata, which stimulated the enormous expansion of chitinozoan studies during this period. They were especially suitable for company purposes in that their simple morphology and definitive evolution meant that minimal training was required before a person could use them for dating samples. In a series of publications, Taugourdeau not only reported them from the Silurian of the Aquitaine basin, France (1961), the Ordovician of the United States (1965) and the Early Palaeozoic of the Algerian Sahara (Taugourdeau & B. de Jekhowsky 1960), but also suggested novel approaches to their description and classification (1966). He and others presented an annotated bibliography of chitinozoans (Taugourdeau *et al.* 1967). Silurian chitinozoans were reported by P. M. Bouché (1965) from northern Nigeria, by Beju and N. Dănet (1962) from Romania and by Cramer (1964, 1967) from Spain. Jeanne Doubinger and Jacques Poncet (1964) recorded Devonian forms from France; Lange discovered them during the search for petroleum in the Brazilian Devonian (1949*b*, 1952); and R. I. Jodry and Donald E. Campau extolled their biostratigraphical value to US petroleum geologists (1961). Charles Collinson of the Illinois Geological Survey (Fig. 35) not only reported them from the Devonian of that state, but also wrote a valuable joint review of North American chitinozoans (Collinson & Schwalb 1955; Collinson & Scott 1958).

Eisenack himself continued to work on the group he had discovered, in a series of papers that sometimes treated them separately, sometimes along with Palaeozoic acritarchs (1955, 1962, and later papers). Ordovician chitinozoans were reported by Frank H. T. Rhodes (1951) from Wales and by Georg Schultz (1967) and Sven Laufeld (1967) from Sweden. P. Richard Evans recorded them from Western Australia (1961), W. Anthony M. Jenkins from the Ordovician of England (1967) and Oklahoma (1969), and Wilson and Robert T. Clarke from the Early Carboniferous of that state (1960). D. L. Dunn described them from the Devonian of Iowa and Michigan (1959; Dunn & T. H. Miller 1964), Roger F. Boneham from the Middle Devonian of Ontario and Ohio, (1967, 1969) and E. L. Gafford and Evan J. Kidson from the Permian of Kansas (1968) – rather doubtfully, since reworking was thought possible.

In an extended study of the chitinozoans, Kozlowski (1963) pointed out that they occurred quite often as straight or spiral chains, linked aperture to base, side by side, or loosely attached within a sac-like cocoon. He noted also that some specimens of *Cyathochitina* have a spongy mass at the base, which perhaps served for attachment. The presence of a sac-like structure (the opisthosome) within the chamber of solitary or colonial forms, and of an apparently contractile structure (the prosome) within the neck, was also noteworthy.

All these features needed to be taken into account when the affinity of the chitinozoans was considered. Eisenack (1962) and others considered that they were gastropod egg-cases (see Sarjeant 1992*b*, p.501). Kozlowski (1963), although noting parallels in arrangement to polychaete and gastropod eggs, considered the structure of those eggs to be too dissimilar from that of chitinozoans to sustain any relationships; he concluded that the affinity of the chitinozoans remained obscure. Taugourdeau (1964) reported an *Ancyrochitina* containing a roughly spherical body, too large to pass out through the aperture; he felt that this indicated an encysted or reproductive stage but favoured the view (expressed earlier by Deflandre 1945) that they were an independent, extinct group. Jenkins (1970) noted the remarkable correspondence in distribution, and in relative diversity per horizon, between chitinozoans and graptolites, suggesting that they might represent the missing prosicular stage in graptolite development. Though this idea was ingenious, it was ultimately to prove incorrect (Cashman 1990; summary in Miller 1996).

The classification of the chitinozoans was considered in a series of papers by Jan Jansonius (1964, 1967, 1969) but this, like their affinity, was destined to remain controversial.

Other palynomorphs

In the early post-War years, the algal genus *Botryococcus* received little notice. In the 1960s, however, it became the focus for increasing attention. Its presence in English Carboniferous rocks was reported by Alan E. Marshall and A. H. V. Smith (1964) and in US Early Tertiary deposits by Traverse (1955*b*). A. C. Brown *et al.* (1969) described the three physiological states: a green, active growth stage with straight-chain olefines; a brown to orange resting state 'of mulberry habit' with high concentrations of unsaturated hydrocarbons; and a dark green, dormant stage with little hydrocarbon. The importance of *Botryococcus* as a source of oil was stressed in a series of papers (Maxwell *et al.* 1968; Brown *et al.* 1969; Cane 1969; Knights *et al.* 1970), while the contribution of bacterial action to the formation of torbanites and other oil-rich sediments was stressed by A. G. Douglas *et al.* (1969). (For an account of subsequent studies, see Batten & Grenfell 1996.)

The colonial genus *Gloeocapsamorpha* was so named by Zalessky (1917) because of its similarity to the modern cyanobacterium *Gloeocapsa*. There are indications that it might be a marine alga and, though the suggestion by Traverse (1955*b*) and others that it was synonymous with *Botryococcus* is no longer accepted, its systematic position remains uncertain. It is an important component of Ordovician marine shales in the Baltic Basin of Estonia, being styled kukersite and mined as a source of fuel (Bekker 1921). It is present also in Baltic Silurian sediments (Eisenack 1960); however, the report by Timofeyev (1966), from the Lower Sinian (Proterozoic) of China, is considered questionable. The explosion of work on this organism occurred after 1970; it is reviewed by Wicander *et al.* (1996).

Another colonial alga, *Pediastrum*, hitherto known only from freshwater deposits, was reported from Cretaceous strata by Evitt (1963*b*); it is attributed to the Chlorococcales. Evitt also published a detailed study of the ophiobolids (1968); however, their affinity remains uncertain.

Following the first report by Deane (1849), the acid-resistant linings of foraminiferal tests received virtually no attention for more than a century. When studies were renewed, they gave rise initially to errors. John F. Grayson (1956) considered them to be composed of calcium fluoride and dismissed them as fortuitous byproducts of the palynological preparation process. This mistake was corrected independently by Otto Wetzel (1957) and Frederik H. van Veen (1957), who both demonstrated their organic composition. However, they were thereupon misinterpreted as a distinct group of foraminifera with small organic-walled tests, 'microforaminifera' (Wilson & Hoffmeister 1959). Edwin D. McKee, John Chronic and Estella B. Leopold (1959), who encountered them in sediments from a Pacific atoll, doubted this, wondering whether the microfossils might be separate species, dwarfs or juveniles of larger species, or the remains of larger forms whose earliest chambers possessed organic linings. Experiments in which foraminiferal shells were dissolved in dilute acid showed the latter alternative to be correct (see Sarjeant 1992*b*, pp. 507, 508).

A first classification of foraminiferal linings was proposed by Ferenc Góczan (1962), who described five coiled types. Stefan Mackó (1963) and M. H. Déak (1964) likewise proposed formal classifications, but this approach was rejected by Helen Tappan and Alfred R. Loeblich Jr (1965), who preferred to place them instead into the existing classification of foraminifera. (For a history of subsequent developments, see R. P. W. Stancliffe 1996.)

Little attention was paid to melanosclerites during this period. Eisenack (1963*c*), who had elevated them to the status of an Order Melanoskleritoitidea *incertae sedis*, described two new genera and reported further discoveries in 1971. New forms were described by Hanna Górka (1971) from the Polish Ordovician and by R. Pichler (1971) from the German Devonian. However, no progress was to be made in their interpretation until the 1990s (see Cashman 1996).

Three other groups characterized during this period – the 'pyritospheres' of Love (1958), the 'anellotubulates' of Otto Wetzel (1967) and the 'linolotypes' of Eisenack (1962) – have been subsequently shown to be pseudofossils, artifacts of bacterial action or chemical processing (see Love 1962; Sarjeant 1992*b*, pp. 513–514; Miller & Jansonius 1996). In contrast, several hitherto undescribed types of microfossils were distinguished for the first time, including arthropod cuticular fragments (Eisenack 1956; W. D. I. Rolfe 1962; Taugourdeau 1967), early growth stages of graptolites (Eisenack 1959, 1971) and possible eggs of polychaetes (Kozlowski 1974).

Studies of Precambrian palynomorphs were begun by Lucien Cayeux (1894), who reported what he believed to be radiolarians from the Brioverian (late Precambrian) of Brittany, France. Deflandre (1949) showed this to be erroneous, considering instead that the Brioverian forms were hystrichospheres. Raimond Hovasse (1956) reported Precambrian forms from the Ivory Coast. Subsequent studies by Maurice J. Graindor (1956, 1957), and by Deflandre himself (1955, 1957) resulted in the recognition of further taxa; all would later be called acritarchs.

It was Timofeyev who discovered the rich Sinian and Riphean (Late Precambrian) microfloras of eastern Europe, western Russia, Ukraine and China (1959, 1966, 1969, 1973; Timofeyev *et al.* 1976). Most of the microfossils he reported were of quite large size (up to 1 mm in cross-measurement), spheroidal to ovoidal, with single or double walls and a reduced ornament. Since their affinity is questionable, they have usually been placed into the acritarch subgroups Sphaeromorphitae and Disphaeromorphitae. Timofeyev's studies were extended by N. A. Volkova (1968, 1969); his work is assessed by Jankauskus & Sarjeant 2001.

A much more diverse palynoflora was reported by J. William Schopf (1968), the son of James Schopf, from the Late Precambrian Bitter Springs Formation of central Australia. This included cyanobacteria and a variety of other types of solitary or chain-forming organisms, as well as solitary forms doubtfully compared with simple dinoflagellates.

The first record of earlier Precambrian micro-organisms came with the examination by Elso S. Barghoorn and M. A. Tyler (1962, 1965) of cherts from the Palaeoproterozoic Gunflint Formation of southern Ontario, Canada. The Gunflint microflora includes filaments (*Gunflintia*), ellipsoidal structures (*Huroniospora*) and a variety of other morphotypes. Subsequently, in co-operation with Schopf, Barghoorn reported 'three billion year old' micro-organisms from the Precambrian of South Africa (Barghoorn & Schopf 1966; Schopf & Barghoorn 1967). Despite the excitement caused by these discoveries (e.g. Cloud 1965), serious work on Precambrian palynofloras was to continue at only a slow pace until the 1980s. (Subsequent discoveries are reviewed by Knoll 1996.)

General developments in palynology (1945–1970)

Before 1945, only two textbooks had been published in palynology, and both of these –

Wodehouse's *Pollen Grains* (1935) and Erdtman's *An Introduction to Pollen Analysis* (1943) – were concerned almost wholly with actuopalynology, as was Kurt Faegri and Johs. Iversen's *Textbook of Modern Pollen Analysis* (1950). Erdtman's *Pollen Morphology and Plant Taxonomy*, published in four volumes (1952–1965; Erdtman & Sorsa 1971), was vastly larger in content, but scarcely broader in scope.

Though there had been earlier newsletters for pollen specialists, there was no journal concentrating on micropalaeontology, let alone on palynology, and there were no societies with a palynological focus. In consequence, papers on palynology were published in a wide variety of journals, mostly with a national, rather than an international, circulation. Illustration was always restricted, because of high costs; far too many published photographs, and even drawings, were so small as to render crucial features of morphology hard to discern. (My own earliest papers suffered badly from this particular blight; see Sarjeant 1959, 1960*b*, *c*). Consequently, when the Palaeontological Association was formed in Great Britain in 1957, a particular aim was to produce a journal with ampler plates of higher quality. The plates in its journal *Palaeontology*, initially produced by the excellent (albeit now outdated) collotype process, were a revelation.

In other regards, changes also did not come quickly. Improvements in microscopic equipment was slow. Eisenack took his photographs using a Leitz monocular microscope, to which he attached a box camera fashioned from a biscuit tin and furnished with glass negatives (see Gocht & Sarjeant 1983, p. 473). The camera which I fitted to the monocular petrological microscope for my own early studies (between 1956 and 1959) used film, but was not in other respects an improvement. The development in the early 1960s of such fine instruments as the various Zeiss photomicroscopes, in combination with improved techniques of palynological preparation (see Wood *et al.* 1996, for discussion), was an enormous advance.

The first journal to deal specifically with microfossils was *The Micropaleontologist*, scarcely more than a newsletter and essentially without illustrations of quality. The launching, by the American Museum of Natural History in 1955, of the successor journal *Micropaleontology* marked a large step forward; however, though papers on palynology have appeared in that journal in increasing numbers, its emphasis has always been on microfossils with mineralized walls. A year earlier, Erdtman had launched in Sweden *Grana Palynologica* (now *Grana*), the first journal truly devoted to palynology; though featuring papers on other groups and themes from time to time, it has always been concerned primarily with pollen and spores and with actuopalynology. The coverage of the French journal *Pollen et Spores*, inaugurated in 1959, was virtually restricted to those themes.

The first textbook in which palynomorphs, other than spores and pollen, gained extensive treatment was Erdtman's *Handbook of Palynology* (1969), to which I contributed on his invitation a 90-page 'Appendix' on other groups of palynomorphs. Yet this was still outside the main text – almost an afterthought. Much more balanced in treatment was a work published almost simultaneously, *Aspects of Palynology* (edited by Tschudy & Richard A. Scott 1969), in which tasmanitids and acritarchs were treated incidentally in several chapters, with a contribution on Precambrian and Palaeozoic microfloras by James M. Schopf and one on dinoflagellates and other marine palynomorphs by Evitt.

The earliest national society was the Palynological Society, formed in India in 1964. It published two journals, the *Palynological Bulletin* and the *Journal of Palynology*; both were started in 1965, combining under the latter title in 1972. Another Indian journal, *The Palaeobotanist*, continues to be published by the Birbal Sahni Institute of Palaeobotany and, in recent years at least, has frequently featured palynological papers.

International gatherings of palynologists began with a semiformal meeting in Stockholm during the VIIth International Botanical Congress, with Erdtman as host. However, not till twelve years later did Kremp organize the First International Conference on Palynology, held in Arizona in 1962 with around 100 participants. At the Second International Conference on Palynology, staged in Utrecht, The Netherlands, in 1966, I was one of some 150 participants who contributed a paper which, we understood, would be published in a special conference volume. Instead, after we had surrendered the rights in our papers to the conference's organizing committee, we were disconcerted to discover that they were to constitute the early parts of a new Elsevier journal, the *Review of Palaeobotany and Palynology* (first published in 1967). Two useful bibliographies, of palaeopalynology by A. A. Manten (1969) and of actuopalynology by O. K. Hulshof & Manten (1971), were among its subsequent contents.

I was also a participant in the gathering of 35 palynologists at Tulsa, Oklahoma, in December 1967, which inaugurated the second palynological society, the American Association of

Stratigraphic Palynologists (AASP; see Traverse & Sullivan 1983; Sarjeant 1998). It held its earliest annual meetings successively at Louisiana State University (LSU), Baton Rouge (1968), Pennsylvania State University (1969) and the University of Toronto (1970), the papers presented being published as volumes of the LSU series *Geoscience and Man*.

In two papers by Manten (1968, 1970), the numbers of papers published in palynology and its subdisciplines were reviewed and the results presented in diagrammatic form. The absolute number had grown from less than 50 in 1916–1920 to around 5750 in 1961–1966; 34% of these papers were in English, 22% in Russian, 15.5% in German and 19.5% in French.

To try to cope with this volume of publications, various compilative series were established. Potonié's seven-volume *Synopsis der Gattungen der Sporae dispersae* was the first (see p. 285). The *Catalog of Spores and Pollen* was begun by Gerhard Kremp and others in 1957; Kremp's *Morphologic Encyclopedia of Palynology* (1965) also remains useful. Deflandre and his wife, Marthe Deflandre-Rigaud (see Sarjeant 1991*b*), produced for many years a *Fichier micropaléontologique générale* which included dinoflagellates and acritarchs in its coverage; and Eisenack inaugurated in 1964 his *Katalog der fossilen Dinoflagellaten, Hystrichosphären und verwandten Mikrofossilien.*

After 1970: changes and prospects

If I had attempted to continue my history of palynology from 1970 to the present, this paper would have been at least thrice its present length. A number of new groups of microfossils have been recognized, in particular of green and blue-green algae. The classification of living and fossil dinoflagellates, long a cause of taxonomic problems and conceptual controversy, seems at last to have stabilized (see Fensome *et al.* 1993). Though there have been immense advances in the understanding of the detailed structure and actions of living pollen and spores, through the work of John Rowley and others, the bases of nomenclature and classification for fossil pollen and spores remain in dispute and, indeed, the names sometimes change according to the level of the geological column which is under study, without any corresponding morphological changes.

Off-shore records of palynomorphs from samples and cores had been published earlier (e.g. Wilson & Hoffmeister 1955; Stanley 1967*b*, 1969; D. B. Williams 1968), but it was during this period that geology truly expanded into the oceans and palynology became a staple means of correlation of submarine sediments. This expansion was presaged by the work of Daniel Habib (1969, 1970). In particular, wide ranging studies resulted from the international Deep Sea Drilling Project, in which Habib was an early participant (1972). Information is now available concerning the sequences of palynomorphs in all the world's oceans.

All in all, this is an exciting period in the history of palaeopalynology; yet there are major problems. The importance of palynomorphs for biostratigraphical correlation and interpretation of past environments is recognized nowadays by oil companies, local and national geological surveys, and a variety of other bodies concerned with geological and environmental matters. This has generated an ever-growing flood of palynological literature, even though some companies and organizations still prefer to keep their results confidential and all too many theses and dissertations lie unpublished on university shelves.

Enhanced processing methods and improved microscopical equipment have facilitated researches on the detailed structure of palynomorphs; phase contrast, Nomarski-interference contrast, confocal laser and scanning-electron microscopy have brought especially major advances. An inevitable corollary is the proliferation of taxa, some of them differentiated on such fine details as to mean that they can only be recognized when specimens are exceptionally well preserved and ideally oriented. (For example, some dinoflagellate generic names are determined entirely by the relative portion of certain small plates, the plates themselves being in most instances visible only with difficulty, if at all.)

To keep abreast with an expanding nomenclature, such compilative works as the glossaries of dinoflagellate terminology (G. L. Williams *et al.* 1973, 2000), the series of indices to dinoflagellate taxa begun by J. K. Lentin and Williams in 1973 and of acritarch and prasinophyte taxa by Fensome *et al.* (1990), and the continuation of the Eisenack 'Katalog' (Fensome *et al.* 1991 and later parts) are truly invaluable. Unfortunately, though a number of databases concerning pollen and spores are available – for example, Kremp's Palynodata and the AASP's Palydisks furnish valuable reference compilations, while the PalSys computer database of the Laboratory of Palaeobotany and Palynology, Utrecht, brings together figures and text of published taxa – no similarly authoritative analytical guides are currently available to taxa of pollen and spores. Though Erdtman's *Handbook* was reissued in an enlarged edition (edited by Nilsson & Praglowski 1992), only one new single volume textbook on paleopalynology has appeared (Traverse 1988).

Late in 1974, my textbook on *Fossil and Living Dinoflagellates* was published, the first on this theme. Subsequently, David L. Spector (1984) and Taylor (1987) published compilations of papers, largely on living dinoflagellates, and Evitt (1985) furnished an extended account of what he termed *Sporopollenin Dinoflagellate Cysts*. A collection of important papers on all aspects of palynology, edited by Marjorie D. Muir and me, appeared in 1977.

In 1998 the AASP produced a comprehensive survey of information on *Palynology: Principles and Practice*, under the editorship of Jansonius and McGregor. The size of this work – three volumes and 1400 pages – is indicative of the growth of the field in the 31 years since the Association was formed.

The circumstances of publication are changing. New journals devoted partially or entirely to palynomorphs have appeared: of these, the *Revue de Micropaléontologie* in France, the *Revista Española de Micropaleontología* in Spain, the *Journal of Micropalaeontology* in the United Kingdom and the AASP journal *Palynology* in the USA are the most important. Unfortunately, declining library budgets in universities and institutions, in combination with a growing tendency of companies to use consultants rather than employing full-time palynologists, has meant wholesale cancellations of journal subscriptions. This is already forcing some journals and serials to cease publication. (The *Catalogue of Spores and Pollen* foundered in 1985, *Pollen et Spores* in 1991, while certain other journals are nowadays appearing with dismaying irregularity.)

Computer accessing of data is certainly an available alternative, but the consequent high investment of funds and of personnel time mean that research by individuals outside large institutions is becoming increasingly difficult. It may be, indeed, that future researches will be done entirely outside the academic milieu. However, I trust not, since company and institutional requirements are inevitably focused so much on the financial bottom-line that little opportunity is afforded for the investigation of such matters as taxonomy and evolution, or even for innovations in technique, unless these are considered likely to yield future profits. Stronger associations between universities and industry may offer a partial solution, even though such arrangements must, to some extent at least, compromise academic freedom.

The development of palynology: an overview

Though the study of the dust that includes spores and pollen grains was begun quite early in the history of microscopy, it assumed importance only during the second half of the twentieth century.

Before 1930, quite a lot had been learned concerning the reproductive function of these minute organic structures. Their significance in plant development and classification had been recognized and it had been realized that the inhalation of pollen could cause medically adverse effects. Spores had been recovered from sediments as ancient as the Devonian, as had prasinophytes (though the latter were not yet distinguished taxonomically). Dinoflagellate cysts, plus some still-mysterious spine-bearing microfossils, had been discovered in Mesozoic and Tertiary sediments. However, though the significance of pollen grains as climatic indices in Quaternary terrestrial sediments had been perceived, palynomorphs were in general receiving little attention from scientists at large.

It was only after 1930 that their true geological potential came to be perceived. Yet progress was slow at first. Researches in Germany, Great Britain and the United States demonstrated the value of pre-Quaternary spores and pollen in the correlation of lignites and coals and showed their usefulness in the tracing of economic deposits underground. Investigations by company geologists were foreshadowing their use in the determination of subsurface structures, and thus in the search for oil and natural gas reservoirs. Even so, it was not until after World War II that their practical application was to become widespread.

The use of pollen in the investigation of Quaternary deposits progressed faster, not merely as a tool for recognizing ancient environments but also for establishing relative dates of sediments and shell-beds. Before World War II, this was being done frequently; after that sad episode, it came to be done routinely. The construction of pollen 'spectra' provided visual references that could be employed by persons with minimal scientific training, facilitating greatly the work of prehistorians and archaeologists. New applications were developed: the allocation of dates to the spread of agricultural practices; elucidation of the diet of extinct animals and ancient humans; determination of the source and purity of honey; the identification of allergens and the demonstration of a link between fossil spore concentrations and silicosis among coal miners; even the use of palynomorphs as evidence in crime investigation.

The study of marine palynomorphs lagged behind that of terrestrial forms. The 'xanthidia' – the spiny bodies that had puzzled Victorian microscopists – came to be renamed 'hystrichospheres', but their nature only began to be comprehended 30 years later. Even after the majority of post-Palaeozoic forms had been shown to be dinoflagellate cysts, the affinity of the residue – the acritarchs – remained long in question. (Indeed, it is only now being elucidated with any confidence). Certain other groups of marine palynomorphs – notably the prasinophytes and the scolecodonts – had been discovered before 1930, but attracted little study until several more decades had elapsed. The chitinozoans were first reported in the 1930s but, even though it now seems clear that they are an independent group of micro-organisms, their affinity is still being questioned. A variety of other groups of palynomorphs were discovered during that period and later, but most of them attract only intermittent study, even today.

The employment of marine palynomorphs for purposes of biostratigraphical correlation really only began in the 1960s. Two factors favoured their use. Their distribution through a broader range of sediment types than those containing calcareous microfossils made them utilizable in samples from which foraminifera and ostracodes could not be extracted. Moreover, their much higher concentration meant that a single gram of sediment might yield in excess of 100 000 specimens, whereas a much larger sample would yield a very much smaller number of those larger microfossils. This was an especial advantage in making subsurface correlations of samples from small-diameter borehole cores or from sidewall cuttings. In consequence, the examination of marine palynomorphs came to be a basic means of dating samples in subsurface investigations by oil companies and consultants.

The presence of palynomorphs of simple character in Early to Middle Proterozoic sediments proved interesting but not stratigraphically helpful, since morphological variation was limited and their evolution relatively slow. However, from the latest Proterozoic to the Late Devonian, the number and variety of acritarch taxa, and their quite rapidly changing morphology, has made them highly suitable for stratigraphical correlation. From earliest Ordovician to Devonian, the information thus gained can be supplemented by study of chitinozoans – a group whose simplicity of morphology and rapid evolution means that even an untrained beginner, if furnished with a correlation chart, can quickly assign dates to samples.

In contrast, marine palynomorphs have, as yet, been only sparsely reported from Carboniferous and Permian strata. At those levels and in the later Devonian, correlation is best done using spores and pollen, and indeed, during those time intervals, continental sediments are both more widely exposed, and more economically important, than marine sediments. That picture does not change in the earliest Triassic. In contrast, from the Middle Jurassic to the present, dinoflagellate assemblages are rich, varied and rapidly changing, making them ideal for surface and subsurface stratigraphical correlation. Moreover, though they do not characterize depth zones in the oceans so clearly as do foraminifera, the dinoflagellates are being regularly used in the interpretation of marine environments.

Even in the 1930s, only a handful of persons worldwide were engaged in palynological studies. By the 1950s, yes, the number had grown, but it remained small. Increasing recognition of the importance of palynology is made apparent by the immense growth in the memberships of the American Association of Stratigraphical Palynologists; this now has over 600 individual members, even though its membership is preponderantly North American and includes few medical practitioners of palynology.

The last 70 years, then, have seen palynology grow from the esoteric pursuits of a few into the day-to-day activity of hundreds – from a scientific backwater into a mainstream of research. Whatever the future holds, the study of palynomorphs will surely continue to be of inestimable value to humanity.

This paper grew out of an invitation from D. R. Oldroyd, to give a historical presentation on palynology at the International Geological Congress in Rio de Janeiro – a meeting in which, for reasons unimportant now, I felt unable to participate. The opportunity to write it came through an accident on fieldwork in Korea, which kept me housebound for several late-summer weeks. During that time, I was aided greatly by the daily visits and other assistance of my research assistant, J. W. C. Sharp. This research, and my other work, has been supported by Operating Grant No. 8,393 of the National Science and Engineering Research Council of Canada.

It should be noted that the portraits contained herein are primarily those of spore-pollen palynologists. Portraits of dinoflagellate/acritarch specialists have been presented by me in an earlier paper (Sarjeant 1998). I should like to have featured more portraits of palynologists working on other groups of palynomorphs, but these were not readily available and time constraints prevented any prolonged search for them.

References

AGELOPOULOS, J. 1967. *Hystrichosphären, Dinoflagellaten und Foraminiferen aus dem eozänen Kieselton von Heiligenhafen, Holstein*. Mathematisch-naturwissenschaftlichen Fakultät der Eberhard-Karls-Universität, Tübingen, Germany, 74 p.

AĞRALI, B. 1963. Étude des microspores du Namurien à Taarlağzi (Bassin houiller d'Amasre, Turquie). *Annales de la Société géologique du Nord*, **83**, 145–159.

ALBERTI, G. 1959*a*. Über *Pseudodeflandrea* n.g. (Dinoflag.) aus dem Mittel-Oligozän von Norddeutschland. *Mitteilungen aus dem Geologischen Staatsinstitut Hamburg*, **28**, 91–92.

ALBERTI, G. 1959*b*. Zur Kenntnis der Gattung *Deflandrea* (Dinoflag.) in der Kreide und im Alttertiär Nord-und Mitteldeutschlands. *Mitteilungen aus dem Geologischen Staatsinstitut Hamburg*, **28**, 93–105.

ALBERTI, G. 1961. Zur Kenntnis mesozoischer und alttertiärer Dinoflagellaten und Hystrichosphaerideen von Nord-und Mitteldeutschland sowie einigen anderen Europäischen Gebieten. *Palaeontographica*, Series A, **116**, 1–58.

ALPERN, B. & LIABEUF, J. J. 1966. Zonation palynologique du bassin houiller Lorrain. *Zeitschrift der Deutschen geologischen Gesellschaft*, **117**, 162–177.

BAKER, G. 1973. Dr. Isabel Clifton Cookson. *In*: GLOVER, J. E. & PLAYFORD, G. (eds) *Mesozoic and Cainozoic Palynology: Essays in Honour of Isabel Cookson*. Geological Society of Australia, Special Publication no. **4**, i–x.

BALME, B. E. 1952. The principal microspores of the Permian coals of Collie. *Bulletin of the Geological Survey of Western Australia*, no. **105**, 164–201.

BALME, B. E. 1970. Palynology of Permian and Triassic strata in the Salt Range and Surghar Range, West Pakistan. *In*: KUMMEL, B. & TEICHERT, C. *Stratigraphic Boundary Problems: Permian and Triassic of West Pakistan*. University of Kansas Department of Geology Special Publication no. **4**. University of Kansas Press, Lawrence, 305–453.

BALTEŞ, N. 1963. Dinoflagellate şi Hystrichosphaeride ćretacice din Platforma moezică. *Petrol şi Gaze*, **14**, 581–597.

BALTEŞ, N. 1964. Albian microplankton from the Moesic Platform, Rumania. *Micropaleontology*, **13**, 324–336.

BALTEŞ, N. 1965. Observaţii asupra microflorei cretacice inferioare din zona R. Biraz. *Petrol şi Gaze*, **16**, 3–17.

BARGHOORN, E. S. & SCHOPF, J. W. 1966. Micro-organisms three billion years old from the Precambrian of South Africa. *Science*, **152**, 758–763.

BARGHOORN, E. S. & TYLER, S. A. 1962. Microfossils from the Middle Precambrian of Canada. *Abstracts, International Conference on Palynology, Tucson, Arizona, 1962*. (Republished in *Pollen et Spores*, **4**, 331).

BARGHOORN, E. S. & TYLER, S. A. 1965. Microorganisms from the Gunflint Chert. *Science*, **147**, 563–577.

BARSS, M. S. 1967. *Illustrations of Canadian fossils: Carboniferous and Permian spores of Canada*. Geological Survey of Canada, Paper, no. **67–11**, 94 pp.

BATTEN, D. J. & GRENFELL, H. R. 1996. *Botryococcus*. *In*: JANSONIUS, J. & MCGREGOR, D. C. (eds) *Palynology: Principles and Practice*, Vol. 1, *Principles*. American Association of Stratigraphic Palynologists, College Station, Texas, 205–214.

BEJU, D. 1969. Jurassic microplankton from the Carpathian foreland of Romania. *Colloquium on the Mediterranean Jurassic, Budapest, 3–8 September 1969*, 1–25.

BEJU, D. & DĂNET, N. 1962. Chitinozoare siluriene din Platforma Moldovenească şi Platforma Moezică. *Petrol şi Gaze*, **13**, 527–536.

BEKKER, H. 1921. The Kuckers stage of the Ordovician rocks of NE Estonia. *Acta et Commentationes Universitatis Dorpatensis*, Series AII, **1**, 1–91.

BELSKY, C. V., BOLTENHAGEN, E. & POTONIÉ, R. 1965. Sporae dispersae der Oberen Kreide von Gabun, Äquatoriales Afrika. *Paläontologische Zeitschrift*, **39**, 72–83.

BENNIE, J. & KIDSTON, R. 1886. On the occurrence of spores in the Carboniferous formations of Scotland. *Proceedings of the Royal Physical Society, Edinburgh*, **9**, 82–117.

BHARDWAJ, D. C. 1962. The miospore genera in the coals of Raniganj stage (Upper Permian), India. *Palaeobotanist*, **9**, 68–106.

BINNEY, E. W. 1871. Observations on the structure of fossil plants found in the Carboniferous strata II. *Lepidostrobus* and some allied cones. *Palaeontographical Society Monographs*, **24**, 33–62.

BLACKBURN, K. B. 1936. *Botryococcus* and the algal coals. I. A re-investigation of the alga *Botryococcus braunii* Kützing. *Transactions of the Royal Society of Edinburgh*, **58**, 841–854.

BOEKEL, N. M. VAN 1963. Uma nova espécie de *Tasmanites* do Devoniano do Pará. *Anais da Academia Brasileira de Ciencias*, **35**, 353–355.

BOLKHOVITINA, N. A. 1961. *Fossil and recent spores of the Schizaeacae*. Trudy Geologicheskogo Instituta, Akademiya Nauk S. S. S. R., **40**, 176 pp (in Russian).

BOLKHOVITINA, N. A. 1968. *The spores of the family Gleicheniaceae and their importance for stratigraphy*. Trudy Geologicheskogo Instituta, Akademiya Nauk S. S. S. R., **186**, 116 pp (in Russian).

BONEHAM, R. F. 1967. Hamilton (Middle Devonian) Chitinozoa from Rock Glen, Arkona, Ontario. *American Midland Naturalist*, **178**, 121–125.

BONEHAM, R. F. 1969. Middle Devonian (Erian) chitinozoan casts from silica, Lucas County, Ohio. *Journal of Paleontology*, **43**, 527–528.

BOSTOCK, J. 1819. Case of a periodical affliction of the eyes and chest. *Medico-Chirurgical Transactions*, **10**, 161–165.

BOSTOCK, J. 1828. On catarrhus aestivus, or summer catarrh. *Medico-Chirurgical Transactions*, **12**, 437–446.

BOUCHÉ, P. M. 1965. Chitinozoaires du Silurien s.l. du

Djado (Sahara Nigerien). *Revue de Micropaléontologie*, **8**, 151–164.

BRADBURY, S. 1967. *The Evolution of the Microscope*. Pergamon Press, London, 357 pp.

BRATSHEVA, G. M. 1969. *Palynological Studies of Upper Cretaceous and Paleogene of the Far East*. Akademiya Nauk S. S. S. R., Ordena Trudovogo Kradnogo Znameni. Trudy Geologikheskii Institut, no. **207**, 56 pp. (in Russian).

BRITO, I. M. 1965. *Nôvos microfosseis Devonianos do Maranhão*. Publicación Avulso, Escola de Geológia, Universidad de Bahia, **2**, 4 pp.

BRITO, I. M. & SANTOS, A. S. 1965. Contribuição do conhecimento dos microfosseis Silurianos e Devonianos as Bacia do Maranhão. *Notas Preliminares e Estudos. Servico Geologico e Mineralogico Brasil*, **129**, 1–21.

BROSIUS, M. & BITTERLI, P. 1961. Middle Triassic hystrichosphaerids from salt-wells Riburg–15 and –17, Switzerland. *Verhandlungen der Schweizerischen Petroleum-Geologen und Ingenieren, Bulletin*, **28**, 33–49.

BROWN, A. C., KNIGHTS, B. A. & CONWAY, E. 1969. Hydrocarbon content and its relationship to physiological state in the green alga *Botryococcus braunii*. *Phytochemistry*, **8**, 543–547.

BRYANT, V. M., Jr 1996. Forensic studies in palynology. *In*: JANSONIUS, J. & MCGREGOR, D. C. (eds) *Palynology: Principles and Practice*, Vol. 3, *New Directions, Other Applications and Floral History*. American Association of Stratigraphic Palynologists Foundation, College Station, Texas, 957–960.

BRYANT, V. M., Jr & HOLLOWAY, R. G. 1996. Archaeological palynology. *In*: JANSONIUS, J. & MCGREGOR, D. C. (eds) *Palynology: Principles and Practice*, Vol. 3, *New Directions, Other Applications and Floral History*. American Association of Stratigraphic Palynologists Foundation, College Station, Texas, 913–918.

BRYANT, V. M., Jr, JONES, J. G. & MILDENHALL, D. C. 1990. Forensic palynology in the United States of America. *Palynology*, **14**, 193–208.

BROZOWSKA, M. & ŽOLDANI, Z. 1958. Uwagio zasięgu stratigraficznym niektorých gatunków megaspor karbońskich. *Kwartalnik Geologiczny*, **2**, 515–531.

BURGER, D. 1966. Palynology of the uppermost Jurassic and lowermost Cretaceous strata in the eastern Netherlands. *Leidse geologische Mededelingen*, 209–276 (also issued separately).

BUTTERWORTH, M. A. 1966. The distribution of densospores. *Palaeobotanist*, **15**, 16–28.

CALANDRA, F. 1964. Sur un présumé dinoflagellé, *Arpylorus* nov. gen. du Gothlandien de Tunisie. *Comptes-rendus hébdomadaires de l'Académie des Sciences, Paris*, **258**, 4112–4114.

CAMERARIUS, R. J. 1692. *De sexu plantarum espistola*. Academiae Caesareo Leopold, Tübingen, 110 pp.

CANE, R. F. 1969. Coorongite and the genesis of oil shale. *Geochimica et Cosmochimica Acta*, **33**, 257–265.

CARRUTHERS, W. 1865. On an undescribed cone from the Carboniferous beds of Airdrie, Lanarkshire: *Flemingites gracilus*. *Geological Magazine*, **2**, 433–440.

CASHMAN, P. B. 1990. The affinity of the chitinozoans; new evidence. *Modern Geology*, **5**, 59–69.

CASHMAN, P. B. 1996. Melanosclerites. *In*: JANSONIUS, J. & MCGREGOR, D. C. (eds), *Palynology: Principles and Practice*, Vol. 1, *Principles*. American Association of Stratigraphic Palynologists Foundation, College Station, Texas, 365–372.

CAYEUX, L. 1894. Les preuves de l'existence d'organismes dans les terrains précambriens. Premier note sur les Radiolaires précambriens. *Bulletin de la Société Géologique de France*, **22**, 197–228.

CHATEAUNEUF, J. J. & GRUAS-CAVAGNETTO, C. 1968. Étude palynologique du Paléogène de quatre sondages du Bassin Parisien; Chaignes, Montjavoult, Le Tillet, Ludes. *In Colloque sur l'Éocène*. Mémoires du Bureau de Recherches Géologiques et Minières, Paris, **59**, 114–152.

CHRISTIANSEN, T. 1962. *Botanik*, Vol. 2, *Systematik Botanik*, No. 2 *Alger*. Munksgaard, Copenhagen, 178 pp.

CHURCHILL, D. M. 1960. Living and fossil unicellular algae and aplanospores. *Nature*, **186**, 493–496.

CHURCHILL, D. M. & SARJEANT, W. A. S. 1962. Fossil dinoflagellates and hystrichospheres in Australian freshwater deposits. *Nature*, **194**, 1094.

CHURCHILL, D. M. & SARJEANT, W. A. S. 1963. Freshwater microplankton from Flandrian (Holocene) peats of south Western Australia. *Grana palynologica*, **3**, 29–53.

CLARKE, R. F. A. 1965. British Permian saccate and monosaccate miospores. *Palaeontology*, **8**, 322–354.

CLARKE, R. F. A. & VERDIER, J. P. 1967. An investigation of microplankton assemblages from the chalk of the Isle of Wight, England. *Verhandelingen der Koninkijke Nederlandse Akademie van Wetenschappen*, 1st Series, **24**.

CLENDENING, J. A. 1962. Small spores applicable to stratigraphical correlation in the Dunkard Basin of West Virginia and Pennsylvania. *Proceedings of the West Virginia Academy of Sciences*, **34**, 133–142.

CLOUD, P. 1965. Significance of the Gunflint (Precambrian) microflora. *Science*, **148**, 27–35.

COCA, A. F., WALZER, M. & THOMMEN, A. A. 1931. *Asthma and Hay Fever in Theory and Practice*. Bailiere, Tindall & Cox, London, 1851 pp.

COLLINSON, C. & SCHWALB, H. 1955. *North American Paleozoic Chitinozoa*. Illinois Geological Survey Report of Investigations, **186**, 33 pp.

COLLINSON, C. & SCOTT, A. J. 1958. *Chitinozoan faunule of the Devonian Cedar Valley Formation*. Illinois Geological Survey Circular, **247**.

COOKSON, I. C. 1947. Plant microfossils from the lignites of Kerguelen Archipelago. *British, Australian and New Zealand Antarctic Research Expedition, 1929–31, Reports*, Series 12, **2**, 127–142.

COOKSON, I. C. 1950. Fossil pollen grains of proteaceous type from Tertiary deposits in Australia. *Australian Journal of Science*, Series B, **3**, 166–177.

COOKSON, I. C. & EISENACK, A. 1958. Microplankton from Australian and New Guinea Upper Mesozoic sediments. *Proceedings of the Royal Society of Victoria*, **70**, 19–79.

COOKSON, I. C. & HUGHES, N. F. 1964. Microplankton from the Cambridge Greensand (mid-Cretaceous). *Palaeontology*, **7**, 37–59.

COOKSON, I. C. & MANUM, S. 1960. On *Crassosphaera*, a new genus of microfossil from Mesozoic and Tertiary deposits. *Nytt Magasin for Botanikk,* **8**, 5–8.

COOPER, C. L. 1942. North American Chitinozoa. *Abstracts of the Paleontological Society, Geological Society of America*, **53**, 1828.

COUPER, R. A. 1953. Upper Mesozoic and Cainozoic spores and pollen grains from New Zealand. *New Zealand Geological Survey, Palaeontological Bulletin*, **22**.

COUPER, R. A. 1958. British Mesozoic microspores and pollen grains. A systematic and stratigraphic study. *Palaeontographica*, Series B, **103**, 75–179.

CRAMER, F. H. 1964. Microplankton from three Palaeozoic formations in the Province of Léon, N.W. Spain. *Leidse Geologische Mededelingen*, **30**, 253–360.

CRAMER, F. H. 1967. Chitinozoans of a composite section of Upper Llandoverian to basal lower Gedinnian sediments in northern Spain. A preliminary report. *Bulletin de la Société Belge de Géologie, de Paléontologie et d'Hydrologie*, **75**, 69–129.

CRAMER, F. H. 1970. Middle Silurian continental movement estimated from phytoplankton-facies transgression. *Earth and Planetary Sciences Letters*, **10**, 87–93

CRAMER, F. H. & DÍEZ DE CRAMER, M. 1968. Consideraciones taxonómicas sobre las acritarcas del Silúrio Medio y Superior del Norte de España. *Boletin Geologico y Minero, 79 Año*, 541–574.

CRAMER, F. H. & DÍEZ DE CRAMER, M. 1970. Acritarchs from the Lower Silurian Neahga Formation, Niagara Peninsula, North America. *Canadian Journal of Earth Sciences*, **7**, 1077–1085.

CRONEIS, C. & SCOTT, H. W. 1933. Scolecodonts. *Bulletin of the Geological Society of America*, **44**, 207.

CROPP, F. W. 1960. Pennsylvanian spore floras from the Warrior Basin, Mississippi and Alabama. *Journal of Paleontology*, **34**, 359–367.

CROPP, F. W. 1963. Pennsylvanian spore succession in Tennessee. *Journal of Paleontology*, **37**, 900–916.

CROSS, A. T. 1947. Spore floras of the Pennsylvanian of West Virginia and Kentucky. *Journal of Geology*, **55**, 285–308.

CROSS, A. T. (ed.) 1964. *Palynology in Oil Exploration; A Symposium*. Society of Economic Paleontologists and Mineralogists, Special Publication no. **11**, 200 pp.

CROSS, A. T. & KOSANKE, R. M. 1995. History and development of Carboniferous palynology in North America during the early and middle twentieth centrury. *In*: LYONS, P. C., MOREY, E. D. & WAGNER, R. H. (eds) *Historical Perspective of Early Twentieth Century Carboniferous Paleobotany in North America*. Geological Society of America, Boulder, Colorado; Memoir no. **185**, 353–387.

CROSS, A. T. & SCHEMEL, M.P. 1951. Representative microfossil floras of some Appalachian coals. *Congrès International de stratigraphie et de Géologie du Carbonifère, 1952*, 123–130.

DAEMON, R. P., QUADROS, L. F. & DA SILVA, L. C. 1967. Devonian palynology and biostratigraphy of the Paraná basin. *In*: BIGARELLA, J. J. (ed.) *Problems in Brazilian Devonian Geology*. Boletin Paraense de Geociências, **21–22**, 99–132.

DAUGHERTY, L. H. 1941. *The Upper Triassic flora of Arizona*. Carnegie Institution of Washington, Publications (Contributions to Paleontology), no. **526**, 108 pp.

DAVEY, R. J. 1969. Non-calcareous microplankton from the Cenomanian of England, northern France and North America. Pt. I. *Bulletin of the British Museum (Natural History), Geology*, **17**, 103–180.

DAVEY, R. J. 1970. Non-calcareous microplankton from the Cenomanian of England, northern France and North America. Pt. II. *Bulletin of the British Museum (Natural History), Geology*, **18**, 335–397.

DAVEY, R. J., DOWNIE, C., SARJEANT, W. A. S. & WILLIAMS, G. L. 1966. Fossil dinoflagellate cysts attributed to *Baltisphaeridium*. *Bulletin of the British Museum (Natural History), Geology*, Supplement **3**, 157–175.

DAVEY, R. J. & WILLIAMS, G. L. 1966*a*. The genera *Hystrichosphaera* and *Achomosphaera*. *Bulletin of the British Museum (Natural History), Geology*, Supplement **3**, 28–52.

DAVEY, R. J. & WILLIAMS, G. L. 1966*b*. The genus *Hystrichosphaeridium* and its allies. *In*: DAVEY, R. J., DOWNIE, C., SARJEANT, W. A. S. & WILLIAMS, G. L. Studies on Mesozoic and Cainozoic dinoflagellate cysts. *Bulletin of the British Museum (Natural History) Geology*, Supplement **3**, 53–106.

DAWSON, J. W. 1871*a*. On spore cases in coals. *American Journal of Science*, Series 3, **1**, 256–263.

DAWSON, J. W. 1871*b*. On spore cases in coals. *Canadian Naturalist*, **5**, 369–377.

DAWSON, J. W. 1886. On rhizocarps in the Erian (Devonian) period in America. *Bulletins of the Chicago Academy of Science*, **1**, 105–118.

DÉAK, M. H. 1964. Les Scytinascias. *Bulletin de la Société Géologique d'Hongrie*, **94**, 95–106.

DEANE, H. 1849. On the occurrence of fossil Xanthidia and Polythalamia in chalk. *Transactions of the Microscopical Society of London*, **2**, 77–79.

DE CONINCK, J. 1965. Microfossiles planctoniques de Sable Yprésien à Merelbeke. Dinophyceae et Acritarcha. *Mémoires de l'Académie Royale de Belgique (Classe des Sciences)*, **36**, 1–64.

DE CONINCK, J. 1969. Dinophyceae et Acritarcha de l'Yprésien du sondage de Kallo. *Mémoires del'Institut Royal des Sciences Naturelles de Belgique*, no. **161**, 5–67.

DE JERSEY, N. J. 1959. Jurassic spores and pollen grains from the Rosewood coalfield. *Queensland Government Mining Journal*, **60**, 346–366.

DE JERSEY, N. J. 1962. *Triassic Spores and Pollen Grains from the Ipswich Coalfield.* Geological Survey of Queensland Publications, **307**, 18 pp.

DEFLANDRE, G. 1933 . Note préliminaire sur un péridinien fossile, *Lithoperidinium oamaruense* n.g., n.sp. *Bulletin de la Société zoologique de France*, **58**, 265–273.

DEFLANDRE, G. 1935. Revue: Les microfossiles des

silex de la craie. *Bulletin de la Société Française de Microscopie*, **4**, 116–120.
DEFLANDRE, G. 1936. Microfossiles des silex crétacés Pt. I: Généralités, flagellés. *Annales de Paléontologie*, **25**, 151–191.
DEFLANDRE, G. 1937. Microfossiles des silex crétacés Pt. II: Flagellés *incertae sedis*, hystrichosphaeridés, sarcodinés, organismes divers. *Annales de Paléontologie*, **26**, 51–103.
DEFLANDRE, G. 1942. Sur les Hystrichosphères des calcaires siluriens de la Montagne Noire. *Comptes-rendus hébdomadaires de l'Académie des Sciences, Paris*, **215**, 475–476.
DEFLANDRE, G. 1945. Microfossiles des calcaires siluriens de la Montagne Noire. *Annales de Paléontologie*, **31**, 41–76.
DEFLANDRE, G. 1946. Radiolaires et hystrichosphaeridés du Carbonifère de la Montagne Noire. *Comptes-rendus hébdomadaires de l'Académie des Sciences, Paris*, **223**, 515–517.
DEFLANDRE, G. 1947*a*. Sur quelques micro-organismes planctoniques des silex jurassiques. *Bulletin de l'Institut Océanographique de Monaco*, no. **921**, 1–10.
DEFLANDRE, G. 1947*b*. Sur une nouvelle Hystrichosphère des silex crétacés et sur les affinités du genre *Cannosphaeropsis* O.We. *Comptes-rendus hébdomadaires de l'Académie des Sciences, Paris*, **224**, 1574–1576.
DEFLANDRE, G. 1947*c*. *Calciodinellum* nov. gen., premier représentant d'une famille nouvelle de Dinoflagellés fossiles à thèque calcaire. *Comptes-rendus hébdomadaires de l'Académie des Sciences, Paris*, **224**, 1781–1782.
DEFLANDRE, G. 1947*d*. Le problème des hystrichosphères. *Bulletin de l'Institut Océanographique de Monaco*, no. **918**, 1–23.
DEFLANDRE, G. 1948. Les Calciodinellidés: dinoflagellés fossiles à thèque calcaire. *Le Botaniste*, **34**, 191–219.
DEFLANDRE, G. 1949. Les soi-disant Radiolaires du Précambrien de Bretagne et la question de l'existence de Radiolaires embryonnaires fossiles. *Bulletin de la Société Zoologique de France*, **74**, 351–352.
DEFLANDRE, G. 1954. Systématique des Hystrichosphaeridés: sur l'acception du genre *Cymatiosphaera*, O. Wetzel. *Comptes-rendus sommaires des Séances de la Société Géologique de France*, no. **12**, 257–258.
DEFLANDRE, G. 1955. *Palaeocryptidium* n.g. *cayeuxi* n. sp., micro-organismes incertae sedis des phtanites briovériens bretons. *Comptes-rendus sommaires de la Société Géologique de France*, **9–10**, 182–184.
DEFLANDRE, G. 1957. Remarques sur deux genres de Protistes du Précambrien (*Arnoldia* Hovasse 1956, *Cayeuxipora* Graindor 1957). *Comptes-rendus hébdomadaires de l'Académie des Sciences, Paris*, **244**, 2640–2641.
DEFLANDRE, G. & COOKSON, I. C. 1954. Sur le microplancton fossile conservé dans diverses roches sédimentaires australiennes s'étageant du Crétacé inférieur au Miocène supérieur. *Comptes rendus hebdomadaires des séances de l'Académie des Sciences*, **239**, 1235–1238.
DEFLANDRE, G. & COOKSON, I. C. 1955. Fossil microplankton from Australian Late Mesozoic and Tertiary sediments. *Australian Journal of Marine and Freshwater Research*, **6**, 242–313.
DETTMAN, M. E. & PLAYFORD, G. 1962. Sections of some spores from the Lower Carboniferous of Spitzbergen. *Palaeontology*, **5**, 679–681.
DEUNFF, J. 1951. Sur la présence de micro-organismes (Hystrichosphères) dans les schistes ordoviciens du Finistère. *Comptes-rendus hébdomadaires de l'Académie des Sciences, Paris*, **233**, 321–323.
DEUNFF, J. 1954*a*. Sur le microplancton du Gothlandien armoricain. *Comptes-rendus sommaires de la Société Géologique de France*, no. **3**, 54–55.
DEUNFF, J. 1954*b*. Sur un microplancton du Dévonien du Canada recélant des types nouveaux d'Hystrichosphaeridés. *Comptes-rendus hébdomadaires de l'Académie des Sciences, Paris*, **239**, 1064–1066.
DEUNFF, J. 1957. Micro-organismes nouveaux (Hystrichosphères) du Dévonien de l'Amérique du Nord. *Bulletin de la Société Géologique et Minéralogique de Bretagne*, **2**, 5–14.
DEUNFF, J. 1961*a*. Quelques précisions concernant les Hystrichosphaeridés du Dévonian du Canada. *Comptes-rendus sommaires de la Société Géologique de France*, no. **8**, 216–218.
DEUNFF, J. 1961*b*. Un microplancton à Hystrichosphères dans le Tremadoc du Sahara. *Revue de Micropaléontologie*, **4**, 37–52.
DEUNFF, J. 1966. Acritarches du Dévonien de Tunisie. *Comptes-rendus sommaires de la Société Géologique de France*, **1**, 22–23.
DIETZ, L. F., SARJEANT, W. A. S. & MITCHELL, T. A. 2000. The dreamer and the pragmatist: A joint biography of Walter Wetzel and Otto Wetzel,with a survey of their contributions to geology and micropaleontology. *Earth Sciences History*, **18**, 4–50.
DIJKSTRA, S. J. 1946. Eine morphographische Bearbeitung des karbonischen Megasporen, mit besonderer Berücksichtigung von Südlimburg (Niederlande). *Mededelingen van de Geologische Stichting*, Series C, **3–1**, 101 pp.
DIJKSTRA, S. J. 1953. Some Brazilian megaspores, Lower Permian in age, and their comparison with Lower Gondwana megaspores from India. *Mededelingen van de Geologische Stichting*, New Series, **9**, 5–10.
DODEKOVA, L. 1967. Les dinoflagellés et acritarches de l'Oxfordien–Kiméridgien de la Bulgarie du nord-est. *Annals of the University of Sofia, Faculty of Geology and Geography*, Series 1, *Geology*, **60**, 9–30.
DODEKOVA, L. 1969. Dinoflagellés et acritarches du Tithonique aux environs de Pleven, Bulgarie central du nord. *Bulgarska Akademiya na Naukite, Izvestiya na geologischeskiya Institut, Seriya Paleontologiya*, **18**, 13–24.
DOUBINGER, J. & PONCET, J. 1964. Présence de nombreux chitinozoaires dans le Dévonien inférieur (Siégénien) du Cotentin. *Comptes-rendus sommaires de la Société géologique de France*, no. **3**, 104–105.
DOUGLAS, A. G., EGLINTON, G. & MAXWELL, J .R.

1969. The hydrocarbons of coorongite. *Geochimica et Cosmochimica Acta*, **33**, 569–577.
DOWNIE, C. 1957. Microplankton from the Kimeridge Clay. *Quarterly Journal of the Geological Society of London*, **112**, 413–434.
DOWNIE, C. 1958. An assemblage of microplankton from the Shineton Shales (Tremadocian). *Proceedings of the Yorkshire Geological Society*, **31**, 331–349.
DOWNIE, C. 1959. Hystrichospheres from the Silurian Wenlock Shale of England. *Palaeontology*, **2**, 56–71.
DOWNIE, C. 1962. So-called spores from the Torridonian: report of demonstration. *Proceedings of the Geological Society of London*, no. **1600**, 127–128.
DOWNIE, C. 1963. 'Hystrichospheres' (acritarchs) and spores of the Wenlock Shales (Silurian) of Wenlock, England. *Palaeontology*, **6**, 625–652.
DOWNIE, C. 1967. The geological history of the microplankton. *Review of Palaeobotany and Palynology*, **1**, 269–281.
DOWNIE, C. 1969. Palynology of the Chuaria Shales of the Grand Canyon. *In*: BASS, D. L. (ed.) *Geology and Natural History of the Grand Canyon Region. Guidebook to the 5th Field Conference*. Four Corners Geological Society, Flagstaff, Arizona, 121–122.
DOWNIE, C., EVITT, W. R. & SARJEANT, W. A. S. 1963. Dinoflagellates, hystrichospheres and the classification of the acritarchs. *Stanford University Publications in Geological Sciences*, **7**, 1–16.
DOWNIE, C. & SARJEANT, W. A. S. 1963. On the interpretation and status of some hystrichosphere genera. *Palaeontology*, **6**, 83–96.
DOWNIE, C. & SARJEANT, W. A. S. 1965. *Bibliography and Index of Fossil Dinoflagellates and Acritarchs*. Memoirs of the Geological Society of America, no. **96** (1964), 180 pp.
DOWNIE, C. & SARJEANT, W. A. S. 1966. The morphology, terminology and classification of fossil dinoflagellate cysts. *In*: DAVEY, R. J., DOWNIE, C., SARJEANT, W. A. S. & WILLIAMS, G. S. Studies on Mesozoic and Cairnozoic Dinoflagellate cysts. *Bulletin of the British Museum (Natural History), Geology*, Supplement **3**, 110–117.
DOWNIE, C., WILLIAMS, G. L. & SARJEANT, W. A. S. 1961. Classification of fossil microplankton. *Nature*, **192**, 471.
DOYLE, J. A. 1969. Cretaceous angiosperm pollen of the Atlantic coastal plain and its evolutionary significance. *Journal of the Arnold Arboretum*, **59**, 1–35.
DRUGG, W. S. 1967. Palynology of the Upper Moreno Formation (Late Cretaceous–Paleocene), Escarpado Canyon, California. *Palaeontographica*, Series B, **120**, 1–74.
DUISBURG, H. von. 1860. Urweltlicher Blütenstaub. *Neues Preussisches Provinzialblatt, III. Folge,* 296 (also in *Verschlag Deutschen Naturforscher u.s.w., Königsberg (1860),*1861, 291 and *Schriften der (Königlichen) Physikalisch-Okonomischen Gesellschaft zu Königsberg,* **31**, 31–32.
DULHUNTY, J. A. 1946. Principal microspore-types in Permian coals of New South Wales. *Proceedings of the Linnean Society of New South Wales*, **70**, 147–157.
DULHUNTY, J. A. 1947. Distribution of microspore types in New South Wales Permian coalfields. *Proceedings of the Linnean Society of New South Wales*, **71**, 239–251.
DULHUNTY, J. A. & DULHUNTY, R. 1949. Notes on microspore types in Tasmanian Permian coalfields. *Proceedings of the Linnean Society of New South Wales*, **74**, 132–139.
DUNN, D. L. 1959. Devonian chitinozoans from the Cedar Valley Formation in Iowa. *Journal of Paleontology*, **33**, 1001–1007.
DUNN, D. L. & MILLER, T. H. 1964. A distinctive chitinozoan from the Alpena Limestone (Middle Devonian) of Michigan. *Journal of Paleontology*, **38**, 725–728.
DURHAM, O. C. 1936. *Your Hay Fever*. Bobbs-Merrill, Indianapolis.
DURHAM, O. C. 1948. Aerobiology: development and technique. *In:* VAUGHAN, W. T. & BLACK, J. H. (eds), *Practice of Allergy*. Mosby, Saint Louis, Missouri, 451–464.
EHLERS, E. 1868*a*. Über eine fossile Funicee aus Solenhofen (*Eunicites aritus*), nebst Bemerkungen über fossile Würmer überhaupt. *Zeitschrift für wissenschaftliche Zoologie*, **18**, 421–443.
EHLERS, E. 1868*b*. Ueber fossiler Würmer aus dem lithographischen Schiefer in Bayern. *Palaeontographica*, **17**, 145–175.
EHRENBERG, C. G. 1837*a*. [Untitled]. *Verhandlungen der Preussischen Akademie der Wissenschaften*, 44.
EHRENBERG, C. G. 1837*b*. [Untitled]. *Königlich Preussische Akademie der Wissenschaften zu Berlin, Bericht über die zur Bekanntmachung geeigneten Verhandlungen*, 61.
EHRENBERG, C. G. 1838. Über das Massenverhältniss der jetzt lebenden Kiesel-Infusorien und über ein neues Infusorien-Conglomerat als Polierschiefer von Jastraba in Ungarn. *Königlich Akademie der Wissenschaften zu Berlin, Abhandlungen (1836)*, **1**, 109–135.
EHRENBERG, C. G. 1840. Über noch jetzt zahlreich lebende Thierarten der Kreidebildung und den Organismus der Polythalamien. *Königlich Preussische Akademie der Wissenschaften zu Berlin, Bericht über die zur Bekanntmachung geeigneten Verhandlungen (1839)*, 81–174.
EHRENBERG, C. G. 1843. Über einige Jura-Infusorien Arten des Corallrags bei Krakau. *Monatsberichte der Akademie der Wissenschaften, Berlin*, 61–63.
EHRENBERG, C. G. 1854. *Mikrogeologie: das Erden- und Felsen-schaffende Wirken des unsichtbaren kleinen selbständigen Lebens auf der Erde.* Leopold Voss, Leipzig, 374 + 31 + 88 pp.
EICHWALD, E. 1854. Die Grauwackenschichten von Lieu-und Esthland. *Bulletin de la Société Impériale des Naturalistes de Moscou*, **27**, 1–111.
EISENACK, A. 1931. Neue Mikrofossilien des baltischen Silurs. *Palaeontologische Zeitschrift*, **13**, 74–118.
EISENACK, A. 1932. Neue Mikrofossilien des baltischen Silurs II. *Paläontologische Zeitschrift*, **14**, 257–277.

EISENACK, A. 1935. Mikrofossilien aus Doggergeschieben Ostpreussens. *Zeitschrift für Geschiebeforschung und Flachlandsgeologie*, **11**, 167–184.

EISENACK, A. 1936*a*. Dinoflagellaten aus dem Jura. *Annales de Protistologie*, **5**, 59–63.

EISENACK, A. 1936*b*. *Eodinia pachytheca* n.g. n. sp., ein primitiver Dinoflagellat aus einem Kelloway-Geschiebe Ostpreussens. *Zeitschrift für Geschiebeforschung und Flachlandsgeologie*, **12**, 72–75.

EISENACK, A. 1938*a*. Die Phosphoritknollen der Bernsteinformation als Überliefer tertiären Planktons. *Schriften der Physikalisch-Ökonomischen Gesellschaft zu Königsberg*, **70**, 181–188.

EISENACK, A. 1938*b*. Hystrichosphaerideen und verwandte Formen im baltischen Silur. *Zeitschrift für Geschiebeforschungen und Flachlandsgeologie*, **14**, 1–30.

EISENACK, A. 1939. Einige neue Annelidenreste aus dem Silur und dem Jura des Baltikums. *Zeitschrift für Geschiebeforschung und Flachlandsgeologie*, **15**, 153–176.

EISENACK, A. 1942. Die Melanoskleritoiden, eine neue Gruppe silurisches Mikrofossilien aus dem Unterstamm der Nesseltiere. *Paläontologische Zeitschrift*, **23**, 157–180.

EISENACK, A. 1948. Mikrofossilien aus Kieselknollen des Böhmischer Ordoviziums. *Senckenbergiana*, **28**, 105–117.

EISENACK, A. 1951. Über Hystrichosphaerideen und andere Kleinformen aus Baltischem Silur und Kambrium. *Senckenbergiana*, **32**, 187–204.

EISENACK, A. 1954. Mikrofossilien aus Phosphoriten des samländischen Unter-Oligozäns und über die Einheitlichkeit der Hystrichosphaerideen. *Palaeontographica*, series A, **105**, 49–95.

EISENACK, A. 1955. Chitinozoen, Hystrichosphären und andere Mikrofossilien aus dem Beyrichia-Kalk. *Senckenbergiana lethaea*, **36**, 157–188.

EISENACK, A. 1956. Beobachtungen an Fragmenten von Eurypteriden-Panzern. *Neues Jahrbuch für Geologie und Paläontologie, Abhandlungen*, **104**, 119–120.

EISENACK, A. 1957. Mikrofossilien in organischer Substanz aus dem Lias Schwabens (Süddeutschland). *Neues Jahrbuch für Geologie und Paläontologie, Abhandlungen*, **105**, 239–249.

EISENACK, A. 1958*a*. Mikroplankton aus dem norddeutschen Apt nebst einigen Bemerkungen über fossile Dinoflagellaten. *Neues Jahrbuch für Geologie und Paläontologie, Abhandlungen*, **106**, 383–422.

EISENACK, A. 1958*b*. *Tasmanites* Newton 1875 und *Leiosphaeridia* n.g. als Gattungen der Hystrichosphaeridia. *Palaeontographica*, Series A, **110**, 1–19.

EISENACK, A. 1959. Einige Mitteilungen über Graptolithen. *Neues Jahrbuch für Geologie und Paläontologie, Abhandlungen*, **107**, 253–260.

EISENACK, A. 1960. Über einige niedere Algen aus dem baltischen Silur. *Senckenbergiana lethaea*, **41**, 13–26.

EISENACK, A. 1962. Mikrofossilien aus dem Ordovizium des Baltikums 2. Vaginatenkalk bis Lyckholmer Stufe. *Senckenbergiana lethaia*, **43**, 349–366.

EISENACK, A. 1963*a*. Hystrichosphären. Cambridge Biological Society, Biological Reviews, **38**, 107–139.

EISENACK, A. 1963*b*. Sind die Hystrichosphären Zysten von Dinoflagellaten? *Neues Jahrbuch für Geologie und Palaeontologie, Monatsschrift*, 1963(5), 225–231.

EISENACK, A. 1963*c*. Melanoskleriten aus anstehenden Sedimenten und aus Geschieben. *Paläontologisches Zeitschrift*, **37**, 122–134.

EISENACK, A. 1963*d*. Über einige Arten der Gattung *Tasmanites* Newton 1875. *Grana Palynologica*, **4**, Erdtman 65th Anniversary Volume, 204–216.

EISENACK, A. 1964. *Katalog der fossilen Dinoflagellaten, Hystrichosphären und verwandten Mikrofossilien*, Vol. 1, *Dinoflagellaten*. Schweizerbart, Stuttgart, 888 pp.

EISENACK, A. 1969. Zur Systematik einiger paläozoischer Hystrichosphären (Acritarcha) des baltischen Gebietes. *Neues Jahrbuch für Geologie und Paläontologie, Abhandlungen*, **133**, 245–266.

EISENACK, A. 1971. Die Mikrofauna der Ostseekalke (Ordovizium). 3. Graptolithen, Melanoskleriten, Spongien, Radiolarien, Problematika ... *Neues Jahrbuch für Geologie und Paläontologie, Abhandlungen*, **137**, 337–357.

EISENACK, A. 1972. Kritische Bemerkung zur Gattung *Pterospermopsis* (Chlorophyta, Prasinophyceae). *Neues Jahrbuch für Geologie und Paläontologie, Monatsschriften N.*, **10**, 596–601.

ELLER, E. R. 1933. An articulated annelid jaw from the Devonian of New York. *American Midland Naturalist*, **14**, 186.

ELSIK, W. C. 1966. Biologic degradation of fossil pollen grains and spores. *Micropaleontology*, **12**, 515–518.

ELSIK, W. C. 1968. Palynology of the Paleocene Rockdale Lignite, Milam County, Texas. *Pollen et Spores*, **10**, 263–314.

ERDTMAN, G. 1943. *An Introduction to Pollen Analysis*. Chronica Botanica, Waltham (Mass), 231 pp.

ERDTMAN, G. 1952–1965. *Pollen Morphology and Plant Taxonomy*. Vols 1–3. Almqvist & Wiksell, Stockholm, 539 + 127 + 131 pp.

ERDTMAN, G. 1967. Glimpses of palynology 1916–1966. *Review of Palaeobotany and Palynology*, **1**, 23–29.

ERDTMAN, G. 1969. *Handbook of Palynology: Morphology–Taxonomy–Ecology: An Introduction to the Study of Pollen Grains and Spores*. Munksgaard, Copenhagen, 486 pp.

ERDTMAN, G. & SORSA, P. 1971. *Pollen Morphology and Plant Taxonomy*, Vol. 4. Almqvist & Wiksell, Stockholm.

EVANS, P. R. 1961. Chitinozoans from Thangoo Nos. 1–1A. *In*: *Thangoo No. 1 and No. 1A Wells, Western Australia*. Commonwealth of Australia, Department of National Development, Bureau of Mineral Resources, Geology and Geophysics, Canberra, Publication no. **14**, 23.

EVITT, W. R. 1961. Observations on the morphology of fossil dinoflagellates. *Micropaleontology*, **7**, 385–420.

EVITT, W. R. 1963*a*. A discussion and proposals concerning fossil dinoflagellates, hystrichospheres and acritarchs. *Proceedings of the National Academy of Science*, **49**, 158–164, 298–302.

EVITT, W. R. 1963*b*. Occurrence of freshwater alga *Pediastrum* in Cretaceous marine sediments. *American Journal of Science*, **261**, 890–893.

EVITT, W. R. 1964. Dinoflagellates and their use in petroleum geology. *In*: CROSS, A. T. (ed.) *Palynology in Oil Exploration*. Society of Economic Paleontologists and Mineralogists, Special Publication no. **11**, pp. 65–72.

EVITT, W. R. 1967. Dinoflagellate studies II. The archeopyle. *Stanford University Publications in the Geological Sciences*, **10**, 1–61.

EVITT, W. R. 1968. The Cretaceous microfossil *Ophiobolus lapidaris* O. Wetzel and its flagellum-like filaments. *Stanford University Publications in the Geological Sciences*, **12**, 1–9.

EVITT, W. R. 1985. *Sporopollenin Dinoflagellate Cysts: Their Morphology and Interpretation*. American Association of Stratigraphic Palynologists Foundation, Dallas, Texas, 333 pp.

EVITT, W. R. & DAVIDSON, S. E. 1964. Dinoflagellate studies I. Dinoflagellate cysts and thecae. *Stanford University Publications in the Geological Sciences*, **10**, 1–12.

FAEGRI, K. & IVERSON, J. 1950. *Textbook of Modern Pollen Analysis*. Munksgaard, Copenhagen.

FEISTMANTEL, O. 1881. The fossil flora of the Gondwana System. *Memoirs of the Geological Survey of India*, Series 12, **3**, 1–77.

FENSOME, R. A., WILLIAMS, G. L., BARSS, M. S., FREEMAN, J. M. & HILL, J. M. 1990. *Acritarchs and Fossil Prasinophytes: An Index to Genera, Species and Intraspecific Taxa*. American Association of Stratigraphic Palynologists, Contributions Series, no. **25**, 771 pp.

FENSOME, R. A., TAYLOR, F. J. R., NORRIS, G., SARJEANT, W. A. S., WHARTON, D. I. & WILLIAMS, G. L. 1993. A classification of living and fossil dinoflagellates. *Micropaleontology* **7** Special Publication no. **7**, 351 pp.

FENSOME, R. A., GOCHT, H., STOVER, L. E. & WILLIAMS, G. L. 1991. *The Eisenack Catalogue of Fossil Dinoflagellates*. New Series, Vol. 1. Schweizerbart, Stuttgart.

FELIX, C. J. & BURBRIDGE, P. P. 1967. Palynology of the Springer Formation of southern Oklahoma, U. S. A. *Palaeontology*, **10**, 347–425.

FENSOME, R. A., RIDING, J. B. & TAYLOR, F. J. R. 1996. Dinoflagellates. *In*: JANSONIUS, J. & MCGREGOR, D. C. (eds) *Palynology: Principles and Practice*, Vol. 1, *Principles*. American Association of Stratigraphic Palynologists Foundation, College Station, Texas, 107–170.

FRESENIUS, G. 1860. Über *Phelonites lignitum*, *Phelonites strobilina* und *Betula salzhausenensis*. *Palaeontographica*, **8**, 158.

FRÜH, J. H. 1885. Kritische Beiträge zur Kenntniss des Torfes. *Geologisches Jahrbuch*, **35**, 677–726.

FUCHS, T. 1905. Über die Natur von *Xanthidium* Ehrenberg. *Centralblatt für Mineralogie, Geologie und Paläontologie*, 340–342.

FUNKHAUSER, J. W. 1959. A survey of non-spore-pollen palynology (Abstract). *Proceedings of the 9th International Botanical Congress*, Montreal, Vol. 2, 126.

GAFFORD, E. L. & KIDSON, E. J. 1968. Probable occurrences of chitinozoans from the Lower Permian of Kansas. *Compass*, **45**, 72–73.

GEINITZ, F. E. 1887. Geologische Notizen aus der Lüneburger Heide. *Jahresheft der Naturwissenschaftlichen Verein der Fürstentum Lüneburg*, **10**, 36.

GERLACH, E. 1961. Mikrofossilien aus dem Oligozän und Miozän Nordwestdeutschlands, unter besonderer Berücksichtigung der Hystrichosphaeren und Dinoflagellaten. *Neues Jahrbuch für Geologie und Paläontologie, Abhandlungen*, **112**, 143–228.

GIJZEL, P. VAN 1967. Autofluorescence of fossil pollen and spores, with special reference to age determination and coalification. *Leidse Geologische Mededelingen*, **46**, 263–317. (Republished in: MUIR, M. D. & SARJEANT, W. A. S. (eds) *Palynology, Part I*. Benchmark Papers in Geology, **46**. Dowden, Hutchinson & Ross, Stroudsburg, 53–93.)

GOCHT, H. 1952. Hystrichosphaerideen und andere Kleinlebewesen aus Oligozänablagerungen Nord und Mitteldeutschlands. *Geologie*, **4**, 301–320.

GOCHT, H. 1957. Microplankton aus dem nordwestdeutschen Neokom I. *Paläontologisches Zeitschrift*, **31**, 163–185.

GOCHT, H. 1970. Dinoflagellaten-Zysten aus dem Bathonium des Erdölfeldes Aldorf (NW.-Deutschland). *Palaeontographica*, Series B, **129**, 125–165.

GOCHT, H. & SARJEANT, W. A. S. 1983. Pioneer in palynology: Alfred Eisenack (1891–1982). *Micropaleontology*, **29**, 470–477.

GÓCZÁN, F. 1956. Pollenanalytische (palynologische) Untersuchungen zur Identifizierung der liassischen Schwarzkohlenflöze von Komló. *Magyar Allami Földtani Intézet, Evkönyve*, **45**, 167–212.

GÓCZÁN, F. 1962. Un microplancton dans le Crétacé de la montagne Bakony. *Annual Reports of the Hungarian Geological Institute* (for 1959), 181–209.

GOLDENBERG, F. 1855. Flora saraepontana fossilis. *Die Planzenversteinerungen des Steinkohlengebirgs von Saarbrücken*, Vol. I. Saarbrücken (copy not located).

GÖPPERT, H. R. 1836. De floribus in statu fossili commentatio. *Nova Acta Academiae Caesarea Leopoldino-Carolinae Germanicum Nature Curiosorus*, 547–572.

GÖPPERT, H. R. 1848. Uber das Vorkommen von Pollen im fossilen Zustände. *Neues Jahrbuch für Mineralogie, Geognosie, Geologie und Petrefaktenkunde*, **11**, 338–340.

GÓRKA, H. 1963. Coccolithophoridés, dinoflagellés, hystrichosphaeridés et microfossiles incertae sedis du Crétacé supérieur de Pologne. *Acta Palaeontologica Polonica*, **8**, 3–90.

GÓRKA, H. 1971. Sur les "melanosclérites" extraits des galets erratiques ordoviciens de Pologne. *Bulletin de la Société Géologique et Minéralogique de Bretagne*, Series C, **3**, 29–40.

GOUBIN, N. 1965. Description et répartition des principaux Pollenites permiens, triasiques et jurassiques des sondages du Bassin de Morondava (Madagascar). *Revue de l'Institut de Pétrôle*, **20**, 1415–1458.

GRAINDOR, M. J. 1956. Note préliminaire sur les

microfossiles du Briovérien. *Comptes-rendus sommaires de la Société geologique de France*, no. **11**, 207–210.

GRAINDOR, M. J. 1957. Cayeuxidae nov. fam., organismes à squelette du Briovérien. *Comptes-rendus hébdomadaires de l'Académie des Sciences, Paris*, **244**, 2075–2077.

GRAYSON, J. F. 1956. The conversion of calcite to fluorite. *Micropaleontology*, **2**, 71–78.

GREW, N. 1682. *The Anatomy of Plants. With an Idea of a Philosophical History of Plants*. Rawlins, London, for the author, 24 + 304 pp.

GRUAS-CAVAGNETTO, C. 1967. Complexes sporopolliniques du Sparnacien du Bassin de Paris. *Review of Palaeobotany and Palynology*, **5**, 243–261.

GUENNEL, G. K. 1963. Devonian spores in a Middle Silurian reef. *Grana Palynologica*, **4**, 245–261.

GUTJAHR, C. C. M. 1966. Carbonization measurements of pollen-grains and their application. *Leidse Geologische Mededelingen*, **38**, 1–29. (Republished in: MUIR, M. D. & SARJEANT, W. A. S. (eds) *Palynology, Part I*. Benchmark Papers in Geology, **46**, Dowden, Hutchinson & Ross, Stroudsburg, 26–52.)

GUY, D. J. E. 1971. *Palynological Investigations in the Middle Jurassic of the Vilhelmsfält Boring, Southern Sweden*. Institutes of Mineralogy, Palaeontology, and Quaternary Geology, University of Lund, Publication no. **168**, 104 pp.

GUY-OHLSON, D. J. E. 1996. Prasinophycean algae. *In*: JANSONIUS, J. & MCGREGOR, D. C. (eds) *Palynology: Principles and Practice*, Vol. I, *Principles*. College Station, Texas: American Association of Stratigraphic Palynologists, College Station, Texas, 181–189.

HABIB, D. 1968. Spore and pollen paleoecology of the Redstone seam (Upper Pennsylvanian) of West Virginia. *Micropaleontology*, **14**, 199–220.

HABIB, D. 1969. Middle Cretaceous palynomorphs in a deep-sea core from the Seismic Reflector Horizon A outcrop area. *Micropaleontology*, **15**, 85–101.

HABIB, D. 1970. Middle Cretaceous palynomorph assemblages from clays near the Horizon Beta deep-sea outcrop. *Micropaleontology*, **16**, 345–379.

HABIB, D. 1972. Dinoflagellate stratigraphy Leg 11, Deep Sea Drilling Project. *In*: HOLLISER, C. D., EWING, J. I. *et al.* (eds), *Initial Reports of the Deep Sea Drilling Project*, **11**, Washington, 367–425.

HACQUEBARD, P. A. 1957. Plant spores in coal from the Horton Group (Mississippian) of Nova Scotia. *Micropaleontology*, **3**, 301–324.

HACQUEBARD, P. A. 1961. Palynological studies of some upper and lower Carboniferous strata in Nova Scotia. *Proceedings, Third Conference on the Origin and Constitution of Coal, Crystal Cliffs, Nova Scotia, June 1956*, Nova Scotia Department of Mines and Nova Scotia Research Foundation, Halifax, 227–256.

HACQUEBARD, P. A. & BARSS, S. 1957. A Carboniferous spore assemblage, in coal from the South Nahanni River area, Northwest Territories. *Geological Survey of Canada Bulletin*, no. **40**, 1–63.

HAJOS, M. 1964. A mecseki miócen diatomaföld rétegek microplanktonja. *Magyar Allami Földtani Intézet Evkönyve*, 139–171.

HAMMEN, T. VAN DER 1954. El desarrollo de la flora Colombiana en los periodos geológicos. I. Maastrichtiano hasta Terciario mas inferior. *Boletin Geologico, Bogotá*, **2**, 49–106.

HAMMEN, T. VAN DER 1956. Nomenclatura palinológica sistemática/A palynological systematic nomenclature. *Boletin Geologico, Bogotá*, **4**, 26–101.

HANSON, B. 1982. Karl Walter Klement (1931–1982). *Bulletin of the American Association of Petroleum Geologists*, **66**, 1156.

HARRIS, T. M. 1938. *The British Rhaetic flora*. British Museum (Natural History), London, 84 pp.

HART, G. F. 1960. Microfloral investigation of the Lower Coal Measures (K2): Ketewaka-Mchuchuma Coalfield, Tanganyika. *Geological Survey of Tanganyika Bulletin*, no. **30**, 1–18.

HART, G. F. 1965. *The Systematics and Distribution of Permian Miospores*. Witwatersrand University Press, Johannesburg, 252 pp.

HEMER, D. O. & NYGREEN, P. W. 1967. Algae, acritarchs and other microfossils *incertae sedis* from the Lower Carboniferous of Saudi Arabia. *Micropaleontology*, **13**, 183–194.

HINDE, G. J. 1879. On annelid jaws from the Cambro–Silurian, Silurian and Devonian Formations in Canada and from the Lower Carboniferous in Scotland. *Quarterly Journal of the Geological Society of London*, **35**, 370–389.

HINDE, G. J. 1880. On annelid jaws from the Wenlock and Ludlow Formations of the west of England. *Quarterly Journal of the Geological Society of London*, **36**, 368–378.

HINDE, G. J. 1882. On annelid remains from the Silurian strata of the Isle of Gotland. *Birand till Kungliga Svensk Vetenskapsakademiens, Hindlingas*, **7**, 3–28.

HINDE, G. J. 1896. On the jaw-apparatus of an annelid (*Eunicites reidiae* sp. nov.) from the Lower Carboniferous of Halkin Mountain, Flintshire. *Quarterly Journal of the Geological Society of London*, **52**, 438–450.

HOEKEN-KLINKENBERG, P. M. J. van. 1964. A palynological investigation of some Upper Cretaceous sediments in Nigeria. *Pollen et Spores*, **6**, 209–231.

HOEKEN-KLINKENBERG, P. M. J. van. 1966. Maastrichtian, Paleocene and Eocene pollen and spores from Nigeria. *Leidse Geologischa Mededelingen*, **38**, 37–48.

HOFFMEISTER, W. S. 1954. *Microfossil prospecting for petroleum*. United States Patent Office, no. **2**, 686, 108, 4 pp.

HOFFMEISTER, W. S. 1959. Lower Silurian spores from Libya. *Micropaleontology*, **5**, 331–334.

HOFFMEISTER, W. S., STAPLIN, F. L. & MALLOY, R. E. 1955. Geologic range of Paleozoic plant spores in North America. *Micropaleontology*, **1**, 9–27.

HOLM, G. 1887. Om Vettern och Visingsöformationen. *Birand till Kungliga Svensk Vetenskapsakademiens, Hindlingas*, Series 2, no. **7**, 1–49.

HOLST, H.O. 1908. Postglaciala tidsbestämnigar. *Sveriges Geologiska Undersökning Årsbok*,

Series C, *Afhandlingar och uppsatser*, no. **2**(216), 94 pp.

HOOKER, J. D. 1848. The vegetation of the Carboniferous period as compared with that of the present day. *Memoirs of the Geological Survey of England and Wales*, **2**, 387–430 (also in *Edinburgh New Philosophical Journal*, **45**, 362–369; **46**, 73–78, 398–400).

HOOKER, J. D. 1852. On the spheroidal bodies, resembling seeds, from the Ludlow Bone Bed. *Quarterly Journal of the Geological Society of London*, **9**, 12.

HOPPING, C. A. 1967. Palynology and the oil industry. *Review of Palaeobotany and Palynology*, **2**, 23–48.

HOVASSE, R. 1956. *Arnoldia antiqua* gen. nov., sp. nov., foraminifère probable du Précambrian de la Côte d'Ivoire. *Comptes-rendus hébdomadaires de l'Académie des Sciences, Paris*, **247**, 2582–2584.

HUGHES, N. F. 1961. Further interpretation of *Eucommiidites* Erdtman 1948. *Palaeontology*, **4**, 292–299.

HUGHES, N. F. 1970. The need for agreed standards of recording in palaeopalynology and palaeobotany. *Paläontologisches Abhandlungen*, Series B, **3**, 357–364.

HUGHES, N. F. 1975. The challenge of abundance in palynomorphs. *Geoscience & Man*, **11**, 141–144.

HUGHES, N. F. & MOODY-STUART, J. C. 1966. A method of stratigraphic correlation using Early Cretaceous miospores. *Palaeontology*, **12**, 84–111.

HUGHES, N. F. & PLAYFORD, G. 1961. Palynological reconnaissance of the Lower Carboniferous of Spitzbergen. *Micropaleontology*, **7**, 27–44.

HULSHOF, O. K. & MANTEN, A. A. 1971. Bibliography of actuopalynology 1671–1966. *Review of Palaeobotany and Palynology*, **12**, 5–241.

HYDE, H. A. & WILLIAMS, D. A. 1944. The right word. *Pollen Analysis Circulars*, **8**, 6.

IBRAHIM, A. C. 1933. *Sporenformen des Aegirhorizonts des Ruhr-Reviers*. Triltsch, Würzburg.

IVERSEN, J. 1941. Land occupation in Denmark's Stone Age. A pollen-analytical study of the influence of farmer culture on the vegetational development. *Danmarks Geologiske Undersøgelse*, **66**, 20–65.

JANKAUSKAS, T. & SARJEANT, W. A. S. 2001. Boris V. Timofeyev (1916–1982): pioneer of Precambrian and Early Paleozoic palynology. *Earth Sciences History* (in press).

JANSONIUS, J. 1962. Palynology of Permian and Triassic sediments, Peace River area, western Canada. *Palaeontographica*, Series B, **110**, 35–98.

JANSONIUS, J. 1964. Morphology and classification of some Chitinozoa. *Bulletin of Canadian Petroleum Geology*, **12**, 901–918.

JANSONIUS, J. 1967. Systematics of the Chitinozoa. *Revue of Palaeobotany and Palynology*, **1**, 345–360.

JANSONIUS, J. 1969. Classification and stratigraphic application of Chitinozoa. *Proceedings of the North American Paleontological Convention*, 789–808.

JANSONIUS, J. & CRAIG, J. H. 1971. Scolecodonts. I. Descriptive terminology and revision of systematic nomenclature. II. Lectotypes, new names for homonyms, index of species. *Bulletin of Canadian Petroleum Geology*, **19**, 251–302.

JANSONIUS, J. & MCGREGOR, D. C. 1996. Introduction. *In*: JANSONIUS, J. & MCGREGOR, D. C. (eds) *Palynology: Principles and Practice*, vol. 1 *Principles*. American Association of Stratigraphic Palynologists Foundation, College Station, Texas, 1–10.

JARDINÉ, S. 1967. Spores à expansions en forme d'élatères du Crétáce moyen d'Afrique Occidentale. *Review of Palaeobotany and Palynology*, **1**, 235–258.

JEFFREY, E. C. 1910. The nature of some supposed algal coals. *Proceedings of the American Academy of Arts and Sciences*, **46**, [273]–290.

JEFFREY, E. C. 1914. On the composition and qualities of coal. *Economic Geology*, **9**, 732–742.

JENKINS, W. A. M. 1967. Ordovician Chitinozoa from Shropshire. *Palaeontology*, **10**, 436–488.

JENKINS, W. A. M. 1969. *Chitinozoa from the Ordovician Viola and Fernvale Limestones of the Arbuckle Mountains of Oklahoma*. Palaeontological Association, London, Special Papers in Palaeontology, **5**.

JENKINS, W. A. M. 1970. Chitinozoans. *Geoscience & Man*, **1**, 1–22.

JODRY, R. I. & CAMPAU, D. E. 1961. Small pseudochitinous and resinous microfossils: new tools for the subsurface geologist. *AAPG Bulletin*, **45**, 1378–1391.

JOHNSON, H. A. & THOMAS, B. W. 1884. The microscopic organisms in the Bowlder Clay of Chicago and vicinity. *Bulletin of the Chicago Academy of Sciences*, **1**, 35–40.

JONES, G. D. & BRYANT, V. M. Jr 1992. Melissopalynology in the United States: a review and critique. *Palynology*, **16**, 63–71.

JONES, G. D. & BRYANT, V. M. Jr 1996. Melissopalynology. *In*: JANSONIUS, J & MCGREGOR, D. C. (eds) *Palynology: Principles and Practice*, Vol. 3, *New Directions, Other Applications and Floral History*. American Association of Stratigraphic Palynologists Foundation, College Station, Texas, 933–938.

JONKER, F. P. 1967. Palynology and The Netherlands. *Review of Palaeobotany and Palynology*, **1**, 31–35.

JUX, U. 1968. Über den Feinbau der Wandung bei *Tasmanites* Newton. *Palaeontographica*, Series B, **124**, 112–124.

JUX, U. 1969. Über den Feinbau den Zystenwandung von *Pachysphaera marshalliae* Parke, 1966. *Palaeontographica*, Series B, **125**, 104–111.

JUX, U. & MOERICKE, V. 1965. *Tytthodiscus suevicus* eine Thekamoebe? *Palaeontographica*, Series B, **115**, 107–116.

KALIBOVÁ-KAISEROVÁ, M. 1951. Megaspory radnické ho jového pásma kladensko-rakovické krame nouhenné pánve. *Sbornik Ustredního Ústavu Geologického*, **18**, 21–92.

KAR, R. K., KIESER, G. & JAIN, K. P. 1972. Permo–Triassic subsurface palynology from Libya. *Pollen et Spores*, **14**, 389–453.

KARKCZEWSKA, A. J. 1967. Carboniferous spores from the Chelm I Boring (eastern Poland). *Acta Palaeontologica Polonica*, **12**, 268–345.

KEDVES, M. 1960. Études palynologiques dans le bassin de Dorog I. *Pollen et Spores*, **2**, 89–118.

KEDVES, M. 1961. Études palynologiques dans le bassin de Dorog II. *Pollen et Spores*, **3**, 101–153.

KEDVES, M. 1962. *Noremia*, a new microfossil genus from the Hungarian Eocene, and systematical and stratigraphical problems about the Crassosphaeridae. *Acta Mineraologica Petrographica Szeged*, **15**, 19–27.

KEDVES, M. 1963. Contribution à la flore Eocène Inférieure de la Hongrie sur la base des examens palynologiques des couches houillières de puits III d'Oroszlány et puits XV/B de Tatabánya. *Acta Botanica. Academiae Scientiarum Hungaricae*, **9**, 31–66.

KHLONOVA, A. F. 1962. Some morphological types of spores and pollen grains from Upper Cretaceous of eastern part of West Siberian lowlands. *Pollen et Spores*, **4**, 297–309.

KIELAN-JAWOROWSKA, Z. 1961. On two Ordovician polychaete jaw apparatuses. *Acta Palaeontologica Polonica*, **6**, 237–260.

KIELAN-JAWOROWSKA, Z. 1962. New Ordovician genera of jaw apparatuses. *Acta Palaeontologica Polonica*, **7**, 241–256.

KIELAN-JAWOROWSKA, Z. 1966. Polychaete jaw apparatuses from the Ordovician and Silurian of Poland and comparison with modern forms. *Acta Palaeontologica Polonica* (special parts).

KLAUS, W. 1960. Sporen der karnischen Stufe des ostalpinen Trias. *Jahrbuch, Geologisches Bundesanstalt* (Wien), *Sonderband*, **5**, 107–184.

KLEMENT, K. W. 1957. Revision der Gattungzugehörigkeit einiger in die Gattung *Gymnodinium* eingestüfter Arten jurassischer Dinoflagellaten. *Neues Jahrbuch für Geologie und Paläontologie Monatshefte*, **9**, 408–410.

KLEMENT, K. W. 1960. Dinoflagellaten und Hystrichosphaerideen aus dem Unteren und Mittleren Malm Südwestdeutschlands. *Palaeontographica*, Series A, **114**, 1–104.

KLUMPP, B. 1953. Beitrag zur Kenntnis der Mikrofossilien des mittleren und oberen Eozän. *Palaeontographica*, Series A, **103**, 377–406.

KNIGHTS, B. A., BROWN, A. C., CONWAY, E. & MIDDLEDITCH, B. S. 1970. Hydrocarbons from the green form of the freshwater alga *Botryococcus braunii*. *Phytochemistry*, **9**, 1317–1324.

KNOLL, A. H. 1996. Archean and Proterozoic paleontology. *In*: JANSONIUS, J. & MCGREGOR, D. C. (eds), *Palynology: Principles and Practice*, Vol. I, *Principles*. American Association of Stratigraphic Palynologists Foundation, College Station, Texas, 51–80.

KOZLOWSKI, R. 1943. The characteristic plant microfossils of the Pittsburgh and Pomeroy coals of Ohio. *American Midland Naturalist*, **29**, 119–132.

KOZLOWSKI, R. 1950. Pennsylvanian spores of Illinois and their use in correlation. *Illinois State Geological Survey Bulletin*, **74**, 1–128.

KOZLOWSKI, R. 1956. Sur quelques appareils masticateurs des Annélides polychètes ordoviciens. *Acta Palaeontologica Polonica*, **1**, 165–205.

KOSANKE, R. M. 1963. Sur la nature des chitinozoaires. *Acta Palaeontologica Polonica*, **8**, 425–449.

KOSANKE, R. M. 1974. Découverte des oeufs de polychètes dans l'Ordovicien. *Acta Palaeontologica Polonica*, **19**, 437–442.

KOZUR, H. 1967. Scolecodonten aus dem Muschelkalke des germanischen Binnenbeckens. *Monatsberichte der Deutschen Akademie der Wissenschaften zu Berlin*, **9**, 842–886.

KOZUR, H. 1970. Zur Klassifikation und phylogenetischen Entwicklung der fossilen Phyllocodida und Eunicida (Polychaeta). *Freiberger Forschungshefte*, Series 1, **260**, 35–81.

KRÄUSEL, R. 1941. Die Sporokarpien Dawsons, eine neue Thallophyten-Klasse des Devons. *Palaeontographica*, Series B, **86**, 113–133.

KREMP, G. O. W. 1952. Sporen-Vergesellschaften und Mikrofaunen-Horizonte in Ruhrkarbon. *Comptes-rendus de Congrès pour l'Avancement des Études de Stratigraphie du Carbonifère*, Vol. 1, 347–357.

KREMP, G. O. W. 1957–1967. *Catalog of Fossil Spores and Pollen* (26 vols). Pennsylvania State University, University Park, Pennsylvania.

KREMP, G. O. W. 1965. *Morphologic Encyclopedia of Palynology*. University of Arizona, Program in Geochronology, Tucson, Contributions no. **100**, 186 pp.

KRIVÁN-HUTTER, E. 1963. Microplankton from the Palaeogene of the Dorog Basin I. *Annales Universitatis Scientiarum Budapestiensis de Rolando Eötvös Budapest, Sectio Geologica*, **6**, 71–91.

KRUTZSCH, W. 1957. Sporen- und Pollengruppen aus der Oberkreide und dem Tertiärs Mitteleuropas und ihre stratigraphische Verteilung. *Zeitschrift für angewandte Geologie*, **3**, 509–548.

KRUTZSCH, W. 1959. Mikropaläontologische (sporenpaläontologische) Untersuchungen in der Braunkohle des Geiseltales. *Geologie (Berlin)*, Beihefte **3**, 425 pp.

KUPRIANOVA, L. A. 1960. Progress of palynology in the USSR during the period 1935–1959. *Pollen et Spores*, **2**, 123–128.

KUPRIANOVA, L. A. & ALESHINA, L. A. 1961. *Address book of palynologists of the Soviet Union*. Akademiya Nauk SSSR, Vsesoiuznoe Botanchoskoe Obstichestro Leningrad.

KÜTZING, F. T. 1849. *Species Algarum*. Brockhaus, Leipzig, 922 pp.

KUYL, O. S., MULLER, J. & WATERBOLK, H. J. 1955. The application of palynology to geology, with reference to western Venezuela. *Geologie en Mijnbouw*, **17**, 49–76.

LAGERHEIM, G. 1895. Uredineae Herbarii Eliae Fries. *Tromsø Museums Aarshefter*, **17**, [25]–132 (also published separately).

LANGE, F. W. 1947. Annelidos poliquetos dos folhelhas Devonianos do Paraná. Arqlivos do Museu Parenaense *Paranaense*, **6**, 161–230.

LANGE, F. W. 1949*a*. Polychaete annelids from the Devonian of Paraná, Brazil. *Bulletins of American Paleontology*, **33**, 5–102.

LANGE, F. W. 1949*b*. Novos microfósseis Devonianos do Paraná. *Arquivos de Museu Paranaense*, **7**, 287–289.

LANGE, F. W. 1952. Chitinozoários do Folhelho Barreirinha, Devoniano do Para. *Dusonia*, **3**, 373–386.

LAUFELD, S. 1967. Caradocian Chitinozoa from Dalarna, Sweden. *Geologiska Föreningens i Stockholm, Förhandlinger*, **8–9**, 275–349.

LEFÈVRE, M. 1932. Sur la présence de Péridiniens dans un dépôt fossile des Barbades. *Comptes-rendus hébdomadaires de l'Académie des Sciences, Paris*, **197**, 2315–2316.

LEFÈVRE, M. 1933. Les Peridinites des Barbades. *Annales de Cryptogamie Exotique*, **6**, 215–229.

LEFFINGWELL, H. A. 1970. Palynology of the Lance (Late Cretaceous) and Fort Union (Paleocene) Formations of the type Lance area, Wyoming. *In*: KOSANKE, R. M. & CROSS, A. T. (eds) *Symposium on Palynology of the Late Cretaceous and Tertiary*. Geological Society of America, Special Paper **127**.

LEJEUNE, M. 1936. L'étude microscopique des silex (1ière Note). *Annales de la Société Géologique Belge*, **59**, *Bulletin* No. 7, 190–197.

LEJEUNE-CARPENTIER, M. 1946. L'étude microscopique des silex (12ième Note). Espèces nouvelles ou douteuses de *Gonyaulax*. *Annales de la Société Géologique Belge*, **69**, *Bulletin* No. 4, B187–B197.

LEJEUNE-CARPENTIER, M. 1951. L'étude microscopique des silex (13ième note). *Gymnodinium* et *Phanerodinium* (Dinoflagellates) de Belgique. *Annales Société Géologique Belge*, **74**, *Bulletin* No. 4, B307–B313.

LENTIN, J. K. & WILLIAMS, G. L. 1973. *Fossil Dinoflagellates: Index to Genera and Species*. Geological Survey of Canada, Paper **73–42**, 176 pp.

LESCHIK, G. 1956. Sporen aus dem Salzton des Zechsteins von Neuhof (bei Fulda). *Palaeontographica*, Series B, **100**, 122–142.

LESCHIK, G. 1959. Sporen aus den "Karru-Sandsteinen" von Norronaub (Südwest Afrika). *Senckenbergiana lethaea*, **40**, 51–95.

LEVET-CARETTE, J. 1963. Étude de la microflore infraliasique d'un sondage effectué dans les soussols de Boulogne-sur-Mer (P.-de-C.). *Annales de la Société géologique du Nord*, **84**, 91–121.

LEVET-CARETTE, J. 1966. Microflore wéaldienne provenant d'un puits naturel à la fosse Vieux-Condé (Groupe de Valenciennes). *Annales de la Société géologique du Nord*, **86**, 153–174.

LEWIS, H. P. 1940. The microfossils of the Upper Caradocian phosphate deposits in Montgomeryshire, North Wales. *Annals and Magazine of Natural History*, Series 11, **5**, 1–39.

LINDBERG, H. 1905. [quoted in MANTEN 1967: reference not located].

LINNARSSON, G. 1880. De äldre Paleozoiska lagren; trakten kring Motala. *Geologiska Föreningens i Stockholm, Förhandlingar,* **5**, 23–30.

LISTER, T. R. 1970*a*. The method of opening, orientation and morphology of the Tremadocian acritarch, *Acanthodiacrodium ubui* Martin. *Proceedings of the Yorkshire Geological Society*, **38**, 47–55.

LISTER, T. R. 1970*b*. *The acritarchs and Chitinozoa from the Wenlock and Ludlow series of the Ludlow and Millichope areas, Shropshire. Part 1.* Palaeontographical Society Monographs, The Palaeontographical Society, London.

LIUBER, A. A. 1938. Spores and pollen from the Permian of the U S S R. *Problemy Sovetskoi Geologii*, **8**, 152–160 (in Russian).

LIUBER, A. A. & VAL'TS, I. E. 1941. *Atlas of Microspores and Pollen Grains of the Palaeozoic of the U. S. S. R.*. Trudy Vsesoiuznogo Nauchno-Issledovatel'skogo Geologicheskogo Instituta, **139**, Gosgeolizdat, Moscow.

LOEBLICH, A. R., III. 1967. Nomenclatural notes on the Pyrrhophyta, Xanthophyta and Euglenophyta. *Taxon*, **16**, 68–69.

LOHMANN, H. 1904. Eier und sogenannte Cysten der Plankton-Expedition. Anhang: Cyphonautes. *Wissenschaftliche Ergebnisse der Plankton-Expedition Humboldt-Stiftung*, New Series, **4**, 25.

LOVE, L. G. 1958. Micro-organisms and the presence of syngenetic pyrite. *Quarterly Journal of the Geological Society of London*, **113**, 429–440.

LOVE, L. G 1960. Assemblages of small spores from the lower Oil-Shale Group of Scotland. *Proceedings of the Royal Society of Edinburgh*, Section B, **67**, 99–126.

LOVE, L. G 1962. Further studies on micro-organisms and the presence of pyrite. *Palaeontology*, **5**, 444–459.

LÜCK, H. 1913. *Beitrag zur Kenntnis des älteren Salzgebirges im Berlepsch-Bergwerk bei Stassfurt, nebst Bemerkungen über die Pollenführung des Salztones*. Doctoral thesis, University of Leipzig.

LUDWIG, R. 1857. Fossile Pflanzen aus der jungsten Welterauer Braunkohle. *Palaeontographica*, **5**, 1–81.

MCGREGOR, D. C. 1960. Devonian spores from Melville Island, Canadian Arctic Archipelago. *Palaeontology*, **3**, 26–44.

MCGREGOR, D. C. 1961. Spores with proximal radial pattern from the Devonian of Canada. *Geological Survey of Canada Bulletin*, **76**.

MCINTYRE, D. S. 1962. Pollen from deeply buried Coal Measures, Taranaki, New Zealand; No. **2**. *New Zealand Journal of Geology and Geophysics*, **5**, 314–319.

MCKEE, E. D., CHRONIC, J. & LEOPOLD, E. B. 1959. Sedimentary belts in the lagoon of Kapingamarangi Atoll. *Bulletin of the American Association of Petroleum Geologists*, **43**, 501–562.

MACKÓ, S. 1957. Lower Miocene pollen flora from the valley of Klodnicka near Gliwice (Upper Silesia). *Travaux de la Société scientifique de Wroclaw*, Series 13, **8**, 313 pp.

MACKÓ, S. 1963. Sporomorphs from Upper Cretaceous near Opole (Silesia) and from the London Clay. *Práce Wroclawskiego Towarzystwa Nauk*, Series B, **106**, 1–136.

MÄDLER, K. A. 1963. Die figurierten organischen Bestandteile der Posidonienscheifer. *Geologisches Jahrbuch, Beihefte*, **58**, 287–406.

MÄDLER, K. A. 1964. Die geologische Verbreitung von Sporen und Pollen in der deutschen Trias. *Geologisches Jahrbuch, Beihefte*, no. **65**, 127 pp.

MÄDLER, K. A. 1967. Hystrichophyta and acritarchs. *Review of Palaeobotany and Palynology*, **5**, 285–290.

MAIER, D. 1958. Zur Gliederung des Tertiärs mit Hystrichosphaerideen. *Neues Jahrbuch für Geologie und Paläontologie, Monatshefte*, no. **10**, 468–472.

MAIER, D. 1959. Planktonuntersuchungen in tertiären und quartären marinen Sedimenten. Ein Beitrag zur Systematik, Stratigraphie und Ökologie der Coccolithophorideen, Dinoflagellaten und Hystrichosphaerideen vom Oligozän bis zum Pleistozän. *Neues Jahrbuch für Geologie und Paläontologie, Abhandlungen*, **107**, 278–340.

MALIAVKINA, V. S. 1949. *Key to spores and pollen: Jurassic–Cretaceous*. Trudy Vsesojuznyi Nauchno-Isslefova tel'skii Geologischeskii Institut (VSEGEI), New Series, **33**, 138 pp. (in Russian).

MALIAVKINA, V. S. 1964. *Spores and Pollen from Triassic Deposits of the West Siberian Lowland*. Trudy Vsesojuznyi Nauchno-Issledovatel'skii Geologischeskii Institut (VNIGRI), **231**, 293 pp. (in Russian).

MALPIGHI, M. 1687. *Opera omnia: seu thesaurus locupletissimus botanico–medico–anatomicus, viginti quatuor tractatus complectens. Et in duos tomos distributus, quorum tractatum seriem videre est dedicatione absoluta* (2 vols). Apud Petrum Vander Aa, Lugduni Batavorum 16 + 170 + (22) pp.: 379 + (37) pp.

MANTELL, G. A. 1845. Notes of a microscopical examination of the Chalk and Flint of southeast England, with remarks on the Animalculites of certain Tertiary and modern deposits. *Annals and Magazine of Natural History*, **16**, 73–88.

MANTELL, G. A. 1850. *A pictorial atlas of fossil remains, consisting of coloured illustrations selected from Parkinson's "Organic remains of a former world" and Artis's "Antediluvian phytology*. Bohn, London, 207 pp.

MANTEN, A. A. 1967. Lennart von Post and the foundation of modern palynology. *Review of Palaeobotany and Palynology*, **1**, 11–22.

MANTEN, A. A. 1968. A short history of palynology in diagrams. *Review of Palaeobotany and Palynology*, **6**, 177–188.

MANTEN, A. A. 1969. Bibliography of palaeopalynology 1836–1966. *Review of Palaeobotany and Palynology*, **8**, 1–572.

MANTEN, A. A. 1970. Statistical analysis of a scientific discipline: palynology. *Earth Science Reviews*, **6**, 181–218.

MARSHALL, A.E. & SMITH, A. H. V. 1964. Assemblages of miospores from some Upper Carboniferous coals and their associated sediments in the Yorkshire Coalfield. *Palaeontology*, **7**, 656–673.

MARTIN, F. 1966. Les acritarches du sondage de la Brasserie Lust, à Kortrijk (Courtrai), (Silurien belge). *Bulletin de la Société belge de Géologie, de Paléontologie et de Hydrologie*, **74**, 354–400.

MARTIN, F. 1967. Les acritarches du parc de Neuville-sous-Huy (Silurien belge). *Bulletin de la Société belge de Géologie, de Paléontologie et de Hydrologie*, **75**, 306–335.

MARTIN, F. 1969*a*. Ordovicien et Silurien belges; données nouvelles apportés par l'étude des acritarches. *Bulletin de la Société belge de Géologie, de Paléontologie et de Hydrologie*, **77**, 175–181.

MARTIN, F. 1969*b*. Les acritarches de l'Ordovicien et du Silurien belges; détermination et valeur stratigraphique. *Mémoires de l'Institut Royal des Sciences Naturelles de Belgique*, no. **160** (1968), 1–175.

MARTINET, J. F. 1779. *Katechismus der Natuur*, Vol. 4. Allart, Amsterdam, 354 pp.

MASSALONGO, A. 1855. *Monografica delle Nereidi fossili del Monte Bolca*. Antonelli, Verona, 55 pp.

MAXWELL, J. R., DOUGLAS, A.G., EGLINTON, G. & MCCORMICK, A. 1968. The botryococcenes-hydrocarbons of novel structure from the alga *Botryococcus braunii* Kützing. *Phytochemistry*, **7**, 2157–2171.

MENÉNDEZ, C. A. 1951. La flora mesozoica de la Formación Llantenes (Provincia de Mendoza). *Revista del Instituto Nacional de Investigaciones de las Ciencias Naturales, Ciencias Botánicas*, **2**, 147–261.

MENÉNDEZ, C. A. & POTHE DE BALDIS, E. D. 1967. Devonian spores from Paraguay. *Review of Palaeobotany and Palynology*, **1**, 161–172.

MERRILL, J. A. 1895. Fossil sponges of the flint nodules in the Lower Cretaceous of Texas. *Bulletin of the Museum of Comparative Zoology, Harvard* (*Geology Section* III), **28**, 1–26.

MILLER, M. A. 1996. Chitinozoa. *In*: JANSONIUS, J. & MCGREGOR, D. C. (eds) *Palynology: Principles and Practice*, Vol. I, *Principles*. American Association of Stratigraphic Palynologists Foundation, College Station, Texas, 307–336.

MILLER, M. A. & JANSONIUS, J. 1996. "Linotolypidae" and cenospheres. *In*: JANSONIUS, J. & MCGREGOR, D. C. (eds) *Palynology: Principles and Practice*, Vol. 1, *Principles*. American Association of Stratigraphic Palynologists Foundation, College Station, Texas, 357–359.

MOORE, L. R. 1946. On the spores of some Carboniferous plants; their development. *Quarterly Journal of the Geological Society of London*, **102**, 251–298.

MOREAU-BENOÎT, A. 1966. Étude des spores du Dévonien inférieur d'Avrille (Le Fléchau), Anjou. *Revue de Micropaléontologie*, **8**, 215–232.

MOREAU-BENOÎT, A. 1967. Premiers résultats d'une étude palynologique du Dévonien de la Carrière de Fours á Chaux d'Angers (Maine-et-Loire). *Revue de Micropaléontologie*, **9**, 219–240.

MORGENROTH, P. 1966*a*. Neue in organischer Substanz erhaltenen Mikrofossilien des Oligozäns. *Neues Jahrbuch für Geologie und Paläontologie, Abhandlungen*, **127**, 1–12.

MORGENROTH, P. 1966*b*. Mikrofossilien und Konkretionen des nordwesteuropaischen Untereozäns. *Palaeontographica*, Series B, **119**, 1–53.

MORGENROTH, P. 1968. Zur Kenntnis der Dinoflagellaten und Hystrichosphaeridien des Danien. *Geologisches Jahrbuch*, **86**, 533–578.

MORGENROTH, P. 1970. Dinoflagellate cysts from the Lias Delta of Lühnde/Germany. *Neues Jahrbuch für Geologie und Paläontologie, Abhandlungen*, **136**, 345–359.

MORRIS, J. 1840. A systematic catalogue of the fossil plants of Britain. *Magazine of Natural History*, **4**, 75–80, 179–183.

MUIR, M. D. & SARJEANT, W. A. S. 1971. An annotated bibliography of the Tasmanaceae and of related living forms (Algae: Prasinophyceae). *In*: JARDINÉ, S. (ed.) *Les acritarches. Microfossiles organiques du Paléozoique*. Éditions du Centre

National de la Recherche Scientifique, Paris, **3**, 52–117. (Republished in: MUIR, M. D. & SARJEANT, W. A. S. (eds) *Palynology, Part II: Dinoflagellates, Acritarchs and other Microfossils*. Benchmark Papers in Geology, **47**. Dowden, Hutchinson & Ross, Stroudsburg, 212–221.)

MUIR, M. D. & SARJEANT, W. A. S. 1977. *Palynology* (2 vols). Benchmark Papers in Geology Series, **47**, **48**. Dowden, Hutchinson & Ross, Stroudsburg, 351 pp: 414 pp.

MULLER, J. 1959. Palynology of recent Orinoco delta and shelf sediments. *Micropaleontology*, **5**, 1–32.

NAGY, E. 1965. The microplankton occurring in the Neogene of the Mecsek Mountains. *Acta Botanica Hungarica*, **11**, 197–216.

NAGY, E. 1968*a*. Moss spores in Hungarian Neogene strata. *Acta Botanica Academiae Scientiarum Hungaricae*, **14**, 113–132.

NAGY, E. 1968*b*. New spore genera from the Mecsek Mountains (Hungary). *Botanica Academiae Scientiarum Hungaricae*, **14**, 357–363.

NATHORST, A. G. 1886. Nagra ord on Visingsöserien. *Geologiska Föreningens i Stockholm Förhandlingar*, **8**, 5–23.

NAUMOVA, S. N. 1939. Spores and pollen of the coals of the U. S. S. R. *Transactions of the 17th International Geological Congress (1937)*, **1**, 353–364.

NAUMOVA, S. N. 1949. Lower Cambrian spores. *Isvestiya Akademiya Nauk, S. S. R.*, Ser. *Geologiya*, 49–56 (in Russian).

NAUMOVA, S. N. 1953. *Spore-pollen assemblages of the Russian Platform and their stratigraphical significance*. Akademiya Nauk S. S. S. R., Trudy Geologicheskogo Instituta, **60**, 204 pp. (in Russian).

NEALE, J. W. & SARJEANT, W. A. S. 1962. Microplankton from the Speeton Clay of Yorkshire. *Geological Magazine*, **99**, 439–458.

NEVES, R. 1958. Upper Carboniferous plant spore assemblages from the *Gastrioceras subcrenatum* Horizon, North Staffordshire. *Geological Magazine*, **95**, 1–18.

NEWTON, E. T. 1875. On "tasmanite" and Australian "white coal". *Geological Magazine*, Decade 2, **2**, 337–342.

NICHOLS, D. J. & TRAVERSE, A. 1971. Palynology, petrology and depositional environments of some Early Tertiary lignites in Texas. *Geoscience & Man*, **3**, 37–48.

NILSSON, S. & PRAGLOWSKI, J. (eds) 1992. *Erdtman's Handbook of Palynology* (2nd edn). Munksgaard, Copenhagen, 580 pp.

NOON, L. 1911. Prophylactic inoculation against hay fever. *Lancet*, **1**, 1572.

NOREM, W. L. 1955. *Tytthodiscus*, a new microfossil genus from the California Tertiary. *Journal of Paleontology*, **29**, 694–695.

NORRIS, G. & SARJEANT, W. A. S. 1965. *A descriptive index of genera of fossil Dinophyceae and Acritarcha*. New Zealand Geological Survey, Paleontological Bulletin, no. **40**, 721 pp.

ORŁOWSKA-ŻWOLIŃSKA, T. 1971. On several stratigraphically important species of sporomorphs occurring in the Keuper of Poland. *Acta Societatis Botanicorum Poloniae*, **40**, 633–651.

O'ROURKE, M. K. 1996. Medical palynology. *In*: JANSONIUS, J. & MCGREGOR, D. C. (eds), *Palynology: Principles and Practice*, Vol. 3, *New Directions, Other Applications and Floral History*. American Association of Stratigraphic Palynologists Foundation, College Station, Texas, 945–956.

OUYANG SU. 1962. The microspore assemblage from the Lungian Series of Changhsing, Chekiang. *Acta Palaeontologica Sinica*, **10**, 76–119 (in Chinese, with English summary).

OWENS, B. 1970. Recognition of the Devonian–Carboniferous boundary by palynological methods. *In*: STREEL, M. & WAGNER, R. H. (eds) *Colloque sur la Stratigraphie du Carbonifère*. Congrès et Colloques d'Université de Liegè, no. **55**, 349–364.

OWENS, B. & BURGESS, I. C. 1965. The stratigraphy and palynology of the Upper Carboniferous outlier of Stainmore, Westmorland. *Bulletin of the Geological Survey of Great Britain*, no. **23**, 17–44.

OWENS, B. & SARJEANT, W. A. S. 1999. Charles Downie 1923–1999. *Geological Society Annual Review*, no. **99**, 28–29.

PACLTOVA, B. 1966. Výsledkey mikropaleobotanických studii chatakvitéinského souvrotvi na Slovensku. *Rozpravy Česke Akademie Ved, Praha*, **76**, 54–65.

PASTIELS, A. 1948. *Contributions à l'étude des microfossiles de l'Éocene belge*. Mémoires du Muséum National d'Histoire Naturelle, Belgique, no. **109**, 77 pp.

PENDLETON, M. W., BRYANT, V. M. Jr & PENDLETON, B. B. 1996. Entomopalynology. *In*: JANSONIUS, J. & MCGREGOR, D. C. (eds) *Palynology: Principles and Practice*, Vol. 3, *New Directions, Other Applications and Floral History*. American Association of Stratigraphic Palynologists Foundation, College Station, Texas, 931–944.

PEPPERS, R. A. & PFEFFERKORN, H. W. 1970. A comparison of the floras of the Colchester (No. 2) coal and Francis Creek Shale. *In*: SMITH, W. H. *et al.* (eds) *Depositional Environments in Parts of the Carbondale Formation – Western and Northern Illinois*. Illinois State Geological Survey Guidebook Series, no. **8**, 61–74.

PFISTER, R. 1895. Versuch ein Mikroskopie des Honigs. *Forschungsberichte Lebensm. Bex. Hyg., Chemie und Pharmacie (München)*, **2**, 1–9, 29–35.

PICHLER, P. 1971. Mikrofossilien aus dem Devon der südlichen Eifeler Kalkmulden. *Senckenbergiana lethaea*, **52**, 315–357.

PIÉRART, P. 1955. Les mégaspores contenues dans quelques couches de houille du Westphalien B et C aux charbonnages Limbourg Meuse. *Publications de l'Association pour l'Étude de la Paléontologie et de la Stratigraphie Houillère*, **21** (*Hors Serié*, **8**), 125–142.

PIÉRART, P. 1958. L'utilisation des mégaspores en stratigraphie houillère. *Bulletin de la Société belge de Géologie, de Paléontologie et d'Hydrologie*, **64**, 587–599.

POCOCK, S. A. J. 1962. Microfloral analysis and age determination of strata at the Jurassic–Cretaceous boundary in the western Canada plains. *Palaeontographica*, Series B, **111**, 1–95.

POKROVSKAYA, I. M. 1956. *Atlas of Oligocene Spore-pollen Complexes of Various Regions of the U. S.*

S. R. Trudy Vsesojuznyi Nauchno-Issledovatel'skii Geologischeskii Institut (VSEGEI), New Series, **16** (in Russian).

POKROVSKAYA, I. M. (ed.). 1966. *Paleopalinologiya* (3 vols). Trudy Vsesojuznyi Nauchno-Issledovatel'skii Geologischeskii Institut (VSEGEI), New Series, **141**, 312 pp. (in Russian).

POTONIÉ, R. 1932. Pollenformen aus tertiären Braunkohle (3. Mitteilung). *Jahrbuch der Preussischen Geologischen Landesanstalt*, **52** (for 1931), 1–7.

POTONIÉ, R. 1934. Zur Mikrobotanik des eozänen Humodils des Geiseltales. *Arbeiten aus dem Institut für Paläobotanik und Petrographie des Brennsteine*, **4**, 25–125.

POTONIÉ, R. 1956. Synopsis der Gattungen der *Sporae dispersae.* I. Sporites. *Geologisches Jahrbuch Beihefte*, **23**, 1–103.

POTONIÉ, R. 1962. Synopsis der Sporae in situ. *Geologisches Jahrbuch, Beihefte*, **52**, 1–104.

POTONIÉ, R. 1967. Versuch der Einordnung der fossilen *Sporae dispersae* in das phylogenetische System der Pflanzenfamilien. *Forschungsberichte des Landes Nordrhein-Westfalen*, No. **1761,** 1–310.

POTONIÉ, R. 1970. *Synopsis der Sporae dispersae.* Beihefte zum Geologischen Jahrbuch, **87**.

POTONIÉ, R. & KREMP, G. O. W. 1954. Die Gattungen der paläozoischen Sporae dispersae und ihre Stratigraphie. *Geologisches Jahrbuch*, **69**, 111–193.

POTONIÉ, R. & KREMP, G. O. W. 1955. Die Sporae dispersae des Ruhrkarbons, ihre Morphologie und Stratigraphie mit Ausblicken auf Arten anderer Gebiete und Zeitabschnitte, I. *Palaeontographica*, Series B, **98**, 1–136.

POTONIÉ, R. & KREMP, G. O. W. 1956*a*. Die Sporae dispersae des Ruhrkarbons, ihre Morphologie und Stratigraphie mit Ausblicken auf Arten anderer Gebiete und Zeitabschnitte, II. *Palaeontographica*, Series B, **99**, 95–181.

POTONIÉ, R. & KREMP, G. O. W. 1956*b*. Die Sporae dispersae des Ruhrkarbons, ihre Morphologie und Stratigraphie mit Ausblicken auf Arten anderer Gebiete und Zeitabschnitte, III. *Palaeontographica*, Series B, **100**, 65–121.

RAISTRICK, A. 1935. The correlation of coal seams by microspore content. Part 1. The seams of Northumberland. *Transactions of the Institute of Mining Engineers*, **88**, 142–149.

RAISTRICK, A. & SIMPSON, J. 1933. The microspores in some Northumberland coals and their use in the correlation of coal seams. *Transactions of the Institute of Mining Engineers*, **85**, 225–235.

RALPH, T. S. 1865. Observations on the microscopical characters presented by a mineral (dysodil) from Tasmania. *Transactions of the Royal Society of Victoria*, **6**, 7.

RAMANUJAM, C. G. K. 1966. Palynology of the Miocene lignite from south Arcot district, Madras, India. *Pollen et Spores*, **8**, 149–203.

REINSCH, P. F. 1881. *Neue Untersuchungen über die Mikrostruktur der Steinkohle des Carbon des Dyas und Trias.* Weigel, Leipzig, 1241 pp.

REINSCH, P. F. 1884. *Micro-Palaeo-Phytologia Formationes Carboniferae* (2 vols). Krische, Erlangen.

REINSCH, P. F. 1905. Die Palinosphärien, ein mikroscopischer vegetabile Organismus in der Mukronatenkreide. *Centralblatt für Mineralogie, Geologie und Paläontologie*, 402–407.

RENAULT, B. 1899–1900. Sur quelques microorganismes des combustibles fossiles. *Bulletin de la Société industrielle et minière de St.-Etienne*, **13**, 865–1169; **14**, 5–159.

RHODES, F. H. T. 1951. Chitinozoa from the Ordovician Nod Glas Formation. *Nature*, **192**, 275–276.

RICHARDSON, J. B. 1962. Spores with bifurcate processes from the Middle Old Red Sandstone of Scotland. *Palaeontology*, **5**, 171–194.

RICHARDSON, J. B. & LISTER, T. R. 1969. Upper Silurian and Lower Devonian spore assemblages from the Welsh borderlands and South Wales. *Palaeontology*, **12**, 201–252.

ROGALSKÁ, M. 1954. *Analiza sporowo-pyłkowa liasowego węgla Blanowickiego z Górnego Śląska/Spore and Pollens* [sic] *analysis of the region of the so-called Blanowice coal in Upper Silesia.* Biuletyn Instytut Geologiczny, Warsawa, **89**. (copy not located).

ROGALSKÁ, M. 1956. *Analiza sporowo-pyłkowa liasowych osadów obszaru mroczków-rozwady w powiecie opoczyńskim.* Biuletyn Instytut Geologiczny, Warsawa, **104**.

ROLFE, W. D. I. 1962. The cuticle of some Middle Silurian ceratiocarid Crustacea from Scotland. *Palaeontology*, **5**, 30–51.

ROSSIGNOL, M. 1962. Analyse pollinique de sédiments marins quaternaires en Israël. II. Sédiments Pleistocènes. *Pollen et Spores*, **4**, 121–148.

ROSSIGNOL, M. 1963. Aperçus sur le developpement des Hystrichosphères. *Bulletin du Muséum National d'Histoire Naturelle*, Series 2, **35**, 207–212.

ROUSE, G. E. 1957. The application of a new nomenclatural approach to Upper Cretaceous plant microfossils from western Canada. *Canadian Journal of Botany*, **35**, 349–375.

SAMOILOVICH, S. R. 1953. Pollen and spores from the Permian deposits of the Cherdyn and Aktiubinok areas, Cis-Urals. *Palaeobotanickeskii Sbornik*, New Series, **75**, 5–57 (in Russian; English translation by M. K. Elias, *Oklahoma Geological Survey Circular*, No. **56**, 1961).

SAMOILOVICH, S. R. 1961. *Pollen and spores of western Siberia: Jurassic–Paleocene.* Trudy Vsesojuznyi Nauchno-Issledovatel'skii Geologischeskii Institut (VNIGRI), New Series, **75**, 92 pp. (in Russian).

SARAUW, G. F. L. 1897. Cromer-skovlaget i Frihavnen og traelevningerne i de ravførende sandlad ved København. *Meddelelser fra Dansk Geologisk Forening*, **4**, 17–44.

SARJEANT, W. A. S. 1959. Microplankton from the Cornbrash of Yorkshire. *Geological Magazine*, **96**, 329–346.

SARJEANT, W. A. S. 1960*a*. The mystery of the hystrichospheres. *Journal of the University of Sheffield Geological Society*, **3**, 161–167.

SARJEANT, W. A. S. 1960*b*. New hystrichospheres from the Upper Jurassic of Dorset. *Geological Magazine*, **97,** 137–144.

SARJEANT, W. A. S. 1960*c*. Microplankton from the

Corallian rocks of Yorkshire. *Proceedings of the Yorkshire Geological Society*, **32**, 389–408.
SARJEANT, W. A. S. 1961*a*. The hystrichospheres: a review and discussion. *Grana Palynologica*, **2**, 102–111.
SARJEANT, W. A. S. 1961*b*. Microplankton from the Kellaways Rock and Oxford Clay of Yorkshire. *Palaeontology*, **4**, 90–118.
SARJEANT, W. A. S. 1963*a*. Fossil algae and modern rock-dating. *New Scientist*, **18**, 668–670.
SARJEANT, W. A. S. 1963*b*. Fossil dinoflagellates from Upper Triassic sediments. *Nature*, **199**, 353–354.
SARJEANT, W. A. S. 1964. The stratigraphic application of fossil microplankton (dinoflagellates and hystrichospheres) in the Jurassic. *Colloque du Jurassique, Luxembourg 1962, vol. Conptes-rendus et Mémoires*, 441–448.
SARJEANT, W. A. S. 1965. The Xanthidia. *Endeavour*, **24**, 33–39 (published also in French, German, Italian, and Spanish).
SARJEANT, W. A. S. 1966*a*. The supposed "sponge spicules" of Merrill, 1895, from the Lower Cretaceous (Albian) of Texas. *Breviora, Museum of Comparative Zoology, Harvard*, **242**, 1–15.
SARJEANT, W. A. S. 1966*b*. Microplankton from the Callovian (*S. calloviense* Zone) of Normandy. *Revue de Micropaléontologie*, **8**, 175–184.
SARJEANT, W. A. S. 1966*c*. Dinoflagellate cysts with *Gonyaulax*-type tabulation. *Bulletin of the British Museum (Natural History), Geology*, Supplement **3**, 107–156.
SARJEANT, W. A. S. 1966*d*. Further dinoflagellates from the Speeton Clay (Lower Cretaceous). *Bulletin of the British Museum (Natural History), Geology*, Supplement **3**, 199–214.
SARJEANT, W. A. S. 1967*a*. The rediscovery of a lost species of dinoflagellate cyst *Hystrichosphaera* (ex: *Spiniferites*) *reginaldi* (Mantell 1844) comb. nov. *Microscopy*, **30**, 241–250.
SARJEANT, W. A. S. 1967*b*. The stratigraphic distribution of fossil dinoflagellates. *Review of Palaeobotany and Palynology*, **1**, 323–343.
SARJEANT, W. A. S. 1968. Microplankton from the Upper Callovian and Lower Oxfordian of Normandy. *Revue de Micropaléontologie*, **10**, 221–242.
SARJEANT, W. A. S. 1970*a*. Xanthidia, Palinospheres and "Hystrix". A review of the study of fossil unicellular microplankton with organic cell walls. *Microscopy*, **31**, 221–253. (Republished in: MUIR, M. D. & SARJEANT, W. A. S. (eds), *Palynology, Part II: Dinoflagellates, Acritarchs and Other Microfossils*. Benchmark Papers in Geology, **47**. Dowden, Hutchinson & Ross, Stroudsburg, 8–40.)
SARJEANT, W. A. S. 1970*b*. Recent developments in the application of fossilized planktonic organisms to problems of stratigraphy and palaeoecology. *Paläobotanik*, Series B, **3**, 669–680.
SARJEANT, W. A. S. 1970*c*. Acritarchs and tasmanitids from the Chhidru Formation, uppermost Permian of West Pakistan. *In*: KUMMEL, B. & TEICHERT, C. (eds) *Stratigraphic Boundary Problems: Permian and Triassic of West Pakistan*. Department of Geology, University of Kansas, Special Publications no. **4**, 277–304.
SARJEANT, W. A. S. 1973. Two great palynologists: Gunnar Erdtman and Georges Deflandre. *Microscopy*, London, **33**, 319–331.
SARJEANT, W. A. S. 1974. *Fossil and Living Dinoflagellates*. Academic Press, London & New York, 184 pp.
SARJEANT, W. A. S. 1978*a*. Hundredth year memoriam: Christian Gottfried Ehrenberg 1795–1876. *Palynology*, **2**, 209–211.
SARJEANT, W. A. S. 1978*b*. *Arpylorus antiquus* Calandra, emend., a dinoflagellate cyst from the Upper Silurian. *Palynology*, **2**, 167–179.
SARJEANT, W. A. S. 1981. A restudy of some dinoflagellate cyst holotypes in the University of Kiel Collections. II. The Eocene holotypes of Barbara Klumpp (1953); with a revision of the genus *Cordosphaeridium* Eisenack, 1963. *Meyniana*, **33**, 97–132.
SARJEANT, W. A. S. 1982. Joseph B. Reade (1801–1870) and the earliest studies of fossil dinoflagellates in England. *Journal of Micropalaeontology*, **1**, 85–93.
SARJEANT, W. A. S. 1983. A restudy of some dinoflagellate cyst holotypes in the University of Kiel Collections. IV. The Oligocene and Miocene holotypes of Dorothea Maier (1959). *Meyniana*, **35**, 85–137.
SARJEANT, W. A. S. 1984*a*. A restudy of some dinoflagellate cyst holotypes in the University of Kiel Collections. V. The Danian (Palaeocene) holotypes of Walter Wetzel (1952, 1955). *Meyniana*, **36**, 121–171.
SARJEANT, W. A. S. 1984*b*. Charles Downie and the early days of palynological research at the University of Sheffield. *Journal of Micropalaeontology*, **3**, 1–6.
SARJEANT, W. A. S. 1985. Alfred Eisenack (1891–1982) and his contribution to palynology. *Review of Palaeobotany and Palynology*, **45**, 3–15.
SARJEANT, W. A. S. 1986. A restudy of Pastiels' (1948) dinoflagellate cysts from the Early Eocene of Belgium. *Bulletin de l'Institut royal des Sciences naturelles de Belgique*, no. **56**, 5–43.
SARJEANT, W. A. S. 1991*a*. Henry Hopley White (1790–1877) and the early researches on Chalk "Xanthidia" (marine palynomorphs) by Clapham microscopists. *Journal of Micropalaeontology*, **10**, 83–93.
SARJEANT, W. A. S. 1991*b*. Sclerites, spicules and systematics: the researches of Marthe Deflandre-Rigaud (1902–1987). *Micropaleontology*, **37**, 191–195
SARJEANT, W. A. S. 1992*a*. Gideon Mantell and the "Xanthidia". *Archives of Natural History*, **19**, 91–100.
SARJEANT, W. A. S. 1992*b*. Microfossils other than spores and pollen in palynological preparations. *In*: NILSSON, S. & PRAGLOWSKI, J. (eds) *Erdtman's Handbook of Palynology* (2nd edn). Munksgaard, Copenhagen, 468–525.
SARJEANT, W. A. S. 1998. From excystment to bloom? Personal recollections of thirty-five years of dinoflagellate and acritarch meetings. *Norges*

teknisk-naturvitenskapelige universitet Vitenskapsmuseet. Rapport botanisk, Series 1998–2, 1–21.

SARJEANT, W. A. S. 1999. Obituary: Charles Downie (1923–1999). *Acritarch Newsletter*, no. **15**, 4–8.

SARJEANT, W. A. S. 2000. Charles Downie (1923–1999) and his work on fossil dinoflagellates. *Stuifmail*, **18**, 33–44.

SARJEANT, W. A. S. & VANGUESTAINE, M. 1999. Maria Lejeune-Carpentier (1910–1995): a memorial. *Journal of Micropalaeontology*, **18**, 137–142.

SCHEURING, B. W. 1970. Palynologische und palynostratigraphische Untersuchungen des Keupers in Bölchentunnel (Solothurner Jura). *Schweizerische Paläontologische Abhandlungen*, **88**.

SCHIMPER, P. W. 1869–1874. *Traité de paléontologie végétale, ou, La flore du monde primitif dans ses rapports avec les formations géologiques et la flore du monde actuel* (3 vols and atlas). Ballière, Paris. (Vol. 2, 1870, relevant).

SCHOPF, J. M. & BARGHOORN, E. S. 1967. Alga-like fossils from the Early Precambrian of South Africa. *Science*, **156**, 508–512.

SCHOPF, J. M., WILSON, L. R. & BENTALL, R. 1944. An annotated synopsis of Palaeozoic fossil spores and the definition of generic groups. *Reports of Investigations, Illinois Geological Survey*, **91**, 1–73.

SCHOPF, J. V. 1938. Spores from the Herrin (No. 6) coal bed in Illinois. *Illinois State Geological Survey, Report of Investigations*, **50**, 1–73.

SCHOPF, J. V. 1968. Microflora of the Bitter Springs Formation, Late Precambrian, central Australia. *Journal of Paleontology*, **42**, 651–688.

SCHULTZ, G. 1967. Mikrofossilien des oberen Llandovery von Dalarne (Schweden). *Sonderveröffentlichungen des Geologischen Institutes der Universität Köln*, **13**, 175–187.

SCHULZ, E. 1962. Sporen-paläontologische Untersuchungen zur Rhät-Lias-Grenze in Thüringen und der Altmark. *Geologie*, **11**, 308–319.

SCHULZ, E. 1965. Sporae dispersae auf der Trias von Thüringen. *Mitteilungen aus dem Zentralen Geologichen Institut*, **1**, 257–287.

SCHULZE, F. 1855. Über das Vorkommen wollerhaltener Zellulose in Braunkohle und Steinkohle. *Bericht über die Verhandlungen der Königlich Preussischen Akademie der Wissenschaften*, 676–678.

SCHWAB, K. W. 1966. Microstructure of some fossil and recent scolecodonts. *Journal of Paleontology*, **40**, 416–423.

SCOTT, A. C. & HEMSLEY, A. R. 1996. Palaeozoic megaspores. *In*: JANSONIUS, J. & MCGREGOR, D. C. (eds) *Palynology: Principles and Practice*, Vol. 2, *Applications*. American Association of Stratigraphic Palynologists Foundation, College Station, Texas, 629–640.

SEGROVES, K. L. 1967. Cutinized microfossils of probable nonvascular origin from the Permian of Western Australia. *Micropaleontology*, **13**, 289–305.

SEGROVES, K. L. 1969. Saccate plant microfossils from the Permian of Western Australia. *Grana Palynologica*, **9**, 174–227.

SELLING, O. H. 1951. Lennart von Post, 16/6 1884–11/1 1951. *Svensk Botanisk Tidskrift, Stockholm*, **45**, 275–296.

SEMENOVA, E. V. 1970. *Spores and Pollen from the Jurassic Deposits Bordering on the Triassic Beds of Donbassa*. Akademiya Nauk S. S. S. R., Ukrainskoi S S R, Instityt Geologicheskikh Nauk, Kiev, 144 pp. (in Russian).

SLATER, L., EVANS, M. M. & EDDY, G. E. 1930. The significance of spores in the correlation of coal seams. *Fuel Research*, **17**, 1–28.

SMITH, A. H. V. 1962. The palaeoecology of Carboniferous peats based on the miospcores and petrography of bituminous coals. *Proceedings of the Yorkshire Geological Society*, **33**, 428–439, 446–465.

SMITH, A. H. V. & BUTTERWORTH, M. A. 1967. *Miospores in the coal seams of the Carboniferous of Great Britain*. Palaentogical Association, Special Papers in Palaeontology, **1**, 324 pp.

SOBOLIK, K. D. 1996. Pollen as a guide to prehistoric diet reconstruction. *In*: JANSONIUS, J. & MCGREGOR, D. C. (eds) *Palynology: Principles and Practice*, Vol. 3, *New Directions, Other Applications and Floral History*. American Association of Stratigraphic Palynologists, College Station, Texas, 927–932.

SOLLAS, W. J. 1901. Fossils in the Oxford University Museum, V. On the structure and affinities of the Rhaetic plant *Naiadita*. *Quarterly Journal of the Geological Society of London*, **57**, 307–312.

SOMMER, F. W. 1953. Os esporomorfos do folhelho de Barreirinha. *Divisão Geológico e Minerio Rio de Janeiro*, no. **140**, 1–59.

SOMMER, F. W. & VAN BOEKEL, N. M. 1963. Some new Tasmanaceae from the Devonian of Pará. *Anais da Academia Brasileira de Ciências*, **35**, 61–65.

SPECTOR, D. L. (ed.) 1984. *Dinoflagellates*. Academic Press, Orlando.

SPINNER, E. G. 1965. Westphalian megaspores from the Forest of Dean Coalfield, England. *Palaeontology*, **8**, 82–106.

STANCLIFFE, R. P. W. 1996. Microforaminiferal linings. *In*: JANSONIUS, J. & MCGREGOR, D. C. (eds) *Palynology: Principles and Practice*, Vol. 1, *Principles*. American Association of Stratigraphic Palynologists Foundation, College Station, Texas, 373–380.

STANLEY, E. A. 1961*a*. A new sporomorph genus from northwestern South Dakota. *Pollen et Spores*, **3**, 155–162.

STANLEY, E. A. 1961*b*. The fossil pollen genus *Aquilapollenites*. *Pollen et Spores*, **3**, 329–352.

STANLEY, E. A. 1965. Upper Cretaceous and Paleocene plant microfossils and Paleocene dinoflagellates and hystrichosphaerids from northwestern South Dakota. *Bulletins of American Paleontology*, **49**, 177–384.

STANLEY, E. A. 1966. The problem of reworked pollen and spores in marine sediments. *Marine Geology*, **4**, 397–408.

STANLEY, E. A. 1967*a*. Cretaceous spore and pollen assemblages from northern Alaska. *Review of Palaeobotany and Palynology*, **1**, 229–234.

STANLEY, E. A. 1967*b*. Palynology of six ocean-bottom

cores from the southwestern Atlantic Ocean. *Review of Palaeobotany and Palynology*, **2**, 195–203.

STANLEY, E. A. 1969. The occurrence and distribution of pollen and spores in marine sediments. *Proceedings of the First International Conference on Planktonic Microfossils, Geneva, 1967*. Brill, Leiden, 640–643.

STAPLIN, F. L., JANSONIUS, J. & POCOCK, S. A. J. 1965. Evaluation of some acritarchous hystrichosphere genera. *Neues Jahrbuch für Geologie und Paläontologie, Abhandlungen*, **123**, 167–201.

STAUFFER, C. R. 1933. Middle Ordovician Polychaeta from Minnesota. *Bulletin of the Geological Society of America*, **44**, 1173–1218.

STAUFFER, C. R. 1939. Middle Devonian Polychaeta from the Lake Erie district. *Journal of Paleontology*, **13**, 500–511.

STOCKMANS, F. & WILLIÈRE, Y. 1960. Hystrichosphères du Dévonien belge (Sondage de l'Asile d'aliénés à Tournai). *Senckenbergiana lethaea*, **41**, 1–11.

STOCKMANS, F. & WILLIÈRE, Y. 1963. Les Hystrichosphères ou mieux les Acritarches du Silurien belge. Sondage de la Brasserie Lust à Courtrai (Kortrijk). *Bulletin de la Société Belge de Géologie, Paléontologie et Minéralogie*, **71**, 450–481.

STREEL, M. 1964. Une association de spores du Givétien inférieur de la Vesdre à Goé (Belgique). *Annales de la Société géologique de Belgique*, **87**, 1–30.

SULLIVAN, H. J. 1958. The microspore genus *Simozonotriletes*. *Palaeontology*, **1**, 125–138.

SULLIVAN, H. J 1962. Distribution of miospores through coals and shales of the Coal Measures sequence exposed in Wernddu Claypit, Caerphilly (south Wales). *Quarterly Journal of the Geological Society of London*, **118**, 353–373.

SZANIAWSKI, H. 1968. Three new polychaete jaw apparatuses from the Upper Permian of Poland. *Acta Palaeontologica Polonica*, **13**, 225–281.

SZANIAWSKI, H. 1970. Jaw apparatuses of the Ordovician and Silurian polychaetes from the Mielnik borehole. *Acta Palaeontologica Polonica*, **15**, 445–478.

SZANIAWSKI, H. 1996. Scolecodonts. *In*: JANSONIUS, J. & MCGREGOR, D. C. (eds) *Palynology: Principles and Practice*, Vol. I, *Principles*. American Association of Stratigraphic Palynologists Foundation, College Station, Texas, 337–354.

TAKAHASHI, K. 1961. Pollen und Sporen des westjapanischen Alttertiärs und Miozäns (Pt II). *Memoirs of the Faculty of Science, Kyushu University*, Series D (*Geology*), **11**, 279–345.

TAPPAN, H. & LOEBLICH, A. R., Jr 1965. Foraminiferal remains in palynological preparations. *Revue de Micropaléontologie*, **8**, 61–63.

TASCH, P. 1963. Hystrichosphaerids and dinoflagellates from the Permian of Kansas. *Micropaleontology*, **9**, 332–336.

TAUGOURDEAU, P. 1961. Chitinozoaires du Silurien d'Aquitaine. *Revue de Micropaléontologie*, **6**, 135–154.

TAUGOURDEAU, P. 1964. Sporulation ou enkystement chez un *Ancyrochitina* (Chitinozoaire). *Comptes-rendus sommaires des Séances de la Société géologique de France*, no. **6**, 238–239.

TAUGOURDEAU, P. 1965. Chitinozoaires de l'Ordovicien des U. S. A.: comparaisons avec faunes de l'ancien monde. *Revue de l'Institut Français du Pétrôle*, **20**, 463–484.

TAUGOURDEAU, P. 1966. Les Chitinozoaires: techniques d'études, morphologie et classification. *Mémoires de la Société géologique de France*, **45**, 1–64.

TAUGOURDEAU, P. 1967. Débris microscopiques d'eurypteridés du Paléozoique Saharien. *Revue de Micropaléontologie*, **10**, 119–127.

TAUGOURDEAU, P. 1968. Les scolécodontes du Siluro-Devonien et du Carbonifère de sondages Sahariens: stratigraphie – systématique. *Revue de l'Institut Français du Pétrôle*, **23**, 1219–1271.

TAUGOURDEAU, P., BOUCHÉ, P., COMBAZ, A., MAGLOIRE, L. & MILLEPIED, P. 1967. *Microfossiles organiques du Paléozoique. I. Les Chitinozoaires: analyse bibliographique illustré*. Editions du Centre National de la Recherche Scientifique, Paris, 96 pp.

TAUGOURDEAU, P. & JEKHOWSKY, B. de. 1960. Répartition et description des Chitinozoaires Siluro-Dévoniens de quelques, sondages de la C. R. E. P. S., de la C. F. I. H. et de la S. N. Réalau Sahara *Revue de l'Institut Français du Pétrole*, **15**, 1199–1260.

TAYLOR, F. J. R. (ed.). 1987. *The Biology of Dinoflagellates*. Botanical Monograph Series, **21**, Blackwell Scientific Publications, Oxford, 785 pp.

TE PUNGA, M. T. 1948. *Nothofagus* pollen from the Cretaceous Coal Measures at Kaitangata, Otago, New Zealand. *New Zealand Journal of Science and Technology*, Series B, **29**, 32–35.

TE PUNGA, M. T. 1949. Fossil spores from New Zealand coals. *Transactions of the Royal Society of New Zealand*, **77**, 289–296.

THIERGART, F. 1944. Die Pflanzenreste der Posidonienschiefer. *In*: BROCKAMP, B., *Zur Paläogeographie und Bitumenführung des Posidonienschiefers im Deutschen Lias*. Archiv für Lagerstattenforschungen, **77**, 1–59.

THIESSEN, R. 1920. *Structure in Palaeozoic bituminous coals*. United States Bureau of Mines Bulletin, No. **117**.

THIESSEN, R. 1925*a*. Microscopic examination of the Kentucky oil shales. *In*: THIESSEN, T. R., WHITE, D. & CROUSE, C. T. (eds) *Oil Shales of Kentucky*, Publications of the Kentucky Geological Survey, Series 6, **21**, 1–47.

THIESSEN, R. 1925*b*. Origin of the Boghead Coals. *US Geological Survey, Professional Papers*, no. **32**, 121–135.

THOMAS, H. H. 1933. On some pteridospermous plants from the Mesozoic of South Africa. *Philosophical Transactions of the Royal Society of London*, Series B, **222**, 193–265.

TIMOFEYEV, B. V. 1955. Finds of spores in Cambrian and Precambrian deposits of eastern Siberia. *Doklady Akademiya Nauk S. S. S. R.*, **105**, 547–550 (in Russian).

TIMOFEYEV, B. V. 1956. Hystrichosphaeridae of the Cambrian period. *Doklady Akademiya Nauk S. S. S. R.*, **106**, 130–132 (in Russian).

TIMOFEYEV, B. V. 1959. The oldest flora of the Prebaltic region and its stratigraphic importance. *Trudy Leningrad (VNIGRI) Geokhimiceskiy Sbornik*, **129**, 1–136, 147 pp. (in Russian).

TIMOFEYEV, B. V. 1965. Fitoplankton pozdnego proterozoya i rannego paleozoya. *Abstracts of Lectures to the First All Soviet Union Paleogeological Conference, Novosibirsk*, 112–114.

TIMOFEYEV, B. V. 1966. *Micropalaeophytological investigation of ancient suites*. Moscow and Leningrad: Nauka (in Russian; English translation 1974: *Micropalaeophytological Research into Ancient Strata*. British Library, Lending Division), 214 pp.

TIMOFEYEV, B. V. 1967. K mikropaleofitologicheskoj harakteristike venda i nizov nizhnego kembriya. *All Soviet Union Conference on Stratigraphy of Marginal Sediments of the Precambrian and Cambrian, May 5–8, Ufa*, 13.

TIMOFEYEV, B. V. 1969. *Sphaeromorphyda of the Proterozoic*. Akademiya Nauka S. S. S. R., Leningradskoe Otdelenie, Izdatelstva Nauka, Leningrad, 145 pp. (in Russian).

TIMOFEYEV, B. V. 1973*b*. *Mikrofitofossilii dokembriya Ukrainy*. Akademia Nauk S. S. S. R., Institut Geologii i Geokhronologii Dokembriya, Leningradskoe Otdelenie, Izdatelstva Nauka, Leningrad.

TIMOFEYEV, B. V., GERMAN, T. N. & MIKHAILOVA, N. S. 1976. *Microphytofossils of the Precambrian, Cambrian and Ordovician*. Akademia Nauk S. S. S. R., Institut Geologii i Geokhronologii Dokembriya, Leningradskoe Otdelenie, Izdatelstva Nauka, Leningrad, 106 pp. (in Russian).

TRALAU, H. 1967. Some Middle Jurassic microspores of southern Sweden. *Geologiska Föreningens Stockholm, Förhandlingar*, **89**, 469–472.

TRAVERSE, A. 1955*a*. *Pollen analysis of the Brandon Lignite of Vermont*. US Department of the Interior, Bureau of Mines, Reports, **5151** (Republished in part in: MUIR, M. D. & SARJEANT, W. A. S. (eds) *Palynology, Part I*. Benchmark Papers in Geology, **46**. Dowden, Hutchinson & Ross, Stroudburg, 167–183).

TRAVERSE, A. 1955*b*. Occurrence of the oil-forming alga *Botryococcus* in lignites and other Tertiary sediments. *Micropaleontology*, **1**, 343–350.

TRAVERSE, A. 1974. Paleopalynology 1947–1952. *Annals of the Missouri Botanical Garden Society*, **61**, 203–236.

TRAVERSE, A. 1988. *Paleopalynology*. Unwin Hyman, Boston, 600 pp.

TRAVERSE, A., CLISBY, K. H. & FOREMAN, F. 1961. Pollen in drilling-mud "thinners", a source of palynological contamination. *Micropaleontology*, **7**, 375–377.

TRAVERSE, A. & SULLIVAN, H. J. 1983. The background, origin, and early history of the American Association of Stratigraphic Palynologists. *Palynology*, **7**, 7–17.

TRYBOM, F. 1888. Bottenprof från svenska insjöar. *Geologiska Föreningens i Stockholm Förhandlingar*, **10**, 489–511.

TSCHUDY, R. H. 1961. Palynomorphs as indicators of facies environments in Upper Cretaceous and lower Tertiary strata, Colorado and Wyoming. *Symposium on Late Cretaceous rocks, Wyoming and adjacent areas, The 16th Wyoming Geological Association Annual Field Conference*, 53–59.

TSCHUDY, R. H. & SCOTT, R. A. (eds) 1969. *Aspects of Palynology*. Wiley, New York.

TURPIN, C. R. 1837. Analyse ou étude microscopique des différents corps organisés et autre corps de nature diverse qui peuve, accidentellement, se trouver envelloper dans le pâte translucide des silex. *Comptes-rendus hébdomadaires de l'Académie des Sciences*, Paris, **4**, 304–314, 351–362.

UPSHAW, C. F. & CREATH, W. B. 1965. Pennsylvanian miospores from a cave deposit in Devonian limestone, Calloway County, Missouri. *Micropaleontology*, **11**, 431–448.

UPSHAW, C. F & HEDLUND, R. W. 1967. Microspores from the upper part of the Cofferville Formation (Pennsylvanian, Missourian), Tulsa County, Oklahoma. *Pollen et Spores*, **9**, 143–170.

VALENSI, L. 1948. Sur quelques micro-organismes planctoniques des silex du Jurassique moyen du Poitou et de Normandie. *Bulletin de la Société Géologique de France*, Series 5, **18**, 537–550.

VALENSI, L. 1953. Microfossiles des silex du Jurassique moyen. Remarques pétrographiques. *Mémoires de la Société Géologique de France*, no. **68**, 100 pp.

VALENSI, L. 1955. Sur quelques micro-organismes des silex du Magdalénien de Saint-Amand (Cher). *Bulletin de la Société Géologique de France*, Series 6, **5**, 35–40.

VALENSI, L. 1960. De l'origine des silex protomagdaléniens de l'Abri Pataud, les Eyzies. *Bulletin de la Société Préhistorique de France*, **57**, 80–84.

VANGUESTAINE, M. 1967. Découverte d'acritarches dans le Revinien supérieur du Massif de Stavelot. *Annales de la Société Géologique de Belgique*, no. **90**, Bulletin 4–6, 585–600.

VANGUESTAINE, M. 1968. Les acritarches du sondage de Grand Halleux. *Annales de la Société géologique de Belgique*, no. **91**, 361–375.

VARIUKHINA, I. M. 1971. *The Spores and Pollen of Red-coloured and Coal-bearing Deposits of the Permian and Triassic in the Northeast Part of Russia*. Akademiya Nauk S. S. S. R., Komi Otdelenie, Institut Geologii, Leningrad, 158 pp. (in Russian).

VARMA, C. P. & RAWAT, M. S. 1963. A note on some disporate grains recovered from Tertiary horizons of India and their potential marker value. *Grana Palynologica*, **4**, 130–139.

VEEN, F. R. VAN 1957. Microforaminifera. *Micropaleontology*, **3**, 74.

VENKATACHALA, B. S. & KAR, R. K. 1967. Palynology of the Kathwai Shales, Salt Range, West Pakistan. 1. Shales 25 ft. above the Talchir Boulder Bed. *Palaeobotanist*, **16**, 156–166.

VENKATACHALA, B. S., KAR, R. K. & RAZA, M. 1968. Carboniferous spores and pollen from the Calareti Zone of the Moesian Platform, Rumania. *Palaeobotanist*, **17**, 68–79.

VENOZHINSKENE, A. I. 1964. Spore assemblages of the Stonishkáy, Sh'ashuvisis and Viesite Formations of Estonia. *Voprosy Stratigrafi i Paleogeografi i Devona Pribaltiki, Vilnius, Issledovaniya Gosgeolkom S. S. S. R.*, 42–45 (in Russian).

VIGRAN, J. O. 1964. Spores from Devonian deposits, Mimerdalen, Spitzbergen. *Norsk Polarinstitutts Skrifter*, no. **132**, 32 pp.
VIRKKI, C. 1937. On the occurrence of winged spores in the Lower Gondwana rocks of India and Australia. *Proceedings of the Indian Academy of Sciences*, Section B, **6**, 428–431.
VISSCHER, H. 1966. Palaeobotany of the Mesophytic III. Plant microfossils from the Upper Bunter of Hergelo, The Netherlands. *Acta Botanica Neerlandica*, **15**, 316–375.
VOLKHEIMER, W. 1968. Esporas y granos de polen del Jurasico de Neuquén (Republica Argentina). I. Descriptiones sistemáticas. *Ameghiniana*, **5**, 333–370.
VOLKOVA, N. A. 1968. Acritarchs from the Precambrian and Lower Cambrian deposits of Estonia. *In*: VOLKOVA, N. A., ZHURAVLEVA, Z. A., ZABRODIN, V. E. & KLINGER, B. (eds) *Problematic Riphean–Cambrian boundary layers of the Russian Platform, the Urals and Kazakhstan.* Trudy Geologikheshogo Instituta, Akademiya Nauk S. S. S. R., **188**, 8–36 (in Russian).
VOLKOVA, N. A. 1969. Acritarchs of the north-western Russian platform. *In*: ROZANOV, A. Y. *et al.* (eds) *The Tommotian Stage and the Cambrian Lower Boundary Problem.* Nauka, Moscow, 224–236 (in Russian).
VON POST, L. 1916. Einige südschwedischen Quellmoore. *Bulletin of the Geological Institute of Upsala*, **15**, 218–278.
VON POST, L. 1918. Skogsträdspollen i sydsvenska torvmosselagerföljder. *Förhandlingar Skandinavisken Naturforskeres, 16 møte, 1916*, 432–465.
VON POST, L. 1927. Myrmarker. *In*: MUNTHE, H., HEDE, J. E. & VON POST, L., *Beskrivning till Kartbladet Hemse.* Sveriges Geologiska Undersökning, Series Aa, no. **164**, 101–138.
VOZZHENNIKOVA, T. F. 1960. Palaeoalgological characteristics of the Mesozoic and Cainozoic beds of the West Siberian lowland. *Akademii Nauk S. S. S. R., Sibirskoi Oldelenyi, Trudy Instituta Geologii i Geografii*, **1**, 7–64 (in Russian).
WALL, D. 1962. Evidence from Recent plankton regarding the biological affinities of *Tasmanites* Newton, 1875 and *Leiosphaeridia* Eisenack, 1958. *Geological Magazine*, **99**, 353–362.
WALL, D. 1965*a*. Microplankton, pollen and spores from the Lower Jurassic in Britain. *Micropaleontology*, **11**, 151–190.
WALL, D. 1965*b*. Modern hystrichospheres and dinoflagellate cysts from the Woods Hole region. *Grana Palynologica*, **6**, 297–314.
WALL, D. & DALE, B. 1967. The resting cysts of modern marine dinoflagellates and their palaeontological significance. *Review of Palaeobotany and Palynology*, **2**, 349–354.
WALL, D. & DOWNIE, C. 1963. Permian hystrichospheres from Britain. *Palaeontology*, **5**, 770–784.
WEBER, C. A. 1893. Über die diluviale Flora von Fahrenberg in Holstein. *Botanisches Jahrbuch*, **18**, Beiblatt, 43.
WEBER, C. A. 1896. Zur Kritik interglacialer Pfanzenablagerungen. *Abhandlungen herausgegeben vom Naturwissenschaftlichen Verein zu Bremen*, **13**, 413–468.
WEICHARDT, J. W. 1905. *Serologische Studien aus dem Gebiete der experimentellen Therapie.* Habilitationsschrift, University of Erlangen, Stuttgart, 60 pp.
WETHERED, E. 1886. On the occurrence of spores of plants in the Lower Limestone Shales of the Forest of Dean Coalfield and in the black shales of Ohio, United States. *Proceedings of the Cotteswold Naturalists' Field Club*, **8**, 167–173.
WETHERELL, E. W. 1892. On the occurrence of Xanthidia (Spiniferites of Mantell) in the London Clay of the Isle of Sheppey. *Geological Magazine*, Decade 3, **9**, 28–30.
WETZEL, O. 1932. Die Typen der Baltischen Geschiebefeuersteine, beurteilt nach ihrem Gehalt an Mikrofossilien. *Zeitschrift für Geschiebeforschung*, **8**, 129–146.
WETZEL, O. 1933. Die in organischer Substanz erhaltenen Mikrofossilien des Baltischen Kreide-Feuersteins. *Palaeontographica*, Series A, **77**, 141–188; **78**, 1–110.
WETZEL, O. 1951. Die Mikropaläontologie des baltischen Kreide-Feuersteins, auch eine Angelegenheit der modernen Paläobotanik. *Svensk Botanisk Tidskrift*, **45**, 249–253.
WETZEL, O. 1957. Fossil 'microforaminifera' in various sediments and their reaction to acid treatment. *Micropaleontology*, **3**, 61–64.
WETZEL, O. 1967. Rätselhafte Mikrofossilien der Oberlias (epsilon); neue Funde von 'Anellotubulaten' O. We. 1957. *Neues Jahrbuch für Geologie und Paläontologie, Abhandlun*gen, **128**, 341–349.
WETZEL, W. 1922. Sediment-petrographische Studien. I. Feuerstein. *Neues Jahrbuch für Mineralogie, Geologie und Paläontologie, Beilagebände*, **47**, 39–92.
WETZEL, W. 1952. Beitrag zur Kenntnis des danzeitlichen Meeresplanktons. *Geologisches Jahrbuch der deutschen geologischen Landesanstalt*, **66**, 391–417.
WETZEL, W. 1955. Die Dan-Scholle vom Katharinenhof (Fehmarn) und ihr Gehalt an Planktonten. *Neues Jahrbuch für Geologie und Paläontologie, Monatschefte*, no. **1**, 30–46.
WHITE, D. 1929. Description of fossil plants found in some "mother rocks" of petroleum from northern Alaska. *AAPG Bulletin*, **13**, 841–848.
WHITE, M. C. 1862. Discovery of microscopic organisms in the siliceous nodules of the Palaeozoic rocks of New York. *American Journal of Science*, Series 2, **33**, 385–386.
WICANDER, R., FOSTER, C. B. & REED, J. D. 1996. *Gloeocapsamorpha*. *In*: JANSONIUS, J. & MCGREGOR, D. C. (eds) *Palynology: Principles and Practice*, Vol. 1, *Principles*. American Association of Stratigraphic Palynologists Foundation, College Station, Texas, 215–225.
WILLIAMS, D. B. 1968. The occurrence of dinoflagellates in marine sediments. *In*: FUNNELL, B. M. & REIDEL, W. R. (eds) *The Micropalaeontology of Oceans*. Cambridge University Press, Cambridge, 23–243.

WILLIAMS, D. B. & SARJEANT, W. A. S. 1967. Organic-walled microfossils as depth and shoreline indicators. *Marine Geology*, **5**, 389–412.

WILLIAMS, G. L. & DOWNIE, C. 1966*a*. The genus *Hystrichokolpoma*. *In*: DAVEY, R.J., DOWNIE, C., SARJEANT, W. A. S. & WILLIAMS, G. L. Studies on Mesozoic and Cainozoic dinoflagellate cysts. *Bulletin of the British Museum (Natural History) Geology*, Supplement **3**, 176–181.

WILLIAMS, G. L. & DOWNIE, C. 1966*b*. *Wetzeliella* from the London Clay. *In*: DAVEY, R.J., DOWNIE, C., SARJEANT, W. A. S. & WILLIAMS, G. L. Studies on Mesozoic and Cainozoic dinoflagellate cysts. *Bulletin of the British Museum (Natural History) Geology*, Supplement **3**, 182–198.

WILLIAMS, G. L. & DOWNIE, C. 1966*c*. Further dinoflagellate cysts from the London Clay. *In*: DAVEY, R.J., DOWNIE, C., SARJEANT, W. A. S. & WILLIAMS, G. L. Studies on Mesozoic and Cainozoic dinoflagellate cysts. *Bulletin of the British Museum (Natural History) Geology*, Supplement **3**, 215–236.

WILLIAMS, G. L., FENSOME, R. A., MILLER, M. A. & SARJEANT, W. A. S. 2000. *A Glossary of the Terminology Applied to Dinoflagellates, Acritarchs and Prasinophytes, With Emphasis on Fossils* (3rd edn). American Association of Stratigraphical Palynologists, Contributions Series, no. **37**, 370 pp.

WILLIAMS, G. L., SARJEANT, W. A. S. & KIDSON, E. 1973. *A Glossary of the Nomenclature of Dinoflagellate Cysts and Amphiesmae, Acritarchs and Tasmanitids*. American Association of Stratigraphical Palynologists, Contributions Series, no. **2**, 222 pp.

WILLIAMSON, W. C. 1871. On the organization of the fossil plants of the Coal-Measures, Part I. *Philosophical Transactions of the Royal Society of London*, **159**, 477–510.

WILLIAMSON, W. C. 1872. On the organization of the fossil plants of the Coal-Measures, Parts II–III. *Philosophical Transactions of the Royal Society of London*, **160**, 197–240, 283–318.

WILLS, L. J. 1910. On the fossiliferous Lower Keuper rocks of Worcestershire, with descriptions of some of the plants and animals discovered therein. *Proceedings of the Geologists' Association*, **21**, 249–331.

WILSON, L. R. 1946. The correlation of sedimentary rocks by fossil spores and pollen. *Journal of Sedimentary Petrology*, **16**, 110–120.

WILSON, L. R. 1962. *Plant microfossils, Flowerpot Formation*. Oklahoma Geological Survey Circular, no. **49**, 50 pp.

WILSON, L. R. 1964. Recycling, stratigraphic leakage, and faulty techniques in palynology. *Grana Palynologica*, **5**, 425–436. (Republished in: MUIR, M. D. & SARJEANT, W. A. S. (eds) *Palynology, Part I*. Benchmark Papers in Geology, **46**. Dowden, Hutchinson & Ross, Stroudsburg, 224–235.)

WILSON, L. R. & CLARKE, R. F. A. 1960. A Mississippian chitinozoan from Oklahoma. *Oklahoma Geology Notes*, **20**, 148–150.

WILSON, L. R. & HOFFMEISTER, W. S. 1955. Morphology and geology of the Hystrichosphaerida. *Society of Economic Paleontologists and Mineralogists, 29th Annual Meeting, Program*, 122–123. (Republished in: *Journal of Sedimentary Petrology*, **25**, 137; *Journal of Paleontology*, **29**, 735).

WILSON, L. R. & HOFFMEISTER, W. S. 1959. Small foraminifera. *Micropaleontologist*, **6**, 26–28.

WILSON, L. R. & URBAN, J. B. 1963. An incertae sedis palynomorph from the Devonian of Oklahoma. *Oklahoma Geology Notes*, **23**, 16–19.

WIMAN, C. 1895. Paläontologische Notizen I: ein Präkambrisches Fossil. *Bulletin of the Geological Institution, University of Uppsala*, **2**, 109–133.

WINSLOW, M. A. 1962. *Plant spores and other microfossils from the Upper Devonian and Lower Mississippian rocks of Ohio*. US Geological Survey, Professional Papers, no. **364**, 93 pp.

WITHAM, H. T. M. of LARTINGTON. 1833. *The Internal Structure of Fossil Vegetables Found in the Carboniferous and Oolitic Deposits of Great Britain, Described and Illustrated*. Black, Edinburgh; Longman, Rees, Orme, Brown, Green & Longman, London, 84 pp.

WITTE, H. 1905. *Stratiotes aloides* L. funnen i Sveriges postglaciala avlafringar. *Geologiska Föreningens i Stockholm Förhandlingar*, **27**, 432.

WODEHOUSE, R. P. 1933. Tertiary pollen II. The oil shales of the Eocene Green River Formation, *Bulletin of the Torrey Botanical Club*, **60**, 479–521.

WODEHOUSE, R. P. 1935. *Pollen Grains. Their Structure, Identification and Significance in Science and Medicine*. McGraw-Hill, New York, 574 pp.

WOLFF-EISNER, A. 1906. *Das Heufieber, sein Wesen und seine Behandlung*. Kehmann, Munich, 19 pp.

WOOD, G. D., GABRIEL, A. M. & LAWSON, J. C. 1996. Palynological techniques – processing and microscopy. *In*: JANSONIUS, J. & MCGREGOR, D. C. (eds) *Palynology: Principles and Practice*, Vol. 1, *Principles*. American Association of Stratigraphic Palynologists Foundation, College Station, Texas, 29–50.

WOODS, R. D. 1955. Spores and pollen – a new stratigraphic tool for the oil industry. *Micropaleontology*, **1**, 368–375.

YAHŞIMAN, K. 1964. Some new megaspores in the Turkish Carboniferous and their stratigraphical values. *Congrès pour l'Avancement des Études de Stratigraphie du Carbonifère, Paris, Comptes-rendus*, Series 5, **3**, 1261–1264.

ZAKLINSKAYA, E. D. 1963. *Angiosperm Pollen and its Significance for Upper Cretaceous and Palaeogene Stratigraphy*. Akademiya Nauk S. S. S. R., Trudy Geologicheskogo Instituta, 256 pp. (in Russian).

ZALESSKY, M. D. 1914. On the nature of *Pila*, the yellow bodies of boghead, and on sapropel of the Ala-Kool Gulf of Lake Balkhash. *Bulletins du Comité Géologique, Pétrograde*, **33**, 495–507 (in Russian).

ZALESSKY, M. D. 1917. On marine sapropelite of Silurian age, formed by blue-green algae. *Izvestiya Imperatroskoi Akademiia Nauk*, Series 4, **1**, 3–18 (in Russian).

ZALESSKY, M. D. 1926. Sur les nouvelles algues découvertes dans le sapropélogène de Lac Beloë. . .et sur une algue sapropélogène, *Botryococcus braunii* Kützing. *Revue générale de Botanique*, **38**, 31–42.

ŽEBERA, K. 1935. Les conodontes et les scolécodontes du Barrandien. *Bulletin international de l'Académie des Sciences de Bohême*, **36**, 88–96.

ZEILLER, R. 1884. Cones de fructification de Sigillaires. *Annales de Sciences Naturelles*, Series 6, *Botanique*, **19**, 256 *et seq.*

ZERNDT, J. 1930. Megasporen aus einen Flöz in Libiąż (Stephanien). *Bulletin International de l'Académie polonaise des Sciences et des Lettres*, Series B, *Sciences Naturelles* (I), No. **7–10 BI**, 39–70.

Collecting, conservation and conservatism: late twentieth century developments in the culture of British geology

SIMON J. KNELL

Department of Museum Studies, 105 Princess Road East, University of Leicester, Leicester LE1 7LG, UK

Abstract: The last three decades of the twentieth century saw a transformation of the place and influence in British society of two cultural themes: environmental conservation and the values of political conservatism. These are here used to examine cultural change in the science of geology at two levels of resolution. First, the micropolitics of the science are revealed through a study of collecting in an era of conservation. Here the scientific hegemony confronted the more populist and commercially driven wings of geology. This was a period of campaign and conflict, leading to the eventual accommodation of opposing views. The second section examines the macropolitics of the science's institutional infrastructure through a study of a science in a period of recession and under the control of an ideologically motivated Conservative government. The challenge for science was to acquire appropriate government patronage. Here patterns of decline and growth in the science are revealed, driven by supposedly 'external' factors. Both perspectives show how the notion of accountability became critical to the science at all levels, and how, in an era which saw the revolutionizing of mass communication, language became fundamental to the political progress of the science.

This paper is an exploration of aspects of the 'sociology' of geology in the last three decades of the twentieth century – a period of rapid cultural change. In terms of 'major developments', it asks how a science adjusts to its changing sociopolitical context. Here, that theme is explored through an examination of two further 'major developments', which act as indicators of change: geological site conservation, and the impact of political conservatism on geology's major institutions (see Fig. 1).

The first half of this paper is a high-resolution study of the micropolitics of geological conservation, as exposed in the tensions which surrounded the issue of fossil collecting. In a recent monograph, I used collecting as a key to understanding the social politics which drove geology forward in its early years (Knell 2000). While no longer a practice at the very heart of the intellectual development of the science, collecting still forms a critical interface between different interest groups: academics, collectors, dealers, amateurs, curators, cognate agencies and the wider public. The development of geological site conservation in the 1970s challenged established patterns of collecting and participation in geology, and thus provides a topic through which changing group relationships and the distribution of power can be understood. In contrast, a macroview is taken of the science's institutions in order to understand the impact of conservative ideology, particularly during the 1980s. A central tenet of this ideology, as espoused by the Conservative governments of the 1980s, was the responsibility of the individual both in terms of action and accountability. This was an ideological opposite to the then socialist Labour party's visions of combined action and the shared responsibilities of society, which were so often embodied in policies of nationalization and central control. (By the end of the century, however, a redefined 'New' Labour party would adopt these 'conservative' values.) In this paper, corporate plans and institutional restructuring become useful indicators of the science's response. Conservation and conservatism were, in their separate ways, key elements in the cultural change that occurred in geology in the last three decades of the century. While these words and concepts have similar roots, they emerged from opposite ends of the political spectrum and not infrequently pull in opposite directions. They are here treated quite separately, and while I allude to the impact of conservatism on conservation it is not the purpose of this paper to deal in depth with the interplay between the two. Here, they become separate centres of focus, for different 'scales' of analysis, the scale and perspective of the two investigations being necessarily contrasting. Both, however, are permeated by desires for possession and control: critical aspects of the culture of geology in the early nineteenth century, which re-emerged as key attributes in the period currently under review.

From: OLDROYD, D. R. (ed.) 2002. *The Earth Inside and Out: Some Major Contributions to Geology in the Twentieth Century*. Geological Society, London, Special Publications, **192,** 329–351. 0305-8719/02/$15.00

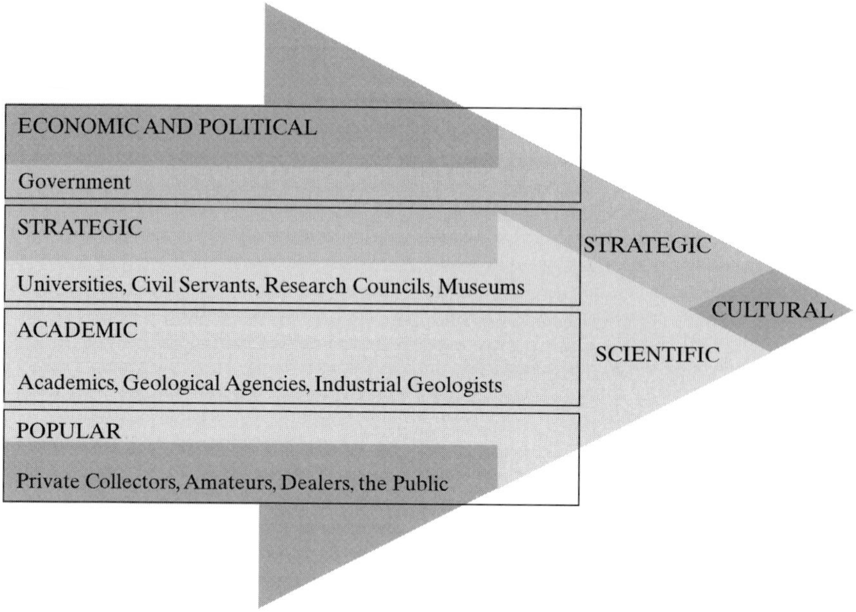

Fig. 1. Simplified diagram of the culture of late twentieth-century geology. Four groups are distinguished and ranked in terms of power and influence within this culture. These are: Economic and Political (government); Strategic (companies, universities, and similar organizations researching in this field and employing geologists); Academic (researchers and communicators); and Popular (essentially the private and amateur geologists and the public at large). The three arrows conceptualize three types of development and bring together their participants. Other types of development may be distinguished. These arrows should not be read as indicating a coincidence of direction or that strategy alone gives cultural change in a particular direction. Each arrow conceals conflicts and agreements that result in change. The two themes explored in this paper can be viewed as operating within these two areas of development: scientific (conservation) and strategic (conservatism). Together they act as indicators of wider social interaction and cultural change.

Set in the context of a volume which principally talks about the development of new approaches, methods, and ideas in geology, the perspective and methodology applied here perhaps demand a little further explanation. Rather than accept development as something entirely within the purview of intellectual ideas and technologies, this paper demonstrates that it is also the result of cultural forces *outside* this core, and indeed that science can be viewed more broadly as a changing cultural entity. Many scientists and some historians of science still reject the role of power, authority and construction in the development of scientific knowledge and, indeed, do not welcome notions of science as a cultural entity. In the 1990s these waters became increasingly muddied as scientists did battle with cultural theorists in the so-called 'Science Wars'. It was a clash which encouraged scientists to adopt polarized views of postmodernist thought, suggesting it to be 'anti-Enlightenment'. I prefer historian Georg Iggers' (1997, p. 16) interpretation: 'The postmodern critique of traditional science and traditional historiography has offered important correctives to historical thought and practice. It has not destroyed the historian's commitment to recapturing reality or his or her belief in a logic of inquiry, but rather it has demonstrated the complexity of both.' This paper also makes use of the thinking of modern philosophers and sociologists of science (such as Barnes 1974; Callon 1986; Latour 1987; Shapin 1992), but I make no attempt to slot this study into a particular school of philosophy.

As Iggers suggests, in dealing with the culture of a science, the historian expects complexity, and rejects simple cause and effect relationships. Key or dominant forces may be located, but they tend to operate in combination, and with subtlety. Cultural change then becomes something nebulous, rooted in reality but also in constructed readings of reality (which I shall demonstrate). Consequently a paper such as this does not simply deal with planned or intended outcomes, or even accidents, but rather with the product of a multitude of forces, sometimes beyond easy articulation or definition. This paper is, then, predicated on the notion that even if science remained unchanged in terms of its beliefs and methods (which it never has or does), change in its cultural setting inevitably results in new meanings, new emphases, new tensions, new methods, new directions, new developments. Our perceptions of a science will develop and change, irrespective of change in that science. This will happen whenever we change cultural setting, whether by moving geographically or temporally. For example, in China, peasant power replaces a Western mechanical excavator, while in India, palaeontologists are, we are told, less inclined to talk to each other (Whybrow 2000). Similarly, if we could travel back to the early nineteenth century, we should find geologists attributing entirely different values and meanings to the science (Knell 2000). Therefore having a specific focus in time and place is essential to the kind of analysis undertaken here (Porter 1977, p. 218; Knell 2000, p. xii). Though this idea has long been central to modern historiography, it runs contrary to many histories of science, which project backwards modern perceptions of a particular science or hold onto a notion that science is a universally agreed concept in time and space. This paper explores cultural change over the period of three decades, from 1970 to 2000. It focuses specifically on the British context, though for the purposes of cultural contrast reference is also made to practice and thinking in other countries.

Science in an era of change

> It is hard enough for someone in his fifties to come to terms with the speed with which his hair has whitened, his face has lined and his frame has thickened; to grasp how utterly almost everything on Earth has changed in only 30 years is a great deal harder still (John Simpson, BBC World Affairs Editor, 1998, p. 16).

The last three decades of the twentieth century saw rapid and striking social change, from which few countries were immune. What Britain in the 1960s saw as 'Americanization' was by the 1990s a process of 'globalization' (Simpson 1998; Scholte 2000). The post-War era was typified by the rise of liberalization and democratization. This was manifest not only in constitutional change but most spectacularly in the unleashing of the voice of youth, of social idealism, of civil rights, sexual equality, nuclear disarmament and antiwar protest, so prevalent in the 1960s and early 1970s. The background to science in this period reflected this social and intellectual idealism–of society undoing the shackles of class, economic inequality, social authority, and so on. However, by the early 1980s these ideals were being tempered by the political conservatism of government. At its birth, geology had known similar social transformation as new wealth created an increasingly powerful middle class for whom the new science was a means of cultural expression and identity (Knell 2000).

Science is not immune to the social changes taking place around it. Prominent in the late 1960s were growing concerns for the environment, which exposed a rift between popular desire and established political agenda. These were awoken by marine biologist Rachel Carson's book *Silent Spring* (1962), which portrayed the extended and unforeseen effects of pesticides. The new environmental reality fully entered the public conscience in December 1968 when NASA's *Apollo 8* looped the moon and beamed back pictures of a finite planet Earth alone in the darkness of space (Ward & Dubos 1972). In 1970, the USA celebrated its first 'Earth Day', when 20 million people rallied to raise awareness of environmental issues. That year also saw the establishment of the US Environmental Protection Agency. Bookshops everywhere became filled with paperbacks exhorting readers to save the planet. The breadth of the perceived problem was enormous: agrochemicals and sustainability, lead levels in children and the proliferation of the car, the consequences of nuclear power and the social costs of coal extraction, industrial disease and the use of such materials as asbestos and heavy metals, population growth and problems of food supply, and much more. All brought a sense of a coming apocalypse which, however, had all but been forgotten by the late 1990s. Such fears created a natural union with protests for nuclear disarmament. In the Cold War tensions of the 1970s many felt that if environmental degradation did not finish mankind, 'the big one' would. There were prophecies of a new millennium of social turmoil and deprivation.

Science was at the heart of these controversies. No longer, it was claimed, 'could the activities of scientists themselves be constructed as floating free above the economic and social base, the abstract accumulation of knowledge about the world, a polite academic interchange of Popperian conjectures and refutations. Science and its practitioners were locked into the social order' (Rose & Rose 1980). Science was now 'incorporated', 'industrialized', 'a factory'. The perceived marriage between social and scientific progress had broken down and geology was implicated, along with the other sciences. A single issue of *New Scientist* in 1980, for example, revealed a public enquiry into the proposed sinking of new coal pits in the unspoilt Vale of Belvoir in Leicestershire, initial test drilling to bury nuclear waste in the rocks of Scotland, and planning for two polluting 'super brickworks' in the Oxford Clay fields of Bedfordshire.

This was not just 'rage against reason' or a 'rage against science and scientists', as some have suggested (Hacking 1999, p. 61). Rather it was about ownership and control, for science was also necessary to locate evidence and arguments against such developments, to determine solutions, to provide an 'alternative'. The establishment's scientists were often portrayed, or liked to portray themselves, as men with exclusive knowledge and consequently exclusive power. Lord Todd, President of the Royal Society, wrote: 'It is, I fear, often the case that the stridency of the protesters fanned by the public media of communication, permit them to exert an undue influence at the expense of the experts' (Todd 1980). Such statements claimed that expertise was value free, that science came without moral or political baggage, that progress would be opened to us through scientific discovery alone. Todd's words also indicated a growing fear, amongst some members of the scientific community, of irrational forces. But, if he was correct, then the fault was also with that same scientific community, as it had failed to communicate effectively with a public that saw such statements as arrogant.

It should be no surprise, then, that in the early 1980s the Royal Society became the parent to a worldwide movement known as 'The Public Understanding of Science'. Its intentions included a desire to ensure that the public used their democratic powers in an informed (i.e. pro-scientific) way. It reflected a line of reasoning more than a century old, which could have been found, for example, among the middle-class supporters of mid-nineteenth-century mechanics' institutes. Whether or not it was an attempt to inform or control (if it is possible to distinguish between the two), the movement was certainly wedded to a notion of cultural orthodoxy sustained by science.

However, by 1980, Britain was beginning one of the most radical economic and political transformations of the twentieth century. The previous decade had seen the first of a succession of post-World War II recessions. The old strategies of Keynesian economics and the goal of full employment, which had formed the mainstay of British politics for 30 years, were, in 1979, replaced by the new, harder, ideological conservatism of Margaret Thatcher. The economics of the 'New Right' turned policy on its head. Now a single political objective, price stability, dominated all thinking. Driven by a philosophical desire for strong authoritarian government and a free marketplace, the Prime Minister committed Britain to suffer the short-term consequences of high unemployment and falling output in a way never before seen as politically acceptable. If campaigners wished for interventionist politics to turn the tide on environmental degradation, they would not find them here. New political desires sought to redirect individual freedoms away from social liberty and into reinvented Victorian self-sufficiency and the capitalist imperative. The decline and chaos of the 1970s were, in the eyes of the new Conservative administration, a product not just of oil-driven recessions but also of permissiveness, weak but interventionist governments, and strong unions (Booth 1995; Green 1989; Healey 1993; Overbeek 1990; Smith 1984). The earlier Conservative administration of Edward Heath had wished to deliver similar hard-line policies in 1970, but its nerve failed. By the time of Thatcher's fall in 1990, the country had been transformed. So much so, that when the Labour party emerged from its wilderness years, to return to government in 1997, socialism no longer appeared on its agenda. Rebranded as 'New Labour', it now preached a new kind of 'caring conservatism'.

Establishing the conservation hegemony

It was against this background, of society shifting from 1960s environmental idealism to 1990s 'conservatism', that geological site conservation can best be understood. A new and narrowly conceived idea in 1970, its journey is one which reflects the cultural changes the science underwent more generally. Nature Conservancy Council (NCC) officer, and one of the subject's chief protagonists, Bill Wimbledon, later referred to it as the 'opinion-ridden art of site conservation' for it exposed a conflict between the desires of a growing and diverse community of private geologists and the perceptions, and actions, of authority (Wimbledon 1988, p. 41). The aim here is not to give a full account of the development of site conservation but rather to examine its cultural implications. Various NCC officers have given 'official' histories of these events (for example, Duff 1979, 1980; Ellis 1996; Wimbledon 1988; NCC 1990).

In the 1960s and 1970s concern for the world's natural resources reflected the egalitarian or 'common ground' politics of ownership and responsibility, which had developed from the political Left. Sustainability became the *raison d'etre* of the conservation movement a quarter century before politicians gave notice to it at the Earth Summit in Rio in 1992. In geology, conservation concerns, which reflect this sense of shared ownership, can be traced back to its earliest years (Knell 2000). However, site selection and protection in Britain really only began around 1945 (Duff 1979, 1980). Inevitably it took on new significance in the 1970s, then under the leadership of the geological section of the Government's independent agency, the Nature Conservancy Council. By 1973, this organization had established several hundred geological Sites of Special Scientific Interest (SSSIs), a small fraction of the 3500 SSSIs then covering 'nature' more generally. In 1977, with the number of geological sites at 1300, the NCC's geological section became the Geological Conservation Review Unit which had the aim of publishing a '"Domesday Book" for geological conservation in Britain' (Wimbledon 1988, p. 42).

The real challenge for the NCC, it seemed, was to change the accepted order, a *way of doing* geology that had evolved over nearly two centuries. In 1970, conservation in geology was barely discussed outside a small body of activists in the mainstream of the science, who could see the damage being done to some classic localities. Inevitably, as the momentum of the movement grew, these rather restricted perceptions of the need for conservation, and of the causes of damage, held sway. Like other aspects of a science that had once been open to anyone who chose to pick up a hammer, collecting and fieldwork now came under bureaucratic scrutiny, management and regulation. Preaching a new kind of evangelical puritanism, the NCC questioned the established and sacred collecting-based culture of a science of leisure, education and commerce. A reader of the geological section's widely distributed *Information Circular* now found that, in the conservation establishment's (i.e. the NCC) view, collecting was synonymous with 'misuse', or so it seemed. The 'god-fearing' amateur worried about a total ban on collecting, knowing that there were many in the conservation fraternity who would welcome such a move (Cotton 1984). The NCC had also roundly castigated educational parties for their unthinking activities, stating that, as a result, access to some sites was now in jeopardy. But here was the nub of the problem. In an era of strident conservationism, and widely adopted values of a shared heritage, the NCC had seen protection as its primary goal. The issue of access was a lower priority and one that it was ill-equipped to address (Doody 1975). However, by 1976, perhaps realizing that its confrontational tone was causing alienation, the organization's *Circular* played down the misuse issue by placing all 'threats' in a broader 'Site News' category. Another section, 'Co-operation in Conservation', indicated an awareness that only by making this a shared issue could a hoped-for consensus be found and NCC targets met. Language here became the key. The NCC was undoubtedly the engine driving forward this 'new way'. It had cultural authority but little direct power over landowners or geological practitioners. In nearly all its dealings its use of language was measured and carefully applied. Although geologists viewed its activities as essentially practical and scientific, its success (or otherwise) relied much on the art of persuasion and ambassadorial diplomacy; ambiguously defined and understood, conservation remained a contested issue.

However, as the 1970s progressed the implications of the NCC's lack of actual power became all too apparent. At the heart of its problems lay the inadequacies of the SSSI system, which merely required planning authorities to *notify* the organization of planning applications affecting sites of scientific importance. This allowed developments to be opposed if the scientific integrity of a site was threatened. However, not all potentially damaging operations required planning permission and, as a consequence, sites were sometimes destroyed by agricultural or other activity. In addition, the NCC's reliance on diplomacy meant that landowners often seemed to be calling the shots, with the agency only too ready to oblige them. There had been successes, such as the prevention of coastal defences at Barton-on-Sea in Hampshire, where property was at direct risk from coastal erosion (for location of key sites, see Fig. 2). But there were also embarrassing failures. One concerned the building of a row of houses across the only entrance to a Permian fossil locality, leaving it virtually inaccessible (Doody 1975, p. 213). However, the great advantage of SSSIs for government was the low cost involved, as it was unnecessary to acquire ownership. But now there were serious doubts about the effectiveness of the system.

By 1979 there was a widespread feeling that geological conservation was in crisis, and a London conference, 'The future development of geological conservation in the British Isles', was organized by the Geological Curators' Group

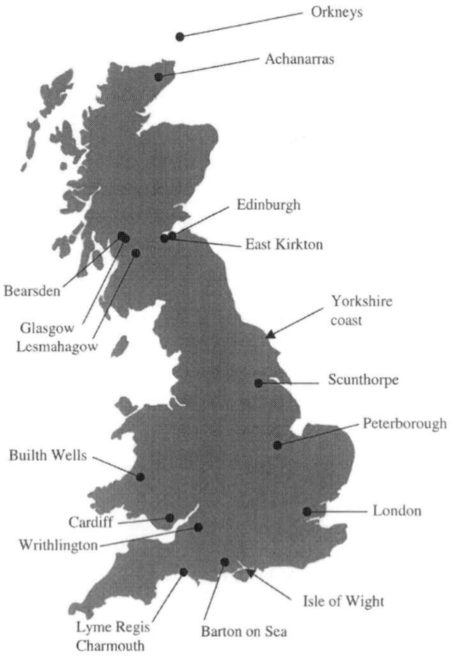

Fig. 2. Map showing locations of key sites mentioned in the text.

(GCG), in the hope of identifying and resolving the key issues (Clements 1984). The meeting exposed a preoccupation with the problem of collecting. Wimbledon (1988, p. 48) later recalled the themes: 'There was then much talk of the need for new legislation: to deter collectors, to prevent the export of "priceless fossils" ', and so on. George Black, head of the NCC's geological section, was unequivocal on this matter of policy: the 'advance of geology', he said, was not served by 'excessive hammering and purposeless collecting', by the use of research localities for educational work, by the use of 'difficult' localities for 'parties of low academic level', or by antagonizing landowners (Black 1984, p. 7; see also Long & Black 1975). The qualifiers used in these arguments were an application of the necessary vagaries of 'political speech'. Who could say what was or was not 'excessive' or 'purposeless'? Certainly, the academic, amateur and commercial collecting communities did not have a common view on the matter. In the NCC's view, the removal of fossils from an SSSI constituted theft. However, it was the owner's responsibility to pursue this, and few had been willing to do so.

A widely perceived cause of destruction was the massive increase in the student population since the 1960s, and the consequent increase in fieldwork programmes. In 1980 it was suggested that the annual number of student field days had risen from nearly 19 000 in 1963–1965, to just over 34 000 in 1971–1972, and almost 55 000 in 1977–1978. NCC research also indicated that the potential threats to some sites might well have been greater than even these figures reveal. The number of sites used had not increased substantially, but in the different survey periods particular localities were favoured, with a temporary move to Scottish sites being detected in the early 1970s (NCC 1980*a*).

But even those universities that were not active in the field found themselves in a difficult position. The Open University, for example, replaced a planned mailing of 4000 real fossils, to its distance-learning students, with plastic replicas, 'as the University is aware that it would otherwise be liable to deplete seriously the country's stock of fossils' (Anon. 1973).

It was clear that without action to increase the number of available localities and control fieldwork at some sites, conflict between the needs of education and conservation could only worsen. In 1975, the NCC had met with the newly formed GCG and soon afterwards the two organizations brought together proposals for what became the National Scheme for Geological Site Documentation. A pilot NCC investigation in Surrey showed some 3500 historical records of sites, of which 30% had survived. It was thought an estimated 100 000 potential field localities existed in Britain (Long & Black 1975). The GCG's scheme was to locate and document these for future educational use. However, tensions subsequently developed between these two organizations when the NCC was seen to have reneged on its promised financial contribution to the 1979 conference, leaving the GCG financially embarrassed. With the country now in recession, the NCC's geologists clearly did not have sufficient financial autonomy, and having been talked into taking on site documentation, the GCG now found grant-aid also drying up. The NCC's own geologists were also showing signs of discontentment with a resource base that was inadequate for the task before them. Outside commentators complained about the slow progress, but there was little that the NCC geologists could do about it.

The organization's difficult position with regard to its powers – too weak for the radical lobby, too interventionist for the more conservative landowners – appeared to be resolved with the Wildlife and Countryside Act (1981) (Smith 1986; but see Caufield 1984). Under this legislation any 'Potentially Damaging Operations', or

PDOs, could be opposed and, if successfully, compensation paid to the landowner. Unfortunately, for the act to be enforced, owners of SSSIs had to be renotified, a process that caused further delay in site identification and designation. By the middle of that decade, with politics tilting firmly towards the Right, the conservation movement in general began to look for new politically acceptable rationales. Thus, the economic argument was increasingly used, even to justify the preservation of species. With the political landscape transformed it was vital that the NCC was seen to be a pragmatic organization with the economic well-being of the country in its sights. To the Thatcher Government, science held no special status. Like every other cost to the taxpayer, it too needed to justify its worth. The NCC's purpose had to be recast, it was now 'cultural', where culture referred to 'the whole mental life of a nation': 'the proper role for NCC ... is to practise nature conservation according to a definition of purpose which is primarily cultural, that is the conservation of wild flora and fauna, geological and physiographic features of Britain for their scientific, educational, recreational, aesthetic and inspirational value' (NCC 1984, p. 75). Dominant amongst these values, however, were the needs of science, which when labelled 'cultural' became imbued with notions of wider social benefit. It was a manipulation of language and meaning that geology had long known, but one in which it needed to become even more adept as the century progressed.

Ownership and the field

For all its debate, the 1979 'Future development of geological conservation' conference did little to change opinions about collecting. Where once the hammer had been an essential tool, without which geology could not be practised, in the new conservation-aware science it became an 'offensive weapon'. This was most profoundly symbolized by the removal of hammers from the badge of the Geologists' Association (GA) in 1990, where they had proudly sat as an emblem of amateur activity for more than a century (Green 1990; Knell 1991; see Fig. 3). In the 1980s, there was a widespread feeling, derived from the views of the conservation establishment, that to collect was wasteful of a limited resource, and that commercial exploitation was repugnant to any right-minded geologist. The amateur and commercial collector saw a scientific elite taking possession of what had been, to all intents and purposes, a public resource. The NCC's group of largely doctoral geologists, who determined policy, had decided that fossils were principally a resource

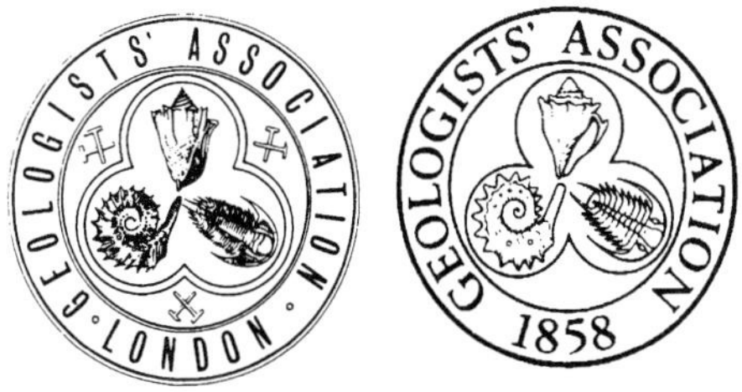

Fig. 3. The Geologists' Association badge, before and after removal of the hammers in 1990 (reproduced by permission of the Geologists' Association).

for science, and that science was to be prosecuted by *bona fide* researchers. The NCC felt it was protecting a scientific heritage.

Tensions were most overt where the amateur or professional came up against the world of the commercial collector. While there was a tacit understanding that commercial collecting was really only a problem where it affected vulnerable and important sites, received wisdom was that such practices should be discouraged. Any possible benefits of commercial extraction were overlooked, despite the reverence paid to Mary Anning and other early practitioners of this art. Rather than see the benefits of mass collecting as a means to reveal rarities, the only product was thought to be shops crammed with data-less and abused specimens, useless to science. The wider cultural implications of constraining collecting or the sale of specimens were overlooked, forgetting that it was here as much as anywhere that members of the public got their first taste for the science. During the 1980s and 1990s geologists frequently complained that theirs was becoming a Cinderella subject, and that the popular media ignored them. It seemed that practical geology was becoming an exclusive activity too. The cultural authority of academic or professional science, which was bound up in the contemporary view of conservation, though understandable, was simply overriding the perceived rights of a much wider group of participants. During the 1980s challenges came from various quarters and ultimately turned these perceptions on their head. The decisive battles took place in Scotland and on the south coast of England.

In 1977, local evidence suggested that German dealers had been using power tools to 'pillage' fish localities in the Orkneys and in Lesmahagow, near Glasgow. Small-scale collecting had been the norm here and not opposed. As Ian Rolfe, geologist at the Hunterian Museum in that city, remarked: 'as a museum man I am not opposed to keen collecting, simply to the illicit collecting currently on the increase' (Rolfe 1977). Commercial dealerships had grown in number through the 1970s as interest in amateur collecting had increased. But at the end of the decade there were still estimated to be only seven or eight professional collectors in Britain, perhaps twelve importers or wholesalers, and a small but diverse group of rockshop operators. There were also a considerable number of amateurs who were not averse to selling fossils (Harker 1984). Contrary to establishment views, the commercial collecting community was not homogenous nor could it be easily defined. Similarly, the amateur community was beyond simple definition in these terms. However, the Lesmahagow incidents alerted the NCC to the risk of future site damage, and in the most publicized case of the decade it oversaw the arrest of two German collectors at the famous Devonian fish locality of Achanarras Quarry, near Thurso, in the far north of the Scottish mainland, in June 1979. In the first conviction of its kind in Britain these two collectors were merely 'admonished', but it was felt at the time that an important warning had been given to others (NCC 1980*b*).

The position of the commercial collector took a new turn in 1981 when the University of Glasgow contracted Stan Wood, a local amateur who had found Namurian (Carboniferous) fish in a stream-bed near the housing estate where he lived, to oversee a fossil dig at the site. The excavation at Bearsden became one of the great British palaeontological stories of the decade, revealing, amongst other things, remarkable new fossil sharks. Partly supported by the NCC, the excavation also delivered fine educational outcomes for the Hunterian Museum. It demonstrated the potential of amateur and educational involvement in a strikingly novel way that seemed to run counter to many NCC preconceptions. A few years later Wood rediscovered the East Kirkton Limestone near Bathgate in West Lothian, a remarkable Lower Carboniferous lacustrine deposit containing terrestrial and amphibious animals, including the famous 'Lizzie' (*Westlothiana*), then thought to be the earliest known reptile (more correctly, 'amniote'; Rolfe *et al.* 1994). Stan Wood's discovery of two new Carboniferous vertebrate localities had, it was claimed, caused a 'quantum leap' in knowledge of this fauna (Unwin 1986). He was given much media coverage when his discoveries toured the country in 1986–1988 in the exhibition, 'Mr Wood's Fossils', and the Scottish 'amateur' soon became, amongst the British public at least, the best-known palaeontologist of the decade. However, in June 1987, finding no opening for 'a fossil hunter' in the academic or museum world, he opened his own fossil shop.

In this same period, the West Dorset District Council in southern England gave consideration to new by-laws to prohibit the removal of fossils from the cliffs around Charmouth and Lyme Regis, the British stronghold of commercial collectors since the birth of the science. There was a local belief that their collecting activity was increasing erosion rates and required control. The NCC and the Geological Society offered to support this move if the local council ensured that *bona fide* geological researchers and educational parties would not be adversely affected. Plans were put in place for licences to control the type of collecting, the size of hammers, and so on. All would pay and be controlled except 'researchers' who would remain completely unregulated beyond the requirement of a free licence. Commercial collectors would retain some access but would require a licence to excavate and might be required to involve a scientist in their activities. Here the NCC had most clearly shown its colours, something the commercial fraternity would long remember. However, it was the collectors who eventually won the day by demonstrating that coastal erosion was not affected by their activities; and the Secretary of State ruled against the local council (NCC 1982, 1983; Taylor 1988).

It was against this background that Stan Wood, in 1985, came out against geological conservation. Desiring a renaissance of interest in palaeontological exploration, and using the Bearsden excavation as a model, he suggested that old sites should be opened up for public participation in collecting, with tools for hire and a caravan on site with a fossil advisor. For him, conservation was the antithesis of this: involving a preservation of the past rather than prospecting for the future, and a 'shading in of no-go areas on geological maps' (Wood 1985). His temper had earlier been aroused when the two Germans, whom he knew, had been prosecuted at Achanarras Quarry. Under the regulations, collectors required a permit and could only take away two fossils despite the presence of fish in their tens of thousands (from 1984 the number that could be collected was raised to ten). In response, Keith Duff (1985) of the NCC stated that only 10% of designated sites (about 150 in all) suffered similar limitations or restrictions, and that future exploration was ultimately a goal of conservation. He quoted Benton & Wimbledon (1985) who had recently expressed an aim: 'to encourage and participate in the systematic use and excavation of sites (but not their total removal) by professionals and responsible amateurs and to promote proper recording of finds and taphonomic information'. Using evidence of the devastation of sites resulting from the commercial emphasis on the perfect and the rejection of the incomplete, he vigorously opposed the encouragement of a commercial market in vertebrate fossils. As so frequently occurred in these kinds of arguments, both sides could offer convincing examples to support their case and both could pounce upon the weaknesses of their opponents. No one felt the need to recognize their opponents' more positive qualities. There was no incentive to compromise.

At this point eleven senior vertebrate palaeontologists weighed in in support of conservation. But here at last was a hint that times were changing, that old assumptions, which had caused so much heartache, were beginning to crumble. Mike Benton, Bill Wimbledon, and others, believed that few sites were non-renewable, that site vandals were a rarity, and that overcollecting was not the threat it was purported to be. Development was the real bogey. In their view, restrictions at Achanarras had been a mistake: 'an over-reaction by some conservation enthusiasts to the threat of foreign collectors pillaging the site. We can hope that such restrictions will never be applied again' (Benton *et al.* 1985). Yet some geologists felt a contradiction in site conservation 'only to have it slowly "destroyed" by fossil collectors' (Cleal 1987).

Still criticized in the geological press for its slow rate of progress and publication, its ears ringing over the Achanarras affair, and with domestic problems arising from the summary transfer of staff from Newbury to Peterborough, the Geological Conservation Review Unit (GCRU) began to break up, and 'a rather shadowy organisation calling itself the Association of GCR Contributors' appeared on the horizon. The watershed came in October 1987, when, in a second London conference organized by the GCG, the Geological Society and the Palaeontological Association, 'The use and conservation of palaeontological sites', the geological community appeared to shift *en masse* to a new consensus which echoed the thoughts of Benton and friends. In the run-up to the conference the GCRU moved to the NCC's new Peterborough headquarters, and following a period of some confusion the team was strengthened to a level comparable with biology, and a final push made towards completing site notification in line with 'corporate objectives'.

During the conference, the former NCC man George Black, and a few others, launched the British Institute for Geological Conservation (BIGC) 'in an atmosphere of unconcealed contempt for the supposed failures of the Nature Conservancy Council.' *Geology Today* reported: 'The plain fact is that geological conservation in Britain is in a shambles, with no general agreement on either aims or priorities' (Anon. 1988*a*). The statement, however, was incorrect, as the conference demonstrated that the geological community now endorsed a more pragmatic (rather than ideological) approach to conservation, which was responsive to, and respected, the needs of other groups. It rested on a notion of responsibility and an increasing emphasis on *use* (Crowther & Wimbledon 1988). Commercial collectors were reclassified as part of the geological community, with a general realization that categorizing and stereotyping had, in practice, done little to advance conservation.

With lines redrawn, a certain amount of repositioning began. What had been entirely acceptable to the conservation establishment prior to the conference now appeared to be a kind of heresy. It was as if the reconstituted culture demanded a witch-hunt for those who had led geological conservation along an erroneous path. Rolfe, for example, who had expressed concerns over the destruction of Silurian fish localities at Lesmahagow, now revealed that he had acted in the interests of pacifying a distraught landowner. Never against collecting, he was now 'in favour of the use of heavy equipment and explosives for controlled excavations'. Techniques once largely the preserve of the commercial collector were now being used by his museum at East Kirkton (comment by Rolfe in Taylor 1988). The NCC team also had to find excuses, though they were inclined to see (or represent) themselves as mere instruments: 'In the past, attempts have been made (by NCC) to restrict collecting at some fossil sites following vocal and written pressure from palaeontologists, only to find that in later years published opinions have almost totally reversed' (Norman *et al.* 1990, p. 92). Staff tried to distance themselves from the recrimination over access agreements and particularly Achanarras. These were now 'historical'. At Lesmahagow and Achanarras, measures had been introduced in 'direct response to pressure from a small number of geologists to curb activities of professional collectors who were thought to be damaging the sites' (Norman & Wimbledon 1988, p. 194). The language was carefully chosen, 'actual' damage had now become 'thought to be', the hammering damage which caused complaints from owners, to which the NCC had responded so quickly and termed 'misuse', was now merely 'perceived'. The Achanarras prosecution was no longer a triumph of conservation but a symbol of Draconian measures. In this new enlightenment, the NCC were to be more cautious, to maintain a watching brief, to discern the 'extent and impact' of collecting. In contradiction to its earlier Dorset stance, it came to the view that the sea did much more damage than the collectors. Most remarkable of all, commercial collectors were now redrawn as 'gifted' and without whose activities academic and museum geologists would be the poorer (Norman & Wimbledon 1988).

As Wimbledon (1988, p. 47) had come to realize:

> Recent years have seen too much attention being paid to the role of the collector and collecting and too little to the real priorities. Arguments have raged over the value of fossiliferous scree, over fossil collecting quotas, the rights of the professional geologist to collect, and whether professional [i.e. commercial] collectors are a "good" or "bad thing"; yet all are insignificant in comparison with the problems of saving sites from the damage and loss that comes from development.

Earlier calls for legislation and control were, it was conceded, based on poor knowledge of the resource. The 'stop collectors' controversy had only served to divert attention from real needs and real threats. 'Fossils, especially invertebrate fossils, are a renewable resource'.

However, while the perspective of the conservation establishment seemingly changed overnight, the mistrust and suspicion that had become polarized into different camps over the previous two decades would not be readily dissipated; 'geological conservation' had been branded. 'The legacy of panic induced by the Caithness [Achanarras] and Lesmahagow experience is still with us', Wimbledon (1988, p. 48) admitted. Remarkably, he also questioned earlier underlying assumptions: 'Geologists should remember that "their" favourite research sites may have other uses, and that scientific use may have no more validity than any other . . . is scientific exploitation the only valid use of the palaeontological resource?' (Wimbledon 1988, p. 41). This was a fundamental shift in thinking: an admission that assumptions concerning the cultural authority of science in relation to the fossil resource could not be universally justified.

The NCC had also been taking an interest in site conservation in other countries, and the 1987 conference provided opportunities to compare practice at home with that elsewhere. Rupert Wild's (1988) explanation of protection in Germany, where fossils could be designated as 'cultural monuments', caused much interest. However, it was developments in the USA that most closely echoed the new British consensus. Here, in 1985, the National Research Council had established the Committee on Guidelines for Paleontological Collecting (CGPC), a panel of 13 individuals from various sections of the geological community, which was to resolve the long-running issue of fossil collecting on public lands. Some 60 federal agencies had responsibilities in this area and a number of cases of quite harmless activity had been pursued in the courts. The same faction-centred issues as affected conservation in Britain were also present in the USA, but the committee saw past them with great clarity of purpose. The report arising from its deliberations was published in 1987 just before the conference 'The use and conservation of palaeontological sites' in London. It came down unequivocally on the side of collecting in all its guises: 'In general, the science of paleontology is best served by unimpeded access to fossils and fossil-bearing rocks in the field . . . . Generally, no scientific purpose is served by special systems of notification before collecting and reporting after collecting because these functions are performed well by existing mechanisms of scientific communication. From a scientific viewpoint, the role of the land manager should be to facilitate exploration for, and collection of, paleontological materials' (Committee on Guidelines for Paleontological Collecting 1987, p. 2; Pojeta 1992). It found that in general the fossil resource was renewable and that 'Fossils are not rare'. These conclusions reasserted a view taken by the Paleontological Society in 1979 (when, it will be recalled, the British were again in conference and then in the depths of a collecting crisis; Clements 1984). The recommendations permitted all groups to participate in fossil collecting while simultaneously ensuring scientific protection. The only need for permits was for commercial extraction where the involvement of scientific oversight was necessary (as the NCC wished to see in Dorset). In the USA the guidelines became a vital working document for many land managers but they did not achieve their stated aim of simplifying and standardizing access arrangements across the country. Amongst its other recommendations was one to establish a National Paleontological Advisory Committee that would identify localities of national significance, much as had been achieved by the GCR.

The era of responsibility

In 1990, at a high-profile launch in the heart of Westminster, London, the NCC revealed its first five-year plan, *Earth Science Conservation in Great Britain: A Strategy*, which showed both an integrated understanding of user needs and a new, tiered, approach to conservation which also recognized that funding for conservation was unlikely to improve. A new Regionally Important Geological/Geomorphological Sites (RIGS) scheme was unveiled, with the intention of democratizing conservation and giving local groups the means to protect and use sites, not just for the research community or to satisfy the requirements of government, but to meet the local needs of educationalists, museums, amateurs and collectors. No longer was conservation a bureaucratic imposition by Government, it was now in the possession of local interest groups; the sense of responsibility placed upon the geological community was being matched by increased opportunity for participation. And with an estimated 1200 active Earth science researchers, and a total of around 6000 working Earth scientists and 3000 geology students in Britain, there was no need to dress geological conservation up as culture in order to sell it to politicians (NCC 1990). Indeed, it had become increasingly important to raise the profile of the science, to talk up its utility and its place in national life. In 1990, the idea of conservation was easier for governments to accept, as it now meant something different. The earnestness of 1970s radicalism had mellowed and conservation was by this time beginning to enter the mainstream politics of even the most conservative thinkers.

The *Strategy* also revealed NCC's ambitious plans to publish its now 2200 geological sites in a 51-volume work (this was later revised to 42 volumes and 3000 sites as publication began). However, as the organization at last began to celebrate progress, it found itself broken up into country-based units: English Nature, the Countryside Council for Wales and Scottish Natural Heritage. The Joint Nature Conservation Committee co-ordinated activity across Britain.

As Wimbledon predicted, collecting as a conservation issue, which seemed to have been resolved a few years earlier, did not go away. Late in 1990 the NCC received the first challenge to its more relaxed attitudes as farmers began to complain about numbers of visitors, including fossil collectors, to Lesmahagow. Its response was to instigate a system of permits but only so as to inform farmers of the timing of visits; this was not regulation. Two years later a commercial excavation for trilobites at Builth Wells, in Wales, met with local opposition whereas the Government's conservation geologists expected the site to be improved by the activity (Kennedy 1993). However, large-scale illegal excavations at a Carboniferous Shrimp Bed in East Lothian reaffirmed old tensions. In the amateur community it would take a while for the new reality to sink in, as one article in the *Proceedings of the Geologists' Association* demonstrated. Here the NCC's 'bureaucracy' and its attempts to control collecting were 'insufferable' (Wright 1989, p. 296). What the writer feared was not overcollecting but undercollecting. The geological staff of the NCC responded *en masse* to defend their activities, listing the threats and benefits in a way that gave little overt indication of how radically the organization had changed. They made this clearer in the *Geologists' Association Circular*: 'If palaeontological sites are to continue to have scientific relevance (rather than becoming a collection of historically interesting locations), further collecting of geological specimens and their study MUST be made possible. . . . Fossil collecting *per se* cannot, in most circumstances, be considered an undesirable activity, whether it is for scientific, educational or commercial purposes' (Norman *et al.* 1990; Norman 1992, p. 255; Knell 1991, p. 106).

The 1990 *Strategy* saw the impact of fossil and mineral collecting as a key area of activity for the NCC's new programme of applied research. However, it clearly stated that for most sites damage could be avoided if collecting was carefully planned and carried out. Even on unique fossil sites: 'In most cases, responsible and scientific collecting for research, education and commerce represents a valuable activity and one of the reasons for conserving the site. In a limited number of cases, however, restrictions and agreements over intensive commercial or education collection may be required' (NCC 1990, p. 41). By 1992, English Nature was ready to publish a fossil collecting code. Now the word 'responsible' had become a universal qualifier for 'collecting', reformulating the fossil resource into something shared and giving the collector a sense of obligation (Knell 1991; Norman 1992; English Nature 1992; Ellis 1996, p. 90; Larwood & King 1996). Here the language returned to terms such as 'fossil heritage' or 'national natural heritage'. This was not to convince Government or the public of a need for support but to promote a sense of responsibility among collectors of all persuasions by imbuing rocks with a shared trusteeship that countered notions of ownership and exploitation, or that fossils were simply the property of the scientific establishment. Indeed by the end of the century geological conservation had been rebranded as 'Earth Heritage'. More than a marketing exercise, the use of language once again became a means to transform perceptions, to distance a largely remodelled activity from the more controversial past which had spawned it.

'Responsible collecting' thus became a linguistic step along this path. The only remaining problem was that of interpretation, for each participant might define the word 'responsible' differently. Certainly an English Nature position-statement of 1996 redefined the term in such a way as to enshrine the rights of science whereas the conference of nine years earlier recognized a larger community: 'Irresponsible collecting delivers no scientific gain and is therefore an unacceptable and irreplaceable loss from our fossil heritage'. Tensions remained between different factions and came to a head in an exchange of views between a few English Nature officers and commercial collectors during the cutting of a bypass at Charmouth in Dorset in 1989–1990. It was a temporary hiccup which did not reflect a change of policy, but the old distrust resurfaced. The problem of the market in fossils was not going away, and no one doubted that it had its 'good' and 'bad' sides. Wright (1989, p. 296) was certainly not alone in his feelings when he wrote: 'Like many others I deplore the idea that fossils have a money value'. It was logical for geologists, particularly those outside the mainstream of conservation, to look for models in species, habitat or archaeological conservation, to desire the exclusion of fossils from the marketplace. But in the eyes of the academic and conservation establishment the resource was now, in the main, renewable. Taylor (1988, p. 129) even went so far as to suggest that the low financial value attributed to fossils affected how they were valued as cultural items and ultimately the care they received in museums.

The arguments of the past 20 years continued to be recycled, but English Nature and the other conservation agencies were embracing a sense of social purpose essential to the survival of public bodies by the 1990s. Collecting remained on the agenda, and two models became frequently cited in the conservation literature. English Nature's excavations of Coal Measure material at Writhlington had given amateurs an opportunity to collect fossil plants and insects and possibly contribute to science, while commercial fossil excavations into the Lias Frodingham Ironstone at Scunthorpe, in collaboration with the local museum, transformed what was known of its fauna and extended access (Robinson 1988; Knell 1990, 1994; Larwood & King 1996; see Figs 4 & 5).

By the end of the decade, local agreements were beginning to resolve longstanding collecting issues. In 1998, the stretch of coast most intensively exploited by commercial collectors – that around Lyme Regis – became the subject of one such development. The language was now more flexible and reflected the realities of the collecting community: it made no distinction between commercial and non-commercial collectors. Collecting was now to be 'responsible' and 'sustainable'. Collectors were to register important finds for which ownership was to be transferred to the collector. No longer was the professional collector ostracized or vilified for needing to make an income. Formulation of the agreement involved many of the same collectors who had negotiated the Scunthorpe agreement, which itself owed much to German practice. It too established two tiers to collecting, ensuring that the needs of science, conservation, leisure and commerce were not in conflict (Jurassic Coast Project 1998).

Some three years earlier, the anonymously authored booklet, *Guidelines for Collecting Fossils on the Isle of Wight*, actually issued by the island's geological museum, had sought to resolve similar local tensions. In 1999, research was commissioned by Scottish Natural Heritage to locate 'consensus fossil collecting sites'. In the same year, negotiations were begun along the Yorkshire coast to establish a policy on collecting as part of the Dinosaur Coast Project.

The 1990s also had its conference on collecting and conservation: 'A future for fossils' in Cardiff in 1998. While this meeting demonstrated that some fundamental tensions remained, this was no rerun of the conference of 1979 or 1987. The fact that the conservation establishment's magazine, *Earth Heritage*, contained a report on this conference that pictured fossil shops as part of the local economy, shows how far the geological establishment had shifted its thinking in 30 years (Anon. 1999).

While the last decade of the century can be viewed as one in which British conservationists built upon the major culture shift of 1987, American opinion on the matter seems to have retreated from its similarly liberal consensus of that same year. Again the issue of commercial collecting formed its most controversial element. The Society of Vertebrate Paleontology (SVP) had, in 1973, adopted a resolution opposing the sale of fossils to the public. In the 1990s it became one of the most influential

Fig. 4. The Scunthorpe collecting agreement enabled commercial collectors to gather fantastic ammonites which were often stripped of their shells and polished for sale as some of the most expensive invertebrate décor fossils on the market. These now scientifically useless specimens fetched prices in excess of £1000 in the early 1990s. However, the agreed by-product of this activity, which was at a site actively exploited for hardcore, was the rescuing of one of the best starfish faunas of the decade (which also proved the presence of obrusion deposits in the Ironstone for the first time). Intensive excavation has long been known to be the best way to locate rarities, and during this period of collecting the first articulated crinoid and fish fossils were recovered from the locality, together with many new ammonite species. These were all placed in the Scunthorpe Museum. In two years, and after 150 years of collecting, this activity made the most significant contribution to our understanding of the fauna of the Lower Jurassic Frodingham Ironstone. Here the commercial collectors Trevor George and David Sole excavate an Upper Lias site at Roxby Mine, Scunthorpe.

Fig. 5. Geologists' Association members and other amateurs take advantage of fossil collecting opportunities in the vast Lias exposures of Crosby Warren Mine, Scunthorpe.

lobbying organizations in the science: 'Worldwide, amateur and professional paleontologists recognize the damage that recent commercialization has done' (Vlamis *et al.* 2000, p. 56). Stimulated by the greatest collecting controversy of the century, that of 'Sue', the South Dakota *Tyrannosaurus* (which left one man in jail, provoked a Government raid and went to a museum for $8.4 million), two bills aimed at regulating collecting on public lands entered Congress. These were the Vertebrate Paleontological Resources Protection Act ('Baucus Bill') of 1992 and the Fossil Preservation Act of 1996 (Pojeta 1992; Catalani 1997, p. 8; Fiffer 2000).

Campaigning under the banner 'Save America's Fossils for Everyone' (SAFE), the SVP successfully opposed the commercial possibilities enshrined in the 1996 bill. In a public poll it believed it could demonstrate that 'the American public are overwhelmingly against commercial collecting on Federal public lands' (Poling 1996). The Association of Science Museum Directors also came out strongly in support of the SVP position. Neither bill became law, but the problem of collecting on Federal Lands did not go away.

The issue culminated in a forum at the US Geological Survey offices in Virginia in 1999. Here the same irreconcilable perceptions were again rehearsed: fossils were to some a renewable resource, while to others they were not; for some a weathered fossil in context was better than one saved in a collection, but others disagreed; many saw fossils as abundant while others thought they were rare; commercial exploration of mineral wealth was fine but of fossils it was not. In some respects the views of particular groups were predictable, but others sat on the fence, and some (such as amateurs) were divided (American Geological Institute 1999). In May 2000, *Fossils on Federal and Indian Lands*, a Report by the Secretary of the Interior, Bruce Babbitt, was published. This recognized the 'complexities' of fossils, their competing interests in science, leisure, commerce and education, and their differing meanings in the setting of Indian and Federal Lands. Here, fossils were, once again, and in contradiction to views across the Atlantic, a non-renewable resource and 'relatively rare'. Commercial collecting activity on public lands had been successfully opposed: 'Two major professional paleontological societies, representing more than 3,000 members, issued a joint statement in October 1999, agreeing that, "because of the dangers of overexploitation and the potential loss of irreplaceable scientific information, commercial collection of fossil vertebrates on federal lands should be prohibited as in current regulations and policies"' (US Department of the Interior 2000, p. 25). The Government, which expressed a sense of custodial responsibility for collected materials, was taking moral possession of the nation's palaeontological resource, re-establishing *Allosaurus, Deinonychus* and their kin as unique and powerful national icons. The 'heritage principle' was now central to the US administration's view of geological conservation, but in pursuing this principle the Americans had, so it seemed, since 1987 travelled in a direction counter to that taken by the conservation movement in Britain where the drift had been towards consensual accommodation and away from strict control by the scientific hegemony. In the USA, the more conservative views of the scientific hegemony prevailed; this was no consensus view. But this report was not the legislation for which many on both sides had been calling: the issue of collecting on Public and Indian Lands remained unresolved and the debate was set to continue (Reed & Wright 2000).

Ownership of the science's material culture

To what extent were the events in geological conservation indicative of wider trends in the culture of geology and in society at large? The debate over commercial collecting centred on sites as fossil repositories with conflicting opinions on their purpose, size, renewability, and rights of access. It should not surprise us that similar beliefs also extended to collections. It does not take a massive leap of argument to see the fear of site pillaging by foreign collectors as also reflecting beliefs that particular individuals, groups or countries have preferred rights of ownership over certain fossils. The NCC had discovered that geologists acquired a sense of ownership over a site that was local or of particular research interest to them. This mirrored the NCC's own early assumptions that science itself had superior rights of ownership over the fossil resource. Many amateurs evidently felt they had a higher 'moral right' to collect than those who exploited fossils for financial profit.

Yet, many stood opposed to any sense of ownership of scientific material (other than the rights of science itself). As university curator Roy Clements told the 1987 conference: 'As a science, palaeontology knows no national boundaries; its materials represent a 'world heritage' and should not be protected on nationalistic boundaries' (comment in Wild 1988, p. 189). This echoed a point enshrined in museum ethics: their role as one of 'trusteeship'; 'rights of ownership' remained problematic. Clements was not alone in his views. David Norman (1992) of English Nature similarly stated: 'The ideal result for the scientist is that the specimens should be adequately curated and available for study in a recognized institution – no matter in which country that might be'. These sentiments were echoed in the USA by Pojeta (1992, p. 11), amongst others: 'In the past few years, a chauvinism, perhaps jingoism extends to smaller and smaller political entities'.

Once again, however, the purity of scientific

ideology was running counter to the cultural makeup of scientific production (i.e. the diversity of factors that determine the outcomes of scientific endeavour). In 1980s West Germany, science benefited from fossils being 'cultural monuments', yet such designations automatically superimposed nationalistic values as Clements detected. These notions were enshrined in international law: a UNESCO Convention sought to protect the material culture of a nation from illegal export; it included palaeontological material within this definition (UNESCO 1970). Nor could science trample over an emerging sense of nationhood as countries and peoples sought to define themselves in what cultural theorists refer to as the post-colonial era (though those colonized object to the term; Green & Troup 1999). In the 1990s, the new National Museum of Australia was asking 'Who are Australians?' The material culture of that country was developing new meanings and increased significance. If Aborigine headdresses were transformed from colonial loot into a means of cultural understanding and bridge building, so fossils provided a link back into the depths of that country's history.

History is critical to nationhood. Collections, as entities which cross time, are not simply products of that history, they also symbolize it. They contribute to identity. They were, in the language of the 1980s, indisputably 'national heritage' (see, for example, Anon. 1996; Stone *et al.* 1998; Taylor 1991). Science was never nationless, it always had nationalistic overtones, and in the conservative sociopolitical settings of the late twentieth century it was vital that this was so. This sense of nationhood, bound up in science, became strongest in those countries that once felt subjugated or colonized. Canada and Scotland possessed desires similar to those that emerged in Australia in the latter decades of the century. The National Museums of Scotland, for example, rushed to acquire Stan Wood's 'Lizzie' not just for its science or for its tourism potential but also because it had become a Scottish icon, a symbol of status (Knell 1999, p. 11; Gagnon & Fitzgerald 1999; Taylor 1999). Similarly, a sense of local ownership coloured those collecting agreements that sought to keep part of the fossil wealth for a local museum (Knell 1994; Taylor 1999).

This sense of ownership was not without its problems, however. Martin (1999) has shown how, since 1970, museums and science have struggled to deal with illegally exported fossils. Chinese dinosaur eggs, containing unhatched young, flowed into Europe and North America from the Southeast Asian black market, while fossil fish from the Santana Formation in NE Brazil found their way into every fossil shop in the West. These two countries had adopted legislation which sought to control fossils as their own 'heritage'; most of those arriving in the marketplace had been illegally exported. The UK was not a signatory to the major international conventions on this illegal trade, but its museums had voluntarily adopted the conventions as an ethical and legal principle. Those specimens which found themselves exported but excluded from public collections were then in a scientific limbo. If they did not enter the public domain they could not be published, despite holding information at the frontier of knowledge (Martin 1999).

Even within nations, geology was indicating ethical and preferred repositories. Driven by the palaeontological research community, the NCC and English Nature frequently made reference to the desirability of placing collected materials in a public museum. This embodied the science's view that such materials must be available for research. It extended a rule which had been in operation for some time: editors of scientific journals required 'published fossils' to be lodged in an appropriate public institution. However, the conservation fraternity visualized museums as extensions to the process of conservation in the field. This had nothing to do with an archive of published fossils or with the process of transfer during publication. It could be applied to just about anything collected and which thus might hold scientific potential. The fear was that important specimens might remain in private ownership and therefore inaccessible to science. Of course, the realities were more complex. The 'responsible collecting' that the wider conservation fraternity (the NCC, amateur societies, academics in charge of field parties, museum curators, and so on) promoted was open to interpretation. Did it mean data-rich collecting from a measured section, collecting restraint, or the gathering of ex-situ material only, as was frequently recommended to amateurs (Knell 1991; Larwood & King 1996)? The latter is usually regarded as being of little use to museums, even though most museums lack geological curators and are thus not in a good position to assess material. Nor could museums collect on the scale that these recommendations seemed to suggest. Indeed, just as site conservation found itself in turmoil so the museum world discovered its own crisis (see Fig. 6).

In 1980, Philip Doughty shook the Museums Association conference with accusations of 'mismanagement' and 'neglect'. His report on the state of geological collections in British

Fig. 6. Lady Anne Brassey and Edward Charlesworth material rescued from the below sea-level basement storeroom of Bexhill Museum, where it had been packed away for some sixty years. This was typical of the rescue curation undertaken in the 1980s.

museums, published a few months later, and his tireless campaigning, pulled geological collections into the professional conscience for the first time in perhaps 50 years. Utilizing a wealth of evidence, it argued that the archive to one of Britain's greatest scientific achievements was rotting, disorganized and unloved in the country's museums (Doughty 1981*a*, *b*; Knell & Taylor 1991; Knell 1996). Doughty was a key member of the highly influential GCG, which sought to reverse decades of neglect. The group also began to pioneer reinvigorated research into the history of these collections, searching for lost specimens, and adding a new dimension to their value. Seeing the attention geology was attracting, other disciplines soon demonstrated a keenness to show that they too had been abused. But what this represented was not recognition of a new problem, for the problem itself was 150 years old (Knell 1996), nor an interdisciplinary battle for resources, but a substantial leap in the professionalization of museum work. Driven by a rapidly expanding and increasingly youthful workforce, in an era when co-operation, direct action, conservation, and a sense of responsibility for heritage were in the public mindset, it too was an important reflection of cultural change. However, as the remaining years of the century passed, with crisis after crisis in public funding, wavering political support, and local government and university reorganization, no professional group had the power to control the fate of museums and collections. While many professionals added new management tools to a previously weak armoury, such as those needed to deal with forward planning and managing change, others saw the only answer in financial autonomy, something welcomed by Thatcherite politicians.

Ownership of the science itself

Like contemporary conservationists, Doughty in 1980 used the term 'culture' so that science could be understood in the bigger picture: 'Government recognition of the place of science in the cultural life of the nation is still awaited' (Doughty 1981*b*, p. 14). Such Government recognition was not to come, at least not in a way scientists wished. The monetarist policies of the Thatcher regime failed to solve the economic difficulties facing the country. A political desire to reduce direct taxation meant inevitable cuts in public spending, which hit the scientific establishment and the museum community hard. With only temporary respite around 1987, further economic failure followed. The sense of crisis continued to deepen.

By the mid-1980s forward planning was widely adopted in the commercial and public sectors. In the form of 'corporate plans' it inevitably involved institutional self-evaluation and redefinition. These plans were more than bureaucratic devices to generate a sense of responsibility: most institutions saw them, literally, as a means to survival. The 1986 plan of the British Museum (Natural History) (BM(NH)) was typical of the period: it was 'permeated with a sense of crisis' (Anon. 1986). Unable to maintain its scientific programme in the current financial year, and forecasting annual cuts of 3% year on year, its future looked bleak. Though director Neil Chalmers took the brunt of the criticism for the changes this report heralded, his predecessor, Ronald Hedley, had already overseen the planning and imposition of that great abhorrence to the British museum profession: the admission charge. By Chalmers' arrival in 1988, the crisis had grown acute. Finding all but 2% of funds spent on salaries, he took drastic action to rescue the institution from what he saw as impending disaster. Some 15% of scientific posts were axed. With expertise consequently lost or redirected, some collections were put on 'care and maintenance' only. Rebranded the 'Natural History Museum', the institution repositioned its research into applied areas: biodiversity, environmental quality, living resources, mineral resources, and human health and human origins. 'But', as one commentator noted, 'pressure to appear "useful" has made research in areas such as palaeobotany and bird systematics all but extinct' (Culotta 1992, p. 1271). The 51 job cuts announced in 1990 caused a furious response from the scientific community, while the apparent repositioning of the institution's research sparked a House of Lords enquiry into the state of systematics. This latter ultimately led to a short-term injection of some additional funding (£5 million over five years) (Anon. 1990*a*, *b*). In the future the Museum was to move to using externally funded postdoctoral workers to undertake much of its research, an approach which led to an overall increase in staff. Its financial position also moved rapidly into the black (Gee 1998).

Museums and university departments around the world endured similar rationalizations. Commentators saw these organizations withdrawing into applied fields, just as the Natural History Museum had done and, in the case of museums, pumping money into profile-improving front-of-house activities (Allman 1992). In Britain, the body responsible for grant-aiding research and research institutions in geology, the Natural Environment Research Council (NERC), was also in crisis. Its five-year corporate plan for 1985 proposed staff cuts of 30%, which were to come mainly from institutes such as the British Geological Survey (BGS). One early casualty of these changes was the demise of the Survey's Geological Museum – in effect British geology's national museum. The building was transferred and incorporated into the BM(NH), while its collections moved with the Survey staff to a rather inaccessible site near Nottingham. Doughty exclaimed to the British Association in Belfast in 1987: 'It is broadly the equivalent of moving the National Gallery to Holmfirth and burying the Rembrandts and Renoirs'. Leaked two months in advance of publication, NERC's plan for reorganization suggested that the Survey's directorate might also be abolished (Anon. 1985).

Having suffered annual funding reductions of 3.5% for the previous four years, NERC was already all too familiar with the current economic and political climate. The central strategy of the present plan was to reposition itself, to shift funding to the university sector. Inevitably, many university geoscientists welcomed the change, but in the main the Survey's cuts were widely condemned. *Geology Today* referred to it as the most severe attack on the geological community in 200 years: correspondent Ted Nield saw support for science as a tottery edifice with geology trapped in its basement (Nield 1986).

Towards the end of the decade, the Nature Conservancy Council also faced cuts and a complete organizational shake-up. Its chairman, Sir William Wilkinson, bemoaning the influence of Government, reflected on a post-war dream of cultural change driven by eminent scientists: 'Science assumed an almost sacred status during these first years. This belief in the power of science within the Conservancy may look naïve now, aware as we are of the way in which scientific understanding can be subservient to political objectives. However, it carried great political clout then and because of global considerations it may again' (Wilkinson 1990, p. 7). Indeed, Wilkinson saw some salvation in being a 'piggy-in-the-middle' organization, for without the loud voices of public protest, the NCC's independent status would surely have been compromised by political interference.

With public spending suppressed, the Thatcher Government pronounced that where industry would benefit, industry would pay. It was a policy that was to affect science profoundly. Where once science was an unquestioned cultural element of national identity, it increasingly became a service industry for the marketplace. With the dawning of biotechnology and other inherently practical, yet new and fashionable, sciences, geology was being pushed to the fringe. Sir Clifford Butler's working group on the future of the BGS reported late in 1987 and reaffirmed the importance of its core survey work. At the suggestion that such a conclusion should be taken as read, Professor James Briden, NERC director of Earth sciences, claimed that this would be a 'dangerously complacent attitude ... we have a commercially-minded government that will need to be fully convinced of the value to the nation of geology survey' (Anon. 1988*b*). A year later NERC was introducing compulsory redundancies and sweeping cuts to its programme, which included the BGS. It planned a cut of more than 100 staff per year for the foreseeable future. Having suffered much criticism from the scientific community for not resisting this erosion to the nation's science base, it, at last, began to make representations to government. In April 1989, NERC removed a further 160 posts, but planned several high priority 'community projects'.

A decade of cuts had also taken its toll on the Survey. Having begun with 40 palaeontologists (mostly micropalaeontologists), by the end it had just one curator and eight palaeontologists (Doughty, unpublished speech, BAAS, Belfast 1987; Doughty 1996, p. 14). One hundred and fifty years earlier the place of the Survey in Government bureaucracy was also much debated. In the late 1980s, it was still not secure, and in the following decade found itself the subject of further cuts and reorganization.

Science in the private sector also suffered similar economic and social upheaval. In the recession of the late 1980s the petroleum industry underwent savage cuts as producers withdrew from exploration to focus on proven reserves. In the USA, the high costs of domestic production combined with greater environmental stringency and overproduction by OPEC countries resulted in the catastrophic combination of low prices for crude oil while debt costs remained high (Leffingwell 1994). The response was 'downsizing' and 'outsourcing' in an attempt to achieve 'increased shareholder value'. A consequence was a collapse in the world population of micropalaeontologists. By 1994, palaeontologists in all institutions seemed to be fighting to justify their positions. Two years later the US Geological Survey axed approximately one-third of its palaeontological staff. Even where the science was applied there was a new emphasis on 'research that produces products that directly fill a societal need' (Brewster-Wingard 1996). The result for the oil industry was to look to collaboration or 'outsourcing' for research, collection management or database support. Conveniently this came at a time when universities and museums were looking for this kind of link-up with industry (O'Neill 1994). Previously, oil companies had attempted to internalize their business and especially their commercially sensitive scientific data.

Museums, which permanently exist in a world of underinvestment, were also realizing their potential as information repositories in the new electronic information age. Such thinking was at the heart of the Natural History Museum's development of a number of consultancy areas. In 2000, these were analytical facilities, environmental assessment, fossil replicas, petroleum, waste management and habitat restoration, biodiversity, collections management, biomedical, mining and training. Such activities had not been part of the core business for museums, or the natural sciences, two or three decades earlier. In similar fashion, the Survey attached itself to the Thatcher Government's belief in the rising promise of new technology and of the saleability of information. In the late 1980s the Survey was resold to Government as the National Geosciences Database and impressive mockups were promoted to gain support from funders and partners. In 1989, the Government took the bait and began to pump in additional funding (Anon. 1989), and 11 years later the Geoscience Data Index went online. But by this time the BGS was once again shedding staff and demonstrating the insecurities that had dogged it since its birth.

In contrast, *Geology Today* claimed American science was much better at selling science to government (Anon. 1991, p. 83). But it too was undergoing transformation. In 1994, reflecting the wishes of the Senate, the National Science Foundation devised key strategic areas for its funding programme: 'advanced material and processing, biotechnology, environment, global change, high performance computing and communications, manufacturing, science, math and engineering education and civil infrastructure' (Bourgeois 1994, p. 2). While basic funding for the Earth sciences remained, palaeontologists were encouraged to think in terms of how to pigeonhole their activities into this framework. It was suggested that 'global change' might provide an opening. Palaeontological researchers were then to ask themselves: 'How will this research help policymakers in understanding and projecting future climate change, on a human timescale (decades to centuries)?' It made clear divisions between what was perceived as cutting edge and vital, and that which simply increased understanding. Reflecting the short-termism of political cycles, it sought to put in place a quantifiable and politically accountable system of expenditure. While the strategy showed pragmatism and responsibility, it also meant that politicians could make statements relating expenditure to socially important outcomes. NERC adopted a similar policy in the early 1990s. Having long ago pursued thematic research, it now chose to remodel this idea and present it in lay terms to politicians. Contemporary reports suggested these themes were only administrative pigeonholes, not intended to influence the areas within which people work. *Geology Today*'s response was to see this as a rather purposeless exercise, but in doing so it had rather missed the point. What NERC was doing was exactly what its counterpart in the USA would do: construct a shared language as an interface between politics and science. Each side could allot its own meanings but the language itself acted as a flexible coupling of two worlds which had not seen eye-to-eye for decades. Now science could achieve its ends while politicians could claim the achievement of their electoral pledges; both were applying quite different interpretations to the same outcome. In this light a US Geological Survey (USGS) press release of 7 February 2000 takes on new meaning. It claimed the budget increases it had been given would enable it to meet the 'critical needs expressed by communities, stakeholders, government agencies and other organizations'. The USGS core programme now had four 'overarching initiatives': 'safer communities', 'livable communities', 'sustainable resources for the future', and 'America's natural heritage'. Social (and therefore political) meaning ensured funds: 'the relevance of USGS science to improved understanding of the changing world'.

Science had learned to talk the language of politicians rather than the jargon of science (though ironically politicians in this period had borrowed many geological terms: 'seismic shift', 'fault lines' and so on). With developments in publishing technologies, science progressively transformed its language and means of communication in these latter decades of the century. Output from the BGS and NCC provide useful examples. In the 1950s, the annual *Report of the Geological Survey Board* looked little different from the sheet memoirs the Survey had been producing for a century. With its formality and assumption of value, it talked the language of science. In the 1980s, the reports took on corporate styling but still talked in technical language. By the late 1990s the technical language was still present but there was now a sense that the organization was demonstrating its worth to a new audience. Through this decade the covers of its reports showed landscapes, then maps, and then buildings; rocks – the stuff of Survey work – were conspicuously absent. In the *Annual Report* for 1989–1990, the director introduced the Survey's work with talk of geochemists, groundwater and radionuclide migration. Six years later his language had changed, and he now talked of 'science and the market economy', 'the public face of the Survey', and 'the public good'. In the report for 1996–1997 much of the Survey's work was framed under 'Geology and the Community'. In 1999, the purpose of the Survey was to 'support the decision making by public and private bodies at national to local levels on broad issues relating to resources, land use, geohazards and the environment. A small, but key element of the Core Strategic Programme is the promotion of the public understanding of science'. In the process its reports had been transformed from technical manuals into full colour expositions of mission. The NCC's geologists followed a similar path. What was formerly a photostat *Circular*, which contained unquestioned assumptions concerning the legitimacy of its conservation perspective, became, through a series of metamorphoses, the full colour magazine, *Earth Heritage*, which in its style, language, title and support spoke of 'shared values'.

The dynamics of late twentieth century cultural change

In the late twentieth century who 'owned' geology? Was it the servant of government, the possession of the academy, or in the ownership of a broader cultural group? The reality was that it was all these things, though few individuals necessarily saw it as such. The field of geological conservation moved from a predominantly academic hegemony of the 1960s to become something much broader, both in terms of concept and participation. This was no simple shift of belief but something driven by campaigns, rhetoric, dispute and debate. But it was also a reflection of its cultural setting as society shifted from the conservation of protest of the late 1960s to the conservation of responsibility and accountability of the late 1980s. In this area of conservation as in others, in 20 years it had crossed the political spectrum, as only by this means could its ends be achieved. In Britain, it was attached to the NCC, a scientific organization that was more successful than most in retaining its funding during the financial and political stringencies of the 1980s. But like other institutions, it too had to change, and to rethink its role, and this too impacted upon the way conservation in Britain developed.

In the wider geological community there were other upsets, as geological provision in universities underwent radical 'rearrangement', and the sector as a whole expanded rapidly at a time of financial stringency. Throughout this period campaigns erupted, whether to save collections, the Survey, or the very status of the science. If Government answered these protests, which it rarely did, it never did so as the protestors wished. There was never a restoration of lost funding. Change inflicted in times of recession, whether due to ideology, mismanagement or the vagaries of world trade, resulted in a new, increasingly 'useful' science. Both research and geology as a broader cultural field (in museums and conservation, for example) became judged by their social relevance. Such pressures changed the very nature of geoscience and its institutions. When the boom returned, society had moved on. Old limbs were not regrown, but new ones budded from the reconceived science. A new period of scientific diversification begins. It was a curious kind of evolution where fitness to survive in terms of intellectual value did not always play a part in selection.

The late-century campaigns concerning the 'ownership' of the science differed from those better-known battles where geology confronted 'scientific creationism', disbelievers of *Archaeopteryx*, or cultural theorists. In these latter debates, science's sense of reality and its application of empiricism became its most valued weapons. But in campaigns about ownership and control these were rarely useful. Here discourse relied upon the vagaries of ideology, meaning and value, which each party constructed. In the year 2000, the official view of a vast nation on one side of the Atlantic was that fossils are rare and irreplaceable; on the other side of this ocean its small and more densely populated neighbour considered them renewable and, in most cases, not rare.

Perhaps the greatest change in all areas of the science, as in wider society in Britain, was the development of a culture of accountability. The pervasive sense of a Britain in decline, which followed the boom years of the 1960s, forced governments to consider financial efficiencies and instil a sense of accountability and responsibility. But these notions went far beyond the performance measurement of industry. Even in the private worlds of the commercial and amateur collector it became a necessary creed, though one brought to them by a government agency. Accountability had entered the mindset of the country and thus became a tool to be applied wherever it had value.

A science which feels unloved by government, which faces on-going cuts, which is situated in an era of accountability, has to shift its focus. It must prove its worth against the reconceived values of the day, even when these have evolved from political ideology. As the opposition parties fell into disarray in the early 1980s, the Conservative Government appeared unstoppable. It had an ideological strength unseen in the post-War period, unfettered power and a desire to use it. To public institutions, and to science, it was clear that a new era had begun and one that was vastly different from anything science had previously known. The institutional establishment rapidly sought tools that would facilitate its survival, but some took even more radical steps, so apparent in the then highly criticized Natural History Museum in London. Having been amongst the most conservative of all institutions, museums were by the early 1990s battling for their very survival. The Natural History Museum underwent radical transformation in an attempt to take greater control of its finances; Government was not going to let it do otherwise. But few museums were able to do this, so they took another course – to increase their public profiles, to 'democratize', to identify with, and respond to, their audiences – and in so doing became some of the most socially adjusted and progressive organizations in the country. There were all kinds of knock-on effects of these cultural changes from 'dumbing down' and 'hyping up' to redefining and restructuring workforces. What seemed at times like culture in chaos was really one undergoing rapid change.

For science, survival in the 1990s relied on communication. Where once universities, research institutes, museums and scientists talked the language of science to their financial masters, they were increasingly learning the language of politics. The language of the Government and its institutions had long been measured, as any slip could result in litigation or unwanted media attention. Thus the NCC's officers could claim late on that they had never come out overtly against collecting, and indeed there was relatively little evidence that this had been the case. Yet the success of such claims relied upon the change that had already taken place, and the forgetting of original context and emphasis. It was a game politicians often played; if it wasn't unambiguously expressed on paper it did not happen. But there was a more radical change in the very words science used – not just in terms of ideas but in the accepted codified language of politics. Politicians had long played with the ambiguity of language: they were notoriously difficult to pin down (Edelman 1977; Pfeffer 1981). Publicly funded organizations learnt that if they used the terms politicians understood – less plate tectonics, more 'safe communities' – both Government and science would understand each other. This understanding was not actual but political. The scientists could continue with their science, with motive and conclusion redrawn, and the politicians could claim their successes and socially relevant decisions. By this means at least, and although radically transformed, scientists could retain ownership of their world.

I should like to thank my two referees, one of whom was P. Doughty (Ulster Museum), for many useful suggestions and comments; and the Geologists' Association for permission to reproduce its badge. I am also grateful to J. Martin (Leicester Museums), R. Clements (University of Leicester) and M. Stanley (former NSGSD co-ordinator) for valuable conversations. However, the text here is entirely my own reading of this period, and is not necessarily theirs.

There are obvious difficulties in writing such a recent history, especially if one has also been an actor in that history, however minor one's role may have been. History, in these circumstances, can be misread as critical commentary. While I recognize that histories are personally constructed I have attempted to discern the assumptions, developments and arguments by understanding the various sides of the debates that I have examined.

References

ALLMAN, W. D. 1992. From the editor. *American Paleontologist*, **1**, 1.

AMERICAN GEOLOGICAL INSTITUTE 1999. Summary of Public Forum on Federal Paleontology Policy (6–21–99), http: //www.agiweb.org/gap/legis106/fossils.html

ANON. 1973. Fossil conservation and the Open University. *Geology and Physiography Section of NCC, Information Circular*, **8**, 3.

ANON. 1985. The NERC five-year plan. *Geology Today*, **1**, 66–67.

ANON. 1986. Corporate planning at the BM(NH). *Geology Today*, **2**, 96–97

ANON. 1988*a*. Splitting the conservationists. *Geology Today*, **4**, 2

ANON. 1988*b*. What future for the BGS? *Geology Today*, **4**, 4–6.

ANON. 1989. The squeeze has now gone on for long enough. *Geology Today*, **5**, 38.

ANON. 1990*a*. What future for British Earth sciences? *Geology Today*, **6**, 8–9.

ANON. 1990*b*. The Natural History Museum: a symposium and taxonomic comment. *Geology Today*, **6**, 142–145.

ANON. 1991. Convincing the paymasters. *Geology Today*, **7**, 83–85.

ANON. 1996. Australians seek to stem fossil thefts. *Science*, **274**, 725.

ANON. 1999. A future for fossils, *Earth Heritage*. **11**, 3–4.

BARNES, B. 1974. *Scientific Knowledge and Sociological Theory*. Routledge and Kegan Paul, London.

BENTON, M. J. & WIMBLEDON, W. A. 1985, The conservation and use of fossil vertebrate sites. *Proceedings of the Geologists' Association*, **85**, 1–6.

BENTON, M. J., CLEAL, C. J., EDWARDS, D., HALSTEAD, L. B., *et al.* 1985, Mothballs? *Geology Today*, **1**, 135–136.

BLACK, G. P. 1984. The geological work of the Nature Conservancy Council. *In* CLEMENTS, R. G. (ed.) *Geological Site Conservation in Great Britain*. Geological Society, London, Miscellaneous Papers **16**, 6–10.

BOOTH, A. 1995. *British Economic Development Since 1945*. Manchester University Press, Manchester.

BOURGEOIS, J. 1994. Climate changes at the National Science Foundation. *American Paleontologist*, **2**, 1–2.

BREWSTER-WINGARD, G. L. 1996. The status of paleontology in the U. S. Geological Survey. *American Paleontologist*, **4**(1), 9–10.

CALLON, M. 1986. Some elements of a sociology of translation: domestication of the scallops and the fishermen of St Brieuc Bay. *In* LAW, J. (ed.) *Power, Action and Belief: A New Sociology of Knowledge?* Routledge and Kegan Paul, London, Sociological Review Monograph **32**, 196–230.

CARSON, R. L. 1962. *Silent Spring*. Riverside, Cambridge, Mass.

CATALANI, J. 1997. An amateur's perspective: public lands – public fossils? *American Paleontologist*, **5**, 6–8.

CAUFIELD, C. 1984. Ministers admit Countryside Act has failed. *New Scientist*, **101** (22 March), 10.

CLEAL, C. 1987 The use and conservation of palaeontological sites. *Geology Today*, **3**, 150–151.

CLEMENTS, R. G. (ed.) 1984. *Geological Site Conservation in Great Britain*. Geological Society, London, Miscellaneous Papers **16**.

COMMITTEE ON GUIDELINES FOR PALEONTOLOGICAL COLLECTING. 1987. *Paleontological Collecting*. National Academy Press, Washington DC.

COTTON, J. A. D. 1984. The amateur collector. *In*: CLEMENTS, R. G. (ed.) *Geological Site Conservation in Great Britain*. Geological Society, London, Miscellaneous Paper **16**, 48–49.

CROWTHER, P. R. & WIMBLEDON, W. A. (eds) 1988. *The Use and Conservation of Palaeontological Sites*. Palaeontological Association, London, Special Papers in Palaeontology, **40**.

CULOTTA, E. 1992. Museums cut research in hard times. *Science*, **256**, 1268–1271.

DOODY, J. P. 1975. Protection and administration of important geological sites. *Newsletter of the Geological Curators' Group*, **1**, 212–215.

DOUGHTY, P. S. 1981*a*. *The State and Status of Geology in UK Museums*. Geological Society, London, Miscellaneous Papers **13**.

DOUGHTY, P. S. 1981*b*. On the rocks. *Museums Association Conference for 1980*, 12–14.

DOUGHTY, P. S. 1996. Museums and geology. *In*: PEARCE, S. M. (ed.) *Exploring Science in Museums*. Athlone, London, 5–28.

DUFF, K. L. 1979. The problems of reconciling geological collecting and conservation. *In*: BASSETT, M. G. *et al.* (eds), *The Curation of Palaeontological Collections*, Palaeontological Association, London, Special Papers in Palaeontology, **22**, 127–135.

DUFF, K. L. 1980. The conservation of geological localities. *Proceedings of the Geologists' Association*, **91**, 119–124.

DUFF, K. L. 1985. Geological conservation – yes please! *Geology Today*, **1**, 103–104.

EDELMAN, J. M. 1977. *Political Language: Words that Succeed and Policies that Fail*. Academic Press, New York.

ELLIS, N. V. (ed.) 1996. *An Introduction to the Geological Conservation Review*. Joint Nature Conservation Committee, Peterborough, Geological Conservation Review Series, **1**.

ENGLISH NATURE. 1992. *Fossil Collecting and Conservation*. English Nature, Peterborough.

FIFFER, S. 2000.*Tyrannosaurus Sue*. Freeman, New York.

GAGNON, J.-M. & FITZGERALD, G. 1999. Towards a national collection strategy: reviewing existing holdings. *In*: KNELL, S. J. (ed.) *Museums and the Future of Collecting*. Ashgate, Aldershot, 166–172.

GEE, H. 1998. Chalmers' choice: make cuts or go under. *Nature*, **395**, 119.

GREEN, A. & TROUP, K. 1999. *The Houses of History: A Critical Reader in Twentieth-Century History and Theory*. Manchester University Press, Manchester.

GREEN, C. P. 1990. The badge of the Geologists' Association: its history on the cover of the proceedings. *Proceedings of the Geologists' Association*, **101**, 97–99.

GREEN, F. 1989. Evaluating structural economic change: Britain in the 1990s. *In*: GREEN, F. (ed.) *The Restructuring of the UK Economy*. Harvest Wheatsheaf, Hemel Hempstead.

HACKING, I. 1999. *The Social Construction of What?* Harvard University Press, Cambridge (Mass.).

HARKER, R. S. 1984. The dealer – friend or foe? *In*: CLEMENTS, R. G. (ed.) *Geological Site Conservation in Great Britain*. Geological Society, London, Miscellaneous Papers **16**, 50–53.

HEALEY, N. M. 1993. From Keynesian demand management to Thatcherism. *In*: HEALEY, N. M. (ed.) *Britain's Economic Miracle: Myth or Reality?* Routledge, London, 1–39.

IGGERS, G. G. 1997. *Historiography in the Twentieth Century: From Scientific Objectivity to the Postmodern Challenge*. Wesleyan University Press, Hanover.

JURASSIC COAST PROJECT. 1998. *Fossil Collecting Code for the West Dorset Coast*, Jurassic Coast Project, Charmouth.

KENNEDY, K. 1993. Fossils for profit – the Gilwern Hill controversy. *Earth Science Conservation*, **32**, 27.

KNELL, S. J. 1990. Working with professional collectors. *Geology Today*, **6**, 112–113.

KNELL, S. J. 1991. The responsible collector. *Geology Today*, **7**, 106–110.

KNELL, S. J. 1994. Palaeontological excavation: historical perspectives. *Geological Curator*, **6**, 57–69.

KNELL, S. J. 1996. The roller-coaster of museum geology. *In*: PEARCE, S. M. (ed.) *Exploring Science in Museums*. Athlone, London, 29–56.

KNELL, S. J. 1999. (ed.) *Museums and the Future of Collecting*. Ashgate, Aldershot.

KNELL, S. J. 2000. *The Culture of English Geology, 1815–1851: A Science Revealed Through Its Collecting*. Ashgate, Aldershot.

KNELL, S. J. & TAYLOR, M. A. 1991. Museums on the rocks. *Museums Journal*, **91**, 23–25.

LARWOOD, J. & KING, A. 1996. Collecting fossils: a responsible approach. *Earth Heritage*, **6**, 11–13.

LATOUR, B. 1987. *Science in Action: How to Follow Scientists and Engineers through Society*. Open University Press, Milton Keynes.

LEFFINGWELL, H. A. 1994. Reinvigorating industrial paleonotology. *American Paleontologist*, **2**, 1–4.

LONG, B. & BLACK, G. P. 1975. The Nature Conservancy Council's viewpoint and experiences. *Newsletter of the Geological Curators' Group*, **1**, 206–211.

MARTIN, J. G. 1999. All legal and ethical? Museums and the international market in fossils. *In*: KNELL, S. J. (ed.) *Museums and the Future of Collecting*. Ashgate, Aldershot, 112–119.

NATURE CONSERVANCY COUNCIL. 1980*a*. Third survey of geological fieldwork. *Earth Science Conservation*, **17**, 12–15.

NATURE CONSERVANCY COUNCIL. 1980*b*. The commercial exploitation of geological sites – the Nature Conservancy Council's response. *Earth Science Conservation*, **17**, 1–3.

NATURE CONSERVANCY COUNCIL. 1982. Public Inquiry into collecting licences at Charmouth. *Earth Science Conservation*, **19**, 19.

NATURE CONSERVANCY COUNCIL. 1983. Public Inquiry into collecting licences at Charmouth. *Earth Science Conservation*, **20**, 40–42.

NATURE CONSERVANCY COUNCIL. 1984. *Nature Conservation in Great Britain*. Nature Conservancy Council, Peterborough.

NATURE CONSERVANCY COUNCIL. 1990. *Earth Science Conservation in Great Britain: A Strategy*. Nature Conservancy Council, Peterborough.

NIELD, T. 1986. Sound and fury, signifying plenty. *Geology Today*, **2**, 37–39.

NORMAN, D. B. 1992. Fossil collecting and site conservation in Britain: are they reconcilable? *Palaeontology*, **35**, 247–256.

NORMAN, D. B. & WIMBLEDON, W. A. 1988. Palaeontology at the Nature Conservancy Council, *Geology Today*, **4**, 194–196

NORMAN, D. B., DOYLE, P., PROSSER, C., DAVEY, N. & CAMPBELL, S. 1990. Comments on C. W. Wright's 'Ideas in palaeontology: prejudice and judgement', *Proceedings of the Geologists' Association*, **101**, 91–93.

O'NEILL, B. J. 1994. Towards a paleontological network. *American Paleontologist*, **2**(3) (unpaginated).

OVERBEEK, H. 1990. *Global Capitalism and National Decline: The 'Thatcher Decade' in Perspective*. Unwin & Hyman, London.

PFEFFER, J. 1981. *Power in Organizations*. Pitman, Marshfield (Mass.).

POJETA, J. 1992. Paleontology and public lands: a status report. *American Paleontologist*, **1**(1), 9–12.

POLING, J. 1996. Analysis of the SAFE poll on fossil collecting. http: //www.dinosauria.com/jdp/law/poll.htm

PORTER, R. 1977. *The Making of Geology: Earth Science in Britain, 1660–1815*. Cambridge University Press, Cambridge.

REED, C. & WRIGHT, L. 2000. The trouble with fossil thieves. *Geotimes*, **45**(10), 22–25, 33.

ROBINSON, J. E. 1988. The interface between 'professional' palaeontologists and 'amateur' fossil collectors. *In*: CROWTHER, P. R. & WIMBLEDON, W. A. (eds) *The Use and Conservation of Palaeontological Sites*. Palaeontological Association,

London, Special Papers in Palaeontology, **40**, 113–121.

ROLFE, W. D. I. 1977. Recent pillage of classic localities by foreign collectors. *Physiography Section of NCC, Information Circular*, **13**, 2–4.

ROLFE, W. D., CLARKSON, E. N. K. & PANCHEN, A. L. (eds) 1994. Volcanism and early terrestrial biotas: proceedings of a conference. *Transactions of the Royal Society of Edinburgh (Earth Sciences)*, **84**, 175–464.

ROSE, H. & ROSE, S. 1980. The rise of radical science. *New Scientist*, **85**, 28–29.

SCHOLTE, J. A. 2000. *Globalization*. St Martin's Press, New York.

SHAPIN, S. 1992. Discipline and bounding: the history and sociology of science as seen through the externalism–internalism debate. *History of Science*, **30**, 333–369.

SIMPSON, J. 1998. *Strange Places, Questionable People*. Pan, London.

SMITH, K. 1984. *The British Economic Crisis*. Penguin, Middlesex.

SMITH, P. J. 1986. The Geological Conservation Review: progress and problems. *Geology Today*, **2**, 153–155.

STONE, M., COUZIN, J. & HUI, L. 1998. Smuggled Chinese fossils on exhibit. *Science*, **281**, 315–316.

TAYLOR, M. A. 1988. Palaeontological site conservation and the professional collector. *In*: CROWTHER, P. R. & WIMBLEDON, W. A. (eds) *The Use and Conservation of Palaeontological Sites*, Palaeontological Association, London, Special Papers in Palaeontology, **40**, 123–134.

TAYLOR, M. A. 1991. Exporting your heritage? *Geology Today*, **7**, 32–36.

TAYLOR, M. A. 1999. What is in a 'national' museum? The challenges of collecting policies at the National Museums of Scotland. *In*: KNELL, S. J. (ed.) *Museums and the Future of Collecting*. Ashgate, Aldershot, 120–131.

TODD, Lord. 1980. The role of scientists in the 1980s. *New Scientist*, **85**, 2.

UNWIN, D. M. 1986. World's oldest terrestrial vertebrates are Scottish. *Geology Today*, **2**, 99–100.

UNESCO. 1970. *Convention on the Means of Prohibiting and Preventing the Illicit Import, Export and Transfer of Ownership of Cultural Property*. UNESCO, Paris.

US DEPARTMENT OF THE INTERIOR. 2000. *Fossils on Federal and Indian Lands*. Report of the Secretary of the Interior, Department of Interior. http://www.doi.gov/fossil/fossilreport.htm

VLAMIS, T. J., FLYNN, J. J. & STACKY, R. K. 2000. Protecting the past: vertebrate fossils as the public trust. *Museum News*, **79**, 56.

WARD, B. & DUBOS, R. 1972. *Only One Earth*. Penguin, Harmondsworth.

WHYBROW, P. J. 2000. *Travels with the Fossil Hunters*. Cambridge University Press, Cambridge.

WILD, R 1988. The protection of fossils and palaeontological sites in the Federal Republic of Germany. *In*: CROWTHER, P. R. & WIMBLEDON, W. A. (eds) *The Use and Conservation of Palaeontological Sites*. Palaeontological Association, London, Special Papers in Palaeontology, **40**, 181–189.

WILKINSON, W. 1990. Chairman's review. *Nature Conservancy Council 16th Report, 1989–1990*, 7–10.

WIMBLEDON, W. A. 1988. Palaeontological site conservation in Britain: facts, form, function, and efficacy. *In*: CROWTHER, P. R. & WIMBLEDON, W. A. (eds) *The Use and Conservation of Palaeontological Sites*, Palaeontological Association, London, Special Papers in Palaeontology, **40**, 41–55.

WOOD, S. 1985. Geological conservation–no thanks! *Geology Today*, **1**, 68–70

WOOD, S. 1988. The value of palaeontological site conservation and the price of fossils: views of a fossil hunter. *In*: CROWTHER, P. R. & WIMBLEDON, W. A. (eds) *The Use and Conservation of Palaeontological Sites*, Palaeontological Association, London, Special Papers in Palaeontology, **40**, 135–138.

WRIGHT, C. W. 1989. Ideas in palaeontology: prejudice and judgement. *Proceedings of the Geologists' Association*, **100**, 293–296.

Index